Heterick Memorial Library
Ohio Northern University
Ada, Ohio 45810

Data Communications Principles

Applications of Communications Theory
Series Editor: R. W. Lucky, *AT&T Bell Laboratories*

Recent volumes in the series:

COMPUTER COMMUNICATIONS AND NETWORKS
John R. Freer

COMPUTER NETWORK ARCHITECTURES AND PROTOCOLS
Second Edition • Edited by Carl A. Sunshine

DATA COMMUNICATIONS PRINCIPLES
Richard D. Gitlin, Jeremiah F. Hayes, and Stephen B. Weinstein

DATA TRANSPORTATION AND PROTECTION
John E. Hershey and R. K. Rao Yarlagadda

DEEP SPACE TELECOMMUNICATIONS SYSTEMS ENGINEERING
Edited by Joseph H. Yuen

DIGITAL PHASE MODULATION
John B. Anderson, Tor Aulin, and Carl-Erik Sundberg

DIGITAL PICTURES: Representation and Compression
Arun N. Netravali and Barry G. Haskell

FIBER OPTICS: Technology and Applications
Stewart D. Personick

FUNDAMENTALS OF DIGITAL SWITCHING
Second Edition • Edited by John C. McDonald

MODELING AND ANALYSIS OF COMPUTER
COMMUNICATIONS NETWORKS
Jeremiah F. Hayes

MODERN TELECOMMUNICATIONS
E. Bryan Carne

OPTICAL CHANNELS: Fibers, Clouds, Water, and the Atmosphere
Sherman Karp, Robert M. Gagliardi, Steven E. Moran, and Larry B. Stotts

PRACTICAL COMPUTER DATA COMMUNICATIONS
William J. Barksdale

SIMULATION OF COMMUNICATION SYSTEMS
Michel C. Jeruchim, Philip Balaban, and K. Sam Shanmugan

A Continuation Order Plan is available for this series. A continuation order will bring delivery of each new volume immediately upon publication. Volumes are billed only upon actual shipment. For further information please contact the publisher.

Data Communications Principles

Richard D. Gitlin
AT&T Bell Laboratories
Holmdel, New Jersey

Jeremiah F. Hayes
Concordia University
Montreal, Quebec

Stephen B. Weinstein
Bell Communications Research
Morristown, New Jersey

Plenum Press • New York and London

Library of Congress Cataloging-in-Publication Data

Gitlin, Richard D.
Data communications principles / Richard D. Gitlin, Jeremiah F. Hayes, and Stephen B. Weinstein.
p. cm. -- (Applications of communications theory)
Includes bibliographical references and index.
ISBN 0-306-43777-5
1. Data transmission systems. 2. Computer networks. I. Hayes, Jeremiah F., 1934- . II. Weinstein, Stephen B. III. Title. IV. Series.
TK5105.G57 1992
621.382--dc20 92-19019
CIP

First Printing—September 1992
Second Printing—January 1994

ISBN 0-306-43777-5

© 1992 Plenum Press, New York
A Division of Plenum Publishing Corporation
233 Spring Street, New York, N.Y. 10013

All rights reserved

No part of this book may be reproduced, stored in a retrieval system, or transmitted in any form or by any means, electronic, mechanical, photocopying, microfilming, recording, or otherwise, without written permission from the Publisher

Printed in the United States of America

621.382
G536d

To BARBARA, RACHEL, and DAVID
for their love, encouragement and patience.

R. D. Gitlin

To MARY, ANN, JEREMIAH, and MARTIN
with special acknowledgement to
MARGIE and NIKI.

J. F. Hayes

To JUDY, BRANT, and ANNA
for their love and understanding

S. B. Weinstein

Preface

This book was written during an era of remarkable development in the art and technology of data communications, and it reflects our experience as designers of techniques and systems for data communication over voice-grade and broadband channels. Electrical data communication is as old as the telegraph, but the modern era began in the 1960s when telephone lines began to be widely used to access and update computer-based information and to broaden the availability of computer resources. Airline reservation, computer timesharing, sales, inventory, credit checking, military communications and other wide-area networks, private and shared, began to grow rapidly. Expecting performance comparable to that available in local computer environments, users demanded better communications equipment, and vendors began a competition to supply higher-speed, more versatile, and more cost-effective modems, multiplexers, and network controllers.

The telephone network, because it was in place and reached almost everywhere, was accepted by the early designers of data communication systems as the principal switching and transmission medium, although its channels often required improvement to meet operational requirements. These improvements ranged from simple line conditioning to creation of computer communication networks through organization of telephone links into networks with routing and error control capabilities. Despite the evolution of high-speed digital networks, voice-grade channels will continue for a long time into the future as components of data communication networks, particularly for remote access from low-traffic locations. Furthermore, the analytical models and practical techniques developed first for voice-grade channels, where severely distorted channels demanded attention and relatively low data rates permitted implementation of advanced communication-theoretic concepts in available digital signal processing, have more recently been applied to high-speed radio and lightwave channels. In the final analysis, all physical channels are analog and physical data communication is the adaptation of digital signals to analog transmission channels. We are confident that the foundation technologies developed in the "telephone era" of data communications will be of significant interest for many years to come.

The revolution in data communication technology can be dated from the invention of automatic and adaptive channel equalization in the late 1960s. These mechanisms make it possible to improve the transmission characteristics of a less than ideal analog channel to meet the requirements of high-performance, and generally high-speed, data communication. Many engineers contributed to this revolution, but the early inventions of Robert W. Lucky, particularly data-driven equalizer adaptation, were the largest factor in realizing higher-speed data communication in commercial equipment. A whole new industry for the manufacture of modems and related equipment resulted from this striking new technology.

Progress has been steady since those early days, with inventive signal processing algorithms and analytical results keeping pace with development in VLSI. All of the functions associated with data transmission, including channel equalization, modulation, echo cancellation, multiplexing, switching, coding, and network routing and control, have been made more efficient and less expensive, and researchers have discovered innovative combinations of techniques, such as the trellis-coded modulation pioneered by Gottfried Ungerboeck. Data communication may become near-universal as personal computers and digital facsimile machines proliferate, and as the Integrated Services Digital Network (ISDN) and its broadband successors reach a large subscribing public. The Golden Age of data communication engineering is not yet at an end.

The quantity of specialized knowledge accumulated in this area is already huge. Rather than attempting to be encyclopedic, we have chosen to offer a view through the filter of our experience, emphasizing basic principles and highlighting the applicability of modulation, equalization, coding, and synchronization techniques to a variety of communications media including the telephone, twisted pair, radio, magnetic recording, and lightwave channels. There are other topics of importance to data communication, but we believe that the material offered here will give the reader a clear and substantive introduction to the state of the art.

Our personal knowledge of data communication comes from many sources; among these are Bennett and Davey's early textbook and the classic reference work of Lucky, Salz, and Weldon. A great deal was learned in discussion over many years with our colleagues at Bell Laboratories and in the industry generally. We would especially like to thank Bob Lucky and Jack Salz for the fruitful associations we have had with them and the strong influence they have had on our careers. We are also heavily indebted to our long-time friends and distinguished colleagues Paul Ebert, David Falconer, Gerard Foschini, Allen Gersho, Edmond Ho, Irving Kalet, Victor Lawrence, Frank Magee, James Mazo,

Howard Meadors, Kurt Mueller, Marcus Mueller, Jean-Jacques Werner, Jack Winters, and the late Tong Lim, for sharing their knowledge with us and working with us on a number of studies. We acknowledge with thanks the direct help we have received in writing this book from many of those colleagues and from Ender Ayanoglu, Vijay Bhargava, Israel Bar-David, Rob Calderbank, John Cioffi, Glenn Golden, Ali Grami, Larry Greenstein, Zygmunt Haas, Mark Karol, David Haccoun, Joseph Lechleider, Johannes Peek, Hemant Thapar, Lee-Fang Wei, and Nicholas Zervos. We further would like to thank our students at Concordia University, McGill University, Princeton University, the Polytechnic University of New York, and UCLA for many helpful suggestions and criticisms. A special thanks is due to Dave Falconer, who read and commented on the entire manuscript. We are indebted to Judy McKelvey for her painstaking text processing work, and to AT&T Bell Laboratories for its generous support of this work. Finally, we are deeply appreciative of the support, encouragement, and infinite patience of our Editor at Plenum Press, Sy Marchand. We are, of course, entirely responsible for any errors or shortcomings of this book.

We hope, then, that the knowledge and insight drawn from our association with so many talented colleagues and from our own efforts will come through in the words and formulas of the next ten chapters. This book can be used on its own as either a text or a reference work, and is a companion to the networking book of one of us (J. F. Hayes, *Modeling and Analysis of Computer Communication Networks*, Plenum Press, 1984).

There are a number of ways that this book could be used in graduate-level courses. The following are cited as illustrative examples for introductory, intermediate, and advanced courses. For students who have had neither detection and estimation theory nor error detecting and correcting codes, the first six chapters would serve as a text for a first-level course. All that is required as prerequisite is a knowledge of probability and random processes at the level covered in Appendix 2A. An intermediate-level course would cover the material in Chapters 4 through 8. In addition to detection and estimation theory, it is essential that the students understand the Viterbi algorithm as presented in Chapter 3. Finally, a course at the advanced level would cover the material in Chapters 6 through 10 in some depth. As is true in most texts, there is more material than can be covered in a single course. According to the interests of the instructor and the background of the students, material can be emphasized or omitted. Exercises have been provided for all but the first and the last chapters.

Suggestions for improvements will always be welcomed, and we expect to develop and change our own views of the world of data communication as it continues its rapid evolution into the Information Age.

Contents

1

Introduction to Data Communications

1.0 A PERSPECTIVE

Data communication has been with us for a long time. Smoke signals, drum beats, and semaphore signals are examples that are commonly given; indeed, semaphore relay may be regarded as the first modern communication network [1]. But the most remarkable example must surely be alphabetical writing. The concept of conveying information by successive choices from a finite alphabet is the very essence of both writing and digital data communication [2]. In fact, many of the ideas of linguistics carry over to information theory, communications, and pattern recognition. It is the purpose of this first chapter, in a book devoted to the principles of data communications, to provide a perspective on the technology of data communications, and to highlight the broad applicability of the *foundation technologies* of modulation/demodulation, equalization, coding, and synchronization. In this text we will demonstrate the communication-theoretic origin and the broad application of these technologies to a variety of communication media, including the telephone channel, twisted pairs, radio, magnetic recording, and optical fiber.

Electrical data communication, including "modern" concepts of multiplexing, such as rudimentary echo cancellation, and efficient coding, first appeared in the telegraph systems of the nineteenth century pioneered by Samuel F. B. Morse. Keyboard text entry, particularly teletype, TWX, and Telex, and visual data communication, such as telephoto and facsimile, appeared in the first half of the twentieth century. Telex, a text messaging service intended for a broad customer base, remains in service, although internetworked electronic mail may make it obsolete. Facsimile machines for document and graphics transmission over telephone lines have been available since the 1920s, but were for decades slow, expensive, and not widely used. Fast and economical machines have now become ubiquitous, stimulated by new scanning, coding, and transmission technologies and, initially, by East Asian demand for convenient exchange of documents containing Chinese characters. Communication among terminals and computers, growing since the 1960s with the spread of computers and the automation of data processing work, is carried over telephone networks and countless data-oriented local and wide-area networks. Broadband communication, at information rates thousands of times larger than those possible on telephone channels, is opening new perspectives on creative interaction among distributed humans and computers. The current explosion in data communication is indeed a hallmark of the dawning Information Age.

Data are numerical values which may or may not be selected from a finite or infinite alphabet. There are communications applications such as telemetry, and applications in signal processing such as filters using charge-coupled delay lines, where the data are analog, i.e., elements from a continuous range of values. These data may be communicated from one point to another (subject to distortion in transit) in their analog form. Ordinarily, however, data communication refers to *digital* data communication, where the data values are elements from a finite (discrete) set. When this set has only two values, the data are called *binary*. Any information to be transmitted that has an alphabetic representation, such as written text or discrete approximations to continuous variables, can be represented as a stream of digits in the binary or any other integral base appropriate for transmission and processing. One of the key contributions of information theory [3–5] is the means to measure the information in an analog signal and represent it in digital form.

The first among many important advantages of digital transmission is the possibility, through pulse regeneration and error control techniques, of information delivery through imperfect channels and over large distances with virtually no degradation in quality. Other advantages include more easily maintained transmission equipment, signal processing and switching in computer-like structures, format convertibility, and convenient interfacing with office and residential information systems. Digital communication has become an important element of both waveform (voice and video) transmission and data-based information systems.

The design of information systems with any significant distance between communicating elements involves *computer communication* [6–9], the field devoted to *networks* of shared data communication links and processors. This book offers some introductory material in this chapter but is not focused on networking issues. It is mainly about digital data transmission on the individual communication links or channels of telephone, computer, and general-traffic networks. In the context of the International Standards Organization's layered architecture for computer communication, this text is primarily at the physical layer. It emphasizes the theory and techniques, including signal-processing technologies, first developed for data transmission through the voice-grade channel of the public telephone network and now applied to other channels as well, including the digital subscriber access line, microwave, optical fiber, and satellite channels. The concept of "channel" should be understood as the performance description of a point-to-point communication link. It may be separate from specification of a transmission medium, since a channel offered to an end user may—and often does—include transmission segments of several different media. This book is largely about such end devices, called "modems" (*mo*dulator–*dem*odulators), that are either incorporated into terminal equipment or placed between terminal equipment and communication circuits, whose purpose is to reliably transport bit streams between distant locations.

1.1 WHO USES DATA COMMUNICATION?

Applications of data communication are found in many areas of contemporary life. Table 1.1 describes several classes of digital traffic, although the first of these, voice and video, have traditionally been regarded as distinct from data. Advances in technology will probably tend to blur this distinction with all information being conveyed in packets. The network of national and international data communication facilities continues to expand at a rapid rate, and discussion of investment, regulation, technical standardization, privacy, interfacing, and economic impact has gone beyond the pages of technical and trade journals and into the general press and the hardware and software of the information age.

Table 1.1: Major Classes of Digital Traffic, Present and Evolving

Class	Characteristics	Applications
Voice and video	Continuous data streams, intolerant to delay (except for broadcasting) and jitter. Rates from several kbps to hundreds of Mbps. Both continuous and dynamically variable capacity demands.	Telephony, teleconferencing, broadcasting, multimedia information retrieval. Video of varying quality levels, up to high-definition television.
Inquiry–response	Central file, remote terminals, intermittent and unbalanced traffic. Largely intolerant to delay. Rates up to tens of kbps.	Reservations, credit checking, electronic funds transfer, order-entry, records and other information retrieval, inventory control.

(Continued)

Table 1.1: Continued

Bulk transfer	Computer-to-computer: large file movements, high rates, relatively tolerant to delay. Rates to hundreds of Mbps.	Transfer of business records and transactional files, remote typography, offline information retrieval "dumps."
Document transmission	Terminal-to-terminal continuous traffic, wide range of rates and acceptable delays.	Facsimile, graphics, computer mail, communicating word processors.
Interactive computing and information retrieval	Networked computers, databases, and terminals, including work stations and supercomputers. Intermittent and unbalanced traffic. Data rates up to Gbps.	Computer-aided remote image processing and design; supercomputer networks, shared multimedia workspaces.
Remote job entry	Central computer and remote peripherals. Bulk traffic. Data rates up to Mbps range.	Payroll, accounting, structural design, and other programs for business and industry.
Process control and telemetry	Computer to controlled or monitored machinery. Generally intermittent and relatively low rate.	Real-time control of industrial processes, and/or monitoring of operation.

Some years ago it was thought that the computer utility, a large computational facility available on a time-sharing basis to human users at remote terminals, would create a large part of the total data traffic [6,10,11]. The original prognosis was altered by the development of cheap and powerful large-scale integrated circuits, greatly reducing the costs of locally available computers, and by the persistence of relatively high transmission costs. The emphasis shifted, toward the use of data networks for exchanging and managing information, with host computers and file servers serving more as switches, database controllers, and service providers than as calculating machines. With these applications in mind, many contemporary networks are designed for peer-to-peer communication with distributed capabilities for data formatting, code conversion, file interrogation and maintenance, information gathering, and message storage and routing. There is a further trend toward multimedia capabilities, the integration of data, voice and image information streams within the same application. The concept of the computer utility is returning, however, with interactive supercomputing made available to scientists, medical personnel, and others at "broadband" communication rates of megabits per second (Mbps) and gigabits per second (Gbps). Broadband communication promises the low latency (fast transfer of bursts of information) required for many applications.

Local area networks (LANs), metropolitan area networks (MANs), and public and private data networks with national and international coverage are all examples of computer communication networks, described in the next section. Data communication as examined in this book is a set of techniques for electrical signaling of digital information through the broad variety of links that make up these networks.

Personal terminals, primarily work stations used by professionals, have become large users of data communication facilities and services. For the general public, videotex services, delivering information as attractive color text and graphics for display on television or color computer screens, have not had the commercial success that some people expected, but simple text-only services have made headway among personal computer users. Low-speed services may be adequate now, but harmonious integration of voice, higher-speed data, and video media will eventually be demanded for satisfaction of personal communication and information needs.

Because of the continual change in the applications environment, and the reduced transmission costs likely to be seen on future fiber-optic networks, the nature of data traffic today may not be representative of what it will be in the future. As direct digital access becomes increasingly available to communication users, by means such as ISDN and Broadband ISDN [12,15], and data traffic mixes with digitized voice and video, which already consume most of the world's digital transmission capacity, data networks as separate entities may become obsolete. A broadband lightwave network may eventually offer such a vast bandwidth that all traffic classes will appear in the same format in the network, much like the present data traffic. As we point out throughout this text, the techniques we describe for efficient communications are applicable to the plurality of media from which current and future networks will be constructed.

1.2 DATA NETWORK PROTOCOLS

We have already suggested that digital signaling through a channel, the "data communication" of this book, is an element of a much larger picture. The brief discussions of network protocols in this section and data network architectures in the next are intended to introduce the reader to that larger picture, as a

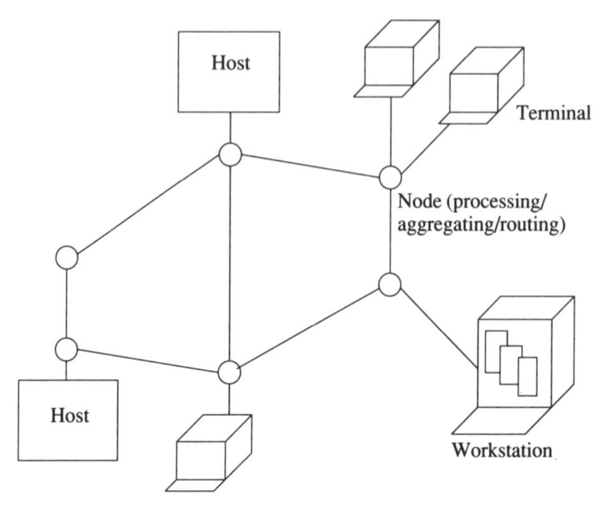

Fig. 1.1 A generic computer communication network.

prelude to and motivation for the discussions of the "physical level" data communication technology that is the main subject of the book.

In broad terms, a network is a connected set of nodes as shown in Figure 1.1. Traffic enters the network at the nodes and travels on the links joining them. The network is designed to share its resources in conveying traffic among users, while meeting service requirements. The key ingredients in this sharing are multiplexing and switching. In this context, multiplexing is the means of sharing the transmission facilities in a link. Switching is choosing and the setting up of a path, consisting of one or more links, between source and destination nodes in the network and routing traffic through the nodes.

When the dramatic growth of data communication began in the late 1960s the principal existing network was the telephone network. The dominant multiplexing techniques in the telephone network were frequency division multiplexing (FDM) and time division multiplexing (TDM). In both FDM and TDM a portion of the capacity of a link is dedicated to a particular source, whether or not it is actively generating traffic. Capacity is assigned in a very few standard sizes, such as a telephone channel or a 64 kbps ISDN channel. With the circuit switching technique used for most traffic in the telephone network, facilities for a continuous end-to-end transmission path are dedicated for the duration of a call.

FDM, TDM, and circuit switching are well suited to ordinary voice traffic, but as already suggested, data traffic, including some forms of "compressed" digitized voice and video, has attributes such as holding times and burstiness that vary significantly from ordinary voice traffic. New forms of communication, such as digitized moving video of varying quality levels or real-time interaction in a distributed computing environment, will require a flexibility in capacity allocation that is inconsistent with the fixed channel sizes of circuit-switched networks. Furthermore, the performance requirements (beyond flexibility of capacity) for data services and new communication types are markedly different from those of voice.

These considerations mandate different approaches for multiplexing and switching of data and of the highly-varied, multimedia traffic of the future. Packet switching, which routes self-contained information units containing encapsulated message segments, is at the heart of these alternative approaches. A packet is simply a fixed or variable-length bit string with a header containing addressing and possibly other control and routing information and a body consisting of user data.

In the remainder of this section, we introduce packet communications, packet-based network architectures, and the layered protocols that allow users to concentrate on applications without worrying about the details of network communications.

1.2.1 PROTOCOL LAYERING

Data communication, or indeed communication generally, can be defined by *protocols* that control relationships among communicating entities, the allocation of network resources, and the orderly flow of information. Modern communication networks are distinguished by their ability to let machines communicate with one another without human intervention. Protocols make it possible to establish and maintain a communication session by conveying appropriate messages among the units carrying out essential tasks. Much of the material in this section follows Comer [13] (see also [14–22]).

These messages might be best understood by analogy to similar signaling among machines and human beings in an ordinary voice call. Dial tone signals that a line is available, prompting the caller to dial, sending a message to switching equipment. A ringing or busy tone indicates the state of the called party. Once the call has been set up, the parties can take an appropriate action, such as repeating a word, if transmission quality is less than perfect; this is a protocol between the human users alone. Finally, the termination of a call is a well-defined procedure with appropriate messages generated by actions such as placing the handset on the switchhook.

For data communication, most, if not all, of the actions associated with session establishment, maintenance, and termination must be carried out by machines. The situation is further complicated by the relatively large number of tasks in comparison with those of a voice call, requiring a cooperating collection of protocols handling functions such as packetization, routing, congestion control, recovery from packet loss damage or delay, and passage from one network to another. Machines involved in a communication session must each be concerned with all of these functions and communicate with one another about them.

In order to avoid the intolerable complexity and chaos of each application program on each machine handling all of these functions, data network designers have devised the concept of *layered protocols*. Each task in setting up and maintaining a communication session is handled by a particular layer, supported by the tasks carried out by lower layers. A signaling message sent from layer n in machine A goes down through the protocol stack in machine A, through the network or collection of networks to machine B, and up through the protocol stack to layer n. Protocol layer n does not have to know anything about how lower-level functions are done; the protocol tasks have been compartmentalized. (This ideal is frequently compromised in the interest of efficiency, e.g., by passing routing information up to higher levels so that packets can be sized for most efficient transport on the actual networks being used.)

In the interest of international standardization to encourage progress toward universal interworking among data communication equipment, the International Standards Organization (ISO) has promulgated a reference model (Figure 1.2) for Open Systems Interconnection (OSI) consisting of seven layers. Each layer is devoted to the fulfillment of a particular task that seems to be a need in most networks.

The lowest level is the *physical* level, providing for the simple exchange of binary digits between two pieces of equipment, such as a terminal and a modem. The basic electrical signaling of user data, as well as control information

7 APPLICATION
Application-dependent services and procedures

6 PRESENTATION
Data formats, representations, and displays

5 SESSION
Control of dialog between processes

4 TRANSPORT
End–to–end control; packet or message assembly and disassembly

3 NETWORK
Routing, network supervision, block, message or packet structure and format

2 LINK
Data flow control on communication links

1 PHYSICAL
Electrical, mechanical, and functional interfaces

Fig. 1.2 The seven layers of the Open Systems Interconnection protocol reference model of the International Standards Organization.

required for data set operation, is treated at the physical level. The RS-232 and the RS-449 interfaces may form part of the physical level. An alternative is the physical layer of X.21* which uses fewer circuits by time division multiplexing data and control information.

The next higher layer, the *link* level, provides structure for the flow of bits over a link between nodes of a network. Typically a transmission frame structure is defined with addressing, control, error control, and data sections. Control of flow over a link takes place at this level. Link level protocols are either *bit* or *character oriented.* A widely encountered example of the former is the *HDLC* protocol which forms level 2 of the *X.25* protocol. *BISYNC* is an example of a character-oriented protocol.

The flow over tandem links is dealt with at the third, or *network* level. Routing through the network is implemented at this level. The structure and format of packets and messages is also a network-level function. Packet multiplexing is the foundation of most computer communication networks. It was devised in response to the inadequacies of conventional multiplexing techniques for bursty data traffic. The essential idea is to fragment messages (or encapsulate whole messages if they are short) into data segments. Addressing, as well as other overhead information, is added to these segments to form a packet. This other overhead might include information on message length and sequencing necessary to reconstruct the original message. Finally, in order to detect the presence of errors, parity check bits are added to the packet. As far as multiplexing and switching are concerned, these packets are self-contained entities. The packetizing technique allows a large number of bursty sources to share the same transmission lines. At the receivers, processors can distinguish packets from different sources on the basis of their addresses.

The packets produced by packet multiplexing must be switched through a computer communication network to their destinations. Two basically different approaches to routing have been advocated, *datagram* and *virtual circuit.* In the datagram technique, packets from the same message are treated as separate entities. Each packet contains a complete definition of the destination. In response to varying traffic conditions, packet routing may change in the middle of a multipacket message, so that packets from the same message may take different routes and arrive out of order. Communication using datagrams is said to be *connectionless,* since no particular circuit (route) is dedicated to the call. In the virtual circuit approach, the unique route that the individual packets of a message follow is fixed at the beginning of a call. A relatively simple lookup table in each node determines the route of a packet once it has been determined to be

---

* In this text, we shall cite several examples of CCITT recommendations, in particular, the X series for networks and the V series for modems. The Comité Consultative Téléphonique et Télégraphique (CCITT), a branch of the International Telecommunications Union (ITU), was renamed the Telecommunications Standardization Sector in 1993, referred to as ITU-T.

part of a particular call. Today, most commercial packet networks use virtual circuit technology. The tradeoff between throughput and delay is seen in the better (shorter) delay performance of datagrams (because of the dynamic routing capabilities), and the better (larger) network throughput when virtual circuit routing is used (because of the simpler routing at each node).

ARPANET, an experimental long-haul network developed by the U.S. Defense Advanced Research Projects Agency (DARPA) beginning in the late 1960s, was one of the largest and widely used packet-switched networks. It was the first of its kind [11], and its technology has been replicated in MILNET, a U.S. defense network. It was the forerunner of the huge Internet, supporting over a million host computers (in 1993) at thousands of locations worldwide. Future research and education networks operating at Mbps and Gbps speeds are being planned [22].

The first three levels of OSI are concerned with components of the path from the originating to the destination node. The fourth layer, the *transport* level, exercises end-to-end control. Flow into the network is controlled and messages are reassembled from packets. The four lowest-level protocols form what is called the *Transport Service*. The Transport Service insures the conveyance of bits from source to destination during a data call or session.

The fifth layer in the hierarchy is the *session* level. As its name implies, users initiate and terminate calls at this level. There may be error control, flow control, or other functions specified at this level. The *presentation* level is concerned with the form in which information is encoded. For example, the coding of data for videotex displays is specified at this level. Data encryption may also take place at the presentation level. The North American Presentation Level Protocol Syntax (NAPLPS) [19] defines character codes and data sequences corresponding to graphical characters and geometric shapes used in videotex displays. The final layer of OSI is the *application* level, which covers those elements of user applications, such as the management of files and data bases, that relate to interaction with distant resources. The creation of electronic mail is done at this level (e.g., the CCITT X.400 Recommendation).

The definition of a layered protocol model does not say anything about the actual protocols written for these layers. Many different protocols are possible to meet different needs. Moreover, when protocols are written, they may not fit exactly into the ISO model. A protocol "suite" (collection of layered protocols) will often have fewer levels, and levels not clearly identified with ISO levels.

The most widely used protocol following the ISO model is X.25, implemented in many public packet-switched networks around the world and standardized by the CCITT. X.25 encompasses the lower four levels, and is notable for its error detection and recovery mechanisms at all levels. At the second (link) level particularly, each frame of data is acknowledged, checked for errors, and retransmitted if necessary. This releases higher protocol levels from such responsibilities, thus supporting "dumb" terminals with little or no data-processing capability, but has a cost in processing complexity and delay within the network. Most X.25 networks operate links at rates of 64 kbps and below.

As data networks move to higher speeds and throughput (total information transferred per unit of time), more "lightweight" protocols are being devised. These faster, "unreliable" protocols [18] leave reliability to the end applications, which in the future, and especially for higher-rate applications, will be running on computers—including powerful work stations—with plenty of processing capability.

The most important layered protocol not conforming to the ISO model, and perhaps the most widely used data communication protocol of all, is TCP/IP (transport control protocol/internet protocol), which facilitates the transparent interconnection of end applications across multiple networks, based on an underlying datagram transport mechanism. The five-layer conceptual model is shown in Figure 1.3. TCP/IP is the basis for the Internet. From the user's perspective, the Internet allows transparent interworking across multiple networks. The combination of the third- and fourth-level protocol layers in the TCP/IP protocol [21] is the powerful basis for the Internet and forms the heart of what is called the Internet Protocol. Under this protocol, a host on one network can establish a virtual circuit connection with a host on another network, via gateway(s) and a series of intermediate networks if necessary (Figure 1.4).

IP is a datagram-based protocol that imposes the least common denominator of network characteristics of the two or more networks involved. As Comer [13] explains, IP provides an unreliable (no error detection or retransmission), connectionless packet transfer. Packets travel through the tandem networks required to join the communicating host computers, from gateway to gate-

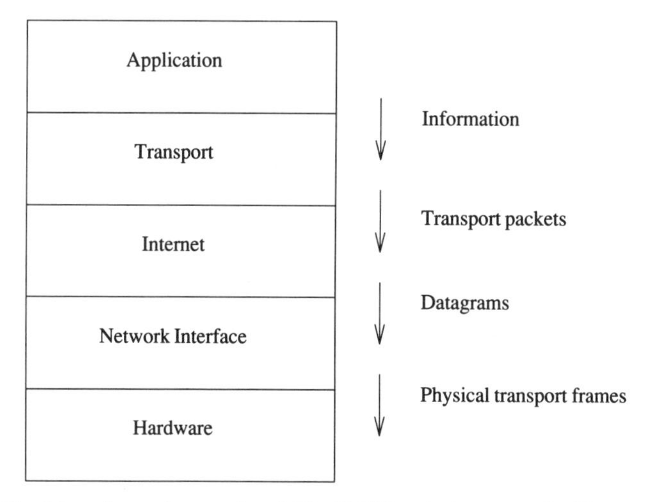

Fig. 1.3 Layered protocol model for TCP/IP and other protocols of Internet.

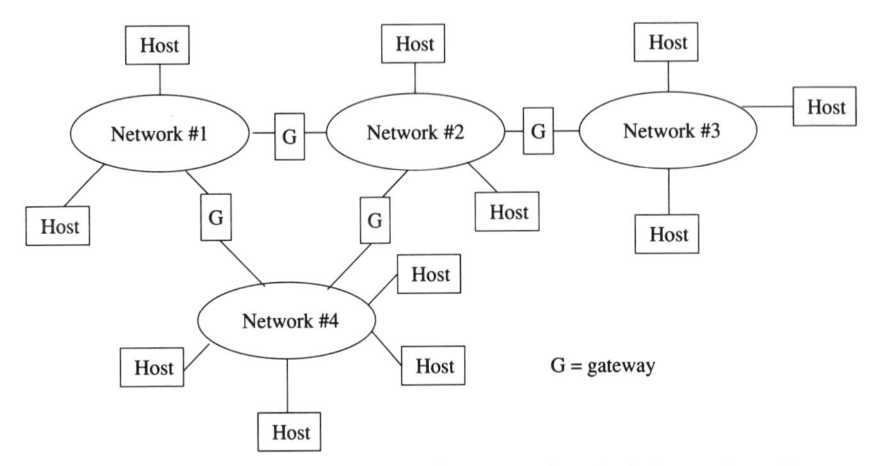

Fig. 1.4 Interconnection of hosts on different networks under the Internet Protocol.

way. A clever universal addressing system is used, with network identification and host identification given separate fields within the address, that allows routing from the source to be made first to the destination network, and then to the particular host. This greatly simplifies the routing tables maintained in hosts and gateways.

1.3 DATA NETWORK ARCHITECTURES

1.3.1 LOCAL AREA NETWORKS

Local area networks (LANs) are computer communication networks that usually span distances of less than a kilometer [23]. They may be found in office buildings, on college campuses, or in other relatively closed environments. LANs are part of the general growth of digital technology, but they have received particular impetus from the advent of distributed processing, including office automation. For the most part, LANs carry data traffic. The usual upper limit to the line transmission rate is 20 Mbps, with most systems operating in the 1–10 Mbps range, but newer fiber-based networks are in the 50–150 Mbps range.

In most applications, LANs are in protected environments. They are usually indoors, and are operated and used by the same organization. The benign environment and the limited area of a LAN means that link reliability is of less concern than in networks covering a larger area. Link reliability has considerable impact on network topology. For example, the original design of the ARPA network, which operated over a wide area, specified that there be at least two paths between any pair of nodes. The effect of this requirement is to mandate a mesh topology. In contrast, because of less concern about link reliability, LANs need only provide a single path between source and destination. A second implication of the limited size of LANs is that bandwidth is not the precious commodity that it usually is in large networks. The transmission media that are available allow transmission at speeds in the range of megabits per second at relatively low cost.

The three most common transmission media used in LANs are twisted pair, coaxial cable, and optical fiber. The twisted pairs of copper wires are the same as those used in the telephone plant. Because of the readily available technology, there is a strong tendency to operate at the same rates as short-haul digital carrier systems, especially at the North American DS1 rate of 1.544 Mbps or the European rate of 2.048 Mbps. The twisted pair has been mainly confined to point-to-point service, while coaxial cable can operate as a multiple-access medium by use of high-impedance taps. Rates up to 100 Mbps are feasible, with 10 Mbps being the standard. Optical fiber offers the highest transmission rate, in the region of 100 Mbps in the simplest configurations using LEDs as sources, but as high as 1–10 Gbps with semiconductor laser sources, single-mode fiber, and other high-performance system components. With currently available technology, optical fiber is basically a point-to-point medium (although point-to-point links with regenerators can be configured in many topologies such as rings). The insertion of passive taps induces appreciable loss and only a limited number can be tolerated, unless optical amplification is provided.

As mentioned above, reliability considerations allow LANs to have simple topologies with a single path between source–destination pairs. The *bus* topology (Figure 1.5a) is appropriate to transmission media such as coaxial cable or radio, which allow what are, in effect, high-impedance taps [24]. These taps do not affect the medium and a large number of stations can be connected. Each of these stations can broadcast simultaneously to the others. The bus topology is particularly appropriate for random-access techniques as means of sharing the transmission medium among the sources. Ethernet, which is so widely deployed as to have become a *de facto* and *de jure* standard, uses carrier sense multiple access with collision detection (CSMA/CD) as an access protocol [25,26]. Stations with messages to transmit sense the line for interfering traffic before transmission. Messages are accompanied by addresses that are recognized by the destination stations. Because of propagation delay within the network, simultaneous transmissions are possible. When collisions of messages from different sources are detected, transmission is aborted and messages are retransmitted after a random time-out interval. If several transmission attempts have failed, the process is handed over to a higher level protocol. The CSMA/CD protocol has been standardized by the IEEE(802.3) [27].

The *ring* topology (Figure 1.5b) is a sequence of point-to-point links with data flow in one direction around the ring (with detection and regeneration at each node); accordingly, any of the three access media mentioned above is appropriate. For reasons of reliability, there are provisions to bypass stations if they become inoperative. The standard medium access technique has been token passing. In this technique, the right to transmit is conveyed by the reception of a unique code sequence called the token. After transmission of its message, a station transmits the token to the next station in a preassigned sequence. Token passing for the ring topology has been standardized [27] in IEEE 802.5 and recently by ANSI (for FDDI) [38]. The token-passing protocol has also been standardized for the bus topology in IEEE 802.4.

The star topology, shown in Figure 1.5c, has been frequently associated with the optical-fiber transmission medium. Users connect to the ends of fibers emanating from a central hub. The fibers carry full-duplex traffic between users, routed through the hub. Variations on the basic theme involve implementation

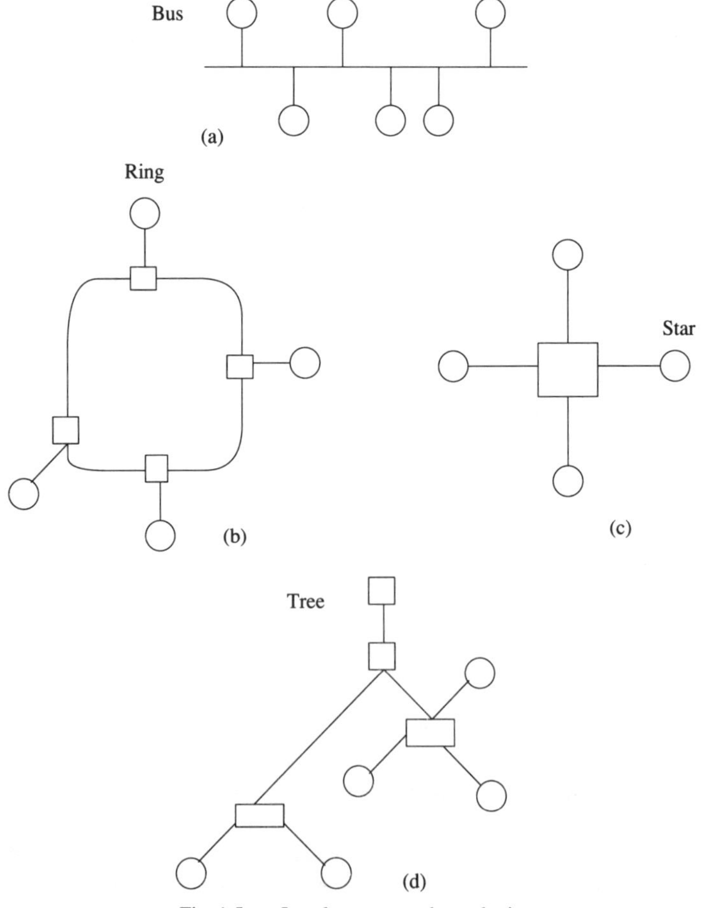

Fig. 1.5 Local area network topologies.

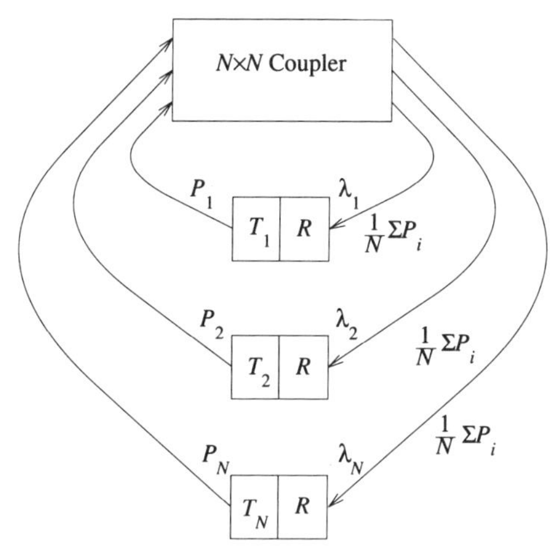

Fig. 1.6 $N \times N$ passive star coupler.

of the hub. Passive star couplers [28–30] are passive broadcast devices that connect N inputs to N outputs (Figure 1.6). The N transmitted signals, which are added together, can be made noninterfering, e.g., by use of different optical wavelengths, as described in Section 1.3.2, although contention systems will often be used, as described below. Since there is no regeneration, there is a 10 log$N$ dB* power loss; however, realistic power budgets would allow up to 1000 users. Since, in essence, every station broadcasts to every other station, the network is similar to the bus discussed above; consequently random-access or token-passing techniques could be used to coordinate traffic flow.

There is a basic problem with many random-access techniques over distributed passive media. Performance (as measured by throughput) deteriorates as the propagation delay in the network increases relative to the message transmission time. This may be understood by observing that even if a station does not detect another active source when it transmits, the message it transmits may collide with a message transmitted by a distant source T_{prop} seconds later (where T_{prop} is the propagation delay). Since stations have to wait T_{prop} seconds to be sure that a collision will not occur, the duty cycle (or the percentage of time that data may be transmitted) is $T_{message}/T_{prop}$, where $T_{message}$ is the duration of a message. This problem is accentuated for optical fiber systems because of their high transmission rates (unless the message duration is increased proportionately). This effect is countered in centralized systems, such as Hubnet [31], by the implementation of an *active* hub. Messages transmitted to the hub are

* Unless otherwise noted, log indicates the logarithm to the base 10. The natural logarithm and the logarithm to the base 2 are denoted by ln and $\log_2$, respectively.

buffered and repeated over all links. The message is accompanied by the address of the destination station. Also the hub chooses one among several nearly simultaneous transmissions for rebroadcast. If its attempted transmission is not echoed, the station retries after a suitable time-out interval. In essence, the hub acts like a zero-length bus. Yet another star configuration with an active hub uses TDM and a form of fast circuit switching to handle a wide range of traffic types [32]. A related approach is also used in Datakit, where a clever centralized contention bus algorithm provides low delay [33]. Finally, switched ATM (Section 1.5.2) is being considered for Local ATM (LATM) networks.

1.3.2 METROPOLITAN AREA NETWORKS

In a certain sense, metropolitan area networks (MANs) are an extension of LANs. They are oriented toward handling traffic in the form of packets; furthermore the topologies and the media-accessing techniques are adaptations of those used in LANs. In terms of basic function, they differ from LANs in geographical extent and, to some extent, in transmission rates. MANs are optimized for operation over distances on the order of 50 km, the size of a large city, with possible operation over larger and smaller distances. Since the medium for MANs is almost exclusively optical fiber, the transmission speeds that are appropriate are 100 Mbps and higher, as explained above.

For the speeds and distances that are relevant to MANs, the propagation delay is not negligible compared to the time required to transmit messages; consequently, random-access protocols that are used in LANs are not appropriate and new ones must be developed. Further, in order to utilize the MAN bandwidth, a wide range of traffic types including data, voice, image, and video, must be handled. The access protocol must be such that the performance requirements for each type can be satisfied. The topologies and protocols that are used in MANs must also be tailored to the physical properties of optical fiber which, as pointed out above, is basically a point-to-point medium allowing a limited number of taps.

The *distributed queue dual bus* (DQDB) technique has been adopted by the IEEE 802.6 Committee as a standard for MANs [34]. It typically operates at rates of 45 or 155 Mbps. The *logical* topology for DQDB is the dual bus shown in Figure 1.7. From purely physical considerations, the taps that the bus topology implies would be limited in number; hence the physical form is likely to be a pair of parallel rings each consisting of point-to-point links with active nodes and an appropriate bypass capability in case of node failure. Customer data flow in fixed-length, contiguous slots on both busses.* Synchronization is the task of the head end. In the slots there is a marker bit that indicates whether or not the slot is filled with customer data. When a station wants to transmit a packet on

* A 53-byte slot (48 bytes data and 5 bytes overhead) has been specified, realizing compatibility with the ATM cell (defined later in this chapter), although the headers differ.

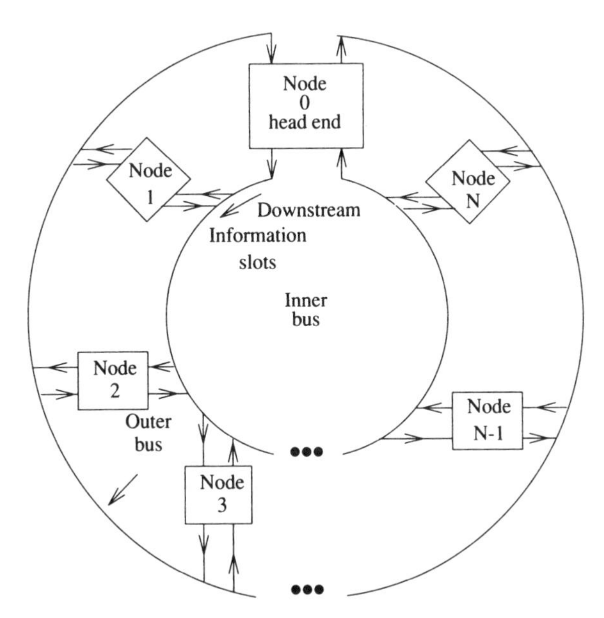

Fig. 1.7 Dual bus MAN as envisioned in draft 802.6 standards.

the inner bus it issues a request on the outer bus. Each of the upstream stations from the requesting station (with respect to the inner bus) increments a counter whenever it observes a request on the outer bus. Each station also monitors the inner bus and decrements the same counter whenever an empty slot is seen. Since the empty slot will be used by a downstream station, the counter at a station indicates the number of downstream stations that have outstanding packets waiting to be sent. Similarly, when a station wants to transmit on the outer bus a request is made on the inner bus.

When a station is ready to transmit a packet, the current contents of its counter is copied into a secondary counter which is associated with the packet. When an empty slot is seen on the bus, this secondary counter is decremented; consequently, its contents indicate the number of outstanding packets submitted by downstream stations prior to this station's request. The packet is put in the first empty slot after the secondary counter reaches zero, i.e., when all prior requests by other stations have, in this station's view, been satisfied.

The DQDB protocol emulates a single distributed queue in which packets are served in their order of arrival. At light loading the delay is small and at heavy loading there are no unused packets. Different classes of traffic may have different delay requirements. For example, a voice signal must be delivered to its destination within a specified time period. These delay requirements can be accommodated by establishing different priority requirements. These priority classes may be served in the DQDB protocol by keeping a counter for each class. A counter increments for each request for its class and higher-priority classes. It decrements when an empty slot is seen. The secondary counter associated with a packet increments when a request for a higher-priority class is received. Again, the packet is put into the first empty slot after the secondary counter becomes empty. One drawback to the DQDB technique is that of fairness [35–37]. Under heavy loads, stations nearer to the head end seem to receive better service. Measures designed to counter this effect are under study.

An alternative to the DQDB protocol is the *fiber distributed data interface* (FDDI) [38], which operates at 100 Mbps and is implemented on an optical fiber dual-ring network. The initial LAN version of FDDI uses multimode fiber and LED transmitters, (and electrical versions on copper twisted pair also exist) but later versions are to use single-mode fibers and laser-diode transmitters, for distances up to 100 km. This technique may be viewed as an augmentation of the token-passing protocol to handle traffic with stringent delay requirements. In this case, there is a time mark that accompanies the token and indicates the time that has elapsed in a cycle through all the stations in the ring. The protocol ensures that there is a maximum duration between service to each node. Each station is allocated a fixed time increment for its traffic. If not all of the stations use their increments there is some slack time that may be used to transmit traffic with more relaxed delay requirements. A further development, FDDI II, provides for the allocation of bandwidth to purely circuit-switched traffic (e.g., voice).

Both DQDB and FDDI may be viewed as incremental progress from existing LAN techniques. In fact, they can be implemented with twisted pair or coaxial cable. Such is not the case for the approach used in *Lambdanet* [39], which utilizes the enormous bandwidth of optical fiber [83]. By either FDM on a single optical carrier, or wavelength division multiplexing (WDM) with numerous lightwave carriers, information may be transmitted in a number of distinct bands. There are estimates of up to 1000 [29] such bands, each operating at 1 Gbps. This capability may be exploited by use of the star coupler shown in Figure 1.6. Each of the receivers in the N stations is tuned to a different fixed wavelength. A transmitter transmits to a particular station by tuning to the appropriate wavelength. A separate control channel coordinates the scheduling and flow of traffic (i.e., to avoid collisions), and the network throughput depends on the traffic pattern.

Lambdanet requires that each of the transmitters be capable of tuning to the wavelengths of each of the receivers. In *ShuffleNet* [40,41] the multiple wavelength capability of optical fiber is exploited by means of fixed wavelength transmitters and receivers. A particular topology, the perfect shuffle network is used to provide connectivity among all stations. Messages are relayed from station to station in the same manner as the ring network. The logical link between stations is a time slot on a particular wavelength channel. Again, the passive-star coupler may be used for the physical implementation.

1.3.3 ISDN and BISDN

Beginning in the 1950s, the digitization of the telephone interoffice network moved forward for a number of good technical and operational reasons. Digital transmission and switching offered the advantages of no degradation in signal quality over unlimited distances, equipment reliability, ease of maintenance, the accommodation in one network of both voice and data traffic, and improved interoffice signaling. The idea of extending digital access to the end user was an obvious one, although very difficult to achieve in practice, that has become imbedded in the concept of the integrated services digital network (ISDN), now a series of CCITT recommendations [42].

By giving the end user direct digital access to the public telecommunications network, at digital transmission rates considerably above what is possible with voiceband data transmission and at reasonable cost, network operators expect a much greater degree of voice/data integration. For example, it would become easy to place a voice/data call in which a spreadsheet, building plan, or other computer-based visual application would be shared by two or more individuals while they discussed it using high-quality (e.g., 7 kHz) audio, possibly encrypted for privacy. The access to the digital signaling network would allow greatly enhanced user control of network services, from various transport options to a number of teleservices and supplementary services such as facsimile and call forwarding. Work station screens would be filled in a fraction of a second, and facsimile transmission of a typical page might take 5 seconds rather than 30.

The fundamental architectural postulates of ISDN are:

— separation of information traffic from signaling traffic
— integrated digital access from subscribers to ISDN switches (the different media are not necessarily carried as integrated traffic within the interoffice network)
— well defined addressing and signaling protocols, giving end users the option of "functional" signaling, in which descriptive messages convey instructions, rather than the "stimulus" signaling that subscribers use in the telephone network, in which on hook-off hook and a limited number of push-button actions have different meanings depending on the status of a call.

The CCITT has defined a family of standard subscriber interfaces. Figure 1.8 shows an ISDN with two such interfaces, the "basic" and "primary" interfaces. The "basic" interface, intended for a large population of individual subscribers, offers two 64 kbps full-duplex "B" transport channels, and one 16 kbps "D" signaling or data channel. Non-ISDN terminals will be supported via terminal adapters. Electrical characteristics for the "basic" interface in the United States, largely conforming to CCITT Recommendation I.430 [42], are an American National Standards Institute (ANSI) Standard [43].

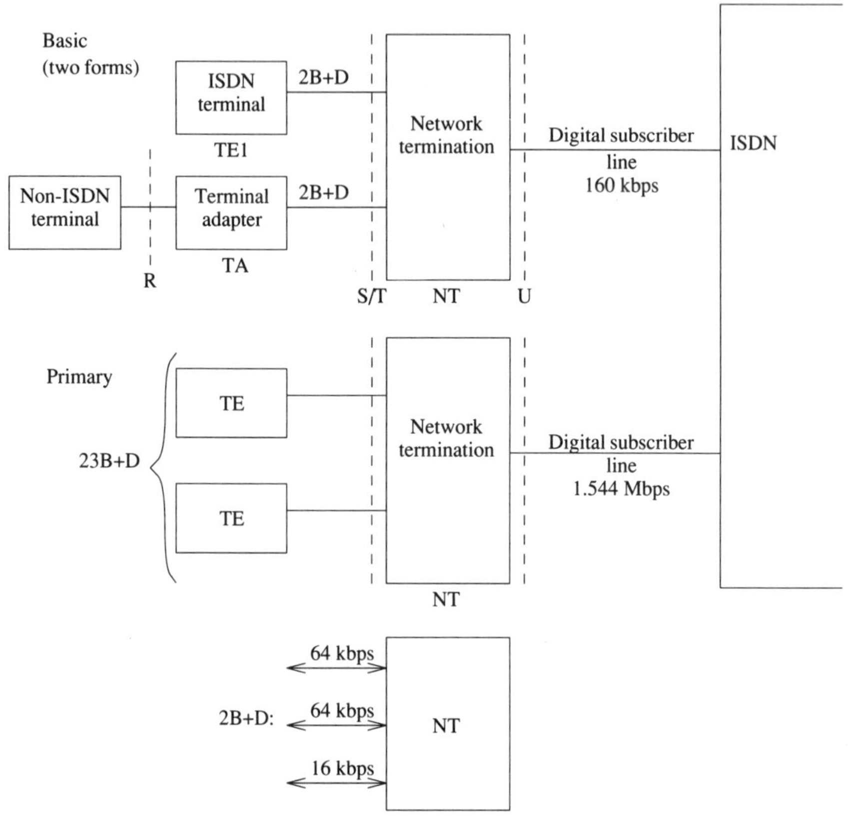

Fig. 1.8 ISDN basic and primary (in North America) subscriber interfaces.

In the application example given above, the 7 kHz voice channel would be accommodated in one of the 64 kbps transport B channels, and the data being shared would be carried in either the other B channel, or in the 16 kbps D channel. Although user-data traffic and signaling can share the D channel, the two data streams are distinguished at the serving office and are logically separate. Signaling traffic need not be restricted to user-network messages; user-to-user signaling is also available as a datagram type of service beyond the usual call processing. Furthermore, data-network transmission modes may be emulated within an ISDN channel with the "frame-relay" option described in the next section.

An ISDN subscriber is linked to the serving office through a digital subscriber line (Section 1.6.2). This line is built from a normal twisted-pair telephone subscriber line with transceivers at the two ends. The transceiver location at the subscriber's end is denoted the "U interface," as shown in Figure 1.8.

With the overhead required for reliable full-duplex transmission, the data rate on the digital subscriber line is 160 kbps full duplex rather than the 144 kbps offered to the subscriber. Chapter 4 briefly describes the U-interface transceiver, emphasizing the line signaling considerations. Other techniques used in the U-interface transceiver are described later in this book, such as channel equalization (Chapter 8) and echo cancellation (Chapter 9).

The "primary" interface is intended more for business offices than individual subscribers. It provides, in the North American digital hierarchy with DS1 facilities operating at 1.544 Mbps, 23 B channels and one D channel. A typical application is connection of a PABX (Private Automatic Branch Exchange, a small private switch) to a serving office.

While ISDN offers many opportunities for enhancements of existing communications applications and realization of new, multimedia applications, it is only a step on the way to an "intelligent, broadband" public telecommunications network [12]. This future Broadband ISDN (BISDN), based on lightwave communications, very-high-speed circuit and packet switching, and pervasive software control, would offer subscribers mixed-traffic access via asynchronous transfer mode transmission (ATM, Section 1.5.2) in multiplexed channels offering users a minimum data rate of about 135 Mbps (Figure 1.9). Higher-rate channels could provide service to the subscriber at about four times this rate and the service could be asymmetric. A great many high-bandwidth applications, including videoconferencing, computer visualization, and high-definition television, will be accommodated. Extensions to ISDN signaling are being made to accommodate the multimedia, multipoint applications of the future BISDN.

1.3.4 FRAME-RELAY NETWORKS [44]

The X.25 suite of protocols has been designed in an era of relatively slow ($\leq$ 64 kbps) transmission facilities and moderately high error rate environment. To take advantage of the higher speed and improved quality of today's (mostly fiber) digital transmission facilities, *frame-relay* networking has been proposed for ISDN to streamline the data transfer phase in virtual circuit networking. Specifically, frame-relay data networking is designed to minimize transit delay and maximize throughput by only performing basic procedures at each network node. In the relay process, incoming LAPD frames are checked only for valid frame-check sequences and address fields; invalid frames are simply discarded, while valid frames are switched ("relayed") toward their destination based on their virtual circuit identity as indicated in the frame address. The remaining protocol procedures and the layer 3 protocol—including, in particular, error control (via retransmission) and flow-control procedures—operate only between end points. In exchange for higher throughput and lower delay, frame-relay networks cannot make use of the network-based congestion control procedures of X.25 networks. Development of efficient congestion control techniques for frame-relay networks is a current research topic of great interest.

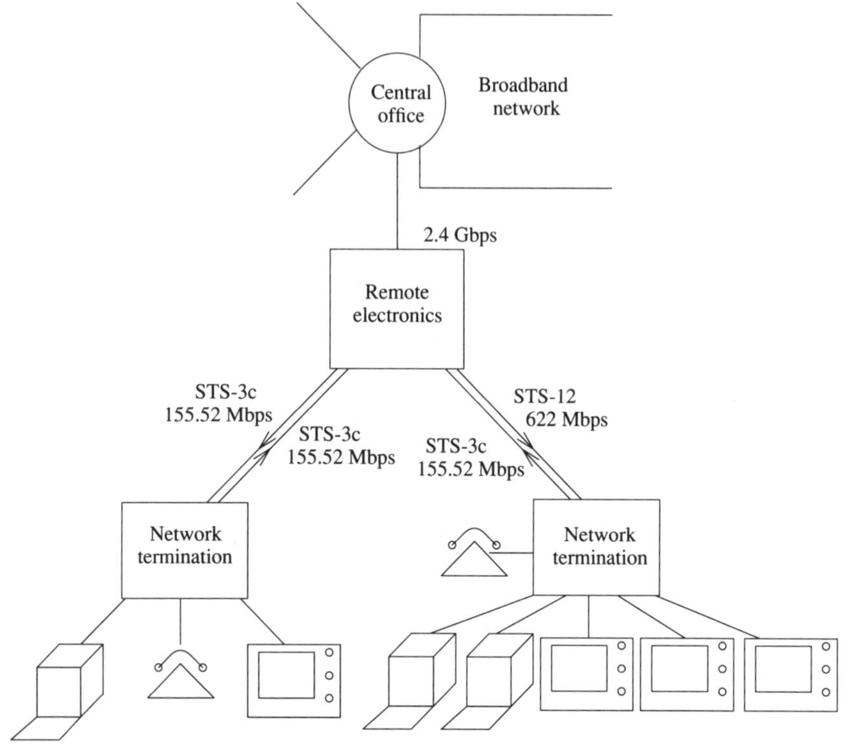

Fig. 1.9 Subscriber access in broadband ISDN, based on STS-3 and higher channel rates and ATM/SONET formats. A usable data rate of about 135 Mbps is provided in the STS-3c channel. The STS-3c channel is used here to transport ATM cells.

1.3.5 APPLICATION-DEDICATED DATA NETWORKS

Unlike the ARPANET and Internet described earlier, there are wide-area private data networks dedicated for specific applications, such as credit authorization or airline reservations. Networks of this kind may operate as random-access, token-passing, or demand multiplexing systems, or have network subsystems of these kinds, but more frequently they are SNA networks that are treelike in topology (Figure 1.5d) and rely on cluster and data concentrators for efficient use of transmission circuits. They originally used conditioned telephone lines even for the backbone links, but today are more likely to use (digital) T1 links for the backbone, with analog telephone line "tails" for accessing multiplexers where lower-rate data streams are combined into higher-rate streams.

Figure 1.10 illustrates a private data network with cluster controllers for control of a group of local terminals, data concentrators (realized here by a local area network) for combining several low-rate data streams into one high-rate

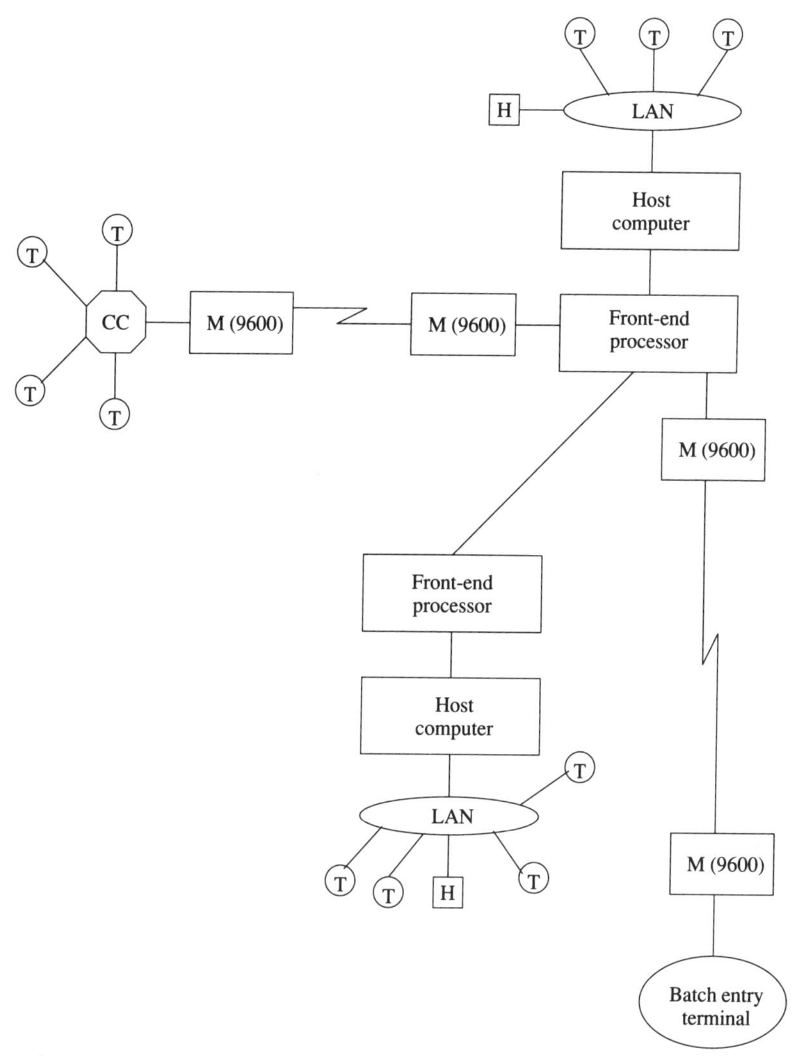

Fig. 1.10 A typical inquiry-type data communication network utilizing cluster controllers concentrators and a front-end processor. Key: T = terminal, CC = cluster controller, M(R) = modem operating at rate R, H = host, M = modem.

stream, and a front-end processor at the host computer site for managing the network, leaving the host computer free for data processing functions. The cluster controller, data concentrator, and front-end processor can each be regarded as a communications processor. The communications processing units are present to (1) maximize efficiency of operation, which in the case of an inquiry-type network usually means minimizing response time, and (2) reduce the total cost for communication links by making the best possible use of each link. When the level of traffic justifies it, the designer will specify high-speed data transmission even on the access links, requiring advanced communications equipment that can get the data through at low error rates despite significant channel impairments. Sophisticated high-speed modems, implementing principles described in this book, are of importance to both shared and application-specific networks.

The polling protocol is commonly used when a large number of terminals are in occasional communication with a central computer. Each terminal is interrogated from time to time by the central processor, possibly but not necessarily, in a regular pattern. *Roll-call* polling is the most direct form. Each station is assigned an address. The central processor broadcasts the addresses in sequence, thereby polling the stations for messages. If a station has an inquiry message to transmit, it responds to the poll with a message. The overhead in polling systems is the time required to poll a terminal, whether or not it has a message.

1.4 DATA COMMUNICATION AT VOICEBAND RATES

With this understanding of the applications and networking implications of data communications, let us now consider the options for data transmission between two points. For the moment, we will concentrate on data transmission at voice-grade channel rates (currently 19.2 kbps and below), since a great deal of the technology described in this book was developed for such channels. It should be noted that 19.2 kbps is approximately 8 bps/Hz (for a nominal 2.4 kHz channel). The techniques for high spectral efficiency are also of great interest for digital radio and the twisted pair ISDN and high-speed digital subscriber line channels introduced later in this chapter.

A user can install privately owned lines, and usually does for a LAN, but for wide-area communication is more likely to lease the lines from a communications carrier. There are a number of facilities; some of them are digital circuits, either private line or one of the packet-switched services cited earlier. A typical accessing arrangement, shown in Figure 1.11, uses a private or dialed analog connection to reach an entrance node of the digital network. A service provider may, for example, arrange for customers to access the services host computer through a public packet-switched network with nodes in many cities. Large users may have nodes of digital networks on their premises, avoiding the problems of access through analog lines.

Alternatively, data can be sent all the way through the analog telephone network. This still leaves a choice between a private line, leased for a fixed monthly fee roughly proportional to distance, and a dialed connection through the switched network (Figure 1.12). The private line, a permanent connection that is not switched, can be purchased in either a four-wire version (a separate channel for each direction) or a two-wire version (signals traveling in both direc-

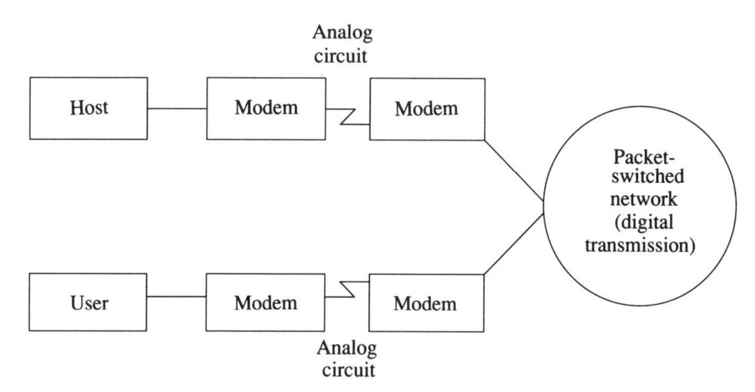

Fig. 1.11 Access to a node of a packet-switched network via a private or dialed analog circuit.

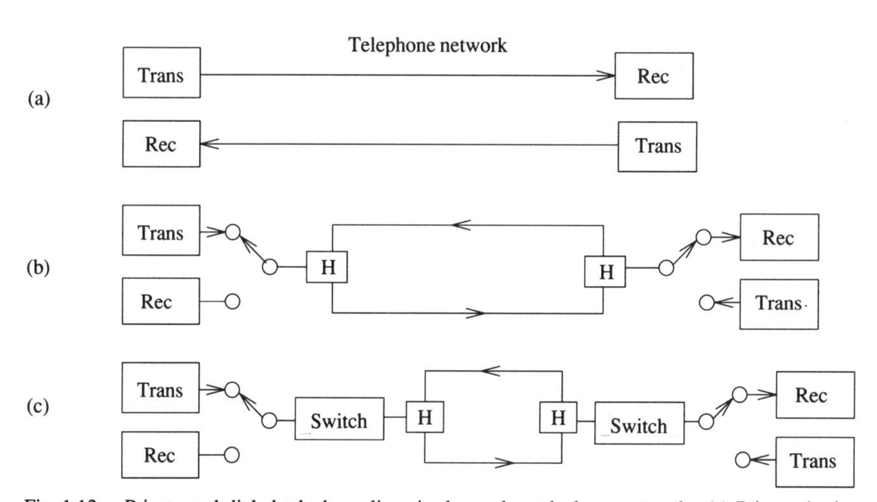

Fig. 1.12 Private and dialed telephone lines in the analog telephone network. (a) Private 4-wire (full duplex). (b) Private 2-wire (half-duplex shown). (c) Dialed 2-wire (half-duplex shown). The "H" boxes are directional couplers, called "hybrid couplers," that provide an interface between 2-wire and 4-wire circuits.

tions on the same pair of wires). The four-wire version is almost always selected, since it allows full-duplex transmission (simultaneous transmission in both directions) without problems of self-interference. The four-wire private line costs very little more than the two-wire private line because long-distance transmission facilities within the telephone plant are already four-wire. The only extra facilities needed for a four-wire private line are additional two-wire subscriber termination circuits.

A dialed line is a two-wire connection. Full-duplex transmission is possible, despite the inherent self-interference problems, by band-splitting, time-compression multiplexing, or echo-cancellation techniques (Chapter 9).

Leasing rates for private lines are high enough so that only high-traffic user locations, or links combining traffic from a number of locations, can justify their use. Banks, credit card companies, airlines, and other large companies often operate private computer communication networks built from leased circuits. The locations with light traffic are left with the dialed network, unless access to a public data network is available, as described earlier. It was, in fact, the limitations of the dialed network for intermittent data traffic, limitations such as transmission impairments, half-duplex operation, long-setup time, and one- or three-minute minimum charging time, that spurred the development of the specialized digital data networks.

There is one compromise analog network solution that meets the needs of many users, and that is the *multipoint line* (Figure 1.13). This is a four-wire private line used in polling applications. By connecting up to 20 terminals to one line, the transmission costs are considerably reduced from what they would be if each terminal had an individual connection to the central processor. Traffic from the master modem is broadcast continuously, while inbound traffic is switched to the tributary modem that is actively responding to the poll.

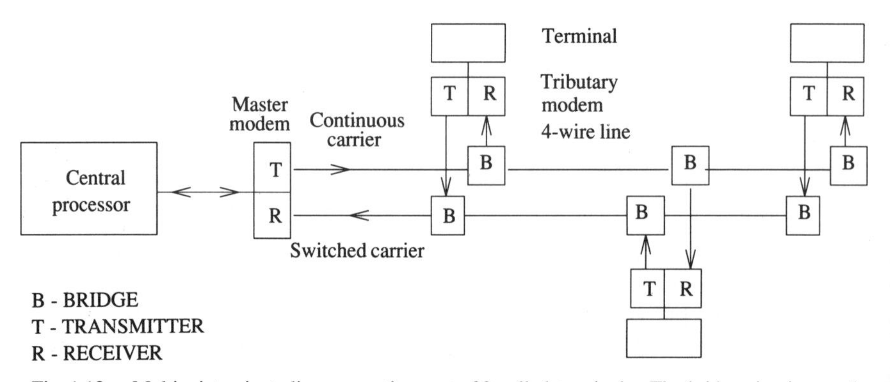

Fig. 1.13 Multipoint private line supporting up to 20 polled terminals. The bridge circuits match impedances. They may, in certain cases, have switched ports or other controlled features.

1.4.1 MODEMS

When data are to be transmitted through an analog circuit, such as one of those illustrated in Figures 1.10–1.13, a *modem* is needed to transform the digital signal produced by a terminal or other data terminating equipment (DTE) into an analog signal that is passed by the analog channel. The discussion of modems takes up most of the book. Modem is a contraction of *mo*dulator–*demo*dulator, and expresses the fundamental purpose of the modem as a device to convert a baseband digital signal into a passband analog signal that can be reliably transmitted over a passband channel. A long-distance telephone channel that does not pass frequencies below about 200 Hz is a passband channel, as are radio and satellite channels.

Many modems do other things as well, including multiplexing of several data streams, conversion of a two-level (binary) digital signal into a multilevel digital signal in order to conserve bandwidth, encoding or scrambling to guarantee signal transitions for timing purposes, and channel equalization to improve the characteristics of the transmission channel. These will be discussed later in this text.

In digital networks, the data channel equipments (DCEs) that terminate the transmission links may not be modems at all because they may not modulate and demodulate signals. But they may well do line signal coding, scrambling, timing recovery, and other functions related to initiation and continuation of data communication. Through incorrect usage, the term modem has become synonymous with DCEs even when transmission is at baseband.

In either case, the working relationship between DTE and DCE, at the physical (level 1) layer of the ISO protocol model, is governed by an interface specification for which a number of standards have been developed by various organizations, most notably the CCITT. The X.21 recommendation partially described earlier, the lowest level of the CCITT X.25 communication protocol, is a widely used definition of this electrical interface.

The characteristics of modems for different operating conditions and data rates are also covered by CCITT recommendations. Table 1.2 lists the major recommendations. These recommendations are respected, if not always followed, and the trend is to greater adherence despite continuing differences between modems at similar rates in different countries. Manufacturers may go beyond the standards in providing additional features.

The telephone voice channel is characterized in Section 1.6.1.

1.5 CARRIER SYSTEMS [46]

The data traffic carried by the telephone network, or on similar facilities, passes through a variety of *carrier systems*. These systems, mostly digital with some analog, largely determine the characteristics of the analog telephone channel, including the impairments which are elements of the analytical models of

Table 1.2: CCITT Recommendations for Modems [45]

Designation	Speed	Signaling format	Application
V.16	(analog)	FM, half duplex, subcarrier frequencies 950 Hz	ECG transmission on dialed circuits
V.19	10 char/sec	Asynchronous, half duplex parallel tones form "A" set (697, 770, 852, 941 Hz) and "B" set (1209, 1336, 1477, 1633 Hz)	Parallel (telephone pushbutton tones) signaling on dialed circuits
V.20	20-40 char/sec	Asynchronous, half duplex parallel tones (As in V.19, but faster transmission and possible third signal set).	As in V.19
V.21	Up to 300 baud (asynchronous)	FSK, full duplex, carrier frequencies in two directions 1080 and 1750 Hz, freq. deviation ± 100 Hz	Dialed network
V.22	1200 bps (optional 600) (synchronous)	Four-phase DPSK, full duplex, carrier frequencies in two directions 1200 and 2400 Hz.	Dialed network and leased circuits
V.22 bis	2400 bps (synchronous)	QAM, 16 signal points, carrier frequencies in two directions 1200 and 2400 Hz. Fixed compromise equalization.	Dialed network and leased circuits
V.23	Up to 2300 baud (plus 75 baud reverse channel) (asynch or synchronous)	FSK, half duplex, carrier 1700 Hz, freq. deviation + 400 Hz.	Dialed network
V.26	2400 bps (plus 75 baud supervisory channel) (synchronous)	Four-phase DCPSK, full duplex, carrier frequency 1800 Hz.	Four-wire leased circuits
V.26 bis	2400 bps plus optional 75 baud reverse channel (1200 bps reduced rate capability) (synchronous)	As V.26, but half duplex.	Dialed network
V.26 ter	2400 bps (1200 bps reduced rate capability)	Four-phase DCPSK, full duplex, compromise or adaptive equalization, echo cancellation to separate directions, carrier frequency 1800 Hz.	Dialed network and leased circuits
V.27	4800 bps (plus 75 baud reverse channel) (synchronous)	Eight-phase DCPSK, full or half duplex, manual equalizer, carrier 1800 Hz.	Four-wire or two-wire leased circuits

(Continued)

Designation	Speed	Signaling format	Application
V.27 bis		As above, with automatic equalizer	
V.27 ter		As above, half duplex	
V.29	9600 bps (synchronous)	QAM, 16 signal points, full duplex, automatic equalization, carrier 1700 Hz, diamond-shaped constellation.	Four-wire leased circuits
V.32	Up to 9600 bps (synchronous)	QAM, full duplex, 32 signal points with channel trellis coding (16 otherwise), echo cancellation, 2400 baud pulsing.	Dialed network and leased circuits
V.32 bis	14,400 bps	As above, 64 signal points	Dialed network
V.33	14,400 bps	AM-PM, 128 signal points.	Four-wire leased circuits

FSK: Frequency-shift keying
DCPSK: Differentially coherent phase-shift keying
QAM: Quadrature amplitude modulation (rectangular array of AM-PM signal points)
AM-PM: Simultaneous amplitude-phase carrier modulation, equivalent to pairwise amplitude modulation on quadrature cosine and sine carrier waveforms.

later chapters. The impairments associated with digital carrier systems arise more from the process of converting incoming analog signals to digital format (and back again) than from digital errors or quantization noise.

The multiplexing hierarchies of the more widely used carrier systems in North America are described in Tables 1.3–1.5 and Figure 1.14. The analog systems use FDM, combining a large number of narrowband channels into one wideband modulation signal by frequency translation of narrowband voice-channel signals to assigned slots in the available frequency spectrum. The L-carrier systems, used since the 1930s, are declining in importance as the digital carrier hierarchy becomes more prevalent.

Table 1.3: Multiplexing Hierarchy of Analog Carrier Systems Used in the North American Telephone Network [45]

Analog	Frequency band	Number of voice channels
Group	60–108 kHz	12
Supergroup	312–552 kHz	60
Mastergroup	60–2788 kHz	600
Jumbogroup	3.1–20.1 MHz	3600

Table 1.4: Digital Multiplexing Hierarchy

Digital	Hierarchy data rate	Equivalent number 64 kbps voice channels
DS0	64 kbps	1
DS1	1.544 Mbps	24
RDT	1.544 Mbps (digital radio)	
DS1C	3.152 Mbps	48 (2 DS1)
DS2	6.312 Mbps	96 (4 DS1)
DS3	44.736 Mbps	672 (28 DS1)
DS3C	89.472 Mbps	1344
DS4	274.176 Mbps	4032
DS5	470 Mbps	

Table 1.5: CCITT Rec. G.702 Digital Multiplexing Hierarchies North America, Europe, and Japan

Level	North America	Europe	Japan
1	1.544 Mbps	2.048 Mbps	1.544 Mbps
2	6.312 Mbps	8.448 Mbps	6.312 Mbps
3	44.736 Mbps	34.368 Mbps	32.064 Mbps
4		139.264 Mbps	97.728 Mbps

The digital hierarchy was designed for an asynchronous-multiplexing scheme, in which no attempt is made to synchronize the tributary data streams that are combined into a higher-rate digital data stream. These tributary streams can have separate clocks. To multiplex them together, the high-speed data stream allows a little extra capacity to accommodate "bit stuffing" [47] to compensate for the slightly varying rates of the tributary streams. Because of these variations, the number of bits stuffed varies from frame to frame in a given multiplexer, and from one multiplexer to another. This solves the multiplexing problem, but at the cost of making it difficult to extract a particular tributary stream from the high-rate stream.

To ease the function of add/drop multiplexing and the transfer of traffic from one high-speed network to another, a synchronous (SONET, Section 1.5.1)

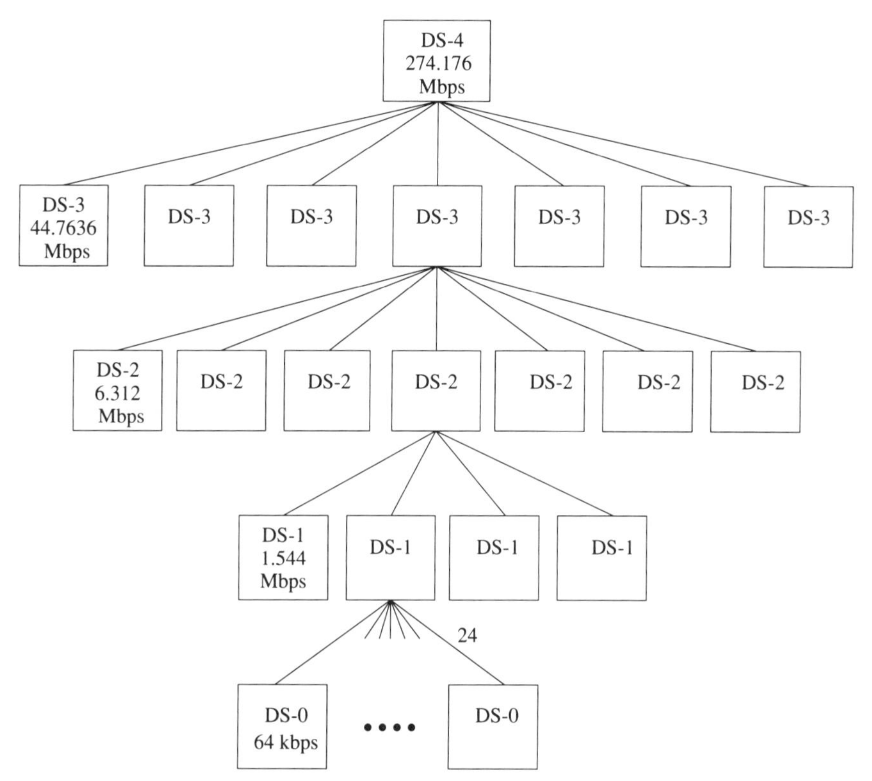

Fig. 1.14 The North American digital hierarchy for the asynchronous digital network (adapted from [67]). Overhead and stuffing bits are added to the sum of tributary data rates.

Table 1.6: The Digital Optical Transmission Hierarchy [48]

OC-1	51.84	Mbps
OC-3	155.52	Mbps
OC-12	622.08	Mbps
OC-24	1244.16	Mbps
OC-48	2.4888	Gbps

multiplexing standard has been promulgated. A new hierarchy of standard digital link transmission rates has been established and is listed in Table 1.6. The optical transmission channels accommodate not only multiplexing (or *framing*) standards such as SONET, but various connection-oriented and connectionless transmission modes. These include the asynchronous transfer mode (ATM, Section 1.5.2), the basis for the future Broadband ISDN. Figure 1.15 illustrates how user data, ATM cells, and SONET frames are related.

DS1, the 1.544 Mbps facility at the bottom of the digital hierarchy, began to appear as a short-haul (less than 300 km) system in the late 1960s. It utilizes two-wire twisted-pair cables with repeaters spaced at roughly 1800 meter intervals. Above the top of the existing hierarchy, and still only a future possibility at the time of writing, is a 470-Mbit DS5 system based on high-performance optical-fiber transmission.

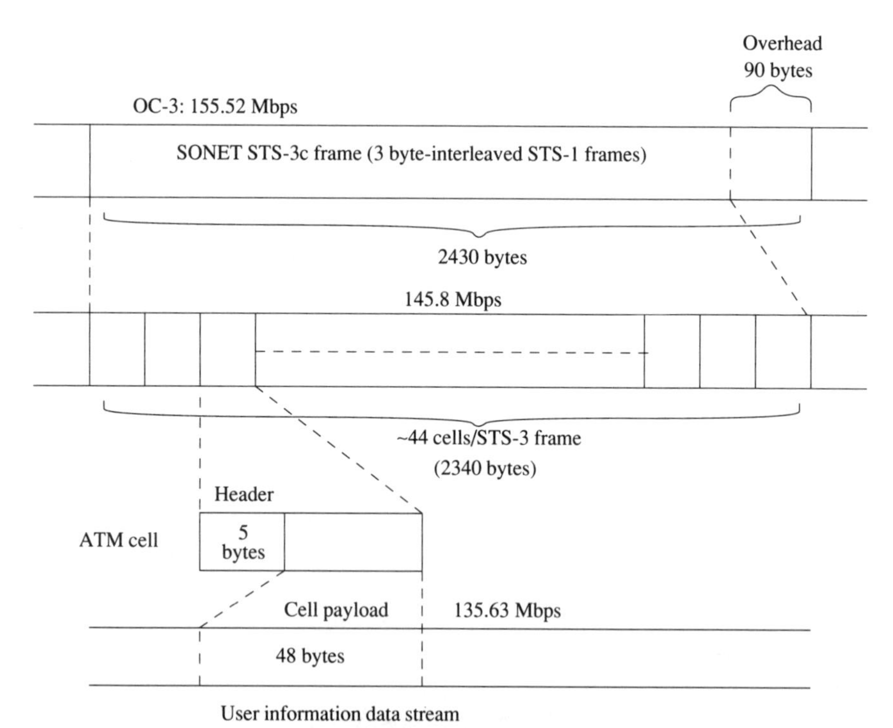

Fig. 1.15 Illustration of encapsulation of user data into cells and cells into SONET frames in the synchronous digital network. In this particular case, the user data rate is limited to 135 Mbps out of the 155.52 Mbps link transmission rate. Note also that an ATM cell may be spread over two SONET frames. See the next section for explanation of the SONET STS-1 and STS-3c frames.

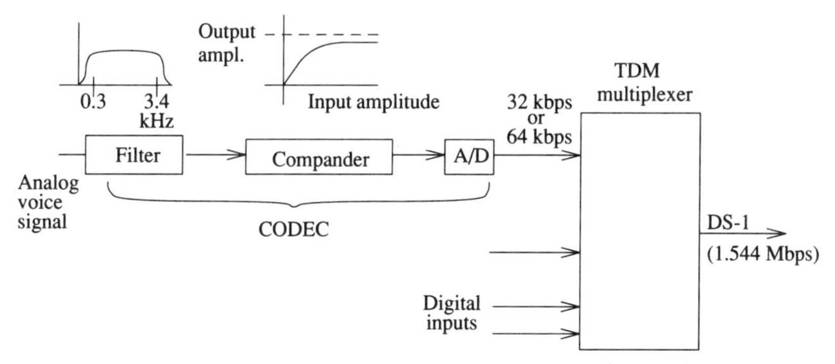

Fig. 1.16 Channel bank at input of a digital carrier system. Some of the digital capacity may be dedicated to digital inputs separate from the digitized analog input channels.

Telephone-network channels are built as tandem connections of local facilities and carrier channels. An end-to-end digital channel will transport bits the whole distance on digital transmission systems, asynchronous or synchronous, while an analog channel may or may not (but probably will) incorporate a digital carrier system somewhere along the way. It is a curious fact that a line signal from a 1200 bps modem on a dialed (analog) telephone line may be PCM-coded into a 64 kbps data stream in a digital carrier system, contributing the longest part of the dialed connection. This waste persists because of the unacceptably high cost of providing digital network access to low-traffic locations, a situation which may exist for some time.*

A digital carrier system fed with analog telephone channels (Figure 1.16) converts each analog signal entering the channel bank into a digital data stream. The *codec* performing this function includes filtering, companding (*com*pression at the sending end of the circuit and exp*anding* at the receiving end), and digital encoding. The compression operation reduces the digital quantization noise for signals with the statistical amplitude distribution characteristic of speech. Some of the distortion in a data signal which has traversed a digital carrier system comes from imperfect filtering and slight mismatches between the compressor in the sending codec and expansion in the receiving codec. For digital encoding, 64 kbps pulse code modulation (PCM) is being gradually replaced by 32 kbps adaptive differential PCM (ADPCM), as described in CCITT Recommendation G.72 [48].

---

* An interim solution would be to recognize and communicate with a plurality of modems at the first digital office encounted by the line signal. If this "generalized modem" at a digital office could be achieved, then the digital transmission system would carry only a 1200 bps data stream from this user (Figure 1.17).

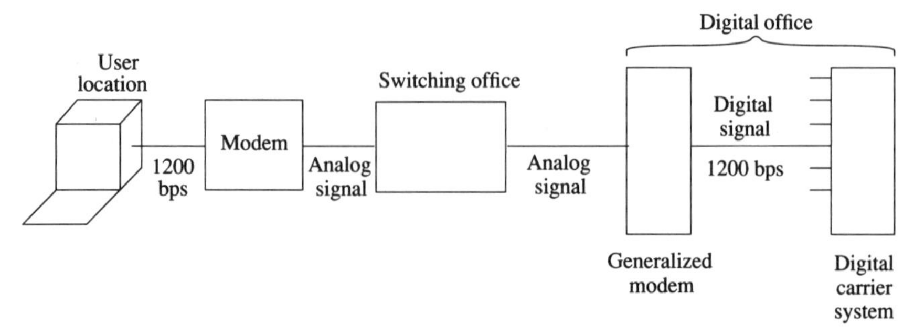

Fig. 1.17 A "generalized modem" at a digital network office converting an analog line signal into a digital stream for further transmission through a digital carrier system.

Carrier systems are found at several levels in the hierarchical network, maintained by the various local exchange and interexchange companies which operate separate parts of the network. This hierarchy of switches is not to be confused with the multiplexing hierarchy described earlier. Most voice-grade telephone channels pass through tandem connections of several carrier systems. Although the end-to-end characteristics of these channels can be measured and described without knowledge of the carrier systems traversed, an understanding of the channel impairments, and of ways to correct or bypass them, is helped by knowledge of basic carrier system features. Some of these are described in Section 1.6.1 in connection with the discussion of specific impairments.

1.5.1 SONET [50]

The deployment of optical fiber in the common carrier network has led to the formulation of a new all-encompassing standard for digital multiplexing, synchronous optical network, SONET. Two basic thrusts are behind SONET. The primary motivation is the increasing data rates that are present in optical networks. The second is that the increasingly international character of telecommunications requires the unification of techniques across national and continental boundaries. This latter point was a severe test of those who formulated the SONET standard since, as shown in Table 1.5, there is significant variation in regional digital multiplexing standards.

The system model and a Synchronous Transport Signal—Level 1 (STS-1) frame are depicted in Figure 1.18. STS-1 is the basic building block and first level of the SONET signal hierarchy. The frame is shown as consisting of 9 rows of 90 bytes (8 bits each), where 3 columns are used for overhead information. The order of transmission of the frame is row by row from left to right. Frames are transmitted at a rate of 8000 frames per second, for a total transmission rate of $9 \times 90 \times 8 \times 8000 = 51.84$ Mbps. With three columns of frame

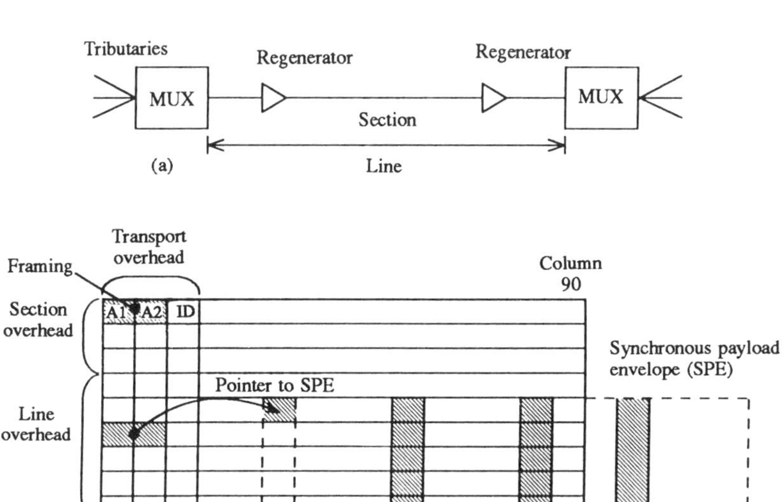

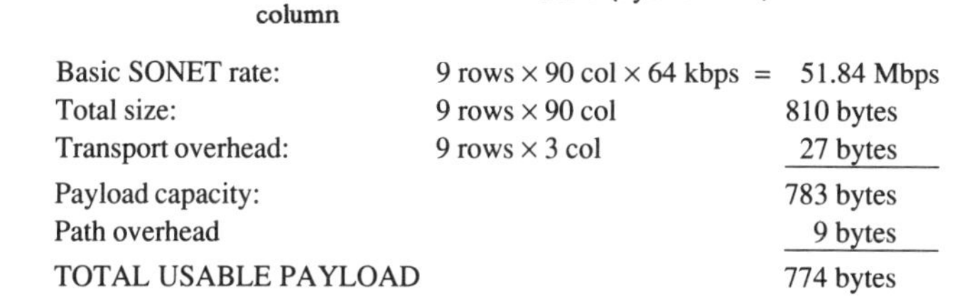

Basic SONET rate:	9 rows × 90 col × 64 kbps =	51.84 Mbps
Total size:	9 rows × 90 col	810 bytes
Transport overhead:	9 rows × 3 col	27 bytes
Payload capacity:		783 bytes
Path overhead		9 bytes
TOTAL USABLE PAYLOAD		774 bytes

Fig. 1.18 SONET STS-1 format. Each square is one byte (8 bits). The rows of 90 bytes are sent sequentially to realize the 810-byte frame. When the Synchronous Payload Envelope is offset, as shown here, it overlaps two frames. Rows wrap around: bytes to the right of the frame are put into the frame to the left of the SPE. For ATM rather than synchronous traffic, the H4 byte points to the beginning of an ATM cell.

overhead and one of path overhead, the payload in terms of user data is 49.536 Mbps. This rate enables the transmission of level three in the North American digital hierarchy, DS-3. Note that a single byte in the Synchronous Payload Envelope (SPE) bears information at a 64 kbps rate.

In the SPE, which may, as shown in Figure 1.18b, overlap two frames (with an address in the frame overhead pointing to where the new SPE begins), either ATM cells or synchronous traffic may be encapsulated. In the first case, a byte in the path overhead points to the beginning of the first new ATM cell. In the second case, overhead bytes indicate which columns are assigned to synchro-

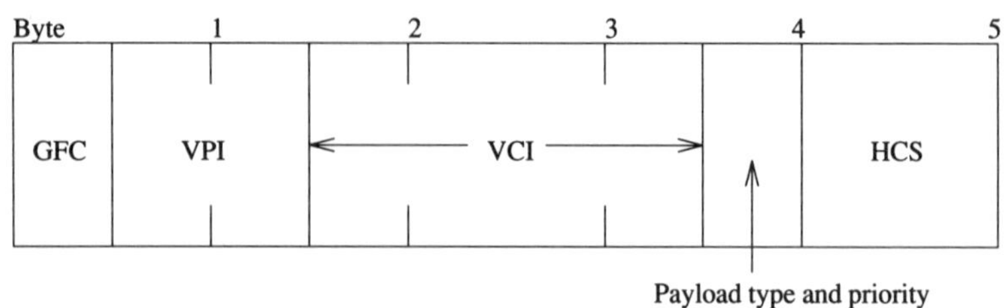

GFC = Generic flow control (4 bits)
VPI = Virtual path identifier (8 bits)
VCI = Virtual circuit identifier (16 bits)
Payload type (3 bits) and priority (1 bit)
HCS = Header-check sequence (8 bits)

Fig. 1.19 The ATM header definition proposed by CCITT Study Group XVIII for the user–network interface. The header for the network–network interface is similar except for elimination of the GFC and expansion of the VPI with these 4 bits.

nous channels. For a DS-1 channel, three columns are assigned. By byte interleaving several STS-1 frames, higher rates may be achieved. The standard rates have been listed earlier in Table 1.6, where OC-N, $N = 1, 3, 12$, and 48, indicates the number of STS-1 frames that are combined. OC-3 (STS-3) and OC-12 (STS-12) are particularly important since they accommodate the two standard access rates for broadband optical networks: 155.52 Mbps and 622 Mbps. The STS-3c frame interleaves three STS-1 frames, as does STS-3, and thus has nine columns of transport overhead, but reduces the three-column path overhead to one. It carries what is effectively one data stream instead of three interleaved data streams.

The overhead bytes carry out the usual functions associated with digital transmission. Framing bytes indicate the start of the STS-1 frame. A significant part of the overhead is allocated for maintenance and network operation. Two bytes are provided for parity checks. The overhead bytes also allow a multiplexing capability that is unique to SONET. The difficulty with multiplexing tributary streams, 64 kbps for example, into the SONET payload is phase misalignment due to small differences in carrier frequencies and phases. If the tributary stream is to occupy the same position in each frame, an elastic store of up to 125 μsec must be provided. However, SONET allows the position of a tributary to drift within certain tolerances by use of *pointing bytes*. These bytes are used to initiate the new position in a frame. The overhead can also be pressed into service to carry a limited amount of data in transition periods.

1.5.2 ATM [51–53]

Optical networks will be required to handle a wide range of traffic types including video, voice, and high-speed data. The need for flexibility in service definition, in particular arbitrary bandwidths and a uniform format for subscriber access, has motivated study and standardization of the asynchronous transfer mode (ATM).

Under ATM all information is repackaged, through the use of adaptation layer protocols, in short, fixed-length packets called cells. A cell consists of 48 bytes of user payload together with 5 bytes of header. The header (Figure 1.19) is concerned with five basic functions. The first is routing. The VCI (virtual channel identification) is a 20-bit label for channel identification, analogous to the assignment of a time slot in a synchronous transmission system. The VCI will typically correspond to an output port on the next switch. Another label, the VPI (virtual path identification) associated, for example, with a server for a group of destinations, may occupy part or all of this field. ATM is associated with connection-oriented (virtual-circuit) transport. A route or virtual circuit is established for a connection between source–destination pairs. All cells generated for the duration of the connection follow this route, with a table at each node in the network defining the next VCI for that route.

A second function is contention resolution in situations where several terminals share a common access medium, done in the GFC field. Four bits are allocated for the third and fourth functions, to identify traffic type and priority. Finally, one byte has been allocated for error detection and correction of the header.

ATM service can be provided at any transmission rate, with BISDN leaning toward use of the OC-3 and OC-12 channels defined earlier (see Table 1.6). There is, at this writing, some debate as to the framing to be used in connection with subscriber access via ATM. It seems most likely that ATM cells will be carried in the user payload envelope in SONET (Figure 1.15). Other alternatives are either no framing or a sequence consisting of a fixed number of information-bearing cells accompanied by a framing cell.

Above this ATM-level protocol is the ATM Adaptation Layer (AAL), which defines uses of part of the 48-byte information field for user needs such as message identification, management of cell loss, and timing control. Five types have been proposed as standards, as illustrated in Figure 1.20. Type 5, also known as Simple and Efficient Adaptation Layer (SEAL), reduces overhead by eliminating the message identifier in each cell.

The advance that ATM represents is in handling a broad speed range of relatively constant-rate traffic such as voice and video by means of a high-speed packet switching mechanism. Such traffic has been handled by circuit switching because of the regularity of the flow and stringent delay requirements. Previously, packet switching had been applied mainly to bursty traffic such as data. The presumption for ATM is that advances in switching technology will enable ATM to emulate the performance of circuit switching while retaining the flexibility of packet switching [54], especially in accommodating new services at arbitrary rates.

1.6 CHANNEL CHARACTERIZATIONS

1.6.1 THE TELEPHONE CHANNEL

Definitive surveys of telephone channel characteristics, including impairments, were carried out in the 1970s [55–57]. Analog channels used for data

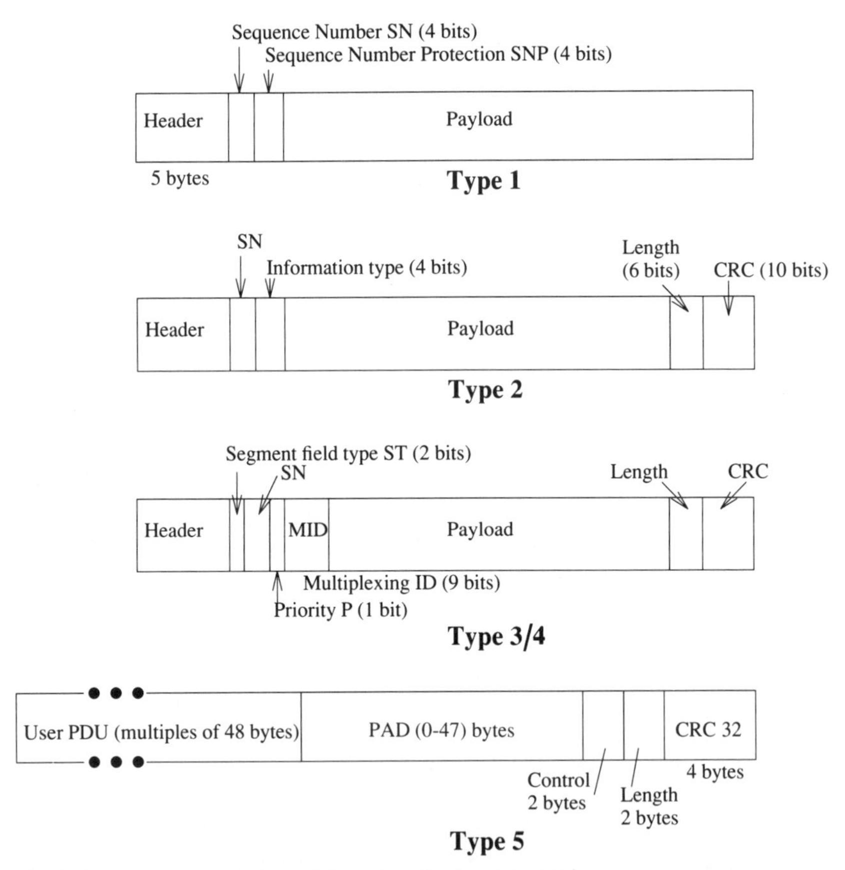

Fig. 1.20 The proposed standard ATM Adaptation Layer protocols.

Type 1: For constant bit rate services. Segmentation and reassembly of user information, cell loss or delay procedures, timing recovery.

Type 2: For variable bit rate video and audio.

Type 3: For connection-oriented data service and signalling. Segmentation and reassembly of variable-length user data sequences.

Type 5: "Simple and efficient" ATM adaptation (SEAL) for bulk data transfer.

transmission are defined by their limitations and impairments. The most significant impairments in telephone voice channels are linear distortion, Gaussian and impulse noise, phase jitter, phase and amplitude "hits," frequency offset, nonlinear distortion, and crosstalk, all defined in the next several pages.

For telephone channels, the channel banks of carrier systems introduce linear and nonlinear distortions while bandlimiting, compressing, and multiplexing the analog signals in preparation for carrier transmission, and the analog signals are perturbed by additive noises of electronic and atmospheric origin. In addition, analog signals traversing radio links are subject to the selective fading (described later) that is characteristic of radio carrier systems. In practice, radio carrier systems incorporate protective switching to replace a faded channel with an operating spare. The switching to the spare channel (and possibly back again) is often perceived in voiceband data transmission as a phase *jump* (a step change in the phase of a transmitted carrier waveform) or a phase *hit* (a rapid forward and reverse phase change). These perturbations naturally cause difficulties in coherent-phase data transmission. They are not apparent in speech since the human ear is not sensitive to phase distortion.

Linear distortion in telephone channels is an impairment arising from the practical limitations of codec filters and transmission lines. Ideally, a channel should have a transfer function which is uniform over all frequencies in amplitude and linear in phase (corresponding to a uniform transmission delay). Figure 1.21 illustrates the linear characteristics of a realistic telephone channel.

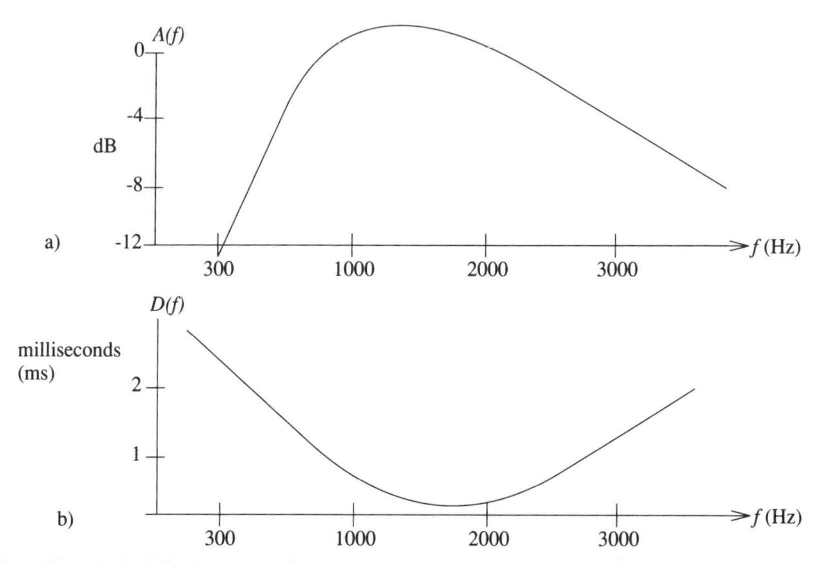

Fig. 1.21 Typical linear characteristics of a telephone channel. (a) Amplitude vs. frequency. (b) Envelope delay distortion vs. frequency.

The transfer function is ordinarily defined in terms of amplitude and relative envelope delay distortion rather than amplitude and phase. *Delay distortion*, the negative derivative of the phase function, is the propagation delay of the envelope of a narrowband signal centered on carrier frequency f within the voice channel. Relative envelope delay distortion is the difference between this delay and that at the frequency exhibiting the least delay.

For most channels, when the signal-to-noise ratio (SNR) is high, linear distortion can be largely compensated by a linear filter. The invention of *adaptive channel equalizers* by R. W. Lucky to perform this function was one of the great achievements of communication engineering, and innovations continue to be made in these complicated structures. More is said about equalization in Section 1.7, and Chapter 8 considers adaptive equalization in considerable detail.

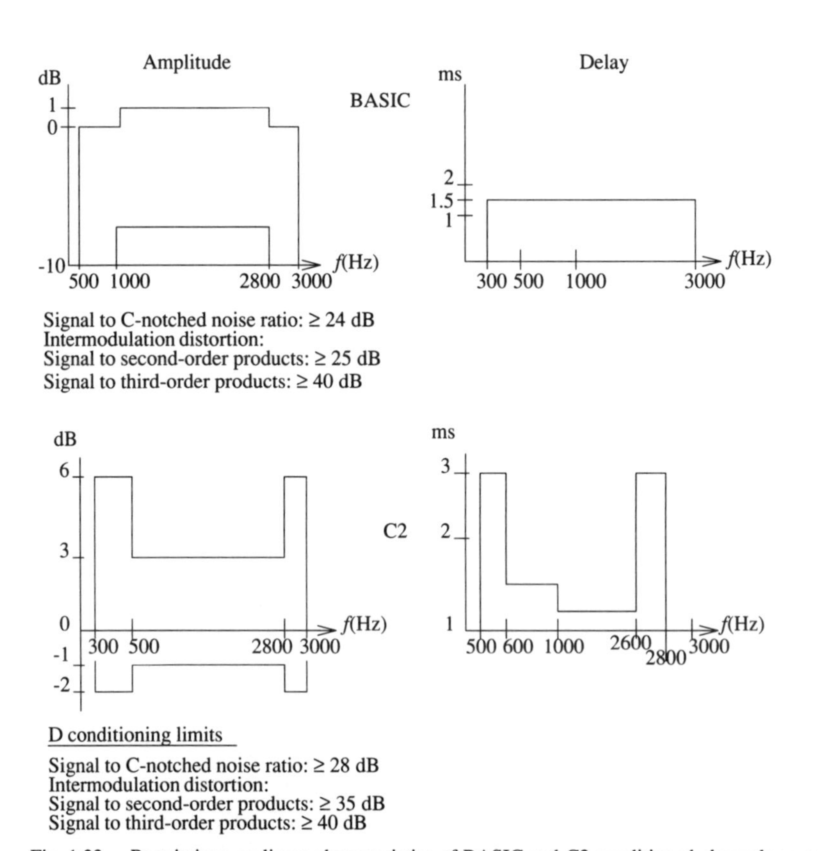

Fig. 1.22 Restrictions on linear characteristics of BASIC and C2 conditioned channels as a function of frequency. The linear characteristics must lie within the "amplitude" and "delay" boxes. D conditioning applies more stringent restrictions on SNR and intermodulation distortions [58].

A severely distorted channel generally cannot be corrected without a cost in signal-to-noise ratio. Common carriers offer private lines in which impairments are held below certain defined and acceptable levels. Channels meeting these specifications are said to be *conditioned.* In the telephone network, the fundamental specification on linear distortion is given by a pair of templates illustrated for BASIC (met by the overwhelming bulk of dialed lines), C2, and D1 conditioning in Figure 1.22. The linear characteristics must lie within these templates, but are not restricted outside of the boxes. Although it would seem to always be advantageous to have a conditioning with less variation in amplitude and envelope delay distortion, the filters used to improve a channel may introduce passband ripple which is more difficult for a modem to compensate than the original distortion.

Modulation techniques are often compared on the basis of performance in Gaussian noise, even though there are many other impairments and the Gaussian assumption is only an approximation. These studies usually give a correct measure of relative performance even if the absolute measures are somewhat inaccurate.

Gaussian noise, which is invariably present on almost all communication channels, is the sum of many small interferences, thermal noises, power-supply ripples, radio noises, and other contributions from natural and man-made sources. It is usually (but not always) at a rather low level, but the important criterion is the ratio of the data signal at the input of a modem to the noise present at that point. For the telephone voice channel, this ratio is usually high enough to correspond to a maximum error-free information transfer rate (the channel capacity, explained in Chapter 2) of 28 kbps, but it has proven difficult to transmit data at rates above 19.2 kbps with low error probability, indicating that the noise and (correctable) linear distortion are not the only significant channel impairments. Nevertheless, it is important, in the design of data receivers, to avoid or minimize noise enhancement in the process of compensating other impairments.* Impulse noise may also be present, but for telephone voice channels is usually dispersed enough to be included in the Gaussian noise without serious error. A fundamental principle of communication theory, developed in Chapter 2, is that signals, or codes, which have dependencies among successive transmitted signals have the potential for providing the most immunity to Gaussian noise. Modern bandlimited communication systems have been revolutionized by codes which combine the coding and modulation process. This technique, called *trellis-coded modulation* is discussed in Chapter 5. Such codes have also been found to provide increased immunity to other impairments.

Phase jitter and frequency offset (or phase roll) arise from imperfect oscillators used for frequency translation in FDM channel banks. Consider, in Fig-

---

* Noise enhancement occurs when a linear compensating filter is used to invert the channel response at all frequencies. At those frequencies where there is a severe attenuation, inversion will enhance the noise.

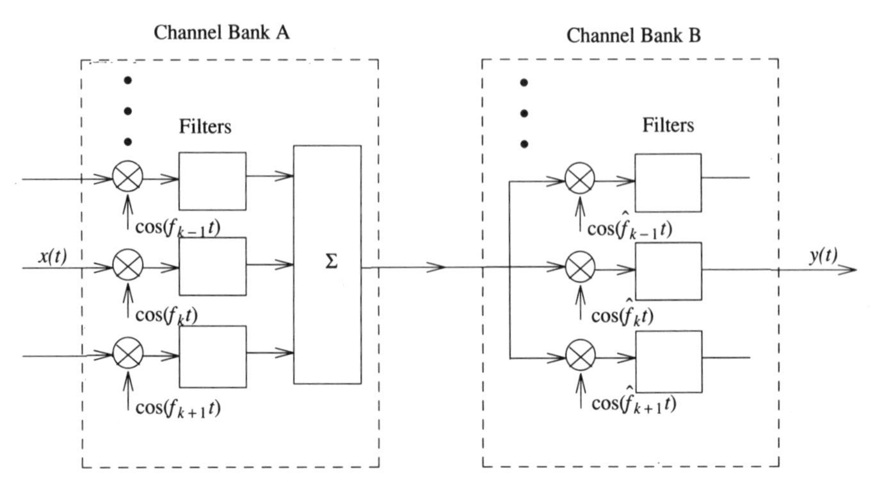

Fig. 1.23 Up- and down-modulation operations in FDM channel banks. The down-modulation reference frequencies $\{\hat{f}_k\}$ are ideally equal to the up-modulation frequencies $\{f_k\}$.

ure 1.23, the modulation and demodulation operations in channel banks A and B, respectively. If a modulated voice channel signal

$$x(t) = m(t) \cos(2\pi f_c t), \tag{1.1}$$

where f_c denotes the carrier frequency, is translated up by $f_k + \Delta$ Hz and down by f_k Hz, the result will be

$$y(t) = m(t) \cos(2\pi f_c + 2\pi\Delta)t, \tag{1.2}$$

containing a frequency offset of Δ Hz or (equivalently) a linearly increasing phase. If the oscillator in channel bank B produces a phase-noisy signal $\cos[2\pi f_k t + \theta(t)]$ instead of $\cos(2\pi f_k t)$, where $\theta(t)$ is a random perturbation called phase jitter, the output signal will be

$$y(t) = m(t)\cos[(2\pi f_c + 2\pi\Delta)t - \theta(t)] \tag{1.3}$$

containing both frequency offset and phase jitter. This is no problem for voice transmission, since the human ear is insensitive to phase, but can be very damaging to data signals.

Frequency offset above 1 Hz, which is the lowest level likely to cause serious trouble in (four-wire) data communication, has become rare* in modern telephone networks but phase jitter is still cause for concern in coherent modulation systems. Figure 1.24 suggests the typical amplitude and spectral content of severe, but not rare, phase jitter. Phase jitter having an amplitude exceeding 10°

* For echo cancellation based on two-wire systems, round-trip frequency offsets exceeding 0.01 Hz can cause severe degradation.

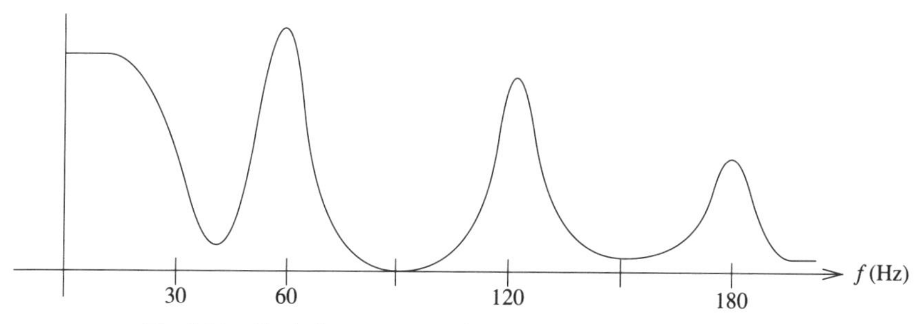

Fig. 1.24 Typical spectrum (Fourier transform) of phase jitter of (1.3).

peak-to-peak is highly unusual in the telephone network. It is apparent from the concentrations at the power-line frequency of 60 Hz (in the U.S.) and its harmonics that power-line and power-supply interferences are present in carrier system channel banks. Some of the lower-frequency components of phase jitter can be traced to the 20 Hz telephone ringing frequency.

Nonlinear distortion is ordinarily modeled as a concatenation of a memoryless nonlinear transformation with a linear transformation of the signal at one or more points within the transmission channel. All electronic equipment is to some extent nonlinear, but severe nonlinear distortion can sometimes be traced to compander mismatches in carrier systems using 64 kbps nonlinear pulse-code modulation, or to distortions in the 32 kbps adaptive differential pulse-code modulation being introduced to increase the number of voice channels without building more transmission facilities.

Figure 1.25 illustrates a typical, but by no means exclusive, model for a coherent nonlinear distortion that occurs in a telephone voice channel—a concatenation of a linear distortion, a memoryless nonlinearity, and a second linear distortion. Empirical studies have determined that second- and third-order harmonic distortions are the most damaging for data transmission. If the input to the nonlinearity is the single sinusoid $\cos f_c t$ and the output (neglecting all harmonics beyond the third) is

$$y(t) = \alpha_1 \cos(2\pi f_c t) + \alpha_2 \cos(4\pi f_c t) + \alpha_3 \cos(6\pi f_c t) \quad , \tag{1.4}$$

then the ratios α_2/α_1 and α_3/α_1 are, respectively, second- and third-order harmonic distortions. This is a simplified model. The test signal used in practice [57] is a collection of sinusoids, usually four, and only those harmonic components falling within the nominal transmission band are included in the distortion.

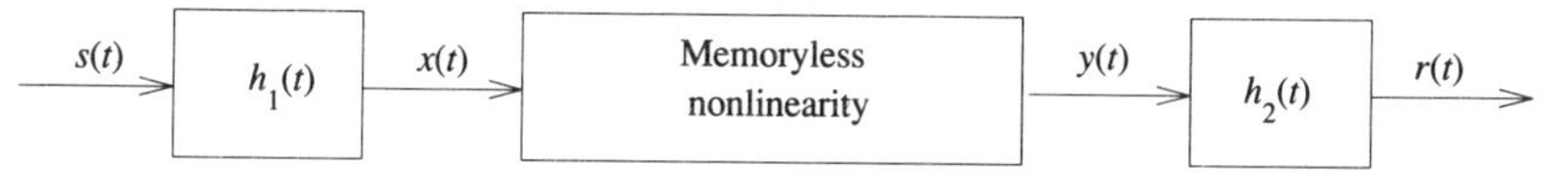

Fig. 1.25 Memoryless nonlinearity imbedded in an otherwise linear transmission channel.

Nonlinear distortions are usually expressed in decibels relative to the fundamental frequency. The second- and third-order harmonic distortions are generally below –27 and –32 dB, respectively. Although telephone channel data transmission at lower rates is rarely degraded by nonlinear distortions, transmission at 9600 bps and above is significantly affected. Furthermore, the transmission, together with an information signal, of one or more pilot tones, which is sometimes desirable for synchronization purposes, can result in interference from harmonics of the pilot tones.

The compensation of nonlinear distortions interspersed with linear distortions is very difficult. In telephone channel transmission, the difficulty can be slightly mitigated by appropriate choice of signaling format, but is usually avoided either by appropriate line conditioning or by keeping the transmission rate below the level at which nonlinearities become significant impairments. Adaptive receiver structures can be derived which compensate for nonlinear distortion, but their complexity has precluded widespread application. Recently, however, progress in feedforward cancellation of nonlinear distortion has been made [59], and practical implementations may be forthcoming. These, and other equalization structures, are treated in Chapter 8.

Crosstalk [46] is the phenomenon of interference from signals in other channels which are supposed to be isolated from the affected channel. For the telephone and assuming proper and economically feasible physical precautions have been taken, the residual crosstalk is usually regarded as a component of uncorrelated additive noise. In circumstances where the interfering channel is accessible, adaptive noise cancellation techniques can be applied to reduce the level of crosstalk interference. For voice-grade telephone lines, crosstalk is generally regarded as part of the background noise, but this is not the case for the digital subscriber line described in the next section.

1.6.2 THE TWISTED-PAIR DIGITAL SUBSCRIBER LINE [60, 62]

The characteristics of the telephone channel impose a theoretical channel capacity of about 28 kbps. In contrast, a digital network should offer a subscriber direct digital access, accepting rectangular baseband pulses at rates commonly produced by present and future terminals and computers, for which rates from 64 kbps to tens of megabits per second will soon be common.

Fundamentally, however, *all* transmission links are analog. Whether or not the digital network operator places a network node on a subscriber's premises, the operator has to design a data communication system for the links of the network. In practice, the network node is usually off subscriber premises, and the subscriber is left with a responsibility for data communication between his or her location and the network node. For example, nodes of the integrated service digital network (ISDN) are located in telephone offices, and "Basic" access at 144 kbps full duplex from a subscriber location is via a twisted-pair loop, the same access line used for plain old telephone service. The data communication

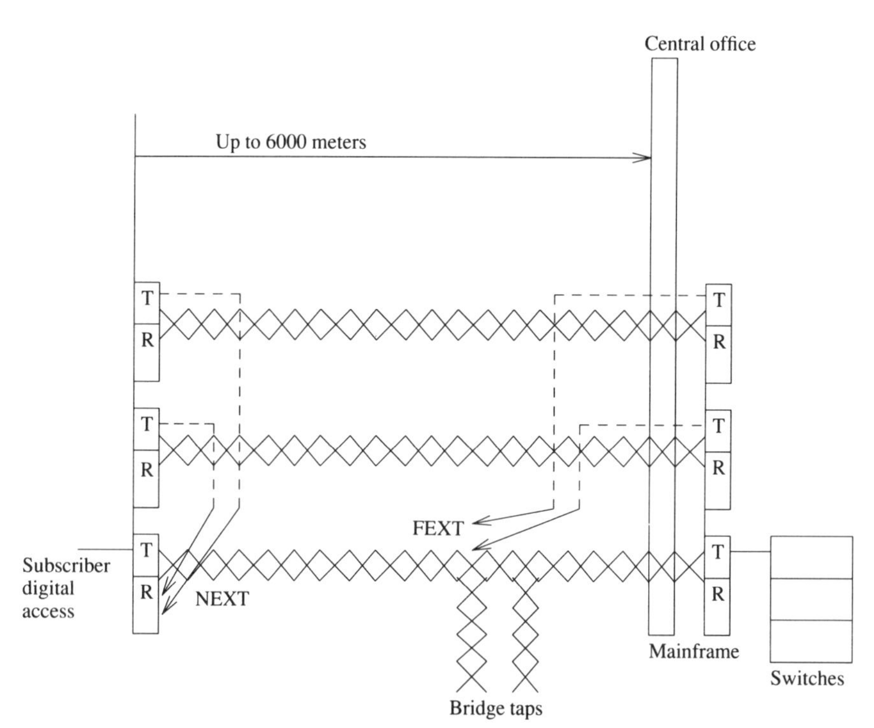

Fig. 1.26 Twisted-pair digital subscriber line. Lines run in bundles, leading to crosstalk interference. From the subscriber's point of view, near-end crosstalk (NEXT) derives from transmitters near the subscriber premises, and far-end crosstalk (FEXT) from transmitters at the central office.

system devised for this link is as sophisticated as that used for 4800 bps full-duplex data communication over a regular bandlimited telephone channel. The parameters of the channel, and the implementation of the data communication system, are significantly but not entirely different, and proven techniques of pulse design, channel equalization, echo cancellation, and channel trellis coding have been successfully transferred from modems to the transceivers used for the twisted-pair digital line.

The twisted-pair subscriber line (Figure 1.26) is a completely passive pair of (mostly) 24- to 26-gauge copper wires, twisted together to minimize noise and running a distance of up to 18 kilofeet (for unloaded circuits) from a subscriber residence to a telephone central office, where it terminates on a mainframe. When used as a digital subscriber line, full-duplex data transceivers are placed at both ends.

Ideally, twisted pairs may be modeled as a uniform transmission line [61]; however serious impairments come from the deviations of the twisted-pair line from the ideal model. The lines often contain impedance discontinuities from

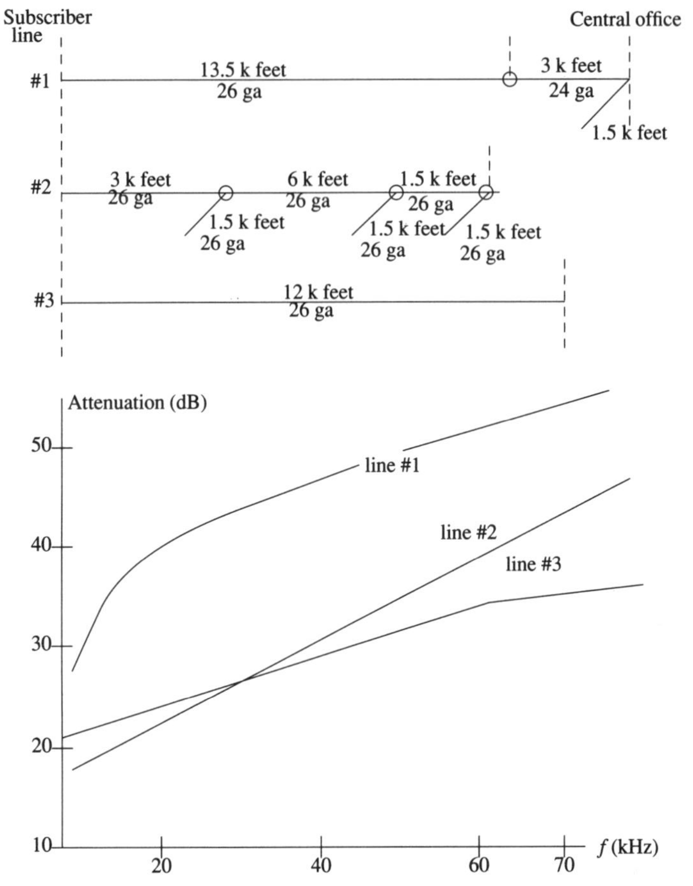

Fig. 1.27 Structure and loss characteristics of three subscriber lines from among the longest 10% subscriber lines in the telephone plant. (From Lechleider [60]).

splicing of different gauge cables and especially from the presence of *bridge taps*, attachments of open-ended cable "spurs" made for a variety of past needs, such as prior routings to other locations. Bridge taps contribute much of the loss on "bad" subscriber lines (Figure 1.27), and near-end crosstalk (NEXT) from nearby transmitters on physically adjacent circuits contributes a high level of interference (Figure 1.28). Frequency-dependent loss leads to intersymbol interference, which must be corrected by channel equalization (Chapters 7, 8), and a high level of frequency-dependent crosstalk reduces the noise margin and forces the use of lower-frequency signaling (Chapter 4) than might otherwise be desirable.

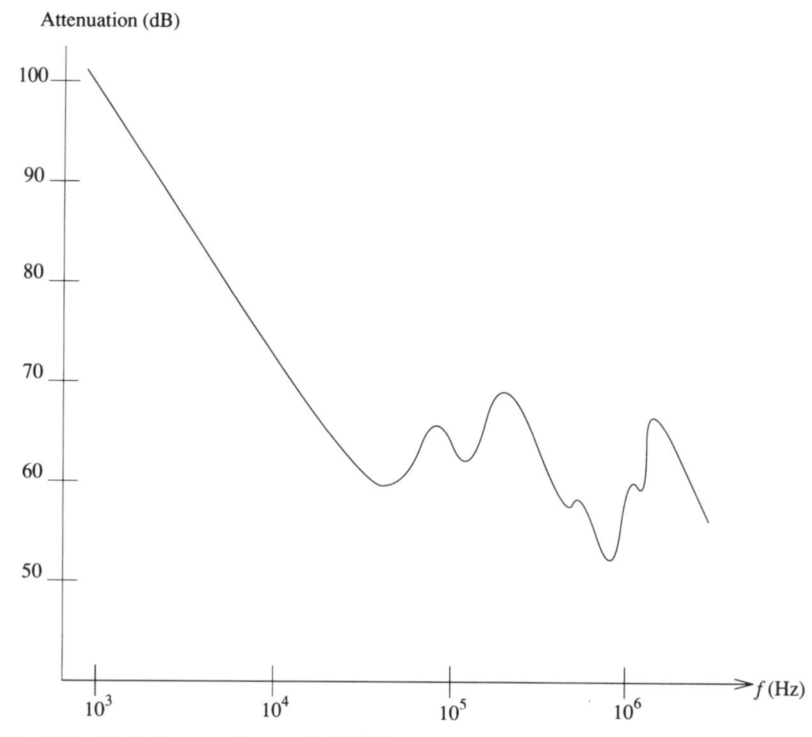

Fig. 1.28 Typical near-end crosstalk (NEXT) loss vs. frequency for a twisted-pair line picking up energy from one other line in the same bundle. The loss is the signal attenuation from one line to another, and does not imply any particular interference level with respect to the desired signal in the recipient line. (Lechleider [60]).

The NEXT power transfer function (from one wire pair to another) is reasonably well modeled* as [62]

$$\left| H_{NEXT}(f) \right|^2 = \beta f^{3/2} . \tag{1.5}$$

In practice, NEXT is much more significant than far-end crosstalk (FEXT) as an interference, and is almost always more significant than random noise. The echos returning from impedance discontinuities, especially hybrid leakage and bridge taps, may be almost as large as the local transmitted signal, which could be as high as 40 dB above the desired distant signal. The echo dispersion is moderate, typically 1–2 ms, and echo cancelers have been designed for digital subscriber line applications (Chapter 9), making it possible to operate full duplex in the same relatively low-frequency band. Ordinary additive noise, from all sources other than crosstalk and echo, is sufficiently below the received

* The model is inaccurate in a mid-frequency region where the cable attenuation flattens.

signal level not to be an important impairment. Very-low-frequency noise is eliminated by highpass filtering (above about 1 kHz).

The unloaded twisted-pair subscriber line has considerable bandwidth. Full-duplex data transmission at rates of 160 kbps, enough to accommodate the ISDN Basic-rate subscriber interface of 144 kbps plus overhead, is possible for reliable transmission over the subscriber line. Two alternative techniques have been considered for use in commercial implementations: time-compression multiplexing (TCM) and echo cancellation. The first technique uses the channel one way at a time, storing data in buffers and releasing them in high-speed transmission bursts exceeding twice the user data rate. It is simple, but has a large bandwidth requirement, exacerbated by the loss of valuable transmission time during the "turn-around" periods between bursts. The echo-cancellation technique permits continual transmission in both directions in the same bandwidth. The self-interference from the locally-generated signal, including echoes from the bridge taps, is canceled by adjusting (or adapting) a filter to emulate the leakage path through the hybrid. The echo-cancellation technique is treated in Chapter 9. Depending on line length and whether one-way or two-way transmission is required, transmission at 1.544 Mbps (and above) is feasible.

1.6.3 DATA TRANSMISSION OVER DIGITAL RADIO CHANNELS [65–69]

Radio carrier systems are the systems, now largely digital, that carry aggregated traffic of hundreds of Mbps or even Gbps and whose parabolic antennas are a common sight on radio relay towers. Their component voice channels are used in telephone circuits and their fading characteristics affect telephone channel characteristics, although to a large extent the portions of telephone channels contributed by digital radio are simple digital pipes. But to those who work with digital radio systems, the engineering design of the radio link between relay points is a complex engineering problem comparable to the design of a high-speed voiceband modem.

Digital-carrier systems use many of the same modulation techniques as telephone channels. These modulation techniques are introduced in the next section. In particular, high spectral efficiency, up to 8 bps/Hz in current systems, is obtained through use of quadrature modulation techniques with hundreds of two-dimensional signal points. This is made possible by the high signal-to-noise ratio (SNR), typically of the order of 30 dB. The tradeoff between SNR and spectral efficiency is a constant theme of data communications and is discussed at some length in Chapter 4.

The drive to higher spectral efficiency was a consequence of the need to squeeze more voice circuits out of the available microwave spectrum. In the United States, the Federal Communications Commission ordered, in late 1974, that radio carrier systems operating in the 4, 6, and 11 GHz bands, have minimum capacities of 1152 voice circuits per channel [63,64]. Channel

Table 1.7: Candidate Modulations for Digital Radio Systems in Europe and North America (Adapted from [6]). The theoretical maximum spectral efficiency, in bps/Hz, is given in parentheses.

Bit Rate	Channel Bandwidth		
	20 MHz	30 MHz	40 MHz
34 Mbps	4-PSK (2)	4-PSK (2)	4-PSK (2)
68 Mbps	16-QAM (4)	8-PSK (3)	4-PSK (2)
140 Mbps	256-QAM (8)	64-QAM (6)	16-QAM (4)
280 Mbps	-	1024-QAM (10)	256-QAM (8)
90 Mbps	64-QAM (6)	16-QAM (4)	8-PSK (3)
135 Mbps	256-QAM (8)	64-QAM (6)	16-QAM (4)
180 Mbps	1024-QAM (10)	256-QAM (8)	64-QAM (6)
270 Mbps	-	1024-QAM (10)	256-QAM (8)

bandwidths were specified as 20, 29.65, and 40 MHz in the 4, 6, and 11 GHz bands respectively.

For consistency with the digital hierarchy (Table 1.4 of Section 1.5), the minimum number of 64 kbps voice circuits must in fact be 1344, corresponding to about 90 Mbps transmission rate. A spectral efficiency of 4.5 bps is therefore demanded in the 4 GHz band. Table 1.7 indicates the modulation formats considered for radio carrier systems in Europe and the United States. The theoretical maximum spectral efficiency is not achievable in practice, where real-pulse filters have an "excess bandwidth" beyond the Nyquist minimum bandwidth (Chapter 4).

The microwave channels used for digital-radio data communications exhibit fading from multipath propagation and distortion from nonlinear power amplifiers in the transmitters [66,67]. If the channel is broad enough so that different frequency components fade independently, the fading is said to be selective, as described further below.

Digital radio systems are line of sight, with relay stations on buildings and hills typically separated by about 40 km. A long-haul radio may have dozens of tandem radio links. With pulse regeneration at relay stations and very low per-link error rates outside of fading intervals, overall performance is excellent. But severe degradation of performance, including a complete outage, can occur with *multipath fading*, the result of signal propagation along several near-parallel paths. The signal fading has a non-frequency-dependent component, which can be largely compensated by an automatic gain control, and a frequency-dependent component that has notches as deep as 40 dB and will cause outages unless compensated by equalization and diversity techniques.

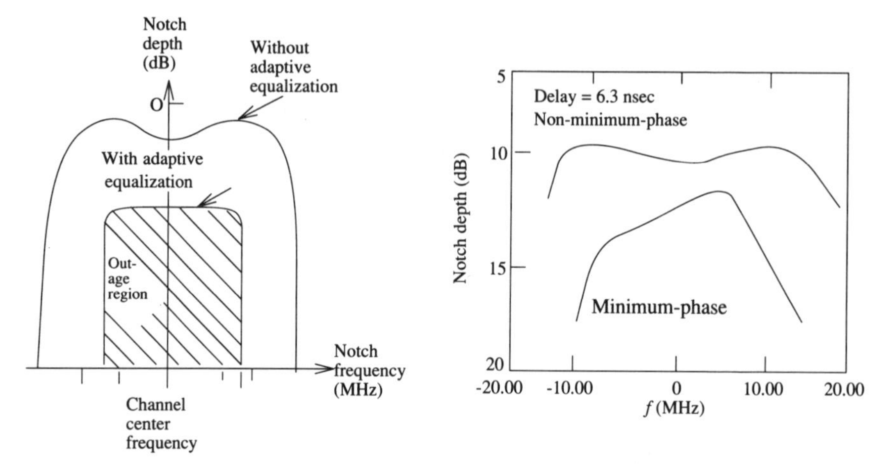

Fig. 1.29 Signature curves showing frequency-selective sensitivity of digital radio systems to fading. (a) General shape, with and without equalization (from [66]). (b) Examples for 16 QAM at 10^{-3} error rate, with minimum phase (occurring for relative amplitude $b < 1$) and non-minimum-phase responses (from [67]).

The sensitivity of a digital radio system to frequency-selective fading is characterized by a *signature curve* such as those shown in Figure 1.29. For specified modulation, equalization, and diversity parameters, a curve can be drawn showing the notch depth vs. frequency that can be sustained without causing an outage. An outage may be characterized by a high bit error rate, such as 10^{-3}. Without adaptive equalization, the curve typically has an "M" shape, showing more susceptibility to outage at frequencies closer to the carrier and at the band edges. With adaptive equalization, the frequency sensitivity is reduced

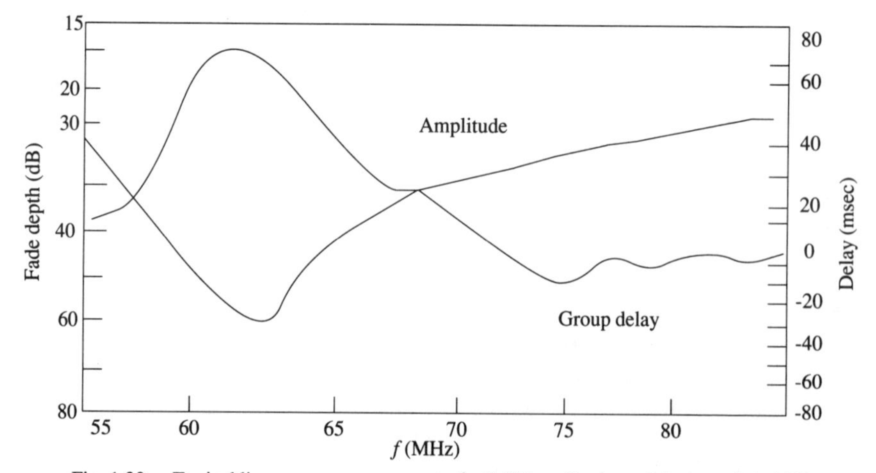

Fig. 1.30 Typical linear response over part of a 6-GHz radio channel during a fade [67].

(the curve flattened) and a much larger notch depth can be sustained without outage. Figure 1.30 shows part of a linear channel characteristic during a fade. The simplified three-path model

$$H(f) = a[1 - b \exp\{-j2\pi(f - f_o)\tau\}] \tag{1.6}$$

offers a good fit to most measured responses of (relatively) narrowband radio channels [68]. Models such as this one (and alternative two-path and polynomial formulations) help to both characterize fading channels and design compensating equalization and diversity structures. The three paths referred to are a direct, unfaded path, a second path close enough in attenuation and delay so that their combination yields a flat amplitude factor a, and a third path at relative delay τ and amplitude b. A fixed τ is often used, with channel responses fitted through choices of a, b, and f_o.

Several adaptive receiver techniques are used to counter multipath effects, which cause severe intersymbol interference [69]. Intersymbol interference is described at length in Chapter 4. Adaptive channel equalization is realized by either a simple frequency-domain system or a more sophisticated time-domain systems of the kind described in Chapter 8. The simplest frequency-domain system is a slope equalizer, compensating for a linear slope in the channel amplitude characteristic across the channel bandwidth. To compensate for the notch characteristic of the two-ray propagation model, a notch equalizer can be devised, which works well for minimum-phase fading but not for non-minimum-phase fading. Time-domain equalization techniques are required for the radio systems with higher spectral efficiency.

For the transmitter, a major consideration is to not interfere with adjacent channels. The FCC has defined sidelobe restrictions, described in Chapter 5, where attention is paid to pulse characteristics, such as minimal shift keying (MSK), appropriate for digital radio applications.

CELLULAR SYSTEMS

The extant digital microwave radio systems described above are becoming overshadowed by mobile and personal portable radio systems. Cellular mobile radio is a system in which users equipped with 0.6–10 watt telephone transceivers communicate with a network of base stations [70]. The base stations may be regarded as serving "cells" (Figure 1.31) with cell base station or stations handling mobile units in its cell. Voice channels are on individual carrier frequencies that are re-used in the system, but not in nearby cells. A “hand-off” procedure transfers a mobile unit to a new base station when the communication channel with that base station becomes better than that with the old base station. In the United States, the 806–947 MHz band used for the mobile telephone service [71] has channels spaced at 30 kHz. Analog (FM) modulation was used at the time of writing, but there is a worldwide move toward digital systems.

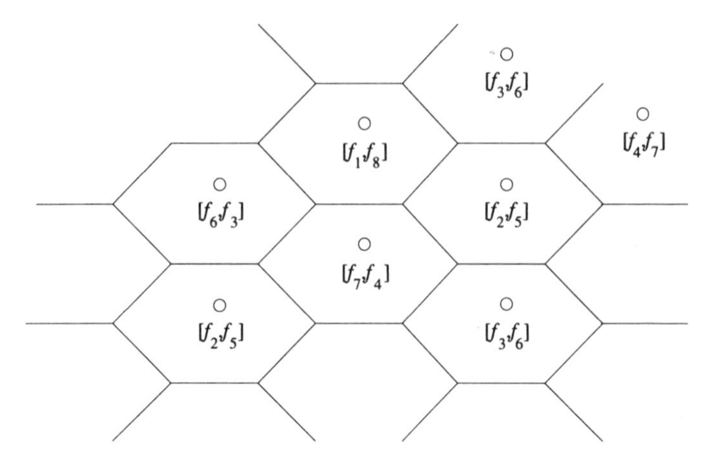

Fig. 1.31 A cellular network of radio base stations, illustrating frequency reuse.

The first introduction of digital mobile systems is in Europe, where the Group Special Mobile (GSM) system has been standardized. It uses shared time-division multiple-access (TDMA) channels operating at 270 kbps [71]. It supports voice channels compressed to 16 kbps or below. Spread spectrum code division multiple access (CDMA), in which all stations share the same wide band, is challenging TDMA for the future. It has realization complexities, but has advantages in spectrum utilization and resistance to selective fading.

Personal portable communications presume a microcellular base station network supporting portable units with 1–10 mW transmitters [72,73]. Microcells might be defined as long, slender cells along major roads, rooms in buildings, or areas around base stations on telephone poles, at about 2000 feet spacing. The microcellular system, just as the high-powered mobile cellular system, might use time-division, frequency-division, or CDMA techniques, individually or combined.

1.6.4 FIBER-OPTIC CHANNEL [74,76]

The extraordinary advances in optical fiber and associated semiconductor technology have resulted in exponential growth in the performance of fiber-optic communication systems. As shown in Figure 1.32, the product of bit rate times unrepeatered distance has grown by a factor of two per year. In the thirty years since the invention of the laser and the fifteen years since the development of low-loss optical fibers, lightwave systems have become the technology of choice for new interoffice and long-haul (including undersea) transmission systems, and are becoming increasingly important contenders for local distribution (i.e., the loop plant) and local area networks. The enormous bandwidth (> 40 THz) of

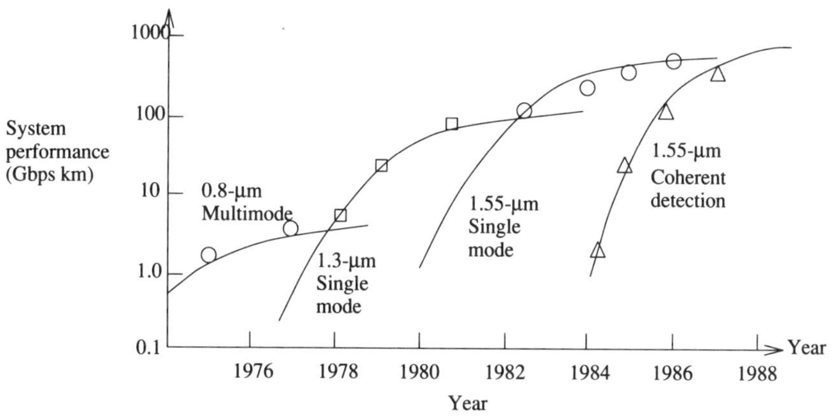

Fig. 1.32 Progress in lightwave transmission expressed in transmission rate–distance product at several wavelengths [74].

optical fibers has the potential to make broadband services, such as video, widely available. The recent advent of optical amplifiers has dramatically altered the architectural alternatives and performance limits associated with fiber-optic networks. In Chapter 10 we present a detailed discussion of signal processing, including coding and equalization, for fiber-optic communication systems. This section is limited to a brief overview of the fiber channel.

Most commercial lightwave systems are well represented by the simple model of Figure 1.33, which consists of a transmitter, a fiber transmission medium, and a receiver. The most common transmitter converts an input binary data stream to on–off light pulses. These pulses are clocked into the fiber, detected, and converted into electrical signals. A minimum acceptable bit error rate for such systems is 10^{-9}. The most important parameter in lightwave systems, except for very short links, is fiber loss. Today, fibers can be made that have extremely low attenuation on the order of 0.2 dB/km. In Figure 1.34 we show the attenuation of high-quality fiber in the range from 0.8 μm to 1.8 μm. An optical fiber uses total internal reflection (see Figure 1.35) to confine light to the fiber. Light can only propagate (i.e., reflect) at a finite number of angles (they are called modes). When the fiber radius is many times larger than the wavelength, there are many modes and this is called a *multimode* fiber. When the radius is the order of a wavelength, only one mode is supported, and the fiber is called a *single-mode* fiber.

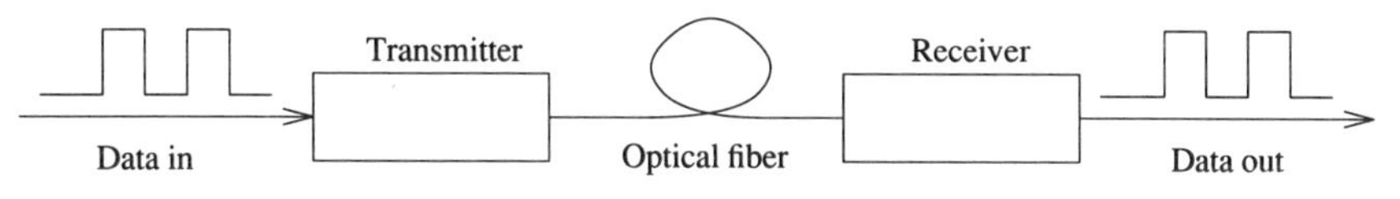

Fig. 1.33 Digital lightwave transmission link [74].

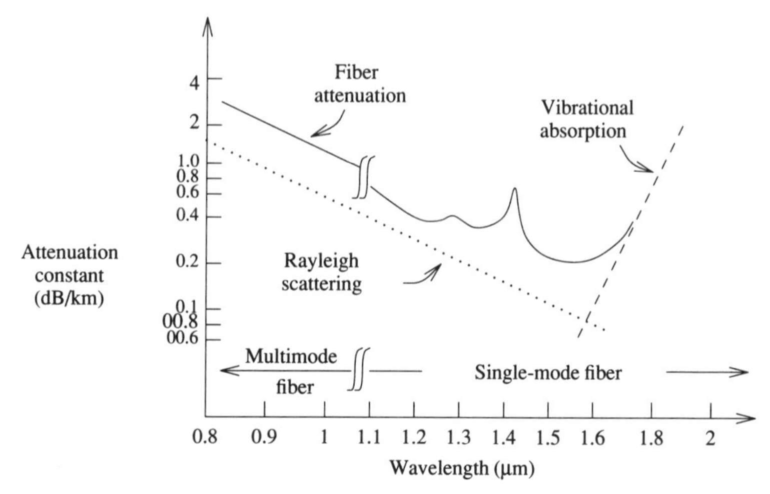

Fig. 1.34 Attenuation of high-quality optical fiber [74].

The extremely low loss of optical fibers becomes quite apparent by comparing•the loss (in dB/meter) for various electromagnetic wave-guiding media. As shown in Figure 1.36, optical fiber has, by far, the lowest loss of any medium and has a bandwidth many orders of magnitude greater than the other media.

When the symbol rate of the system is increased, dispersive effects in the fiber may affect performance. As the symbol rate increases, the transmitted pulses spread out in time and overlap each other at the detector. The deleterious effects of dispersion can be compensated for by an equalizer (Section 1.7 and Chapters 8 and 10), which boosts the gain at the attenuated higher frequencies.

There are two types of light sources: lasers and light-emitting diodes (LEDs). LEDs are *incoherent* sources of light (i.e., like a lightbulb) that emit energy over a range of wavelengths much greater than the data rates typically transmitted over optical fibers, typically 20–30 nm. Consequently, LEDs can only be modulated in intensity (typically, on–off). A laser, usually in the form of a laser diode, provides a *coherent* light source (narrowband relative to the signal symbol rate). Laser diodes have a spectral width in the range 1–5 nm and

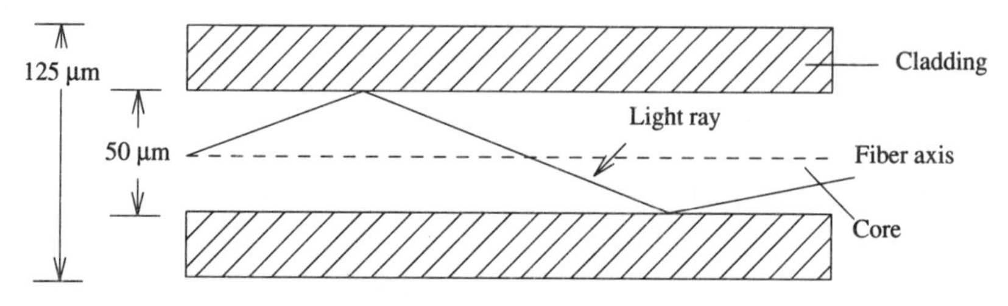

Fig. 1.35 Internal reflection in single-mode and multimode fibers.

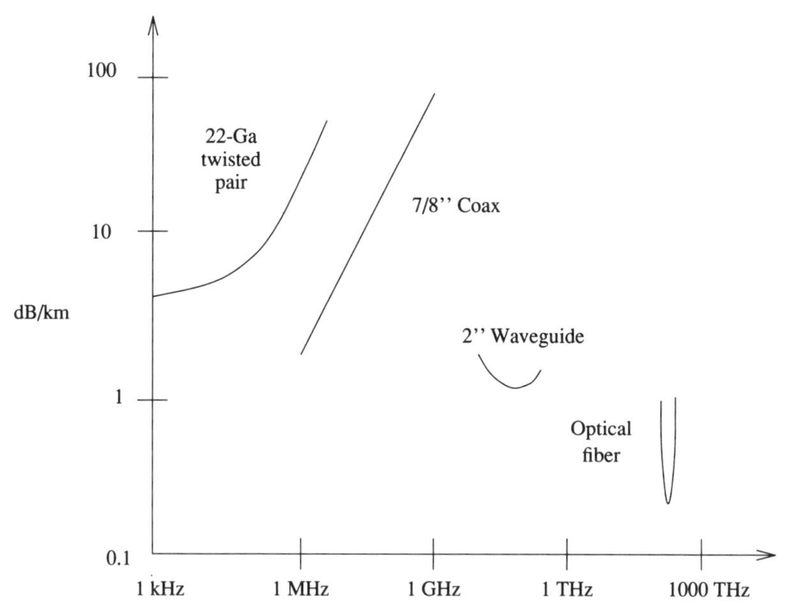

Fig. 1.36 Relative losses of different media [74]. The huge optical fiber bandwidth is obscured by the logarithmic scale of frequency.

single-mode laser diodes less than 0.2 nm. Even though a laser can be coherently modulated and detected, current commercial systems almost always use intensity modulation, even with a laser source. Coherent modulation of a laser enables the construction of device wavelength division multiplexed systems [75,76]. The output of a laser can be coupled into a fiber with very little loss (<3dB) and typically operates with a 1-mW output. An LED emits light over a much larger solid angle than a laser, so the coupling loss is much greater than for a laser. The LED is less expensive than the laser, and may remain the source of choice for LANs, while the laser will be used for long-distance systems.

At the output of the fiber, optical energy is converted to an electrical signal by a photodetector that produces a current proportional to the incident optical power. There are two types of detectors in use: the PIN photodiode for rates ~100 Mbps and the avalanche photodiode (APD) for speeds $\geq$ 1 Gbps.

1.7 SIGNAL PROCESSING FOR DATA COMMUNICATIONS

As we have seen, the function of a modem is to convert the digital data signal into a form which is suitable for transmission over the transmission medium. At the receiver, the various impairments introduced by the network are compensated for to recover the original data. In the implementation of modem transmitters and receivers, there are three important components: modulation, synchronization, and equalization.

1.7.1 MODULATION TECHNIQUES

Aside from the twisted-pair subscriber line, all of the channel models offered above are *passband*, with no significant power transmission below a lower band-edge frequency. Conventional data signals are, however, *baseband*, and must be passed through modulators to generate the passband signals that the passband channels will accommodate. These modulators are either linear, causing little or no bandwidth expansion, or nonlinear, often, but not necessarily, causing substantial bandwidth expansion. The choice of modulation format depends on the data rate and the channel, with inexpensive FSK (frequency-shift keying) used when spectral efficiency is not important, and more complicated linear quadrature modulators when it is. These choices are reflected in the CCITT modem recommendations of Table 1.2.

FSK is the accepted standard on telephone channels at rates below 1200 bps. FSK modems are usually asynchronous. At the receiver, timing information is not provided by the modem to the data terminal and must be rederived by the terminal from a character each time one is received. The bandwidth expansion of FSK modulation and the presence or absence of lines in the spectrum is critically dependent on the ratio of modulation index (the frequency deviation per unit of input amplitude) to (maximum) pulsing rate. For binary FSK, a value of this ratio in the region of 0.5 to 0.8 has been found to produce signals with continuous and reasonably compact spectra.

At 2400 bps on telephone channels and in some coherent lightwave communication systems at rates of hundreds of megabits per second, four-phase is used. Since each choice among four phases conveys two bits of information, the pulsing (symbol) rate, or *baud*, is half the data rate, e.g., 1200 for a 2400 bps modem. It will be shown in Chapter 5 that the spectral efficiency is 2 bps/Hz. There is, of course, a price paid in SNR, and thus in bit error probability, as the separation in signal points is reduced for constant average transmitted power.

Although eight-phase modulation and higher may be used for higher spectral efficiency, a rapidly increasing price is paid in sensitivity to phase jitter as signal points are brought closer together in phase. Nevertheless, eight-phase modulation is used in some telephone line modems at 4800 bps and digital radio systems at 90 Mbps. *Adaptive channel equalization* becomes desirable to reduce the effects of channel distortion.

For higher spectral efficiencies, a more uniform distribution of signal points, one tending to maximize the minimum distance between signal points [77,78], is preferred. At higher data rates on telephone lines, 16-point AM-PM, defined by some relatively uniform allocation of signal points within the signal space of Figure 1.37, was for years the standard modulation format.

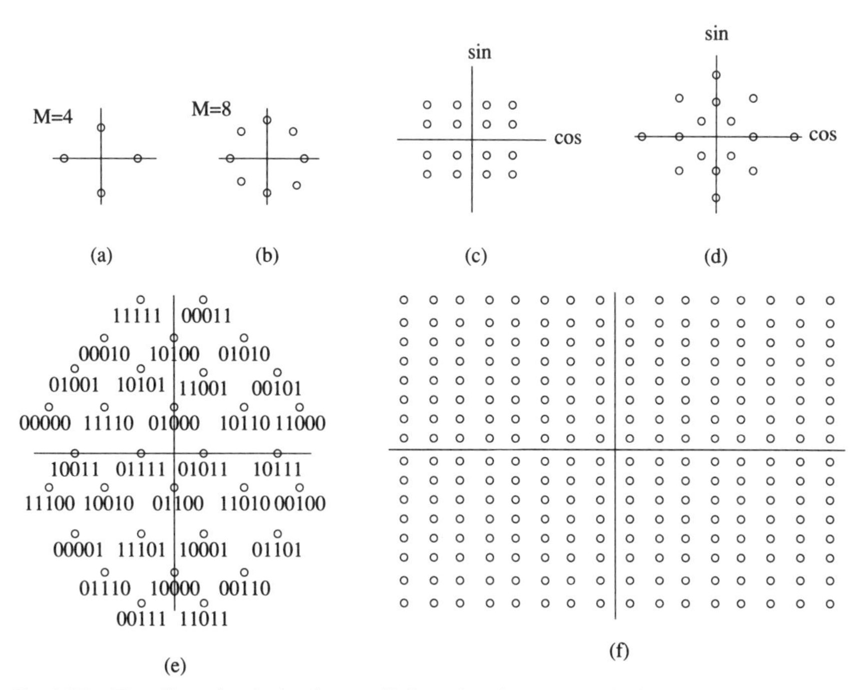

Fig. 1.37 Two-dimensional signal constellations for data communication. Each signal point defines the amplitude and phase of a sinusoidal carrier waveform. (a) 4-phase, (b) 8-phase, (c) 16-point QAM, (d) 16-point CCITT V.29, (e) 32-point CCITT V.32, (f) 256-point QAM.

The spectral efficiency is 4 bps/Hz. Two 16-point signal constellations that have been widely used are quadrature amplitude modulation (QAM) and CCITT Recommendation V.29. More recently, the opportunity to realize performance improvements through use of channel trellis coding (Chapter 5) has resulted in modems using larger signal constellations, particularly the 32-point constellation of CCITT Recommendation V.32. We have already noted that digital radio systems, which enjoy relatively undistorted and low-noise channels except during fades, have been pressing for higher spectral efficiency and already use 256-point signal constellations, achieving 8 bps/Hz. In general, these are all coherent modulation systems, with carrier phase "recovered" by the data-directed tracking loops, as described in Chapter 6.

For data transmission on satellite channels, in which the satellite-borne transponders are strictly power limited, the preferred modulation technique is a constant-envelope format, such as QPSK (quadrature phase-shift keying, the same as 4-phase PSK), MSK, or OQPSK (offset QPSK, where "offset" means that the cosine and sine carrier waveforms are not pulsed at the same time). The use of such formats is of little help in random-access systems in which several signals can add so that the total signal power fluctuates widely anyway, and for this reason there is growing use of TDM random-access systems with guaranteed constant power.

Coherent lightwave communication systems still use relatively simple binary or four-phase modulation, because phase noise is a serious problem and bandwidth is relatively plentiful. With technology improvements, the application of more complex modulation formats can be expected.

1.7.2 SYNCHRONIZATION REQUIREMENTS

Data communication requires, to an extent determined by the particular application, synchronization of operations at the transmitting and receiving ends of the communication system. Coherent modulation systems need carrier synchronization (or recovery), and correct sampling of the receiving signal requires timing synchronization (or recovery), both illustrated in Figure 1.38. These functions are handled by fixed or adaptive algorithms implemented in modem receivers. Carrier and timing recovery techniques must strike a compromise between tracking rate and noise enhancement, where noise enhancement means an increase in the random fluctuations of the parameter estimates due to an increase in the tracking rate or (equivalently) the tracking loop bandwidth.

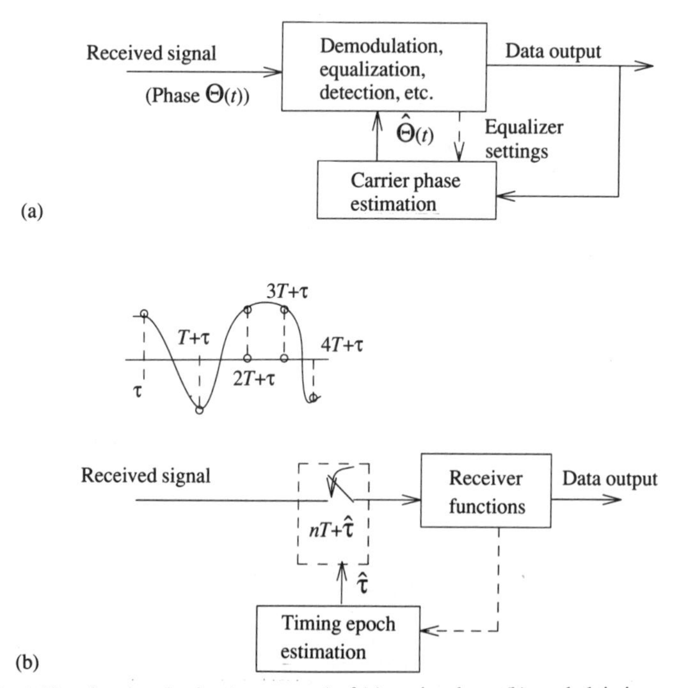

Fig. 1.38 Synchronization (or recovery) of (a) carrier phase, (b) symbol timing.

Tracking capabilities are needed because of channel phase perturbations, oscillator misadjustment and aging, and noise perturbations. Carrier and timing recovery are covered in Chapter 6.

In addition to these transmission synchronization requirements, which relate to receiver operation, there can be data stream synchronization requirements imposed by the use of digital frames, block codes, digital scramblers, data encryption, and other data formats which presume knowledge of where various initialization and transition points are in the data stream. (Initialization of receiver functions may also require a known data sequence, but the synchronization problem is usually avoided by use of a short repetitive data pattern, as discussed in Chapter 6.) We shall not elaborate on these matters here; they are discussed in later chapters.

1.7.3 CHANNEL EQUALIZATION

We have already called attention to the need to correct the impairments and deficiencies of the transmission channel in order to achieve reliable data communication. In the voiceband telephone channel the most troubling impairment is the intersymbol interference (ISI) in the received data signal induced by linear channel distortion. ISI is described in Chapter 4. The essential concept is that the edges, or tails, of a pulse misshapen in transit will interfere with neighboring pulses. A sample taken at the center of a pulse will contain components from the tails of other pulses. If the sum of these undesired components is large enough and in the wrong direction, it can lead to an incorrect decision (see Figure 1.39a).

Advantages in spectral shaping, such as reducing signal power near band edges to avoid interfering with neighboring frequency channels, can be obtained by intentionally introducing intersymbol interference into a transmitted data sequence. As usual, there is a price in signal-to-noise ratio. The encodings in this class of *partial-response* techniques (Chapter 4) are widely used in digital carrier systems, where their inherent redundancy provides some error-detection capability in addition to spectral smoothing.

To reduce ISI, an *equalizer* can be provided to reshape each received pulse into something more like its transmitted form. The extent of this reshaping is limited by the frequency distribution of the received noise, since a substantial reshaping will (in some circumstances) increase the noise level. The equalizer can be either a fixed compromise filter, intended to achieve satisfactory performance over a broad range of channels, or an adaptive structure [79,80] designed to configure itself into a filter which is particularly effective for the channel being used. An adaptive equalizer, using data decisions in an adjustment mechanism, can track changes in the channel during the course of normal data transmission without damaging or interfering with this transmission. Voiceband modems operating at 4800 bps and above (i.e., ≥2 bps/Hz) usually incorporate adaptive equalization.

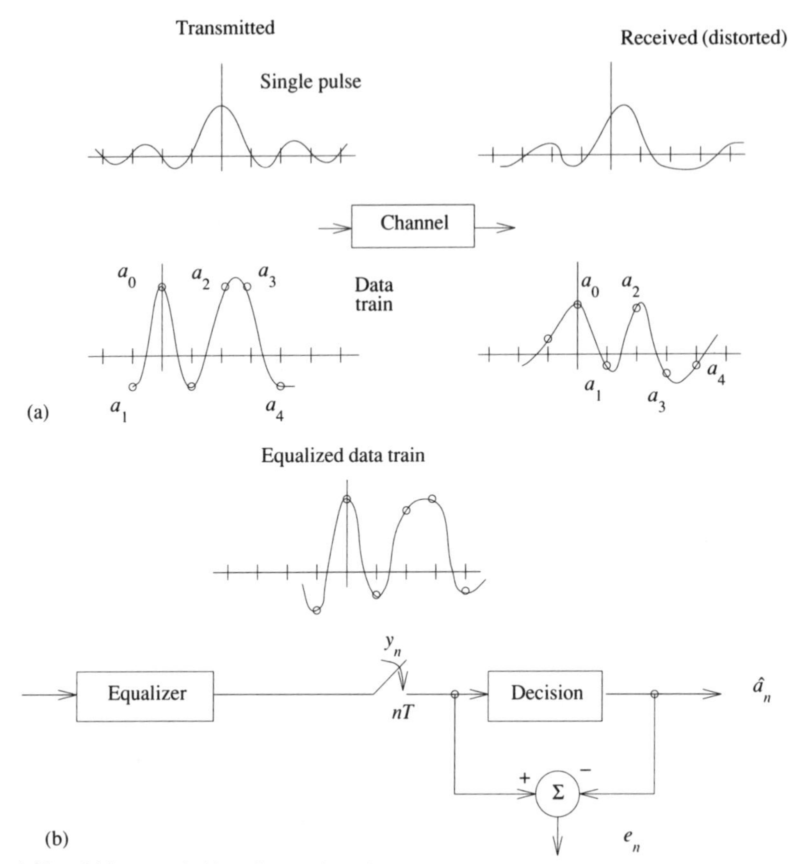

Fig. 1.39 (a) Intersymbol interference in a channel with linear distortion. (b) Correction of intersymbol interference in an equalizer.

The conventional structure for adaptive equalization is a *transversal filter* — a tapped delay line with tap weights adjusted so as to satisfy a given performance criterion (shown in Figure 1.39b). This structure is not an arbitrary choice, but may be derived from the optimum linear receiver structure derived in Chapter 7. Various nonlinear structures [81], discussed in Chapter 8, can improve performance somewhat, but at costs in complexity and sensitivity to the channel model.

1.8 ORGANIZATION OF THIS BOOK

Throughout this introductory chapter, references have been made to other chapters that explain and analyze the concepts introduced here. It is our intention that the book progress in a logical fashion from one chapter to the next, although the book can serve as a reference work for those already working in the field. Each chapter ends with a group of exercises intended to reinforce understanding rather than stump or confuse the student. We hope we have been successful.

Chapter 2 provides a background in information theory and in detection and estimation that we believe is important for real understanding of data communication. The derivations in later chapters assume a reasonable knowledge in these areas, as well as in basic probability theory, stochastic processes, and Fourier analysis.

Chapter 3 is a brief review of coding theory, introducing block and convolutional codes and the Viterbi algorithm for convolutional decoding. The material on the Viterbi algorithm is pertinent to the derivation of the optimum (nonlinear) receiver in Chapter 7, to channel trellis coding in Chapter 5, and to a number of other topics covered elsewhere in the book.

Chapter 4 covers baseband data transmission, introducing the basic Nyquist theory and several important pulse-coding techniques, examining trade-offs among spectral efficiency, SNR, and complexity, and presenting useful error-rate bounds.

Chapter 5 describes the modulation techniques used for passband data transmission and their appropriate applications. It introduces complex representations for modulated signals and bandpass channels, channel trellis coding, and spectrally efficient modulation such as minimal-shift keying (MSK) and multipoint AM-PM formats.

Chapter 6 describes the carrier acquisition and tracking techniques appropriate to synchronous passband data communications, including phase-locked loops and predictive phase trackers. It also covers timing acquisition and tracking.

Chapter 7 offers several "optimum" solutions to the problem of design of a baseband data communication system. The two principal solutions are the nonlinear maximum likelihood sequence estimation receiver, implemented in the Viterbi algorithm, and the optimum linear receiver, which is a matched filter followed by a tapped delay line. The decision feedback structure is also described.

Chapter 8, a particularly important part of this book, is devoted to adaptive equalization. It shows the applicability of the least mean squared error (LMS) algorithm [79] for tap-weight adjustment in a transversal equalizer, describes the transient and steady-state performance of the synchronous (symbol interval) equalizer, and then demonstrates the superior capabilities of the fractionally-spaced equalizer. Sections are also included on the Kalman, lattice, and decision-feedback equalizers and on cyclic equalization and other fast start-up techniques.

Chapter 9 examines techniques for echo cancellation as a means for implementing full-duplex operation on two-wire channels such as the twisted-pair subscriber line and the dialed telephone line.

Chapter 10 gives selected examples of the interrelationship between the choice of system architecture and digital implementation technology. We also describe the potential application of advanced modulation and equalization technologies to fiber-optic communication systems.

Although the organization of this book and many of its discussions are consciously associated with problems of current interest in the field of data communication, we believe its analytical content to be of broad and long-term interest. We hope that readers will find the book a useful and accessible source of basic knowledge and an inspiration for their creative work.

REFERENCES

[1] *From Semaphore to Satellite, The Story of the ITU,* International Telecommunications Union, 1965.

[2] R. W. Lucky, *Silicon Dreams: Information, Man, and Machine,* St. Martins Press, 1989.

[3] N. Abramson, *Information Theory and Coding,* McGraw-Hill, 1963.

[4] R. Gallager, *Information Theory,* Wiley, 1968.

[5] R. E. Blahut, *Principles and Practice of Information Theory,* Addison-Wesley, New York, 1987.

[6] A. S. Tanenbaum, *Computer Networks,* Prentice-Hall, 1981.

[7] M. Schwartz, *Telecommunications Networks*, Addison-Wesley, New York, 1986.

[8] J. F. Hayes, *Modeling and Analysis of Computer Communication Networks*, Plenum Press, New York, 1984.

[9] D. Bertsekes and R. G. Gallager, *Data Networks,* Prentice-Hall, 1987.

[10] R. E. Kahn, "Resource Sharing Computer Communication Networks," *Proc IEEE,* Vol. 60, November 1970.

[11] L. G. Roberts and B. D. Wessler, "Computer Network Development to Achieve Resource Sharing," AFIPS Conference Proceedings, Vol. 36, 1970.

[12] S. B. Weinstein, "Communication in the Coming Decades," *IEEE Spectrum*, Vol. 24, No. 11, pp. 62–67, November 1987.

[13] D. Comer, *Internetworking with TCP/IP, Principles, Protocols and Architecture*, Prentice-Hall, 1988.

[14] W. Stallings, *Data and Computer Communications*, Macmillan, New York, 1985.

[15] P. Verma, Editor, *ISDN Systems*, Prentice-Hall, 1990.

[16] G. D. Schultz, "Anatomy of SNA," *Computerworld*, Vol. 15, No. 11a, pp. 35–38, March 1981.

[17] V. L. Hoberecht, "SNA Function Management," *IEEE Trans on Comm.*, Vol. COM-28, No. 4, April 1980, Reprinted in P. E. Green, Jr., *Computer Network Architectures and Protocols*, Plenum Press, New York, 1982.

[18] H. R. Rudin, Ed., *Special Issue of IEEE Communications Magazine on High-Speed Network Protocols*, Vol. 27, No. 6, June 1989.

[19] North American Presentation Level Protocol Syntax (NAPLPS), American National Standard X3.110-1983, American National Standards Institute, New York, 1983.

[20] J. E. White, "ASN.1 and ROS: The Impact of X.400 in OSI," *IEEE J. Selected Areas in Communications*, Vol. 7, pp. 1060–1072, September 1989.

[21] Information Sciences Institute, Transmission Control Protocol NIC-RFC 793. DDN Protocol Handbook, Vol. 2, pp. 2.179–2.198, September 1981.

[22] *Toward a National Research Network,* Computer and Science and Technology Board, National Research Council, National Academy Press, Washington, 1989.

[23] J. F. Hayes, "Local Distribution in Computer Communications," *IEEE Communications Magazine*, Vol. 19, No. 2, pp. 6–14, March 1981. Reprinted in IEEE Communication Society, *Tutorials in Modern Communications,* V. B. Lawrence, J. L. LoCicero, and L. B. Milstein, Editors, Computer Science Press, 1983. A revised version appears in *Advances in Local Area Networks,* K. Kummerle, J. O. Limb, and F. Tobagi, Editors, IEEE Press, 1987.

[24] J. F. Hayes and M. Mehemet-Ali, "Random Access Systems - ALOHA's Progeny," *Canadian Journal of Electrical Engineering*, Vol. 14-1, pp. 3–10, January 1989.

[25] R. M. Metcalfe and D. R. Boggs, "Ethernet: Distributed Packet Switching for Local Computer," *Communications of the ACM,* Vol. 19, pp. 395–404, July 1976.

[26] J. F. Shock *et al.*, "Ethernet," in *Advances in Local Area Networks,* K. Kummerle, J. O. Limb, and F. A. Tobagi, Editors, IEEE Press, 1987.

[27] ANSI/IEEE Standards 802.1 to 802.6 for Local Area Networks, Wiley Interscience, 1985.

[28] M. R. Finley, Jr., "Optical Fiber in Local Area Networks," in *Advances in Local Area Networks,* op. cit.

[29] I. P. Kaminow, "Non-Coherent Photonic Frequency — Multiple Access Networks," *IEEE Network,* Vol. 3, No. 2, pp. 4–13, March 1989.

[30] R. A. Linke, "Frequency Division Multiplexed Optical Networks Using Hetrodyne Detection, ibid.

[31] S. Lee and P. P. Boulton, "The Principles and Performance of Hubnet: a 50 Mbps Glass Fiber Local Area Network," Special Issue on Local Area Networks, *IEEE J. Selected Areas Commun.,* Vol. SAC-1, pp. 711–721, November 1983.

[32] M. Mehmet-Ali and J. F. Hayes "An Optical Fiber Based Local Backbone Network," *Canadian Journal of Electrical Engineering*, Vol. 14, No. 4, pp. 127–133, 1989.

[33] A. G. Fraser, "Towards a Universal Data Transport System," in *Advances in Local Area Networks*, K. Kummerle, J. O. Limb, and F. Tobagi, Editors, IEEE Press, 1987.

[34] R. M. Newman *et al.*, "The QPSX Man," *IEEE Communications Magazine,* Vol. 26, No. 6, pp. 20–28, April 1988.

[35] E. L. Hahne, A. K. Choudhury, and N. F. Maxemchuk, "Improving the Fairness of Dual Queue Dual Bus Networks," Proc. IEEE Infocom '90, San Francisco, June 1990.

[36] J. W. Wong, "Throughput of DQDB Networks Under Heavy Load," EROL/LAN 89, Amsterdam.

[37] M. H. Huber, K. Sawer, and W. Scheodl, "QPSX and FDDI-II Performance Study of High Speed LANS," *EFOC/LAN 88*, Amsterdam, June 1988.

[38] F. E. Ross, "An Overview of FDDI: The Fiber Distributed Data Interface," *IEEE Communications Magazine,* Vol. 7, No. 7, pp. 1043–1051, September 1989.

[39] M. Goodman, C. Brackett, C. Lo, H. Kobrinski, M. Vecchi, and R. Bulley, "Design and Demonstration of the LAMBDANET™ system: a multiwavelength optical network," Conference Record *IEEE Globecom* '87, Tokyo, June 1987.

[40] A. S. Acampora, M. J. Karol, and M. G. Hluchyj, "Terabit Lightwave Networks: The Multihop Approach," *AT&T Technical Journal,* Vol. 66, No. 6, November/December 1987.

[41] M. G. Hluchyj and M. J. Karol, "ShuffleNet: An Application of Generalized Perfect Shuffles to Multihop Lightwave Networks," *INFOCOM* '88, New Orleans, March 1988.

[42] CCITT Blue Book, Vol. III, Fascicle III-5, *Integrated Services Digital Network (ISDN),* Recommendations of the Series I, IXth Plenary Assembly, Geneva, 1989.

[43] ANSI T1.605-1989, Integrated Services Digital Network (ISDN — Basic Access Interface for S and T Reference Points (Layer 1 Specification), American National Standards Institute, Washington, 1989.

[44] B. T. Doshi and H. Q. Nguyen, "Congestion Control in ISDN Frame Relay Networks," *AT&T Technical Journal,* Vol. 67, No. 6, November/December 1988.

[45] *Data communication over the telephone network*, CCITT Red Book, vol. VIII, Fascicle VIII.1, International Telecommunications Union, Geneva, 1985.

[46] Bell Telephone Laboratories, Inc., *Engineering & Operations in the Bell System,* third printing, 1978.

[47] B. Fleury, "Asynchronous High Speed Digital Multiplexing," *IEEE Communications Magazine*, Vol. 24, No. 8, pp. 17–25, August 1986.

[48] CCITT Blue Book, Vol. III, Fascicle III.3, October 1988, "7 KHz: audio coding within 64 Kbps."

[49] *Telecommunications Transmission Engineering,* Vol. 2, Third Edition, Bellcore, 1990.

[50] R. Ballart and Y. C. Ching, "SONET: Now Its the Standard Optical Network," *IEEE Communication Magazine,* Vol. 29, No. 3, pp. 8–15, March 1989.

[51] B. Schaffer, "Synchronous and Asynchronous Transfer Modes in the Future ISDN," *Conference Record, ICC '88.*

[52] M. de Prycker, *Asynchronous Transfer Mode*, Ellis Horwood, 1991.

[53] S. Minzer, "Broadband ISDN and Asynchronous Transfer Mode (ATM)," *IEEE Communications Magazine*, Vol. 27, No. 9, pp. 17–24, September 1989.

[54] L. Wu and M. Kerne, "Emulating Circuits in a Broadband Packet Network," *Proc. IEEE Globecom '88.*

[55] F. P. Duffy and T. W. Thatcher, "Analog Transmission Performance on the Switched Telecommunications Network," *Bell System Tech. J.*, Vol. 50, pp. 1311–1347, April 1971.

[56] T. C. Spang, "Loss-noise-echo study of the direct distance dialog network," *Bell System Tech. J.*, Vol. 55, pp. 1–36, January 1976.

[57] F. P. Duffy *et al.*, "Echo performance of toll telephone connections in the United States," *Bell System Tech. J.*, Vol. 54, pp. 229–243, February 1975.

[58] Bell System Technical Reference PUB 41008, *Transmission Parameters Affecting Voiceband Data Transmission — Description of Parameters,* AT&T, 1974.

[59] E. Biglieri, A. Gersho, R. D. Gitlin, and T. Lim, "Adaptive cancellation of nonlinear intersymbol interference," *IEEE J. Selected Areas of Communications,* SAC-2, No. 5, 1984.

[60] J. W. Lechleider, "Line codes for digital subscriber lines," *IEEE Communications Magazine*, Vol. 27, No. 9, pp. 25–32, September 1989.

[61] W. H. Hayt, *Engineering Electromagnetics*, McGraw-Hill, 1989.

[62] H. S. Lin and C-P. J. Tzeng, "Full Duplex Data Over Local Loops," *IEEE Communications Magazine*, Vol. 26, No. 1, January 1988.

[63] Bell Laboratories, *Transmission Systems for Communications*, Fifth Edition, 1982.

[64] FCC docket 19311, FCC 74-985, September 19, 1974, revised January 25, 1975.

[65] T. Noguchi, Y. Daido, and J. A. Nossek, "Modulation Techniques for Digital Radio," *IEEE Communications Magazine*, Vol. 24, No. 10, pp. 21–30, October 1986.

[66] D. P. Taylor and R. P. Nartreann, "Telecommunications by Digital Radio," *IEEE Communications Magazine*, Vol. 24, No. 8, pp. 11–16, August 1986.

[67] W. D. Rummler, R. P. Coutts, and M. Liniger, "Multipath Fading Channel Models for Digital Microwave Radio," *IEEE Communications Magazine*, Vol. 24, No. 11, pp. 30–42, November 1986.

[68] L. J. Greenstein and M. Shafi, "Outage Calculation Methods for Microwave Digital Radio," *IEEE Communications Magazine*, Vol. 25, No. 2, pp. 30–39, February 1987.

[69] J. K. Chambelain, F. M. Clayton, H. Sari, and E. Vandamme, "Receiver Techniques for Microwave Digital Radio," *IEEE Communications Magazine*, Vol. 24, No. 11, pp. 43–54, November 1986.

[70] V. H. MacDonald, "The Cellular Concept," *Bell System Tech. J.*, Vol. 58, No. 1, pp. 15–49, January 1979.

[71] E. Lee and D. Messerschmitt, *Digital Communication,* Kluwer Academic Publishers, 1988.

[72] R. Steele, "The cellular environment of lightwave handheld portables," *IEEE Communications Magazine*, Vol. 27, No. 7, pp. 20–29, July 1989.

[73] D. C. Cox, "Portable digital radio communications — an approach to tetherless access," *IEEE Communications Magazine*, Vol. 27, No. 7, pp. 30–40, July 1989.

[74] P. S. Henry, R. A. Linke, and A. J. Gnauck, "Introduction to Lightwave Systems," in *Optical Fiber Telecommunications II,* edited by S. E. Miller and I. P. Kaminow, Academic Press, 1988.

[75] J. Salz, "Modulation and Detection for Coherent Lightwave Communications," *IEEE Communications Magazine*, Vol. 24, No. 6, June 1986.

[76] S. D. Personick, *Fiber Optics Technology and Applications,* Plenum Press, New York, 1985.

[77] G. J. Foschini, R. D. Gitlin, and S. B. Weinstein, "Optimization of Two-Dimension Signal Constallations in the Presence of Gaussian Noise," *IEEE Trans on Communications,* Vol. COM-22, No. 1, January 1974.

[78] G. J. Foschini, R. D. Gitlin, and S. B. Weinstein, "On the selection of Two-Dimensional Signal Constellations in the Presence of Phase Jitter and Gaussian Noise," *Bell System Tech. J.*, Vol. 52, pp. 927–965, July-August 1973.

[79] B. Widrow and S. Stearns, *Adaptive Signal Processing,* Prentice-Hall, 1985.

[80] R. W. Lucky, J. Salz, and E. J. Weldon, Jr., *Principles of Data Communication,* McGraw-Hill, New York, 1968.

[81] D. D. Falconer, "Adaptive Equalization of Channel Nonlinearities in QAM Data Transmission," *Bell System Tech. J.*, Vol. 57, pp. 2589–2611, September 1978.

[82] E. F. O'Neill, Editor, *A History of Engineering and Science in the Bell System,* AT&T Bell Laboratories, 1985.

[83] P. E. Green, Jr., *Fiber Optic Networks*, Prentice Hall, New York. 1993.

2

Theoretical Foundations of Digital Communications

2.0 INTRODUCTION

The generic communications system of interest in this book is depicted in Figure 2.1. This model contains most of the elements which will be discussed in this and subsequent chapters. The output of an information source* is first encoded into a digital, usually binary, data stream. The encoding operation may involve several steps. If the source is analog, sampling and quantization by an analog-to-digital (A/D) converter are involved. The next step may be source coding in the form of redundancy removal. Huffman coding and the Lempel-Ziv algorithm, discussed in Section 2.5, are examples of this operation. Redundancy in the form of parity bits may be added in order to guard against channel errors. We shall be discussing redundancy encoding in Chapter 3. Once a digital stream is available, the operation of the encoder may also involve *encryption* for security. The digital stream may also be scrambled to provide randomization of the transmitted signal to facilitate adaptive equalization and timing recovery in the receiver. Scrambling is discussed in Section 6.7. The function of the modulator is to put the digital data stream into a form which is suitable for transmission over the physical channel. This may involve simply translating a sequence of binary digits into a sequence of pulses. This step may also involve baseband pulse encoding to combat intersymbol interference (see Section 4.6). In the case of passband systems, the data sequence may be used to modulate a carrier, thereby placing the signal into the passband of the channel (see Chapter 5). The receiver reverses the operations performed at the transmitter.

* In Section 2.3 below, we define information in a quantitative way.

The output of the channel is translated into a digital format by the demodulator. This would ordinarily involve carrier and timing recovery (see Chapter 6) and equalization (see Chapter 8). The decoder places the digital data sequence into a form which is suitable for delivery to the information destination. If the source is analog, we may have digital-to-analog (D/A) conversion. Error correction and detection can be carried out in the decoder. This is also the point where the data are decrypted and descrambled, i.e., the original message is derived from the text.

The partitioning of the transmitter and receiver into distinct functions in Figure 2.1 is for pedagogical and historical reasons (i.e., this typifies the design of most extant digital communications systems). It has been recognized that combining modulation and coding, as well as joint demodulation and decoding, can improve system performance—in fact, trellis coding, described in Section 5.7 of the text, is based on a novel combination of these functions.

The two foundation disciplines for communication systems are *information theory* and *statistical detection and estimation theory*. The former quantifies the information content of analog and digital sources and relates this quantity to the ultimate information-carrying ability of physical channels. As we have seen in Section 1.6 of Chapter 1, a fundamental barrier to communications is *noise*, which may be broadly defined as random disturbance introduced by the channel. We use detection and estimation theory to derive statistical estimates of signal parameters that have been corrupted by noise. Since it is less abstract than information theory, we begin our discussion with detection and estimation theory.

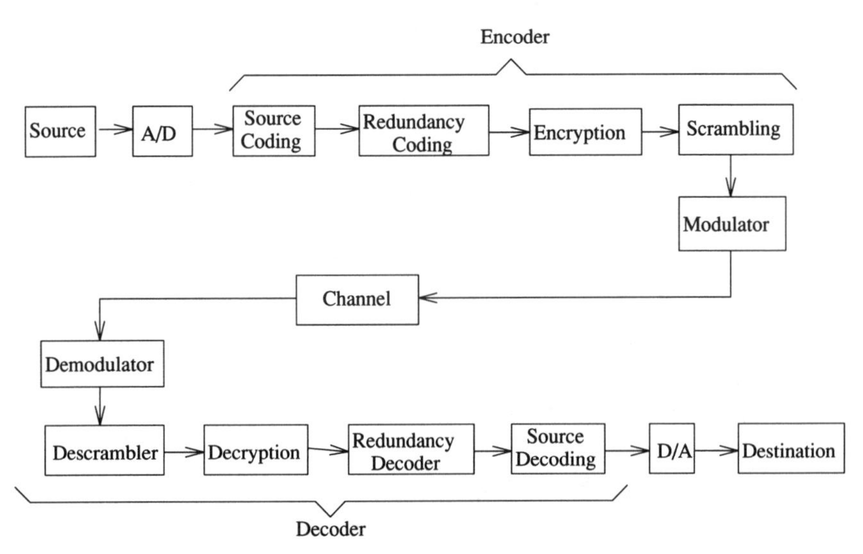

Fig. 2.1 Generic communications system.

We consider the problem of making decisions about the state of a system based upon observations which are tempered by a knowledge of the underlying statistical structure. A decision rule which minimizes the probability of making an erroneous decision will be derived. This decision rule is applied to the case of transmitting signals over channels which are disturbed by additive white Gaussian noise. (See Appendix 2A for definitions.) The derivation leads to the first important communications, as distinguished from mathematical statistics, result, the *matched-filter* receiver. In the case of nonwhite noise, the matched filter must be supplemented by a noise-whitening filter. The basic theory is then applied to pulse amplitude modulation (PAM), where the amplitude of a pulse is modulated by a number of levels greater than or equal to two. The performance in terms of probability of error is then derived for the binary and higher-order modulation, and the central roles of bandwidth and signal-to-noise ratio in the reliable transmission of information are explored.

In the material discussed above, the focus was upon deciding which of two or more signals was transmitted over a channel. Our focus then shifts to the estimation of analog quantities. As we shall see in Chapter 6, this material is directly applicable to the estimation of the phase of a carrier. The estimation results are derived, for the most part, under the mean-square error (MSE) criterion. As we shall see, the guiding principle is a geometrically intuitive result that the error produced using the MSE estimate is orthogonal to the observed signal. Throughout this book, we will emphasize geometrical interpretations of results since this view often remains when the equations have become a distant memory.

The final broad topic treated in the chapter is information theory. The first topic covered is the definition and derivation of the *entropy* of a source, which quantifies the information supplied by source outputs. The entropy of a source gives the minimum average number of binary digits needed to encode a source output. The Huffman code gives a practical way to achieve this bound. Finally, the concept of mutual information is used to define both the information in an analog source and information-carrying *capacity* of a channel. The capacities of the additive Gaussian channel and the binary symmetric channel (BSC) among others are derived.

2.1 INTRODUCTION TO DECISION THEORY*

2.1.1 OPTIMUM DECISION REGIONS

Decision theory is a subset of detection and estimation theory concerning choices between statistical hypotheses. In the typical decision-theoretic context,

---

* The reader is also referred to several of the excellent texts which apply decision theory to communication systems [1]–[4].

one is presented with statistical data whose probability distribution depends on a discrete set of underlying parameters. Decision theory offers tools for processing the data to enable one to choose among these parameters in a statistically optimal fashion. Specifically, we shall be applying elements of the theory to the detection of signals in a digital communication system in which one of a finite set of symbols is transmitted over a channel.

The derivation of optimum decision rules begins with the definition of what is called the *observation space*. This is the raw data that are to be processed by the receiver. The dimensions of this space depend upon the particular application at hand, and the points in the space are the observed output of the channel. Consider the following simple one-dimensional example. A zero or a one is conveyed over a pair of wires by voltage levels of $+s$ or $-s$, respectively. A single observation of the voltage is made. We assume that this observation is disturbed by additive noise, giving

$$Y = X + N, \tag{2.1}$$

where $X = +s$ or $-s$. Assume that the noise voltage, N, is a zero mean Gaussian random variable.* The observation space in this case is the range of values assumed by the random variable Y.

Because of the random nature of the channel, different transmitted symbols may be mapped into the same point in the observation space. In general, all that distinguishes these different transmitted symbols at the channel output are different conditional probability distributions. In the example of (2.1), conditioned on the transmitted symbol, Y is a Gaussian random variable with either mean $+s$ or mean $-s$. The essential task of detection theory is to decide among the conditional distributions based on the observed values of Y. In the above example this is done by partitioning the observation space into two nonoverlapping regions (Figure 2.2) which cover the whole space. More generally, when one of L symbols is transmitted, the observation space is partitioned into L regions. Each region is identified with a transmitted symbol; a channel output falling in a particular region leads to deciding that the corresponding symbol was transmitted.

The choice of the criterion for partitioning the space led to a great deal of controversy in the formative years of detection theory. Part of the difficulty was in the assignment of *a priori* probabilities to the parameters of the distributions. Fortunately, matters are considerably simplified in the communications context, since the *a priori* probabilities of the transmitted symbols are known. In communications systems, there is also a convenient criterion for partitioning the

---

* The Gaussian random variable, which is defined in Appendix 2A, along with other probabilistic concepts, is often a remarkably robust assumption. Many physical processes can be modeled as Gaussian. In other cases where the noise is, strictly speaking, not Gaussian, it can sometimes be assumed Gaussian for the purposes of comparing alternative systems and techniques, considerably simplifying the mathematical analysis. Furthermore, the results obtained for a Gaussian channel often provide a lower bound on the performance of systems operating on a non-Gaussian channel.

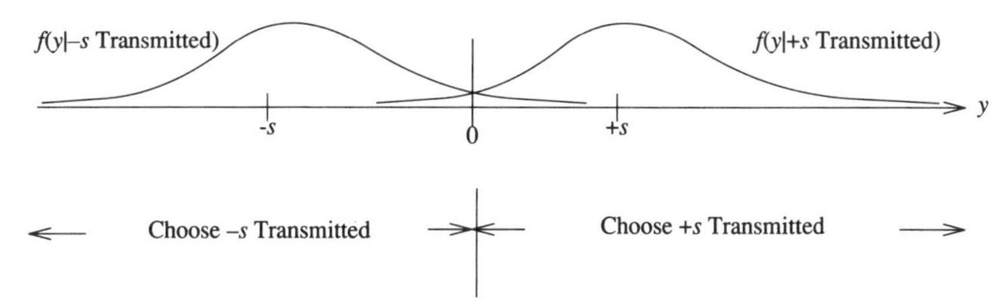

Fig. 2.2 Decision regions for one-dimensional binary transmission example.

observation space. One attempts to minimize the probability of making an error in deciding upon the transmitted symbol.

The derivation of the optimal decision regions is simplified if we assume the binary case, i.e., two transmitted symbols. The results of this analysis form the basis of the general case. The derivation of the detection rule involved partitioning the observation space into two disjoint regions; $\mathbf{R}_0$ and $\mathbf{R}_1$, each region being identified with one or the other signal. A decision is made as to whether the output of the channel falls in $\mathbf{R}_0$ or $\mathbf{R}_1$, respectively. The problem is to find the partition which minimizes probability of error. We shall assume that the observation space is of dimension M, i.e., all of the variables of equation (2.1) are vectors rather than scalars. We shall see later how signal and noise wave forms can be represented by such vectors. We generalize from $\pm s$ of the example above by supposing that the symbols s_0 and s_1 are transmitted with probabilities P_0 and P_1, respectively. If the received sample falls in region $\mathbf{R}_i$, the decision is made that s_i $(i=0,1)$ was transmitted. A detection error is made if s_0 is transmitted and the observed random variable Y falls in the region $\mathbf{R}_1$ or vice versa. The probability error given that s_0 is transmitted is denoted by

$$Pr[Error|\mathbf{s}_0\ Transmitted] = Pr[\mathbf{Y}\in\ \mathbf{R}_1|\mathbf{s}_0\ Transmitted]$$

and, similarly,

$$Pr[Error|\mathbf{s}_1\ Transmitted] = Pr[\mathbf{Y}\in\ \mathbf{R}_0|\mathbf{s}_1\ Transmitted].$$

From the law of total probability, we may write the probability of error as

$$\begin{aligned}Pr[Error] &= Pr[Error,\ \mathbf{s}_0\ Transmitted] + Pr[Error,\ \mathbf{s}_1\ Transmitted]\\ &= P_0\ Pr[\mathbf{Y}\in\ \mathbf{R}_1|\mathbf{s}_0\ Transmitted] + P_1\ Pr[\mathbf{Y}\in\ \mathbf{R}_0|\mathbf{s}_1\ Transmitted].\end{aligned} \tag{2.2}$$

We now derive the rule for choosing $\mathbf{R}_0$ and $\mathbf{R}_1$ which minimizes the probability of error. In case of a continuous channel output, we have

$$Pr[Error] = P_0 \int_{\mathbf{R}_1} f(\mathbf{y}|\mathbf{s}_0 \ Transmitted) \ d\mathbf{y} + P_1 \int_{\mathbf{R}_0} f(\mathbf{y}|\mathbf{s}_1 \ Transmitted) \ d\mathbf{y} \ , \tag{2.3}$$

where $f(\mathbf{y}|\mathbf{s}_i \ Transmitted)$, $i = 0,1$ is the probability density function of the channel output conditioned on the transmitted signal. Integration is over the multidimensional regions $\mathbf{R}_1$ and $\mathbf{R}_0$, respectively.

In the case of a discrete channel output, the probability of error may be written as:

$$Pr[Error] = P_0 \sum_{\mathbf{y}_j \in \mathbf{R}_1} Pr(\mathbf{Y}=\mathbf{y}_j|\mathbf{s}_0 \ Transmitted)$$

$$+ P_1 \sum_{\mathbf{y}_j \in \mathbf{R}_0} Pr(\mathbf{Y}=\mathbf{y}_j|\mathbf{s}_1 \ Transmitted) \ , \tag{2.4}$$

where $\mathbf{y}_j$, $j = 1,2...,M$, are the points in the observation space and $Pr(\mathbf{Y}=\mathbf{y}_j|\mathbf{s}_i \ Transmitted)$, $i = 0,1$, are the appropriate conditional probabilities.

Since $\mathbf{R}_0$ and $\mathbf{R}_1$ partition the observation space, $\mathbf{R}_0 \cap \mathbf{R}_1 = \phi$ and $\mathbf{R}_0 \cup \mathbf{R}_1 = \Omega$, where ϕ denotes the null space and Ω denotes the whole space. Based on these relations, and the fact that

$$\int_{\mathbf{R}_0} f(\mathbf{y}|\mathbf{s}_0 \ Transmitted) \ d\mathbf{y} + \int_{\mathbf{R}_1} f(\mathbf{y}|\mathbf{s}_0 \ Transmitted) \ d\mathbf{y}$$

$$= \int_{\Omega} f(\mathbf{y}|\mathbf{s}_0 \ Transmitted) \ d\mathbf{y} = 1 \ ,$$

we may express (2.3) in the following form which makes the minimization problem straightforward:

$$Pr[Error] = P_0 \int_{\Omega} f(\mathbf{y}|\mathbf{s}_0 \ Transmitted) \ d\mathbf{y} + \int_{\mathbf{R}_0} \{-P_0 \ f(\mathbf{y}|\mathbf{s}_0 \ Transmitted)$$

$$+ P_1 \ f(\mathbf{y}|\mathbf{s}_1 \ Transmitted)\} \ d\mathbf{y}$$

$$= P_0 + \int_{\mathbf{R}_0} \{P_1 f(\mathbf{y}|\mathbf{s}_1 \ Transmitted) - P_0 \ f(\mathbf{y}|\mathbf{s}_0 \ Transmitted)\} \ d\mathbf{y}. \tag{2.5}$$

The only unknown quantity in (2.5) is the set of points which compose $\mathbf{R}_0$. Observe that $Pr[Error]$ can be minimized by placing in $\mathbf{R}_0$ only those points for which the integrand is negative. The proof that this leads to a minimum is by contradiction. Suppose that $\mathbf{R}_0$ contains a set of points for which the integrand is positive. The probability of error can be decreased simply by placing these points in $\mathbf{R}_1$. Points which are exactly equal to zero can be placed in either region without changing the probability of error.

Since they are the product of positive quantities, a probability and a probability density, the two terms in the integrand of (2.5) are positive and the integrand is negative at a point y if

$$\frac{P_0 f(\mathbf{y}|\mathbf{s}_0\ Transmitted)}{P_1 f(\mathbf{y}|\mathbf{s}_1 Transmitted)} = \frac{Pr(\mathbf{s}_0\ Transmitted|\mathbf{Y}=\mathbf{y})}{Pr(\mathbf{s}_1\ Transmitted|\mathbf{Y}=\mathbf{y})} > 1\ . \quad (2.6a)$$

Thus the decision rule that minimizes the probability of error is "decide s_0 was transmitted if (2.6a) is true and s_1 otherwise."

The decision rule in (2.6a) chooses the signal which maximizes the *a posteriori probability* of the transmitted symbol, given the observation; accordingly, it is called the maximum *a posteriori* (MAP) detector. As we shall see when we treat source coding in Section 2.4.2, for efficient transmission the *a priori* probabilities should be equal, $P_0 = P_1$. In this case the rule is to declare that s_0 was transmitted if

$$f(\mathbf{y}|\mathbf{s}_0\ Transmitted) > f(\mathbf{y}|\mathbf{s}_1\ Transmitted)\ , \quad (2.6b)$$

otherwise, choose $\mathbf{s}_1$ as the transmitted symbol.

EXAMPLE 1 We now apply the MAP detector to the one-dimensional Gaussian example that we considered in (2.1) above. Since the Gaussian noise has zero mean and mean square value σ^2, the probability density of Y is

$$f(\mathbf{y}|\mathbf{s}_i\ Transmitted) = \exp(-(\mathbf{y}\pm\mathbf{s})^2/2\sigma^2)/\sqrt{2\pi\sigma^2}\ .$$

Applying (2.6), we have the decision rule "choose the transmitted signal as s if

$$\exp(-(\mathbf{y}-\mathbf{s})^2/2\sigma^2) > \exp(-(\mathbf{y}+\mathbf{s})^2/2\sigma^2)."$$

Taking logarithms of both sides, we have, after some manipulation, the threshold rule, which is to choose $+\mathbf{s}$ as the transmitted signal if

$$\mathbf{y} > 0.$$

The decision space is shown in Figure 2.2, along with the conditional probability density functions. A similar development can be followed in the case of a discrete observation space. Again, the probability of error is minimized by choosing the maximum *a posteriori* estimate (MAP)

$$\frac{P_0 Pr(\mathbf{Y}=\mathbf{y}_j|\mathbf{s}_0\ Transmitted)}{P_1 Pr(\mathbf{Y}=\mathbf{y}_j|\mathbf{s}_1\ Transmitted)} = \frac{Pr(\mathbf{s}_0\ Transmitted|\mathbf{Y}=\mathbf{y}_j)}{Pr(\mathbf{s}_1\ Transmitted|\mathbf{Y}=\mathbf{y}_j)} \quad (2.7a)$$

In the case of equal *a priori* probabilities, it can be shown that the probability of error is minimized for each $\mathbf{y}_j$, $i = 1,2,...,M$, in the observation space by choosing $\mathbf{s}_0$ as the transmitted signal if

$$Pr(\mathbf{Y} = \mathbf{y}_j|\mathbf{s}_0\ Transmitted) > Pr(\mathbf{Y} = \mathbf{y}_j|\mathbf{s}_1 Transmitted)\ . \quad (2.7b)$$

Otherwise, choose $\mathbf{s}_1$ as the transmitted signal.

EXAMPLE 2 Consider the following example for the discrete case. Suppose that two signals are transmitted with equal probability over an optical fiber line. At the receiver, we assume that the optical detector emits electrons at a Poisson rate whose mean value depends upon the transmitted signal. For the two possible transmitted signals, s_0 and s_1, we denote these averages as λ_0 and λ_1 electrons per second, respectively. We decide which of the two signals was transmitted by counting the number of electrons emitted in a t second interval. Let Y_t denote the number of electrons emitted by the optical detector over the observation interval $(0,t)$; then the receiver decides in favor of s_0 over s_1 if the actual number of observed electrons, n, is such that

$$Pr(Y_t = n | s_0 \; Transmitted) > Pr(Y_t = n | s_1 \; Transmitted).$$

Since the number of emitted electrons is a Poisson* random variable, we have

$$Pr[Y_t = n | s_i \; Transmitted] = \exp(-\lambda_i t) \, (\lambda_i t)^n / n!, \; i = 0,1, \; n = 0,1,2,...$$

Since Y_t conditioned on the transmitted signal is a Poisson random variable, we choose s_0 over s_1 as the average rate if

$$\exp(-\lambda_0 t) \, (\lambda_0 t)^n / n! > \exp(-\lambda_1 t) \, (\lambda_1 t)^n / n!$$

or

$$n > (\lambda_0 - \lambda_1) t / (\ln \lambda_0 - \ln \lambda_1).$$

The equation says to choose the decision rule that says to select s_0 if the number of electrons is greater than the threshold on the right-hand side.

There is a related approach to decision making which results in the same decision rule as (2.6b) when the *a priori* probabilities are identical. The technique is called *maximum likelihood.* Given the received signal y one chooses the input which maximizes the likelihood $f(\mathbf{y} | \mathbf{s}_i \; Transmitted)$, $i = 0$ or 1. Maximum likelihood is appropriate when there is no way of determining the *a priori* probabilities P_0 and P_1. For example, in radar, where one attempts to detect the presence or absence of a reflected signal, the *a priori* probabilities generally have no meaning. In this case decisions are made on the basis of costs associated with each kind of error.

2.1.2 *L*-ARY TRANSMISSION

In many applications the channel can be utilized more efficiently by using L-ary transmission, where one of $L > 2$ symbols, s_i, $i = 0,1...,L-1$ is transmitted. It is almost always true that L is a power of two, $L = 2^l$, so that

* See Appendix 2A for the definition of a Poisson random variable.

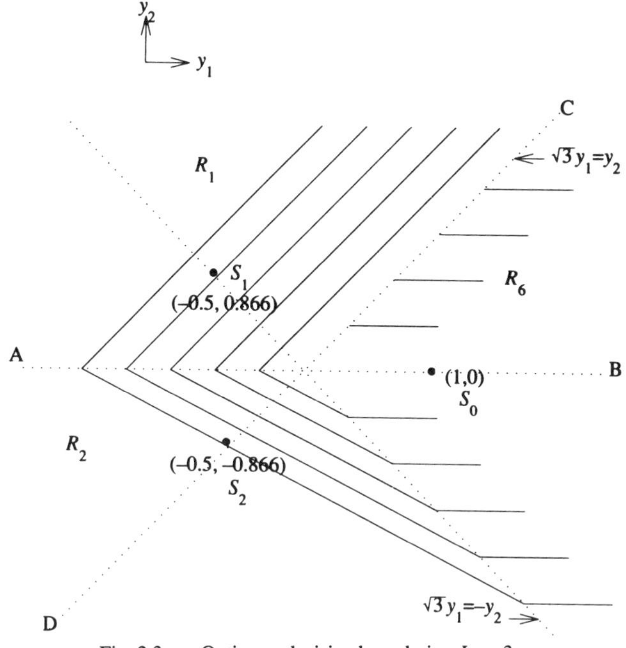

Fig. 2.3 Optimum decision boundaries: $L = 3$.

encoding a binary data stream is straightforward; each symbol conveys $\log_2 L$ bits.*

The results that we have obtained for the detection of binary signals apply directly to L-ary transmission. We imagine that the search for the most probable transmitted signal is conducted by comparisons of all possible pairs of transmitted signals. In the continuous case, the transmitted signal is chosen as s_i, where i is the value of the index such that $P_i\ f(\mathbf{y}|\mathbf{s}_i\ Transmitted)$ is maximum over $i = 0,1,...,L-1$ and where P_i is the probability that s_i is transmitted. In the discrete case, for a channel output i, s_i is chosen such that $P_i\ Pr(Y = \mathbf{y}_j|\mathbf{s}_i\ Transmitted)$ is maximum over $i = 0,1,...,L-1$. In the usual communications context, where all transmitted symbols are equally probable, the above decision rule is equivalent to maximum likelihood.

The partitioning of the two-dimensional observation space for the simple case of $L = 3$ is shown in Figure 2.3. Let the transmitted symbols be represented by points on the plane, $s_0 = (1,0)$, $s_1 = (-0.5, 0.866)$, and $s_2 = (-0.5, -0.866)$. (In Chapter 5, we shall see that these points might represent the phases of a sinusoidal carrier.) Let us assume that each of the two

---

* In speaking of transmitting in bits per second, we are using the expression in its commonly used sense; however, as we shall see in Section 2.4, there is a mathematically rigorous definition of the information-bearing content of a signal in bits.

components of the signal point representing a symbol is disturbed independently by additive Gaussian noise with mean square value σ^2. We also assume that the signals are equally probable. Since the noise components are independent, the joint conditional densities are the product of the marginals. The conditional densities of the channel outputs are:

$$f(\mathbf{y}|\mathbf{s}_0\ Transmitted) = \exp(-((\mathbf{y}_1 - 1)^2 + \mathbf{y}_2)/2\sigma^2)/2\pi\sigma^2$$

$$f(\mathbf{y}|\mathbf{s}_1\ Transmitted) = \exp(-((\mathbf{y}_1 + 0.5)^2/2 + (\mathbf{y}_2 - 0.866)^2/2\sigma^2)/2\pi\sigma^2$$

$$f(\mathbf{y}|\mathbf{s}_2\ Transmitted) = \exp(-((\mathbf{y}_1 + 0.5)^2/2 + (\mathbf{y}_2 + 0.866)^2/2\sigma^2)/2\pi\sigma^2$$

The derivation now proceeds in the same fashion as that of the one-dimensional Gaussian example considered above. The signal $\mathbf{s}_0$ is chosen over $\mathbf{s}_1$ if

$$||\mathbf{Y} - \mathbf{s}_0||^2 < ||\mathbf{Y} - \mathbf{s}_1||^2,$$

where $||\cdot||$ denotes the familiar Euclidean distance. It is interesting to note that the latter relationship describes a detector that chooses the transmitted signal that is *closest* to the received signal vector. The concept of a receiver which decides in favor of the signal that is at the *minimum distance* from the received signal is an important attribute of many of the receivers we will discuss later in this book. Returning to the above example, we find that after some manipulation, this leads to the criterion choose s_0 over s_1 transmitted if $\sqrt{3}\ \mathbf{y}_1 \geq \mathbf{y}_2$. The remaining regions are found the same way, with the result in the form of optimum decision regions shown in Figure 2.3. As we see from the figure, the boundaries of the decision regions are the perpendicular bisectors of the lines connecting the signal points.

2.1.3 PERFORMANCE—THE UNION BOUND

It is essential to describe the performance of these optimum detectors. In principle, the calculation of performance, i.e., the probability of making an error, is quite direct for the binary case. Having derived the optimum decision regions R_0 and R_1, one may evaluate (2.3) and (2.4) for the continuous and discrete cases, respectively. In this chapter we shall consider calculations of performance for two types of channels, the additive white Gaussian noise (AWGN) channel and the binary symmetric channel (BSC).

The calculation of performance is somewhat more complicated in the case of L-ary transmission, $L > 2$. From the law of total probability, we may write

$$Pr[Error] = 1 - Pr[Correct] = 1 - \sum_{i=0}^{L-1} P_i\ Pr[Correct|\mathbf{s}_i\ Transmitted]$$

$$= 1 - \sum_{i=0}^{L-1} P_i\ Pr[\ \mathbf{Y} \in \mathbf{R}_i|\mathbf{s}_i\ Transmitted]. \tag{2.8}$$

The key to evaluating the probability of error is the calculation of the term

$$Pr[\mathbf{Y} \in \mathbf{R}_i | \mathbf{s}_i \ Transmitted] = \int_{\mathbf{R}_i} f(\mathbf{y} | \mathbf{s}_i \ Transmitted) \ d\mathbf{y} \tag{2.9a}$$

in the continuous case, and

$$Pr[\mathbf{Y} \in \mathbf{R}_i | \mathbf{s}_i \ Transmitted] = \sum_{\mathbf{y}_i \in \mathbf{R}_i} Pr(\mathbf{Y} = \mathbf{y}_i | \mathbf{s}_i \ Transmitted) \tag{2.9b}$$

in the discrete case.

There are many situations in which the calculation of the quantities in (2.9a–b) is not straightforward. We shall see several examples later in the text. In these situations, we can obtain a useful upper bound on the error probability by applying the *union bound,* which is composed of errors involving only pairs of signals.

Suppose that the transmitted signal is s_i. Let E_{ij} denote the event that the observed point is such that s_j rather than s_i would be chosen. For example, in Figure 2.3, the event E_{12} would occur if the observation falls below the line A–B when s_1 is transmitted. The actual error event consists of the union of these events over all $j \neq i$, and we may write

$$Pr[Error | s_i \ Transmitted] = Pr[\bigcup_{i \neq j} E_{ij}]. \tag{2.9c}$$

It is important to observe that the events E_{ij} and E_{ik}; $i \neq j \neq k$ are not disjoint. This is obvious in the example in Figure 2.3 where E_{12} and E_{10} have considerable overlap since E_{10} consists of points to the right of C–D. An application of the standard laws of probability to (2.9c) yields the *union bound*

$$Pr[Error | s_i \ Transmitted] \leq \sum_{i \neq j} Pr[E_{ij}]. \tag{2.10}$$

The calculation of the pairwise errors in (2.10) is the same as we have considered for binary transmission. For example, for the case depicted on Figure 2.3, one would integrate the density conditioned on s_1 over the half plane below A–B and the half plane to the right of C–D if the transmitted signal were s_1 [see (2.3) and (2.4)].

2.2 THE ADDITIVE WHITE GAUSSIAN NOISE (AWGN) CHANNEL

2.2.1 THE MATCHED-FILTER RECEIVER

In physical channels the information symbols are translated into time-varying waveforms or pulses whose form is suitable for transmission over the

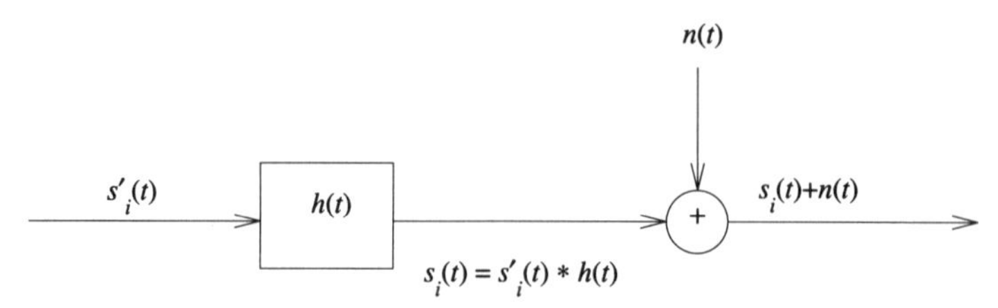

Fig. 2.4 The additive, linear noisy, communications channel.

channel. In a general sense, this process is called *modulation*.* We begin with the binary case, where the waveforms $s_0'(t)$ and $s_1'(t)$ are transmitted for a zero and a one, respectively. There are many practical examples of modulation techniques. Perhaps the most basic, known as non-return to zero (NRZ), is the encoding of a zero into a positive pulse and a one into its negative. In frequency shift keying† (FSK), data bits are represented by bursts of carrier at different frequencies. Over a short time interval we have $s_i'(t) = \sin(2\pi f_i t)$, $i = 0,1;\ f_0 \neq f_1$.

The signal is transmitted over the channel, where it is subject to the impairments discussed in Section 1.6 of Chapter 1. The model of the channel is depicted in Figure 2.4. The transmitted signal suffers delay and attenuation distortion represented by the channel which has impulse response $h(t)$ and the corresponding transfer function $H(f)$. In this chapter, we shall assume that $h(t)$ or, equivalently, $H(f)$ is known at the receiver. Random impairments are presented by additive noise. We also assume that there is no intersymbol interference, where energy from one symbol spills over into adjacent symbol intervals. Under this assumption, we can consider the detection of a single signal in isolation.**

Under these assumptions, the output of the channel may be written

$$y(t) = s_i(t) + n(t), \quad i = 0,1 , \tag{2.11}$$

where $s_i(t)$ is the convolution of the transmitted signal with the channel impulse response, $s_i(t) = s_i'(t) * h(t)$, $i = 0,1$.‡ The detection problem is to decide upon the transmitted signal, $s_0'(t)$ or $s_1'(t)$, based upon the output of the channel as represented in (2.11). Since the impulse response of the channel is known at

---

* There are often two levels of modulation in digital communication systems: the first level maps customer information, e.g., a binary stream, into a discrete set of signals, and the second level translates the signals into a form which is suited to the transmission media, modulation into a passband channel.

† See Section 5.6 for a discussion of FSK.

** In Chapter 4 intersymbol interference and the effect of channel shaping and of transmitter and receiver filtering are considered.

‡ The symbol * indicates convolution $s_i'(t) * h(t) = \int_{-\infty}^{\infty} s_i'(\tau)h(t-\tau)\,d\tau = \int_{-\infty}^{\infty} s_i'(t-\tau)h(\tau)\,d\tau$.

the receiver, this is equivalent to choosing between $s_0(t)$ and $s_1(t)$ transmitted over an undistorted channel. We shall adopt this approach throughout the chapter. For the moment, we shall assume that the noise is white,* with double-sided power spectral density that is constant with frequency

$$N(f) = N_0/2; \; -\infty < f < \infty \, . \tag{2.12a}$$

(In a practical system, the noise density need only be constant over the range occupied by the transmitted signal.) The autocorrelation function of the process $n(t)$ is the inverse Fourier transform of the spectral density $N(f)$ and is given by

$$R_N(t) = (N_0/2) \; \delta(t), \tag{2.12b}$$

where $\delta(t)$ is the Dirac delta function.

The detection of continuous time signals is a straightforward application of the detection principles that we have already developed. Suppose that the received signal is contained in the interval $(0,T)$. (We shall say more about the parameter T when we discuss the realization of the matched filter.) The derivation of the optimum receiver for continuous signals in additive Gaussian noise begins with the representation of $y(t)$ in $(0 \le t \le T)$, by a vector, $\mathbf{y} = (y_1, y_2, ..., y_M)$. Let $\phi_n(t)$ $n = 1,2,...$ be a set of orthonormal basis functions over the interval $(0, T)$ such that

$$\int_0^T \phi_m(t) \; \phi_n^*(t) \; dt = \begin{cases} 1, & m = n \\ 0, & m \neq n \end{cases} \; . \tag{2.13a}$$

where $\phi_n^*(t)$ is the complex conjugate of $\phi_n(t)$. The waveform $y(t)$ can be represented over $(0,T)$ by the expansion

$$y(t) = \sum_{n=1}^{\infty} y_n \; \phi_n(t) \tag{2.13b}$$

where

$$y_n = \int_0^T y(t) \; \phi_n^*(t) \, dt \, . \tag{2.13c}$$

Because it is so well known, perhaps the best example of a set of orthonormal basis functions is the Fourier series, where the orthogonal basis functions are the complex exponentials

$$\phi_n(t) = \exp\,(j2\pi nt/T)/\sqrt{T}, \qquad n = 0 \pm 1, \pm 2, \, ...$$

and the coefficients of the vector are

$$y_n = 1/\sqrt{T} \int_0^T y(t) \; \exp\,(-j2\pi nt/T) \; dt.$$

---

* The concept of white noise is explained in Appendix 2A.

In theory, an infinite number of coefficients may be required to represent a signal; however, one can represent the signal to any desired degree of accuracy with a truncated series. We assume that the number of coefficients, M, to represent a signal is large enough so that the mean square error,

$$E[(y(t) - \sum_{i=1}^{M} y_n \phi_n(t))^2] ,$$

is small, where $E[\cdot]$ indicates expectation over all realizations of the process. If this expression approaches zero as $M \to \infty$, then it is said that the series is *mean square convergent* to the process $y(t)$.

The received signal $y(t)$ is the sum of a signal component $s_i(t)$, $i = 0,1$ and a noise component. These components may also be expressed as vectors, relative to the same set of basis functions

$$\mathbf{s}_i = (s_{i1}, s_{i2}, \ldots, s_{iM}),\ i = 0,1$$

and

$$\mathbf{n} = (n_1,\ n_2, \ldots, n_M)$$

such that

$$\mathbf{y} = \mathbf{s}_i + \mathbf{n},$$

where each coefficient is given by

$$y_j = s_{ij} + n_j. \tag{2.14a}$$

The coefficients of the vectors defined above are given by

$$s_{ij} = \int_0^T s_i(t)\ \phi_j(t)\ dt,\ i = 0,1, \quad j = 1,2,\ldots,M \tag{2.14b}$$

and

$$n_j = \int_0^T n(t)\ \phi_j(t)\ dt,\ j = 1,2,\ldots,M. \tag{2.14c}$$

Expressing the received signal as a vector allows us to use the results of the previous section. Consider the projection of the white Gaussian noise on a set of orthonormal vectors as indicated in (2.14c). Since this operation is linear, n_j is also a Gaussian random variable. Further, since the noise has zero mean

$$E(n_j) = \int_0^T E[n(t)]\ \phi_j(t)\ dt = 0. \tag{2.15a}$$

Second moments can be found from the following considerations (see (2.12b)).

$$E[n_i n_j] = E\int_0^T dt_1 \int_0^T dt_2 n(t_1)n(t_2)\phi_i(t_1)\phi_j(t_2)$$

$$= \int_0^T dt_1 \int_0^T dt_2 E[n(t_1)n(t_2)]\phi_i(t_1)\phi_j(t_2)$$

$$= N_0/2 \int_0^T dt_1 \int_0^T dt_2 \delta(t_1 - t_2)\phi_i(t_1)\phi_j(t_2).$$

From the properties of the Dirac delta function we have

$$E[n_i n_j] = N_0/2 \int_0^T \phi_i(t)\ \phi_j(t)\ dt = \begin{cases} N_0/2, & i = j \\ 0, & i \neq j \end{cases}. \tag{2.15b}$$

From (2.15b) we see that $(n_1, n_2, \ldots, n_M)$ are uncorrelated. Since they are Gaussian, this implies that the noise coefficients are independent. We see also that each of the random variables $n_1, n_2, \ldots, n_M$ has zero mean and variance $N_0/2$. The joint Gaussian density function of these M independent random variables is

$$f_N\ (n_1, n_2, \ldots, n_M) = \prod_{i=1}^{M} (\exp(-n_i^2/N_0)/\sqrt{\pi N_0}). \tag{2.16}$$

Returning to processing the received vector, we have

$$\mathbf{y} = \mathbf{s}_i + \mathbf{n},\ \ i = 0,1.$$

Conditioned on the transmitted signal (either $s_0(t)$ or $s_1(t)$), the received signal y has a Gaussian distribution with mean s_i, $i = 0$ or 1 and variance determined by the noise variance. The probability density of the received signal plus noise conditioned on the transmitted signal, is given by

$$f_{\mathbf{Y}}(y_1, y_2, \ldots, y_M | s_i(t)\ \mathit{Transmitted}) = \prod_{j=1}^{M} \exp(-(y_j - s_{ij})^2/N_0) \Big/ \sqrt{\pi N_0},\ j = 0,1. \tag{2.17}$$

Substituting (2.17) into (2.6b), we have the decision criterion, choose $s_0(t)$ as transmitted if

$$\prod_{j=1}^{M} \exp\left[-(y_j - s_{0j})^2/N_0\right] > \prod_{j=1}^{M} \exp\left[-(y_j - s_{1j})^2/N_0\right]. \tag{2.18}$$

Since the logarithm is a monotonic function of positive arguments, taking the logarithm of both sides of (2.18) preserves the inequality. Rearranging terms, we derive the following decision rule:

Given the received vector $y_1, y_2, \ldots, y_M$, choose $s_0(t)$ as the transmitted signal if

$$\sum_{j=1}^{M} (y_j - s_{0j})^2 < \sum_{j=1}^{M} (y_j - s_{1j}) . \tag{2.19}$$

On the right- and left-hand sides of (2.19) we have the *Euclidean* distances in M-dimensional space between the received signal plus noise and s_0 and s_1 respectively. Our decision rule is therefore quite reasonable; we take the transmitted signal to be the one that is *closest* to the received signal. As we shall see, the idea of distance is a quite general criterion for choosing and discriminating between signals to transmit information; simply put, we desire signals which are as far apart as possible. We cannot simply increase the transmitted levels to values which are arbitrarily far apart, since the transmitted power, which is generally limited by system considerations, is a function of these levels. Therefore, it is a design problem to choose signals which make best use of the transmitted signal space and available power. We shall return to this point presently.

In theory, the foregoing derivation leads to an implementable receiver. For any received signal $y(t)$, we find the components $y_1, y_2, \ldots, y_M$ on a set of orthogonal basis functions. Then we find the distances given in (2.19) and compare. This is the way things might be implemented in a digital realization of a system. There is an alternative analog implementation which may be realized by means of time-invariant linear filters. This involves expressing the summations of coefficients in (2.19) by integrals of time functions. If $\phi_i(t)$, $i = 1,2,\ldots,M$ are a set of orthonormal basis functions*, then for

$$x(t) = \sum_{i=1}^{M} x_i \, \phi_i(t)$$

we have the following relationship between the coefficients and the continuous time signals:

$$\int_0^T x^2(t) \, dt = \int_0^T \sum_{i=1}^{M} x_i \, \phi_i(t) \sum_{j=1}^{M} x_j \, \phi_j(t) \, dt$$

$$= \sum_{i=1}^{M} \sum_{j=1}^{M} x_i \, x_j \int_0^T \phi_i(t) \, \phi_j(t) \, dt = \sum_{i=1}^{M} x_i^2 . \tag{2.20}$$

Since y_j and s_{ij}, $i = 0,1, j = 1,2,\ldots,M$ are coefficients of an orthogonal set of

* In Appendix 2B we consider a particular set of orthogonal basis functions, the Karhunen–Loève expansion, that provides a more rigorous basis for the derivation of the optimum receiver.

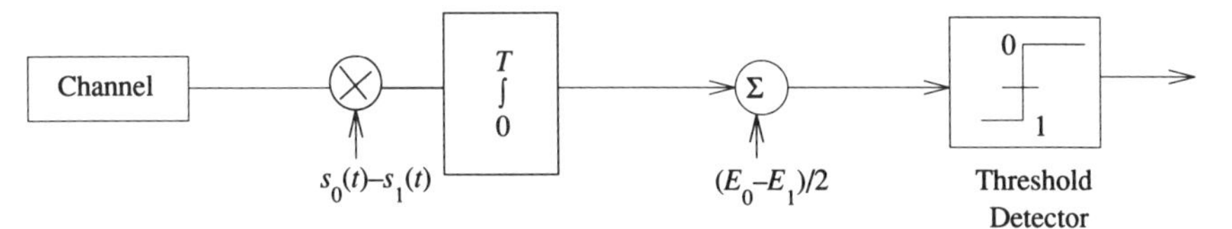

Fig. 2.5 Correlation receiver for binary transmission.

basis functions, we may express the terms in (2.19) as

$$\int_0^T [y(t) - s_0(t)]^2 \, dt < \int_0^T ([y(t) - s_1(t)]^2) \, dt \tag{2.21}$$

(see (2.14 a-b) and (2.15b)). By rearranging terms in (2.21) we have the following equivalent expression: choose $s_0(t)$ as the transmitted signal if

$$z = \int_0^T y(t)[s_0(t) - s_1(t)] \, dt > (E_0 - E_1)/2, \tag{2.22}$$

otherwise, choose $s_1(t)$ as the transmitted signal. In (2.22), $E_i = \int s_i^2(t) \, dt$ is the energy in signal $s_i(t)$ and (2.22) may be implemented as a *correlation receiver* whereby the received signal, $y(t)$, is correlated with $s_0(t)$ and $s_1(t)$. The correlation receiver is shown in Figure 2.5. There is an alternative implementation that is central to most of the receiver structures described in this text and those used in practice. Suppose we have a filter whose impulse response, $c(t)$, is such that

$$c(t) = s_0(T - t) - s_1(T - t). \tag{2.23}$$

The filter is said to be *matched* to $s_0(t) - s_1(t)$. Note that the filter derived in this fashion may be noncausal, inasmuch as $c(t)$ is not necessarily zero for $t < 0$. In practical realizations this can be remedied by choosing T large enough so that any response prior to $t = 0$ is negligible. The output of the matched filter sampled at time $t = T$, with input $y(t)$, is the desired correlation

$$z = \int_0^T y(t) \, c(T - t) \, dt = \int_0^T y(t) \, [s_0(t) - s_1(t)] \, dt.$$

The *matched filter** implementation involving $c(t)$, given by (2.23), is shown in Figure 2.6. The reason for the name matched filter can be seen by reference to

* The matched filter has also been called the North filter after its inventor [6]. See also [7] for a classic tutorial on the subject.

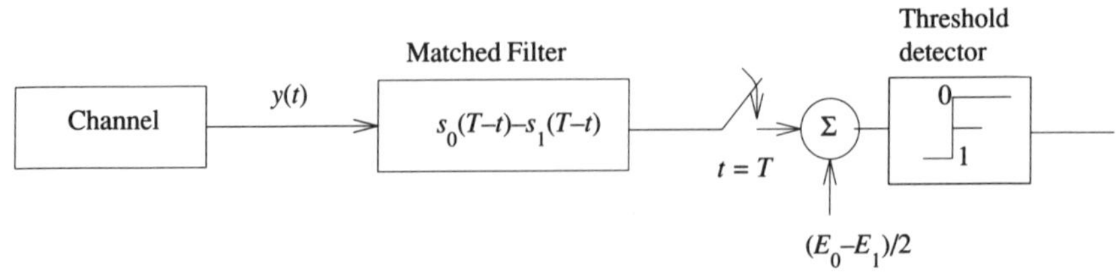

Fig. 2.6 Matched filter receiver: binary case.

the frequency domain. A filter with impulse response $c(t) = s_i(T-t)$ has a transfer function of the form $C(f) = S_i^*(f)\exp(-j2\pi fT)$, where $S_i(f)$ is the Fourier transform of $s_i(t)$ and the asterisk denotes the complex conjugate signal. In Figure 2.7 we sketch the absolute value of a typical transfer function, as well as the white noise spectrum with spectral density $N_0/2$. As indicated, the matched filter emphasizes the received signal over the frequency region where it is strong; since the noise is white, it is treated the same at all frequencies. It can be shown that the matched filter, with the impulse response given by (2.23), also maximizes the signal-to-noise ratio at time $t = T$, when $s_0(t)$ and $s_1(t)$ are the possible transmitted signals. (See Exercise 2.11 at the end of the chapter.)

2.2.2 NONWHITE NOISE

The preceding analysis was developed for the case of white Gaussian noise. Because of this assumption, the projections of the noise on the set of orthogonal basis functions resulted in a set of coefficients that are uncorrelated, and, since the joint density depends only on correlations, the noise samples are independent. The joint density of the elements of $\mathbf{y}$ is the product of the conditional marginal densities (see (2.16)). If the noise is colored, then the uncorrelatedness of the coefficients will be true only for the coefficients of what is known as the *Karhunen–Loève* expansion of the noise process.

A standard rigorous development of the receiver for colored noise which is based on the Karhunen–Loève expansion is given in Appendix 2B to this chapter; however, we can arrive at the same result by a heuristic argument. The essence of the argument is the realization that any *invertible* operation may be used at the receiver front end. If this operation were to degrade performance, it could be removed by a subsequent inverse operation. With this in mind, let us

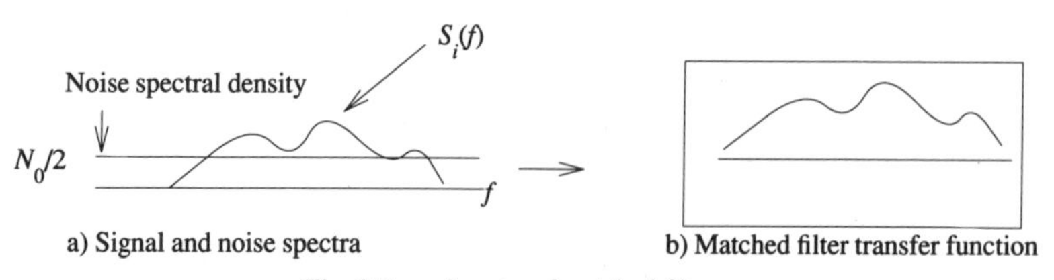

Fig. 2.7 Spectra of matched filter.

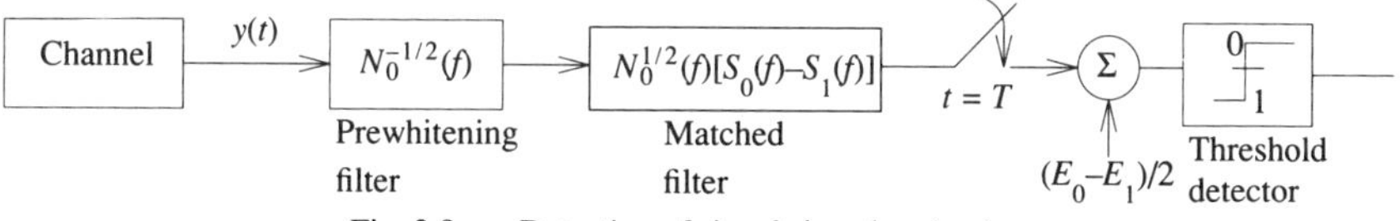

Fig. 2.8 Detection of signals in colored noise.

suppose that the power spectral density of the noise, is $N_0(f)$, which is not flat, and let us assume that there is no interval in the frequency domain for which the noise is zero. This is a reasonable assumption, since if there were such an interval, no matter how small, we could achieve an arbitrarily high transmission rate by placing a multilevel signal in this zero-noise region. Suppose we pass the channel output $y(t)$ through an invertible filter with transfer function $N_0^{-1/2}(f)$ (see Figure 2.8). Recall from Appendix 2A that the output spectral density is the product of the input spectral density and the magnitude *squared* of the filter transfer function. Thus the noise at this filter output will be white. Since $N_0(f)$ is everywhere nonzero, except for isolated points, the filter provides an invertible operation so that the signal $y(t)$ can always be fully recovered and we have not lost any information.

If the signal component of the input to this filter is $s_i(t)$, $i = 0,1$ then the signal component of the output is the convolution of $s_i(t)$ with the impulse response of the filter. We denote this output signal component as $r_i(t)$, $i = 0,1$. Now the noise component of the output is the colored noise passed through the $N_0^{-1/2}(f)$ filter and has power spectral density $N_0(f)|N_0^{-1/2}(f)|^2 = 1$, which is white. For obvious reasons, this filter is called a *noise-whitening filter*.

After passing through the noise-whitening filter we have a signal $r_i(t)$ immersed in white Gaussian noise. Since there has been no loss of information, the detection process is exactly the same as before with $s_i(t)$ replaced by $r_i(t)$, $i = 0,1$. Note that now the matched filter is matched to the line signal passed through the noise whitening filter. Thus the optimum receiver for colored noise is the matched-filter (or correlation receiver) preceded by a noise whitening filter.

2.2.3 *L*-ARY SIGNALING — PULSE AMPLITUDE MODULATION*

The results we have obtained for binary transmission over the AWGN channel are directly applicable to L-ary transmission. Anticipating material in Section 2.4, transmitting $L = 2^l$ signal levels is equivalent to transmitting l binary signals. In general, we say that selecting one of L equally probable levels conveys $\log_2 L$ bits of information.

---

* In Section 4.2, PAM is also treated. The distinction is that in the present case a single signal is considered, whereas in the sequel a sequence of signals is studied.

Suppose that the transmitted symbols are $s_i(t)$, $i = 0, 1,2,...,L-1$; $L > 2$. Suppose also that we wish to minimize the probability of error, i.e., the probability of choosing an output different than the true input. We can conceive of the detection process in this case as making a choice between all possible pairs of signals. Since the noise is white and Gaussian, the same derivation which led to (2.21) applies and we have the rule, choose $s_k(t)$ if

$$\int_0^T [y(t) - s_k(t)]^2 \, dt \tag{2.24a}$$

is minimum over all possible transmitted signals

$$s_i(t), \; i = 0,1,...,L-1.$$

This is equivalent to choosing the signal that maximizes the term

$$\int_0^T [s_k(t) \, y(t) - s_k^2(t)/2] \, dt. \tag{2.24b}$$

The correlation receiver and the matched-filter receiver for L-ary detection are shown in Figures 2.9 a and b, respectively.

The results in the preceding paragraph apply irrespective of the form of the transmitted signals $s_0(t), s_1(t),..., s_{L-1}(t)$. The case of L-ary signaling most often used is *pulse amplitude modulation* (PAM), where the information is conveyed by the amplitude of the same basic pulse, $p(t)$. Thus, $s_i(t) = a_i p(t)$, $i=0,1,...,L-1$. We assume, with no loss of generality that $a_0 < a_1 < \cdots < a_{L-1}$. In this case from (2.24) we choose $s_k(t)$ over $s_m(t)$ if

$$a_k \int_0^T y(t) \, p(t) \, dt - 1/2 \, a_k^2 \, E_p > a_m \int_0^T y(t) \, p(t) \, dt - 1/2 \, a_m^2 \, E_p \tag{2.25}$$

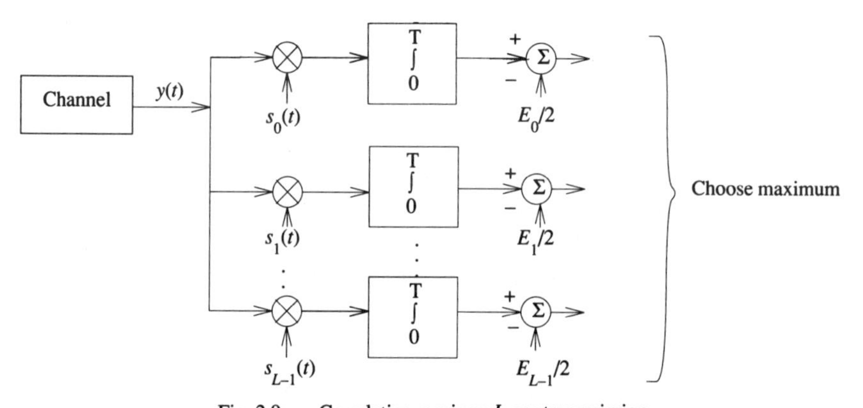

Fig. 2.9a Correlation receiver: L-ary transmission.

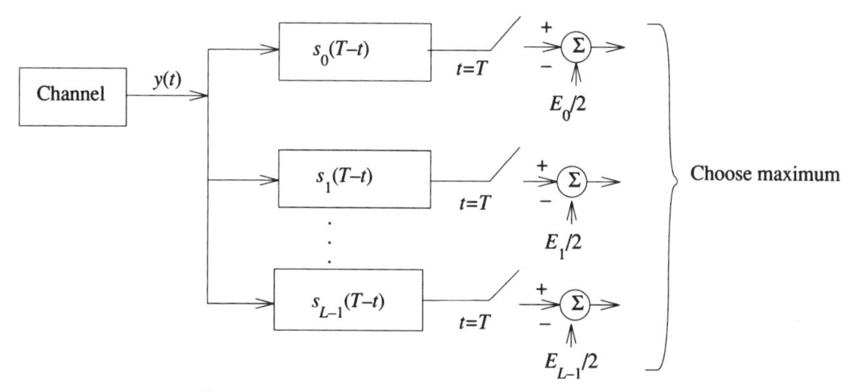

Fig. 2.9b Matched filter receiver: L-ary transmission.

where E_p is the pulse energy $\int_0^T p^2(t)\ dt$. Rearranging terms, we have the decision rule, choose $s_k(t)$ over $s_m(t)$ if

$$\int_0^T y(t)\ p(t)\ dt > (a_k + a_m)\ E_p/2\ . \tag{2.26}$$

Thus a threshold between two pairs lies midway between their respective pulse amplitudes. Considering all of the possible transmitted signals, we have the following decision rules:

(1) Choose $s_0\ (t)$ if $\int_0^T y(t)\ p(t)\ dt \leq (a_0 + a_1)\ E_p/2$

(2) Choose $s_i(t),\ i = 1,2,...,L-2,$ if

$$(a_{i-1} + a_i)\ E_p/2 < \int_0^T y(t)\ p(t)\ dt \leq (a_i + a_{i+1})\ E_p/2 \tag{2.27}$$

(3) Choose $s_{L-1}(t)$ if $(a_{L-2} + a_{L-1})\ E_p/2 < \int_0^T y(t)\ p(t)\ dt$

If we assume* that there is equal spacing of 2 units of signal amplitudes between levels and that the levels are odd integers that are symmetrical about zero, we have $a_i = [2i + 1 - L],\ i = 0,1,...,L-1$. The detection regions, which lie between the *slicing levels*, are as shown in Figure 2.10, where $E_p = 1$ and $R_{(L-1+i)/2}$ is the decision region for the ith level.

* Since there is asymmetry between outer and inner decision regions, there may be some slight advantage in nonequally spaced signal levels.

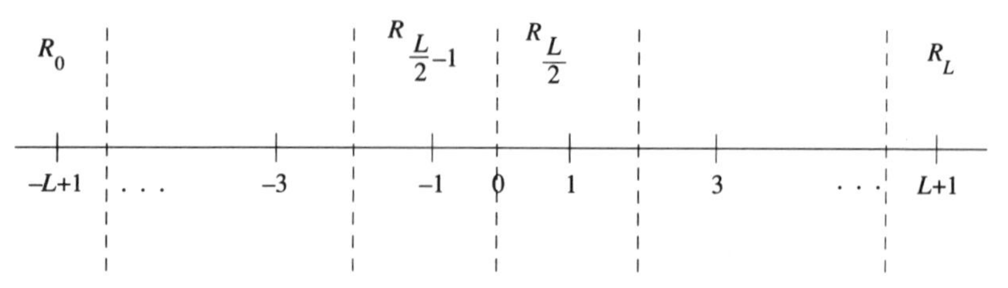

Fig. 2.10 Decision regions for equally spaced multilevel PAM.

2.2.4 CALCULATION OF PERFORMANCE — BINARY SIGNALS

From (2.22) we observe that the decision on the actual transmitted signal is based on a single number, z, the output of the matched filter sampled at time $t = T$. This number, which is called a *sufficient statistic* in the language of statistical decision theory, is distilled from the received waveform over a whole time interval. This quantity, z, being a function of the transmitted signal and channel noise, is a random variable. The probability distribution of z determines the value of our performance measure, the probability of error.

We begin with the binary case. A transmission error is made if the transmitted signal is $s_0(t)$, and $z < (E_0 - E_1)/2$, or if the transmitted signal is $s_1(t)$ and $z \geq (E_0 - E_1)/2$ [see (2.22)]. Using the law of total probability, the probability of error is given by

$$Pr(Error) = P_0 \, Pr(z < ((E_0 - E_1)/2) | s_0(t) \; Transmitted)$$

$$+ P_1 \, Pr(z \geq ((E_0 - E_1)/2) | s_1(t) \; Transmitted) \; . \tag{2.28}$$

Conditioned on the transmitted signal, $s_0(t)$ or $s_1(t)$, z is a Gaussian random variable since it is simply the result of the linear filtering of a Gaussian process. The distribution of a Gaussian random variable is determined by its mean and variance. The conditional expectation of z is

$$\begin{aligned} E[z|s_i(t) \; Transmitted] &= \int_0^T [s_0(t) - s_1(t)] \, E[y(t)|s_i(t) \; Transmitted] \, dt \\ &= \int_0^T [s_0(t) - s_1(t)] \, s_i(t) \, dt \\ &= \begin{cases} E_0 - \rho, & i = 0 \\ \rho - E_1, & i = 1 \end{cases} \end{aligned} \tag{2.29}$$

where the cross correlation between the signals is defined as

$$\rho = \int_0^T s_0(t) \, s_1(t) \, dt \, .$$

The conditional variance of z is

$$Var\ (z|s_i(t)\ Transmitted) = E\{[\int_0^T [y(t)-s_i(t)][s_0(t)-s_1(t)]\,dt]^2\}$$
$$= \int_0^T dt_1 \int_0^T dt_2\ [s_0(t_1) - s_1(t_1)]\ [s_0(t_2) - s_1(t_2)]E[n(t_1)n(t_2)]\ . \tag{2.30}$$

Since the channel noise is white, we have

$$Var\ (z|s_i(t)\ Transmitted) = \int_0^T dt_1 \int_0^T dt_2\ (s_0(t_1) - s_1(t_1))$$
$$(s_0(t_2) - s_1(t_2))\ (N_0/2)\ \delta\ (t_1-t_2) = (N_0/2) \int_0^T dt\ (s_0(t) - s_i(t))^2$$
$$= N_0(E_0 + E_1 - 2\rho)/2, \qquad i = 0,1\ . \tag{2.31}$$

From the mean and variance of z, the conditional error probabilities can be calculated as

$$Pr[z < (E_0 - E_1)/2|s_0(t)\ Transmitted]$$
$$= \int_{-\infty}^{(E_0-E_1)/2} \frac{\exp(-(z - (E_0 - \rho))^2/N_0(E_0 + E_1 - 2\rho))}{\sqrt{\pi N_0\ (E_0 + E_1 - 2\rho)}}\, dz\ .$$

After the change of variable $w = -(z - (E_0 - \rho))/\sqrt{N_0(E_0 + E_1 - 2\rho)/2}$, we have

$$Pr[z < (E_0 - E_1)/2|s_0(t)\ Transmitted] = Q([(E_0 + E_1 - 2\rho)/2N_0]^{1/2})\ , \tag{2.32a}$$

where* $Q(x) = \int_x^\infty \frac{\exp(-t^2/2)}{\sqrt{2\pi}}\, dt$. Similarly, it can be shown that

$$Pr[z \geq (E_0 - E_1)/2\ |\ s_1(t)\ Transmitted]$$
$$= \int_{(E_0-E_1)/2}^{\infty} \frac{\exp(-(z - (\rho - E_1))^2/N_0(E_0 + E_1 - 2\rho))}{\sqrt{\pi N_0(E_0 + E_1 - 2\rho)}}\, dz$$
$$= Q([(E_0 + E_1 - 2\rho)/2N_0]^{1/2})\ . \tag{2.32b}$$

---

* An alternative to $Q(x)$, often used to express Gaussian probabilities, is the complementary error function, which is defined

$$\text{erfc}(x) = \frac{2}{\sqrt{\pi}} \int_x^\infty \exp\ (-t^2)\ dt\ .$$

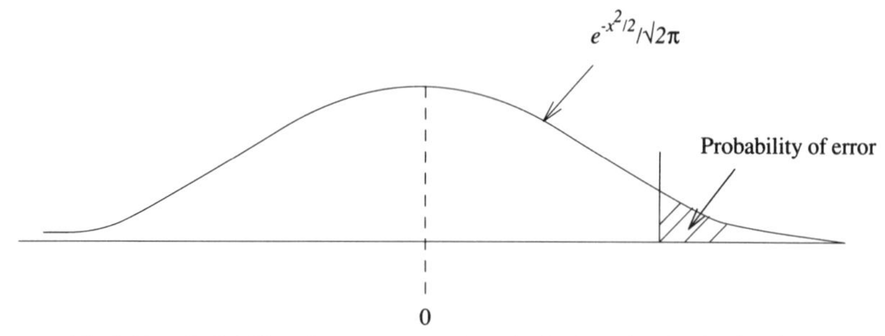

Fig. 2.11 Probability of error as area under tail of a Gaussian density function.

Thus we have

$$Pr[Error] = Q([(E_0 + E_1 - 2\rho)/2N_0]^{1/2}) , \qquad (2.33)$$

and the probability of error is simply the area under the tail of a Gaussian density function which has mean zero and variance one. (See Figure 2.11.) When $x > 2.5$, $Q(x)$ may be bounded fairly closely.[2] To see this, we integrate $Q(x)$ by parts to obtain, for $x > 0$,

$$Q(x) = \frac{1}{\sqrt{2\pi}} [\exp(-x^2/2) / x - \int_x^\infty \exp(-t^2/2) / t^2 dt],$$

but

$$0 < \int_x^\infty \exp(-t^2/2)/t^2 \, dt < \frac{1}{x^3} \int_x^\infty t \exp(-t^2/2) \, dt = \exp(-x^2/2)/x^3 \, ;$$

consequently,

$$(1 - 1/x^2) \exp(- x^2/2) / x\sqrt{2\pi} < Q(x) < \exp(-x^2/2) / x\sqrt{2\pi} \qquad (2.34a)$$

Another bound which is useful, albeit looser, is given by

$$Q(x) < \exp(- x^2/2)/2 \, . \qquad (2.34b)$$

The bounds given in (2.34a and b) are plotted in Figure 2.12, where P_e in equation (2.33) is shown as a function of $10\log[E_0 + E_1 - 2\rho)/2N_0]$ This is in accordance with the usual practice of referring to the signal-to-noise ratio in units of decibels (dB); we use a logarithmic scale for the abscissa. It is standard to divide both terms in the energy-to-noise density ratio, E/N_0, by the time duration (or equivalently the inverse of the bandwidth), T. This ratio is then referred to as the signal-to-noise (power) ratio. As we see, the upper and lower bounds of (2.34a) are reasonably close to one another. The bound of (2.34b) is often used because of its simple form.

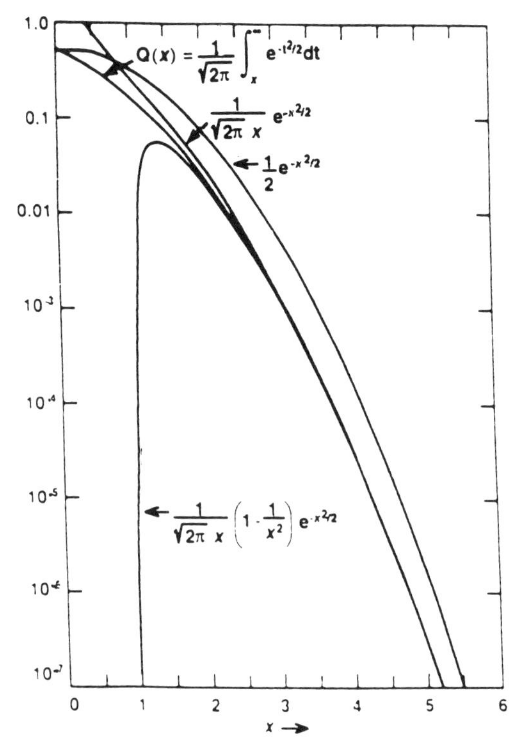

Fig. 2.12 Plot of error rate bounds.

2.2.5 SIGNAL DESIGN—BINARY CASE

In the foregoing development, it was not necessary to specify the actual shapes of the transmitted signals $s_0(t)$ and $s_1(t)$. However, through the parameters E_0, E_1, and ρ, the transmitted signals affect the performance. It is clear from (2.33) that the probability of error is minimized when $(E_0 + E_1 - 2\rho)/2N_0$ is maximized. From the *Schwartz inequality*, we have

$$\rho^2 = \{\int_0^T s_0(t)s_1(t)\, dt\}^2 \le \int_0^T s_0^2(t)\, dt \int_0^T s_1^2(t)\, dt = E_0\, E_1 ,$$

with equality if $s_0(t) \sim s_1(t)$. With this in mind, it is clear then that the probability of error is minimized when ρ is as negative as is possible. This is obtained by *antipodal* signals, i.e., $s_0(t) = -s_1(t)$. In this case,

$$\rho = -E = -\int_0^T s_i^2\, (t)\, dt,\ i = 0,1 ,$$

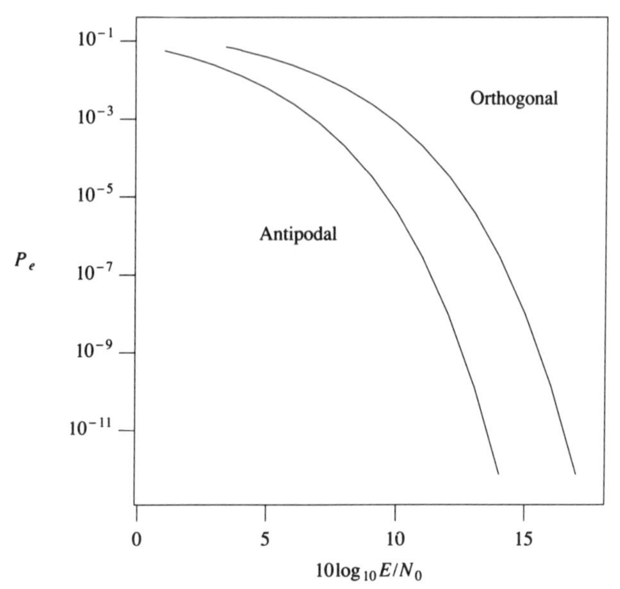

Fig. 2.13 Probability of error vs. signal-to-noise ratio for binary antipodal and binary orthogonal signaling.

and the error rate is

$$Pr[Error] = Q([2E/N_0]^{1/2}) \,. \tag{2.35}$$

The fact that antipodal signals are optimal is not surprising. In Gaussian noise, the detection process depends on distances and, for a given average power, antipodal signals are maximally distant from one another.

There are situations where it may not be convenient or possible to transmit antipodal signals. For example, in a passband channel, phase coherence is needed to transmit signals of opposite signs (see Chapter 5). Thus, if it is not possible to achieve phase coherence (e.g., the carrier phase is jittering at a very high rate), then antipodal signaling may not be possible. Another example is direct-detection fiber optic communications, where the most common form of signaling is intensity modulation of the optical source. Since intensity is a positive quantity, antipodal signaling is not possible. A suboptimum, but often used alternative is *orthogonal* signaling, where $\rho = 0$. Frequency shift keying (FSK), which is discussed in Chapter 5, is a good example of orthogonal signaling. From (2.33) the probability of error for binary orthogonal signaling is

$$Pr[Error] = Q([E/N_0]^{1/2}) \,, \tag{2.36}$$

under the assumption that $s_0(t)$ and $s_1(t)$ have equal energy.

In Figure 2.13, the probability of error is shown as a function of signal-to-noise ratio, E/N_0, for both antipodal and orthogonal signaling. As is evident from (2.35) and (2.36), antipodal signaling achieves the same error rate with a factor of 2 or 3 dB, less signal-to-noise ratio than orthogonal signaling. Thus, in order to achieve the same probability of error, orthogonal signals must have twice as much energy as antipodal signals. However, sometimes designers will specify orthogonal signaling either because coherence cannot be achieved at the receiver or because of the simplicity of the detectors for orthogonal signals.

2.2.6 PERFORMANCE — PAM

Because of the linear structure of the signals and the decision regions, it is not difficult to calculate the probability of error for a multilevel PAM signal. The same formula for $Pr[Error]$ used in the binary case applies; we integrate a conditional Gaussian density function over an interval. Again, we focus on the sufficient statistic

$$z = \int_0^T y(t)\ p(t)\ dt\ .$$

For L-ary PAM, the transmitted signal is $s_k(t) = [2k + 1 - L]\ p(t)$, where $k = 0,1,...,L-1$. The conditional mean of z is

$$E(z|s_k(t)\ Transmitted) = [2k + 1 - L] \int_0^T p^2(t)\ dt = [2k + 1 - L]E_p\ ,$$
$$k = 0,1,...,L-1 \tag{2.37a}$$

and the conditional variance is

$$\begin{aligned} var(z|s_k(t)\ Transmitted) &= E \int_0^T dt_1 \int_0^T dt_2\ p(t_1)\ p(t_2)\ n(t_1)\ n(t_2) \\ &= N_0\ E_p/2 \end{aligned} \tag{2.37b}$$

An error will occur when z falls outside the decision boundaries for $s_k(t)$. Suppose that a_k is an inner level, i.e., $k = 1,2,...,L-2$, then the probability of an error is the probability that z is outside the adjacent boundaries, $[2k - L]E_p$ and $[2k + 2 - L]E_p$. For the conditional probability of error we have

$$\begin{aligned} &Pr(Error|s_k(t)\ Transmitted) \\ &= Pr(z<(2k-L)E_p \text{ or } z>(2k+2-L)E_p|s_k(t)\ Transmitted) \\ &= 1-Pr[|z-(2k+1-L)E_p|<E_p|s_k(t)\ Transmitted]\ . \end{aligned} \tag{2.38}$$

For the two outer levels, a_0 and a_{L-1}, the calculation of probability of error is slightly different and is given by

$$Pr[Error|s_0(t)\ Transmitted] = Pr[z > (-L + 2)E_p|s_0(t)\ Transmitted] \ , \tag{2.39a}$$

and

$$Pr[Error|s_{L-1}(t)\ Transmitted] = Pr[z < (L - 2)E_p|s_{L-1}(t)\ Transmitted] . \tag{2.39b}$$

There are $L-2$ terms of the same form as (2.38). With the conditional means and variance of the Gaussian random variable, z, given by (2.37a) and (2.37b), the conditional error rate is

$$Pr(Error|s_k(t)\ Transmitted) = 2\ Q([2E_p/N_0]^{½}),\ k = 1,2,...,L-2 . \tag{2.40a}$$

For each of the outer terms in (2.39a-b) we have

$$Pr(Error|s_0(t)\ Transmitted) = Pr(Error|s_{L-1}(t)) = Q([2E_p/N_0]^{½}) \ , \tag{2.40b}$$

and the probability of error averaged over all transmitted signals, assuming that they are equally probable, is then

$$Pr(Error) = 2(1-1/L)\ Q([2E_p/\ N_0]^{½}) \ , \tag{2.41}$$

As a check, we notice that for $L=2$ (binary antipodal signaling) we have the correct formula for probability of error (see 2.35). The formula in (2.41) is in terms of energy in the pulse $p(t)$. We can express the average transmitted signal energy in terms of this quantity* by

$$E = (1/L)\sum_{i=0}^{L-1} (2i + 1 - L)^2\ E_p = [L^2 - 1]E_p/\ 3 \ , \tag{2.42}$$

---

* In deriving (2.42), the identity

$$\sum_{i=1}^{I} i^2 = I(2I+1)(I+1)/6$$

is quite useful. This may be shown by the following steps. From a simple change of variables, we have

$$\sum_{i=1}^{I}(i+1)^3 = \sum_{i=1}^{I} i^3 + (I+1)^3 - 1 .$$

Expanding terms, we have

$$\sum_{i=1}^{I}(i+1)^3 = \sum_{i=1}^{I} i^3 + 3\sum_{i=1}^{I} i^2 + 3\sum_{i=1}^{I} i + I .$$

Equating the RHSs and eliminating common terms, we prove the identity.

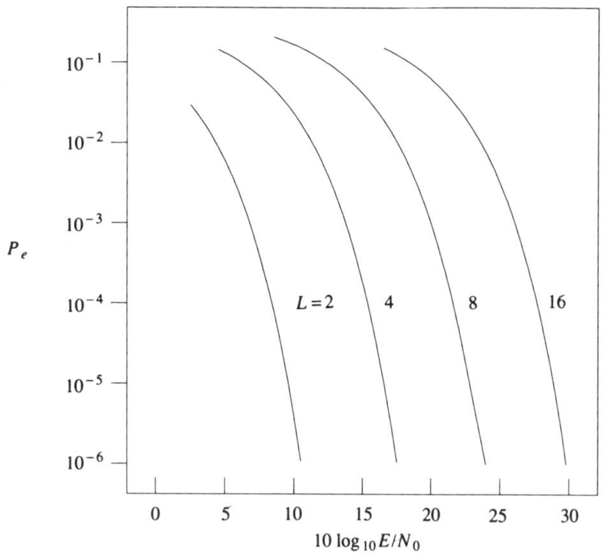

Fig. 2.14 Probability of error P_e vs. signal-to-noise ratio for L-level PAM, where E/N_0 is the signal-to-noise ratio.

and the probability of error becomes

$$Pr[Error] = 2(1 - 1/L)\,Q\,([6E/(L^2 - 1)N_0]^{1/2})\,. \qquad (2.43)$$

It is common practice to *Gray code* signals so that there is a change of only a single bit between adjacent signal levels. Hence, for high signal-to-noise ratios, the probability of error is the probability of a single bit error with this coding. Equation (2.43), which is plotted in Figure 2.14, indicates one of the basic characteristics of communications engineering—a tradeoff between bit rate and signal-to-noise ratio. If the energy per transmitted symbol is held constant, and the number of levels is increased, thereby increasing the number of bits per transmitted symbol, the probability of error increases. Thus, to increase the number of transmitted bits per symbol and maintain performance, it is necessary to increase the transmitted energy. For example, in order to double the number of signal levels, say from 4 to 8, we need 6 dB more signal-to-noise ratio. Notice that doubling the number of signal levels only increases the number of bits per symbol by *one bit*.

2.2.7 BANDWIDTH AND TRANSMISSION RATE

In the immediately preceding section, the effect of signal-to-noise ratio on the transmission of information over a channel of arbitrarily wide bandwidth was discussed. For practical systems of limited bandwidth, the other important factor is the dimensionality of a signal which is embodied in its time-bandwidth product. The dimensionality of a signal is the number of its linearly independent components. It can be shown that a signal having bandwidth W and of duration T has the approximate dimensionality $2WT$. Thus a signal with wider bandwidth can convey more information than a signal with the same duration and narrower bandwidth. To illustrate—from the time-domain sampling theorem we know that a signal which is strictly bandlimited to W Hz may be described by $2W$ samples per second. If the duration of the signal were T seconds, it could be specified, approximately, by $2WT$ samples. The approximation here lies in the fact that a bandlimited signal cannot be strictly time limited and vice versa.

In practical communication systems, user data is transmitted as a sequence of signals. In this process, the channel bandwidth, as well as the signal-to-noise ratio, comes into play. If a PAM signal having $L = 2^K$ possible levels is transmitted every T seconds, this is a data rate of $R = K/T$ bits per second. Suppose we wish to increase R. There are two ways that this can be done, either by increasing the signaling rate, $1/T$, or increasing the number of levels per signal. An increase in either of these quantities will require an increase in the signal-to-noise ratio for the performance to remain constant.

If the signaling (or symbol) rate is increased, the bandwidth occupied by the signal is also increased, since the bandwidth of a signal is proportional to the reciprocal of its duration. As indicated in (2.43), for PAM the probability of error is a function of the ratio of the signal energy, and the noise power spectral density, E/N_0, irrespective of the signal duration. However, if the duration of a signal is halved while keeping its power constant, the average received noise power, N_0/T, is doubled. Nevertheless, there is a gain associated with using the bandwidth, if it is available. From Figure 2.14 we see that doubling the data rate by increasing the number of levels requires at least 6 dB in signal-to-noise ratio; whereas doubling the signaling rate requires only 3 dB more signal power. Assuming that the channel has enough bandwidth so that the signals remain undistorted, then increasing the symbol rate is a much more efficient means of increasing the data rate than increasing the number of levels. For channels that either distort the signals or have frequency-dependent noise, a detailed evaluation must be made.

We shall further explore the role of bandwidth in data transmission by presenting two contrasting examples of signaling. Suppose we wish to transmit K bits in a T-second interval. The first approach we consider is binary pulse-by-pulse signaling in which K antipodal-nonoverlapping signals, each of duration T/K, are transmitted over an AWGN channel. The signals transmitted in each time slot of T/K sec are detected independently. The probability of error for each bit, P_e, is given by (2.35) with parameter E_b/N_0, where E_b is the energy of the antipodal signal. The energy per bit here is simply E_b. To determine the probability of at least one error in the entire K-bit block, we note that since the binary signals occupy separate intervals and the noise is white, errors in each interval are independent events and the probability of a block error is given by

$$P_K = 1 - (1 - P_e)^K . \qquad (2.44)$$

It is not difficult to see that as $K \to \infty$, $P_K \to 1$. Thus, unless the signal-to-noise ratio is made arbitrarily large, the probability that an error will be made approaches unity. As far as the block error rate is concerned, it doesn't matter if K is increased by increasing T while keeping the pulse rate constant, or by increasing the pulse rate. In the latter case, it is necessary to increase the bandwidth and the transmitted power in proportion. As K increases, the probability of an individual bit error stays the same, while the number of places where an error can occur increases.

Consider now block orthogonal signaling, which is a form of *pulse position modulation* (PPM). The interval T is segmented into $L = 2^K$ slots. The transmission of K bits is accomplished by transmitting a pulse in *one* of these slots during the T-second interval. In this example, the transmission and reception of the k bits are *not* done independently of each other. Since any pair of signals occupy nonoverlapping slots, they are orthogonal. Suppose that the pulses all have energy E_s, then the probability of any pairwise error is given by (2.36) with parameter E_s/N_0. This pairwise error can be inserted into union bound, (2.10), to obtain the upper bound on the error rate. The error in comparing any pair of orthogonal signals is

$$P_K \leq (L-1)Q([E_s/N_0]^{1/2}).$$

We apply the bound of (2.34b) in order to simplify the expression thereby obtaining

$$P_K < L \exp(-E_s / 2N_0). \qquad (2.45)$$

Since we are sending K bits in T seconds, the transmission rate is $R = K/T$ bps, so $L = 2^{RT}$. Using (2.45) we can write

$$P_K < 2^{RT} \exp(-P_s T/2 N_0) = \exp(-T [(P_s/2N_0) - R\ln 2]), \qquad (2.46)$$

where $P_s = E_s/T$ is the signal power and where ln denotes the natural logarithm. For (2.46), we see that the probability of error goes to *zero*, at an *exponential* rate, with increasing T, as long as the rate over the channel satisfies

$$R < P_s/(2N_0 \ln 2) . \qquad (2.47)$$

As we shall see later in this chapter, this result is a particular example of a profound concept developed by Claude Shannon in the 1940's, the information-bearing *capacity* of a channel. In deriving (2.47), we were fairly crude. As we shall see presently, it is theoretically possible to transmit at *twice* this information rate with a zero error rate.

For orthogonal signaling, the signals have equal energy and signal space here can be viewed as points on a sphere in 2^K dimensions, with each signal point occupying a dimension. However, as the space grows in dimension, the distance between signals remains constant at $\sqrt{2\, E_s}$. Since the energy per bit is $E_b = E_s/K$, the distance in terms of E_b increases with $\sqrt{K}$.

Another difference between the two signaling schemes lies in their bandwidth. In the case of bit-by-bit signaling, the bandwidth of the signals is proportional to K/T, whereas for block orthogonal signaling the bandwidth is proportional to $2^K/T$. Block orthogonal signaling is primarily of theoretical interest, since it is quite profligate in its use of bandwidth. There are more efficient ways of trading bandwidth for performance. As we shall see in the next chapter, bandwidth may be used to transmit redundant binary digits in the form of error correcting codes.

2.3 THE BINARY SYMMETRIC CHANNEL

In certain applications, the communications channel can be characterized as a digital pipe with known input/output performance. In this case, it is useful to make an abstraction of the communication system by looking at the purely digital level of end-to-end performance. Suppose that we regard the channel as consisting of the modulator and the demodulator, as well as the physical channel (see Figure 2.1). For the case of L-ary transmission over the channel, we regard the binary to L-ary encoder as being part of the modulator. The input and output of the augmented channel would be a binary sequence. As we shall see in Chapter 3, this abstraction is particularly useful in dealing with error correcting and detecting codes.

The binary channel may be characterized by output probabilities conditioned on the input symbols, which are given by

$$Pr\,[\text{output } 0 \,|\, \text{input } 0] = P_0,\; Pr\,[\text{output } 1 \,|\, \text{input } 0] = 1 - P_0$$

$$Pr\,[\text{output } 0 \,|\, \text{input } 1] = 1 - P_1,\; Pr\,[\text{output } 1 \,|\, \text{input } 1] = P_1$$

If the physical channel were the AWGN Channel, these error probabilities would be given by (2.33). In most cases of interest there is symmetry and $P_0 = P_1 = P$ and $P < 1/2$. We have then the binary symmetric channel (BSC), which may be represented by the diagram shown in Figure 2.15. A variation on the BSC is the *binary erasure channel* (BEC). In this case there is a third output, the erasure symbol indicating indecision on the channel output. The input–output relations for the BEC are shown in Figure 2.16.

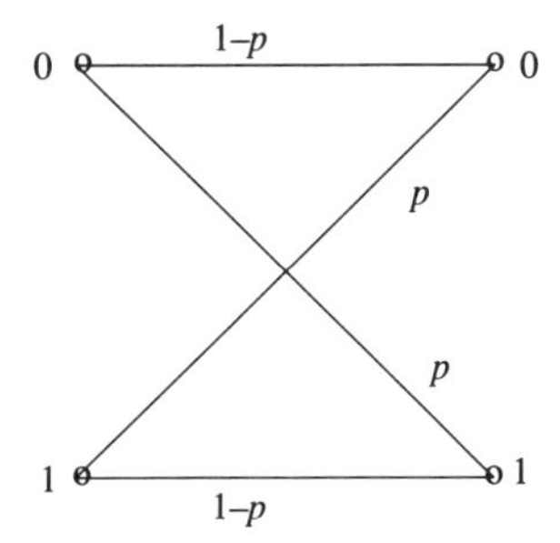

Fig. 2.15 Representation of the binary symmetric channel.

As we shall see in Chapter 3, error correcting and detecting codes improve performance by adding redundant binary digits to a binary information sequence. One of the two ways of doing this is by means of *block codes*. In these codes a block of k information bits is encoded into one of 2^k codewords which are n bits in duration. The redundancy lies in the fact that $n > k$, so that of the 2^n possible received sequences only one of a much smaller number, 2^k, was actually transmitted.

In Section 2.5, we saw that protection against channel noise is obtained by designing signals to be maximally distant from one another in the sense of Euclidean distance. (Recall antipodal vs orthogonal signaling.) An analysis of codeword detection in the BSC shows that this same concept carries over to digital block codes. (Here, the signals are restricted to binary vectors of length n.) Information symbols are encoded into n-bit codewords and transmitted over the channel. The n-bit blocks at the output of the channel may be corrupted by noise. This is referred to as hard *decision* decoding in that the demodulator makes symbol-by-symbol decisions without using knowledge of the redundan-

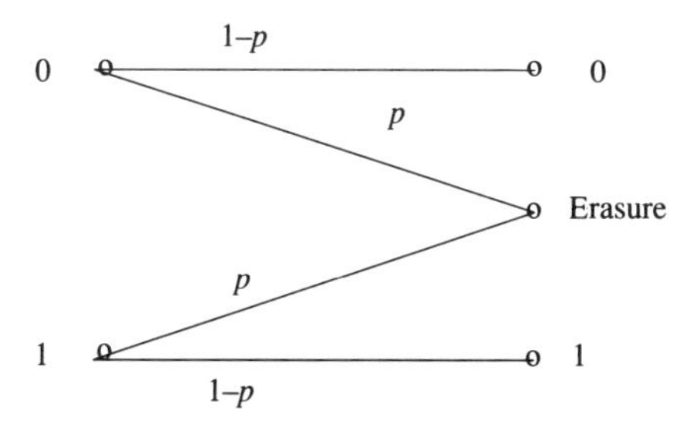

Fig. 2.16 Representation of the binary erasure channel.

cies built into the code.* (Again, the demodulation is assumed to be part of the channel, so corruption by noise implies replacing "0 bits" by "1 bits" and vice versa). It is the function of the decoder to decide which of the 2^k codewords were transmitted based on the n-bit blocks. As in the case of L-ary transmission above, the decision is made by choosing the codeword with the largest *a posteriori* probability. As we have seen, a MAP receiver will make decisions by deciding between all possible pairs of code words. We denote the pair of codewords being compared as c_0 *and* c_1. Under the assumption of equiprobable codewords, we choose c_0 over c_1 if

$$Pr(z|c_0) > Pr(z|c_1) \ , \tag{2.48}$$

where z is the n-bit block out of the channel and $Pr(z|c_i)$, $i = 0,1$ is the probability of the output conditioned on the input. (See (2.6b) and (2.7).) For errors caused by white Gaussian noise, the assumption that channel errors are independently made on successive bits is often justified. Note that, for a *given* signal c_i, knowledge of the error pattern,† $z \oplus c_i$, is sufficient to describe the state of the system. Under this assumption, the probability of a *particular* pattern of i errors in a block of n binary digits is given by

$$Pr[Particular\ pattern\ of\ i\ errors\ in\ n\ bits] = p^i(1-p)^{n-i} \tag{2.49}$$

Let d_i indicate the number of places in which z differs from c_i. From the properties of the BSC, we have

$$Pr(z|c_0) / Pr(z|c_1) = p^{d_0}(1-p)^{n-d_0} / p^{d_1}(1-p)^{n-d_1} = (p\ /\ (1-p))^{d_0-d_1} \ . \tag{2.50}$$

From (2.48) and (2.50) we have the decision rule, choose c_0 over c_1 if $d_0 < d_1$. (Recall that $p < 1/2$). From this it follows that error correcting and detecting codes should be designed so that codewords are maximally distant from one another in terms of the number of places in which they differ. This measure of the distance between pairs of codewords is called the *Hamming distance*. For example, the Hamming distance between the codewords 010011 and 100110 is four. It is easy to verify that the Hamming distance between codewords c_i and c_j, $D(i, j)$, satisfies the following criteria for a true distance measure:

(1) $D(i, j) \geq 0$ with $D(i, j) = 0$ if and only if $c_i = c_j$

(2) $D(i, j) = D(j, i)$

(3) $D(i, k) \leq D(i, j) + D(j, k)$ (the triangle inequality) .

---

* In Section 5.6 we shall discuss a recent breakthrough, trellis coding, which tightly couples the coding and modulation processes as well as the modulation and decoding processes, to arrive at powerful, efficient codes for the bandlimited AWGN channel.

† The expression $z \oplus c_i$ denotes the modulo 2 sum of z and c_i.

The BSC is called the *hard limited* channel and we speak of *hard decisions* on each bit coming out of such a channel. If bits passing through such a channel are independent of one another, the performance of the receiver is optimum. Error correcting and detecting codes (as well as most physical channels) introduce correlations among the bits in a sequence, and hard-decision decoding is not optimum—even in white noise. Because of these dependencies, each bit interval contains information about adjacent bit intervals. When hard decisions are made, this information is destroyed. The alternative is *soft decisions*, where the collection and processing information from the received waveform is processed in order to make later decisions about individual bit sequences.

In the next chapter on error correcting and detecting codes, we shall consider hard and soft decisions in detail when we examine the *Viterbi algorithm*; however, a simple example will serve to illustrate the difference. Consider the simple code with a zero encoded into three positive pulses and a one into three negative pulses. In the case of hard limiting, each of the three pulses is detected individually and a majority logic decision is made about the data bit from the three individual decisions on the pulses. An error is made if two or three detection errors are made on individual pulses, and it follows that the probability of a data bit error is

$$Pr[Error] = 3p^2 (1 - p) + p^3 , \tag{2.51}$$

where p is the probability of an individual pulse being in error. If the channel is disturbed by additive white Gaussian noise, p is given by (2.35), since we have antipodal signals. For example, if the signal-to-noise ratio is $10 \log_{10} E/N_0 = 6.0$ dB then $p = 2.4 \times 10^{-3}$. Substituting into (2.51), we find that $Pr[Error] = 1.73\times10^{-5}$.

Soft decision or analog decoding can also be applied to the received signal. In this case the three pulses are treated as one contiguous waveform. In the sample calculation, since the contiguous pulse is three times as long as the individual pulses, the signal power is tripled and the signal-to-noise ratio is 10.77 dB with the resulting probability of bit error 5.47×10^{-7}, an improvement in error rate of almost two orders of magnitude. This example illustrates two fundamental principles of communication theory:

(1) Adding redundancy to a message stream will generally provide performance improvement.

(2) When the received bits are correlated, detecting an entire message, rather than detecting on a bit-by-bit basis, will also provide significant performance improvement.

One of the challenges for data communications designers is to balance their particular system requirement with the gains promised by theory. For example, detecting an entire message generally implies considerable delay at the receiver. The engineer will have to decide, on a case-by-case basis if this delay is tolerable when viewed from the overall user perspective, as well as the needs of other subsystems for reliable decisions with minimal delay.

2.4 ELEMENTS OF ESTIMATION THEORY

In contrast to detection theory, *estimation theory* deals with a continuous rather than a discrete unknown quantity. For example, in Chapter 6, we shall use the theory to estimate the phase angle of a carrier in an analog communication channel. We shall consider only the estimation of a parameter, but *Wiener filtering* and *Kalman filtering* have been used to estimate random waveforms. (We consider Wiener and Kalman filtering in our discussion of equalization in Chapters 7 and 8.)

2.4.1 BAYESIAN ESTIMATE WITH MEAN SQUARE ERROR CRITERION

As in the case of detection in the preceding section, we have an observation space whose elements, Y, are used to obtain the estimate; one calculates $\hat{X}(Y)$ where $\hat{X}(\cdot)$ is a function chosen so that the error between the quantity to be estimated X and $\hat{X}(Y)$ is minimum, in some sense. To begin we consider *Bayesian* estimation where the *a priori* probability of the quantity to be estimated is known. A loss function or estimation criterion which has the virtue of being mathematically tractable is the mean–square error (MSE)

$$L(X, \hat{X}) = E(X - \hat{X}(Y))^2 . \tag{2.52}$$

As we shall see, the estimate obtained from the mean-square error criterion is robust since it is obtained for a wide range of criteria and probability distributions.

Averaging over X and Y we find that the MSE can be expressed as

$$\begin{aligned} E(X - \hat{X}(Y))^2 &= \int_{\Omega_X} dx \int_{\Omega_Y} dy\; f_{XY}(x,y)\; (x - \hat{X}(y))^2 \\ &= \int_{\Omega_Y} dy\; f_Y(y) \int_{\Omega_X} dx\; f_{X|Y}(x|y)\; (x - \hat{X}(y))^2 , \end{aligned} \tag{2.53}$$

where Ω_X and Ω_Y indicate, respectively, the input and the observation space. Since $f_Y(y) \geq 0$, $E(\hat{X} - X(Y))^2$ is minimized by minimizing the inner integral in (2.53) for each point y on the observation space. This minimum is found by differentiating the inner integral with respect to $\hat{X}(y)$ and setting the result equal to zero. Thus, $2\hat{X}(y) - 2\int dx\; x\; f_{X|Y}(x|y) = 0$. Solving for $\hat{X}(y)$, we find that the optimum estimate is

$$\hat{X}(y) = \int_{\Omega_x} dx\; x\; f_{X|Y}(x|y) = E(X|Y{=}y). \tag{2.54}$$

Thus, from (2.54) the minimum mean-squared estimate is the conditional expectation of the parameter X given the observation Y. It may not be apparent from our concise derivation, but this is a powerful and general result which is applicable to vector variables and even to random processes. We direct the reader's attention to the intuitive appeal of the result; the estimate is the average of X conditioned the observation of Y.

EXAMPLE 3 The Bayes estimate of a parameter can be illustrated by means of Gaussian random variables. Suppose that X, a Gaussian random variable with mean zero and variance σ_X^2, is immersed in Gaussian noise which has mean 0 and variance σ_N^2. We observe

$$Y = X + N$$

and form the minimum mean-square estimate of X. From Bayes's rule, we may write

$$f_{X|Y}(x|y) = f_X(x)\ f_{Y|X}(y|x)/f_Y(y)\ , \tag{2.55}$$

where

$$f_{Y|X}(y|x) = \exp(-(y - x)^2/2\sigma_N^2)\ /\sqrt{2\pi\sigma_N^2}$$

$$f_X(x) = \exp(-x^2/2\sigma_X^2)\ /\ \sqrt{2\pi\sigma_X^2}$$

$$f_Y(y) = \exp(-y^2/2(\sigma_X^2 + \sigma_N^2))\ /\ \sqrt{2\pi\sigma_X^2 + (\sigma_N^2)}\ . \tag{2.56}$$

For this last density we remind the reader that Y is the sum of two independent Gaussian random variables, X and N. Substituting into (2.55) and completing the square, we find

$$f_{X|Y}(x|y) = \exp(-(x - y/(\ 1 + \sigma_N^2/\ \sigma_X^2))^2/\ 2\sigma_T^2)\ /\ \sqrt{2\pi\sigma_T^2}\ , \tag{2.57}$$

where $\sigma_T^2 = \sigma_X^2\ \sigma_N^2\ /\ (\ \sigma_X^2 + \sigma_N^2)$.

We see from (2.57) that $f_{X|Y}(x|y)$ is Gaussian with mean value

$$E[X|Y=y] = y/(\ 1 + \sigma_N^2\ /\ \sigma_X^2)\ . \tag{2.58}$$

Note the weighting of the estimate by the noise and signal variances; e.g. as the noise vanishes, $\sigma_N^2 \to 0$, consequently $\hat{X} \to y$, and as the signal decreases $\hat{X} \to (\sigma_X^2\ /\ \sigma_N^2)y$.

The fact that the conditional mean is a minimum mean-square estimate in this case can be illustrated graphically. In Figure 2.17, we plot the cost function $(X - \hat{X})^2$ and the conditional probability density function $f_{X|Y}(x|y)$. Since the Gaussian density function has a maximum at its mean (see Figure 2.17) and since both curves are symmetric, it follows that the average is a minimum when $\hat{X}$ is selected to coincide with the maximum of $f_{X|Y}(x|y)$. It is of interest that the optimum estimate in (2.58) is a *linear* function of the observation. This is

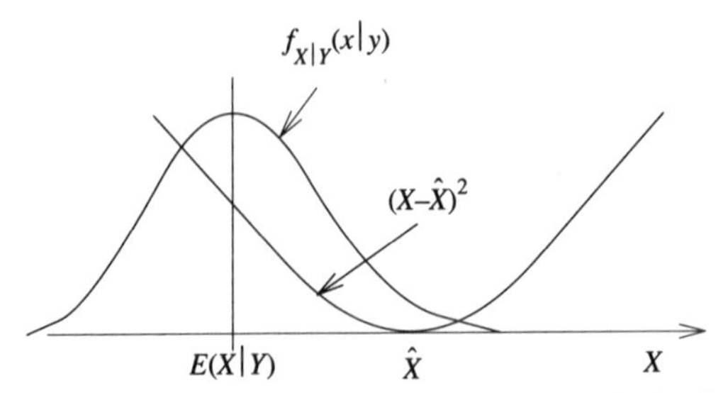

Fig. 2.17 Cost function and conditional probability density.

generally true for Gaussian random variables. We demonstrate this for the case of multiple observations. Again suppose that we have a Gaussian random variable, X, disturbed by Gaussian noise, N. In this case, we make L independent measurements on the observation space, $\mathbf{y} = (y_1, y_2, ..., y_L)$. Conditioned on $X = x$, each observation is Gaussian with mean x and variance σ_N^2. In this case we have

$$f_{X|\mathbf{Y}}(x|\mathbf{y}) = f_X(x)\ f_{\mathbf{Y}|X}(\mathbf{y}|x)/f_{\mathbf{Y}}(\mathbf{y})\ , \tag{2.59}$$

where

$$f_{\mathbf{Y}|X}(\mathbf{y}|x) = \prod_{i=1}^{L} \exp(-(y_i - x)^2/2\sigma_N^2)\Big/\sqrt{2\pi\sigma_N^2}$$

$$f_X(x) = \exp(-x^2/2\sigma_X^2)\Big/\sqrt{2\pi\sigma_X^2}$$

$$f_{\mathbf{Y}}(\mathbf{y}) = \prod_{i=1}^{L} \exp(-y_i^2/2(\sigma_X^2 + \sigma_N^2))\Big/\sqrt{2\pi(\sigma_X^2 + \sigma_N^2)}\ .$$

Substituting into (2.59), we find

$$f_{X|\mathbf{Y}}(x|\mathbf{y}) = \frac{1}{\sqrt{2\pi\sigma_X^2}\left[1 + \sigma_X^2/\sigma_N^2\right]^{L/2}} \exp\left\{-\sum_{i=1}^{L}\frac{(y_i - x)^2}{2\sigma_N^2} - \frac{x^2}{2\sigma_X^2} + \sum_{i=1}^{L}\frac{y_i^2}{2(\sigma_N^2 + \sigma_X^2)}\right\} \tag{2.60}$$

Of course we could proceed as in the one-dimensional case by completing the square; however, there is a simpler way. Since $f_{X|\mathbf{Y}}(x|\mathbf{y})$ is Gaussian, it has a maximum at its mean value. We find the maximum of the density in (2.60) by finding the minimum of the quadratic form in the exponent. Differentiating with respect to x and setting the result equal to zero, we find

$$\hat{X} = \left[1 / (L + \sigma_N^2/\sigma_X^2)\right] \sum_{i=1}^{L} y_i \,. \tag{2.61}$$

As expected, the estimate is a linear function of the observations.

The results that we have obtained for the Gaussian distribution and the mean-square error criterion carry over to a number of other cases. We state, without proof, the following theorems concerning estimates [2,5]. These theorems bear witness to the robustness of the estimate given in (2.54).

THEOREM 2.1 If the performance criterion, $L(X, \hat{X})$, has the properties of being an even, nondecreasing function of $| X - \hat{X} |$ and $f_{X|Y}(x|y)$ is unimodal and symmetric about this mode, then the optimum estimator is $\hat{X} = E(X|Y)$.

THEOREM 2.2 If $L(X, \hat{X})$ is an even and convex function of $|X - \hat{X}|$ and $f_{X|Y}(x|y)$ is symmetric about its mean, then the optimum estimator is $\hat{X} = E(X|Y)$.

As in the case of detection, there are occasions where the idea of an *a priori* distribution on X is inappropriate. In this case we use the maximum likelihood estimate. We take as the estimate of X that value for which $f_{Y|X}(y|x)$ is maximum. From (2.55), it follows that the maximum likelihood estimator is the same as the Bayes estimate when X has a uniform distribution and the loss function is symmetric and nondecreasing. In the multidimensional case that we just considered, we have

$$f_{\mathbf{Y}|\mathbf{X}}(\mathbf{y}|\mathbf{x}) = \exp\left(-\sum_{i=1}^{L} (y_i - x_i)^2 / 2\sigma_N^2\right) \Big/ \sqrt{2\pi\sigma_N^2} \,, \tag{2.62}$$

and the maximum likelihood estimate is

$$\hat{X} = \frac{1}{L} \sum_{i=1}^{L} y_i \,. \tag{2.63}$$

Note that this estimate is not weighted by σ_N^2/σ_X^2, as is (2.58), since *a priori* information is not utilized.

ORTHOGONALITY CONDITION

From (2.54), the minimum mean-square estimate is the conditional mean of the random variable, X, given the observation Y. An important and frequently used property of minimum mean-square estimates is the *orthogonality condition* between the estimation error, $X - \hat{X}$, and the received sample, Y. Before demonstrating this condition, we introduce the following conditional expectation notation. For any function $g(x,y)$, we have

$$E[g(X,Y)] = E_X[E_{Y|X}[g(X,Y) \mid X]]$$

$$= E_Y[E_{X|Y}[g(X,Y) \mid Y]] , \tag{2.64}$$

where $E_{X|Y}[g(X,Y) \mid Y]$ denotes the average of $g(X,Y)$ over X for a fixed value of Y. With this notation, we can write the correlation between the estimation error and the received sample as

$$E[(X - \hat{X})Y] = E[XY] - E[\hat{X}Y] , \tag{2.65}$$

where the second term may be written

$$E[\hat{X}Y] = E[E_{X|Y}(X|Y)Y] = E_Y[E_{X|Y}(XY|Y)] = E[XY] . \tag{2.66}$$

Combining (2.65) and (2.66) gives the desired result

$$E[(X - \hat{X})Y] = 0 , \tag{2.67}$$

which states that the estimation error is orthogonal to (i.e., uncorrelated with) the received sample for the optimum estimate. This property will be very useful in our discussion of adaptive equalization and adaptive echo cancellation. Note that (2.67) can be generalized to include a linear function of the received samples, when X and Y are replaced by vectors of samples.

2.5 FUNDAMENTALS OF INFORMATION THEORY

In the preceding sections of this chapter we have seen examples relating the rate of transmitting information and the performance in terms of probability of error, (2.47). In this section, we shall explain this relationship in a systematic fashion by presenting fundamentals of information theory,* which emerged fully formed with the publication of Claude Shannon's landmark paper in 1948 [8]. This theory has recently assumed renewed relevance as information rates in several applications approach the bounds that it has derived. Heretofore, the bounds were thought to be so loose as to be unattainable.

2.5.1 ENTROPY OF A DISCRETE SOURCE

In the preceding section, we saw examples of a relationship between the number of signals that are transmitted and performance. For example, in Fig-

* There are a number of good texts on the subject treated in this section including [9–13]. For an interesting, personal view of information theory written for a general audience, see [14].

ure 2.14 we saw that the performance of PAM deteriorates as the number of levels is increased, unless the signal-to-noise ratio is increased. Further, as shown in (2.47), the rate at which binary digits can be transmitted over a channel is limited by the signal-to-noise ratio.

The profound relationships between performance in terms of probability of error and the information rate exemplified in the previous paragraph are the subject of the science of information theory. The three fundamental contributions of this subject which are relevant to data communications are: 1) the quantification of the information content of the source, 2) the characterization of the information-bearing capacity of the communication channel in terms of its physical characteristics, and 3) establishment of the relationship between the information content of the source and the capacity of the channel. The importance of the last contribution is that it ties the first two together.

2.5.2 ENTROPY OF A DISCRETE MEMORYLESS SOURCE

The *entropy* of a source denotes its information content. The origin of the term itself is in thermodynamics, where it is a measure of the randomness or disorder in a system. We also use, synonymously, the term *uncertainty* of a source. It is the measure of our lack of knowledge before the source output is revealed to us. We begin with a discrete, memoryless source.* Such a source has L possible output symbols, $X_0, X_1, \ldots, X_{L-1}$, each occurring independently with probability $P(X = x_i) = P_i$, $i = 0,1,\ldots,L-1$. The source entropy per symbol is defined to be

$$H(P_0,P_1,\cdots,P_{L-1}) = -\sum_{i=0}^{L-1} P_i \log_2 P_i \ . \tag{2.68}$$

In (2.68), the logarithm to the base 2 is used, in which case the entropy is measured in bits. If the natural logarithm is used, the corresponding measure is the *nat*. In (2.68) we may consider the source entropy to be the average over all of the symbols of the quantity $\log_2 P_i$, $i = 0,1,\ldots,L-1$. This quantity may be considered as the amount of information obtained when the output of the source is the ith symbol.

EXAMPLE 4 For a binary source with equiprobable outputs, $P_0 = P_1 = ½$, the source entropy is one bit per symbol.

EXAMPLE 5 For the English language, the frequency of occurrence of each of the letters is given in Table 2.1. The entropy of a source producing letters with probabilities equal to these frequencies is 4.03 bits/symbol.

---

* After suitable definitions have been given, we shall treat other kinds of sources in a limited fashion.

Table 2.1 Frequency of Occurrences in English Text (From Reza [17])

Symbol	Probability	Symbol	Probability
space	0.1859	N	0.0574
A	0.0642	O	0.0632
B	0.0127	P	0.0152
C	0.0218	Q	0.0008
D	0.0317	R	0.0484
E	0.1031	S	0.0514
F	0.0208	T	0.0796
G	0.0152	U	0.0228
H	0.0467	V	0.0083
I	0.0575	W	0.0175
J	0.0008	X	0.0013
K	0.0049	Y	0.0164
L	0.0321	Z	0.0005
M	0.0198		

That $H(P_0,P_1,\ldots,P_{L-1})$ is an appropriate measure of information may be deduced* from the fact that it is the only function satisfying the following properties of a measure of information which we would consider to be reasonable from an intuitive point of view:

1. It is a continuous function of each of the P_i.

2. When all outputs are equally probable, $P_i = 1/L$, the information measure is monotonic increasing with L. This is in keeping with intuitive notions of information: the wider the choice, the more information is gained when the source output is revealed.

3. If the source outputs are revealed in stages, the entropy of the total should be the sum of the entropies for each stage. To illustrate, suppose that a source outputs the symbols A, B, or C with the respective probabilities $P_A = 1/2$, $P_B = 1/3$, $P_C = 1/6$, respectively. We are told first that the output is either A or is either B or C. The entropy of this "first stage source" is $H(1/2,1/2)$. If the output is either B or C, we are told the final choice in the second stage. Note that the second stage occurs with probability 1/2 and that the conditional probabilities for B and C are 2/3 and 1/3 respectively. Clearly, as far as information content is concerned, the order in which the output is revealed should be irrelevant and we have $H(1/2,1/3,1/6) = H(1/2,1/2) + 1/2\,H(2/3,1/3)$.

* An axiomatic foundation for information theory has been established by Khinchin [15].

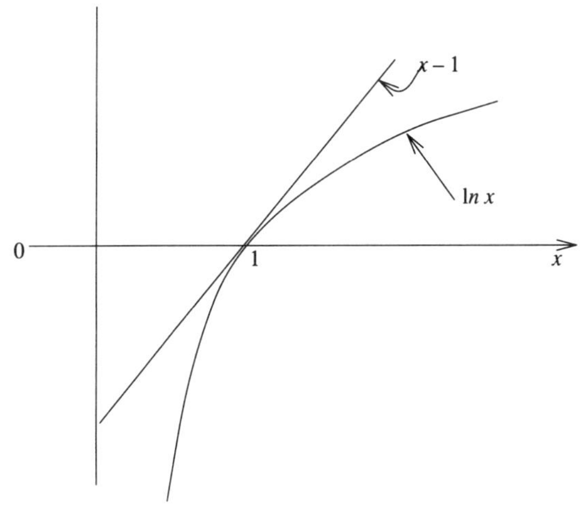

Fig. 2.18 Plots of $x - 1$ and $\ln x$.

The following theorem follows directly from the properties of the entropy.

THEOREM 2.3 The entropy of a source is maximum when all source outputs are equally probable, i.e.,

$$H(P_0, P_1, \ldots, P_{L-1}) \leq \log_2 L \; . \tag{2.69}$$

The proof of (2.69) is based on the inequality $x - 1 \geq \ln x$, with equality for $x = 1$ (see Figure 2.18). Let $x_i = Q_i / P_i$ where Q_i and P_i are valid discrete probabilities, i.e.,

$$P_i,\, Q_i \geq 0 \quad \text{and} \quad \sum_{i=0}^{L-1} P_i = \sum_{i=0}^{L-1} Q_i = 1 \; .$$

Then since $x_i - 1 \geq \ln x_i$ we have

$$Q_i / P_i - 1 \geq \ln(Q_i / P_i) \; , \quad \text{or} \quad Q_i - P_i - P_i \ln Q_i \geq -P_i \ln P_i \; .$$

Summing over i gives

$$\sum_{i=0}^{L-1} (Q_i - P_i - P_i \ln Q_i) \geq - \sum_{i=0}^{L-1} P_i \ln P_i \; .$$

But $\sum_{i=0}^{L-1} Q_i - \sum_{i=0}^{L-1} P_i = 1 - 1 = 0$, and we have

$$- \sum_{i=0}^{L-1} P_i \ln Q_i \geq - \sum_{i=0}^{L-1} P_i \ln P_i = H(P_0, P_1, \ldots, P_{L-1}) \; . \tag{2.70}$$

Note that equation (2.70) applies to any probability distribution, in particular,

$Q = 1/L$. Substituting this into (2.70), and noting that $\ln x_i = \log_2 x_i \ln 2$, (2.69) follows immediately.

EXAMPLE 6 A source producing four symbols has the following output probabilities $P_0 = 1/2$, $P_1 = 1/4$, $P_2 = 1/8$, and $P_3 = 1/8$. The entropy of this source is 1.75 bits per symbol. If the source had equiprobable outputs, the entropy would be 2 bits per symbol.

Now that the preliminaries are completed, we find it convenient to change notation slightly. We will let $H(X) = H(P_0, P_1, \ldots, P_{L-1})$, where X is a discrete random variable taking on values $x_0, x_1, \ldots, x_{L-1}$ with probabilities $P_0, P_1, \ldots, P_{L-1}$, respectively.

2.5.3 JOINT ENTROPY AND EQUIVOCATION

In the preceding we defined an entropy function of a single random variable which is the output of a discrete information source. We now extend the concept to two random variables which are correlated with one another. This extension will allow us to extend the entropy concept to sources whose output is a sequence of dependent variables, i.e., sources with memory. It will also enable us to define and compute the information-carrying capacity of a channel. Suppose that we have two random variables X and Y. We define the *joint entropy* of X and Y as

$$H(X,Y) = H(Y,X) = -\sum_{i,j} P_{XY}(x_i,y_j) \log_2 P_{XY}(x_i,y_j) \quad , \tag{2.71}$$

where $P_{XY}(x_i,y_j)$ is the joint probability for X and Y, $Pr(X = x_i, Y = Y_j)$. This is not very different from the previous definition of entropy. It can be shown that

$$H(X,Y) \le H(X) + H(Y) \ , \tag{2.72}$$

with equality if and only if X and Y are independent random variables. Equation (2.72) indicates that correlation between X and Y removes some of the randomness of the joint events associated with X and Y. Thus if a source has memory, successive outputs are correlated and the entropy of a sequence is reduced.

The conditional entropy of X given that $Y = y_j$ is defined as

$$H(X|Y = y_j) = -\sum_{i} P_{X|Y}(x_i|y_j) \log_2 P_{X|Y}(x_i|y_j) \ , \tag{2.73}$$

where $P_{X|Y}(x_i|y_j)$ is the conditional probability $Pr(\, X = x_i | Y = y_j)$.

Averaging (2.73) over Y, we have

$$H(X|Y) = -\sum_{i,j} P_{XY}(x_i,y_j) \log_2 P_{X|Y}(x_i|y_j) \ , \tag{2.74a}$$

where $H(X|Y)$ is the average uncertainty of X given Y. It is also called the *equivocation* of X with respect to Y. In a similar fashion, we can define

$$H(Y|X) = -\sum_{i,j} P_{XY}(x_i,y_j) \log_2 P_{Y|X}(y_j|x_i) . \tag{2.74b}$$

Note from (2.72) and (2.74a–b) we have

$$H(X|Y) \leq H(X) \tag{2.75a}$$

$$H(Y|X) \leq H(Y) , \tag{2.75b}$$

and it is not difficult to show the following relations between the joint uncertainty and the equivocation

$$H(X,Y) = H(X) + H(Y|X) = H(Y) + H(X|Y) . \tag{2.76}$$

In order to illustrate the meaning of the equivocation, let us assume that X and Y are independent random variables. Consequently, we have $P_{Y|X}(y_j|x_i) = P_Y(y_j)$. Substituting into (2.76) we have

$$\begin{aligned} H(Y|X) &= -\sum_{i,j} P_{XY}(x_i,y_j)\log_2 P_Y(y_j) \\ &= -\sum_{j} P_Y(y_j) \log_2 P_Y(y_j) = H(Y) . \end{aligned} \tag{2.77}$$

Under the same assumption $H(X|Y) = H(X)$. These relations conform to intuition, since knowing one of these random variables gives no information about the other.

The final definition of this section is the *mutual information* between X and Y,

$$I(X|Y) = H(X) - H(X|Y) . \tag{2.78}$$

It is left as an exercise (see Exercise 2.19) to show that

$$I(X|Y) = I(Y|X) = H(X) + H(Y) - H(X,Y) \tag{2.79}$$

The quantity $I(X|Y)$ is the information about X conveyed by Y and vice versa. Once again, we illustrate this concept by means of extreme examples. If X and Y are independent random variables, then $I(X|Y) = 0$, no information is conveyed. On the other hand, if $X = Y$, then $I(X|Y) = H(X) = H(Y)$. The relationships among the various quantities that we have defined are given in Figure 2.19.

The quantities defined in this section can be defined for an analog source in a similar fashion. Suppose we have two continuous random variables X and Y with joint density function $f_{XY}(x,y)$ and marginal density functions $f_X(x)$ and $f_Y(y)$. We define the equivocation as

$$H(X|Y) = -\int_{\Omega_X} dx \int_{\Omega_Y} dy\, f_{XY}(x,y) \log_2 \left[f_{XY}(x,y)/f_Y(y)\right] \tag{2.80a}$$

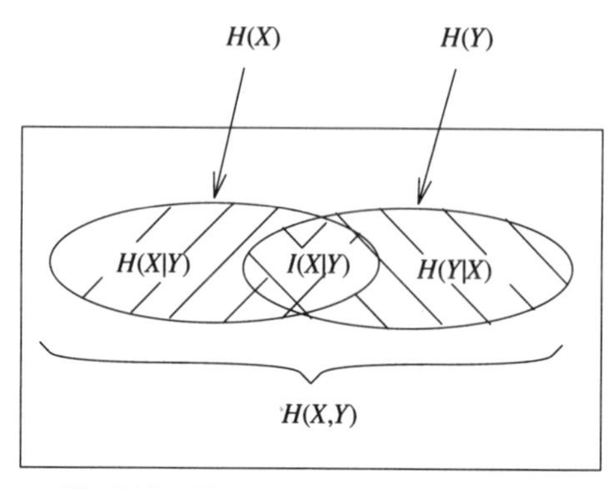

Fig. 2.19 Illustration of entropy concepts.

and

$$H(Y|X) = -\int_{\Omega_X} dx \int_{\Omega_Y} dy\ f_{XY}(x,y)\ \log_2 \left[f_{XY}(x,y)/f_X(x) \right] . \quad (2.80b)$$

We also define the mutual information as

$$I(X|Y) = H(X) - H(X|Y) \quad (2.81a)$$

$$I(Y|X) = H(Y) - H(Y|X) , \quad (2.81b)$$

where $H(X)$ and $H(Y)$ are defined as

$$H(X) = -\int_{\Omega_X} f_X(x)\ \log_2 f_X(x)\ dx \quad (2.82)$$

and

$$H(Y) = -\int_{\Omega_Y} f_Y(y)\ \log_2 f_Y(y)\ dy. \quad (2.83)$$

Note that there is a difference between the definitions of (2.80a–b) and (2.74), (2.75); here $f_X(x)$ and $f_Y(y)$ are not probabilities but probability densities.

It is not difficult to show that reversible transformations on X or Y have no effect on this definition of the mutual information. As in the discrete case, it can be shown that if X and Y are independent then $I(X|Y) = 0$. The concepts of joint entropy and equivocation will be used to find the capacity of a channel.

2.5.4 ENTROPY OF A DISCRETE MARKOV SOURCE

We have considered single outputs of sources, whereas in data communications we are interested in dealing with sequences of information symbols. The simplest case is that of a memoryless source where successive symbols are independent of one another. In this case, the application of the results that we have so far is straightforward. At this point, it is convenient to again change notation slightly. From the previous section, it is not difficult to show that the entropy of a block of n successive outputs of a discrete memoryless source is given by

$$H(X^n) = nH(X) \ , \tag{2.84}$$

where X^n denotes the n symbol output sequence. As we shall see presently, correlations between symbols reduces this entropy. Such correlation will be present in natural languages. For example, Shannon estimated that the entropy of English text is between 0.6 and 1.3 bits per symbol. This stands in sharp contrast with the 4.03 bits per symbol entropy that we computed for the (presumed independent) 26-symbol source which we computed earlier.

Most sources of information are not memoryless. In English text, for example, the letter q is almost always followed by the letter u. Furthermore, these dependencies among source outputs reduce the information content of the source. The concepts of joint entropy and equivocation allow us to calculate the entropy of sources with memory.

We deal with the simplest source with memory, the sources which can be modeled as a stationary first-order Markov chain. The ideas carry over to more general sources in a straightforward fashion. Consider a sequence of source output symbols, $X_0, X_1, \ldots, X_n$. Each of these symbols may take on values with probabilities $P_0, P_1, \ldots, P_{L-1}$ respectively. Such a source is first-order *Markov* if the statistical dependence of the current output depends only upon the previous output, i.e.,

$$Pr(X_i = x_i | X_0 = x_0, X_1 = x_1, \ldots, X_{i-1} = x_{i-1}) = Pr(X_i = x_m | X_{i-1} = x_{i-1}). \tag{2.85}$$

As a consequence of (2.85), we can write

$$Pr(X_0 = x_i, X_1 = x_1, \ldots, X_n = x_n) = Pr(X_0 = x_0) \prod_{i=1}^{n} P_r(X_i = x_i | X_{i-1} = x_{i-1}). \tag{2.86}$$

We define the transition probability

$$P_{jk} = P_r(X_i = x_k | X_{i-1} = x_j), \ j,k = 1,2,\ldots,L \ , \tag{2.87}$$

and it can be shown [16] that the steady state probabilities satisfy the equation

$$P_k = \sum_{j=0}^{L-1} P_j \, P_{jk}, \ k = 0,1,\ldots,L-1 \ . \tag{2.88}$$

Now consider the joint entropy of the sequence of output symbols, $X_0, X_1, \ldots, X_{n-1}$.

$$H(X_0, X_1, \ldots, X_{n-1})$$
$$= -\sum P(X_0 = x_0, X_1 = x_1, \ldots, X_n = x_n) \log_2 P(X_0 = x_0, X_1 = x_1, \ldots, X_n = x_n) \tag{2.89}$$

From (2.86) and (2.89), we have

$$H(X_0,X_1,\ldots,X_{n-1}) = Pr(X_0=x_0)\log_2 Pr(X_0=x_0) - \sum_{i=1}^{n} Pr(X_{i-1}=x_{i-1},X_i=x_i)\log_2 Pr(X_i=x_i|X_{i-1}=x_{i-1}) \tag{2.90}$$

which can be written

$$H(X_0,X_1,\ldots,X_{n-1}) = \sum_{i=1}^{L} -P_i \log_2 P_i - \sum_{i,j=1}^{L} P_i P_{ij} \log_2 P_{ij}. \tag{2.91}$$

In terms of the equivocation we have

$$H(X_0,X_1,\ldots,X_n) = H(X) + (n-1)\, H(X_i|X_{i-1}) , \tag{2.92}$$

where $H(X_i|X_{i-1})$ indicates the information contained in a symbol based on the previous symbol. From (2.75) we have

$$H(X_i|X_{i-1}) \leq H(X) . \tag{2.93}$$

Thus, in principle the correlations between successive source symbols imply a reduction in the information content of a source.

2.5.5 SOURCE CODING—THE HUFFMAN CODE [18]

In this section, we shall relate the entropy of a discrete source to the number of bits actually needed to encode the source. We shall consider block encoding for which a symbol or sequence of symbols is mapped into a unique codeword, a block of binary digits. These codewords need not all be of the same length. If b_i binary digits are required to encode the ith symbol, the average number of binary digits needed to encode the output of a source is

$$L(X) = \sum_{i=0}^{L-1} b_i P_i . \tag{2.94}$$

In this section, we show that the entropy of the source is a tight lower bound on the minimum average codeword length.

The simplest case is that of a memoryless source in which all symbols are equally probable and the number of symbols is a power of two, $L = 2^k$. The entropy of this source is k bits per symbol. This figure gives guidance on encoding. Possible outputs are divided in two equal groups of 2^{k-1} symbols. The question is asked, "To which group does the output belong?" This group is divided in two and the question is repeated. In each case, one bit of information is obtained in response to a question since each subgroup is equiprobable. The question is repeated $k-1$ times. Now if the successive splits of the group were designated "upper" and "lower" and if a zero and a one, respectively, were assigned to each answer, we would have a code which is of minimum length. This is illustrated in Figure 2.20 for $L=8$.

Symbol	Probability	Codeword
0	1/8	0 0 0
1	1/8	0 0 1
2	1/8	0 1 0
3	1/8	0 1 1
4	1/8	1 0 0
5	1/8	1 0 1
6	1/8	1 1 0
7	1/8	1 1 1

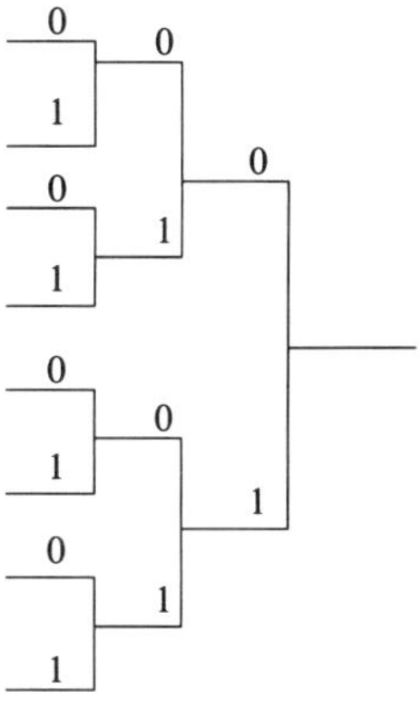

Fig. 2.20 Coding of equiprobable L-ary source.

Consider now the case where the symbol probabilities are not equal but are still powers of $1/2$. An example is the case where the probabilities are $P_i = (1/2)^{i+1}$, $i = 0,1,\ldots,L-2$; $P_{L-1} = (1/2)^{L-1}$. As in the previous case, the output is revealed in stages. The symbols are ordered with the highest probability symbol at the top, and the symbols are split into groups of equal probability. In the present example, the first symbol is in one group and all of the other symbols in the other. (See Figure 2.21 for an example.) Each question yields one bit of information. If the output is symbol i, the number of questions is i for $i = 1,2,\ldots,L-1$ and $L-1$ for $i = L$. Again the sequence of yes–no answers can be translated into a sequence of binary digits specifying the output

Symbol	Probability	Codeword
0	1/2	0
1	1/4	10
2	1/8	110
3	1/16	1110
4	1/16	1111

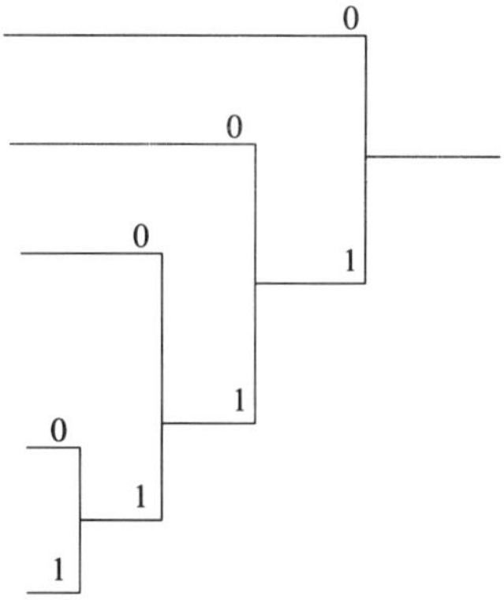

Fig. 2.21 Coding symbols with unequal probability.

symbol. Note that in this case the length of the codeword for the ith symbol is given by $b_i = -\log_2 P_i$.

We turn now to the general case where output probabilities are arbitrary. We assume that, as in the preceding paragraph, code words are of lengths b_i and they satisfy

$$-\log_2 P_i \leq b_i < -\log_2 P_i + 1 , \tag{2.95}$$

and averaging (2.95) over P_i we find

$$H(X) \leq L(X) < H(X) + 1 . \tag{2.96}$$

It can be shown that there exists an uniquely decipherable code whose word lengths conform to (2.95); in fact, later in this section we shall give one, the *Huffman code*.

By averaging over blocks of symbols this bound can be made tighter. A block of n symbols is regarded as source with L^n possible outputs. Codewords are assigned in the same manner as in (2.95). Averaging over all possible symbols, we find

$$H(X^n) \leq L(X^n) < H(X^n) + 1 . \tag{2.97}$$

But, for the memoryless source, we have from (2.84) $H(X^n) = nH(X)$. Substituting into (2.97) and dividing by n, we have

$$H(X) \leq L(X^n)/n < H(X) + 1/n . \tag{2.98}$$

In (2.98), $L(X^n)/n$ is the number of binary digits per symbol. As we see, in the limit

$$\lim_{n\to\infty} L(X^n)/n = H(X) . \tag{2.99}$$

As we code over longer blocks, the bound becomes tighter, and (2.99) is known as the *noiseless coding theorem* of Shannon, which says, that as we use the Huffman coding algorithm over longer and longer blocks of symbols, the average number of bits required to encode each symbol approaches the entropy of the source.

The Huffman code gives a method of constructing a code which has the minimum average number of binary digits per symbol. It is a member of the class of codes called *prefix codes*, in that no codeword is the prefix of another codeword. For this class of code, sequences of binary digits can be decoded uniquely and instantaneously. The codes obtained in the examples of Figures 2.20 and 2.21 are prefix codes.

The *Huffman coding* algorithm begins by listing the output symbols in decreasing order of probability. The two lowest probability symbols are combined into a group and a new list is formed with this group as a single entry. Again the list is in decreasing order of probability. The first step is repeated to form another list. At each step larger groups are formed and treated as a single entry. The process continues until the list contains only two entries. A zero is assigned to each member of one group and a one to each member of the other.

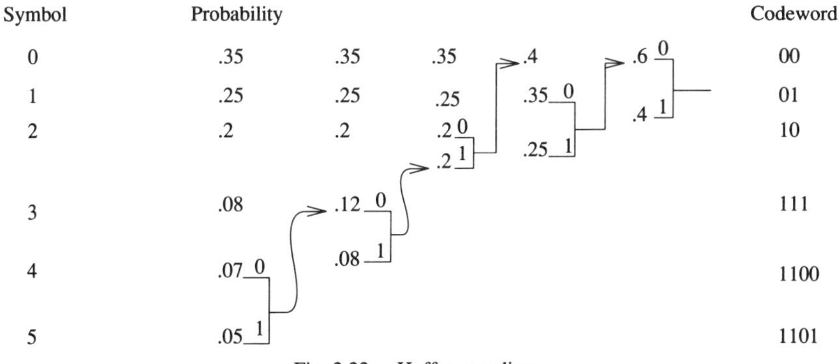

Fig. 2.22 Huffman coding.

The processes of grouping is then retraced; each of the groups is split and a zero and a one are assigned. The process is similar to the sequence of questions seen earlier. The Huffman coding algorithm is illustrated in Figure 2.22.

Beyond its theoretical implications, Huffman coding is used in some commercial data compression units to reduce the number of transmitted bits and thus either reduce the time for transmission of customer data or allow multiplexing with other data streams for a fixed user data rate. In the standard for digital facsimile, run lengths are encoded using a modified version of the Huffman code [19].

2.5.6 MORE ON HUFFMAN CODING

Despite its ubiquity, the Huffman code does have certain drawbacks. The most obvious is that the source statistics must be known. This is no handicap when general text is being handled since the probabilities given in Table 2.1 would suffice. However, these probabilities would not lead to efficient coding in the case of specialized text such as stock market quotations, airline reservations, or baseball box scores. This suggests an adaptive approach in which the probabilities are estimated as the text is transmitted and the coding changed appropriately. A compression of approximately 43% has been reported by use of this approach [14].

A second problem lies in the fact that, for ordinary text only single characters are encoded. For example, the obvious correlation between q and u is ignored, or as many times as "the" occurs in the text, the three separate letters are encoded. Of course, one could expand the symbol alphabet to include all digrams or even trigrams; however, the resulting complexity would be prohibitive. For example, there are 26×26 possible digrams of ordinary English letters. Furthermore, it is frequently the case that a source produces long strings of the same characters. Huffman coding has no special facility for handling runs.

Huffman coding also poses problems from the point of view of implementation.* For example, in the case of data transmission, there is a mismatch of source and channel rates since sources product symbols at a constant rate which are then encoded into variable-length code words. In order to implement Huffman codes it is necessary to provide buffering at the transmitter. The size of this buffer is a tradeoff between overflow and coding efficiency [20].

2.5.7 THE LEMPEL–ZIV ALGORITHM [21,22]

An alternative source coding technique, the *Lempel–Ziv* (L–Z) algorithm, remedies some of the deficiencies of the Huffman coding algorithm. The algorithm is intrinsically adaptive and can encode frequently occurring groups of source symbols.

The L–Z algorithm is simplicity and elegance itself. A source output sequence is searched for a sequence of symbols which have not been seen before. The new sequence is encoded in terms of previously seen sequences which have been compiled in a code book. We illustrate by means of a simple example involving the output of a binary source. Suppose that the output of the source is

$$0011001000110101 0001111 .$$

Let us assume that the symbols "0" and "1" are already in the code book; accordingly, the first previously unseen symbol is "00". This new symbol is transmitted as two 0's and the new symbol "00" is put in the code book. The string "11" which is next in the sequence is handled in the same way. The string "001" is new. It may be transmitted as the concatenation of the symbols "00" and "1" both of which are already in the code book. The working of the algorithm in segmenting the remainder of the output sequence is shown below.

$$00/11/001/000/110/10/100/01/111/ .$$

The encoded symbols may be transmitted simply by transmitting their position in the code book. The position is the concatenation of the positions of the two code book entries that are used to represent the new entry. In the above example, the code book at the end of the sequence would be

$$0,1,00,11,001,000,110,10,100,01,111 ,$$

and the sequence would be transmitted by the sequence of numbers

$$11,22,32,31,41,21,81,12,42 .$$

In the above example, the sequence 00 is recognized as the concatenation of the

* Reference [19] contains a discussion of the problems encountered in implementing the Huffman encoding algorithm for data storage applications.

first code book entry, 0 with itself; hence 11 is transmitted. Also, the sequence 100 is recognized as the concatenation of the eighth and the first code words. The commas here are only for ease of presentation. The decoder would look at pairs of symbols.

The integers representing the code words would themselves be represented by a block of binary digits whose size depends upon the eventual size of the code book. In practice [23] twelve-bit blocks, implying 4096 entries in the code book, are used. Note that, in the foregoing, a new sequence always ends in a "0" or a "1" which are one and two, respectively, in the book; consequently, only one bit is required for the ends of the code words and code words can be represented by thirteen bits. Note also that in the Lempel–Ziv algorithm fixed-length code words represent a variable number of source output symbols. This is the inverse of the Huffman coding algorithm and is well suited to synchronous transmission.

The algorithm may be used to encode text from any digital source. As an illustration of the Lempel–Ziv algorithm encoding English text, we encode the following beautiful passage by James Joyce.

> Yes, the newspapers were right, snow was general all over Ireland. It was falling on every part of the dark central plain, on the treeless hills, falling softly on the Bog of Allen, and farther westward, falling into the dark mutinous Shannon waves.

We assume that the code book already contains all of the letters of the alphabet, both upper and lower case, space, and punctuation marks. The segmentation of the text is as follows:

> Ye/s,/ t/he/ n/ew/sp/ap/er/s /we/re/ r/ig/ht/; /sn/ow/ w/as/ g/en/era/l /al/l o/ve/r /Ir/el/an/d./ I/t /wa/s f/all/in/g /on/ e/ver/y /pa/rt/ o/f /th/e /da/rk/ c/ent/ra/l p/la/in,/ on/ th/e t/ree/le/ss/ h/il/ls/, /fa/ll/ing/ s/of/tl/y o/n /the/ B/og/ of/ A/lle/n a/nd/, f/ar/ther/ we/st/war/d,/ f/alli/ng/ i/nt/o /the /dar/k /mu/ti/no/us/ S/ha/nn/on /wav/es/./

Again, each new sequence is encoded into two words already in the code book. The latter of these two are chosen from a smaller set, the initial code book. Note that even in this short passage, the common words "on" and "the" are compiled in the code book. We may also observe the action of the algorithm in encoding the word "falling" which is repeated three times in the text.

It has been shown that the Lempel–Ziv algorithm achieves a compression of approximately 55% on ordinary English text [14,23}. This corresponds roughly to what would be obtained by encoding pairs of letters. For more structured text such as COBOL files, higher compression can be achieved. In contrast, as has been mentioned above, it has been demonstrated that Huffman coding provides a compression of approximately 43% on English text.

It is clear that adaptivity is built into the algorithm; only strings that appear in the text are put in the code book. Moreover, it there is an abrupt change in the nature of the text being transmitted, this will be evident from the new entries in the code book. A new code book can be compiled in order to accommodate this change.

2.5.8 RATE DISTORTION THEORY

We have been dealing with sources which may be characterized as periodically generating a discrete set of symbols. However, many information sources are inherently analog, for example speech. The information content of such sources may be determined by means of *rate distortion theory* [24]. It is recognized that infinite information would be required to specify a wave form with infinite precision; consequently, a certain amount of error is tolerated. In general, smaller error means that more bits are required to specify the waveform. This corresponds to the usual experience in sampling and quantizing analog waveforms.

The minimum amount of information needed to specify the output of a source within a certain specified error is called the *rate distortion function* of the source. More precisely, the rate distortion function is the minimum mutual information (see (2.78)) between a source output and its approximation under a constraint on the error. For example, consider a Gaussian process which is limited to the band $(-W, W)$ and has mean-squared value P_{AV}. If a mean square error D is allowed, then a rate distortion function for the process is

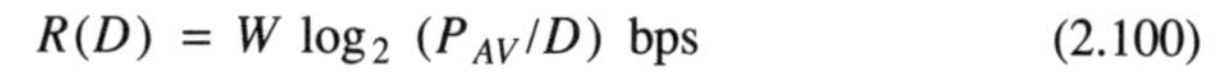

$$R(D) = W \log_2 (P_{AV}/D) \text{ bps} \tag{2.100}$$

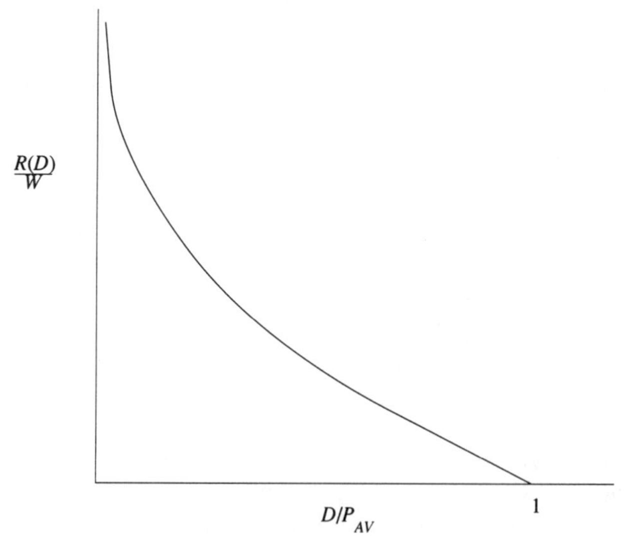

Fig. 2.23 Rate distortion function.

(See Exercise 2.23.) An illustration of the relationship between rate and error $R(D)/W = \log_2 (P_{AV}/D)$ is shown as a function of D/P_{AV} in Figure 2.23. It has been shown that one can come close to the bound by practically realizable techniques [25].

2.6 CHANNEL CAPACITY

The mutual information can be used to define the amount of information which can be transmitted through a channel. Let X be the input to a channel and Y the output. Since $I(X|Y)$ is the number of bits conveyed about X given knowledge of Y, it should be clear that $I(X|Y)$ is the rate at which information can be transmitted through a channel for a particular source encoding. In an information theoretic sense this encoding can be characterized by the probability distribution of the source symbols, $P(X)$. Note that sources with the same entropy may have very different probability distributions of output symbols. The information-carrying capacity of the channel is defined as the maximum of $I(X|Y)$ over all possible source encodings.* The *channel coding theorem* states that, if the rate of information produced by a source, as measured by the entropy, is less than the capacity of the channel, then the source output can be transmitted through the channel without error.

We will derive the capacity of the two channels that we studied in Sections 2.2 and 2.3, the analog additive Gaussian noise channel, either white or colored, and the discrete binary symmetric channel. We begin the derivation for the Gaussian channel by considering the one-dimensional case of a voltage or current level, X, transmitted over a channel which is disturbed by Gaussian noise. The signal level X may be, for example, a sample of a waveform which modulates the amplitude of a pulse. The output of the channel may be represented as

$$Y = X + N ,$$

where N is a Gaussian random variable with zero mean and various σ^2. Since X and N are assumed to be independent random variables we may express the information transmitted through the channel as

$$I(Y|X) = H(Y) - H(Y|X) = H(Y) - H(N) . \tag{2.101}$$

Note that the uncertainty about Y given X is simply the uncertainty due to the noise, $H(N)$. We maximize $I(Y|X)$ over all possible input distributions for a given noise level in the channel and with a constraint on the mean square value

---

* The capacity is often called the *Shannon bound* after the founder of information theory.

or power of the transmitted signal*, $E(X^2) \leq P_{AV}$. If for example, X modulates the amplitude of a pulse $p(t)$, we assume the normalization $\int_{-\infty}^{\infty} p^2(t)\, dt = 1$. The noise level in the channel specifies $H(N)$. The probability distribution is

$$f_N(n) = \exp(-n^2/2\sigma^2)/\sqrt{2\pi\sigma^2} \ , \tag{2.102}$$

and using (2.102) and (2.82) we have for

$$H(N) = 1/2 \ \ln 2 + (1/2)\log_2(2\pi\sigma^2) \ . \tag{2.103}$$

Thus maximizing $I(Y|X)$ comes down to maximizing $H(Y)$ over the distribution functions of the input. This is equivalent to optimizing over the probability distribution of Y, since X and N are independent random variables. The power constraint on X is equivalent to the constraint† $E[Y^2] \leq P_{AV} + \sigma^2$. Thus, the optimization problem comes down to finding $f_y(y)$ such that

$$H(Y) = -\int_{-\infty}^{\infty} f_Y(y)\ \log_2 f_Y(y)\, dy \tag{2.104a}$$

is maximized, subject to the constraint

$$\int_{-\infty}^{\infty} y^2\ f_Y(y)\, dy \leq P_{AV} + \sigma^2 \ . \tag{2.104b}$$

Since $f_y(y)$ must be a valid probability density, we have the additional constraint

$$\int_{-\infty}^{\infty} f_Y(y)\, dy = 1 \ . \tag{2.104c}$$

The derivation is a straightforward exercise in the calculus of variations. We form the Lagrangian

$$\int_{-\infty}^{\infty} [-f_Y(y)\ (\log_2 f_Y(y) + a_1 + a_2 y^2)]\ dy \ , \tag{2.105}$$

where a_1 and a_2 are Lagrange multipliers whose values are chosen to satisfy the constraints. Taking the variation of (2.105) with respect to the function $f(x)$ shows that the optimum must satisfy the equation

$$\log_2 f_Y(y) + \frac{1}{\ln 2} + a_1 + a_2\ y^2 = 0 \ , \tag{2.106a}$$

or equivalently

$$f_Y(y) = A_1\ \exp(-a_2 y^2) \ , \tag{2.106b}$$

---

* For a derivation based on the peak rather than the average power of a transmitted signal see [25].

† We are assuming, without loss of generality, that X has mean value zero.

where A_1 is a composite of several constants, including a_1. Substituting into (2.104b) and (2.104c) we find that the optimum is

$$f_Y(y) = \exp\left[-y^2/2(P_{AV} + \sigma^2)\right] / \sqrt{2\pi(P_{AV} + \sigma^2)} \,, \qquad (2.107a)$$

and the maximum of $H(Y)$ is

$$H(Y) = 1/2 \ln 2 + 1/2 \log_2 \left[2\pi(P_{AV} + \sigma^2)\right] \quad \text{bits/sample} \,. \qquad (2.107b)$$

Substituting (2.103) and (2.107b) into (2.101), we have for the simple, one-dimensional channel, the capacity

$$C = \frac{1}{2} \log_2 \left[1 + \frac{P_{AV}}{\sigma^2}\right] \text{ bits/sample} \,. \qquad (2.108)$$

2.6.1 BANDLIMITED CHANNEL

We now consider a channel which transmits a sequence of information symbols. Now suppose that the channel is strictly bandlimited such that

$$H(f) = \begin{cases} 1 & |f| \le W \\ 0 & |f| > W \,. \end{cases} \qquad (2.109)$$

Strictly speaking, this filter characteristic is unrealizable but it forms a useful approximation to many physical situations. There is no point in transmitting signal energy outside this band; accordingly, the receiver would filter out all received signal and noise outside the band $(-W, W)$. If the noise is white, the *received* noise has power density spectrum

$$S_N(f) = \begin{cases} \dfrac{N_0}{2}, & |f| \le W \\ 0, & |f| > W \,, \end{cases} \qquad (2.110a)$$

with autocorrelation function given by the inverse Fourier transform

$$R_N(\tau) = F^{-1}\left[S_N(f)\right] = N_0 \frac{\sin 2\pi W \tau}{2\pi\tau} \,. \qquad (2.110b)$$

Since the received waveforms are limited to the band $(-W, W)$, they may be represented by samples spaced $1/2W$ seconds apart.* The ith sample can be

* This is proved in Theorem 4.1 in Chapter 4.

written

$$Y_i = X_i + N_i \ , \quad i = 0, \pm 1, \pm 2, \ldots \ ,$$

where X_i and N_i are samples of the transmitted signal and noise, respectively. From the autocorrelation function of the noise, we see that

$$E[N_i N_j] = R_N \left[(i-j)/2W \right] = \begin{cases} N_0 W, & i = j \\ 0, & i \neq j \ . \end{cases} \tag{2.111}$$

Thus the noise samples are independent and have variance $N_0 W$. From the previous derivation we see that each sample can convey $1/2 \ \log_2 (1 + P_{AV}/N_0 W)$ bits of information. Since there are $2W$ samples per second, the capacity of the channel is given by the celebrated formula

$$C = W \log_2 \left(1 + \frac{P_{AV}}{N_0 W}\right) \text{ bps.} \tag{2.112}$$

Notice that the capacity of this channel can be increased by increasing either the bandwidth or the signal-to-noise ratio, $P_{AV}/N_0 W$. As the bandwidth of the channel increases we have, in the limit

$$\begin{aligned} C_\infty &= \lim_{W \to \infty} W \log_2 \left[1 + \frac{P_{AV}}{N_0 W} \right] \\ &= P_{AV}/(N_0 \ln 2) \text{ bps.} \end{aligned} \tag{2.113}$$

We see that this rate is exactly double that which we derived for the case of orthogonal L-ary signaling over the AWGN channel in Section 2.3 (see equation (2.47)). As was stated in connection with the former derivation, more sophisticated arguments would produce (2.113). Finally it should be noted that (2.112) and (2.113) give upper bounds on transmission rates. In Section 2.7, we shall calculate the channel capacity of several well-known channels.

The results of (2.112) and (2.113) may be used to establish other interesting bounds on the transmission rate of information. Suppose that the actual rate of information transmission is R bps. From the channel coding theorem this transmission can be made to be error free, provided R is less than the channel capacity, i.e.,

$$R \leq W \log_2 (1 + P_{AV}/N_0 W) \ . \tag{2.114}$$

This can be expressed in terms of energy per bit, $E_b = P_{AV} \ / \ R$, as

$$\frac{R}{W} \leq \log_2 \left[1 + \frac{E_b}{N_0} \left[\frac{R}{W} \right] \right] ,$$

or

$$\frac{E_b}{N_0} \geq \frac{2^{(R/W)} - 1}{R/W} \; . \tag{2.115}$$

For the case of infinite bandwidth this becomes

$$\frac{E_b}{N_0} \geq \ln 2 \tag{2.116a}$$

or

$$10 \log \frac{E_b}{N_0} \geq -1.6 \text{ dB} \; . \tag{2.116b}$$

Equations (2.115)–(2.116a,b) give the lower bound on the amount of energy per transmitted bit that is required for reliable communications through the additive white Gaussian noise channel. If the signal-to-noise ratio per bit is less than –1 dB, it is not possible to transmit error free no matter how much bandwidth is used.

2.6.2 COLORED NOISE CHANNEL

While the strictly bandlimited channel gives some insight into the transmission of information, we are interested in the capacities of more practical channels, where the attenuation characteristics may assume an arbitrary shape and where the noise may not be white. Let the power density spectrum of the noise be $N(f)$ and the channel transfer function be $H(f)$. We find it convenient to represent the channel as shown in Figure 2.24. This representation is entirely equivalent to that of Figure 2.4 from the point of view of the received signal plus noise. The equivalent noise power density spectrum,

$$N'(f) = \frac{N(f)}{|H(f)|^2} \; , \tag{2.117}$$

allows the noise and the channel characteristics to be represented in a compact fashion. It is possible to derive the capacity of this channel in a mathematically rigorous fashion by use of the Karhunen–Loève expansion; however, the level of mathematics required is beyond the scope of the text. We proceed with a heuristic derivation.

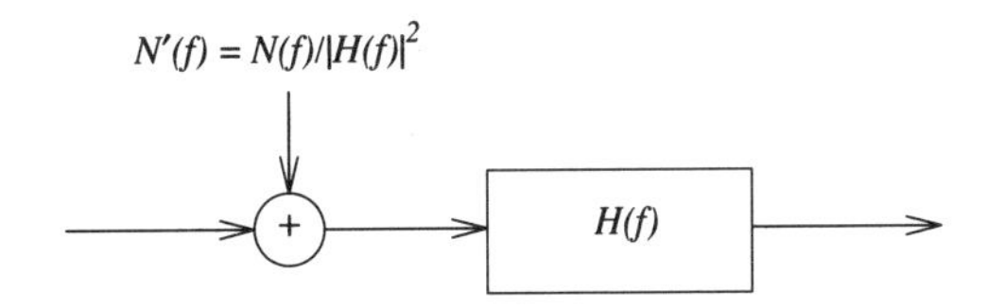

Fig. 2.24 Equivalent additive noise channel.

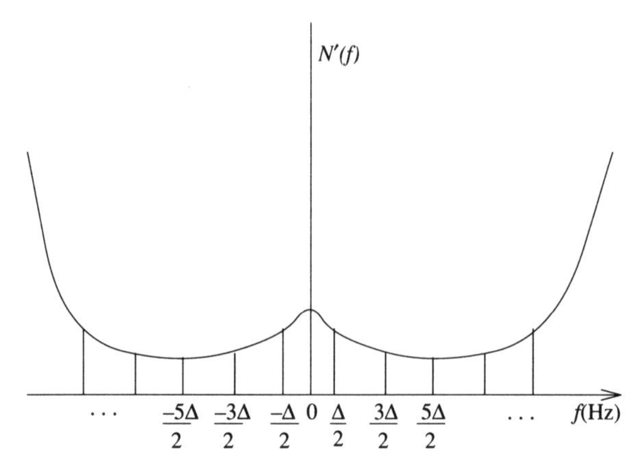

Fig. 2.25 Equivalent noise power spectral density.

We assume that the channel may be represented by a set of nonoverlapping narrowband channels each having bandwidth Δ Hz. (See Figure 2.25.) We may approximate the noise as being flat over each of the subbands, with the mean-square value of the noise in the ith subband given by $N'(i\Delta)\,\Delta$, $i = 0, \pm 1, \ldots$. The received signal plus noise in the ith subband may be represented as

$$Y_i = X_i + N'(i\Delta), \qquad i = 0, \pm 1, \pm 2, \ldots .$$

The fact that Y_i is seen through the filter $H(f)$ is of no consequence since the filtering operation can be reversed. The amount of information that can be transmitted through each of these channels is, from (2.108),

$$C_i = \frac{\Delta}{2} \log_2 \left(1 + P_i/(N'(i\Delta)\Delta)\right) \text{ bps.}$$

where P_i is the average power in the ith channel. Since all of these subchannels are independent of one another, the total information rate is

$$C_\Delta = \sum_i C_i = \frac{1}{2} \sum_i \Delta \log_2 \left(1 + P_i/(N'(i\Delta)\Delta)\right), \qquad (2.118a)$$

and the total average power in the channel is

$$P_\Delta = \sum_i P_i . \qquad (2.118b)$$

Now suppose that we wish to allocate the available transmitted power among the subchannels so as to maximize the total capacity. We form the Lagrangian

$$C_\Delta + \lambda P_\Delta = \frac{1}{2} \sum_i \left[\Delta \log_2 \left(1 + P_i/(N'(i\Delta)\Delta)\right) + \lambda P_i\right], \qquad (2.119)$$

where λ is a Lagrange multiplier whose value is chosen so as to satisfy the

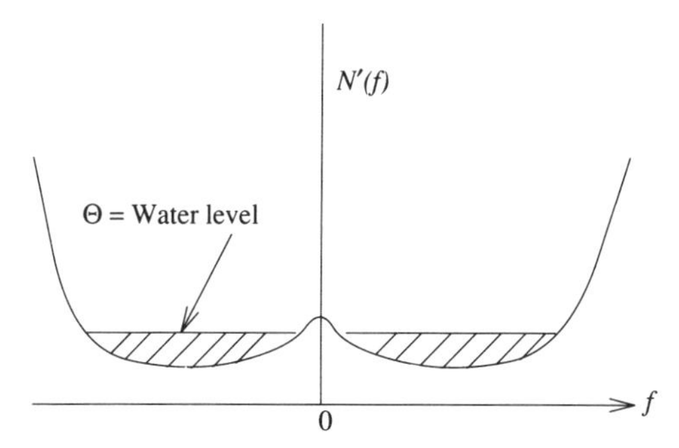

Fig. 2.26 Water-filling argument to determine the transmitted spectrum (which is given by the shaded area).

power constraint. Differentiation shows that the optimum is obtained for all i when $P_i/(N'(i\Delta)\Delta)$ is a constant for all i. If $N'(i\Delta)$ is independent of the transmitted signal, the optimum transmitted level is

$$E_i = \begin{cases} \Theta - N'(i\Delta)\Delta, & \text{for } \Theta - N'(i\Delta)\Delta \geq 0 \\ 0, & \text{otherwise}, \end{cases} \tag{2.120}$$

where Θ is a constant which is chosen to satisfy the average power constraint. It is important to note that Θ is the *same* for all i.

The optimum channel capacity is then

$$C^*_\Delta = \frac{1}{2} \sum_i \Delta \max\,(0, \log_2 \Theta/(N'(i\Delta)\Delta)) \tag{2.121a}$$

with the average power

$$E^*_\Delta = \sum_i \max\,(0, \Theta - N'(i\Delta)\,\Delta)\;. \tag{2.121b}$$

This can be represented by the so-called water-filling model (see Figure 2.26). The channel characteristic $N'(f)$ forms a trough since it will always increase for large enough frequency as $H(f) \to 0$. The parameter Θ gives the level of water in the trough. If Θ is larger than $N'(i)\Delta$ then the difference $\Theta - N'(i\Delta)\Delta$ gives the amount of power in a particular channel, which is shown as the dashed area. If the allowed transmitted power is increased, the water level rises and the capacity of the total channel increases. (See (2.120a) and (2.120b).)

We now make the final step by letting the width of the subchannel go to zero. Defining the parameter

$$\theta = \frac{\Theta}{\Delta}\,,$$

we that the channel capacity is given by

$$C = \frac{1}{2} \lim_{\Delta \to \infty} \sum_i \Delta \max \left(0, \log_2(\theta/N'(i\Delta)\right)$$
$$= \frac{1}{2} \int_{-\infty}^{\infty} df \max \left[0, \log_2 \left[\theta |H(f)|^2 / N(f)\right]\right], \qquad (2.122a)$$

and for the average transmitted power by

$$P_{AV} = \int_{-\infty}^{\infty} df \max \left(0, \theta - N(f)/|H(f)|^2\right). \qquad (2.122b)$$

From (2.122a-b), we see that channel capacity is attained when energy is placed in the frequency band where noise is lowest (see also Figure 2.26). As we have seen, below a certain signal-to-noise ratio, transmission of information is not possible and placing signal energy in a high-noise band is a waste of transmitted power.

2.7 CALCULATIONS OF CHANNEL CAPACITY FOR SELECTED CHANNELS

2.7.1 VOICE-BAND TELEPHONE CHANNEL

The most widely available communications channel is the voice telephone channel. The attenuation characteristics for this channel are shown in Figure 1.33. It is reasonable, as a first approximation, to assume that the attenuation is constant between 200 and 3000 Hz giving a bandwidth of 2800 Hz. We may assume that the noise disturbing the channel is white. In the usual telephone channel, the signal-to-noise ratio is ~30 dB. Since we are seeking an upper bound on capacity, we shall ignore the other impairments discussed in Section 1.6 of Chapter 1. One would expect that, with the advance of telecommunications technology, the effect of these impairments would diminish. From (2.112) we have that the channel capacity is $C = 28$ kbps. For several years, the commercially available limit for transmission over the voiceband channel remained at 96 kbps and the capacity bound seemed unattainable. However, the recent breakthrough in trellis coding, which will be discussed in Chapter 5, has increased the viable rate to 19.2 kbps and it seems that the Shannon bound is in sight. Because of its ubiquity, the voiceband channel is the most widely studied. With the advance of technology, one would expect that attainable performance would also approach the Shannon bound for other channels which now seem out of reach.

2.7.2 TWISTED-PAIR CHANNEL [26]

We can use the results for the colored noise channel to transmission over a twisted-wire pair, which is the predominant medium between the telephone handset and the rest of the telephone plant. For RC-type lines of less than 1000 feet, the transfer function of this channel can be modeled as having the $\sqrt{f}$ attenuation characteristic

$$|H(f)|^2 = e^{-\alpha\sqrt{f}} \quad . \tag{2.123a}$$

The constant α has the following make-up

$$\alpha = kl/l_0 \quad , \tag{2.123b}$$

where k is a constant representing such factors as wire gauge, l_0 is a reference line length, and l is the actual length of the line under study. For example, for 24-gauge wire, $k = 1.158$ for $l_0 = 18{,}000$ feet.

We shall begin by assuming that the noise disturbing the twisted-pair channel is Gaussian noise. While the white noise assumption gives a bound for transmission over an isolated twisted pair, the reality is that a single pair is usually part of a bundle of pairs; consequently, over short lengths, the dominating impairment is *near-end crosstalk* (NEXT). In this case the Gaussian assumption is more difficult to justify, since crosstalk is only coupled over five to seven adjacent pairs. Nevertheless, an analysis under the Gaussian assumption is useful, since it can be shown that the results furnish a lower bound on channel capacity. The analysis is considerably simplified since it can be shown that the equivalent noise power density spectrum is a function of the input power spectrum, and, as a consequence, there is no need to optimize over the input power density spectrum. If it is assumed that the power density spectrum is the same for all twisted pairs, the power density spectrum of the noise (i.e., the crosstalk) may be taken as

$$N(f) = S(f)\ |H_c(f)|^2 \ , \tag{2.124a}$$

where $S(f)$ is the power density spectrum of the transmitted signal and $H_c(f)$ represents the transfer function in coupling from adjacent wire pairs. It has been shown that the coupling has the form

$$|H_c(f)|^2 = \beta f^{3/2} \ , \tag{2.124b}$$

where β is a parameter which depends upon the particular cable type. For example, for a typical twisted pair used in a local area network, $\beta = 10^{-9}$. The equivalent noise power spectral density is then (see (2.117))

$$N'(f) = \frac{S(f)|H_c(f)|^2}{|H(f)|^2} \quad .$$

As we have seen, the optimum input spectrum is such that $E_i/(N'(i\Delta)\Delta)$ is a constant where E_i is the energy in an incremental band of width Δ. (See (2.120).) But $E_i = S(i\Delta)\Delta$ and $E_i/(N'(i\Delta)\Delta) = |H(f)|^2/|H_c(f)|^2$ for any $S(f) > 0$. Thus substituting into (2.103a) and letting $\Delta \rightarrow 0$, the capacity of the

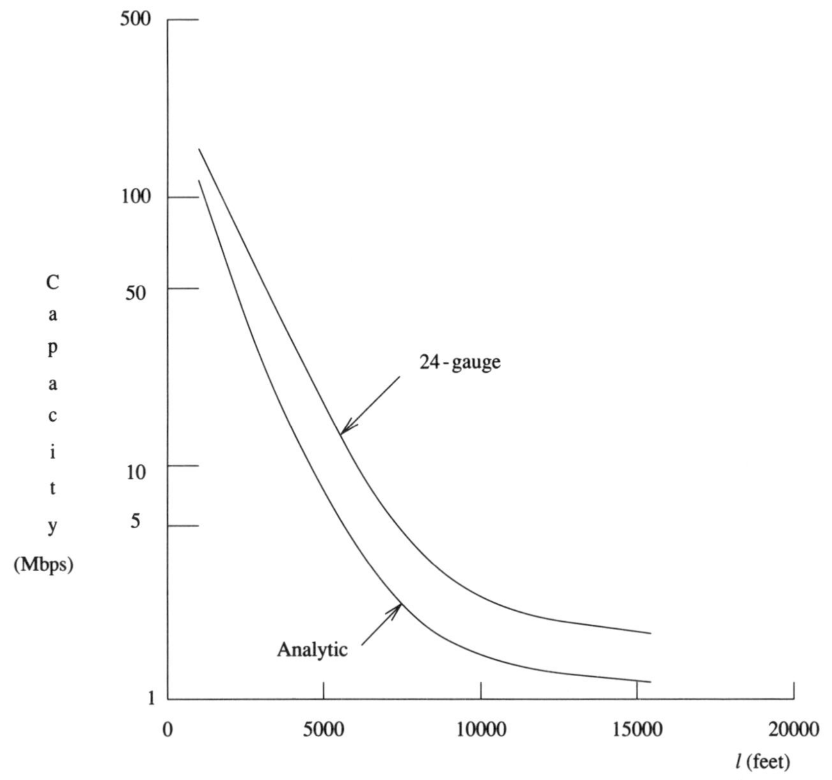

Fig. 2.27 Channel capacity as function of cable length (for analytic channel model and 24-gauge physical model).

crosstalk-dominated twisted-pair channel is given by

$$C_{NEXT} = \int_{\Omega} \log_2 \left[1 + \frac{|H(f)|^2}{|H_c(f)|^2} \right] df \ , \tag{2.125}$$

where Ω is the set of positive frequencies for which $S(f) > 0$. Under the assumption that $\Omega = (0,\infty)$, we have from equations (2.124a), (2.124b), and (2.125)

$$C_{NEXT} = \int_0^{\infty} \log_2 \left[1 + \frac{e^{-\alpha\sqrt{f}}}{\beta f^{3/2}} \right] df \ . \tag{2.126}$$

This integral is not analytically tractable and must be evaluated numerically. It is plotted in Figure 2.27 as a function of cable length for both the above analytical model and for an actual cable characteristic. Note that Mbps capacity exists from many premises to the central office. The capacity falls from

515 Mbps at 150 feet to less than 100 Mbps at 700 feet. At 18,000 feet the capacity is only 1.20 Mbps. At the T1 repeater spacing, 6000 feet, the capacity is 6 Mbps. The channel capacity of the NEXT-dominated channel is independent of the spectrum of the transmitted signal so long as the spectrum is nonzero over the entire frequency range. As the length of line increases, additive, white, Gaussian noise becomes a factor. In this case, there is an optimum transmitted power spectral density which is bandlimited [27].

2.7.3 PAM SIGNALING [12]

It is often the case that inexpensive transmitter and receiver implementations are more important than optimum use of the channel. It is of interest to determine the potential of such simple schemes by comparing their capacity to the optimal. An example of such a simple scheme is pulse amplitude modulation (PAM) discussed in Section 2.2.3. Recall that the amplitude of pulses were modulated by the amplitudes $a_i = [2i + 1 - L]$, $i = 0,1,2,\ldots,L - 1$. We assume that the encoding is such that all signal levels are equally probable. Assuming that the channel noise is white, the output of the channel is passed through a filter matched to the pulse shape thereby maximizing the signal-to-noise ratio. From (2.37a-b) we have that the probability distribution of the output of the matched filter conditioned on the transmitted signal is

$$f_{Z|X}(z|X = a_i) = \frac{\exp(-(z - a_i E_p)^2 / N_0 E_p)}{\sqrt{\pi N_0 E_p}}, \tag{2.127a}$$

where $E_p \triangleq \int_0^T p^2(t)\,dt$, the energy in the pulse. The marginal distribution of z is given by

$$f_Z(z) = \frac{1}{L}\sum_{i=0}^{L-1} \exp(-(z - a_i E_p)^2 / N_0 E_p) \Big/ \sqrt{\pi N_0 E_p}\,. \tag{2.127b}$$

Now the mutual information which indicates the amount of information which is transmitted through this channel is given by (2.81a-b) and (2.82a-b). Since probabilities and signal format, are fixed, the channel capacity is simply the mutual information in this particular case, which is given by

$$C_{PAM} = H(Z) - H(Z \mid X)$$

$$= -\frac{1}{L}\sum_{j=0}^{L-1}\int_{-\infty}^{\infty} dz \frac{\exp\left[-(z - a_j E_p)^2 / N_0 E_p\right]}{\sqrt{\pi N_0 E_p}} \log_2\left[\frac{1}{L}\sum_{i=0}^{L-1}\exp\left[-(z - a_i E_p)^2 / N_0 E_p\right] / \sqrt{\pi N_0 E_p}\right]$$

$$+\frac{1}{L}\sum_{j=0}^{L-1}\int_{-\infty}^{\infty} dz \frac{\exp\left[-(z - a_j E_p)^2 / N_0 E_p\right]}{\sqrt{N_0 E_p}} \log_2 \exp\left[\left[-(z - a_j E_p)^2 / N_0 E_p\right] / \sqrt{N_0 E_p}\right] \tag{2.128}$$

The substitution $w = (z - a_j E_p)/\sqrt{N_0 E_p}$ allows simplification for computational purposes, giving

$$C_{PAM} = \log_2 L - \frac{1}{L} \sum_{j=0}^{L-1} \int_{-\infty}^{\infty} dw \frac{e^{-w^2/2}}{\sqrt{2\pi}} \log_2 \sum_{i=0}^{L-1} \exp\left[-\left(b_j - b_i\right)^2 /2 - w\left(b_j - b_i\right)\right], \tag{2.129}$$

where $b_i = a_i \left(\frac{E_b}{N_0} \right)^{1/2}$. As we have seen in (2.42), the average signal energy normalized to the noise is given by

$$E/N_0 = \frac{1}{L} \sum_{i=0}^{L-1} b_i^2 = \frac{1}{L} \left(\frac{E_p}{N_0} \right) \sum_{i=0}^{L-1} [2i + 1 - L]^2$$

$$= \frac{(L^2 - 1)}{3} \frac{E_p}{N_0} .$$

Equation (2.129), evaluated numerically, is plotted in Figure 2.28 as a function of E/N_0. This gives the information carried by each symbol when the input to

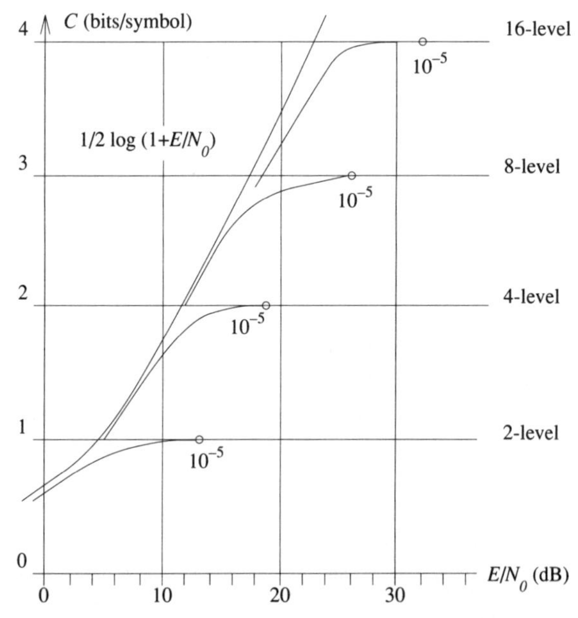

Fig. 2.28 A plot of the information conveyed by a multilevel symbol over a channel with additive white Gaussian noise as a function of SNR. Also shown is the channel capacity of continuous-valued input signals. The points labeled 10^{-5} indicate the SNR at which a probability of error of 10^{-5} is achieved with direct techniques (no coding) [12].

the Gaussian channel is the PAM signal. For high signal-to-noise ratio, the curves are asymptotic to $\log_2 L$. For purposes of comparison we also plot $(1/2)\log_2(1+E/N_0)$, which is the information conveyed through the channel when optimization is carried out over the source distribution. As we see from Figure 2.28, as the signal-to-noise ratio increases it is advisable to increase the number of signaling levels in order to increase the information-bearing capacity of the channel. Thus by choosing a discrete signaling alphabet we lose very little in capacity—a powerful rationalization for digital communications.

2.7.4 PSK AND QAM [12]

This same calculation can be carried out for the bandpass channel, which will be studied in Chapter 5. In this case, there is an additional factor, the phase of the carrier; consequently, signals are represented as points on a plane rather than on a line as in the foregoing. The coordinates are projections on the in-phase and quadrature components of a reference carrier. Typical in-phase and quadrature (or complex) *signal constellations* are shown in Figure 1.37. If information is conveyed by the phase of the carrier only, this technique, known as phase shift keying (PSK), has the advantage of being easier to implement. However, by modulating amplitude as well as phase, more information can be transmitted. The resulting signals are called quadrature amplitude modulated (QAM) signals.

Assuming that the phase of the carrier is known at the receiver, the bandpass channel can often be treated as two independent channels, in-phase and quadrature. (For details, see Chapter 5.) The analysis of the information-carrying capacity of the bandpass channel proceeds in very much the same way as that of the PAM channel. The probability density of the received signal out of the matched filter conditioned on the transmitted signal, is

$$f_{Z|W}(z|W = a_i) = \frac{\exp(-|z - a_i E_p|^2/N_0 E_p)}{\sqrt{\pi N_0 E_p}}, \tag{2.130}$$

where a_i is now a point in the complex plane rather than a point on a line. Note that (2.130) represents a two-dimensional density since it represents in-phase (real) and quadrature (imaginary) components. Following the same steps that lead to (2.128), we find

$$\begin{aligned} C_{QAM} &= \log_2 L - \frac{1}{L}\sum_{j=0}^{L-1}\int_{-\infty}^{\infty} du \int_{-\infty}^{\infty} dv\, \frac{e^{-|z|^2/2}}{\sqrt{2\pi}} \\ &= \log_2 \sum_{i=0}^{L-1} \exp\left[-|b_j - b_i|^2/2 - \operatorname{Re} z^*(b_j - b_i)\right], \end{aligned} \tag{2.131}$$

where $b_i = a_i(E_p/N_0)^{1/2}$ and $z^* = u - iv$ is the complex conjugate of z. Again E_p is the pulse energy and $N_0/2$ is the noise power spectral density in the

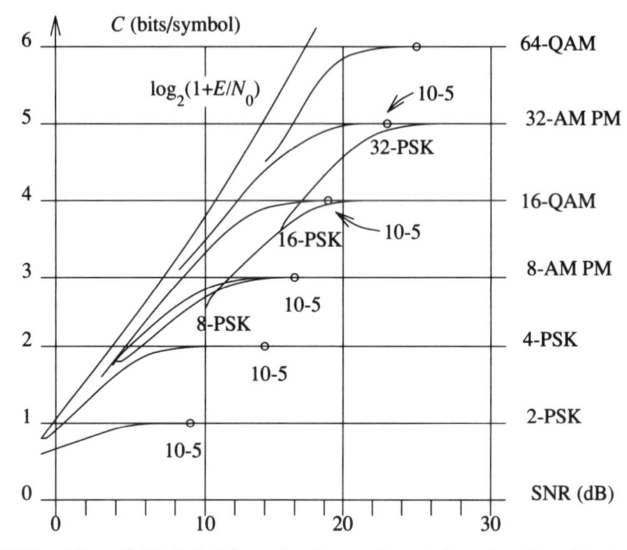

Fig. 2.29 Figure 2.28 plotted for an in-phase and quadrature signaling alphabet [12].

channel. The results of a numerical calculation for the various constellations is shown in Figure 2.29. As we see, for the same number of points in the plane, the QAM symbols convey more information than the PSK. Again, for high signal-to-noise ratio, the curves approach $\log_2 L$. We also compare these curves to the capacity for an unconstrained signal alphabet to $\log_2 (1 + S/N)$. In keeping with the fact that QAM signaling may be treated as independent signaling over different channels, we see that the QAM curves are simply double those of PAM signaling.

APPENDIX 2A BASIC CONCEPTS OF PROBABILITY THEORY

In this appendix we review the basic concepts of probability theory. This review is intended as a summary of material presented in one of any number of excellent texts on the subject.*

2A.1 AXIOMS OF PROBABILITY

Basic set theory forms the foundation for the definition of probability. We define Ω as the set of all possible outcomes of an experiment, e.g., 1, 2, 3, 4, 5, 6 in the toss of a die. We define the set of events $\boldsymbol{A}$. An *event* is defined to be a subset of the set of all experimental outcomes, e.g., odd numbers in the toss of a

* References [28] and [29] are two classical treatments of the subject.

die. The elements of **A** are subsets of Ω. On the set of events we define the following operations:

1. Complementation: A^c is the event that event A does not occur.
2. Intersection: $A \cap B$ is the event that events A and B both occur.
3. Union: $A \cup B$ is the event that events either (or both) A or B occur.
4. Inclusion: $A \subset B$ means that A occurring implies event B occurs.

The set of events, **A**, is closed under the first three of these operations, i.e., performing the operations on one or two of these sets, as appropriate, produces another member of the set. Thus if A and B are events, then A^c, B^c, $A \cup B$, $A \cap B$,... are also events. For example, if in the toss of a die, $\{1, 4, 5\}$ and $\{1\}$ are events, so are $\{2, 3, 6\}$, or $\{1, 2, 3, 6\}$, $\{2, 3, 4, 5, 6\}$, $\{4, 5\}$; $\varnothing$ (empty or null set), and Ω. Two sets of events are said to be disjoint if they have no common elements, i.e., events A and B are disjoint if $A \cap B = \varnothing$. Let $Pr(A)$ denote the probability of event A. The probability of event A is a function on the set of events which satisfies the following axioms:

2A.I There is a nonempty set of experimental outcomes, Ω, and **A**, a set of events defined over the set of outcomes.

2A.II For any event $A \in \mathbf{A}$, $Pr(A) \geq 0$.

2A.III For the set of all possible experimental outcomes, Ω, $Pr(\Omega) = 1$.

2A.IV If the events A and B are disjoint then $Pr(A \cup B) = Pr(A) + Pr(B)$.

2A.V For countably infinite sets A_1, A_2, A_3,..., such that $A_i \cap A_j = \varnothing$ for $i \neq j$ we have

$$Pr\left(\bigcup_{i=1}^{\infty} A_i\right) = \sum_{i=1}^{\infty} Pr(A_i)$$

From these axioms all of the properties of the probability can be derived. A partial list of these properties is as follows:

1. $Pr(A) \leq 1$ for any event $A \in \mathbf{A}$.
2. $Pr(A^c) = 1 - Pr(A)$, where A^c denotes the complement of A.
3. $Pr(A \cup B) = Pr(A) + Pr(B) - Pr(A \cap B)$, for arbitrary A and B.
4. $Pr(A) \leq Pr(B)$ if $A \subset B$.

2A.2 CONDITIONAL PROBABILITY

The idea of the probability of an event can be extended to establish relationships between events. We define the *conditional* probability

$$Pr(A|B) = \frac{Pr(A \cap B)}{Pr(B)} \tag{2A.1}$$

We term $Pr(A|B)$ the probability of event A conditioned on the occurrence of event B. It can be shown that $Pr(A|B)$ is a valid probability in as much as it satisfies the axioms. The events A and B are said to be *independent* if $Pr(A \cap B) = Pr(A)\ Pr(B)$. From (2A.1) we find that this implies $Pr(A|B) = Pr(A)$, meaning that the occurrence of event B has no bearing on the probability of event A.

Consider now a set of events which partition the set of experimental outcomes. If

$$\bigcup_{i=1}^{n} A_i = \Omega \tag{2A.2a}$$

and

$$A_i \cap A_j = \varnothing, \qquad i \neq j\ , \tag{2A.2b}$$

then we say that the sets $A_1, A_2, \ldots, A_n$ *partition* the space. We can write any event B in terms of the disjoint events $A_1, A_2, \ldots, A_n$;

$$B = \bigcup_{i=1}^{n} (A_i \cap B)\ . \tag{2A.3a}$$

From axiom 2A.IV and (2A.1) we have

$$Pr(B) = \sum_{i=1}^{n} Pr(A_i \cap B) = \sum_{i=1}^{n} Pr(B|A_i)\ Pr(A_i)\ . \tag{2A.3b}$$

This is the *law of total probability,* which is of considerable utility in the text. The application of this law leads to *Bayes's rule,* which states that

$$Pr(A_i|B) = \frac{Pr(A_i)Pr(B|A_i)}{\sum_{i=1}^{n} Pr(A_i)Pr(B|A_i)}\ . \tag{2A.4}$$

2A.3 RANDOM VARIABLES—PROBABILITY DISTRIBUTIONS AND DENSITIES

A *random variable* is a function on the set of experimental outcomes mapping the experimental outcomes onto the real line. *Discrete random variables* are mapped onto a countable set of points on the real line, possibly infinite. *Continuous random variables* are mapped into an interval on the real line. Discrete or continuous random variables denoted as X may be characterized by their *probability distribution functions,* which are defined as

$$F_X(x) \triangleq Pr(X \leq x)\ . \tag{2A.5}$$

Thus the probability distribution function evaluated at x is the probability that a set of experimental outcomes will map into the interval $(-\infty, x)$. In the sequel we use the convention of denoting random variables by upper-case letters.

From the basic axioms of probability the following properties for the probability distribution function may be shown: $F_X(-\infty) = 0$, $F_X(\infty) = 1$, and $F_X(x_1) \le F_X(x_2)$ if $x_1 \le x_2$. From the probability distribution function one can calculate the probability of an event lying in an interval, i.e.,

$$Pr(x_1 < X \le x_2) = F_X(x_2) - F_X(x_1), \quad x_1 \le x_2 . \tag{2A.6}$$

For a *discrete* random variable the probability distribution function is a sequence of steps. The height of each step is the probability of the random variable assuming a particular value (see Figure 2A.1). From (2A.5) and (2A.6), we have

$$Pr(X = x) = F_X(x) - F_X(x^-) , \tag{2A.7}$$

where

$$x^- = \lim_{\varepsilon \to 0} x - \varepsilon, \quad \varepsilon > 0 .$$

The probabilities $Pr(X = x_i)$, $i = 1,2,...,N$, where $x_1, x_2, \ldots, x_N$ are all possible values of X, serve as an alternative characterization of the random variable. From the axioms of probability we have

1. $Pr(X = x_i) \ge 0, \qquad i = 1,2,...,N$
2. $Pr((X = x_i) \cup (X = x_j)) = Pr(X = x_i) + Pr(X = x_j), \qquad i \ne j$
3. $\sum_{i=1}^{N} Pr(X = x_i) = 1$

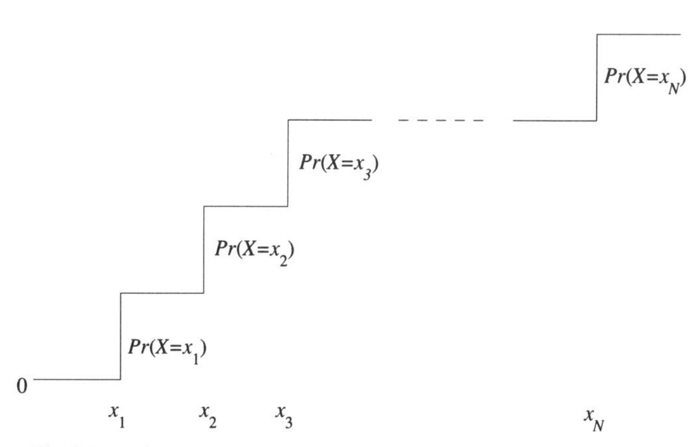

Fig. 2A.1 Probability distribution function for a discrete random variable.

There are three important examples of discrete random variables. For the *binomial* distribution

$$Pr(X = k) = \begin{bmatrix} n \\ k \end{bmatrix} p^k (1-p)^{n-k}, \qquad k = 0,1,...,n\,, \qquad (2A.8a)$$

where $0 \le p \le 1$ and the positive integer n are parameters. The *Poisson* distribution is given by

$$Pr(X = k) = \frac{\lambda^k e^{-\lambda}}{k!}, \qquad k = 0,1,2,...\,, \qquad (2A.8b)$$

where $\lambda \ge 0$ is the average value of X, to be defined presently. Finally we have the *geometric* distribution in which

$$Pr(X = k) = q^k(1-q), \qquad k = 0,1,...\,, \qquad (2A.8c)$$

where $0 \le q \le 1$. It is an easy exercise to demonstrate that the probabilities in each of these cases sum to one.

For a *continuous* random variable, the probability distribution function is a continuous monotonically nondecreasing function. In this case the random variable may be characterized by the *probability density function,* which is defined as

$$f_X(x) \triangleq \frac{dF_X(x)}{dx} \qquad (2A.9a)$$

or

$$F_X(x) = \int_{-\infty}^{x} f_X(t)\,dt\,. \qquad (2A.9b)$$

The properties of the probability distribution function imply the following properties for the density function: $f_X(x) \ge 0$ for all x and $\int_{-\infty}^{\infty} f_X(x)\,dx = 1$. Although the distribution function and the density function may be defined for discrete random variables, they find greater use in connection with continuous random variables. By way of illustration we cite three prominent examples of continuous random variables. First of all, we have the Gaussian or normally distributed random variable which has density function

$$f_X(x) = \frac{e^{-(x-\mu)^2/2\sigma^2}}{\sqrt{2\pi}\,\sigma}, \qquad -\infty < x < \infty \qquad (2A.10a)$$

where μ and σ^2 are the mean and the variance of x respectively. These quantities will be defined presently. Integrating (2A.10a) from $-\infty$ to ∞ and after a change of variable, $y = (x - \mu)/\sigma$, we obtain a standard integral which may be found in any number of integral tables to be equal to 1

$$\int_{-\infty}^{\infty} \frac{e^{-y^2/2}}{\sqrt{2\pi}}\,dy = 1\,. \qquad (2A.10b)$$

The *exponentially* distributed random variable has density function

$$f_X(x) = \begin{cases} 0, & x < 0 \\ \lambda\, e^{-\lambda x}, & x \geq 0, \end{cases} \tag{2A.11}$$

where λ is a parameter. Finally given x_1 and x_2, with $x_1 \leq x_2$, as parameter, the *uniform* distribution is

$$f_X(x) = \begin{cases} 1/(x_2 - x_1), & x_1 \leq x \leq x_2 \\ 0, & \text{otherwise} . \end{cases} \tag{2A.12}$$

The probability distribution function of these random variables can be found from (2A.9b). It is left as exercises to show that the two densities in (2A.11) and (2A.12) integrate to 1 over the appropriate limits.

There are applications where a mixed random variable is encountered. The probability is spread over an interval as for a continuous random variable but there are points of concentration as for a discrete random variable. In this case the probability distribution has jumps at the points of concentration represented by unit steps

$$F_X(x) = \sum_{i=1}^{N} P_i U(x - x_i) + G_X(x) ,$$

where P_i are the probabilities and $U(x)$ is the unit step, and $G_x(x)$ is a continuous function. The probability density may be represented in terms of Dirac delta functions

$$f_x(x) = \sum_{i=1}^{N} P_i \delta(x - x_i) + g_X(x) ,$$

where $g_X(x)$ is the derivative of $G_X(x)$.

2A.4 JOINT DISTRIBUTIONS OF RANDOM VARIABLES

In this section we consider relations between two or more random variables. We define the *joint* probability distribution function of two random variables X and Y as

$$F_{XY}(x,y) \triangleq P_r(X \leq x,\, Y \leq y), \qquad -\infty < x,y < \infty , \tag{2A.13}$$

where x and y are real numbers. The random variables may represent different mappings onto the real line from the same space of experimental outcomes, and the properties of the joint density function are

$$F_{XY}(-\infty, -\infty) = 0, \qquad F_{XY}(\infty, \infty) = 1,$$

$$F_{XY}(x_1, y) \leq F_{XY}(x_2, y) \qquad \text{for } x_1 \leq x_2$$

and

$$F_{XY}(x,y_1) \le F_{XY}(x, y_2) \qquad \text{for } y_1 \le y_2 .$$

The joint probability density function of the random variables X and Y is defined as

$$f_{XY}(x,y) \triangleq \frac{\partial^2 F_{XY}(x,y)}{\partial x \, \partial y}, \qquad -\infty < x,y < \infty . \tag{2A.14}$$

By analogy with the one-dimensional case we have the following properties for the joint density function:

$$f_{XY}(x,y) \ge 0, \quad -\infty < x,y < \infty \quad \text{and} \quad \int_{-\infty}^{\infty} dx \int_{-\infty}^{\infty} dy \, f_{XY}(x,y) = 1 .$$

The *marginal distribution* and *density functions,* when they exist, may be calculated from the joint functions. We have

$$F_X(x) = F_{XY}(x,\infty) , \tag{2A.15a}$$

$$F_Y(y) = F_{XY}(\infty, y) , \tag{2A.15b}$$

$$f_X(x) = \int_{-\infty}^{\infty} f_{XY}(x,y) \, dy , \tag{2A.15c}$$

$$f_Y(y) = \int_{-\infty}^{\infty} f_{XY}(x,y) \, dx . \tag{2A.15d}$$

Two random variables are independent if

$$F_{XY}(x,y) = F_X(x) \, F_Y(y) , \tag{2A.16a}$$

or equivalently in terms of probability density functions,

$$f_{XY}(x,y) = f_X(x) \, f_Y(y) . \tag{2A.16b}$$

For random variables which are not independent, the concepts of conditional distribution and density functions come into play. We define the conditional distribution function as

$$F_{X|Y}(x|y) = \frac{Pr(X \le x, Y = y)}{Pr(Y = y)} = \frac{F_{XY}(x,y)}{f_Y(y)} . \tag{2A.17a}$$

The event here is the probability that $X \le x$ is given that $Y = y$. We may also define the conditional density of the random variable X as

$$f_{X|Y}(x|y) = \frac{f_{XY}(x,y)}{f_Y(y)} , \tag{2A.17b}$$

which is the probability density of X conditioned on the event $Y = y$.

Although we have discussed joint density functions, joint distribution functions, and independence in connection with two random variables, these same concepts apply in a straightforward way to three or more random variables.

2A.5 EXPECTATION OF A RANDOM VARIABLE—MOMENTS

The distributions of random variables are characterized by a number which is called the *expectation* or the *mean value* of a random variable, which is defined to be

$$E[X] \triangleq \sum_{i=}^{\infty} x_i Pr(X = x_i) \tag{2A.18a}$$

for discrete random variables and

$$E[X] \triangleq \int_{-\infty}^{\infty} x\, f_X(x)\, dx \tag{2A.18b}$$

for continuous random variables. The expectation is often written $\overline{X}$.

EXAMPLE 7 Since it is so important to the material presented in the text, we calculate the mean of a Gaussian random variable. Substituting (2A.10a) into (2A.18b), we have

$$E[X] = \int_{-\infty}^{\infty} x \frac{e^{-(x-\mu)^2/2\sigma^2}}{\sqrt{2\pi}\,\sigma} dx\,.$$

A change of variable $y = \dfrac{x-\mu}{\sigma}$ yields

$$E[X] = \int_{-\infty}^{\infty} y \frac{e^{-y^2/2}}{\sqrt{2\pi}} dy + \mu \int_{-\infty}^{\infty} \frac{e^{-y^2/2}}{\sqrt{2\pi}} dy\,. \tag{2A.19}$$

Since it is the integral of an odd function over symmetric limits, the first term on the right-hand side of (2A.19) is equal to zero. From (2A.10b) we have

$$E[X] = \mu\,.$$

The expectation of a random variable which can be expressed as a function of another random variable, X, i.e., $E[Y]$, where $Y = g(X)$, can be found from the distribution function of X. In the discrete case we have

$$E[Y] = E[g(X)] = \sum_{i=1}^{\infty} g(x_i) P[X = x_i] \tag{2A.20a}$$

and in the continuous case

$$E[Y] = E[g(X)] = \int_{-\infty}^{\infty} g(x)\, f_X(x)\, dx\,. \tag{2A.20b}$$

The most useful functions in terms of characterizing random variables are of the form $Y = X^k$. The kth moment of a random variable is $E[X^k]$ and the kth central moment is $E[(X - E[X])^k]$. The *variance* of a random variable is the second central moment:

$$Var(X) = E(X - E[X])^2 = EX^2 - (E[X])^2 . \tag{2A.21}$$

The square root of the variance is a quantity called the *standard deviation* of a random variable.

EXAMPLE 8 To calculate the variance for a Gaussian random variable,

$$Var(X) = \int_{-\infty}^{\infty} \frac{(x-\mu)^2 \, e^{-(x-\mu)^2/2\sigma^2}}{\sqrt{2\pi}\,\sigma} .$$

(Recall that $E(X) = \mu$.)

We again perform the change of variable $y = \dfrac{x-\mu}{\sigma}$, to obtain

$$Var(X) = \sigma^2 \int_{-\infty}^{\infty} \frac{y^2}{\sqrt{2\pi}} e^{-y^2/2} \, dy .$$

After integration by parts we have

$$Var(X) = \sigma^2 \left[-y \, e^{-y^2/2} \Big|_{-\infty}^{\infty} + \int_{-\infty}^{\infty} \frac{e^{-y^2/2}}{\sqrt{2\pi}} \, dy \right] = \sigma^2 . \tag{2A.22}$$

Note that in all of these cases if the form of the distribution, e.g., Gaussian, is known, then knowledge of the mean and the variance allows the complete distribution to be found. In the cases of the Poisson, geometric, and exponential distributions only the mean is required. From the properties of expectation the mean and the variance of a linear transformation of a random variable follow immediately:

$$E[aX + b] = aE[X] + b \tag{2A.22a}$$

$$Var(aX + b) = a^2 Var(X) \tag{2A.22b}$$

EXAMPLE 9 Consider the linear transformation $Y = aX + b$ where X is a random variable with a Gaussian distribution. With a change of variable in the density function given by (2A.10a) we have

$$f_Y(y) = \frac{\exp(-(y - a\mu - b)^2/2a^2\sigma^2)}{\sqrt{2\pi}\; a\sigma} ; \tag{2A.23}$$

consequently, the random variable Y is itself Gaussian with $E[Y] = a\mu + b$

and $Var(Y) = a^2\ Var(X)$. In dealing with linear transformations of a Gaussian random variable, it is not necessary to go through these steps. We are assured that the resulting random variable has a Gaussian distribution which is characterized by its mean and variance. It is only necessary to compute these two quantities. The means and the variances for the distributions that we have discussed in the forgoing are given in Table 2A.1.

Table 2A.1. Means and Variances

Distribution	Mean	Variance
Binomial (A.8a)	np	$np(1-p)$
Poisson (A.8b)	λ	λ
Geometric (A.8c)	$q/(1-q)$	$q/(1-q)^2$
Gaussian (A.10a)	μ	σ^2
Exponential (A.10b)	$1/\lambda$	$1/\lambda^2$
Uniform (A.12)	$(x_2 - x_1)/2$	$(x_2 - x_1)^2/3$

If the joint probabilities of two random variables are known, joint moments may be calculated. In the discrete and the continuous cases we have, respectively,

$$E[XY] = \sum_{i=1}^{\infty} \sum_{j=1}^{\infty} x_i y_j P(X = x_i,\, Y = y_j) \tag{2A.24a}$$

and

$$E[XY] = \int_{-\infty}^{\infty} dx \int_{-\infty}^{\infty} dy\ xy f_{XY}(x,y)\ . \tag{2A.24b}$$

This joint first moment is called the *correlation* or the *autocorrelation* of the random variables X and Y. The *covariance* or the *autocovariance* is defined as

$$\begin{aligned} Cov(X,Y) &= E[(X-E[X])(Y-E[Y])] \\ &= \sum_{i=1}^{\infty} \sum_{j=1}^{\infty} (x_i - E[X])(y_j - E[Y]) \cdot Pr[X = x_i,\, Y = y_j] \quad \text{(discrete case)} \\ &= \int_{-\infty}^{\infty} dx \int_{-\infty}^{\infty} dy\ (x - E[X])(y - E[Y]) \cdot f_{XY}(x,y) \quad \text{(continuous case)} \end{aligned} \tag{2A.25}$$

We say that two random variables are uncorrelated if $E[XY] = E[X]E[Y]$ or equivalently $Cov\ (X,Y) = 0$. When random variables are independent they are of necessity uncorrelated. Since taking the expectation of a random variable is a linear operation, the expected value of the sum of two random variables is simply the sum of the expected values:

$$E[X_1 + X_2] = E[X_1] + E[X_2] \; . \tag{2A.26}$$

This holds even if the random variables are not independent. When two random variables are uncorrelated it is evident from (2A.21) and (2A.25) that the variance of their sum is equal to the sum of the variance:

$$Var(X_1 + X_2) = Var(X_1) + Var(X_2) \; . \tag{2A.27}$$

The distribution of sums of independent random variables can be found by means of convolution. Let $Y = X_1 + X_2$. If X_1 and X_2 are discrete random variables, then

$$Pr[Y = k] = \sum_{i=-\infty}^{\infty} Pr[X_1 = i] Pr[X_2 = k - i] \; , \tag{2A.28a}$$

and for continuous random variables we have

$$f_Y(y) = \int_{-\infty}^{\infty} dx f_{X_1}(x) f_{X_2}(y - x) \tag{2A.28b}$$

when $f_Y(y)$, $f_{X_1}(x)$, and $f_{X_2}(x)$ are the appropriate probability density functions.

2A.6 THE JOINT DISTRIBUTION OF GAUSSIAN RANDOM VARIABLES

The Gaussian distribution has properties which simplify analysis in many situations. This is fortunate since many physical systems can be modeled using Gaussian random variables. Furthermore, it has been a serendipitous experience that analysis based on the assumption of Gaussian random noise has even given insight into systems for which it is known that the disturbances do not have a Gaussian distribution.

A key property of Gaussian random variables is that the sum of independent Gaussian random variables is itself Gaussian. Let X_1 have a Gaussian distribution with mean μ_1 and variance σ_1^2 and let X_2 have a Gaussian distribution with mean μ_2 and variance σ_2^2. The random variable $Y = X_1 + X_2$ is Gaussian with mean $\mu_1 + \mu_2$ and variance $\sigma_1^2 + \sigma_2^2$. In order to show this we substitute the Gaussian density function given by (2A.10a) into the convolution operation, (2A.24b), giving

$$f_Y(y) = \int_{-\infty}^{\infty} \frac{e^{-(x-\mu_1)^2/2\sigma_1^2}}{\sqrt{2\pi}\;\sigma_1} \frac{e^{-y-x-(\mu_2)^2/2\sigma_2^2}}{\sqrt{2\pi}\;\sigma_2} \, dx \; . \tag{2A.29}$$

The changes of variables

$$t_1 = x - \mu_1 \tag{2A.30a}$$

and

$$t_2 = y - \mu_1 - \mu_2 \tag{2A.30b}$$

gives

$$f_Y(y) = \int_{-\infty}^{\infty} \frac{\exp(-t_1^2/2\sigma_1^2 - (t_2 - t_1)^2/2\sigma_2^2)}{2\pi\, \sigma_1 \sigma_2} dt_1 \ .$$

We focus on the exponent, and completing the square gives

$$\frac{t_1^2}{\sigma_1^2} + \frac{(t_2 - t_1)^2}{\sigma_2^2} = \frac{\sigma_1^2 + \sigma_2^2}{\sigma_1^2 \sigma_2^2} \left(t_1 - \frac{t_2 \sigma_1^2}{\sigma_1^2 + \sigma_2^2} \right)^2 + \frac{t_2^2}{\sigma_1^2 + \sigma_2^2} \ . \tag{2A.31}$$

Notice that the second term on the right-hand side of (2A.31) can be factored out of the integral. Another change of variable in the integral

$$z = \left(t_1 - \frac{t_2 \sigma_1^2}{\sigma_1^2 + \sigma_2^2} \right) \Bigg/ \frac{\sigma_1 \sigma_2}{\sqrt{\sigma_1^2 + \sigma_2^2}} \ ,$$

gives the expression

$$f_Y(y) = \frac{\exp(-t_2^2/2\sqrt{\sigma_1^2 + \sigma_2^2}\,)}{\sqrt{2\pi(\sigma_1^2 + \sigma_2^2)}} \int_{-\infty}^{\infty} \frac{e^{-z^2/2}}{\sqrt{2\pi}} dz = \frac{\exp(-(y - \mu_1 - \mu_2)^2/2\sqrt{\sigma_1^2 + \sigma_2^2}\,)}{\sqrt{2\pi(\sigma_1^2 + \sigma_2^2)}} \ . \tag{2A.32}$$

Comparison of (2A.32) with (2A.10a) proves the assertion.

This relationship is easily extended; the sum of any number of Gaussian random variables is Gaussian with mean equal to the sum of the means and variance equal to the sum of the variances.

The two random variables X and Y have a joint Gaussian distribution when their joint density function is

$$f_{XY}(x,y) = \frac{\exp\left\{ -\left[\frac{(x - \mu_1)^2}{\sigma_1^2} - \frac{2r(x - \mu_1)(y - \mu_2)}{\sigma_1^2} + \frac{y - \mu_2)^2}{\sigma_2^2} \right] /2(1 - r^2) \right\}}{2\pi\sigma_1\sigma_2\sqrt{1 - r^2}} \tag{2A.33}$$

where $r = \dfrac{E[(X - \mu_1)(Y - \mu_2)]}{\sigma_1 \sigma_2}$. From equations (2A.15c) and (2A.15d), it can be shown that the marginal distributions of X and Y are Gaussian with means and variances of μ_1, σ_1^2 and μ_2, σ_2^2, respectively. If the random variables are uncorrelated, $r = 0$ and (2A.33) decomposes into

$$f_{XY}(x,y) = f_X(x)\, f_Y(y) = \frac{\exp\left[-\frac{(x-\mu_1)^2}{2\sigma_1^2}\right]}{\sqrt{2\pi}\,\sigma_1} \; \frac{\exp\left[-\frac{(y-\mu_2)^2}{2\sigma_2^2}\right]}{\sqrt{2\pi}\,\sigma_2} \tag{2A.34}$$

We see from (2A.34) that uncorrelated Gaussian random variables are independent. In general, uncorrelated random variables are not independent.

The conditional distribution of Y given X is given by

$$f_{Y/X}(y/x) = \frac{\exp\left[-\left[y-\mu_2-\frac{r\sigma_2}{\sigma_1}(x-\mu_1)\right]^2 / 2\sigma_2^2(1-r^2)\right]}{\sigma_2\sqrt{2\pi(1-r^2)}} \tag{2A.35}$$

We see that this is a Gaussian random variable with mean $\mu_2 - r\,\frac{\sigma_2}{\sigma_1}(x-\mu_1)$ and variance $\sigma_2^2(1-r^2)$.

2A.7 PROBABILITY-GENERATING FUNCTIONS AND CHARACTERISTIC FUNCTIONS

Frequently it is more convenient to express distributions in terms of transforms. The *probability generating function* of the discrete random variable X is defined as

$$X(z) \triangleq E[z^X] = \sum_{i=0}^{\infty} z^i Pr[X = i] \ , \tag{2A.37}$$

where z is a complex variable. The probability-generating function has an obvious relationship to the z transform of the sequence $Pr[X = i]$, $i = 0,1,2,\ldots$. The probability-generating function contains all of the information that the probability distribution does. One can be obtained from the other. A noteworthy property of the probability-generating function is that it is analytic on the unit disk, which implies that it is bounded. Note that

$$|X(z)| \le \sum_{i=0}^{\infty} |z|^i Pr[X = i] \le \sum_{i=0}^{\infty} P_r[X = i] = 1 \ . \tag{2A.38}$$

Given the probability-generating function of a random variable, the moments of the random variable can be easily found. For the first moment we have

$$E[X] = \frac{dX(z)}{dz}\Big|_{z=1} = \sum_{i=0}^{\infty} iz^{i-1} Pr[X = i]\Big|_{z=1} = \sum_{i=0}^{\infty} iPr[X = i] \ .$$

Higher moments can be found from successive differentiation and manipulation.

For example, we find

$$E[X^2] = \frac{d^2X(z)}{dz^2}\Big|_{z=1} + \frac{dX(z)}{dz}\Big|_{z=1}$$

From (2A.28a) and the definitions of generating functions it can be shown that the generating function of the sum of independent random variables is the product of the individual generating function; i.e., if $Y = X_1 + X_2$ then

$$Y(z) = X_1(z)X_2(z) , \tag{2A.39}$$

where $X_1(z)$ and $X_2(z)$ are the probability-generating functions of X_1 and X_2, respectively. For the continuous random variable, X, we define the *characteristic function* to be

$$\phi_X(\omega) = E[e^{j\omega x}] = \int_{-\infty}^{\infty} dx\; e^{jx\omega} f_X(x) \tag{2A.40a}$$

and

$$j = \sqrt{-1} .$$

The properties of the characteristic function are similar to those of the probability-generating function. The characteristic function and the probability density functions are Fourier transform pairs. Given the characteristic function, the density function is given by

$$f_X(x) = \frac{1}{2\pi} \int_{-\infty}^{\infty} \phi_X(\omega)\; e^{-j\omega x}\; \alpha\omega . \tag{2A.40b}$$

The moments of the random variable X can be found in an even more straightforward fashion:

$$\frac{d^k \phi_X(\omega)}{d\omega^k}\Big|_{\omega=0} = \int_{-\infty}^{\infty} dx\; x^k f_X(x) = j^k E[X^k] . \tag{2A.41}$$

If the random variables X_1 and X_2 are independent, then the characteristic function of $Y = X_1 + X_2$ is

$$\phi_Y(\omega) = \phi_{X_1}(\omega)\; \phi_{X_2}(\omega) , \tag{2A.42}$$

where $\phi_{X_1}(\omega)$ and $\phi_{X_2}(\omega)$ are the characteristic functions of X_1 and X_2, respectively. The characteristic function of a Gaussian random variable with mean μ and variance σ^2 is

$$\phi(\omega) = e^{j\omega\mu - \frac{1}{2}\sigma^2\omega^2} . \tag{2A.43}$$

2A.8 BOUNDS AND LIMIT THEOREMS

It is often the case that exact expressions for quantities are difficult to obtain, but certain inequalities give insight. The best known of these is the *Che-*

bychev inequality, which states that

$$Pr[X \geq a] \leq \frac{E[g(x)]}{g(a)} , \tag{2A.44a}$$

where $g(x)$ is a non-negative even function which is increasing on $[0, \infty]$. A case which is of particular interest is

$$Pr(|X - E[x]| \geq \varepsilon) \leq \frac{Var\ X}{\varepsilon^2} . \tag{2A.44b}$$

The Chebychev inequality may be used to say something about sums of independent random variables. Let $Y_n = (1/n) \sum_{i=1}^{n} X_i$, where X_i, $i = 1,2,...,$ are independent identically distributed random variables with mean μ and variance σ^2. Since

$$E[Y_n] = \mu \quad \text{and} \quad Var\ Y_n = \sigma^2/n ,$$

the Chebychev inequality gives

$$\lim_{n \to \infty} Pr(|Y_n - \mu| > \varepsilon) = 0.$$

This result shows that the arithmetic average approaches the mean of the distribution with probability 1. This is known as the *strong law* of *large numbers.* A related result is the *central limit theorem:* Let $S_n \triangleq \sum_{i=1}^{n} X_i$. The distribution of S_n approaches Gaussian as $n \to \infty$.

The *Chernoff* bound finds application in bounding tail probabilities in a simple analytically tractable fashion. If $f_x(x)$ is the probability density of the random variable X, we may write

$$Pr(X > x) = \int_x^\infty f_x(t)\ dt = \int_{-\infty}^\infty U(t - x) f_x(t)\ dt. \tag{2A.45}$$

Where the unit step function is defined as

$$U(x) = \begin{cases} 1 & \text{for } x \geq 0 \\ 0 & \text{for } x < 0 \end{cases} .$$

Now for any value of $a > 0$ we have

$$U(t - x) \leq \exp\left[a(t - x)\right] ; \ -\infty < t \ \infty . \tag{2A.46}$$

Substituting (2A.46) into (2A.45) yields

$$Pr(X > x) \leq \int_{-\infty}^\infty \exp\left[a(t - x)\right] f_x(t)\ dt . \tag{2A.47a}$$

The right-hand side is seen as the expected value of $\exp\left[a(t - x)\right]$, and we have

$$Pr(X > x) \leq \exp(-ax)\ E\left[\exp(aX)\right] . \tag{2A.47b}$$

In terms of the characteristic function of X, we have

$$Pr(X > x) \leq \exp(-ax)\ \phi_x(a/j) . \tag{2A.47c}$$

The parameter $a > 0$ can be chosen so as to minimize the right-hand side of (2A.47b) or (2A.47c).

2A.9 RANDOM PROCESSES

Frequently in the text we deal with functions of time which exhibit random properties; noise is probably the most frequently encountered example. This kind of function can be modeled as a *random process*. A random process can be defined in a fashion similar to the definition of a random variable. Suppose that a function of time is associated with each outcome of a random experiment. The *ensemble* (or collection) of these functions is a random process. Two simple examples will serve to illustrate the concepts. A fair coin is flipped. To the outcome "head," we associate the constant voltage +1 volts and to the outcome "tail," the voltage –1 volts. The ensemble here consists of the two constant functions of time. A second example involves an experiment with a wheel, with marked angles, spun relative to a constant point. The outcome of the experiment is an angle, θ, which is uniformly distributed between $-\pi$ and π. This angle is mapped into the phase of a sinusoid with a specified carried frequency, f_0. The ensemble of functions

$$X(\Omega,t) = \cos\left[2\pi f_0\ t + \theta(\omega)\right] , \qquad -\infty < t < \infty,\ -\pi < \theta(\omega) < \pi , \tag{2A.48}$$

forms the random process. The dependence on the experimental outcome is indicated by the ω in (2A.48). In the rest of this section, we suppress this variable to conform to the usual practice.

It should be realized that the idea of an underlying experiment is a mathematical construct which allows one to deploy the concepts of probability, random variables, probability distributions, etc. In many examples of random processes, the underlying experiment would neither be so simple nor so direct.

Time samples of a random process are themselves random variables since they represent mapping from a space of an experimental outcome onto the real line. For example, the sample of the process derived from the coin-flipping example described above is a random variable with distribution $Pr(X(t) = -1) = Pr(X(t) = +1) = 1/2$. The moments of the samples can be calculated. For the coin-flipping process we have

$$E(X(t)) = 1/2 - 1/2 = 0 \quad \text{and} \quad E(X^2(t)) = 1/2 + 1/2 = 1.$$

For the second example of a random process, a sinusoid with a random phase, the first two moments are

$$E(X(t)) = \frac{1}{2\pi} \int_{-\pi}^{\pi} \cos\ (2\pi f_0 t + \theta)\ d\theta = \frac{1}{2\pi} \sin\ (2\ \pi\ f_0 t + \theta)\ \Big|_{-\pi}^{\pi} = 0$$

and

$$E(X^2(t)) = \frac{1}{2\pi} \int_{-\pi}^{\pi} \cos^2(2\pi f_0 t + \theta)\ d\theta = \frac{1}{2\pi} [\frac{\theta}{2} - \frac{1}{4}\ \sin(4\pi f_0 t + 2\theta)] \Big|_{-\pi}^{\pi} = \frac{1}{2}$$

Samples of the random process which are taken at different times are correlated random variables, in general. The *autocorrelation function* is a particularly important measure of this correlation. This function is defined as

$$R(t_1, t_2) = E(X(t_1)\ X(t_2))\ , \qquad (2A.49)$$

where $X(t_1)$ and $X(t_2)$ indicate samples taken at times t_1 and t_2, respectively. From the Schwartz inequality, we have

$$R(t_1, t_2) \leq \{E(X^2(t_1))\ E(X^2(t_2))\}^{1/2}\ . \qquad (2A.50)$$

For the two processes exemplified above, the autocorrelation functions are, respectively,

$$R(t_1, t_2) = 1/2(-1)(-1) + 1/2(+1)(+1) = 1 \qquad (2A.51a)$$

$$R(t_1, t_2) = \frac{1}{2\pi} \int_{-\pi}^{\pi} \cos(2\pi f t_1 + \theta)\ \cos(2\pi f t_2 + \theta)\ d\theta \qquad (2A.51b)$$

$$= \int_{-\pi}^{\pi} \left[\frac{1}{2}\ \cos(2\pi f_0 (t_1 - t_2)) + \frac{1}{2}\ \cos(2\pi f_0 (t_1 + t_2) + 2\theta) \right] d\theta$$

$$= \frac{1}{2}\ \cos(2\pi f_0 (t_1 - t_2))\ .$$

2A.10 STATIONARITY AND ERGODICITY

In connection with the two example processes, we notice that there is a certain time invariance. The means are constant with time and the correlation between samples is a function of time difference and not on the absolute time. Processes with these properties are called *wide-sense stationary*. If this time invariance were true for the whole distribution of the process, hence, all of its moments, we would have a *strict-sense stationary* process.

For a wide sense stationary process $X(t)$, we may write the correlation function as a single variable.

$$R(t_1, t_2) = R_X(t_1 - t_2)\ . \qquad (2A.52a)$$

Note that, from (2A.46),

$$R_X(\tau) \le R_X(0) = E(X^2(t)) . \tag{2A.52b}$$

Generally speaking, we may observe a single realization of a random process. A reasonable question would be "Is the realization typical of the other realizations in the ensemble?" We may, for example, compute *time averages* on the realization of the process that is available, $x(t)$. The time-average value is given by

$$\overline{X} = \lim_{T\to\infty} 1/T \int_{-T}^{T} x(t)\ dt \tag{2A.53a}$$

and the mean-square value by

$$\overline{X^2} = \lim_{T\to\infty} 1/T \int_{-T}^{T} x^2(t)\ dt . \tag{2A.53b}$$

Higher moments are computed in a similar fashion. Correlations between different sample points can be computed as

$$\overline{X(t_1)X(t_2)} = \lim_{T\to\infty} 1/T \int_{-T}^{T} x(t-t_1)x(t-t_2)\ dt . \tag{2A.54}$$

When a process is what is called *ergodic*, all of the time-average moments are equal to the moments computed by ensemble average. In an approximate way, an ergodic process is one where every realization of the process passes through the same set of states. The importance of the ergodic property lies in the fact that the particular realization of the process that is obtained from an observation is in some sense typical. We are assured that the time average will not vary from realization to realization. The coin-flipping example above is an example of a process which is not ergodic. The voltage +1 never passes through –1. The time average is +1 or –1 volt, depending upon the realization; whereas, the ensemble average is 0 volts. In contrast, the above example of the sinusoid with the uniformly distributed phase is an ergodic process, where the time average is computed as follows:

$$X = \lim_{T\to\infty} 1/T \int_{-T}^{T} \cos(2\pi f_0 t+\theta)\ dt = \lim_{T\to\infty} \frac{1}{2}\ \frac{\sin 2\pi f_0 t}{2\pi f_0}\Bigg|_{-T}^{T} = 0$$

We notice that the time average equals the ensemble averages for this process. Similar results can be shown for the higher-order moments.

A *Gaussian process* is one for which any arbitrarily chosen set of time samples, $X(t_1)$, $X(t_2)$,..., $X(t_n)$, have a joint Gaussian distribution. Each pair of samples has the joint density function given by (2A.33). The correlation between $X(t_i)$ and $X(t_j)$ is given by

$$\begin{aligned} r_{ij} &= E(X(t_i)) - E(X(t_i))\ (X(t_j)) - E(X(t_j))/\sigma_1\sigma_2 \\ &= [R(t_i,t_j) - E(X(t_i))E(X(t_j))]/\sigma_i\sigma_j , \end{aligned} \tag{2A.55}$$

where σ_i^2 and σ_j^2 are the variances of $X(t_i)$ and $X(t_j)$, respectively.

From (2A.33) and (2A.55), it is clear that the probability distribution of the process is governed by the mean and the correlation function of the process. If a Gaussian process is wide-sense stationary, then it is strict-sense stationary as well. The Gaussian process, particularly the stationary Gaussian process, is important in the modeling of communications systems. Thermal noise, which is an underlying random disturbance present in all systems, is a stationary Gaussian process.

2A.11 POWER DENSITY SPECTRA

We now focus exclusively on wide-sense stationary processes. As we have seen, for such processes, the mean is a constant with time and the autocorrelation function is a function only of the time difference. We express this as

$$E(X(t)) = m_X$$

and

$$R(t_1,t_2) = R_X(t_1-t_2) \ .$$

Notice that the autocorrelation function is an even function, i.e.,

$$R_X(t_1-t_2) = R_X(-t_1+t_2) \ .$$

We define the *power density spectrum* of a wide-sense stationary process as the Fourier transform of the autocorrelation function.

$$S_X(f) = 5F1\{R_X(\tau)\} = \int_{-\infty}^{\infty} \exp(-2j\pi f\tau)\ R_X(\tau)\ d\tau \ . \quad \text{(2A.56a)}$$

Since $R_X(\tau)$ is an even function, we have

$$S_X(f) = \int_{-\infty}^{\infty} \cos(-2\pi f\tau)\ R_X(\tau)\ d\tau \ ; \quad \text{(2A.56b)}$$

consequently, $S_X(f)$ is a real function of f. Given the power density spectrum of a process, its autocorrelation function may be found from the inverse Fourier transform.

$$R_X(\tau) = (1/2\pi) \int_{-\infty}^{\infty} S_X(f)\exp(2\pi jf\tau)\ df \ . \quad \text{(2A.57a)}$$

For $T = 0$, we have

$$R_X(0) = E(X^2(t)) = (\ 1/2\pi) \int_{-\infty}^{\infty} S_X(f)\ df \ . \quad \text{(2A.57b)}$$

Thus the integral of the power density spectrum integrated over all frequencies is equal to the mean-square value of the process. In the sequel, we show that the power density spectrum is the power in the process in a narrow band; accordingly, the power density spectrum must be a non-negative quantity. From this point of view (2A.57b) is eminently reasonable.

We now consider the output process when a wide-sense stationary process is inserted into a time-invariant linear filter with impulse response $h(t)$. If the input to the filter is a realization of the process, $x(t)$, its output is

$$y(t) = \int_{-\infty}^{\infty} h(\tau)x(\tau - t)\, d\tau .$$

Since the linear operations of convolution and ensemble averaging are independent, the mean value of the output is

$$m_y = E(Y(t)) = \int_{-\infty}^{\infty} h(\tau)\, E(X(t-\tau))\, d\tau = m_X \int_{-\infty}^{\infty} h(\tau) d\tau . \qquad (2A.58)$$

The autocorrelation function of the output process is

$$R_Y(\tau) = E(Y(t)Y(t-\tau)) = E(\int_{-\infty}^{\infty} h(t_1)x(t_1 - t)\, dt_1 \int_{-\infty}^{\infty} h(t_2)x(t_2 - t+\tau)\, dt_2)$$

$$= \int_{-\infty}^{\infty} \int_{-\infty}^{\infty} h(t_1)\, h(t_2) R(t_1 - t_2 - \tau)\, dt_1\, dt_2 . \qquad (2A.59)$$

As is evident from (2A.56) and (2A.57) the output process is wide-sense stationary. It can be shown that if $X(t)$ is a Gaussian process, then so is $Y(t)$. Intuitively, we might expect such a result. As we have seen, the sum of independent Gaussian random variables is Gaussian. We can regard linear filtering as a kind of summation.

Let us take the Fourier transform of the autocorrelation function of the output process. We find

$$5F2(R_Y(\tau)) = \int_{-\infty}^{\infty} d\tau \exp(2\pi jf\tau) \int_{-\infty}^{\infty} dt_1\, h(t_1) \int_{-\infty}^{\infty} dt_2\, h(t_2) R(t_1 - t_2 - \tau) .$$

From the change of variable $v = t_1 - t_2 - \tau$ and rearranging terms we have

$$S_Y(f) = \int_{-\infty}^{\infty} dt_1 \exp(2\pi jt_1) h(t_1) \int_{-\infty}^{\infty} dt_2 \exp(-2\pi jft_2) h(t_2) \int_{-\infty}^{\infty} dv \exp(-2\pi jfv) R_X(v)$$

$$= S_X(f)\, |H(f)|^2 , \qquad (2A.60)$$

where $|H(f)|$ is the transfer function of the filter, the Fourier transform of $h(t)$. This shows the important relation between the spectra of the input and output of linear time invariant filters.

We are now in a position to show the power density characteristic of the spectrum. Suppose that a process $X(f)$ is passed through a filter with a very narrow band centered at f_0

$$H(f) = \begin{cases} 1, & |f_0 - f| < \Delta/2 \\ 0, & \text{otherwise.} \end{cases}$$

From (2A.60) and (2A.59) the mean-square value of the output process is

$$E(Y^2(t)) = \Delta\, S_X(f) \geq 0 . \qquad (2A.61)$$

Equation (2A.61) shows that $S_X(f)$ must be a non-negative quantity; further, regarding it as the power of the process in a narrow frequency band is justified.

A process which is of particular importance in communications studies is the white noise process. It has a zero mean and, in analogy with white light, it has a constant power density spectrum at all frequencies; i.e.,

$$N(f) = N_0/2; \quad -\infty < f < \infty . \tag{2A.62}$$

This is called the *double-sided power density spectrum,* since it is defined for both positive and negative frequencies. The correlation function is found from the inverse transform to be

$$R_N(t) = (N_0/2)\ \delta(t) , \tag{2A.63a}$$

where $\delta(t)$ is the Dirac delta function. White noise is an abstract concept which cannot exist in a physical system since its definition implies infinite power. However, it is useful in modeling wide band processes, thermal noise, for example. Such processes are always seen through a filter or some sort, even if the filter consists of parasitic capacitance. In almost all cases of interest, the filter cuts off, while the background is still constant and the spectrum of the output of a filter with white noise input is

$$S_Y(f) = (N_0/2)\ |H(f)|^2 . \tag{2A.63b}$$

Equations (2A.60) and (2A.63a-b) allow one to calculate the mean square value of the output of a time-invariant linear filter with white noise input. Since it is noise we may take its mean to be zero. If the noise is Gaussian as well, the probability distribution of the output is known as well. Such considerations play a large role in data communications.

APPENDIX 2B DETECTION OF SIGNALS IN COLORED NOISE

In Section 2.2.2 of the text we showed in a heuristic fashion that the detection of a signal in Gaussian colored noise could be carried out in an optimal fashion by a noise-whitening filter followed by a filter matched to the line signal passed through the noise-whitening filter. In this appendix we justify this approach by a more formal derivation employing the Karhunen–Loève expansion. This derivation, which is drawn from [1] and [2], also serves to introduce the student to this expansion, which is a useful tool in a number of areas.

Consider the derivation of the matched filter receiver for white noise in Section 2.2.1. The approach was to express the received signal plus noise in terms of a set of orthonormal basis functions [see (2.13a–c) and (2.14a,b)]. The likelihood ratio was based on the coefficients of this expansion [see (2.16) and (2.17).] Considerable simplification is possible in the white noise case because for any set of orthonormal basis functions, the coefficients for the noise are independent Gaussian random variables.

In the case of colored noise, a quite similar analysis carries through, provided a particular set of basic functions are used to represent the signal and the noise. We assume that the channel noise is stationary with an autocorrelation function represented as $R_N(\tau)$. The functions $\phi_i(t)$, $i = 1,2,...$, which solve the equation

$$\lambda_i \phi_i(t) = \int_0^T R_N(t-\tau)\, \phi_i(\tau)\, d\tau \; , \; i = 1,2,... \tag{2B.1}$$

are called the *eigenfunctions* of the linear transformation on $(0,T)$ with $R_N(\tau)$ as kernel. The real numbers λ_i, $i = 1,2,...$, are the called *eigenvalues.* Since $R_N(\tau)$ is positive definite, it can be shown that $\phi_i(t)$, $i = 1,2,...,N$, form a complete set of orthonormal basis functions on the space of square integrable functions on $(0,T)$. Furthermore, from Mercer's theorem we have that the kernel can be expressed as the following expansion

$$R_N(t_1-t_2) = \sum_{i=1}^{\infty} \lambda_i \, \phi_i(t_1)\, \phi_i(t_2) \; . \tag{2B.2}$$

Note that for white noise, (2B.1) holds for any set of basis functions and $\lambda_i = N_0/2$, $i = 1,2,...$, so that

$$N_0/2 \; \delta(t_1-t_2) = N_0/2 \sum_{i=1}^{\infty} \phi_i(t_1)\phi_i(t_2).$$

If we express the received noise in terms of the eigenfunctions of equations (2B.1) and (2B.2) we have the Karhunen–Loève expansion.

$$n(t) = \sum_{i=1}^{\infty} n_i \, \phi_i(t) \; . \tag{2B.3}$$

Since we are dealing with random quantities we understand by (2B.3) that the series converges in the mean-squared sense. The salient feature of this expansion is that the coefficients of the Karhunen–Loève expansion for the noise are independent random variables. To demonstrate this we only need to show that the coefficients are uncorrelated. Let

$$n_i = \int_0^T n(t)\, \phi_i(t)\, dt \; , \tag{2B.4}$$

and from (2B.1) we have

$$E[n_i n_j] = E\int_0^T dt_1 \int_0^T dt_2\, n(t_1)n(t_2)\phi_i(t_1)\phi_j(t_2)$$

$$= \int_0^T dt_1 \int_0^T dt_2\, E\left[n(t_1)n(t_2)\right] \phi_i(t_1)\phi_j(t_2)$$

$$= \int_0^T dt_1\, \phi_i(t_1) \int_0^T dt_2\, R_N(t_1-t_2)\, \phi_j(t_2)$$

$$= \lambda_i \int_0^T dt\, \phi_i(t)\phi_j(t) = \begin{cases} \lambda_i & i=j \\ 0 & i\neq j. \end{cases} \tag{2B.5}$$

Since (2B.4) is a linear operation on a Gaussian process, the coefficients are Gaussian random variables. From (2B.5), they have variance λ_i, $i = 1,2,\ldots$.

Now suppose that binary signaling $s_i(t)$, where $i = 0$ or 1 is transmitted through the colored channel noise. The output of the channel is $y(t) = s_i(t) + n(t)$, and the joint density function for the received signal conditioned on the transmitted signal is

$$f(Y|s_i\ Transmitted) = \prod_{j=1}^{\infty} \exp(-(y_i - s_{ij}))^2/2\lambda_i/\sqrt{2\pi\lambda_i}\ ,\ i = 0,1\ . \tag{2B.6}$$

Following the steps which led to (2.19) for equiprobable transmitted signals, we have the decision rule, choose s_0 over s_1 if

$$\sum_{j=1}^{\infty} (y_j - s_{0j})^2/\lambda_j < \sum_{j=1}^{\infty} (y_j - s_{1j})^2/\lambda_j\ . \tag{2B.7}$$

Let us look at each of these terms separately. We may write

$$\sum_{j=1}^{\infty} (y_j - s_{ij})^2 / \lambda_i = \sum_{j=1}^{\infty} \left[\int_0^T \phi_j(t)\left[(y(t) - s_i(t)\right] dt\right]^2 / \lambda_i$$

$$= \int_0^T dt_1 \int_0^T dt_2\, [y(t_1) - s_i(t_1)]\, [y(t_2) - s_i(t_2)] \sum_{i=1}^{\infty} \frac{\phi_i(t_1)\, \phi_i(t_2)}{\lambda_i}$$

$$= \int_0^T dt_1 \int_0^T dt_2\, [y(t_1) - s_i(t_1)][y(t_2) - s_i(t_2)]\, Q_N(t_1, t_2)\ . \tag{2B.8}$$

The term $Q_N(t_1, t_2) \triangleq \sum_{i=1}^{\infty} \phi_i(t_1)\ \phi_i(t_2)/\lambda_i$ is recognized as the *inverse kernel* of $R_N(t_1 - t_2)$. Formally, the inverse relation is demonstrated by noting that

$$\int_0^T R_N(t_1 - t)\, Q_N(t_2, t)\, dt = \sum_{i=1}^{\infty} \phi_i(t_1)\, \phi_i(t_2) = \delta\,(t_1 - t_2)\,. \tag{2B.9}$$

Now we would like to show that $Q_N(t_1, t_2)$, the inverse kernel can also be written as

$$Q_N(t_1, t_2) = \int_0^T w(t - t_1) w(t - t_2)\, dt\,, \tag{2B.10a}$$

where $w(t)$ is the impulse response of the *noise-whitening* filter found as the inverse of Fourier transform of $N^{-1/2}(f)$

$$w(t) = F^{-1}[N^{-1/2}(f)]\,. \tag{2B.10b}$$

If we assume that is true, then because of the noise-whitening effect of this filter

$$E \int_0^T d\tau_1 n(\tau_1) w(t_1 - \tau_1) \int_0^T d\tau_2 n(\tau_2) w(t_2 - \tau_2)$$

$$= \int_0^T d\tau_1\; w(t_1 - \tau_1) \int_0^T d\tau_2\; w(t_2 - \tau_2),\; E\Big[n(\tau_1) n(\tau_2)\Big]$$

$$= \int_0^T d\tau_1\; w(t_1 - \tau_1) \int_0^T d\tau_2\; w(t_2 - \tau_2)\; R_N(\tau_1 - \tau_2)$$

$$= \delta\,(t_1 - t_2)\,. \tag{2B.11}$$

Now we multiply both sides of (2B.11) by $w(t_2 - t)$ and integrate with respect to t_2 to obtain

$$w(t_1 - t) = \int_0^T d\tau_1\; w(t_1 - \tau_1) \int_0^T d\tau_2\; R_N(\tau_1 - \tau_2) \int_0^T dt_2\; w(t_2 - t)\; w(t_2 - \tau_2)\,. \tag{2B.12}$$

The identity operation in (2B.12) can be expressed as a Dirac delta function. We have

$$\int_0^T d\tau_2\; R_N(\tau_1 - \tau_2) \int_0^T dt_2\; w(t_2 - t)\; w\,(t_2 - \tau_2) = \delta\,(\tau_1 - t)\,. \tag{2B.13}$$

By comparison with (2B.9) we have the verification of (2B.10).

The proof of (2B.10) allows us to show that the statistics in (2B.8) can be generated by means of the noise-whitening filter followed by a matched filter. Let the output of the noise-whitening filter be written

$$y^*(t) = s_i^*(t) + n^*(t),\; i = 0,1\,. \tag{2B.14}$$

(The asterisk here denotes passage through the noise-whitening filter.) Since $n^*(t)$ is white noise, the detection of the signals, $s_i^*(t)$, $i = 0,1$ follows the derivation in Section 2.2.1. From (2.21) we choose $s_0(t)$ as the transmitted signal if

$$\int_0^T (y^*(t)-s_0^*(t))^2 dt < \int_0^T (y^*(t)-s_1^*(t))^2 \, dt \, . \tag{2B.15}$$

Since

$$y^*(t) = \int_0^T w(t-\tau)\, y(\tau)\, d\tau$$

and

$$s_i^*(t) = \int_0^T w(t-\tau) s_i(\tau)\, d\tau \, ,$$

we may express the terms in (2B.15) as

$$\int_0^T dt\, (y^*(t)-s_i^*(t))^2 = \int_0^T dt \left[\int_0^T d\tau_1 \; w(t-\tau_1)(y(\tau_1)-s_1(\tau_1)) \right]^2$$

$$= \int_0^T dt \int_0^T d\tau_1 \int_0^T d\tau_2 w(t-\tau_1)[y(\tau_1)-s_i(\tau_1)]\, w(t-\tau_2)[y(\tau_2)-s_i(\tau_2)]$$

$$= \int_0^T d\tau_1 \int_0^T d\tau_2 [y(\tau_1)-s_i(\tau_1)][y(\tau_2)-s_i(\tau_2)] \int_0^T dt \; w(t-\tau_1) w(t-\tau_2), \; i = 0,1 \tag{2B.16}$$

From (2B.10a) we have

$$\int_0^T dt (y^*(t)-s_i^*(t))^2 = \int_0^T d\tau_1 \int_0^T d\tau_2 \; [y(\tau_1)-s_i(\tau_1)] \; [y(\tau_2)-s_i(\tau_2)] Q_N(\tau_1,\tau_2) \tag{2B.17}$$

A comparison of equations (2B.8) and (2B.17) shows that $\int_0^T [y^*(t)-s_i^*(t)]\, dt$ is indeed the optimum statistic. The receiver is then the whitening filter followed by a filter matched to $s_i^*(t)$, $i = 0,1$.

REFERENCES

[1] H. L. Van Trees, *Detection Estimation and Modulation Theory*, Wiley, 1968.

[2] J. M. Wozenkraft and I. M. Jacobs, *Principles of Communication Engineering*, Wiley, 1967.

[3] J. B. Thomas, *Statistical Communication Theory*, Wiley, 1969.

[4] G. L. Turin, *Notes on Digital Communications,* Van Nostrand Reinhold, 1969.

[5] S. Sherman, "Non-Mean Square Error Criteria," *IRE Trans. Information Theory* IT-4, No. 13, pp. 125-126, 1958.

[6] D. O. North, "An Analysis of Factors Which Determine Signal/Noise Discrimination in Pulse Carrier Systems," RCA Report PTC-6C, 1943.

[7] G. L. Turin, "An Introduction to Matched Filters," *IRE Trans. Information Theory*, IT-6, pp. 311–329, 1960.

[8] C. E. Shannon, "A Mathematical Theory of Communications," *Bell System Technical Journal*, Vol. 27, Part I, pp. 379–423, Part II, pp. 623–656.

[9] N. Abramson, *Information Theory and Coding,* McGraw-Hill, 1963.

[10] F. M. Ingels, *Information and Coding Theory*, Intx Educational Publishers, 1971.

[11] R. G. Gallager, *Information Theory and Reliable Communications,* Wiley, 1968.

[12] R. E. Blahut, *Principles and Practice of Information Theory,* Addison-Wesley, 1987.

[13] R. J. McEliece, *The Theory of Information and Coding,* Addison-Wesley, 1977.

[14] R. W. Lucky, *Silicon Dreams,* St. Martin's Press, 1989.

[15] A. Khinchin, *Mathematical Foundations of Information Theory,* Dover, 1957.

[16] W. Feller, *An Introduction to Probability and Its Applications,* Vol. 1, Wiley, 1950.

[17] F. M. Reza, *An Introduction to Information Theory*, McGraw-Hill, 1961.

[18] D. A. Huffman, "A Method for Constructing Minimum Redundancy Codes," *Proc. IRE*, Vol. 40, pp. 1098–1101.

[19] R. Hunter and A. H. Robinson, "International Digital Facsimile Coding Standards," *Proc IEEE*, Vol. 68, No. 7, pp. 854–867, July, 1980.

[20] F. Jelinek, "Buffer Overflow in Variable Length Coding of Fixed Rate Sources," *IEEE Trans. on Information Theory*, Vol. IT-14, No. 3, pp. 490–501, May, 1968.

[21] J. Ziv and A. Lempel, "A Universal Algorithm for Sequential Data Compression," *IEEE Trans. on Information Theory*, Vol. IT-23, No. 3, pp. 337–343, May, 1977.

[22] J. Ziv and A. Lempel, "Compression of Individual Sequences Via Variable-Rate Coding," *IEEE Trans. on Information Theory*, Vol. IT-24, No. 5, pp. 530–536, September, 1979.

[23] T. A. Welch, "A Technique for High Performance Data Compression," *IEEE Computer*, Vol. 17, No. 6, pp. 8–19, June, 1984.

[24] T. Berger, *Rate Distortion Theory,* Prentice-Hall, 1971.

[25] T. J. Goblick and J. L. Holsinger, "Analog Source Digitization Measure: A Comparison of Theory and Practice," *IEEE Trans. on Information Theory*, Vol. IT-13, pp. 323–326.

[26] S. Shmai (Shitz) and I. Bar-David, "Upper Bounds on Capacity of a Constrained Gaussian Channel," *IEEE Trans. on Information Theory*, Vol. 35, No. 9, pp. 1079–1084, September, 1989.

[27] I. Kalet and S. Shmai (Shitz), "On the Capacity of the Twisted-Wire Pair: Gaussian Model," *IEEE Trans. on Information Theory*, Vol. 38, No. 3, pp. 379–383, March, 1990.

[28] W. Feller, *Introduction to Probability and Its Applications,* Wiley, 1950.

[29] A. Papoulis, *Probability, Random Variables and Stochastic Processes,* McGraw-Hill, 1965.

EXERCISES

2.1: In connection with equation (2.1) of the text we considered the signals $+s$ and $-s$ immersed in Gaussian noise. Now suppose that the noise is uniformly distributed between $-2s$ and $+2s$.

(a) Find the optimum receiver when the signals are equiprobable.

(b) Repeat part (a) when $Pr(+s\ Transmitted) = 2\ Pr(-s\ Transmitted)$

2.2: An experiment consists of n Bernoulli trials. The trials are independent and have the same probability of success. Suppose that there are two possible underlying states which determine the probability of success on each trial.

(a) Assuming that the underlying states are equally probable, find the test on the outcomes of the trials which finds the correct underlying state with minimum probability of error.

(b) Repeat part (a) for states which are not equally probable.

2.3: A narrow-band process may be represented as $Z(t) = X(t)\cos(2\pi f_c t) + Y(t)\sin(2\pi f_c t)$, where $X(t)$ and $Y(t)$ are stationary random processes whose bandwidth is small compared to f_c. In polar coordinates this can be written $Z(t) = R(t)\cos(2\pi f_c t + \theta(t))$, where $R(t) = [X(t)^2 + Y(t)^2]^{1/2}$ and $\theta(t) = \arctan(X(t)/Y(t))$. If $X(t)$ and $Y(t)$ are independent zero mean Gaussian processes with the same variances, samples of the envelope follow the Rayleigh distribution with probability density

$$f_Z(z) = (z/\sigma^2)\exp(-z^2/2\sigma^2) ,$$

where σ^2 is the variance of $X(t)$ and $Y(t)$. Now suppose that there are two possible underlying states each of which produce difference variances for $X(t)$ and $Y(t)$. Find the optimum test for deciding between the underlying state.

2.4: Find the probability of error for the tests in each part of Exercise 2.1.

2.5: Find the probability of error for the tests in each part of Exercise 2.2.

2.6: Find the probability of error for the test in Exercise 2.3.

2.7: Suppose that the four signals in the complex plane $S_0 = (+s,+s)$, $S_1 = (-s,+s)$, $S_2 = (+s,-s)$, and $S_3 = (-s,-s)$ are transmitted over a channel. Each component of these vectors is disturbed independently by Gaussian noise which has the mean-square value of one for each component.

(a) Find the optimum decision boundaries assuming that the four signals are equally probable.

(b) Repeat part (a) under the conditions that
$Pr(S_0\ Transmitted) = 1/2$, $Pr(S_1\ Transmitted) = 1/4$,
$Pr(S_2\ Transmitted) = 1/8$, $Pr(S_3\ Transmitted) = 1/8$

2.8: For each of the optimum decision boundaries found in Exercise 2.7,

(a) Find the probabilities of pairwise error.

(b) Calculate the union bound when S_0 is the transmitted signal.

2.9: The derivation of the matched filter receiver in the binary case assumed equiprobable transmitted signals.

(a) Repeat the derivation for the case where $s_1(t)$ is three times as likely to be transmitted as $s_0(t)$.

(b) Sketch the matched filter in this case.

2.10: Suppose that binary digits are transmitted using the signals shown in Figure E2.1. Assume that the signals are equally probable and the channel is disturbed by white Gaussian noise with power density spectrum $N_0/2$ watts/Hz.

(a) Show the form of the matched filter receiver which minimizes the probability of error.

(b) Repeat (a) for the correlation receiver.

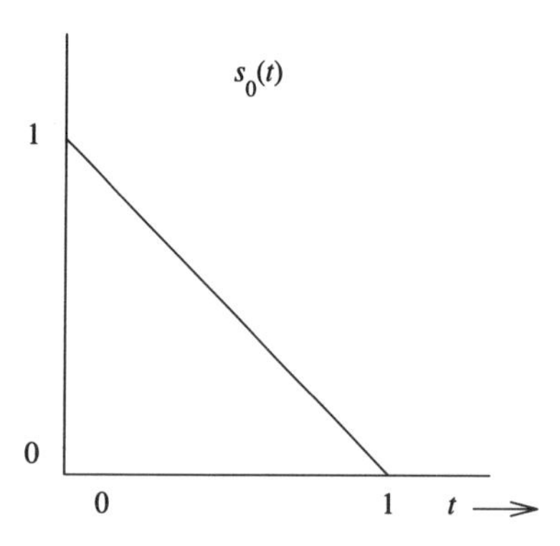

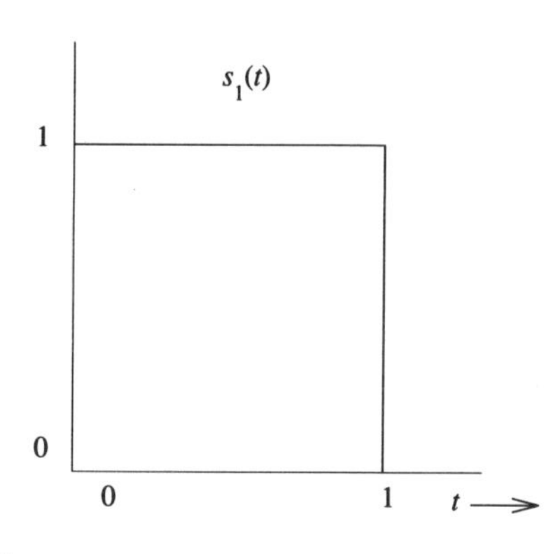

Fig. E2.1.

2.11: Show that the filter matched to $s_i(t)$ maximizes the ratio of the average signal power to the average noise power at the sample instant when $s_i(t)$ is the signal component of the input. (Hint: you may find that the Schwartz inequality is useful.)

2.12: Information is transmitted over an optical fiber channel by means of *pulse position modulation*, which means that there are $L = 2^m$ different positions for a positive pulse in a signal interval.

(a) Assuming that the signal is disturbed by white Gaussian noise in the receiving equipment, find the optimum receiver.

(b) Find an expression for the probability of error.

2.13: Consider the estimation of a Gaussian random variable immersed in Gaussian noise. Assume that the noise is not stationary so that its mean-square value varies from sample to sample. Find the minimum mean-square estimate.

2.14: Find the entropy of the source having L output symbols with the probabilities $p_i = (1/2)^i, i = 1,2,\ldots,L-1,\ p_L = (1/2)^{L-1}$.

2.15: Two zero memory sources S_1 and S_2 have L_1 and L_2 symbols respectively. The probabilities of source outputs for S_i are p_i, $i = 1,2,...,L_1$, and the outputs for S_2 are q_i, $i = 1,2,...,L_2$. A new zero-memory source, $S(w)$, with $L_1 + L_2$ symbols is formed. The first L_1 symbols of $S(w)$ have probabilities wp_i, $i = 1,2,...,L_1$, and the second L_2 have probabilities $(1 - w)q_i$, i, $1,2,...,L_2$. Show that the entropy of $S(w)$ is given by

$$H(S(w)) = w\,H(S_1) + (1-w)H(S_2) - w\log_2 w - (1-w)\log_2(1-w).$$

2.16: Show that the entropy of a geometrically distributed random variable is $H(q, 1 - q)/(1 - q)$.

2.17: Assume a binary source with output symbol probabilities 0.1 and 0.9.

(a) What is the entropy of this source?

(b) Find the minimum average number of binary digits per symbol encoding over blocks of 1, 2, and 3 symbols using Huffman coding.

2.18: Prove equation (2.79).

2.19: Show that $I(X/Y) \geq 0$ with equality if and only if X and Y are independent random variables.

2.20: Suppose that a binary, stationary first-order Markov source has the state transition matrix

$$P_{11} = 3/4, \qquad P_{12} = 1/4$$

$$P_{21} = 1/3, \qquad P_{22} = 2/3$$

Find the entropy per symbol over a long sequence.

2.21: Calculate the entropy and find a Huffman code for a source with the following output symbol probabilities. $P_1 = 0.08$, $P_2 = 0.02$, $P_3 = 0.1$, $P_4 = 0.4, P_5 = 0.05, P_6 = 0.05, P_7 = 0.3$.

2.22: In Section 2.5.5, the Lempel–Ziv algorithm was used to encode a passage from the writings of James Joyce. The next sentence in the passage is

> It was falling too upon every part of the lonely churchyard where Michael Furey lay buried.

Encode this sentence as a new transmission not connected with the passage in the text.

2.23: The objective of this exercise is to derive equation (2.100). In order to lighten this task, we provide some guidance. The derivation is quite similar to that leading to the channel capacity formula of (2.112).

(a) Show that the proof of (2.100) is equivalent to showing that the rate distortion function for a single Gaussian random variable with variance σ^2 is

$$r(D) = \frac{1}{2} \log_2 (\sigma^2/D).$$

If the random variable X is used to approximate the Gaussian random variable Z, we write $Z = X + N$ where N is the error of the approximation.

(b) Show that $I(X/Z) = I(Z/X)$ is minimized if N is a Gaussian random variable with zero mean and variance D. Since X and N are independent, X is Gaussian with variance $\sigma^2 + D$.

(c) Show that the rated distortion function for Z is as shown above.

2.24: A channel has transfer function $|H(f)|^2 = 1/(1 + (f/f_0)^2)$ and is disturbed by white Gaussian noise.

(a) Derive the water filling expression for the capacity of this channel.

(b) Calculate and plot the capacity as a function of the average transmitted power.

2.25: Show that the capacity of the binary symmetric channel is

$$C = - P\log_2 P - (1 - P) \log_2(1 - P) .$$

2.26: Find the capacity of the L-ary uniform channel described by the following matrix of conditional probabilities.

$$Pr(Y = i/X = j) = \begin{cases} p, & i = j \\ q, & i \neq j \end{cases} \qquad i,j = 0,1,\dots,L-1 .$$

3

Error Correcting and Detecting Codes*

3.0 INTRODUCTION

As we have seen in Section 2.5, information theory provides theoretical upper bounds on the information rates that can be obtained over physical channels. While these bounds can be computed for a wide range of channels, the theory gives little indication of how they may be attained. It is fortuitous that, about the same time information theory was conceived, a theory of error correcting and detecting codes emerged. By the systematic injection of redundant bits into the encoding of information, the reliability of transmission could be improved. Hopes that codes achieving the bounds could be easily obtained proved to be illusionary; however, the theory has steadily developed so that codes have become indispensable components of many communications systems yielding considerable performance gains. The remarkable series of pictures that have been received from deep-space probes attest to the power of coding. A recent breakthrough, trellis coding, which is a direct outgrowth of the theory and which will be covered in Section 5.8, has put the Shannon bound within reach for bandlimited channels.

A large part of the work on error correcting and detecting codes has been concerned with implementation. While this chapter is of necessity a survey of the field, we hope that the reader gains some appreciation of how the sophisticated mathematics that is involved has yielded results which provide guidance in the implementation of digital encoding and decoding.

---

* In writing this chapter significant use was made of the reference texts [1]–[4], particularly the first of these. In addition to these, we benefited from insightful review articles [5]–[8]. In keeping with the focus of the text, we shall consider codes in a communication context; however, a significant application of the technique lies in data storage and recording. (See [9] as an example of the latter.)

The basic idea of error correcting and detecting codes is to increase the *distance* between information-bearing signals by the addition of redundancy. We have seen a rudimentary example of this in Section 2.3. There are two complementary and fundamentally different approaches to adding this redundancy: *block codes* and *convolutional codes*. In our presentation we shall focus almost exclusively on binary codes in which the elements are 0's or 1's. Much of the theory carries over to p-ary codes where the elements are $0, 1, \ldots, p-1$, where p is a prime number. We shall discuss the role of p-ary codes in connection with *Reed–Solomon* codes. We begin our study with block codes which consist of a fixed-length block of information and redundant bits. A particular class of these codes, the parity-check codes, is easily implemented and, consequently, of particular interest. Within the class of parity-check block codes, we narrow our focus to a particular subclass, cyclic codes in which codewords are cyclic shifts of other codewords. Again the motivation is implementation. Cyclic codes may be generated and decoded by means of linear sequential circuits which are combinations of shift registers and modulo 2 adders. An important part of our treatment of cyclic codes is their performance, both for hard and soft decision decoding.

The second part of the chapter is devoted to convolutional codes which are obtained by passing the data sequence through a linear shift register circuit. The redundant data stream is derived by modulo 2 addition of the sample contents of the shift register. The redundancy lies in the fact that for every input bit there is more than one output bit from the encoder. Decoding of convolutional codes is carried out by variations of the *Viterbi algorithm;* a technique which is well suited to both hard- and soft-decision decoding. Performance bounds for hard and soft decisions with convolutional codes are derived.

The final section of the chapter is concerned with error detection and retransmission systems, commonly called automatic repeat request systems (ARQ). Three varieties of ARQ are described and analyzed, stop and wait, go back N, and selective repeat. Finally, hybrid systems which employ both error detection and correction are considered.

3.1 BLOCK CODES

Block codes are composed of a set of fixed-length codewords. In standard terminology, we speak of (n,k) codes meaning that codewords are n binary digits long and contain k bits of information. One can view the encoding process as mapping a block of k information bits into one of 2^k codewords which are n bits in duration. The redundancy lies in the fact that, generally, $n > k$. An alternate characterization of a code is the rate $R_c = k/n$ in recognition of the proportion of information bits in a codeword. A (6,3) (rate 1/2) code discussed below is shown in Table 3.1. The first three binary digits encode eight possible symbols. For example, the word 101011 encodes the block 101. The method of

Table 3.1 - A (6,3) Code

	0	0	0	0	0	0	
	0	0	1	1	0	1	
	0	1	0	0	1	1	
	0	1	1	1	1	0	
Codewords	1	0	0	1	1	0	Parity bits c4 = c1 ⊕ c3
	1	0	1	0	1	1	Parity bits c5 = c1 ⊕ c2
	1	1	0	1	0	1	Parity bits c6 = c2 ⊕ c3
	1	1	1	0	0	0	
	Information bits			Parity bits			

constructing this particular code is discussed in the next subsection on parity-check codes.

3.1.1 PARITY-CHECK CODES

We may view the words of an n-bit binary code as occupying the vertices of a cube in an n-dimensional space. Choosing a good set of codewords for a code consists of choosing a set of vertices which have good distance properties in the n-dimensional space. As we have seen in section 2.3.1, the probability of error between two blocks of binary digits is reduced by increasing the *Hamming distance* between the words.

One might imagine that it may be possible to pack an n-dimensional space with a code in a very efficient manner, but the resulting code may be difficult to implement. Accordingly, the efforts of coding theorists have concentrated on codes that are easy to implement. Fortunately, it can be shown that the performance of a large number of such codes comes close to the upper bounds that can be established for codes in general. (We shall present some of these bounds in the next section of this chapter.) A basic code having such a structure is the *parity-check code.* The modulo 2 sum of a subset of the information bits to be transmitted is formed. To this sum is added a parity-check bit such that the modulo 2 sum over the information bits and the parity-check bit is 0. The redundant bits accompanying the information bits in a codeword are referred to as *parity-check bits.* What is probably the most widely encountered example of parity-check codes is contained in the American Standard Code for Information Interchange (ASCII). Information is conveyed in the form of seven binary digits. The 128 possible messages that can be represented by seven bits make possible the transmission of control characters, such as ACK for positive acknowledgement, as well as alphanumeric characters. ASCII characters are almost always accompanied by an eighth parity bit which enables errors in transmission to be detected. This eighth bit is chosen so that the total number of ones is odd, if odd parity is used, or even, if even parity is used. For example, if the letter W is to be transmitted using even parity, the codeword is 11101011. If a single bit error is made, the resulting eight-bit block corresponds to no legitimate codeword and it is evident that an error has been made. Note that there is not enough redundancy to correct errors. Suppose, for example, that there is an error in the second bit to produce 10101011. (Note that the number of ones is odd.) This could have been produced by transmitting U with an error in the eighth bit or an R with an error in the first bit. The limited capability of the ASCII code is due to the relatively small Hamming distance* between codewords. Two errors in a codeword moves it closer to another codeword.

A more powerful parity-check code, which can correct errors, is the (6,3) shown in Table 3.1. The three information bits are accompanied by three parity-check bits over the information bits as shown. We observe that the minimum Hamming distance between any pair of codewords in this code is three; consequently, single errors may be *corrected.* For example, 101110 can be decoded as 100 assuming that only a single error has occurred. If this assumption cannot be made and double errors are possible, then the code can *detect* double errors. Notice that, in the case of ASCII codes, double errors resulted in a legitimate codewords which resulted in a character error.

3.1.2 GENERATOR AND PARITY-CHECK MATRICES

As is evident from Table 3.1, a matrix formulation can be used to characterize a particular code. Let the row vector ($X = x_1 x_2 \cdots x_k$) denote the information bits of a codeword and $C = (c_1, c_2, \ldots, c_n)$ the entire codeword. Each of the elements of the codeword may be expressed as

$$c_i = \sum_j x_j \, g_{ij} \ , \quad j = 1,2,\ldots,n \ , \tag{3.1a}$$

where g_{ij} are binary numbers and Σ indicates addition modulo 2. (See Table 3.1 for an illustration.) This relation may be written in matrix notation as

$$C = XG \ , \tag{3.1b}$$

where G is the (kxn) generator matrix with elements g_{ij}. For example, for the code given in Table 3.1 we have

$$G = \begin{bmatrix} 1\,0\,0\,1\,1\,0 \\ 0\,1\,0\,0\,1\,1 \\ 0\,0\,1\,1\,0\,1 \end{bmatrix} \ .$$

In general, the row vectors in G span a space of k dimensions. If the generator matrix for a linear block code can be put in what is called the *systematic*

* Recall that the Hamming distance between sequences was defined in Section 2.3.

form in which the first k columns form an identity matrix, the code is a member of the class of *systematic codes*. The remaining $n-k$ columns define the parity checks which specify the code. We have

$$G = [I_k : P] = \begin{bmatrix} 1\ 0 \ldots 0 & p_{11} & p_{12} & \cdots & p_{1n-k} \\ 0\ 1 \ldots 0 & p_{21} & p_{22} & \cdots & p_{2n-k} \\ \cdot & \cdot & & & \\ \cdot & \cdot & & & \\ \cdot & \cdot & & & \\ \cdot & \cdot & & & \\ 0\ 0 \ldots 1 & p_{k1} & p_{k2} & \cdots & p_{k,n-k} \end{bmatrix}, \tag{3.2}$$

where I_k is the $(k \times k)$ identity matrix and P is the $k \times (n-k)$ matrix indicating parity-check bits.

We define the *parity-check* matrix of a systematic code as the $(n-k) \times n$ matrix

$$H = [P^T : I_{n-k}] , \tag{3.3}$$

where the superscript T denotes transpose. By inspection we see that the element of the ith row and the jth column of the matrix GH^T is

$$p_{ij} + p_{ij} = 0 \bmod 2 , \tag{3.4a}$$

and we have the important relationship

$$GH^T = \Phi \tag{3.4b}$$

where Φ represents the $(kx(n-k))$ matrix, all of whose elements are zero. For example, the (6,3) code in Table 3.1 has the parity-check matrix

$$H = \begin{bmatrix} 1\ 0\ 1\ 1\ 0\ 0 \\ 1\ 1\ 0\ 0\ 1\ 0 \\ 0\ 1\ 1\ 0\ 0\ 1 \end{bmatrix} .$$

Thus,

$$G \times H^T = \begin{bmatrix} 1\ 0\ 0\ 1\ 1\ 0 \\ 0\ 1\ 0\ 0\ 1\ 1 \\ 0\ 0\ 1\ 1\ 0\ 1 \end{bmatrix} \begin{bmatrix} 1\ 1\ 0 \\ 0\ 1\ 1 \\ 1\ 0\ 1 \\ 1\ 0\ 0 \\ 0\ 1\ 0 \\ 0\ 0\ 1 \end{bmatrix} = \begin{bmatrix} 0\ 0\ 0 \\ 0\ 0\ 0 \\ 0\ 0\ 0 \end{bmatrix} .$$

Table 3.2 (7,4) Hamming Code

Information bits				Codeword						
0	0	0	0	0	0	0	0	0	0	0
0	0	0	1	0	0	0	1	0	1	1
0	0	1	0	0	0	1	0	1	1	0
0	0	1	1	0	0	1	1	1	0	1
0	1	0	0	0	1	0	0	1	1	1
0	1	0	1	0	1	0	1	1	0	0
0	1	1	0	0	1	1	0	0	0	1
0	1	1	1	0	1	1	1	0	1	0
1	0	0	0	1	0	0	0	1	0	1
1	0	0	1	1	0	0	1	1	1	0
1	0	1	0	1	0	1	0	0	1	1
1	0	1	1	1	0	1	1	0	0	0
1	1	0	0	1	1	0	0	0	1	0
1	1	0	1	1	1	0	1	0	0	1
1	1	1	0	1	1	1	0	1	0	0
1	1	1	1	1	1	1	1	1	1	1

The generator matrix for an important (7,4) Hamming code, which is shown in Table 3.2, is

$$G = \begin{bmatrix} 1\,0\,0\,0\,1\,0\,1 \\ 0\,1\,0\,0\,1\,1\,1 \\ 0\,0\,1\,0\,1\,1\,0 \\ 0\,0\,0\,1\,0\,1\,1 \end{bmatrix}, \tag{3.5a}$$

and the parity-check matrix for this code is

$$H = \begin{bmatrix} 1\,1\,1\,0\,1\,0\,0 \\ 0\,1\,1\,1\,0\,1\,0 \\ 1\,1\,0\,1\,0\,0\,1 \end{bmatrix}. \tag{3.5b}$$

Again, it is easy to verify that $GH^T = \Phi$.

The value of the parity-check matrix lies in decoding. From (3.1b) and (3.4b), we have

$$CH^T = XGH^T = \Phi \tag{3.6}$$

In (3.6) Φ represents an $n - k$ element row vector all of whose elements are zero. Thus valid code words are mapped into zero vectors by the parity-check matrix. Now suppose that a codeword, C, is corrupted by an error, E, the resulting received word is $R = C + E$. We have

$$RH^T = CH^T + EH^T = EH^T .$$

The result of this calculation is the *syndrome* of the error in the form of a $(n-k)$ row vector. If there is enough redundancy in the code, the syndrome will uniquely correspond to the error vector and a correction can be made. For the (7,4) code suppose that the error vector is 0100000. Multiplication by the parity-check matrix gives 111. If the error vector is 1000000, multiplication by H gives 101. It is not difficult to show that in this particular case, each error pattern produces a unique syndrome which is not a codeword. A basic assumption is that an error pattern cannot be a codeword; otherwise detection or correction is not possible. The proof is by contradiction. Let E_1 and E_2 be error patterns such that $E_1 \neq E_2$ but $E_1 H^T = E_2 H^T$. But this implies that $(E_1 + E_2)H^T = \Phi$, which, in turn, implies that E_1 and E_2 are codewords, a contradiction of our assumption.

The parity-check codes are also *linear* or *group* codes. It is easily shown that element-by-element modulo 2 addition of two codewords results in another codeword. The sums of the k-bit information blocks results in another k-bit information block. Modulo 2 sums of bits that add to zero also add to zero. The all-zero sequence must also be a codeword since parity checks over k zero bits yield zero bits. It follows then that the words of a parity-check code form a group* under element-by-element modulo 2 addition.

As we have seen, the Hamming distance between codewords indicates the error detecting and correcting capabilities of a code. For linear codes there is an alternative characterization of the distance properties of a code. The *weight* of a codeword is the number of its nonzero elements. Since the modulo 2 sum of a pair of codewords is a codeword, the Hamming distance of two codewords from one another is the weight of the codeword formed by their sum. Since the all zero sequence is a codeword, the minimum distance is equal to the minimum weight. Thus the distance properties of a linear code are determined by the weights of the codewords. If all the words in a code have the same weight we have a *fixed-weight code.*

3.1.3 BURST ERRORS AND INTERLEAVING

It is often the case in practical systems that errors occur in bursts rather than at random. In a burst-error situation the occurrence of a bit in error means that the next bit is more likely to be in error. A rare burst of errors can cause output errors even though the probability of bit error is low. This stands in contrast to random errors, which are independent of one another and are modeled by the BSC.

As we shall see in Section 3.2.2, codes can be devised which detect burst errors; however, the simple device of *interleaving* causes the burst errors to be

---

* The recognition of the group property of parity-check codes is due to Slepian [10].

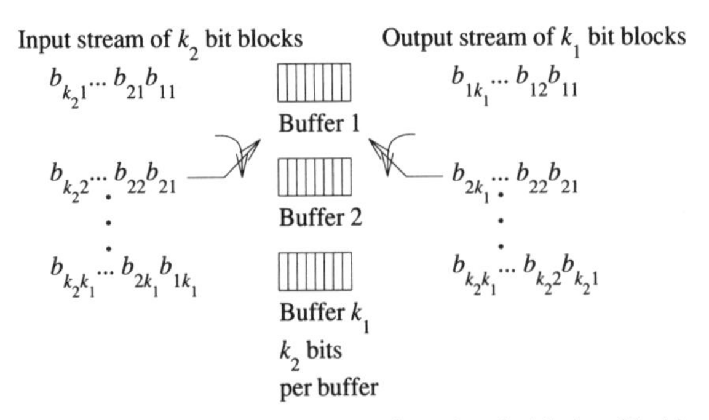

Fig. 3.1 Bit interleaver that groups input into k_1 blocks of k_2 bits.

randomized over a sequence (see Figure 3.1). Suppose, for example, that data is segmented in k_1 successive blocks each consisting of k_2 bits. The data can be arranged in a $k_1 \times k_2$ matrix, and a (n_1, k_1) block code can be used on each column of the matrix and a (n_2, k_2) block code on each row. Burst errors tend to occur on single rows; consequently, they look like random errors to the codes over the columns. The (n_1, k_1) block codes on each row serve to detect and/or correct random errors. The obvious drawback to interleaving is delay. Up to $k_1 \times k_2$ bits must be buffered at the transmitter before transmission.

3.1.4 HAMMING DISTANCE AND ERROR CORRECTION

At this point we have seen enough of the structure of block codes to be able to evaluate their error correcting and detecting capability. Recall that the detector will output the (allowed) codeword closest in Hamming distance to the received sequence. The ability of a code to correct and detect errors is characterized by the minimum Hamming distance between any pair of its codewords, since this distance determines the maximum number of channel errors that can be sustained without incurring errors in the decoded information bits. The relationship between distance and immunity to errors is illustrated in Figure 3.2. If the Hamming distance between codewords is one, a single error will result in a true code word. No error detection or correction capability is possible. When the distance is two, a single channel error will give an n-bit word which is not a codeword and the channel error is detected. A single channel error can be corrected if the Hamming distance between codewords is at least three. In this case the channel error leads to an n-bit sequence which is closer to the correct codeword than any other. Note that for distance three, a double error may look like a single error in terms of distance, from a true codeword (but the wrong one); accordingly, if double errors are possible the code may be used only to detect double errors. As the distance between codewords increases, the code's capability increases. For example, at distance four, a single error will be corrected and a double error detected.

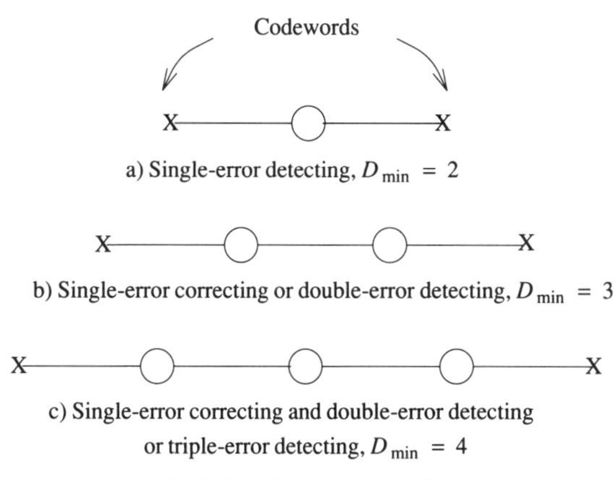

a) Single-error detecting, $D_{min} = 2$

b) Single-error correcting or double-error detecting, $D_{min} = 3$

c) Single-error correcting and double-error detecting or triple-error detecting, $D_{min} = 4$

Fig. 3.2 Distance properties.

These distance properties are illustrated further in Figures 3.3a-b. In both figures, codewords are represented as the vertices of a cube. In Figure 3.3a the codewords are 000, 101, 011, and 110. The distance between codewords is two. The codewords in Figure 3.3b are 000 and 111 and we have distance three.

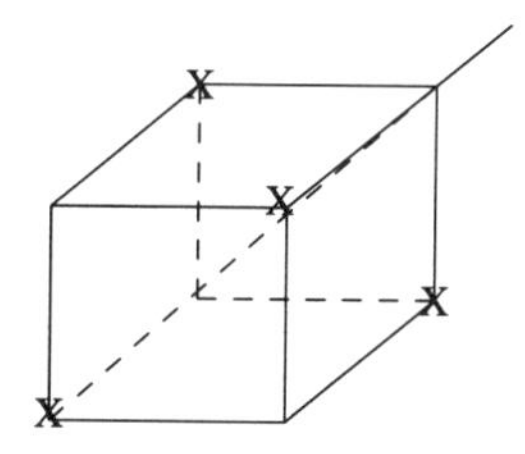

a) Distance between codewords = 2
($k = 2$)

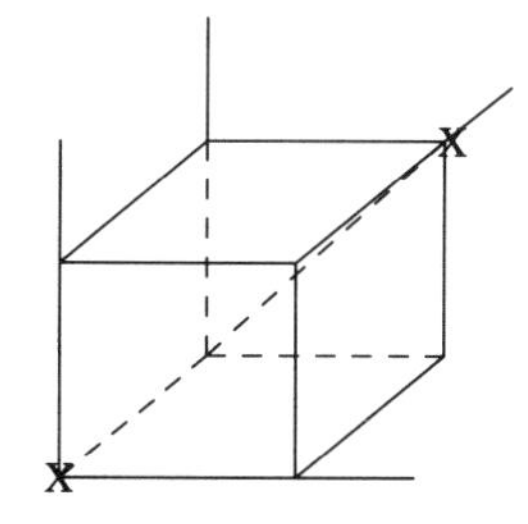

b) Distance between codewords = 3
($k = 1$)

Fig. 3.3 An example code for $n = 3$.

From these considerations, a general relation between codeword minimum distance and error correcting and detecting capability can be derived. Let $D_{\min}$ indicate the minimum Hamming distance between any pairs of words in a code. In order for the code to correct c errors, the following relation must hold:

$$D_{\min} \geq 2c + 1 \tag{3.7a}$$

Thus in Figure 3.2 we see that a code with distance three allows all one-bit errors to be corrected. A minimum distance of four allows double-bit errors to be corrected as well. If only error detection is considered, the minimum distance to detect d errors is

$$D_{\min} \geq d + 1 . \tag{3.7b}$$

3.1.5 CODE STRUCTURE—BOUNDS

As mentioned earlier, an (n,k) linear code consists of 2^k points on an n-dimensional hypercube. In order for the code to be effective, it is necessary that these points be as far apart from one another as possible. Basically the problem is one of the class of *sphere-packing* problems,* where the idea is to place as many as possible spheres of a given size into a larger sphere in an n-dimensional space. The codewords are at the centers of the spheres, each with the same diameter. A dense packing corresponds to more codewords and a higher data rate for the same-dimensional space. Furthermore, in a digital communication context, it is convenient, if not necessary, that the number of points (little spheres) be a power of two. For example, it can be shown that 240 points (little spheres of radius 1) can be packed into the surface of a sphere of radius $\sqrt{12}$ centered at the origin in eight-dimensional space [11]. This could form the basis of a code for which each n-bit word conveys $\log_2 240$ bits of information. However, while this is a very dense packing, the number 240 is inconvenient for practical implementation since it is not a power of two.

The most tangible characterization of a block code is by means of the parameters n and k. The ratio k/n gives the proportion of information to total number of bits. The error correcting and detecting properties of a code are given by the minimum distance between codewords, $D_{\min}$. In this section, we consider general relations between these quantities in the form of the Hamming, Plotkin and Varsharmov–Gilbert bounds. The position of a particular code relative to these bounds is a measure of the efficiency and effectiveness of the code.

THE HAMMING BOUND

We begin with the *Hamming* or *sphere-packing* bound. Consider a single word in the code, represented by a point in an n-dimensional space. There are

---

* For a review of sphere-packing problems from a broad perspective, see Reference [11].

$\binom{n}{i}$ ways that a pattern of up to i errors can occur in an n-bit word. Suppose that a code can correct up to c errors in an n-bit word. There are then $\sum_{i=1}^{c} \binom{n}{i}$ ways that a correctable error can occur for each codeword. Moreover, these points are associated with a particular codeword. Since there are 2^k codewords, the total number of points associated with codewords is

$$2^k \sum_{i=0}^{c} \binom{n}{i} \quad .$$

Clearly, this number must be less than the total number of points in the space, 2^n, and we have

$$\sum_{i=0}^{c} \binom{n}{i} \leq 2^{n-k} \quad . \tag{3.8}$$

The connection between distance and error correction is given by (3.7a). Equation (3.8) gives an upper bound to the number of errors that can be corrected in terms of the number of redundant binary digits $n-k$. We note in passing that if we have equality in equation (3.8), we have what is called a *perfect code*. In this case, c or less errors can be corrected for every codeword and each point in the space is associated with a codeword. As we shall see in Section 3.1.7, the Hamming codes are perfect codes. An example of a Hamming code is the (7,4) code given in Table 3.2.

THE PLOTKIN BOUND

A second useful upper bound is the Plotkin bound. Recall that the minimum distance between the words of a code is equal to the minimum weight of the nonzero codewords. It also follows that the minimum weight must be less than the average weight of all of the other codewords. In order to find this average, we find the total number of ones in all of the codewords. We begin by forming the $(2^k \times k)$ matrix whose rows are simply the binary numbers 0 to $2^k - 1$, e.g., for $k = 3$, the matrix is simply

$$\begin{bmatrix} 0 & 0 & 0 \\ 0 & 0 & 1 \\ 0 & 1 & 0 \\ 0 & 1 & 1 \\ 1 & 0 & 0 \\ 1 & 0 & 1 \\ 1 & 1 & 0 \\ 1 & 1 & 1 \end{bmatrix} \quad .$$

If this matrix, which contains all possible data sequences of length k, is multiplied by a generator matrix, we have the $(2^k \times n)$ matrix whose rows are all of the codewords. Each of the columns of the $(2^k \times n)$ matrix are linear combinations of the columns of the $(2^k \times k)$ matrix. Now in all of the columns of the $(2^k \times k)$ matrix, zeros and ones appear in equal numbers. Since the parity check is simply modulo 2 addition, zeros and ones appear in equal numbers in each of the columns of the $(2^k \times n)$ matrix as well, i.e., 2^{k-1} times each. Since there are n columns, $n2^{k-1}$ ones appear in all of the code words. Finally, we find the average by dividing by the number of nonzero codewords, $2^k - 1$. The result is the Plotkin bound.

$$D_{\min} \leq n\ 2^{k-1}/(2^k - 1)\ . \tag{3.9}$$

This bound is effective for $n >> k$.

THE VARSHARMOV–GILBERT BOUND

Consideration of the properties of the parity-check matrix leads to a lower bound on the minimum distance of codes. From (3.6) we have

$$\sum_{j=1}^{n} h_{ij}\, c_j = 0; \qquad i = 1,2,\ldots,n-k\ , \tag{3.10}$$

where c_j is the jth element of the codeword C. From this relation it is evident that by summing all $(n-k)$ equations in (3.10) the columns corresponding to nonzero elements of the codeword C sum to 0. For example, if c_2 and c_5 are one while all other elements are zero, then $h_{12} + h_{22} + h_{(n-k),2} + h_{15} + h_{25} + h_{(n-k),5} = 0$. Thus, if the minimum-weight codeword is $D_{\min}$, then at least one set of $D_{\min}$ columns must sum to zero. Furthermore, no set of $D_{\min} - 1$ or fewer columns may sum to zero, otherwise there would exist a codeword with less than the assumed minimum weight.

Guided by these observations, we construct a parity-check matrix for a code with a certain minimum distance, $D_{\min}$. Recall that the parity-check matrix has $n-k$ rows and n columns. From the parity-check matrix, the generator matrix and the code follow immediately. We choose as the first column of the parity check, matrix H any nonzero $(n-k)$-tuple. The second and succeeding columns are $(n-k)$-tuples which are not linear combinations of $D_{\min} - 2$ or fewer previous columns. In this way we are assured that no linear combination of $D_{\min} - 1$ or fewer previous columns can sum to zero. Assume that the algorithm is at the point where there are $j-1$ columns in the matrix. The number of linear combinations of $D_{\min} - 2$ or fewer columns is $\sum_{i=1}^{D_{\min}-2} \binom{j-1}{i}$. A jth column can be chosen provided the linear combinations have not exhausted all of the nonzero $(n-k)$-tuples, of which there are $2^{n-k} - 1$. Assuming that each of the linear combinations produces a distinct word, a jth column can be added provided that

$$\sum_{i=1}^{D_{\min}-2} \binom{j-1}{i} < 2^{n-k} - 1 . \qquad (3.11)$$

The adding of columns continues until (3.11) is not satisfied. Recall that the number of columns in the parity-check matrix is equal to the length of the codeword, n: accordingly, n is the largest value of j for which the inequality in (3.11) holds, giving the *Varsharmov–Gilbert bound*

$$\sum_{i=0}^{D_{\min}-2} \binom{n-1}{i} \le 2^{n-k} . \qquad (3.12)$$

An (n,k) code with minimum distance $D_{\min}$ can always be constructed provided the inequality in (3.12) is satisfied. Actually, the minimum distance between codewords could be higher than the bound indicated in (3.12) because of the assumption that all of the linear combinations of $D_{\min}-2$ or fewer columns are distinct. If certain combinations were the same, more terms could be included in the summation since the same $(n-k)$ tuple would not be counted twice.

In Table 3.3 the bounds that we have derived are compared to the distance properties of known, easily implemented codes. As is the usual notation: n, the codeword length, and k, the number of bits per codeword, are indication by (n,k). The abbreviations are H for the Hamming code, BCH for the Bose–Chaudhuri–Hocquenghem code, and MLSR for the Maximal-Length Shift Register code. The single Golay (23,12) code is also listed. (We shall be taking a more detailed look at these codes in the sequel.) For each of the codes the minimum distance is shown. As k and the rate of transmitting information increases, the minimum distance decreases. From the calculations, we see that the two upper bounds, the Hamming and the Plotkin, are complementary in a certain sense. The former is effective for large values of k, while the latter is tight for small values of k. For the most part, the codes are close to the upper bound. For large n and small k we see that the BCH code is far from the upper bound and indeed may fall below the V–G bound ($n = 127$, $k = 8$). This is not a contradiction since the V–G bound means that for a given n and k, one may construct a code with a better minimum. Indeed, the maximal-length shift register codes achieve the upper bound, in this case the Plotkin bound, for small k.

3.1.6 PROBABILITY OF UNDETECTED ERRORS

In a wide range of applications, codes are used for error correction either exclusively or in conjunction with error detection in a hybrid scheme. Even though there is a certain distance between the words of a code, it is evident from Figure 3.2 that the noise in the channel can cause errors which are undetected.

Table 3.3 - Bounds for Various Codes (H=Hamming, BCH=Bose-Chaudhuri-Hocquenghem, MLSR=Maximal Length Shift Register)

Code	D_{min}	Hamming bound	Plotkin bound	Varsharmov–Gilbert bound
H(7,4)	3	3	3	3
MLSR(15,4)	8	9	8	6
BCH(15,7)	5	5	7	4
H(15,11)	3	3	7	3
Golay(23,12)	7	7	11	5
MLSR(31,5)	16	19	16	11
BCH(31,6)	11	19	15	11
BCH(31,16)	7	7	15	5
H(31,26)	3	3	15	3
MLSR(63,6)	32	45	32	24
BCH(63,45)	7	7	31	5
H(63,57)	3	3	31	3
MLSR(127,7)	64	99	64	51
BCH(127,8)	35	97	63	50
BCH(127,22)	31	71	63	28
BCH(127,50)	23	41	63	19
BCH(127,92)	11	13	63	8
H(127,120)	3	3	63	3

The probability of an undetected error can be quantified as a function of the code parameters. Note that an error in a bit position is equivalent to modulo 2 addition of a one. Since adding the error vector to a true codeword produces another codeword undetected, errors occur when errors in the channel exactly match a nonzero codeword. Let A_i; $i=0,1,\ldots,n$ be the number of words in a code which have weight i. The probability of an undetected error is given by

$$P(UE) = \sum_{i=1}^{n} A_i p^i (1-p)^{n-i} , \tag{3.13}$$

Table 3.4 Weight Distribution for Various Codes

Code	Weight distribution
Hamming(7,4)	$A_3 = 7, A_4 = 7, A_7 = 1$
Hamming(15,11)	$A_3 = 35, A_4 = 105, A_5 = 168, A_6 = 280$ $A_7 = 435, A_8 = 435, A_9 = 280, A_{10} = 168$ $A_{11} = 105, A_{12} = 35, A_{15} = 1$
Golay(23,12)	$A_7 = 253, A_8 = 506, A_{11} = 1288,$ $A_{12} = 1288, A_{15} = 506, A_{16} = 253,$ $A_{23} = 1$

where p is the probability of error in a BSC. (See (2.49).) Given the weight distributions of a code, the probability of undetected error can be calculated from (3.13). The weight distribution of several codes which will be discussed in the sequel are given in Table 3.4. In most cases of interest $p << 1$; consequently, the minimum weight codewords determine the probability of undetected error. For example, for the Hamming (7,4) code in Table 3.1, $A_3 = 7$, $A_4 = 7$ and for $p = 10^{-4}$, $P(UE) = 7 \times 10^{-12}$. Note for this simple code the probability of undetected error is proportional to p^3.

3.1.7 THE HAMMING CODE

Hamming codes illustrate the role of the parity-check matrix in error correcting. These codes are block codes with $n = 2^m - 1$ bit words having $k = 2^m - 1 - m$ information bits, where m is a parameter; $m = 1, 2, \ldots$.. The parity-check matrix for this code has m rows and $2^m - 1$ columns which are all of the possible m bit sequences, excluding the all-zero sequence. For any value of m, the minimum distance between codewords is three, since three columns can be found which sum to zero and no two sum to zero. As we have seen, with minimum distance three, single errors can be corrected. The (7,4) code shown in (3.5a-b) is an example of a Hamming code for $m = 3$. Note that the columns parity-check matrix given in (3.5b) are all combinations of three-bit words.

It can be shown that Hamming codes are perfect in that they satisfy the Hamming bound with equality. Since the minimum distance between any pair of codewords is three, there is a sphere with radius one around each codeword, and the total number of points at distance one from a code word is $n+1$. Since there are 2^k codewords we have $(n+1)2^k = 2^m 2^{2^m - m - 1} = 2^{2^m - 1} = 2^n$ points associated with codewords, the whole n-space. The simple code depicted in Figure 3.3b is a Hamming code for $m = 2$.

The distance properties of the Hamming code can be improved by a simple change in the parity-check matrix. As mentioned in the preceding paragraph, the columns of this matrix consist of all possible nonzero m-bit sequences. Now the element-by-element modulo 2 sum of any two columns with an even number of ones yields a column with an even number of ones which must be one of the other columns of the matrix; consequently, the sum of all three gives the all-zero column. This is why words of weight three satisfy equation (3.6) and are valid codewords. (Recall that the minimum weight is equal to the minimum distance between pairs of codewords.) Now suppose that the parity-check matrix is altered by deleting all of the columns with an even number of ones. A combination of at least four columns must be added before the all-zero column is obtained. This is simply because a sum of two columns with an odd number of ones yields a column with an even number of ones. The matrix obtained by this deletion has m rows and 2^{m-1} columns. It may be used to generate a code with words of length $n = 2^{m-1}$ containing $k = 2^{m-1} - m$ information bits. The number of parity-check bits is the same as before, m. In this case codewords must have weight four in order to satisfy (3.6). The code obtained by this technique is the *shortened Hamming code*. Notice that the improved distance properties are obtained at the expense of the information-bearing capabilities of the codewords.

3.2 CYCLIC BLOCK CODES

3.2.1 RELATION TO GALOIS FIELDS

The most important class of linear block codes is the class of cyclic codes in which each codeword is a cyclic shift of another codeword, i.e., if $c_{n-1}, c_{n-2}, \ldots, c_0$ is a codeword, then so is $c_{n-2}, c_{n-3}, \ldots, c_0, c_{n-1}$. For example, if 1110 is a codeword then so are 1101, 1011, and 0111. Notice that all the words in a code are not shifts of the same codeword. The importance of these codes lies in their ease of encoding and decoding. As we shall demonstrate shortly, these operations are carried out by means of linear sequential circuits [12], which are combinations of shift registers and modulo 2 adders, components allowing high-speed digital implementation.

In order to describe the properties of cyclic codes, they are represented by polynomials whose coefficients are the code digits. For example, the codeword $c_{n-1} c_{n-2}, \ldots, c_1, c_0$ is represented by the polynomial in the variable, d,

$$C(d) = c_{n-1} d^{n-1} + c_{n-2} d^{n-2} + \cdots + c_1 d + c_0 .$$

Thus the codeword 1110 is represented by $d^3 + d^2 + d$.

The polynomials which represent cycle codewords are elements of what is called a Galois field.* It is useful at this point to review some of the concepts of

* Any of the textbooks that we have referenced may be consulted for more detail. We consider other aspects of the theory in Section 6.7.3 in connection with self-synchronizing scramblers.

this theory. A field consists of p elements where p is a prime number. The Galois field with p elements is denoted GF(p). From GF(p) a Galois field of p^k elements can be constructed. This is called an extended field which comes into play for Reed–Solomon codes.

The operations of addition and multiplication are defined on these elements. The field is closed under these operations in that multiplication or addition of two elements in the field produces an element already in the field. The identity under addition, 0, and the identity under multiplication, 1, are elements of the field. Reciprocals under addition, $-a$, and under multiplication, a^{-1}, are also elements of the field. Finally, the usual associative, distributive, and commutative laws prevail.

Since, in practice, we are most interested in the binary case, we illustrate the concepts with GF(2) whose elements are 0 and 1; all of the arithmetic is modulo 2. Thus, for example, $d^m + d^m = 0$. Consider the set of polynomials whose coefficients are elements of GF(2). Addition and multiplication of such polynomials are performed in the usual way with modulo 2 arithmetic on the coefficients. We pause to illustrate the arithmetic that is involved on the Galois field of polynomials. The sum is carried out element by element in modulo 2 arithmetic. Thus, for example, the sum of the polynomials $d^3 + d^2 + d$ and $d^3 + d + 1$ is $d^2 + 1$. Multiplication is carried out in the usual fashion. For example, the product of the polynomials $d + 1$ and $d^3 + d^2 + 1$ is $d^4 + d^2 + d + 1$. Division of polynomials in the field is also carried out in the usual way. It is left as an exercise to show that $(d^7 + 1)/(d^3 + d + 1) = d^4 + d^2 + d + 1$. Note that the examples of multiplication and division serve to show that the polynomial $d^7 + 1$ has factors $d^3 + d + 1$, $d^3 + d^2 + 1$, and $d + 1$ since $(d^3 + d + 1) \times (d^3 + d^2 + 1)(d + 1) = (d^3 + d + 1)(d^4 + d^2 + d + 1) = d^7 + 1$. We shall find this particular factorization useful in later examples. In the preceding example of division the polynomials were chosen so that one goes into the other evenly with no remainder. As the following example indicates, this is not always the cas.

$$
\begin{array}{r|l}
 & d^2 + 1 \\
\hline
d^3 + d + 1 & d^5 + 1 \\
 & d^5 + d^3 + d^2 \\
\cline{2-2}
 & d^3 + d^2 + 1 \\
 & d^3 + d + 1 \\
\cline{2-2}
 & d^2 + d
\end{array}
$$

Clearly, the remainder is $d^2 + d$, and we thus have $d^5 + 1 = (d^2 + 1) \times (d^3 + d + 1) + d^2 + d$. We can use the same simple calculation to show that $d^3 + d + 1$ does not divide $d^6 + 1$ or $d^4 + 1$ either. This will be used in the next paragraph as an important example.

We consider the properties of polynomials which describe the words of a cyclic code. The degree of a polynomial, $Q(d)$, is the highest power of d in $Q(d)$. A polynomial is *irreducible* if it cannot be factored into polynomials of lower degree. For example, d^3+d+1 is an irreducible polynomial. (For this simple case, this can be seen by attempting to divide by all lower-degree polynomials.) The exponent of a polynomial, $Q(d)$, is the minimum value of ℓ such that $Q(d)$ divides $d^\ell+1$, i.e., $(d^\ell+1)/Q(d)$ is a polynomial. For example, the exponent of d^3+d+1 is 7, since $(d^7+1)/(d^3+d+1) = d^4+d^2+d+1$. Furthermore, d^3+d+1 does not divide d^n+1 for $n < 7$. An irreducible polynomial of degree m is called a *primitive polynomial* if its exponent is 2^m-1. For example, d^3+d+1 is a primitive polynomial. A table of irreducible polynomials, many of which are primitive, is given in [3].

These properties of polynomials are directly applicable to cyclic codes. It can be shown that a cyclic shift to another codeword can be accomplished by first multiplying the polynomial representing a codeword, C(*d*), by *d* and then dividing by (d^n+1), i.e.,

$$dC(d)/(d^n+1) = c_{n-1} + \left[c_{n-2}d^{n-1}+c_{n-3}d^{n-2}+...+c_0d+c_{n-1}\right]/(d^n-1) \; . \tag{3.14a}$$

Note that the numerator of the right-hand side is the polynomial of the codeword produced by a cyclic shift of one. This suggests that we may describe the shift to produce a codeword $C'(d)$ from the codeword $C(d)$ as

$$C'(d) = c_{n-1}(d^n-1) + dC(d)$$

or

$$C'(d) = dC(d) \quad \text{modulo}(d^n+1) \quad . \tag{3.14b}$$

We conclude that all of the codeword polynomials for codewords which are cyclic shifts of one another may be generated in this fashion. We have

$$C_i(d) = d^iC(d) \quad (\text{modulo}(d^n+1)), \quad i = 1,2,...,2^k-1 \quad , \tag{3.15a}$$

alternatively

$$C_i(d) = d^iC(d) + Q(d)(d^n+1), \quad i = 1,2,...,2^k-1 \quad , \tag{3.15b}$$

where $Q(d)$ is a polynomial of degree less than *n*. Polynomials representing valid codewords satisfy (3.15-a-b).

We now show that cyclic codes may be generated by multiplying a polynomial representing information bits by a generator polynomial. As we shall see, this operation is easily carried out by means of linear sequential circuits. Suppose that the information bits to be transmitted are represented by a $(k-1)$ degree polynomial

$$X(d) = x_{k-1}d^{k-1} + x_{k-2}d^{k-2} + \cdots + x_1 d + x_0 ,$$

where $x_{k-1}, x_{k-2}, \ldots, x_1, x_0$ are the (binary) information bits. These are 2^k possible polynomials, which we denote by $X_i(d)$, $i = 1,2,\ldots,2^k$. Now suppose that the $n-k$ degree polynomial $G(d)$ is a factor of the polynomial d^n+1, i.e., $H(d)G(d) = d^n+1$ where $H(d)$ is a k degree polynomial code. Codes generated by the product $G(d)X(d)$ where $X(d)$ represents a particular set of information bits satisfy the cyclic property. In (3.15) assume the $C(d) = X(d)G(d)$, then $G(d)$ is a factor of $C_i(d)$ since it divides both terms on the right-hand side of (3.15b). We may write

$$C_i(d) = d^i X(d)\, G(d) + Q(d)(d^n+1) = [d^i X(d) + Q(d)H(d)]\, G(d)$$

or

$$C_i(d) = X_i(d)G(d) , \quad i = 1,2,\ldots,2^k . \tag{3.16}$$

The polynomial $G(d)$ is called the *generator polynomial* of the cyclic code. Note that the degree of the generator polynomial is $n-k$ and $G(d) = g_{n-k}d^{n-k} + \cdots + g_1 d + g_0$.

It is not difficult to see that, in general, each codeword generated by (3.16) is unique, i.e., $C_i(d) = X_i(d)G(d) \neq X_j(d)G(d) = C_j(d)$ for $X_i(d) \neq X_j(d)$. Further, it follows that cyclic codes are linear group codes in that the modulo 2 sum of codewords produces another codeword, i.e.,

$$X_i(d)G(d) + X_j(d)G(d) = [X_i(d)+X_j(d)]G(d) = X_k(d)G(d),$$

which satisfies (3.15b); hence it is a codeword. Since cyclic codes are linear block codes, they may also be described by means of the generator matrix discussed in Section 3.1.2. From the generator polynomial the generator matrix may be constructed as

$$G = \begin{bmatrix} g_0 & g_1 & \cdots & g_{n-k} & 0 & 0 & \cdots & 0 \\ 0 & g_0 & \cdots & g_{n-k-1} & g_{n-k} & 0 & \cdots & 0 \\ & \vdots & & \vdots & & & & \\ 0 & 0 & \cdots & 0 & g_0 & g_1 & \cdots & g_{n-k} \end{bmatrix} .$$

We illustrate these concepts by means of an example. Consider the cyclic code of length $n=7$. As we have seen, the factors of d^7+1 are $d+1$, d^3+d^2+1 and d^3+d+1. We take $G(d) = d^3+d+1$ to be the generator polynomial for the code. (Either of the other two factors could have been chosen as well.) The four-bit information sequence 0110 has the polynomial $X(d) = d^2+d$. Multiplying by the generator yields $X(d)G(d) = d+d^3+d^4+d^5$. (Addition is modulo 2). The corresponding codeword is 010111. The generator matrix for this code is given by

$$G = \begin{bmatrix} 1 & 1 & 0 & 1 & 0 & 0 & 0 \\ 0 & 1 & 1 & 0 & 1 & 0 & 0 \\ 0 & 0 & 1 & 1 & 0 & 1 & 0 \\ 0 & 0 & 0 & 1 & 1 & 0 & 1 \end{bmatrix}.$$

3.2.2 DECODING CYCLIC CODES—BURST CAPABILITIES

As indicated earlier, the motivation for cyclic codes and the attendant mathematics is ease of implementation of decoding in systems where the channel may be modeled by the binary symmetric channel (see Section 2.3). In this case, the input to the decoder is a sequence of binary digits containing information and parity-check bits.

In order to show this, we begin by showing that a received codeword is decoded by dividing by the generator polynomial for the code. According to (3.16), this division will yield the information word. However, if there is an error pattern which does not have $G(d)$ as a factor in its polynomial representation, this division will not go evenly and there will be a remainder. Let $E(d)$ represent the error sequence. The division by $G(d)$ produces

$$\begin{aligned} R(d) &= (X(d)G(d) + E(d))/G(d) \\ &= X(d) + E(d)/G(d) \end{aligned}$$

and the remainder $E(d)/G(d)$ serves as a syndrome to the error.

The generator polynomial determines the characteristics of the code, hence the kind of error patterns that can be detected and corrected. The following properties can be shown:

1. All single error patterns can be detected. A single bit error pattern has the polynomial $E(d) = d^i; 0 \le i < n$. Clearly this is not divisible by a factor of $1 + d^n$, since these factors must have at least two terms, a one and at least one power of d. (Recall the factors of d^7+1.)

2. If $G(d)$ has $d + 1$ as a factor, all error patterns with an odd number of terms can be detected. This is because no polynomial with an odd number of terms, which represents an odd number of errors, may have $d+1$ as a factor. To see this, assume the opposite, i.e., an error pattern such that

$E(d) = (d+1)F(d)$. Now evaluate the polynomial at $d = 1$, $E(1) = (1+1)F(0) = 0$ a contradiction, since if there were an odd number of terms, we should have $E(1) = 1$.

3. If $G(d)$ contains a primitive polynomial as a factor, then all double errors can be detected. A double error can be represented as $d^i + d^j = d^i(1 + d^{j-i})$, $i < j$. As we have seen, $G(d)$ cannot divide d^i. Further, since $G(d)$ contains a primitive polynomial, it cannot divide $1 + d^l$ for $l < n$.

A burst of errors can be defined as any sequence beginning with an error and ending with an error with any number of errors in between. It can be represented by the polynomial $E(d) = d^j B(d)$, $0 \leq j \leq n-1$, where $B(d)$ represents the particular pattern of errors and j its position within the n-bit block. If the duration of the burst of errors is $n-k$ or less, then $E(d)$ is not divisible by $G(d)$. The maximum degree of $B(d)$ is $n-k-1$ and the degree of $G(d)$ is $n-k$. We have already pointed out that d^j is not divisible by $G(d)$. This observation leads to the conclusion:

4a. All bursts of duration $(n-k)$ or less can be detected.

Now suppose that the duration of a burst is $n-k+1$. The burst will be undetectable only if it is of the form $E(d) = d^i G(d)$. Now there are $2^{-(n-k-1)}$ possible bursts with an error in position i and an error in position $i + n - k$ only one of which may be represented by $d^i G(d)$ and is therefore undetectable. We then have that:

4b. The fraction of undetectable errors of length $n-k=1$ is $2^{-(n-k-1)}$.

Now consider bursts which are longer then $n-k+1$ bits. There are $2^{-(l-2)}$ possible bursts, $l > n-k+1$, with an error in the ith position and an error in the $(i+l-i)$th position. Among these bursts, those which are undetectable have the form $E(d) = d^i A(d) G(d)$. The number of bursts of this form are $2^{l-(n+k)+2}$, consequently:

4c. The fraction of undetected bursts of duration $l > n-k+1$ is $2^{-(n-k)}$.

3.2.3 EXAMPLES OF CYCLIC CODES

There are several cyclic codes which are international standards under the name *cyclic redundancy check (CRC)* codes. These are generated by the following polynomials:

$$G_{CRC-12}(d) = d^{12} + d^{11} + d^3 + d^2 + d + 1 ,$$

$$G_{CRC-16}(d) = d^{16} + d^{15} + d^2 + 1 ,$$

$$G_{CRC-CCITT}(d) = d^{16} + d^{12} + d^5 + 1 .$$

The simple division illustrated in the preceding section shows that these polynomials can be factored as follows:

$$G_{CRC-12}(d) = d^{12}+d^{11}+d^3+d^2+d+1 = (d+1)(d^{11}+d^2+1) \ ,$$

$$G_{CRC-16}(d) = d^{16}+d^{15}+d^2+1 = (d+1)(d^{15}+d+1) \ ,$$

$$G_{CRC-CCITT}(d) = d^{16}+d^{12}+d^5+1$$

$$= (d+1)(d^{15}+d^{14}+d^{13}+d^{12}+d^4+d^3+d^2+d+1) \ .$$

As we see, $d + 1$ is a factor for all three polynomials; therefore, all odd error patterns can be detected. By consulting the appropriate tables [3], it can be shown that the other factors are primitive polynomials. As we have seen, this property allows all double errors to be detected. The polynomial $G_{CRC-12}(d)$ is used when data are in the form of six-bit words since the degree of the generator polynomial, $n-k = 12$. Errors in two successive characters can be detected. As we have seen in 4a and 4b above, a high percentage of longer strings can be detected. The same is true of $G_{CRC-16}(d)$ and $G_{CRC-CCITT}(d)$ when information is transmitted by eight-bit characters.

The foregoing are practical examples of codes which are in common use. There are a number of classes of codes which form a theoretical basis for practical codes. If the generating polynomial of a cyclic code is primitive, it can be shown [2-4] that the code is Hamming code. Since they are primitive polynomials, by definition the generator polynomials are factors of d^n+1. Recall that Hamming codes are perfect codes which can correct single errors. For example, the polynomial d^3+d+1 generates the [7,4] Hamming code. Further, the primitive polynomial d^4+d+1 generates the [15,11] Hamming code. The weights of the [7,4] and the [15,11] Hamming codes are shown in Table 3.4.

The *[23,12] Golay code* is also a perfect code, but it has distance 7; accordingly it may correct up to 3 errors. The generator polynomial for this code is

$$G(d) = d^{11}+d^9+d^7+d^6+d^5+d+1 \ .$$

Since the Golay code is perfect, it satisfies the Hamming bound with equality. The distribution of weights for the Golay code is shown in Table 3.4. The extended [24,12] Golay code with minimum distance 8 is obtained by adding an overall parity bit to each codeword. Thus, for example, all codewords of weight 7 become codewords of weight 8.

The obvious advantage of the Golay codes over the Hamming code is the ability to correct multiple errors. The difficulty is the limited range of block

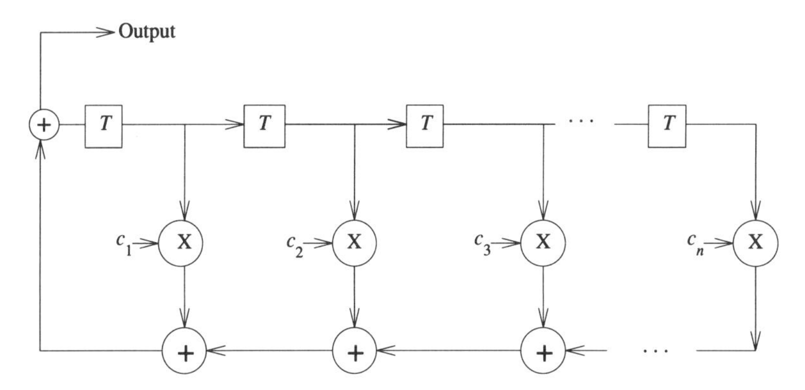

Fig. 3.4 Maximal-length shift register encoder.

lengths, $n = 23$ and $n = 24$. It was not until the invention of the *Bose–Chaudhuri–Hocquenghem (BCH)* code that flexible multiple error correcting codes were available. The BCH codes are generalizations of the Hamming code which correct an arbitrary number of errors. Their generating polynomials are factors of $d^M + 1$ where $M = 2^m - 1$. The code length is $n = 2^m - 1$ and $n - k \leq mt$ where m and t are arbitrary integers. It can be shown that [2–4] the minimum distance between codewords is $D_{\min} \geq 2t + 1$; consequently at least t errors can be corrected. For example, the [63,45] BCH code has minimum distance 7; consequently it can correct up to 3 errors. (See Figure 3.2.) As pointed out in connection with Table 3.3, these codes compare favorably to the upper bound except for large n and k.

The *Reed–Solomon* (*RS*) codes are a subclass of BCH codes that apply to the nonbinary case. RS codes are used to encode symbols consisting of m ($m \geq 1$) binary digits into blocks consisting of $n = 2^m - 1$ symbols (or $m\,(2^m - 1)$ binary digits). There are $n - k = 2t$ parity symbol and the distance between codewords is $2t + 1$ *symbols*. RS codes find wider application because of good distance properties and ease of decoding. RS codes are often used in *concatenated coding* schemes, which will be discussed in Section 3.8.

A final class of cyclic codes which are of interest are the *maximum-length shift register codes*. Consider the feedback shift register circuit shown in Figure 3.4. We assume that the polynomial which characterizes this circuit, $C(z)$, is a primitive polynomial of degree m.* It can be shown that with any nonzero initial state, the circuit will shift through all $2^m - 1$ nonzero states. The output sequence will have period $2^m - 1$ with 2^{m-1} ones.

---

* We shall consider maximal-length shift register circuits in some detail in Section 6.7 when we treat self-synchronizing scramblers.

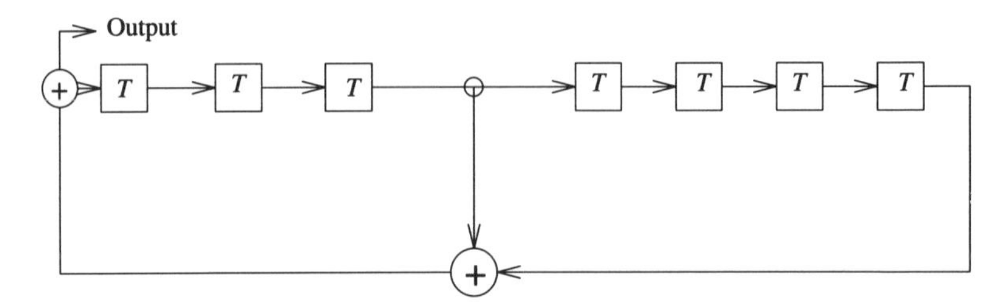

Fig. 3.5 Maximum-length sequence generator for a (127,7) code.

The maximum-length shift register circuit can be used to generate a $(2^m - 1, m)$ systematic code. The m information bits to be transmitted are loaded into the shift register to form the initial state. The codeword is produced by allowing $2^m - 1$ shifts. Note that the codewords are cyclic shifts of the same word, hence all of the codewords have the same weight, 2^{m-1}. Since the code is linear, this is the minimum distance between codewords. Maximum-length shift register codes provide flexibility with respect to length together with excellent distance properties. Their disadvantage is low information rates. These codes are complementary to Hamming codes which are flexible with respect to length, have a high information rate, but which have limited error correcting capability. An example of a maximal-length shift register code is the [127,7] code with minimum distance 64 generated by the primitive polynomial $C(z) = z^7 + z^3 + 1$, shown in Figure 3.5.

3.2.4 LINEAR SEQUENTIAL CIRCUITS

Linear sequential circuits over *GF(p)* are composed of modulo p adders, multipliers, and delay elements connected according to a few elementary rules. As the name implies, such circuits are *linear* over modulo p arithmetic. The laws of commutativity, associativity, and superposition apply. For example, the response of a circuit to the sum of two inputs is the sum of the responses to each input separately. The summations are carried out term-by-term modulo p on the input and output sequences.

The importance of cyclic codes lies in the fact that they may be simply generated and decoded by means of linear sequential circuits. As we have seen, the two basic operations that must be implemented are polynomial multiplication and division. An essential tool for the analysis of linear sequential circuits is the *z-transform* of a time series.† Consider the sequence of binary digits $x_0, x_1, \ldots$. The z-transform of this sequence is defined as

† We use the variables d and z to distinguish between polynomial descriptions of codes and transforms of time series, respectively.

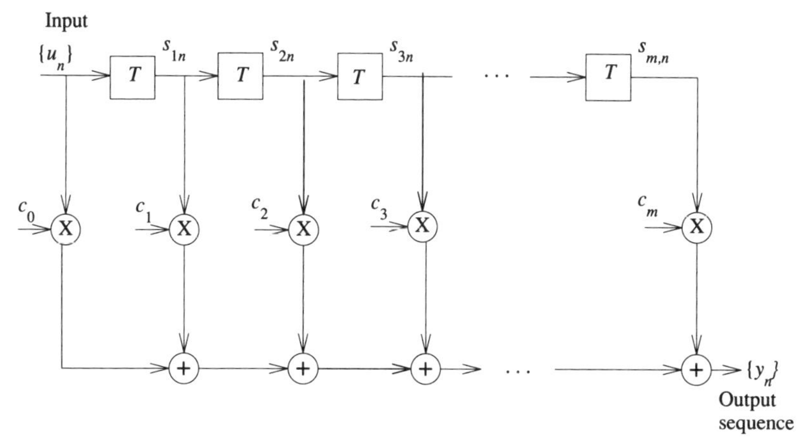

Fig. 3.6 Polynomial multiplier circuit.

$$X(z) = \sum_{i=0}^{\infty} x_i z^i \ , \tag{3.17}$$

and we assume that $x_i = 0$ for $i < 0$. For a finite time sequence, $X(z)$ is simply a polynomial in z. The only property of the z-transform we shall use is the delay operation. If $A(z)$ is the z-transform of the sequence $a_n, n = 0,1,\ldots,$ then the z-transform of a_{n-1} is

$$\sum_{i=0}^{\infty} a_{i-1} \ z^i = \sum_{j=0}^{\infty} a_j z^{j+1} = zA(z) \ .$$

(We assume that $a_i = 0$ for $i < 0$.)

Consider the *multipler circuit* shown in Figure 3.6. The elements of this circuit are modulo 2 adders and delay elements. In the general case of modulo p arithmetic, we would have multipliers as well; however, here the terms c_i, $i = 0,1,\ldots,m$ are binary digits, so they are simply either present or absent. The input to the circuit is denoted as u_n, $n = 0,1,\ldots$, and the output as y_n, $n = 0,1,\ldots$. We also define the sequence s_{1n} as the output of the ith delay element. The vector $(s_{1\,n}, s_2, \ldots, s_{m\,n})$ is the state of the system. Together with the input, the state determines the output.* The input–output relations of the multiplier, under the assumption that the initial state is zero, are given by

$$y_n = \sum_{i=1}^{m} c_i s_{in} + c_0 u_n, \qquad n = 0,1,\ldots \tag{3.18a}$$

* See Section 6.7 for more detail on the properties of linear sequential circuits.

$$s_{1n} = u_{n-1} \tag{3.18b}$$

$$s_{in} = s_{i-1\,n-1}, \qquad i = 0,1,\ldots,m-1 \tag{3.18c}$$

By taking z-transforms of both sides of (3.18a-c) we have

$$Y(z) = \sum_{i=0}^{m} c_i S_i(z) + c_0 U(z) \tag{3.19a}$$

$$S_1(z) = zU(z) \tag{3.19b}$$

$$S_i(z) = z\, S_{i-1}(z)\ , \tag{3.19c}$$

and by substitution

$$Y(z) = U(z)\ C(z)\ , \tag{3.20}$$

where $C(z) = \sum_{i=1}^{n} c_i z^i$ is the transfer function of the circuit. Thus, the output $Y(z)$ is the product of $C(z)$ and $U(z)$.

EXAMPLE 1 The multiplier shown in Figure 3.7 has the input sequence 011001. By direct computation we find that the output sequence is 010110101. The transform of the input is $U(z) = z+z^2+z^5$ and the transfer function of the circuit is $C(z) = 1+z+z^3$. The product is $Y(z) = z+z^3+z^4+z^6+z^8$, which corresponds to the output sequence that we have already found.

Now if the transfer function, $C(z)$, which is a polynomial of degree m, is a factor of z^n, then the circuit can be used to generate $(n,n-m)$ cyclic codes. In the immediately preceding example the transfer function $C(z) = 1+z+z^3$ is a factor of z^7-1; consequently, it can act as the generating polynomial for a (7,4) code.

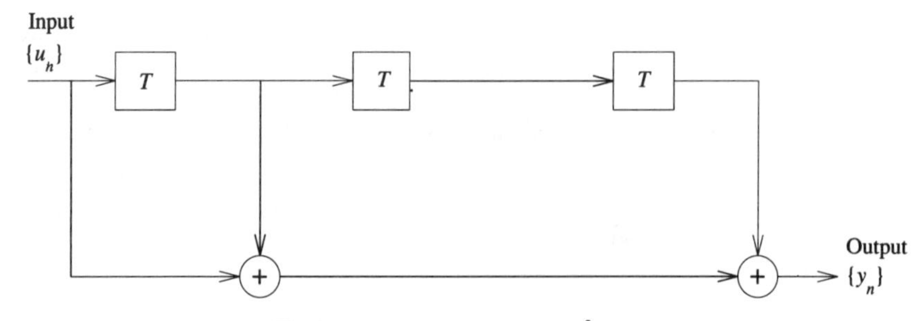

Fig. 3.7 Multiplier circuit for z^3+z+1.

A divider may be used to generate codes and to decode. The generic divider is shown in Figure 3.8. Again the multipliers are simple connections since we focus on the binary case. We may write the following set of equations to describe the system

$$
\begin{aligned}
s_{1\,n} &= u_{n-1} \oplus c_m\, y_{n-1} \\
s_{2\,n} &= s_{1\,n-1} \oplus c_{m-1}\, y_{n-1} \\
&\vdots \\
s_{m-1\,n} &= s_{m-2\,n-1} \oplus c_2 y_{n-1} \\
s_{m\,n} &= s_{m-1\,n-1} \oplus c_1 y_{n-1} \\
y_n &= c_0\, s_{m\,n}\ .
\end{aligned}
\tag{3.21}
$$

Again, we assume that the system is initially at rest so the state vector is zero. We take transforms of both sides of (3.21):

$$
\begin{aligned}
S_1(z) &= zU(z) \oplus c_m\, zY(z) \\
S_2(z) &= zS_1(z) \oplus c_{m-1}\, zY(z) \\
&\vdots \\
S_m(z) &= zS_{m-1}\,(z) \oplus c_1 z\, Y(z) \\
Y(z) &= c_0 S_m(z)\quad .
\end{aligned}
\tag{3.22}
$$

Solving these equations by eliminating $S_1(z)$, $S_2(z)$,..., $S_m(z)$, we find

$$
Y(z) = z^m U(z)/C(z)\ , \tag{3.23}
$$

where $C(z)$ is the previously defined transfer function. (Note that $c_0^{-1} = c_0$ since all operations are modulo 2.)

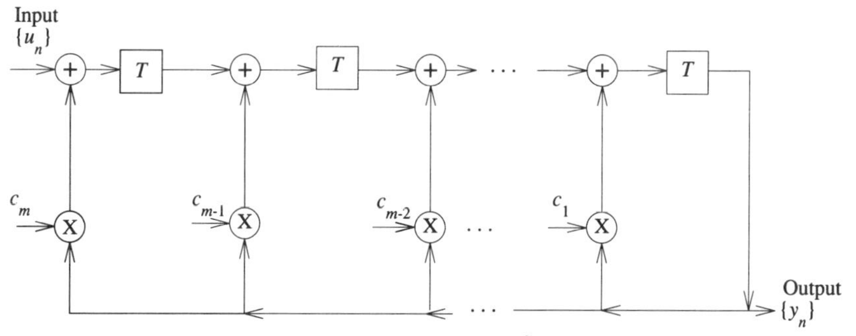

Fig. 3.8 Divider circuit for z^3+z+1.

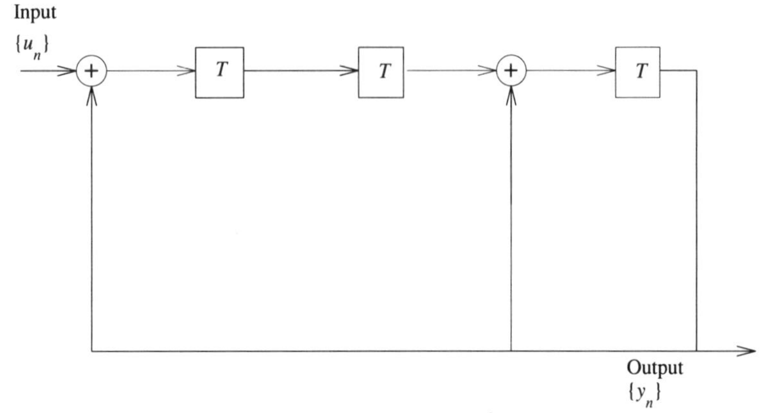

Fig. 3.9 Divider circuit for z^3+z+1.

EXAMPLE 2 Let the input to the divider circuit shown in Figure 3.9 be 010110101. (Note that this is the output of the multiplier in the previous example.) By hand calculation it can be shown that the output sequence is 0000110010, which is the original input to the multiplier with delay. The same result is obtained from polynomial division: $(z^8+z^6+z^4+z^3+z)/(z^3+z+1) = z^5+z^2+z$.

The following derivation shows a divider can be used for code generation. The *Euclidean division algorithm* states that

$$\text{Dividend} = \text{Quotient} \times \text{Divisor} + \text{Remainder}$$

Let the dividend be a polynomial of degree k multiplied by d^{n-k}, whose coefficients are the information bits to be transmitted, and let the divisor be a generator polynomial. We have

$$d^{n-k}X(d) = D(d)\ G(d) \oplus R(d) \tag{3.24a}$$

or

$$D(d)G(d) = d^{n-k}X(d) \oplus R(d)\ , \tag{3.24b}$$

where $D(d)$ and $R(d)$ are to be determined. Equating powers on both sides of (3.24a) we see that the degree of $D(d)$ must be k since $G(d)$ is of degree $n-k$. Moreover, the degree of the remainder, $R(d)$, must be less than that of the divisor, $G(d)$. Since $D(d)$ is an arbitrary polynomial, the right-hand side on (3.24b) is a codeword; moreover the codeword is in standard form since the coefficients of $d^{n-k}, d^{n-k+1}, \ldots, d^{n-1}$ in the polynomial $d^{n-k}X(d)$ are the information bits. The coefficients of $R(d)$, the remainder, correspond to the parity check digits. (3.24a) shows that $R(d)$ may be generated by dividing $d^{n-k}X(d)$ by $G(d)$ and adding a shifted version of the information polynomial, $X(d)$.

The divider can be used to decode by dividing by the generator polynomial. The problem with this straightforward approach is that the decoder complexity grows linearly with the length of the code. However, cyclic codes have considerable structure which can be used to implement simpler decoders. In fact, much research on block codes involves implementation of decoders by exploiting this structure.*

3.3 PERFORMANCE

3.3.1 HARD-DECISION DECODING

In Section 3.1.5 we derived bounds on the minimum distance between codewords. These bounds translate directly into performance bounds in terms of probability of error for both hard- and soft-decision decoding. We begin with hard-decision decoding; consequently, the binary symmetric channel model is appropriate. (See Section 2.3 of Chapter 2.) If, for example, binary-antipodal signaling is used, the probability of bit error is given by (2.35), which we repeat here for convenience.

$$p = Q([2E/N_0]^{1/2}) , \tag{2.35}$$

where E is the energy in each transmitted pulse. Now if a code has minimum distance $2c+1$, it can correct all patterns of up to c errors and the probability of a codeword being in error (i.e., the probability of an undetected error) is bounded by

$$p_e \le \sum_{i=c+1}^{n} \binom{n}{i} p^i \, (1-p)^{n-i} . \tag{3.26a}$$

(See (2.49) and (3.7a).) In most applications p is small enough that the first term in the sum dominates and we have

$$p_e \le \binom{n}{c+1} p^{c+1} \, (1-p)^{n-c-1} . \tag{3.26b}$$

If the code is not a perfect code, some patterns of $c+1$ errors may be corrected, hence the bound in (3.26b). In Figure 3.10 the upper bound in (3.26a) is shown for several codes as a function of E/N_0 assuming binary-antipodal signaling. For signal-to-noise ratios of interest, there is not much difference between

---

* See, for example, Chapters 7 and 8 of Reference [2].

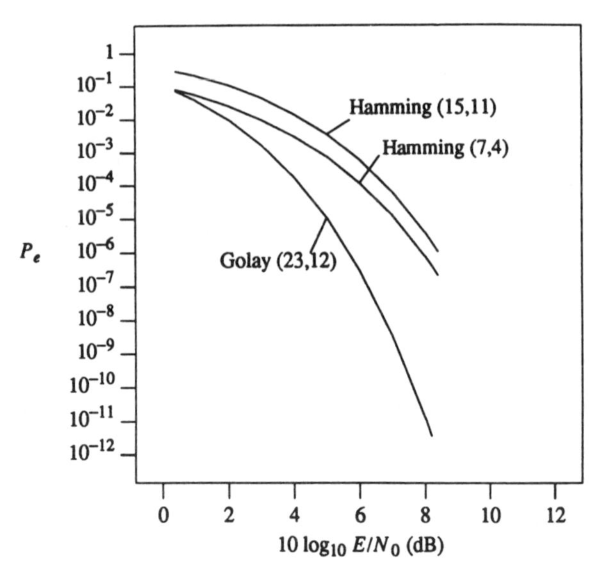

Fig. 3.10 An upper bound on the probability of error as a function of signal-to-noise ratio for several codes.

(3.26a) and (3.26b). The codes we have chosen are Hamming codes, for which $c=1$, and the (23,12) Golay code, for which $c=3$. Note that these are both perfect codes; consequently, we have approximate equality in (3.26b) for large signal-to-noise ratio.

In (3.26b) we have a bound in terms of the maximum number of errors. By application of the *Chernoff bound* one can derive an upper bound in terms of the weights of codewords, a more tangible quantity. Recall that this bound was derived in Appendix 2A of Chapter 2.

Repeating (2A.47b), we have

$$Pr[X \geq x] \leq \exp(-ax)E[\exp(aX)] \ , \tag{3.27}$$

where $E[\cdot]$ denotes expectation and $a \geq 0$ is chosen to minimize the bound.

Now assume that the all-zero codeword is transmitted. This will be mistaken for a codeword, C_i of weight w_i, if there are $w_i/2$ or more errors in places where c_i has ones. Define the independent identically distributed random variables X_j, $j = 1,2,\ldots,w_i$, which correspond to the places where c_i has ones. The probability distribution of these random variables is given by

$$Pr\ [X_j = 1] = p$$

$$Pr[X_j = -1] = 1-p \ .$$

The probability of $w_i/2$ or more errors is simply the probability that the sum of the X_i are positive. By application of the Chernoff bound for a particular error pattern, we have

$$P_{ei} = Pr(\sum_{j=1}^{w_i} X_j \geq 0) \leq E(\exp(a \sum_{j=1}^{w_i} X_j))$$
$$= E(\exp(aX))^{w_i} = [(1-p)\exp(-a) + p \exp(a)]^{w_i} \quad . \tag{3.28}$$

We minimize the bound by differentiating with respect to a and setting the result equal to 0. The optimum value of a is found to be

$$a^* = \ln([(1-p)/p]^{1/2}) \quad .$$

Substituting into (3.28), we find

$$P_{em} \leq \{p[(1-p)/p]^{1/2} + (1-p)[p/(1-p)]^{1/2}\}^{w_m} \leq [4p(1-p)]^{w_m/2} . \tag{3.29}$$

The event whose probability we have bounded is only one way that an error can take place. The union bound furnishes a bound for all possible error events. From (2.10) we have

$$P_e \leq \sum_{i=1}^{2^k-1} [4p(1-p)]^{w_i/2} , \tag{3.30a}$$

where the summation is over all nonzero code words. By rearranging terms in the right-hand side of (3.30a), the bound can be written as

$$P_e \leq \sum_{j=1}^{n} A_j [4p(1-p)]^{j/2} , \tag{3.30b}$$

where A_j is the number of codewords of weight j. The weight distributions for many codes have been tabulated [2]. In Table 3.4 we show the weights for the (23,12) Golay code, the (7,4) Hamming code, and the (15,11) Hamming code. These weights can be substituted into (3.30b) to find the upper bound. The results of this computation are shown in Figure 3.11, where the right-hand side of (3.30) is plotted as a function of E/N_0 for several codes for binary-antipodal signaling.

A looser but simpler bound can be found by substituting the minimum weight of a codeword for w_i in (3.30a), giving

$$P_e \leq (2^k - 1) [4p(1-p)]^{D_{\min}/2} . \tag{3.31}$$

As pointed out above, the minimum weight is equal to the minimum distance, $D_{\min}$. It can be shown that these bounds are reasonably tight for large signal-to-noise ratios. For example, calculations by Proakis [1] for the (23,12) Golay code show that the bounds are within 1/3 and 1/2 dB of the exact probability of

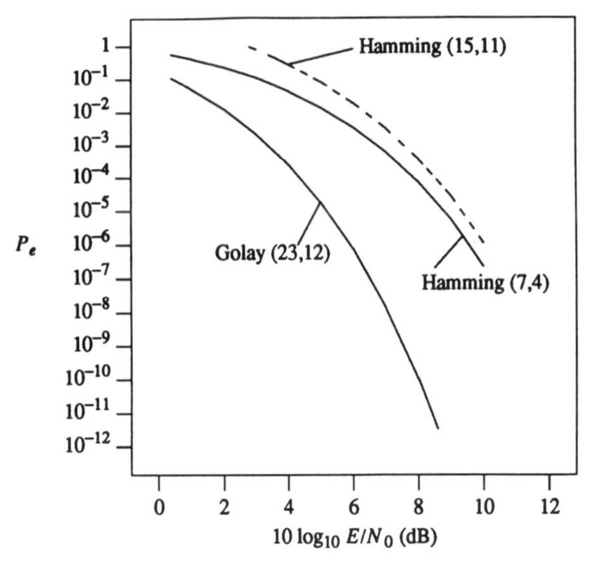

Fig. 3.11 Bounds on probability of error as calculated for various codes with hard-decision decoding.

error for signal-to-noise ratios for which the probability of error is in the neighborhood of 10^{-3} or lower.

3.3.2 SOFT-DECISION DECODING

As we have seen in Section 2.3, hard decisions give the optimum performance when there is no correlation between the binary digits; however, coding introduces correlation and hard limiting destroys some of this information. For purposes of illustration we assume binary-antipodal signaling. The received signal for a single codeword, C_i, may be written as

$$y(t) = \sum_{j=1}^{n} (1 - 2c_{ij})p(t - (ni + j)T) + n(t) , \tag{3.32}$$

where c_{ij}, $j = 1,2, \ldots, n$ are the binary digits in the codeword C_i, $p(t)$ is the transmitted pulse, and $n(t)$ is white Gaussian noise. (3.32) indicates n pulses, one for each digit in the codeword, immersed in white Gaussian noise. We can regard the n pulses as one signal and we can calculate the probability of error for pairs of codewords by applying the results of Chapter 2. Without loss of generality, we assume that the all-zero codeword is transmitted and we calculate the probability of confusing it with the codeword C_j. All that is required for this calculation is the signal energies and their correlation with one another (see (2.33)). The energy in the signals generated by each codeword is nE, where E is

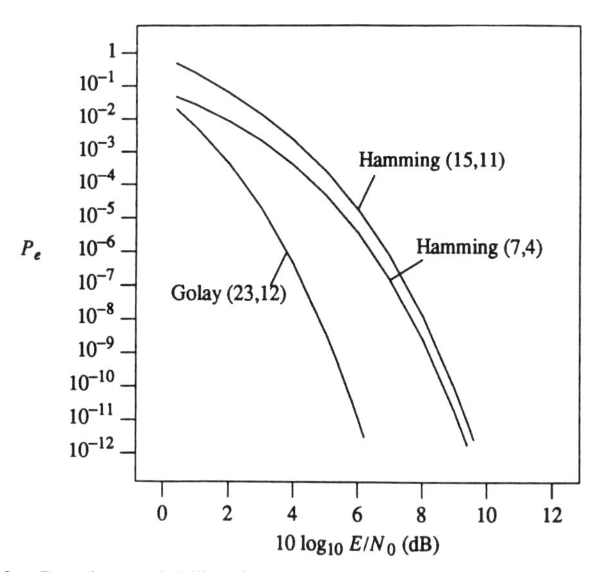

Fig. 3.12 Bounds on probability of error for various codes and soft decision decoding.

the energy in the pulse $p(t)$. The correlation between the codewords is $\zeta_i = E(n - 2w_j)$, where w_j is the weight of the codeword C_j. From (2.35) we find that the probability of error in comparing pairs of codewords is

$$P_{ij} = Q([2Ew_j/N_0]^{1/2}) . \tag{3.33}$$

We apply the union bound (2.10) to find

$$P_e \leq \sum_{j=1}^{2^k-1} Q([2Ew_j/N_0]^{1/2}) , \tag{3.34a}$$

where the summation is over all nonzero codewords. Again, the summation can be rearranged in the form

$$P_e \leq \sum_{k=1}^{n} A_k \; Q([2E \cdot k/N_0]^{1/2}) . \tag{3.34b}$$

From (2.34) we can find a looser but more convenient bound. We apply (2.34b) to (3.34a-b) giving

$$P_e \leq (2^k - 1) \exp(-ED_{\min}/N_0)/2 , \tag{3.35}$$

where $D_{\min}$ is the minimum weight of a codeword. For the codes whose weights are tabulated in Table 3.3, the bounds of (3.34b) are plotted in Figure 3.12.

From these figures we see that soft-decision decoding enjoys a 1 to 2 dB advantage over hard-decision decoding. In view of the fact that a 1 dB increase in signal-to-noise ratio translates to an order of magnitude decrease in probability of error, this advantage is significant. However, in order to deploy soft-decision decoding, increased processing complexity is necessary.

In the bounds we have considered in this and the preceding section the focus was upon probability of an error in decoding a word containing k information bits. Finding the corresponding bit error probability is a difficult problem; however, a reasonable estimate can be obtained by assuming that no more than half of the k bits will be in error when there is a symbol error. The probability of a bit error is then

$$P_b \lesssim P_e/2\,. \tag{3.36}$$

3.3.3 CODING GAIN

The effectiveness of a code may be measured by the *coding gain,* which as its name implies, indicates the efficiency of error correcting codes. Consider an (n,k) block code which corrects up to c errors. In order to compensate for the extra parity check bits, it is necessary to increase the transmission rate by a factor of n/k, which is the inverse of the code rate, R_c. We can take this into account by expressing the probability of error in terms of the energy that is transmitted per bit in the uncoded bit stream, $E_b = nE/k = E/R_c$. ($E_b > E$ since $n > k$.) Substituting into (3.34), for soft-decision decoding we have

$$P_e \leq \sum_{j=1}^{2^k-1} Q([2R_cE_bw_j/N_0]^{1/2}) \leq \sum_{k=1}^{n} A_k\, Q([2R_cE_bk/N_0]^{1/2}) \tag{3.37}$$

Applying the bound of (2.34), we have

$$P_e \leq 2^k \exp(-R_c\, E_b w_{\min}/N_0)/4 = \exp(-R_cE_bw_{\min}/N_0 + k\ \ln 2)/4. \tag{3.38}$$

This is to be compared with the probability of error for uncoded single pulses which is bounded by $\exp(-E_b/N_0)$. The gain in coding is approximately $10\log(R_c\ w_{\min} - N_0k\ln 2/E_b)$ dB. We ignore the factors not appearing in the exponent (see (3.36)).

A more exact calculation of coding gain can be performed for a particular code. For example, consider hard decoding for the (23,12) Golay code which can correct up to 3 errors. The probability of symbol error is given by (3.26a) with $c = 3$. We assume that the probability of bit error is small enough so that the expression in (3.26b) may be used for calculations. Now if a block of k bits were transmitted uncoded with the same energy as the coded case, the energy in each bit would be $23E/12$. The probability of error is given in (2.35). The probabilities of error for the coded and uncoded cases are plotted as a function of the signal-to-noise ratio E_b/N_0, in Figure 3.13.

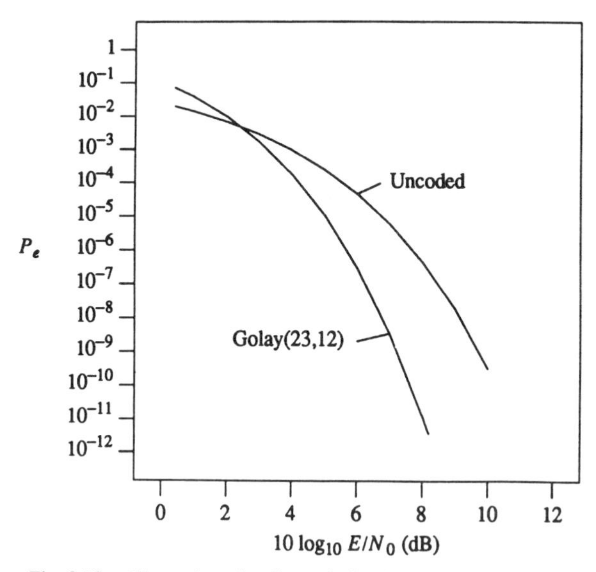

Fig. 3.13 Illustration of coding gain for the (23,12) Golay code.

3.4 CONVOLUTIONAL CODES

3.4.1 GENERAL FORM

The second form of coding that we shall consider is convolutional coding.* In this case the data stream is fed into a linear sequential circuit in the form of a shift register.† The output consists of one or more bits for each input bit formed from linear combinations of the bits in the shift register. A simple example of a sequential encoder is shown in Figure 3.14, where there are three output bits for each input bit. Because of the memory inherent in the shift register, all the output bits in a sequence are functionally related to one another. This stands in contrast to block codes, where the codewords are independent of each other.

The general form for the convolutional encoder is shown in Figure 3.15. It consists of M stages each containing k bits. The input consists of blocks of k information bits which are inserted into the first stage. The previous contents of the first stage are shifted to the second and so on. Blocks of k bits are shifted from stage-to-stage in sequence. Linear combinations of the stored bits are used to form n output bits. The former contents of the last stage are discarded

* We shall postpone, until Section 3.8, a discussion of the relative merits of block and convolutional codes.

† Convolutional codes may be considered to be a subclass of *sliding-block codes* in which blocks of data may be combined in a general fashion, linearly or nonlinearly.

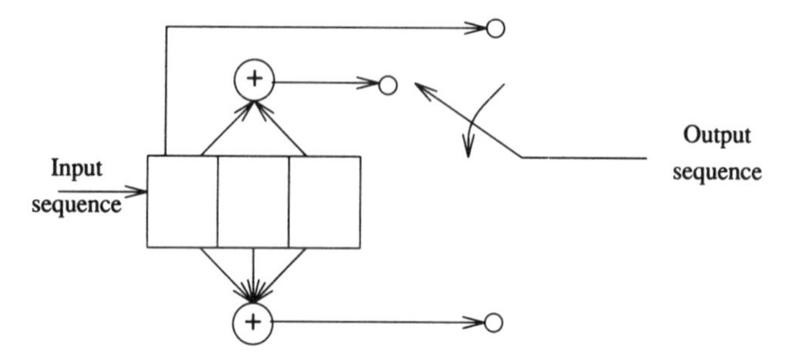

Fig. 3.14 Simple convolutional encoder: 3 outputs are produced for each input.

entirely. The process is repeated after the next block of k information bits is generated.

The latency in the system is embodied in the M k-bit stages. The quantity, M, is called the *constraint length* of the code. The coding rate in this system is k/n. In the simple example cited above in Figure 3.14, $k=1$, $M=3$, and $n = 3$. For the somewhat more complex encoder shown in Figure 3.16, the parameters are $k=2$, $M=3$, and $n=4$.

The input-output relation of the convolutional encoder can be described by means of a generator matrix. The elements of the matrix are binary digits indicating whether a particular digit contributes to a output digit; thus, if element g_{ij} is one, the output of delay i contributes to output j. In vector notation, the input-output relationships can be written

$$\mathbf{y} = \mathbf{S}\, G \tag{3.39}$$

where $\mathbf{S}$ is a kM-dimensional row vector indicating the contents of the shift

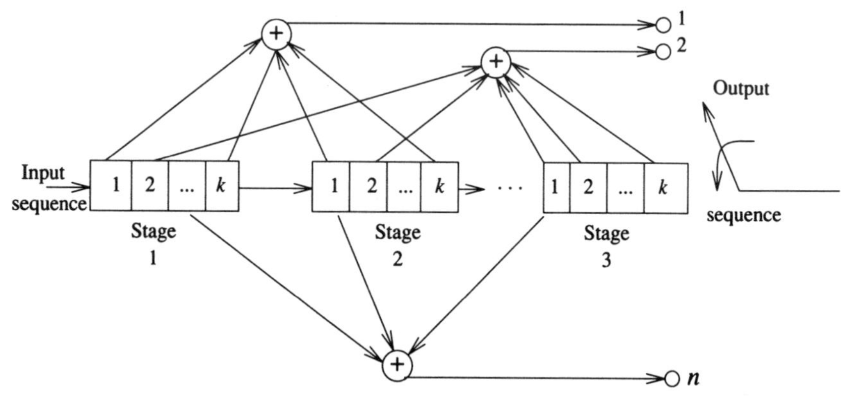

Fig. 3.15 Generic convolutional encoder.

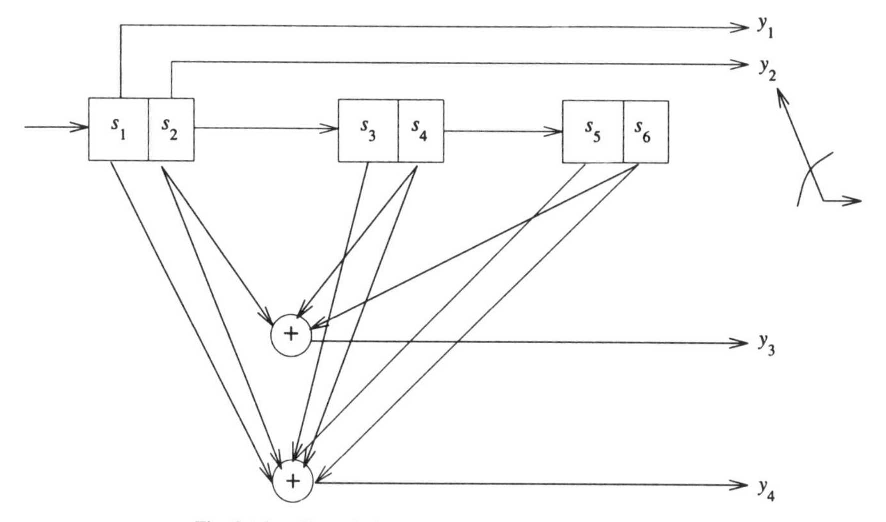

Fig. 3.16 Convolutional encoder: $k=2$, $M=3$, and $n=4$.

register and $\mathbf{y}$ is the n-dimensional output vector. For the convolutional encoder shown in Figure 3.14 the generator matrix is

$$G = \begin{bmatrix} 111 \\ 001 \\ 011 \end{bmatrix},$$

and the generator matrix for the convolutional shown in Figure 3.16 is given by

$$G = \begin{bmatrix} 1001 \\ 0111 \\ 0001 \\ 0011 \\ 0001 \\ 0011 \end{bmatrix}.$$

The generator matrix is essentially a static representation of the operation of the convolutional encoder. However, as we shall see presently, the virtue of convolutional encoding lies in its dynamics, i.e., the way in which the output process unfolds with the input process. Clearly the output of the encoder is a function of the bits stored in the shift register. Accordingly, we define the *state* of the system to be the bit sequence stored in the shift register. The state is the past input sequence. For the convolutional encoder shown in Figure 3.14, the input sequence 110100 with the initial state 000 gives the output sequence 111 110 010 100 001 011.

3.4.2 TREE AND TRELLIS REPRESENTATIONS

A convolutional code can be described by a *tree* in which branches from a node show the states and output which result from the inputs. The tree diagram for the convolutional encoder of Figure 3.14 is shown in Figure 3.17. In this case, each node has two branches. We use the convention that the upper branch gives the output for a 0 input. Each branch is labeled by the n-bit output and by the internal state of the system, in parentheses. Note that the output is determined by the input, which resides in the first stage, plus the contents of the other two stages. Given the tree for a particular encoder, the output sequence for a particular input sequence can be found by tracing the input sequence through the tree. For encoders which operate on k-bit input blocks, there are 2^k branches emanating from each node. As is evident from the tree representation, any state can be reached from any initial state after M–1 steps. Note that the output is a function only of the initial state and the current input.

The difficulty with the tree representation is the exponential growth of the size of the tree with the length of the input sequence. However, since the state

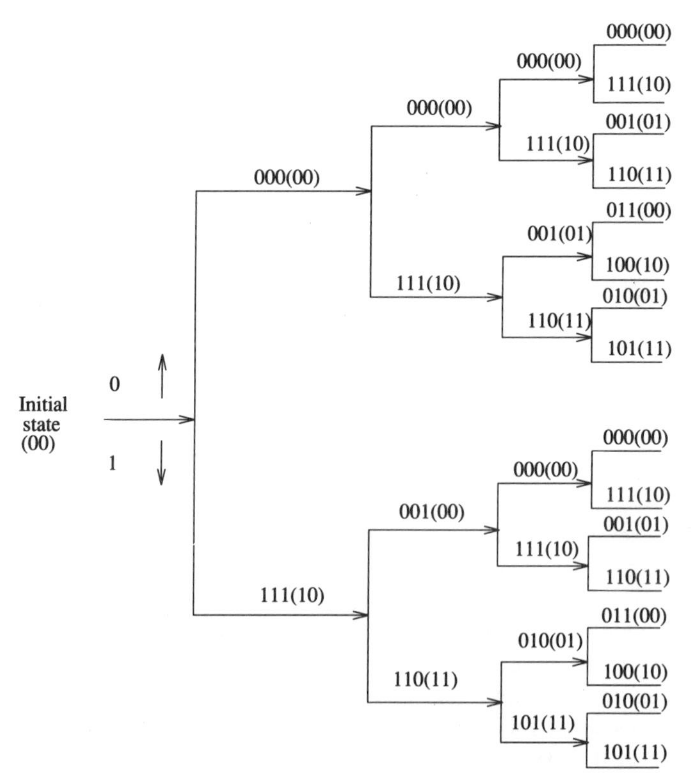

Fig. 3.17 Tree diagram for convolution encoder of Figure 3.14.

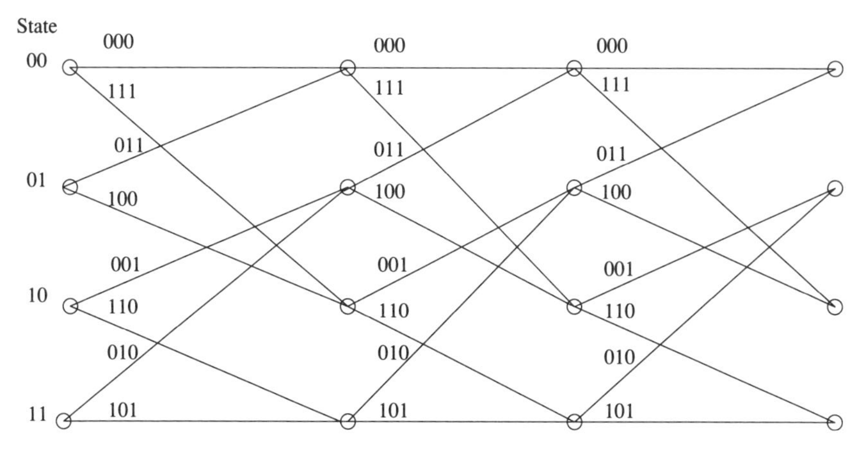

Fig. 3.18 Trellis diagram for convolution encoder of Figure 3.14.

of the system consists of only $k(M-1)$ binary digits, patterns are repeated after penetrating kM branches into the tree. This suggests an alternative representation of the convolutional encoder—the *trellis*. The trellis for the encoder of Figure 3.14 is shown in Figure 3.18. Each node in the trellis corresponds to a state and has two outgoing branches where the output is labeled on each branch. We use the convention that the upper branch indicates the output for zero input. As in the case of the tree representation, the output can be found by tracing through the trellis. Note that the number of states is 2^{M-1}, where M is the number of shift-register stages.

For the case of inputs which are k-bit blocks, each node of the trellis has 2^k outgoing branches. The startup interval allows a total of $(M-1)\cdot k$ input bits. After the startup interval, the trellis shows $2^{(M-1)\cdot k}$ levels at each stage.

3.5 DECODING CONVOLUTIONAL CODES—THE VITERBI ALGORITHM

Decoding for convolutional codes proceeds in the same general fashion as for all of the other signals that we have considered. In this case, we begin by considering the entire sequence of information. Each possible information sequence will be translated into a different transmitted sequence by the convolution encoder. At the receiver, the actual received sequence signal plus noise, is compared with the received sequence for each of the possible outputs of the encoder. The closest hypothetical sequence indicates the optimum information sequence. The power of convolutional codes lies in the fact that an input sequence is mapped into an encoded sequence which is correlated over the duration of transmission. The ability of a convolutional code to correct errors lies in this correlation. This is manifested in the increased distance between output sequences, relative to uncoded transmission. Ideally, the changing of a single input bit should result in large distances between output sequences, which implies tolerance to channel errors. As we shall see, the distance that is important is the Euclidean* as well as the Hamming distance.

The action of the convolutional encoder insures that correlation extends through the whole received sequence. Thus if one knew for certain any portion of the sequence, this could be used in the detection of the remainder of the sequence. Because of this correlation, the optimum receiver should contain a detector which detects the entire received sequence—we cannot treat separate portions independently. Suppose that a transmitted sequence contains N binary pulses. By applying the results of the previous chapter in a brute-force fashion, we obtain a receiver which contains 2^N matched filters, one for each possible received sequence. Clearly this realization is impractical since it requires exponential growth in complexity as the duration of the received sequence grows. In the next section, we shall present an alternative realization of the optimum receiver which is far easier to realize since its growth in complexity is exponential with the constraint length independent of the length (which is proportional to time) of the sequence. This realization is based on the *Viterbi algorithm* [13,14].

3.5.1 DYNAMIC PROGRAMMING

The Viterbi algorithm is a form of *dynamic programming*[15,16] the key element of which is the *principle of optimality.* Dynamic programming problems can be put in the form of finding a minimum distance path between two locations. The principle of optimality states that any portion of the overall optimum path must be the minimum distance path between the end points of that portion. Stated in this manner the principle is obvious, since if it were not true the subpath could be replaced yielding a lower overall path length, a contradiction.

The principle of optimality can be illustrated by the following, rather pedestrian, example.† A certain Professor X walks each day from his office in the Electrical Engineering and Computer Science Building to the faculty club for lunch (see Figure 3.19). Between the two buildings lie two small streams, christened by some campus wag as the Publish and the Perish. Each stream runs from north to south and is crossed by two foot bridges. In our example, these bridges shall be designated by the stream that they cross and by the appellation north or south. One day our scholarly friend decides to find the shortest path to the faculty club. Of course, he could simply calculate the lengths of all possible paths and simply choose the shortest. However, sensing that a deeper principle

---

* The Euclidean distance between signals was treated in Chapter 2; see (2.19).

† The example is taken from Reference [17], a tutorial exposition of the Viterbi algorithm.

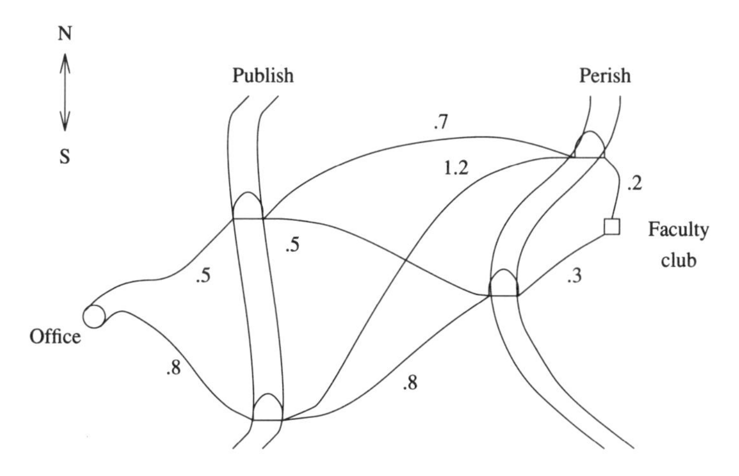

Fig. 3.19 A pedestrian example of dynamic programming.

is involved, Professor X eschews the brute force approach. He first writes down the distance between his office and Publish. He then *postulates* that the optimum path is via the north bridge across the Perish. Under this assumption, he calculates the minimum path from his office to *this bridge* by comparing the two paths over the Publish.

The same procedure is repeated for the south bridge across the Perish. At this point the professor notes that for the purpose of further calculation, he need only keep track of the *shortest* path to the north bridge on the Perish and the *shortest* path to the south bridge on the Perish. He may discard all other possible paths. In observing this simplification the good professor has utilized the principle of optimality. The optimal total path must coincide with the optimum path from his office to either the north or the south bridge across the Perish. The final step is a comparison of the total distance to the faculty club via the north and south bridges across the Perish, showing a slight advantage for the route via the south bridge on the Perish. The step-by-step procedure is shown in Table 3.5.

Table 3.5

Minimum distance from office to club	Publish	Perish	Faculty club
Via north bridge	0.5	1.2	1.4
Via south bridge	0.8	1.0	1.3

In carrying out these calculations, six additions are necessary. Brute force enumeration of all possible would require eight additions. Now, if Professor X had to cross N streams, each with two bridges, dynamic programming would require $4(N-1)+2$ additions whereas straight enumeration would require $(N+1)2^N$ additions. Note the difference between linear and exponential growth with the number of stages that is evident here.

3.5.2 THE VITERBI ALGORITHM AND HARD-DECISION DECODING

The Viterbi algorithm applies (backward) dynamic programming to decoding convolutional codes. This technique gives the maximum likelihood estimate of an entire sequence of transmitted symbols. Consider the system depicted in Figure 3.20. The output of a convolutional encoder is used to modulate a sequence of pulses. Without loss of generality, we shall take these pulses to be binary antipodal. We begin by considering a hard-decision decoder in which the received pulses are passed through a threshold detector. The sequence out of the threshold detector is then fed into the Viterbi decoder, which estimates the binary sequence which was the the original input to the convolutional encoder.

As we shall see presently, soft-decision decoding can also be applied in this case. As we discovered in Section 2.3.2, soft-decision decoding offers performance advantage since no information is discarded. However, hard-decision decoding does offer a certain simplicity of operation. We shall be carrying out the derivation of hard-decision Viterbi decoding in more detail than is really necessary; however, we wish to lay the groundwork for wider applications of the algorithm in subsequent chapters.

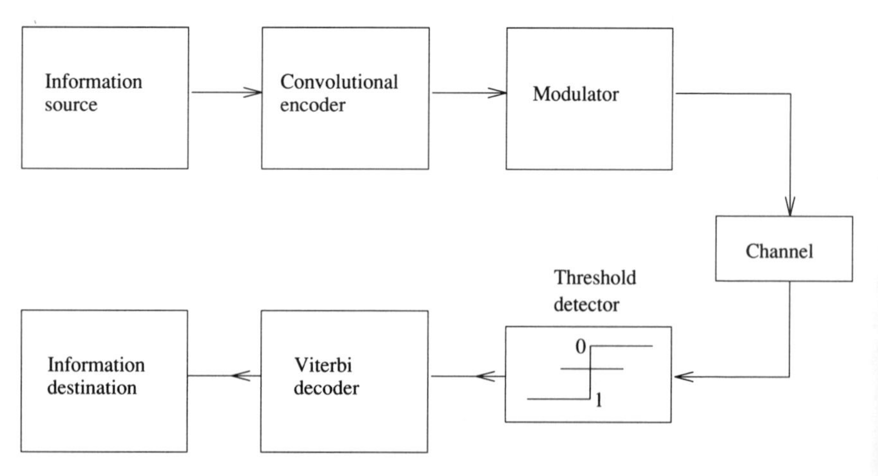

Fig. 3.20 System using hard decisions with a Viterbi decoder.

Suppose that a sequence consisting of N k-bit words is to be transmitted. We denote this sequence as $\boldsymbol{a}_1, \boldsymbol{a}_2, \ldots, \boldsymbol{a}_N$, where $\boldsymbol{a}_i = [a_{i1}, a_{i2}, \ldots, a_{ik}]$ is the ith k-bit word to be transmitted. As we have seen in our discussion of convolutional codes, the output is a function of the state of the encoder, $\boldsymbol{s}_i = \boldsymbol{a}_{i-M+1}, \boldsymbol{a}_{i-M+2}, \ldots, \boldsymbol{a}_i$, $i = 1,2, \ldots, N$. The output of both the convolutional encoder in the transmitter, and the threshold detector, which makes hard decisions, in the receiver is a sequence of N n-bit words. We denote the output of the threshold detector as, $\boldsymbol{b}_1, \boldsymbol{b}_2, \ldots, \boldsymbol{b}_N$ where $\boldsymbol{b}_i = [b_{i1}, b_{i2}, \ldots, b_{in}]$ and $n \geq k$.

Now there are 2^{kN} possible transmitted sequences. Detecting the correct transmitted sequence is, in principle, the same as the signal detection that we have studied in Chapter 2. The only difference, which is considerable from the point of view of computation, is the size of the signal space. We begin by calculating the conditional probability $Pr[\boldsymbol{b}_1, \boldsymbol{b}_2, \ldots, \boldsymbol{b}_N | \boldsymbol{a}_1, \boldsymbol{a}_2, \ldots, \boldsymbol{a}_N]$, which is simply the probability of a particular channel output sequence conditioned on the input sequence. The task of decoding is to find the input sequence which maximizes this *a posteriori* probability.

The fact that we have a binary symmetric channel considerably simplifies the derivation of a useful expression for this probability. If there were no channel errors the output of the channel for each possible input sequence would be known. If we condition on the input sequence, the difference in actual output sequence is due to channel errors. Let $D(\boldsymbol{B}^N, \boldsymbol{A}^N)$ be the *Hamming distance* between the actual received sequence $\boldsymbol{B}^N = \boldsymbol{b}_1, \boldsymbol{b}_2, \ldots, \boldsymbol{b}_N$ and the output if the transmitted sequence had been derived from, $\boldsymbol{A}^N = \boldsymbol{a}_1, \boldsymbol{a}_2, \ldots, \boldsymbol{a}_N$. The conditional probability which is of interest is given by

$$Pr[\boldsymbol{b}_1, \boldsymbol{b}_2, \ldots, \boldsymbol{b}_N / \boldsymbol{a}_1, \boldsymbol{a}_2, \ldots, \boldsymbol{a}_N] = p^{D(\boldsymbol{B}^N, \boldsymbol{A}^N)} (1-p)^{Nn - D(\boldsymbol{B}^N, \boldsymbol{A}^N)} , \tag{3.40}$$

where p is the probability of a channel error (see (2.49)). By taking the logarithm of the *a posteriori* probability, we have the objective function,

$$C = D(\boldsymbol{B}^N \boldsymbol{A}^N) \cdot \log(p/(1-p)) + Nn \cdot \log(1-p) , \tag{3.41}$$

which is to be optimized with respect to the input sequence.

Since it is reasonable to assume that $p < 1/2$, this objective function is maximized by minimizing the Hamming distance $D(\boldsymbol{B}^N, \boldsymbol{A}^N)$. Optimum sequence detection is then simply the computation of the Hamming distance between each of the possible output sequences and the actual output sequence. The problem is that there are 2^{Nk} possible output sequences and the complexity of the receiver grows *exponentially* with the length of the input sequence, an impossibility as far as implementation is concerned.

The virtue of the Viterbi algorithm is that it does the comparison with only a *linear* increase in the complexity of the receiver with the length of the input sequence. This economy is achieved through the application of dynamic programming. The first step is to decompose the total distance into a set of smaller distances each of which is a function of a single internal state of the convolutional encoder. Recall that each internal state generates an n-bit output word from a k-bit input word. These words are then compared with n-bit segments of the output of the threshold detector to generate the distances. Accordingly we may write

$$D(\boldsymbol{B}^N, \boldsymbol{A}^N) = \sum_{i=1}^{N} d(\boldsymbol{b}_i, \boldsymbol{a}_i) \ , \tag{3.42}$$

where $d(\boldsymbol{b}_i, \boldsymbol{a}_i)$ is the Hamming distance between the n-bit words $\boldsymbol{a}_i$ and $\boldsymbol{b}_i$. This segmentation allows us to apply dynamic programming. We proceed in the same manner as in the pedestrian example above. Corresponding to bridges in the example are the internal states of the encoder. In fact, the reader may wish to compare the trellis representation of the convolutional encoder, Figure 3.18, with Figure 3.19 of the pedestrian example. We attempt to find the sequence of states which minimizes the distance. This corresponds to finding an optimum path through the trellis. Again the principle of optimality applies. At each point in the sequence there are 2^{Mk} possible internal states. We proceed by assuming that the optimum path goes through each of these states, in turn. We can then compute the optimum predecessor state. This optimization procedure can be expressed as follows. From (3.42), the optimization problem can be written

$$\min_{\boldsymbol{s}_1, \boldsymbol{s}_2 \cdots \boldsymbol{s}_N} D(\boldsymbol{B}^N, \boldsymbol{A}^N) = \min_{\boldsymbol{s}_1, \boldsymbol{s}_2 \cdots \boldsymbol{s}_N} \{ D(\boldsymbol{B}^{N-1}, \boldsymbol{A}^{N-1}) + d(\boldsymbol{b}_N, \boldsymbol{s}_N) \} \ . \tag{3.43}$$

We begin breaking the minimization into two steps. First we condition on the state $\boldsymbol{s}_N$ and optimize over the sequence of states which lead to this state. We then optimize over the state $\boldsymbol{s}_N$. These steps may be written

$$\min_{\boldsymbol{s}_1, \ldots, \boldsymbol{s}_N} D \ (\boldsymbol{B}^N, \boldsymbol{A}^N) = \min_{\boldsymbol{s}_N} \ \min_{\boldsymbol{s}_1, \ldots, \boldsymbol{s}_{N-1} | \boldsymbol{s}_N} \{ D(\boldsymbol{B}^{N-1}, \boldsymbol{A}^{N-1}) + d(\boldsymbol{b}_N, \boldsymbol{a}_N) \} \tag{3.44}$$

Where $\boldsymbol{s}_1, \ldots, \boldsymbol{s}_{N-1} | \boldsymbol{s}_N$ indicates that the optimization is to be carried out over $\boldsymbol{s}_1, \ldots, \boldsymbol{s}_{N-1}$ with $\boldsymbol{s}_N$ fixed. We proceed by making the following observation

$$\min_{\boldsymbol{s}_1, \boldsymbol{s}_2, \ldots, \boldsymbol{s}_{N-1} | \boldsymbol{s}_N} \{ D(\boldsymbol{B}^{N-1}, \boldsymbol{A}^{N-1}) + d(\boldsymbol{b}_N, \boldsymbol{a}_N) \} = d(\boldsymbol{b}_N, \boldsymbol{a}_N)$$
$$+ \min_{\boldsymbol{s}_1, \boldsymbol{s}_2, \ldots, \boldsymbol{s}_{N-1} | \boldsymbol{s}_N} \{ D(\boldsymbol{B}^{N-1}, \boldsymbol{A}^{N-1}) \} \tag{3.45}$$

Equation (3.45) simply states that the distance $d(\boldsymbol{b}_N, \boldsymbol{a}_N)$ depends only on the final state, $\boldsymbol{s}_N$ and not on the predecessor states as long as we are conditioning on the final state. If the minimum distance to all of the states $\boldsymbol{s}_{N-1}$ is known, the minimum distance to a particular state $\boldsymbol{s}_N$ is found by searching over the two predecessor states. This allows decoding by iteration. Continuing to the next step, we condition on the state $\boldsymbol{s}_{N-1}$,

$$\min_{s_1,\ldots,s_{N-1}|s_N} D(\boldsymbol{B}^{N-1},\boldsymbol{A}^{N-1}) = \min_{s_{N-1}|s_N} \{d(\boldsymbol{b}_N,\boldsymbol{a}_{N-1})$$

$$+ \min_{s_1,\ldots,s_{N-2}|s_{N-1},s_N} D(\boldsymbol{B}^{N-2},\boldsymbol{A}^{N-2})\} . \quad (3.46)$$

A final observation allows the iteration to go forward. The conditioning on state $\boldsymbol{s}_N$ in (3.46) is not necessary since the distance $D(\boldsymbol{B}^{N-2},\boldsymbol{A}^{N-2)}$ does not depend on this state. Thus (3.46) simplifies to

$$\min_{s_{N-2}|s_{N-1},s_N} D(\boldsymbol{B}^{N-2},\boldsymbol{A}^{N-2}) = \min_{s_{N-2}|s_{N-1}} D(\boldsymbol{B}^{N-2},\boldsymbol{A}^{N-2}) \quad (3.47)$$

Combining (3.46) and (3.47) gives the iterations that are the heart of the Viterbi algorithm:

$$\min_{s_1,\ldots,s_{i-1}|s_i} \{D(\boldsymbol{B}^{i+1},\boldsymbol{A}^{i+1})\} = \min_{s_{i-1}\,|\,s_i} \{d(\boldsymbol{b}\cdot\boldsymbol{a}_{i-1})$$

$$+ \min_{s_1,s_1\ldots,s_{i-1}|s_{i-1}} D(\boldsymbol{B}^{i-2},\boldsymbol{A}^{i-2})\} . \quad (3.48)$$

Finding the optimal transmitted path can be seen as finding a minimum distance path through a maze. In this case the maze can be thought of as the trellis diagram for the convolutional code (see Figure 3.18). The basic step is embodied in (3.48). Suppose that the optimum path goes through a *particular state* $\boldsymbol{s}_i$. We search over all of the predecessor states, $\boldsymbol{s}_{i-1}$ to find the minimum distance to $\boldsymbol{s}_i$. This minimum distance is composed of an incremental distance $d(\boldsymbol{b},\boldsymbol{a}_{i-1})$ and the minimum distance to the predecessor state

$$\min_{s_1\cdots s_{i-1}\,|\,s_{i-1}} D(\boldsymbol{B}^{i-2},\boldsymbol{A}^{i-2}) .$$

The incremental distance is the Hamming distance between the output of the convolutional encoder and the received signal. (There is an obvious analogy between the states and the bridges across the streams in the pedestrian example.)

We summarize the operation of the Viterbi algorithm by means of an example. As we have seen, when the sequence 110100 is fed into the rate 1/3 convolutional encoder shown in Figure 3.14 with initial state 00, the output sequence is 111 110 010 100 001 011. This sequence is used to modulate binary-antipodal pulses which are, in turn corrupted by noise in the channel. We assume that the sampled output of the matched filter is the sequence 1.5,1.3,0.8, 0.9,1.2,–1.1, –0.9,1.5,0.08, 1.2,–0.75,–0.85, –1.2,–1.03,0.85, –.85,1.1,1.2. The output of the threshold detector is: 111 110 011 100 001 011. Note that there is one error in this sequence.

The operation of the Viterbi algorithm on the output of the hard limiter is shown in Figure 3.21. The minimum distances are shown in boldface next to the states. The incremental distances between states are as shown on the arcs between the states. To begin the iteration, we assume that the system is in the zero state; accordingly, the distance is zero for this state and infinity for the oth-

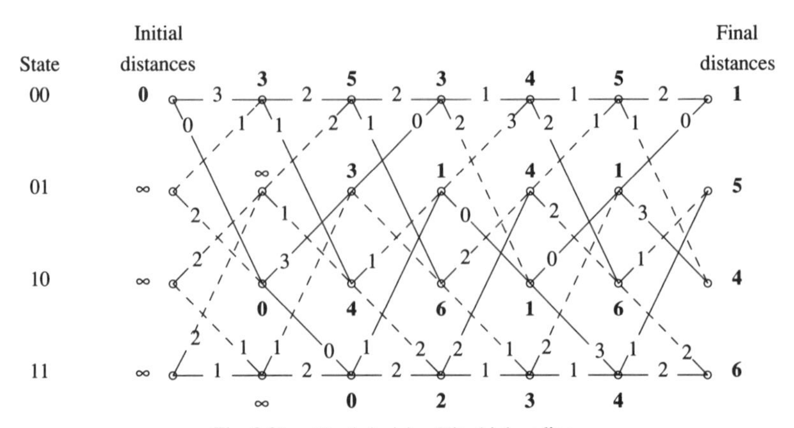

Fig. 3.21 Hard-decision Viterbi decoding.

ers. The next step is to compute the incremental distance between the actual received sequence 111 and the outputs of the various states for zero or one input. For example, with 0 into state 0 the output is 000 (see Figure 3.18) which is distance 3 from 111. The next step is to compute the minimum distances to each of the second states. For each of the states in this example there are two possible predecessor states. The minimum distance predecessor is shown as a solid line. The process continues in this fashion. As an example, consider the sixth step. Assume the optimum path goes through state 01. The two possible predecessors are 10 and 11. The received sequence for this stage is 011. The outputs of 10 and 11 that lead to 11 are 110 and 101, respectively, (see Figure 3.18) and the incremental distance is two, in both cases. The minimum distance to state 11 in this stage is 6, as shown in Figure 3.21. The distances to the final states are compared to find the minimum. In this case, it is 00 with distance 1. The sequence of states that led to this final state are 00 10 11 01 10 01 00. The corresponding output sequence is 111 110 010 100 001 011, and the errors are corrected.

Examination of Figure 3.21 discloses that it is not necessary to wait until the final state has been reached before making a decision. All of the paths after step 6 go through state 10 on step 4—this is called a *merge*; consequently a decision can be made on the first four information bits. The probability that a *merge* does not occur in a sequence decreases exponentially with its length. However, in a practical system it is not feasible to wait too long without making a decision. Therefore, merges are forced, so to speak. Periodically one compares the distances to the set of states at a certain number of bit intervals in the past. The sequence leading to the minimum distance state is taken to be the transmitted sequence; all others are discarded.

3.5.3 THE VITERBI ALGORITHM AND SOFT–DECISION DECODING

In this section we consider soft-decision decoding of a convolutionally encoded sequence. As stated above, the output of the convolutional encoder modulates a sequence of antipodal pulses. We assume that a zero is transmitted as a negative pulse and a one as a positive. The amplitudes of the pulses may then be written $A_{ij} = 2a_{ij} - 1$, and the transmitted sequence can be written as

$$x(t) = \sum_{i=1}^{N} \sum_{j=1}^{n} A_{ij}\, p(t - [i(n-1) + j]T) \ , \tag{3.49}$$

where N is the number of k bit information blocks to be transmitted and n is the number of output bits for each block. Furthermore, $p(t)$ is the pulse whose amplitude is modulated and T is the interval between successive pulses. We assume that the channel introduces no distortion so that its output is

$$y(t) = \sum_{i=1}^{N} \sum_{j=1}^{n} A_{ij} p(t - [i(n-1) + j-1]T) + n(t) \ , \tag{3.50}$$

where $n(t)$ is the Gaussian noise in the channel.

The channel output in (3.50) has the same form that we have considered in the previous chapter, a signal plus noise. The only difference is in terms of quantity since the signal, $\sum_{i=1}^{N} \sum_{j=1}^{n} A_{ij} p(t - [i(n-1) + j-1]T)$, may assume a large number of realizations. Each one of these realizations represents a different sequence $\boldsymbol{a}_1, \boldsymbol{a}_2, \ldots, \boldsymbol{a}_N$. However, the basic problem is the same as that considered previously (see Section 2.3). We decide that the transmitted signal is that which is minimum distance from the received signal. Of the 2^N possible transmitted sequences, we find the transmitted sequence $\boldsymbol{a}_i$, $i = 1,2,\ldots,N$, for which

$$\int_0^{NT} \{y(t) - \sum_{i=1}^{N} \sum_{j=1}^{n} A_{ij}\, p(t-[n(i-1) + j-i]T)\}^2 dt$$

is minimum. This is equivalent to finding the transmitted sequence which minimizes

$$\int_0^{NT} \{-2 \sum_{i=1}^{N} \sum_{j=1}^{n} A_{ij}\, y(t)\, p([t - [n(i-1) + j-1]])T + \\ [\sum_{i=1}^{N} \sum_{j=1}^{n} A_{ij}\, p(t-[n(i-1) + j-1]T)]^2\} dt \, . \tag{3.51}$$

If we assume that there is no memory in the channel, i.e., $\int_0^T p(t-iT)$ $p(t - jT)\, dt = 0$ for $i \neq j$, then (3.51) becomes

$$D(\boldsymbol{B}^N, \boldsymbol{A}^N) = -2 \sum_{i=1}^{N} \sum_{j=1}^{n} A_{ij} z_{in+j} + E_p \sum_{i=1}^{N} \sum_{j=1}^{n} A_{ij}^2 \ , \tag{3.52}$$

where $z_{in+j} = \int_0^T y(t)p(t - [n(i-1)+j-1]T)\,dt$ and E_p is the energy in a pulse $E_p = \int_0^T p(t)^2\,dt$. Notice that the sequence z_{in+j}, $i = 1,2,\ldots,N$; $j + 1,2,\ldots,n$ represents the output samples of a filter matched to the pulse $p(t)$.

The optimization now proceeds in much the same way as in the hard-decision case. We want to find the sequence of $\boldsymbol{a}_i$, $i = 1,2,\ldots,N$, for which $D(\boldsymbol{B}^N,\boldsymbol{A}^N)$ in (3.52) is minimum. In the case of binary-antipodal signaling further simplification of (3.52) is possible since $A_{ij}^2 = 1$. We can therefore simplify the cost function by eliminating the second term on the right-hand side of (3.52), and removing the factor of 2 so that

$$D(\boldsymbol{A}^N,\boldsymbol{B}^N) = \sum_{i=1}^{N}\sum_{j=1}^{n} -A_{ij}z_{in+j}\,. \tag{3.53}$$

Note that the A_{ij}, $i = 1,2,\ldots,N$, $j = 1,2,\ldots,n$, are uniquely specified by the sequence of internal states; $\boldsymbol{s}_1,\ldots,\boldsymbol{s}_N$. In both (3.42) and (3.53) we have the sum of a sequence of terms which are governed by the transitions of an underlying state. Equations (3.42) to (3.48) apply directly with $d(z^i,\boldsymbol{s}^i)$ replaced by $-A_{ij}z_{in+j}$, $i = 1,2,\ldots,N$.

In order to illustrate the operation of the Viterbi algorithm in the case of soft-decision decoding we consider the following example. The seven-bit sequence 11010 is fed into the convolutional encoder shown in Figure 3.14. The output sequence is 111 110 010 100 001 011. This sequence modulates a binary antipodal sequence and is corrupted by noise resulting in the sequence 1.5,1.3,0.8,0.9,1.2,–1.1,–0.9,1.5,0.08,1.2,–0.75,–0.85–1.2,–1.03,0.85–0.85,1.1,1.2. This is the same received sequence that we have seen previously in connection with hard decision decoding. This sequence is processed by a Viterbi decoder to produce the detected sequence. This trellis diagram representation of this is shown in Figure 3.22. In order to simplify the diagram only the lowest distance path into each state is shown in Figure 3.22. Again, we start with the initial state 0. The distance 0 is assigned to this state and ∞ to the others. If the input to the convolutional encode is 0 or 1, the outputs are 000 and 111, respectively, which are transmitted as pulses with amplitudes $A_{ij} = 1$, $j = 1,2,3$ or $A_{ij} = -1$, $j = 1,2,3$. The incremented distances as measured by (3.53) are 3.6 and –3.6, respectively, and are the minimum distance to the states in the first stage. The process continues in this fashion. Consider, for example, the calculation for the minimum distance to state 00 in stage 5. The outputs of the matched filter are –1.2, –1.03, 0.85. The possible predecessor states are 00 and 01 with outputs 000 and 011, respectively. The incremental distances implied by these outputs are –1.38 and –1.02, respectively. (Note that distances are Euclidean implying real numbers.) Adding the incremental distances to the minimum distance to the predecessor states, we find that the minimum distance to the state 00 is –7.7. At the end, the distances to the final states are compared and the minimum length sequence chosen. It turns out that the detected sequence is the

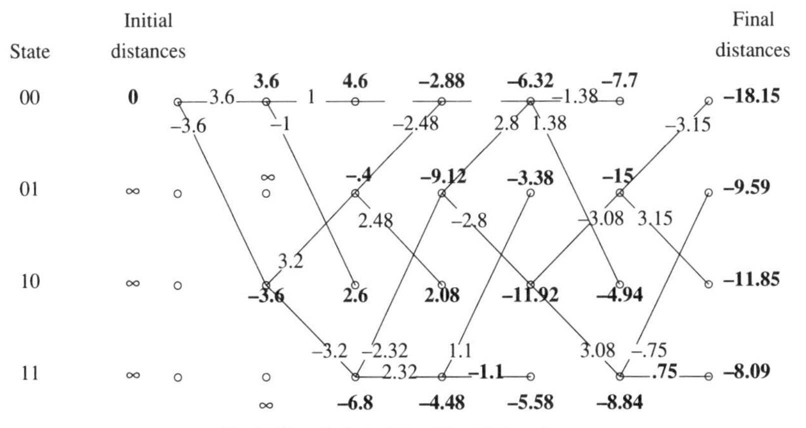

Fig. 3.22 Soft-decision Viterbi decoding.

same as in the hard-decision case. This would not be the case in general. As we shall see presently, soft-decision decoding enjoys an advantage overall and we would expect significant differences in systems with lower signal-to-noise ratios.

3.6 PERFORMANCE OF CONVOLUTIONAL CODES

The performance of convolutional codes for both hard and soft decisions depends upon their distance properties. These properties can be found by means of the third way of characterizing convolutional codes—the *state diagram*. This diagram shows the transition from one state to another as a function of the input. The state diagram for the rate 1/3 convolutional encoder shown in Figure 3.14 is shown in Figure 3.23. In this diagram we follow the convention that a transition caused by a one is indicated by a solid line while a zero transition is indicated by a dotted line. In this case there are two output words from each state. Also

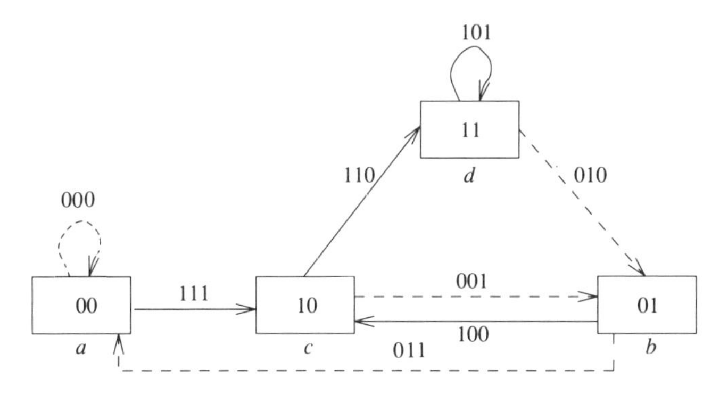

Fig. 3.23 State diagram for the convolutional encoder of Figure 3.14.

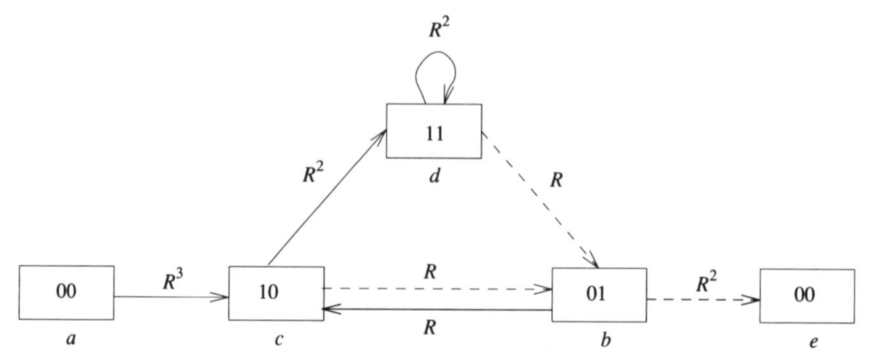

Fig. 3.24 Calculation of the transfer function for convolutional codes.

shown are the n-bit output words which result from the input bit and the internal state.

Since the convolutional code is linear, its distance properties for all sequences can be found by calculating the distances to the all-zero sequence. To calculate the distance we consider paths which diverge from the all-zero sequence and remerge at a later time. This process is indicated by the modified state transition diagram shown in Figure 3.24 in which the initial and final all-zero states are indicated as a and e, respectively. The other states are designated b, c, and d for 01, 10, and 11, respectively. The path information is indicated by the *transfer function* between these two states. Each of the arcs of the graph indicates a particular output. We weight each arc by R^i where R is a dummy variable and where i is the Hamming distance of the output from the all-zero sequence. The distance properties of the code are summarized by equations for the distances between adjacent nodes. For example, the transition into state $b(01)$ is via $c(10)$ or $d(11)$. In either case the output is distance one from the all-zero sequence. We represent the transition to state b from c or d by the equation

$$X_b = RX_c + RX_d \ , \tag{3.54a}$$

The transitions to the other states may be written

$$X_c = R^3 X_a + RX_b \tag{3.54b}$$

$$X_d = R^2 X_c + R^2 X_d \tag{3.54c}$$

$$X_e = R^2 X_b \ . \tag{3.54d}$$

Note, for example, in going from state $a(00)$ to $c(10)$ the output is 111 which is Hamming distance 3 from 000. This is indicated by the term $R^3 X_a$ in (3.54b). The transfer function is defined as the transition from state a to e, $T(R) = X_e / X_a$. This indicates the total distance of the output from the all-zero

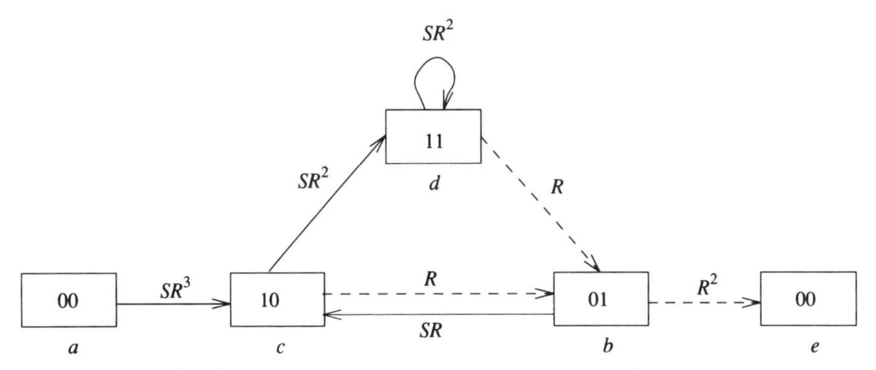

Fig. 3.25 Calculation of bit error rate using the transfer function of convolutional codes.

sequence. From (3.54), we have

$$T(R) = R^6/(1-2R^2) = R^6 + 2R^8 + 4R^{10} + \cdots = \sum_{d=6}^{\infty} A_d R^d , \qquad (3.55)$$

where A_d indicates the number of paths of distance d from the all zero state. As we shall see presently, $T(R)$ enables us to bound the sequence error.

A further embellishment on the state diagram, which enables us to bound bit errors, is counting the number of input ones which cause a transition. The variable S is factored into a transition if there is a one in the input. The state transition diagram is shown in Figure 3.25 and the equations for transition between adjacent states are

$$X_b = RX_c + RX_d \qquad (3.56a)$$

$$X_c = SR^3X_a + SRX_b \qquad (3.56b)$$

$$X_d = SR^2X_c + SR^2X_d \qquad (3.56c)$$

$$X_e = R^2X_b , \qquad (3.56d)$$

and solving for $T(S,R) = X_e/X_a$, we find that

$$T(S,R) = SR^6/(1-2SR^2) = SR^6 + 2S^2R^8 + 4S^3R^{10} + \cdots \qquad (3.57)$$

The exponents of S represent the number of information bits that are in error when a path is incorrectly chosen over the all-zero path. By differentiating $T(S,R)$ with respect to S and setting S=1 gives the number of errors in a path which is a given distance from the all-zero path. For example, for the rate 1/3 code we have from (3.57) $\partial T(S,R)/\partial S|_{S=1} = R^6 + 4R^8 + 12R^{10} + \cdots$. Thus, for a path distance eight from the all-zero sequence, there are four bit errors.

3.6.1 HARD-DECISION DECODING

As in the case of block codes, we shall begin by computing the probability of pairwise errors between paths. Suppose that the all-zero path is correct. Consider a particular erroneous path which deviates from and eventually returns to the all-zero path. Suppose that the distance to this erroneous path is D. If D is an odd number, then the probability of error is

$$P_2(D) = \sum_{k=(D+1)/2}^{D} \binom{D}{k} p^k (1-p)^{D-k}, \tag{3.58a}$$

where p is the probability that a single bit is in error. If the distance is an even number, the expression changes slightly in order to take ties into account. We have

$$P_2(D) = \sum_{k=\frac{D}{2}+1}^{D} \binom{D}{k} p^k (1-p)^{D-k} + 1/2 \binom{D}{D/2} p^{D/2} (1-p)^{D/2}. \tag{3.58b}$$

Applying the union bound yields the following bound for the symbol error

$$P_{es} \le \sum_{d=D_{\min}} A_d \, P_2(d). \tag{3.59}$$

Recall that A_d is the number of paths distance d from the all-zero path and D_{min} is the smallest possible divergence from the all-zero path.

We can also bound the probability of bit error. Suppose we differentiate $T(S,R)$ with respect to S and set $S=1$. The coefficients of the resulting polynomial in R are the number of bit errors corresponding to a given path distance. We designate these coefficients as B_d and write

$$P_{eb} \le \sum_{d=D_{\min}} B_d P_2(d). \tag{3.60}$$

This bound is plotted in Figure 3.26 for the rate 1/3 encoder shown in Figure 3.14.

3.6.2 PERFORMANCE OF SOFT-DECISION DECODING

In the case of soft-decision decoding for binary-antipodal signals, the metric is given by (3.53). The difference in the distance between different paths is due to terms of the form $r_{ij} = -A_{ij} \, z_{in+j}$ which are independent identically distributed Gaussian random variables. If A_{ij} matches the transmitted pulse amplitude, the mean of this random variable is $-E_p$; otherwise it is E_p. In either case the variance is the variance of the noise, $E_p N_0/2$. If the distance between a path and the all-zero path is D binary digits, the difference in the mean values of the paths is $2DE_p$. Thus an error is made if the noise is greater than DE_p. Since the D noise samples have variance DE we have

$$P_2(D) = \int_{DE_p}^{\infty} \exp(-x^2/N_0E_p) \Big/ \sqrt{\pi N_0 E_p}\, dx = Q([2DE_p/N_0]^{1/2}) \; . \tag{3.61}$$

This probability of error is then applied to the union bound to produce the symbol error rate

$$P_{es} \leq \sum_{d=D_{\min}}^{\infty} A_d Q([2dE_p/N_0]^{1/2}) \; , \tag{3.62}$$

where $D_{\min}$ is the the same quantity that was defined in conjunction with (3.59). As in (3.59), the summation is over all path distances. This can be simplified by applying the exponential bound of (2.34b) to the function $Q(\cdot)$. Using (3.55) we can write the symbol error probability as

$$P_{es} \leq 1/2 \sum_{d=D_{\min}}^{\infty} A_d \exp(-2dE_p/N_0) = T(2E_p/N_0) \; . \tag{3.63}$$

As in the case of hard-decision decoding, we can also bound the probability of bit error as well as the probability of sequence error. The expression is the same as in (3.60) with the difference that $P_2(D)$ given by (3.61). Again the coefficients B_d are found by differentiating $T(B,D)$, giving

$$P_{eb} \leq \sum_{d=D_{\min}}^{\infty} B_d \; Q([2dE/N_0]^{1/2}) \; . \tag{3.64}$$

By substituting the bound of (2.34b) into (3.64) we find

$$P_{eb} \leq \sum_{d=D_{\min}}^{\infty} B_d \exp(-2Ed/N_0)/2 = \left. \frac{\partial T(S,R)}{\partial S} \right|_{\substack{S = 1 \\ R = e^{-2E/N_0}}} \tag{3.65}$$

The bound for soft-decision decoding for the circuit of Figure 3.14 is shown in Figure 3.26.

As we see from (3.59), (3.60), and (3.62) through (3.65), the bounds on performance depend upon the distance properties of the convolutional codes. For the high signal-to-noise ratios of most operational systems, the crucial parameter is the quantity $D_{\min}$. By means of computer search, convolutional codes having optimum distance properties have been found [18]–[21].*

3.7 SEQUENTIAL DECODING OF CONVOLUTIONAL CODES

It is evident from the foregoing that the Viterbi algorithm requires a fixed amount of computation irrespective of the noise level in the channel; moreover this level of computation grows exponentially with the constraint length of the

---

* For a tabulation see Chapter 5 of reference [1].

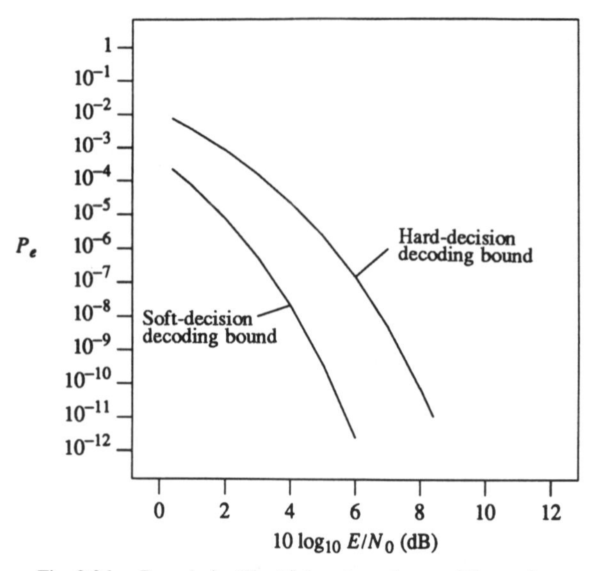

Fig. 3.26 Bounds for Viterbi decoding: for rate 1/3 encoder.

code. This is a disadvantage which is not present in an alternative approach, *sequential decoding*. In this case the computational effort is a function of the uncertainty introduced by the channel. The most common version of sequential decoding is the algorithm due to Fano [22]. In this algorithm, a single most likely path is traced through the tree describing the convolutional code (see Figure 3.17). At each node a metric, the likelihood tempered by the code rate, is computed. If the metric is above a certain threshold, the path continues; otherwise, the path is retraced and an alternative sought. If no better alternative path is found, the original path is continued at a lowered threshold. This process continues until the terminal node of the tree.

An alternative to the Fano algorithm which is faster but which requires more storage is the stack algorithm [23,24]. In this case, several of the most-likely paths are stored, in decreasing order of likelihood, in a stack. The continuations of the current most likely paths are computed and the stack is updated with reordering if necessary. This continues until all of the received data has been treated. At the end, the most likely path is accepted as the decoded path.

In general, the difficulty with sequential-decoding algorithms is the highly variable nature of the computations that are involved. A burst of noise could cause a burst of computations along with buffer overflow. Further, a variable delay in delivering information bits is incurred. It has been shown that, for typical sequential-decoding algorithms, the variance of the number of decoding operations per bit could be infinite in certain cases. However, newer techniques, which may be viewed as hybrids of the Viterbi algorithm and sequential-decoding algorithms, may provide the best of both techniques, thereby reducing computer complexity [25].

3.8 BLOCK AND CONVOLUTIONAL CODES CONCATENATED

The question naturally arises as to the relative merits of block and convolutional codes. The comparison is difficult, indeed controversial [7,8]. Convolutional codes do not seem to have the mathematical depth of block codes, a depth that is, in its way, beautiful.* This may be in a state of flux because of development of *trellis codes* which may be viewed as an extension of convolutional codes. (These codes will be discussed in Chapter 5.) The situation is further clouded by the rapid advance of the technology that is employed to implement codes.

This caveat notwithstanding, it is possible to make general statements about these classes of codes as they are currently deployed. Convolutional codes tend to work best against randomly occurring errors. Codes with short constraint lengths, together with a Viterbi decoder, are quite effective against this kind of disturbance. The advantage of convolutional codes lies in the ability of the Viterbi decoder to make soft decisions in an entirely natural manner. As we have seen, this advantage translates into a signal-to-noise ratio advantage of approximately 2 dB or a decrease of two orders of magnitude in the probability of error.

A problem with convolutional codes is that errors tend to occur in bursts; that is, if an error occurs another is likely. This can be countered simply by increasing the constraint length of the code, thereby introducing memory into the system over a longer span. However, as we have seen, the complexity of the Viterbi decoder increases exponentially with an increase in the constraint length. Sequential decoding suffers buffer overflow and variable delay in dealing with this kind of impairment. As we mentioned earlier in the text, bursts of errors can be dealt with by interleaving. A burst of channel errors is redistributed to appear as random errors after deinterleaving, and can be handled by conventional coding/decoding techniques. The obvious drawback is increased delay and storage requirements.

Reed–Solomon codes, which were considered in Section 3.2.1, are block codes that have a natural immunity to burst errors since they operate at the symbol level. Since the number of symbols in a codeword can be made quite large, bursts occurring within a long codeword can be corrected. To illustrate, if the symbols are eight-bit bytes ($m = 8$, see Section 3.2.1), the block length is $n = 2^8 - 1 = 255$ bytes. If the number of parity symbols per codeword is 4, the

---

* Indeed, the wonders of block codes have inspired one of the authors to write a poem on the subject. [26]

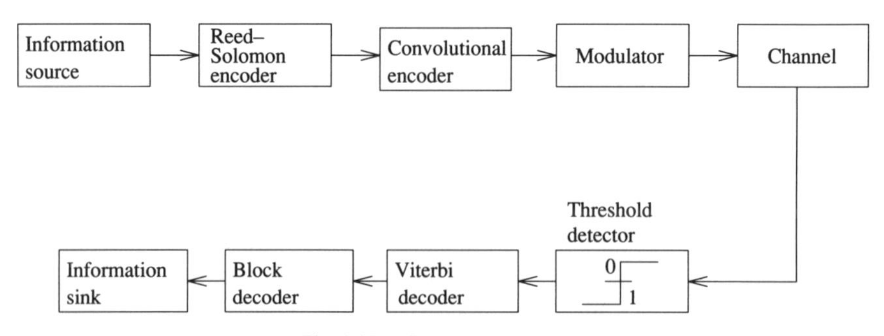

Fig. 3.27 Concatenated coding.

minimum distance between codewords is 5 and thus two symbol errors can be corrected. In general, it is difficult to gain the advantage of soft-decision decoding for block codes when n and k are large because of implementation difficulties. However, the mathematical elegance of block codes allow relatively easy implementation of hard-decision decoding.

The strengths of both block and convolutional codes can be best deployed by the *concatenated coding* scheme shown in Figure 3.27. The outer-level is a Reed–Solomon encoder which encodes blocks consisting of m bits. The resulting binary data stream is then encoded by a convolutional encoder. At the receiver, a Viterbi decoder using soft decisions unravels the convolutional code. This is followed by an algebraic decoder for the Reed–Solomon code. The idea is that soft-decision decoding in the inner level reduces uncorrelated errors while bursts of error are handled by the outer-level Reed–Solomon code.

3.9 AUTOMATIC REPEAT-REQUEST SYSTEMS

As we stated in the introduction, codes can be used for either error detection in automatic repeat-request (ARQ) systems for for error correction in forward error correction (FEC) schemes. As we shall see there is also the possibility of hybrid techniques which combine correction and detection. The appropriateness each of these techniques, ARQ, FEC, or hybrid, depends upon the particular application at hand. As we shall see in the sequel, round trip delay has an important effect on the throughput in ARQ systems. Accordingly, on terrestrial links where round trip delay is small, ARQ is widely applied. FEC which is unaffected by delay is appropriate to satellite communications systems or other systems that are propagation delay limited.

The operation of ARQ systems is depicted in general terms in Figure 3.28. Data is transmitted in the form of fixed-length blocks containing redundant bits for error detection. The general mode of operation for ARQ systems suggests that block codes are more appropriate than convolutional codes in this application. The basic premise of ARQ systems is the presence of a feedback channel

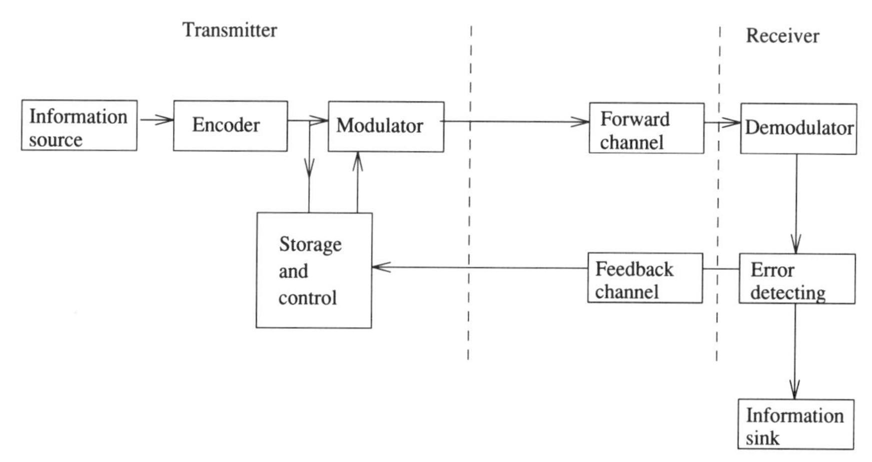

Fig. 3.28 ARQ system.

from the receiver to the transmitter. Over this channel the receiver transmits positive or negative acknowledgments, designated ACK and NACK, respectively, according to the condition of a received block of data. An ACK indicates no detected error and a NACK the opposite. The transmitter responds to an unsuccessful transmission by retransmitting the block.

Within the basic framework there are three varieties of ARQ systems whose names are descriptive of the techniques: 1) *stop-and-wait* (SAW) 2) *go-back-N* (GBN), and 3) *selective repeat* (SR). After the transmission of a message in the stop-and-wait scheme, the transmitter waits for an ACK before discarding a transmitted block. If either a NACK is received or a fixed timeout interval has elapsed, the block is retransmitted. This procedure is repeated until an ACK is received within the time-out interval. The drawback to this technique lies in the fact that the transmitter is idle while waiting for a message acknowledgment. As we shall see in the sequel, in systems with round trip delay comparable to the message transmission time, the stop-and-wait protocol suffers from low throughput. Nevertheless, stop-and-wait has the virtue of simplicity; consequently, it was the first to be implemented in widely available systems. The bisync protocol discussed in Section 1.2 uses a form of stop-and-wait.

In the go-back-N protocol the transmitter is in continuous operation. Again the transmitter saves all unacknowledged blocks. Now, if an error is detected at the receiver, the block which is in error is specified in the NACK and this block, along with all subsequent blocks of data, is retransmitted. The transmitter and the receiver have only a modest increase in complexity. For example, the receiver has the same storage requirements since all blocks after an erroneous block can be discarded. The obvious penalty is the repeated transmission of correct blocks.

In the final ARQ technique to be considered, selective repeat, the transmitter continuously sends blocks of data which are held until acknowledged by the receiver. The NACK specifies the block that is in error and only this block is retransmitted. Selective repeat improves performance since only blocks that had been in error are retransmitted. The price of this improvement is increased complexity in the receiver. Since data must be delivered in the proper sequence, correct blocks must be stored until all preceding erroneous blocks have been correctly received.

3.9.1 PERFORMANCE

As we have seen in Section 3.1.2, block codes can detect errors more easily than they correct them. Therefore, in situations where ARQ is appropriate, relatively few redundant bits need to be transmitted in order to achieve the same performance as FEC. An analysis of performance of the varieties of ARQ furnishes guidance on the utility of ARQ.

When an n-bit block of data is transmitted over a noisy channel three mutually exclusive events may ensue — correct reception, undetected errors, or detected errors. The probability of correct reception is the probability

$$P(C) = (1 - p)^n \quad , \tag{3.66}$$

where p is the probability of a bit error.

As we have seen in Section 3.15, the probability of an undetected error corresponds to a pattern of channel errors which match a codeword and is given by (3.13). The probability of a detected error on each transmission is $P(DE) = 1 - P(UE) - P(C)$, where $P(UE)$ and $P(C)$ are respectively the probabilities of undetected error and correct transmission on a single attempt. Since ACKs and NACKs are short, hence easily protected, it is reasonable to assume that no errors are made in the feedback channel and $P(DE)$ is the probability of retransmitting a block. If errors are made, independently from block to block, the probability of k transmissions (initial + retransmissions) of a block are

$$Pr\ (k\ transmissions) = (1 - P(DE))\ P(DE)^{k-1},\ k = 1,2,... \tag{3.67a}$$

From (3.67a) the average number of transmissions is given by

$$TR = E[No.\ transmissions] = 1/(1 - P(DE)) = 1/(P(C) + P(UE))\ . \tag{3.67b}$$

The distribution here is geometric implying the possibility of an unbounded number of retransmissions. Certainly, a higher-level protocol would intervene at some point.

As (3.67a) implies, the transmission of a block terminates when no error is detected. As we have seen, there may be an error which may not be detected.

The probability of this event is

$$Pr(Undetected\ error) =$$
$$\sum_{k=1}^{\infty} P(UE)\ P(DE)^{k-1} = P(UE)\ /\ (1-P(DE)) =$$
$$P(UE)/(P(C) + P(UE)) \leq \frac{P(UE)}{P(C)} . \tag{3.68}$$

Notice the distinction between an undetected error on a single transmission, with probability P(*UE*) and the probability of accepting a block which contains an undetected error after repeated transmissions.

If the proper codes are used, the probability of an undetected error is negligible for the range of probability of bit errors which are likely to be encountered, $p < 10^{-3}$. For example, suppose that the (23,12) Golay code is used and the probability of bit error is $p = 10^{-3}$. From (3.13) and Table 3.4, we have $P(UE) \cong 2.5\times10^{-19}$. The resulting probability of an undetected error is, from (3.66) and (3.68), *Pr* (*Undetected error*) $\cong 2.5\times10^{-19}$. This contrasts with FEC for the same code and the same probability of bit error. Since the Golay (23,12) code corrects up to three bit errors, four or more errors will not be corrected. We have (see (2.49))

$$Pr(Undetected\ error) = \sum_{i=4}^{23} \begin{pmatrix} 23 \\ i \end{pmatrix} p^i\ (1-p)^{23-i} = 8.7\times10^{-9} . \tag{3.69}$$

As we have indicated in the beginning of this section, the potential problem with ARQ techniques is reduced throughput. For each of the techniques that we have described above, we shall calculate throughput as a function of the system and code parameters. In addition to *p*, the probability of a bit error, the system parameters that are relevant are *DEL*, the round trip delay of the system, and *R*, the rate at which bits are transmitted in bits per second. For the stop-and-wait (SAW) system the probability distribution for the number of transmissions is given by (3.67a). Since each transmission lasts $n/R + DEL$ seconds the average time that the channel is engaged in transmitting a *n*-bit block is, from (3.67b)

$$T_{SAW} = (n/R + DEL)\ TR = (n/R + DEL)\ /\ (P(C) + P(UE)) \tag{3.70}$$

Since each *n*-bit codeword contains *k* information bits, the rate of transmitting correct bits is

$$TP_{SAW} = k/T_{SAW} \text{ bps}$$
$$= (k/n)\ R\ (P(C) + P(UE))/(1+W) , \tag{3.71}$$

where *W* is the number of codewords that can be transmitted in a round-trip delay time, *DEL*. Note that, for this case as well as the succeeding, the delivered bits may contain errors. The rate of correct bits may be found by use of (3.68).

In the case of go-back-N (GBN), each retransmission involves w blocks when a transmission error is detected. The usual case is for $W = R\ DEL/n$ as in stop-and-wait. In contrast to the stop-and-wait protocol, the channel is always occupied. If there are no errors, the time to transmit a block is n/R. The time to transmit an n-bit block, T_{BGN}, is

$$T_{GBN} = n/R + DEL \sum_{k=1}^{\infty} (k-1)\ Pr\ [k\ retransmissions]$$

$$= n/R + DEL\ P(DE)/(P(C) + P(UE))\ . \qquad (3.72)$$

The throughput is

$$TP_{GBN} = k/T_{GBN}\ \text{bps}$$

$$= \frac{k}{n} R(P(C) + P(UE))/(1 + (W-1)P(DE))\ . \qquad (3.73)$$

In the case of selective repeat (SR), only one block need be retransmitted when an error is detected. The time required for this is

$$T_{SR} = (n/R)\ /\ (1 - P(DE)) = (n/R)/(P(C) + P(UE))\ , \qquad (3.74)$$

with the resulting throughput

$$TP_{SR} = k/T_{SR} = (kR/n)(P(C) + P(UE))\ \text{bps}\ . \qquad (3.75)$$

The three ARQ techniques are compared in Figure 3.29 for the (23,12) Golay code. The throughput normalized to the transmission rate, R, is shown as a function of the probability of bit error for several values of W. Note that for $W = 1$, selective repeat and go-back-N have the same performance. Further, the performance of selective repeat is insensitive to W.

Hybrid techniques employing both error correction and error detection have proved useful in system having long delay. There are two levels of sophistication in these hybrid coding techniques. In situations where the noise level is constant with time, what is called the Type I hybrid is appropriate. In this case enough redundancy is put into the code to correct the great majority of the errors. The code also has the ability to detect certain unlikely error patterns. If such error patterns occur, retransmission is requested. The receiver discards the erroneous packet and attempts to correct the errors in the retransmitted packets. The process repeats until no error is detected.

The second approach, Type II, adapts to changing conditions in the channel. No attempt is made to correct errors on the first transmission; however, when an error is detected in the first transmission the retransmission contains extra parity bit for error correction. The original erroneous packet is saved, so

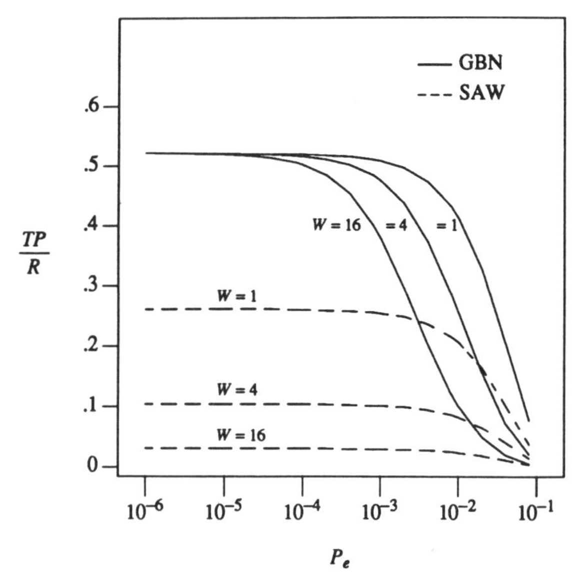

Fig. 3.29 Throughput as a function of probability of bit error for stop-and-wait, go-back-N, and selective repeat.

that it along with the second and subsequent parity bits can be used to retrieve the original message.

REFERENCES

[1] J. G. Proakis, *Digital Communications*, McGraw-Hill, New York, 1983.

[2] S. Lin and D. J. Costello, *Error-Control Coding*, Prentice-Hall, 1983.

[3] W. W. Peterson and E. J. Weldon, *Error-Correcting Codes*, Second Edition, MIT Press, Cambridge, Mass., 1972.

[4] G. C. Clark and J. B. Cain, *Error-Correction Coding for Digital Communications,* Plenum, 1981.

[5] V. J. Bhargava, "Forward-Error Correction for Digital Communications," *IEEE Communications Society Magazine*, Vol. 21, No. 1, pp. 11–19, January 1983.

[6] S. Lin et al., "Automatic-Repeat-Request Error Control Schemes," *IEEE Communications Society Magazine*, Vol. 22, No. 12, pp. 5–16, December 1984.

[7] E. R. Berlekamp et al., "The Application of Error Control to Communications," *IEEE Communications Society Magazine*, Vol. 25, No. 4, pp. 44–57, April 1987.

[8] A. J. Viterbi, "Letter to the Editor," *IEEE Communications Society Magazine*, Vol. 25, June 1987.

[9] B. H. Peek, "Communications Aspects of the Compact-Disk Digital Audio System," *IEEE Communications Society Magazine*, Vol. 23, No. 2, pp. 7–15, February 1985.

[10] D. Slepian, "A Class of Binary-Signaling Alphabets," *BSTJ*, Vol. 35, pp. 203–234, 1956.

[11] N. J. A. Sloane, "The Packing of Spheres," *Scientific American*, January 1984.

[12] A. Gill, *Linear Sequential Circuits,* McGraw-Hill, 1967.

[13] A. J. Viterbi, "Error Bounds for Convolutional Codes and an Asymptotically Optimum Decoding Algorithm," *IEEE Transactions on Information Theory*, Vol. IT-13, pp. 260–269, April 1967.

[14] J. K. Omura, "Optimal Receiver Design for Convolutional Codes and Channels with Memory Via Control Theoretical Concepts," *Information Science*, Vol. 3, pp. 243–266, July 1971.

[15] R. E. Bellman, *Dynamic Programming,* Princeton University Press, 1957.

[16] S. E. Dreyfus, *Dynamic Programming and the Calculus of Variations*, Academic Press, 1965.

[17] J. F. Hayes, "The Viterbi Algorithm Applied to Digital-Data Transmission," *IEEE Communications Society Magazine*, Vol. 13, No. 2, pp. 5–16, March 1975.

[18] J. P. Odenwalder, "Optimal Decoding of Convolutional Codes," Ph.D. dissertation, Department of Systems Science, University of California, Los Angeles, 1970.

[19] K. J. Larsen, "Short Convolutional Codes with Maximal-Free Distance for Rates 1/2, 1/3 and 1/4," *IEEE Transactions on Information Theory*, Vol. IT-19, pp. 371–372, May 1973.

[20] E. Parke, "Short Binary Convolutional Codes with Maximal-Free Distance for Rates 2/3 and 3/4," *IEEE Trans Information Theory*, Vol. IT-20, pp. 683–689, September 1974.

[21] D. G. Dant, J. W. Modestino, and L. D. Wismer, "New Short Constraint Length Convolutional Code Construction for Selected Rational Rates,"

IEEE Transactions on Information Theory, Vol. IT-28, pp. 793–799, September 1982.

[22] R. M. Fano, "A Heuristic Discussion of Probabilistic Decoding," *IEEE Transactions on Information Theory*, Vol. IT-9, pp. 64–74, April 1963.

[23] K. Zigangirov, "Some Sequential-Decoding Procedures," *Probl. Peredachi Inf.*, Vol. 2, pp. 13–25, 1966.

[24] F. Jelinek, "A Fast Sequential-Decoding Algorithm Using a Stack," *IBM Journal of Research and Development*, Vol. 13, pp. 675–685, November 1969.

[25] D. Haccoun, "A Branching Process Analysis of the Average Number of Computations in the Stack Algorithm," *IEEE Transactions on Information Theory*, Vol IT-30, No. 3, pp. 497–508, May 1984.

[26] S. B. Weinstein, "In Galois Fields," *IEEE Transactions on Information Theory*, Vol. 17, p. 220, March 1971.

EXERCISES

3.1: Consider the $(2m+1,1)$ code for which the binary digits are transmitted by codewords consisting of all zeros or ones respectively.

(a) What is the rate of transmitting information?

(b) What is the Hamming distance between codewords?

(c) Assuming a binary symmetric channel with probability of error less than ½, what is the optimum detection algorithm?

(d) What is the minimum bit error probability?

(e) Is this a perfect code?

3.2: Consider the (9,5) code defined by
$(a_1, a_2, a_3, a_4, a_5, a_1 \oplus a_2 \oplus a_4 \oplus a_5, a_1 \oplus a_3 \oplus a_4 \oplus a_5, a_1 \oplus a_2 \oplus a_3 \oplus a_5, a_1 \oplus a_2 \oplus a_3 \oplus a_4)$, where a_1, a_2, a_3, a_4, a_5 are the information digits to be transmitted.

(a) What is the generator matrix for this code?

(b) What is the parity check matrix for this code?

(c) Find the codeword for 10110 and show how it is mapped into the all-zero sequence by the parity check matrix.

(d) What are the Hamming Plotkin and the V-G bounds for this code?

3.3: Find the probability of an undetected error for the (15,11) Hamming code when $p = 10^{-2}$.

3.4: The polynomial $d^4 + d + 1$ is the generator polynomial for (15,11) Hamming code.

(a) Find the generator matrix for this code in systematic form.

(b) Find the feedback shift register circuit which will generate this code.

3.5: Shorten the (15,11) Hamming code to produce a (8,4) code. Find the generator matrix for this code.

3.6: Find the weight distribution for the extended (24,12) Golay.

3.7: Consider the maximum-length shift register code generated by the primitive polynomial $d^5 + d^3 + 1$. Find all of the codewords of this code.

3.8: Find the upper bound on the probability of error for hard-decision decoding when the code is the maximal-length shift register code of Exercise 3.9.

3.9: Repeat Exercise 3.8 for soft-decision decoding.

3.10: For binary-orthogonal signaling find the upper bound on probability of error for both hard and soft decision decoding for the (23,12) Golay code. (These bounds were found in the text for binary-antipodal signaling.)

3.11: Repeat Exercise 3.10 for the extended (24,12) Golay code.

3.12: For a channel SNR of 10 dB and binary-orthogonal signaling, find the coding gain for the (23,12) Golay code. In order to simplify the calculations, you should use the approximation of Equation (2.34b).

3.13: Find the output sequence and the sequence of internal states for the input sequence 101111 when the initial state is 00 in the rate 1/3 encoder shown in Figure 3.14.

3.14: Find the output sequence and the sequence of internal states when the input to the encoder in Figure 3.16 is 10 11 11 00 10 11 00 10 10 10. Assume that the initial state is 00 00.

3.15: For the convolutional encoder shown in Figure 3.16 derive the tree and trellis diagrams.

3.16: Suppose that the rate 1/3 code of Figure 3.14 is being used. Further suppose that the sequence out of a hard-decision decoder is 011 001 110 010 101 101. Find the most likely transmitted sequence.

3.17: Suppose that the rate 1/3 convolutional encoder of Figure 3.14 is used in conjunction with a soft-decision Viterbi decoder. The output of the matched filter is 0.95, 0.90, 1.02, –1.1, –0.96, 0.1, 1.1, 0.97, 0.99, 1.00, –1.1, –1.02, 0.99, 1.02, –0.96, 1.02, –1.03, 0.96, 1.03, –0.98, 1.02, 0.97, –1.01, –0.3. Run this through the Viterbi decoder to find the most likely transmitted sequence. (These are tedious calculations by hand and a computer should be used.)

3.18: In order to carry out the following exercise without excessive computation, we simplify the convolutional encoder shown in Figure 3.16 by eliminating the last stage, s_5 and s_6. We have taken back $y_1 = s_1, y_2 = s_2$, $y_3 = s_2 \oplus s_4$, and $y_4 = s_1 \oplus s_2 \oplus s_3 \oplus s_4$. For the simplified convolutional encoder, find the state diagram and find the transfer function from the all-zero sequence back to the all-zero sequence.

3.19: Use the results of Exercise 3.18 to bound the probability of error.

3.20: For $W = 4$ and $P_e = 10^{-3}$ find the normalized throughput for the (7,4) Hamming code for all three ARQ techniques.

4

Baseband Pulse Transmission

4.0 INTRODUCTION

The foregoing chapters have given an overview of data communications, and reviewed topics in statistical communication theory, coding, and computer communication relevant to data communications analysis and systems design. Chapter 2 in particular described detection techniques for isolated pulse signaling. This chapter begins a detailed examination of signaling techniques for pulse trains. Such techniques constitute the art of conveying digital information through analog channels, which means essentially all channels, since every meaningful physical channel is analog. They include telephone channels, twisted-pair subscriber access lines, magnetic recording channels, shared coaxial media, and optical fiber channels. Not all of the discussion is analytical, for signal design is to some extent a collection of clever techniques developed over time, but there are some unifying theoretical foundations.

The main themes are:

(1) A data stream can be efficiently *coded* into a sequence of analog pulses, usually identical except for amplitude, for transmission through a band-limited baseband channel.

(2) Desirable spectral characteristics can be realized through *shaping* the frequency spectrum of the pulse, shaping the frequency spectrum of the amplitude sequence (by coding the data stream into a correlated amplitude sequence), or via both methods.

(3) Bandwidth can be reduced at the expense of noise immunity. This is consistent with the channel capacity formula (2.112) derived from information theory.

(4) Although pulses may overlap in time, they can be designed to not interfere with one another, or to interfere in controlled ways.

(5) Intersymbol interference caused by linear channel distortion can be largely compensated by adaptive adjustment (equalization) of pulse shaping filters in the communication equipment. Adaptive compensation of other channel impairments is also possible to some extent.

(6) Timing information, for definition of boundaries between pulses for detection purposes, can be derived from the data signals themselves, often at no cost in other desirable properties.

To varying degrees, these ideas will be pursued in this chapter. Themes 5 and 6 will be treated exhaustively in later chapters. Criteria for performance of a data communication system will be developed and some "good" pulse designs will be introduced, but optimum systems will not be derived until Chapter 7. Although the emphasis here is on bandwidth-efficient signaling, bandwidth is not always a constraint, and useful non-bandwidth-efficient signals are also described. Many current textbooks, including [1–7], contain excellent material on pulse design.

4.1 DIRECT-BASEBAND TRANSMISSION

In practice, data transmission may be either *baseband*, with data pulses not modulated onto a carrier waveform, or *passband*, with a recognizable carrier waveform and negligible power near DC. The study of baseband systems is relevant to passband systems also. In particular, linearly modulated passband systems can be analyzed in terms of an equivalent complex baseband model, derived in Chapter 5, which extends virtually all of the baseband concepts and techniques described in this chapter. Many fundamental principles of data communication can be explained in the relatively simple context of baseband data transmission.

Baseband transmission is used directly in a number of commercial communication systems. Examples include T1 carrier systems, "limited distance" wire transmission such as subscriber circuits in the integrated services digital network (ISDN), and coaxial cable and fiber local area networks (LANs). In these and other applications, the signal must carry *timing* information, i.e., have frequent enough transitions between signal levels so that the receiver can locate the pulse boundaries in order to realize near-optimal detection. It may also be desirable to achieve bandwidth efficiency (at a cost in noise immunity) by the use of *multilevel* signaling, or to minimize transmitted energy near an upper band edge to reduce noise and adjacent channel interference. Elimination of energy near DC is required in many amplifiers and transmission systems. All of these considerations receive attention in signal design.

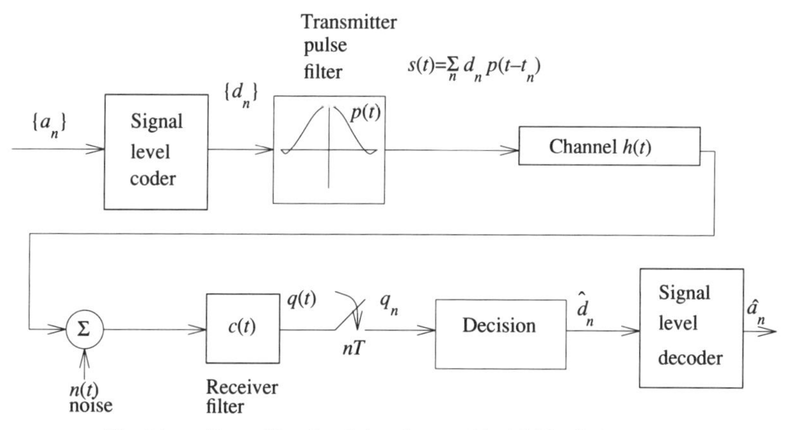

Fig. 4.1 General baseband signaling model with identical pulses $p(t)$.

A generic baseband transmission system is depicted in Figure 4.1. The raw data sequence $\{a_n\}$ is encoded into the sequence $\{d_n\}$. An example of such encoding, discussed in Section 4.6 in connection with partial response signaling, prevents error propagation. The sequence $\{d_n\}$ modulates the amplitudes of pulses which are represented in the time and frequency domains as $p(t)$ and $P(f)$ respectively. The transmitted signal is

$$s(t) = \sum_n d_n p(t-t_n) , \tag{4.1}$$

where the amplitudes, pulse shapes and pulse locations may vary from one pulse to the next, and information may be carried by any of these variables.* But in many applications, especially if bandwidth is a consideration, there is no advantage in using anything other than identical pulses $p(t)$ spaced at uniform *symbol intervals* T, with all information carried in the amplitude sequence $\{d_n\}$. As we have seen, if d_n may assume L equiprobable levels, the rate of transmitting information is $R = (\log_2 L)/T$ bps (see Section 2.2.7). (The Miller code, described later, is an exception to the identical pulse rule.)

As we have already suggested, the coding of the original data stream into this sequence, and the choice of the pulse $p(t)$, can each contribute to defining spectral characteristics and optimizing overall system error rate performance. The channel is represented as a time-invariant linear filter with impulse response $h(t)$ (see Figure 4.1) followed by additive noise. The received signal plus noise are passed through a receiver filter which has impulse response $c(t)$. The sam-

---

* Here, as throughout this book, a summation over n will be assumed to run from $n = -\infty$ to $n = \infty$ unless otherwise noted.

pled output of this filter is fed into a decision circuit whose output, $\{\hat{d}_n\}$, is an estimate of $\{d_n\}$. The estimated sequence is then fed into a decoder. The overall transfer function of the system is given by

$$X(f) = P(f)\, H(f)\, C(f)\ , \tag{4.2}$$

where $H(f)$ and $C(f)$ are the Fourier transforms of $h(t)$ and $c(t)$, respectively. As we shall see in Section 4.2, it is the form of $X(f)$ which governs the extent of *intersymbol interference (ISI)*, i.e., the amount of energy spilling from one symbol interval to another.

4.1.1 SPLITTING THE PULSE SHAPE BETWEEN TRANSMITTER AND RECEIVER

We described in Chapter 2 how a matched filter receiver minimizes the probability of error for isolated pulse signaling in the white Gaussian noise channel. Since that idealized channel has unlimited bandwidth, the pulse shapes are irrelevant; only their energies and cross-correlations matter. In this chapter, where we consider not isolated pulse signaling but rather the transmission of pulse trains, we will impose the additional condition of minimal intersymbol interference. For the white Gaussian noise channel, this means (as we will show in Section 4.3) specifying the overall pulse shape, including transmitter, channel, and receiver.

Before doing that, we will demonstrate that for a data communication system with a *distortionless* channel perturbed by additive white Gaussian noise, with an overall real pulse spectrum $X(f)$, the output SNR is maximized (and hence the error rate is minimized) if the transmitter and receiver characteristics $P(f)$ and $C(f)$ are each equal to $\sqrt{X(f)}$.

Assume that $X(f)$ (shown split between transmitter and receiver in Figure 4.2) is specified so that pulses do not interfere with one another at the receiver sampling instants. Assume also that the channel has enough bandwidth to be transparent, i.e., $H(f) = 1$. We can then restrict our attention to only one of the transmitted pulses, say $d_0 p(t)$ and show that $C(f) = P(f) = \sqrt{X(f)}$ minimizes the probability of detection error for a given average power.

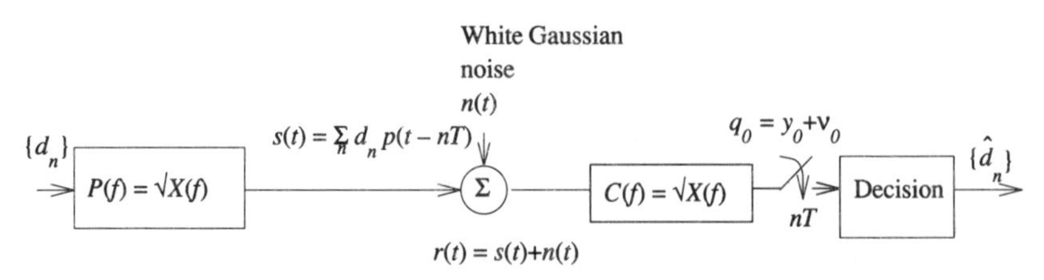

Fig. 4.2 Splitting pulse characteristic between transmitter and receiver.

The proof that $C(f) = P(f) = \sqrt{X(f)}$ minimizes the probability of error is simply that a receiver filter *matched* to the transmitter filter maximizes the output signal-to-noise ratio, as shown in Chapter 2. The filter matched† to $p(t)$ is $c(t) = p(-t)$, so that, with real $p(t)$

$$C(f) = \int_{-\infty}^{\infty} p(-\tau) e^{-j2\pi f\tau} d\tau = \left[\int_{-\infty}^{\infty} p(t) e^{-j2\pi ft} dt \right]^* = P^*(f) \, , \quad (4.3)$$

which is the matched filter derived in Chapter 2, and

$$X(f) = P(f)C(f) = |P(f)|^2 \, . \quad (4.4)$$

The reader will recognize this as the matched filter solution of Section 2.2.1. Since $X(f)$ is nonnegative real,

$$P(f) = \sqrt{X(f)} \, , \quad C(f) = P^*(f) = \sqrt{X(f)} \, . \quad (4.5)$$

This is what we wanted to prove. It is conventional practice to split the pulse characteristic in this way between transmitter and receiver. If the channel filter $H(f)$ is anticipated and at least approximately known, then $P(f)H(f) = \sqrt{X(f)}$.

4.1.2 LINE SIGNAL CODINGS (NRZ, AMI, MILLER, MANCHESTER) AND POWER SPECTRA

A number of line signal codings—translations from binary data stream to transmitted pulse sequence—are in common use. They have their individual compromises, but all strive for the general objectives for line signal codings:

(1) Desirable spectral characteristics, particularly no DC.

(2) Numerous level transitions, or equivalently, significant power near $f = 1/2T$ or its harmonics, to determine symbol interval boundaries and enable reliable timing recovery.

(3) High noise immunity, leading to low bit error rate. As part of this property we might include some form of redundancy which would indicate an obvious error. We shall see examples presently.

BINARY ANTIPODAL (NRZ) SIGNALING

Assume for the moment that the channel is distortionless, with the only perturbation being white Gaussian noise. *Binary antipodal* signaling, in which ones and zeros in the data stream are *independently* coded into positive and negative

---

† For simplicity, $c(t)$ has been defined as a noncausal filter. An appropriate delay t' can be inserted for a causal $c(t) = p(t' - t)$, as suggested in Section 2.2.1.

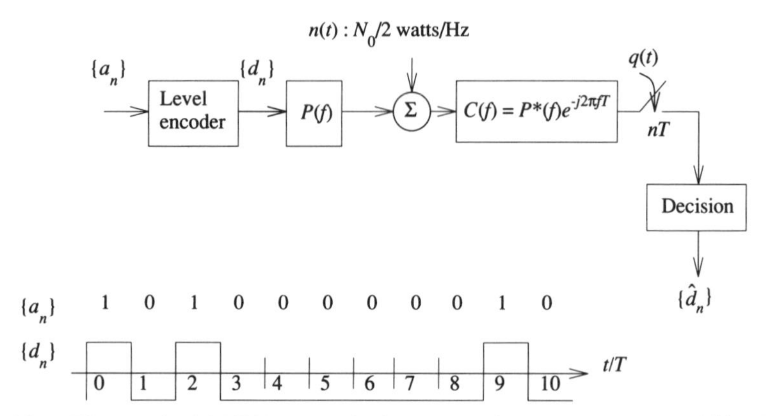

Fig. 4.3(a) Binary antipodal (NRZ) communication system with rectangular pulses, additive Gaussian noise, a distortionless channel, and optimum matched filter receiver.

pulses respectively (Figure 4.3), is perhaps the most straightforward signaling technique. It is also known as one form of nonreturn to zero (NRZ) signaling because one of the alternative nonzero pulse levels is always being sent.

For an isolated pulse in a distortionless, white Gaussian noise channel, or for pulses in an end-to-end system designed to have no intersymbol interference, binary-antipodal signaling yields the lowest possible probability of error. As derived in Chapter 2, probability of error is independent of the pulse shape and is given by (2.35). For convenience, equation (2.35) is repeated here:

$$P_e = Q\,[\sqrt{2\,E_0/N_0}\,]\ ,$$

where E_0 is the energy of the transmitted pulse, $N_0/2$ is the (two-sided) spectral power density of the white Gaussian noise, and $Q(\cdot)$ is the "tail probability" of the Gaussian probability density function defined in Chapter 2. The receiver output is (referring to the system model of Figure 4.3a)

$$q(t) = \sum_n d_n x(t-nT) + \nu(t)\ , \tag{4.6}$$

in which the end-to-end signal pulse is

$$x(t) = p(t) * c(t) \tag{4.7}$$

and the noise component is

$$\nu(t) = n(t) * c(t), \tag{4.8}$$

where $p(t)$ is the transmitter impulse response and $c(t)$ is the receiver impulse response. By the linear, time-invariant nature of the system and the lack of any correlation among neighboring pulse levels (we are assuming for the time being that data are independently coded into pulse levels), the spectrum of the signal component,

$$\sum_n d_n x(t-nT) \ ,$$

is the same as that of the individual pulse $x(t)$. We shall see later that correlated pulse levels change the spectrum of the signal.

Note that the end-to-end spectrum $X(f)$ given by (4.2) is split evenly between transmitter (plus channel) and receiver in order to match the receiving filter to the incoming pulse and maximize the output signal-to-noise ratio, as described earlier. Thus $C(f) = P^*(f)$ (from (4.5)) is matched to the transmitter pulse filter $P(f)$, and

$$X(f) = |P(f)|^2 \tag{4.9}$$

is the overall pulse spectrum.

If $x(t)$ is a rectangular pulse, its power density spectrum (shown in Figure 4.3b and derived later in this section) is

$$S(f) = T^2 \left[\frac{\sin(\pi f T)}{\pi f T} \right]^2 . \tag{4.10}$$

This spectrum, with first null at $1/T$ Hz, is much broader than the minimum bandwidth spectrum for the same signaling rate. But the spectrum is narrower than the spectra of other pulses that sacrifice bandwidth for the sake of more timing information through more pulse-level transitions.

The matched-filter performance for an isolated pulse, or for a pulse train with no intersymbol interference, is often used as a reference for other signaling schemes. However, in considering isolated pulses, an important aspect of data communication—synchronization—is ignored. Consider binary antipodal signaling. A long string of zeros or a long string of ones has no transitions to define symbol interval boundaries. Also, a DC level would be generated by such a string. The lack of timing transitions for long strings of ones or zeros calls for

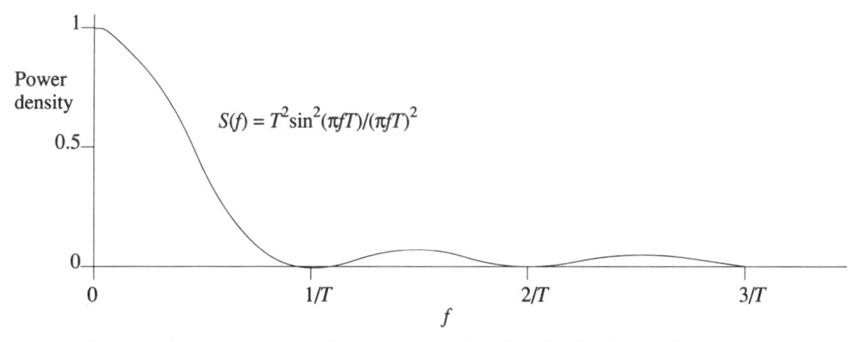

Fig. 4.3(b) Power spectrum for the rectangular pulse of unit height and duration T.

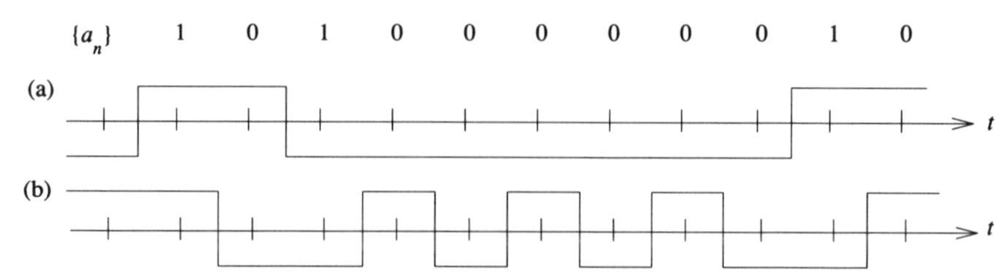

Fig. 4.4 Differential NRZ codings that change line signal level on (a) a data "1", (b) a data "0" (but not on the complementary data value).

special measures to introduce transitions. One approach is to break up the data stream into blocks that are packaged into transmission frames that include extra synchronization bits, guaranteeing at least a few level transitions each frame. Another approach, described in Chapter 6, is to scramble the data stream into one that is guaranteed to go no more than a small fixed number of symbol intervals without a transition. This latter approach also takes care of the DC problem.

Differential forms of NRZ signaling can eliminate the need for an absolute voltage reference level and possible ambiguities in identifying which of a twisted pair of wires is the "plus" wire. As Figure 4.4a illustrates, the line signal level changes when a "1" is sent but not when a "0" is sent. An analogous coding (Figure 4.4b) can, of course, be defined that changes line signal level when a "0" is sent but not when a "1" is sent. The signal power spectrum is the same in these cases as in the standard NRZ coding of Figure 4.3.

ALTERNATE MARK INVERSION (AMI) SIGNALING

A second common line signal coding is *alternate mark inversion* (AMI) (Figure 4.5), also known as *bipolar* signaling. Here we intentionally *do* correlate adjacent pulse levels in order to achieve a desired spectral shaping. We shall derive the power spectrum and evaluate the error-rate performance of AMI. It is used in T1 (1.544 Mbps) and T2 (4×T1) digital transmission systems.* A "1" codes into the opposite polarity pulse from the last "1", and "0" codes into a zero line signal level, yielding a 50% duty cycle.† The zero DC value of the AMI level sequence, even with long data strings of zeros or ones, eases the design of repeater amplifiers. It does not, by itself, solve the timing problem. Note that a three-leveled signal can carry log 3 bits/symbol of information (see (2.70)). AMI, with its one bit/symbol, sacrifices this advantage for advantages in spectral shaping, timing recovery, and performance monitoring.

---

* T1 and T2 conform, respectively, to DS-1 and DS-2 in the North American Digital Hierarchy. See Figure 1.14.

† For an implementation of AMI, see Figure 4.18.

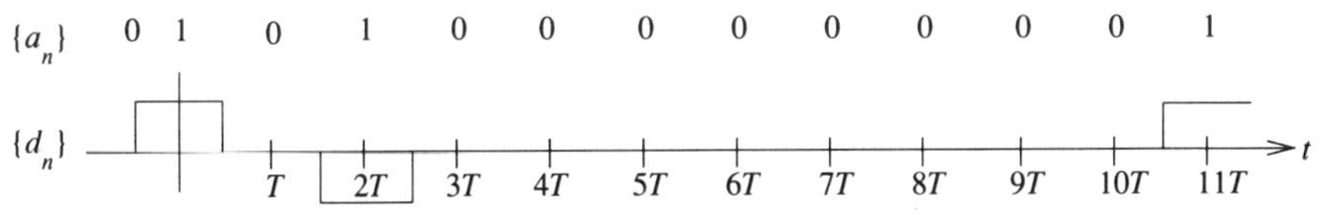

Fig. 4.5(a) Alternate mark inversion (AMI) or bipolar coding.

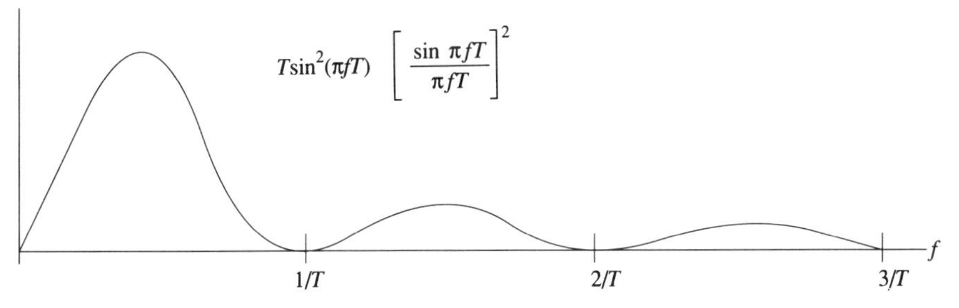

Fig. 4.5(b) Power spectrum for an AMI-encoded random data sequence (rectangular pulses).

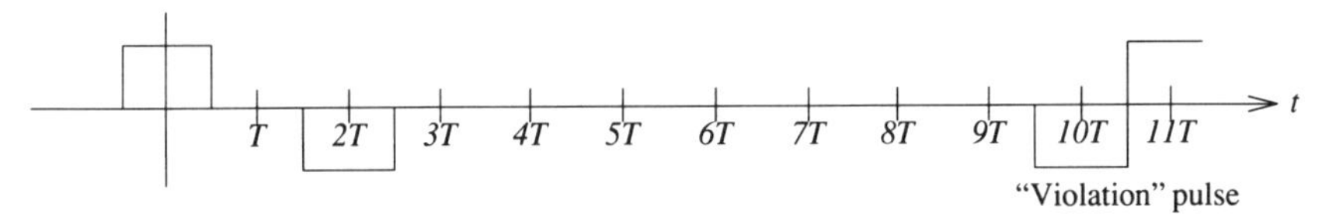

Fig. 4.5(c) Modification in T1 system; eighth consecutive zero is replaced by a "violation," a nonzero pulse in the same direction as the last nonzero pulse.

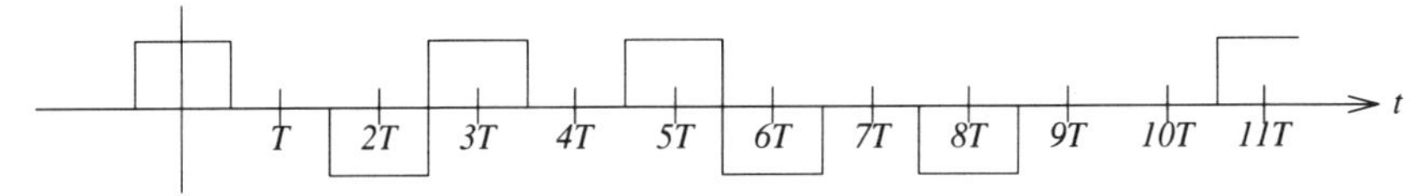

Fig. 4.5(d) Modification in T2 system; sequence of six zeros is replaced by a special violation sequence.

The rectangular pulse shape used in T-carrier systems is not spectrally efficient, but digital repeaters in the cables maintain pulse integrity despite bandwidth limitations. This relative lack of concern about spectral spread is typical of many, but not all, baseband signaling systems.

The T1 system handles the timing problem by replacing the eighth consecutive zero in a data stream with a bipolar "violation" (Figure 4.5c) that is simply interpreted as a zero by the receiver. A violation is two consecutive pulses of the same polarity. This strategy presumes an error-free channel, which is a satisfactory assumption in an environment characterized by long error-free intervals between error bursts. For T2 systems, long sequences of zeros are broken into groups of six which are replaced by a special violation sequence (Figure 4.5d) which has numerous level transitions.

The power density spectrum of the alternate mark inversion line signal is found from the general formula

$$S_s(f) = S_d(f)\ |P(f)|^2/T \tag{4.11}$$

derived in Appendix 4A. Here $P(f)$ is the pulse spectrum, and $S_s(f)$ is the discrete-time Fourier transform of the correlation function of the data level sequence:

$$S_d(f) = \sum_{m=-\infty}^{\infty} r_d(m)\ \exp[-j2\pi mfT]\ . \tag{4.12}$$

Note that this transform differs from the discrete Fourier transform (DFT) in which both time and frequency are quantized.

To evaluate (4.12) for an AMI pulse train, we first derive the autocorrelation function $r_d(m)$ of the amplitude sequence $\{d_n\}$, assuming an input data sequence $\{a_n\}$ of independent random variables with equal probability of being one or zero. It is:

$$r_d(m) = E\left[d_n d_{n+m}\right] = \begin{cases} -1/4, & m=-1 \\ +1/2, & m=0 \\ -1/4, & m=+1 \\ 0, & |m|>1 \end{cases} \tag{4.13}$$

These results were computed as follows for each value of the index m:

For $m = 0$:

d_n	$Pr(d_n)$	$Pr(d_n)\times d_n^2$
0	1/2	0
±1	1/4	1/4

Thus $E\left[d_n^2\right] = 1/2$.

For $m = 1$:

d_n	$Pr(d_n)$	d_{n+1}	$Pr(d_{n+1} \mid d_n)$	$Pr(d_n,d_{n+1})d_n d_{n+1}$
0	1/2	–1	1/4	0
		0	1/2	0
		+1	1/4	0
+1	1/4	–1	1/2	–1/8
		0	1/2	0
–1	1/4	0	1/2	0
		+1	1/2	–1/8

Thus $E\left[d_n d_{n+1}\right] = -1/4$.

For $m = -1$ the results are the same as for $m = 1$.

For $|m| > 1$:

$E[d_n d_{n+m}] = 0$, since +1 and –1 values of d_{n+m} are equally likely regardless of d_n. For example, if $d_n = +1$, then data $a_{n+1} = 0$, $a_{n+2} = 1$ lead to $d_{n+2} = -1$, and data $a_{n+1} = 1$, $a_{n+2} = 1$ lead to $d_{n+2} = +1$. Data pairs (1,0) and (0,0) lead to $d_{n+2} = 0$.

Substituting (4.13) into (4.12), the spectral shaping introduced by the correlated level sequence is

$$S_d(f) = 1/2[1 - \cos 2\pi fT] = \sin^2 \pi fT . \tag{4.14}$$

The Fourier transform of the rectangular pulse used in this example,

$$p(t) = 1, \ -T/2 < t < T/2, \quad \text{and 0 elsewhere}$$

is

$$P(f) = \int_{-T/2}^{T/2} \exp[-j2\pi ft]\,dt = T[\sin \pi fT/\pi fT] . \tag{4.15}$$

This defines the spectral shaping contributed by the pulse itself. Squaring and substituting into (4.11),

$$S_s(f) = 1/T \ S_d(f) \ |P(f)|^2 = T \sin^2 \pi fT \ [\sin \pi fT/\pi fT]^2 . \tag{4.16}$$

The spectrum (4.16), shown in Figure (4.5b), has a null at DC (distinguishing it from the peak at DC for binary antipodal signaling) and a peak at $f = 1/2T$ Hz. AMI thus has the desirable properties of no DC, peak power at $f = 1/2T$ for timing recovery (one has only to filter out a small band of frequencies in the vicinity of $f = 1/2T$ to define symbol interval boundaries by the zero crossings of the filtered signal), and redundancy so that bipolar violations indicate that something is wrong. Violations can be intentionally introduced to enhance timing recovery.

PERFORMANCE OF AMI SIGNALING

The performance of AMI signaling in the distortionless, white Gaussian noise channel (Figure 4.6) will be evaluated here for independent symbol by symbol detection. Actually, there is information in the correlation among signal levels in neighboring pulses that can aid the detection of errors and thus their correction. For example, the sequence of detected level decisions 1, 0, –1, 0, –1 could be corrected to 1, 0, –1, 0, 0, since it is highly improbable that the violation (second –1 in a row) of the last level decision could have derived from a transmitted +1 level. This would allow correct decoding to the data sequence 1, 0, 1, 0, 0. However, we have already noted that violation of the alternating polarity rule is practiced in some systems to break up long sequences of zero

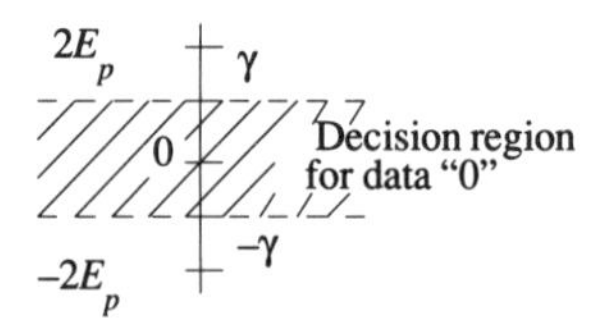

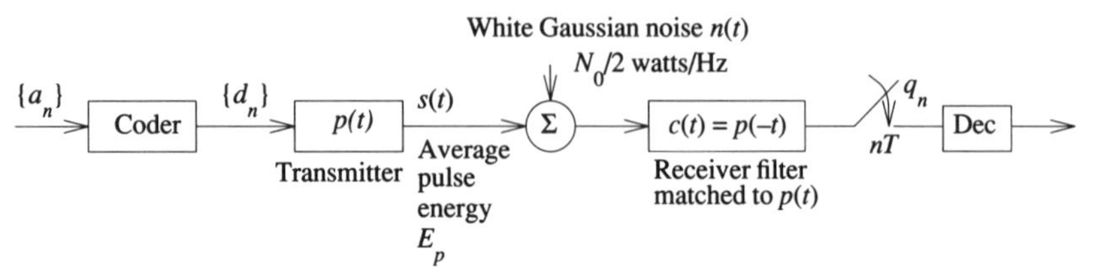

Fig. 4.6 Model for AMI signaling through a distortionless, white Gaussian noise channel, where

$$d_n = \begin{cases} \pm 1, & a_n = 1 \\ 0, & a_n = 0 \end{cases} .$$

levels, so that this error correction advantage is traded for the advantage of more level transitions. We shall, therefore, neglect the correction in probability of error that could be achieved by using information about neighboring pulse levels.

The probability of error derivation is similar to that of Chapter 2 for binary-antipodal and binary-orthogonal signals. In fact, the probability of error is identical, except for a multiplicative constant, to that of binary-orthogonal signals, since the positive and negative waveforms used to represent a data "1" are orthogonal to the zero waveform used to represent a data "0". In the absence of intersymbol interference, the derivation is made, as in Chapter 2, for a single isolated pulse.

Note, from Figure 4.6, that the overall pulse characteristic is split evenly between transmitter and receiver, i.e., the receiver filter $c(t)$ is matched to the received pulse impulse response $p(t)$ as discussed earlier. For the rectangular pulse $p(t)$ as in Figure 4.5a, $c(t) = p(t)$. The output sample of the receiver filter at time $t=0$,

$$q(0) = m(0) + \nu(0) ,$$

contains a Gaussian noise component with variance

$$\sigma^2 = E\left[\nu^2(0)\right] = \int_{-\infty}^{\infty}\int_{-\infty}^{\infty} E[n(t)n(\tau)]\, p(-t)p(-\tau)\, dt\, d\tau$$

$$= \int_{-\infty}^{\infty} [N_0/2]p^2(t)\, dt = N_0 E_p , \tag{4.17}$$

where

$$E_p = 1/2 \int_{-\infty}^{\infty} p^2(t)\, dt \tag{4.18}$$

is the *average* energy in a transmitted pulse, noting that half of the time a zero level is sent and the other half of the time a level of one or minus one is sent.

When a data "1" is applied to the transmitter, a pulse of energy $2E_p$ is sent and the signal component of the receiver output is

$$m(0) = \pm p(t) * p(-t)\big|_{t=0} = \pm \int_{-\infty}^{\infty} p^2(\tau)\,d\tau = \pm 2E_p \,. \tag{4.19}$$

When a data "0" occurs, the transmitted energy and the signal component of the receiver are both zero. These output levels are shown in Figure 4.6.

The receiver's maximum-likelihood decision strategy is to choose "0" when the output sample $q(nT)$ is between the decision region boundaries $(-\gamma, \gamma)$, and "1" otherwise. (Exercise 4.6 determines the optimum value of γ.) As an approximation, we will assume $\gamma = E_p$, midway between the zero and $2E_p$ output signal levels. As a further (and very good) approximation, we will assume that a noise sample ν will never be large enough to carry a receiver output sample from the $+2E_p$ decision region into the $-2E_p$ decision region (or the reverse); if noise causes a movement across a decision region boundary, it will only be to the neighboring decision region.

Exploiting the symmetry of the Gaussian distribution of the noise sample ν, we can write

$$P_e = 3/2\, Pr(\nu > E_p) = 3/2 \int_{E_p}^{\infty} \left[1/\sqrt{2\pi\sigma^2}\right] \exp\left[-z^2/2\sigma^2\right] dz$$

$$= 3/2\, Q\left[\sqrt{E_p/N_0}\right] . \tag{4.20}$$

Only the asymmetrical error probabilities conditioned on a data "0" and a data "1" prevent this expression from being exactly identical to (2.36) for binary orthogonal signals. The exponential expression which the complementary error function asymptotically becomes is dominant for small error probabilities, so the performance is essentially identical for these two systems. There is a loss in performance, however, since the channel has the capability of antipodal signaling which has a 3 dB advantage.

OTHER LINE CODES

There are many other popular baseband pulse codings. The *Manchester* or diphase code, which maps a data "1" into a pulse containing both a positive and a negative level and a data "0" into the same pulse with inverted polarity (Figure 4.7), is very effective for timing recovery because *every* symbol interval includes a level transition. Moreover, there is always energy on the line.

The penalty is paid in bandwidth. The power density spectrum for a Manchester encoded sequence with random binary data input is

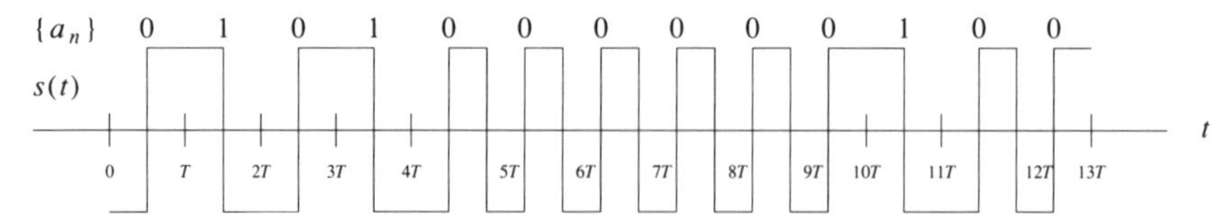

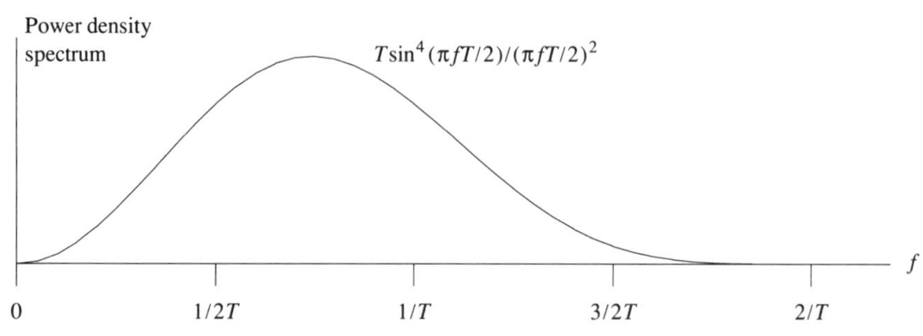

Fig. 4.7 The Manchester (or diphase) code, and the power spectrum for a random data sequence.

$$S_m(f) = T\frac{\sin^4(\pi fT/2)}{(\pi fT/2)^2}. \tag{4.21}$$

As Figure 4.7 shows, the energy peak is considerably higher than for binary antipodal or AMI signals, and significant energy is placed beyond $1/T$ Hz. This tradeoff is acceptable in many cases. The Manchester code is widely used in coaxial cable-based Ethernet local area networks (Chapter 1), and as the electrical signal used to drive optical transmitters in fiber optic communication sys-

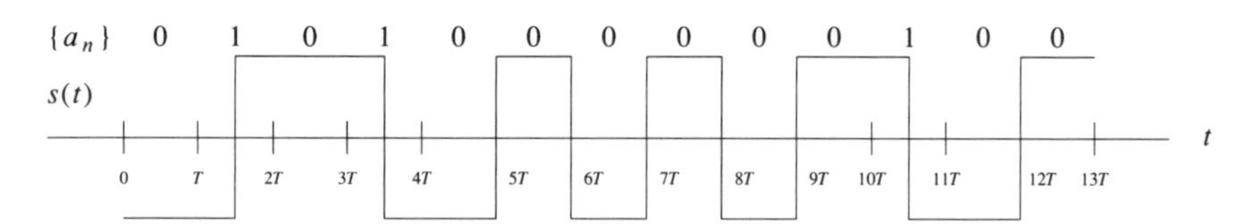

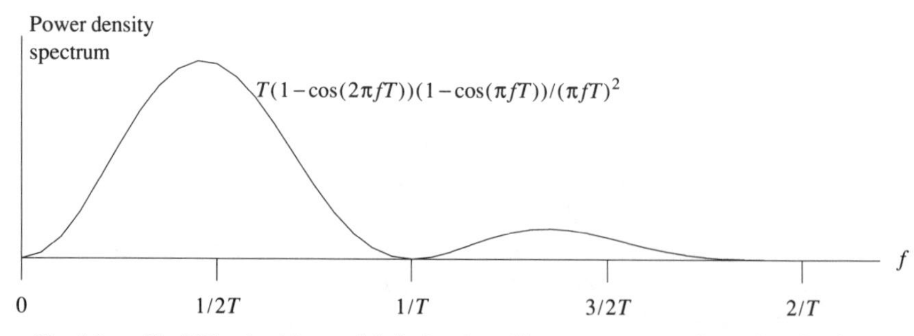

Fig. 4.8 The Miller (or delay modulation) code and its power spectrum for random data input.

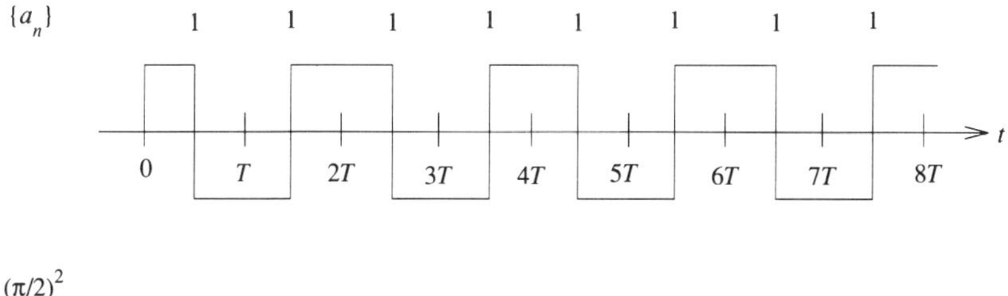

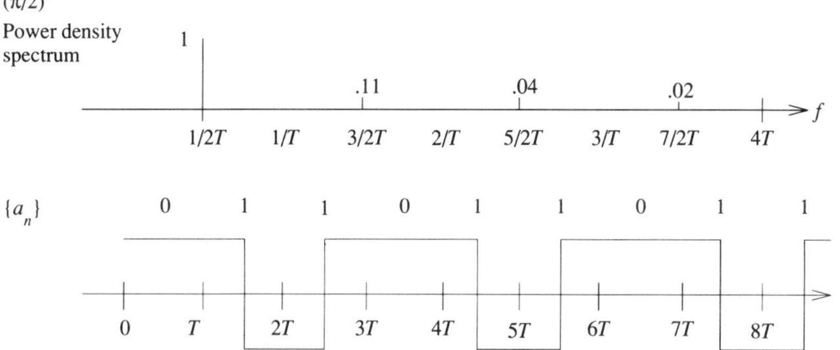

Fig. 4.9 Power spectra of Miller encodings of certain regular (periodic) data sequences.

tems. Since it is an example of binary-antipodal signaling, its error rate is (2.35), derived in Chapter 2.

Another familiar code is the *Miller* or delay modulation code (Figure 4.8). A data "1" codes into the Manchester pulse, or its negative, maintaining continuity from the previous level, and a data "0" is represented by either a positive or negative level over the whole symbol interval, continuing the previous level if that came from a Manchester pulse, corresponding to a data "1", or flipping to the other level if the previous pulse was constant level (corresponding to a data "0"). There are thus two different pulse shapes, Manchester and constant level, and four different possible signaling waveforms in any symbol interval.

The Miller code guarantees at least one level transition every two symbol intervals. Exactly one level transition occurs every two symbol intervals with a "dotting pattern" of alternate input data ones and zeros. For a random input data sequence, the Miller code has a power density spectrum (see Exercise 4.6) that is sharply peaked just below $0.4/T$ Hz and looks very attractive for bandwidth conservation. However, for an input sequence of all ones or all zeros the power spectrum is spread out, and for an input sequence of a zero alternating with a pair of ones there is considerable DC content. This can be seen from a Fourier series analysis of these waveforms, which are periodic with period $2T$. The results of this analysis are indicated in Figure 4.9.

4.2 PULSE AMPLITUDE MODULATION (PAM) IN A DISTORTED, NOISY, BANDLIMITED CHANNEL*

The pulse codings examined in the last section all used binary signaling and pulse shapes not particularly intended to conserve bandwidth. In fact, it was assumed that the channel was largely transparent, $H(f) = 1$. For example, a rectangular pulse shape was often presumed. When the baseband signal is to be used in a bandwidth-restrictive channel, and particularly when it is to be used as the modulation waveform in a modem for voiceband channels, pulse shaping to ameliorate the effect of limited bandwidth is essential. But cutting down on bandwidth causes a smearing out in time, so that a major consideration is to limit the effect of intersymbol interference among neighboring pulses caused by the limited bandwidth.

Pulse spectral shaping is not the only step to be taken. PAM uses *multilevel transmission*, i.e., coding the data into one of L possible amplitudes $(L > 2)$ for each pulse, to achieve an information rate that is faster than the pulsing rate, unlike the binary signals discussed so far. The bandwidth, determined by the pulsing rate and pulse shape $p(t)$, is kept down but for constant average signal power the levels are closer together, resulting in reduced noise immunity. Varying the number of signal levels thus represents the tradeoff between bandwidth and noise immunity suggested in theme 3 at the beginning of this chapter.

Specifically, the data rate for an L-leveled signal is

$$R = \text{symbol rate (baud)} \times \log_2 L = l/T \text{ bps}, \tag{4.22}$$

where $l = \log_2 L$, as described in Chapter 2, Section 2.2.7. For example, the data rate of a PAM signal with 2400 symbols/sec and four levels is 4800 bps. A binary signal with the same 4800 bps data rate would have a baud of 4800 symbols/sec requiring twice as much bandwidth, and a separation between levels $\sqrt{5}$ times as large (Exercise 4.5) giving much better performance against noise.

If bandwidth is freely available for a channel with white Gaussian noise, it is *always* better to use more bandwidth than to increase the number of levels (see Section 2.2.7). A numerical comparison is made at the end of Section 4.6. For voiceband and microwave data communication, unfortunately, there are severe bandwidth constraints that usually force payment of the noise penalty.

---

* The reader may wish to review the discussion in Section 2.2.6 of the performance of PAM in the absence of intersymbol interference.

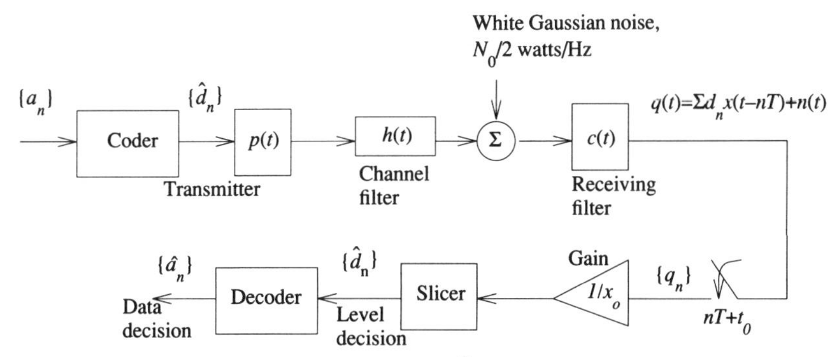

Fig. 4.10 Baseband PAM system. The outputs $\{\hat{d}_n\}$ are estimates of the input levels $\{d_n\}$.

The PAM signal is represented as

$$s(t) = \sum_n d_n p(t-nT), \tag{4.23}$$

where d_n is chosen from an L-letter alphabet and $p(t)$ is the pulse shape generated in the transmitter. The usual assumption is that the data levels are uncorrelated. As we have seen in Section 2.7.5, the values of d_n can be expressed as $[2k+1-L]$ $k=0,1,\ldots,L-1$. The task of the receiver shown in Figure 4.10 is to correctly classify the amplitudes $\{d_n\}$, chosen from the set illustrated in Figure 4.11 for $L=8$, from observation of the received signal $r(t)$. Note the greater generality of the communication system model from those we

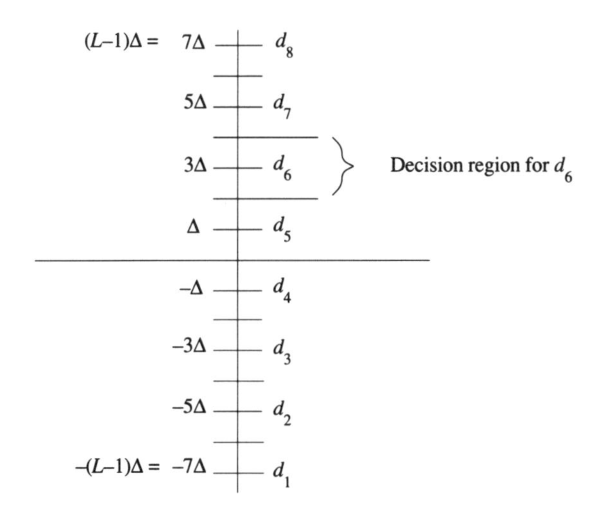

Fig. 4.11 Pulse amplitude levels d_n chosen from eight-level alphabet.

have described before, with a channel filter $H(f)$ and an overall pulse spectrum $P(f)H(f)C(f)$ that may not be ideal.

In linear passband data transmission, as discussed in the next chapter, multilevel PAM signals are frequently the modulation waveforms. Following demodulation in the receiver, detection is done on noisy baseband signals, albeit in two dimensions (complex variables) rather than the one dimension described in this chapter. PAM signaling has become the standard technique in voiceband, digital radio, and optical communication systems.

We will shortly analyze the performance of a PAM data transmission system under the assumption that the channel distortion is known exactly and can be compensated in $P(f)$ and $C(f)$. We will presume, in other words, that the *end-to-end* pulse shape does not introduce intersymbol interference. First, however, we need to understand the conditions that define a pulse shape without intersymbol interference, and furthermore, whether there are such pulse shapes that are also spectrally efficient (which the rectangular pulse is not).

Our pulse design problem concerns the overall system impulse response (Figure 4.10)

$$x(t) = p(t) * h(t) * c(t) . \tag{4.24}$$

The problem is: How should $x(t)$ be chosen so that bandwidth is conserved and one pulse $x(t-nT)$ does not interfere with another $x(t-mT)$? Evaluation of intersymbol interference for imperfect $x(t)$ will be described later in this chapter.

We attack this question by examining the *samples* $q(nT+t_0)$ of the receiver output waveform

$$q(t) = \sum d_n x(t-nT) + \nu(t) \tag{4.25}$$

on which level decision are made. Here $x(t)$ is the overall pulse shape, $\nu(t)$ is noise at the output of the receiver filter, and t_0 is the *sampling time*, or *epoch*, with respect to some clocking waveform. The reason for (possibly) sampling at some time other than the instants $t=nT$ assumed until now is that delay and distortion in the channel filter $h(t)$ might make some other sampling instants

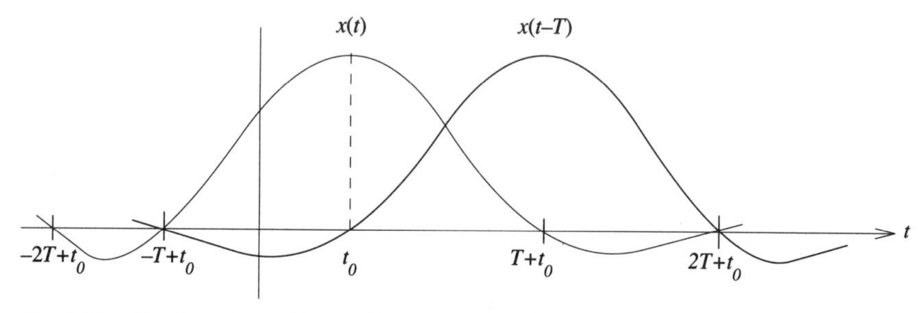

Fig. 4.12 Condition for avoidance of intersymbol interference is that pulses equal zero at all sampling times except their own.

preferable. For example, if $h(t)$ introduces a flat delay of $T/4$ sec, the choice of $t_0 = T/4$ will yield the same signal samples as sampling the transmitted signal at times nT. The clocking waveform (Chapter 6) may be derived from transitions in a received waveform or, for modulated waveforms, from filtering a nonlinear function of the received waveform.

Consider a particular output sample q_0. From (4.25), q_0 is a combination of $d_0 x(t_0)$, the main contribution we would like to detect, and (possible) intersymbol interference contributions from all the other pulses of the form $d_m x(-mT+t_0)$. Note that $x(t_0)$ is the gain of the entire system, which in past work has often been assumed to be unity. The noninterference condition that we would like to achieve occurs when *the tails of all the other pulses pass through zero at time* t_0, i.e.,

$$x(kT+t_0) = 0, \quad k = \pm 1, \pm 2, \ldots \quad . \tag{4.26}$$

This is illustrated in Figure 4.12. Since some intersymbol interference is inevitably present, (4.26) is never exactly satisfied and what we actually get at the sampler output is

$$q(kT+t_0) = \sum_n d_n x[(k-n)T+t_0] + \nu(kT+t_0) \ . \tag{4.27}$$

The following abbreviated notation is henceforth used for convenience:

$$q_k = q(kT+t_0), \quad x_k = x(kT+t_0), \quad \nu_k = \nu(kT+t_0) \ . \tag{4.28}$$

The receiver output sample is therefore

$$q_k = \sum_n d_n x_{k-n} + \nu_k = x_0 [d_k + (1/x_0) \sum_{n \neq k} d_n x_{k-n} + \nu_k / x_0] \ . \tag{4.29}$$

Assume there is a $1/x_0$ gain just after the sampler, as shown in Figure 4.10, so that the input to the threshold detector (slicer) is q_k/x_0. The slicer establishes decision region boundaries at levels 0, $\pm 2\Delta$, $\pm 4\Delta$, ..., $\pm(L-2)\Delta$ (i.e., midway between the signal levels, as shown in Figure 4.11) in order to determine which of the L values of d_n is closest to q_k/x_0.

For all but the outermost levels, an error occurs when

$$(1/x_0) \left| \sum_{n \neq k} d_n x_{k-n} + \nu_n \right| > \Delta \ . \tag{4.30}$$

The problem is to design transmitting and receiving filters $P(f)$ and $C(f)$ to counteract the combined effects of intersymbol interference and noise in order to minimize the probability of error.

The resulting overall pulse spectrum $X(f) = P(f)H(f)C(f)$ is usually an approximation to one meeting the *Nyquist criterion*, derived in the next section, for simultaneous satisfaction of the requirements for narrow bandwidth and no intersymbol interference. The Nyquist criterion will be used to define a large class of signal shapes which have the very useful property, in the time domain, of passing through zero at the peaks of other pulses in a pulse train. It is intuitively reasonable that minimizing the intersymbol interference will maximize the output signal-to-noise ratio at the pulse sampling times, but this is rarely strictly true. When there is noise and significant amplitude distortion, the compensation of transmission nulls by high gains at certain frequencies may increase the noise level at the receiver output more than it reduces the intersymbol interference. This question is addressed directly in Chapter 7, where a linear receiver structure maximizing the mean-square error is derived. But the Nyquist criterion is an important concept, and determines transmitter and receiver filter design in many practical systems, including a default starting point for adaptive systems. We will return to PAM after a discussion of the Nyquist criterion.

4.3 THE NYQUIST CRITERION

The Nyquist criterion [1] specifies the conditions on $X(f)$, the Fourier transform of the overall pulse shape $x(t)$ of the PAM system introduced in the last section, necessary for

$$x(nT) = \begin{cases} 1, & n=0 \\ 0, & n \neq 0 \end{cases} \cdot \tag{4.31}$$

With no loss of generality, a timing epoch $t_0 = 0$ is assumed, since a pulse satisfying (4.31) can be time-shifted by any amount without losing the fundamental property that all but one of its samples at T-second intervals are zero. If (4.31) is satisfied, a pulse *train* at T-second intervals will not exhibit intersymbol interference.

To derive the Nyquist criterion, we need the *sampling theorem* for bandlimited functions:

THEOREM 4.1

If a time function $x(t)$ is bandlimited to frequencies in the range $-W < f < W$ Hz, then

$$x(t) = \sum_n x(n/2W) \frac{\sin 2\pi W(t-n/2W)}{2\pi W(t-n/2W)} \tag{4.32}$$

and

$$X(f) = \begin{cases} (1/2W) \sum\limits_n x(n/2W) \exp[-j2\pi nf/2W], & |f| \leq W \\ 0, \text{ elsewhere}. \end{cases} \tag{4.33}$$

This means that a signal bandlimited to $\pm W$ Hz can be fully described by time samples taken every $1/2W$ seconds.

PROOF

Since $X(f)$ is limited to $|f| < W$, we can represent $X(f)$ by a Fourier series for a periodic function, with period $2W$, that equals $X(f)$ in the interval $-W < f < W$:

$$X(f) = \begin{cases} \sum_n c_n \exp[-j2\pi nf/2W], & |f| \le W \\ 0, \text{ elsewhere} . \end{cases} \tag{4.34}$$

This expansion can be equivalently written

$$X(f) = \sum_n c_n \exp[-j2\pi nf/2W] \; U_W(f) \;, \tag{4.35}$$

where

$$U_W(f) = \begin{cases} 1, & |f| < W \\ 0, & \text{elsewhere} . \end{cases} \tag{4.36}$$

Before proceeding, we remind the reader of the following Fourier transform pairs, $F[\sum_n \delta(t-n/2W)] = \sum_n \exp(-j2\pi nf/2W)$. Then, with F^{-1} designating the inverse Fourier transform and recalling that multiplication in the frequency domain implies convolution in the time domain,

$$\begin{aligned} x(t) &= F^{-1} \, [\sum_n c_n e^{-j2\pi nf/2W}] * F^{-1} \; U_W(f) \\ &= \sum_n c_n \, \delta(t-n/2W) * 2W \frac{\sin 2\pi Wt}{2\pi Wt} \\ &= 2W \sum_n c_n \frac{\sin[2\pi W(t-n/2W)]}{2\pi W(t-n/2W)} \; . \end{aligned} \tag{4.37}$$

The $\{c_n\}$ are simply the Fourier coefficients and are evaluated from

$$\begin{aligned} x(m/2W) &= \int_{-\infty}^{\infty} X(f) \; e^{j2\pi f(m/2W)} \; df \\ &= \sum_n c_n \int_{-W}^{W} e^{j2\pi f(m/2W)} \; df = 2Wc_m \;, \end{aligned} \tag{4.38}$$

where (4.35) has been substituted for $X(f)$, and it has been noted that all integrals are zero except that for $n = m$. Thus

$$c_m = x(m/2W)/2W \;, \tag{4.39}$$

and substitution into (4.35) and (4.37) yields the desired results.

4.3.1 THE NYQUIST CHANNEL

The time interval $T = 1/2W$ sec between samples of $x(t)$ in (4.32) and (4.33) is the *Nyquist interval*. Equivalently, if the sampling interval is T seconds then the frequency $W = 1/2T$ Hz (or the angular frequency π/T radians/sec) is the *Nyquist frequency*. These terms are named for Harry Nyquist, a Bell Laboratories scientist who in 1928 derived the conditions for noninterference of pulses transmitted at regular intervals [9].

The *sampling theorem* establishes that a function bandlimited to the Nyquist frequency is completely specified by samples at Nyquist intervals. Consider such a function $x(t)$ for which

$$x(nT) = \begin{cases} 1, & n=0 \\ 0, & n \neq 0 . \end{cases} \tag{4.40}$$

This is precisely the condition for no intersymbol interference introduced earlier. By (4.33),

$$X(f) = \begin{cases} T, & |f| < 1/2T \\ 0, & \text{elsewhere} . \end{cases} \tag{4.41}$$

This is the *Nyquist channel*. The time function

$$x_{Nyquist}(t) = \sin(\pi t/T)/(\pi t/T) \tag{4.42}$$

and its spectrum are shown in Figure 4.13.

From (4.41), the Nyquist channel is the only channel bandlimited to W Hz that has the samples specified by (4.40). That is unfortunate, because this channel has some disadvantages. It is, strictly speaking, unrealizable, requiring "brick wall" filtering. More significantly, the tail of the time function falls at a

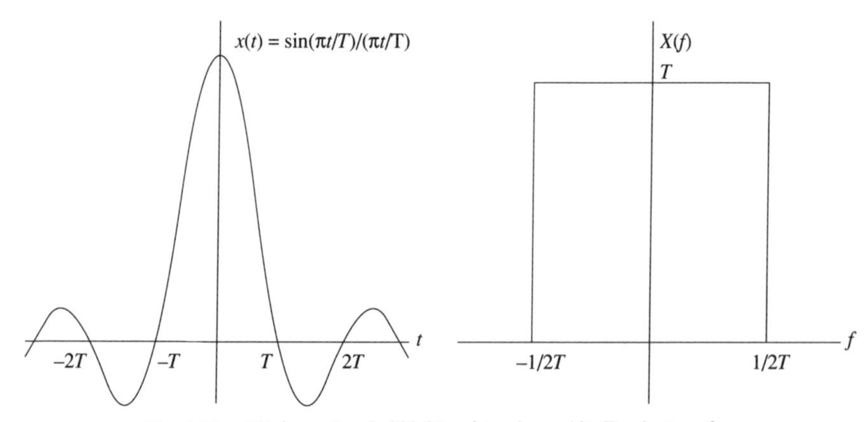

Fig. 4.13 Minimum bandwidth Nyquist pulse and its Fourier transform.

slow $1/t$ rate. Slight errors in the timing epoch of the samples of the equalized (receiver output) pulse in an actual communications system might add up to unacceptably large intersymbol interference. The astute reader may notice that $x_{Nyquist}(t)$ is *noncausal.* It exists for $t < 0$ and cannot be generated by any real-life filter. However, the delayed Nyquist pulse $x_{Nyquist}(t-NT)$ can be generated in a causal filter to any desired degree of accuracy (losing an insignificant left tail) by choosing N large enough.

4.3.2 THE EQUIVALENT NYQUIST CHANNEL (EXCESS BANDWIDTH PULSE)

In order to avoid the unrealistic filtering requirements and stringent timing epoch requirements associated with using $x_{Nyquist}(t)$ for signaling, a certain amount of extra (or *excess*) bandwidth is generally used. It can be shown that the same time samples (4.40) can be obtained from transfer functions $X(f)$ that are not limited to $\pm 1/2T$ Hz ($\pm\pi/T$ radians/sec). The amount of bandwidth beyond $1/2T$ Hz is the *excess bandwidth.* We shall now derive this class of transfer functions.

We desire all $X(f)$ such that

$$x_n = x(nT) = \int_{-\infty}^{\infty} X(f) \exp(j2\pi nfT)\, df = \delta_{n0} , \tag{4.43}$$

where δ_{n0} equals one for $n=0$ and zero for $n \neq 0$. To define this class, break the integral into the sum of integrals over $1/T$ segments of the frequency range.

$$\delta_{n0} = x_n = \sum_k \int_{(2k-1)/2T}^{(2k+1)/2T} X(f) \exp[j2\pi nfT]\, df =$$

$$\sum_k \int_{-1/2T}^{1/2T} X(f+k/T) \exp(j2\pi nfT)\, df = \int_{-1/2T}^{1/2T} X_{eq}(f) \exp(j2\pi nfT)\, df , \tag{4.44}$$

where

$$X_{eq}(f) = \begin{cases} \sum_k X(f+k/T), & |f| \le 1/2T \\ 0, & |f| > 1/2T . \end{cases} \tag{4.45}$$

That is, $X_{eq}(f)$ is constructed by folding segments of $X(f)$ on top of one another. Although the summation on the right-hand side of (4.45) is periodic, $X_{eq}(f)$ is restricted to the interval $[-1/2T,\ 1/2T]$ Hz.

It was already shown that

$$X_{eq}(f) = \begin{cases} \text{constant}, & |f| \le 1/2T \\ 0, & \text{elsewhere} \end{cases} \tag{4.46}$$

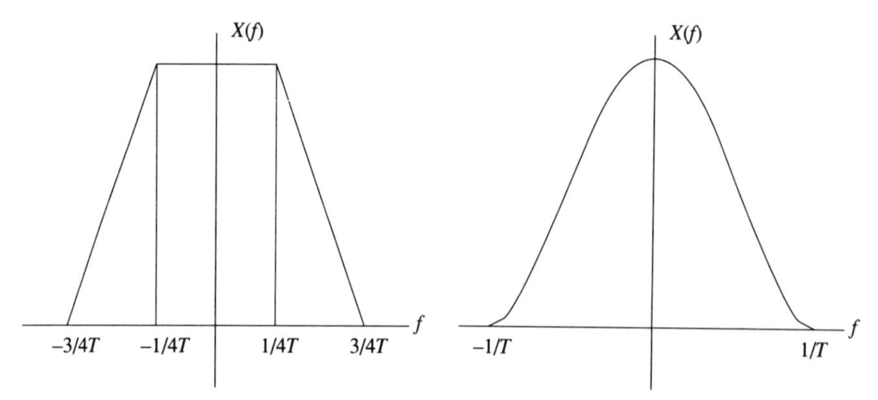

Fig. 4.14 Examples of spectra of excess-bandwidth pulses that satisfy the condition for elimination of intersymbol interference.

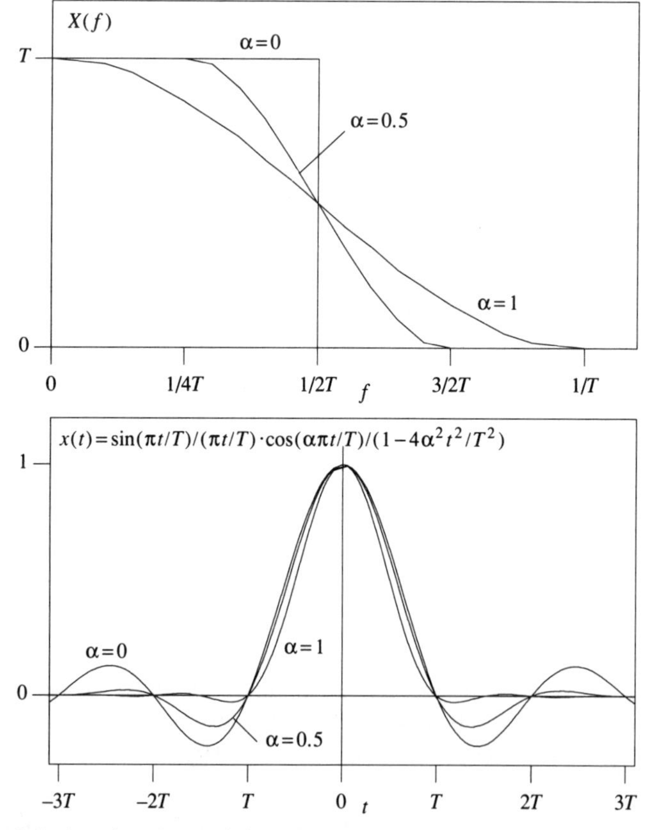

Fig. 4.15 Raised cosine characteristics. The parameter α is the rolloff factor, with $\alpha = 1$ corresponding to 100% excess bandwidth. (From [1].)

is the Fourier transform of the only function bandlimited to $[-1/2T, 1/2T]$ which has a single nonzero sample among samples taken at the time intervals T. Thus, we have the *Nyquist criterion for pulse shaping: X(f) eliminates intersymbol interference for samples at intervals T if and only if*

$$\sum_k X(f+k/T) = \text{constant}, \qquad |f| \leq 1/2T \,. \tag{4.47}$$

Recall that $X(f)$ is an overall transfer function incorporating transmit and receive filters and the channel (see (4.2) and Figure 4.1). Some examples of generalized Nyquist characteristics are illustrated in Figure 4.14.

If $X(f) = 0$ for $|f| > 1/T$, an alternative way to state the Nyquist criterion is that the real part of $X(f)$ must consist of the rectangular function plus any arbitrary function odd about $1/2T$, while the imaginary part of $X(f)$ may be any arbitrary even function about $1/2T$. The proof is left to Exercise (4.11).

4.3.3 RAISED COSINE PULSES

The impulse response for a generalized Nyquist characteristic $X(f)$, although having zeros at $t = nT$ (except for $t = 0$), is, of course, different from the $\sin(\pi t/T)/(\pi t/T)$ impulse response of the classical Nyquist channel. One desirable class of characteristics—not the only one, but widely used—is the *raised cosine* class (Figure 4.15). The transfer function of a member of this class is described as

$$X(f) = \begin{cases} T, & 0 \leq |f| < \dfrac{(1-\alpha)}{2T} \\[2ex] \dfrac{T}{2}\left\{1 - \sin\left[\dfrac{(2\pi|f|T - \pi)}{2\alpha}\right]\right\}, & \dfrac{(1-\alpha)}{2T} \leq |f| \leq \dfrac{(1+\alpha)}{2T} \\[2ex] 0, & |f| > \dfrac{(1-\alpha)}{2T} \,. \end{cases} \tag{4.48a}$$

A visual check will verify that superimposing components of $X(f)$ from the three 1/T intervals $(-3/2T, -1/2T)$, $(-1/2T, 1/2T)$, $(1/2T, 3/2T)$ yields a rectangular characteristic. The parameter α indicates the excess bandwidth over the Nyquist band. It is usually expressed as a percentage, equal to 100α.

The corresponding impulse response is

$$x(t) = \frac{\sin(\pi t/T)}{\pi t/T} \cdot \frac{\cos(\alpha\pi t/T)}{1 - 4\alpha^2 t^2/T^2} \,. \tag{4.48b}$$

It can be seen that $x(t)$ decreases asymptotically as $1/t^3$, in comparison with the $1/t$ of the regular Nyquist pulse (4.42).

It is not difficult to show that

$$\int_{-\infty}^{\infty} X(f)\,df = 1 \tag{4.49}$$

for all values of α. This will be used in the sequel.

4.3.4 PULSE SPECTRUM AND TAIL BEHAVIOR

In general, the tail characteristic of any pulse, including the Nyquist pulse, is determined by the frequency-domain characteristic of the pulse. The exact relationship is given in the following theorem [21]:

THEOREM 4.2

Let $X(f)$ be the Fourier transform of a pulse $x(t)$. If $X(f)$ and its first $k-1$ derivatives are continuous but its kth derivative is not, then $x(t)$ decays asymptotically as $1/|t|^{k+1}$.

PROOF

Assume the $k=1$ case where $X(f)$ is continuous and its first derivative, $X'(f)$, has a discontinuity at $f=f_0$. Using integration by parts, and noting that $X(f)$ and its derivatives are all zero at $\pm\infty$,

$$\begin{aligned}
x(t) &= \int_{-\infty}^{\infty} X(f)e^{j2\pi ft}df \\
&= X(f)(e^{j2\pi ft}/j2\pi t)\Big|_{-\infty}^{\infty} - (1/j2\pi t)\int_{-\infty}^{\infty} X'(f)e^{j2\pi ft}df \\
&= -\,(1/j2\pi t)\,[\int_{-\infty}^{f_0} X'(f)e^{j2\pi ft}df + \int_{f_0}^{\infty} X'(f)e^{j2\pi ft}df\,]\,.
\end{aligned} \tag{4.50}$$

Using integration by parts again,

$$\begin{aligned}
x(t) &= -(1/j2\pi t)\{X'(f)(e^{j2\pi ft}/j2\pi t)\Big|_{-\infty}^{f_0^-} - (1/j2\pi t)\int_{-\infty}^{f_0} X''(f)e^{j2\pi ft}df \\
&\qquad + X'(f)(e^{j2\pi ft}/j2\pi t)\Big|_{f_0^+}^{\infty} - (1/j2\pi t)\int_{f_0}^{\infty} X''(f)e^{j2\pi ft}df\} \\
&= -\,(1/j2\pi t)\{(e^{j2\pi ft}/j2\pi t)\,[X'(f_0^-) - X'(f_0^+)] \\
&\qquad - (1/j2\pi t)[\int_{-\infty}^{f_0} X''(f)e^{j2\pi ft}df + \int_{f_0}^{\infty} X''(f)e^{j2\pi ft}df\,]\}
\end{aligned}$$

$$= (1/2\pi t)^2 e^{j2\pi ft} \left[X'(f_0^-) - X'(f_0^+)\right] + (1/2\pi t)^2 \left[\int_{-\infty}^{f_0} X''(f) e^{j2\pi ft} df + \int_{f_0}^{\infty} X''(f) e^{j2\pi ft} df \right] \tag{4.51}$$

The first term goes to zero as $1/|t|^2$, and the second term declines asymptotically at least as fast as $1/|t|^2$ because the integrals yield square-integrable functions of time that cannot be asymptotically increasing with t. The theorem is thus proved for $k=1$, and can be similarly proved for high values of k by successive integration by parts.

For class 1 and class 4 partial response pulses, which are cases of practical interest described in Section 4.6, the discontinuities occur at the band edges, $\pm 1/2T$. The spectrum $X(f)$ in each case is continuous but there is a discontinuity in $X'(f)$ at $f = \pm 1/2T$. Consequently, the tails of these pulses fall as $1/|t|^2$. Higher order discontinuities are demonstrated in the next example.

EXAMPLE: RAISED COSINE PULSES

$$X(f) = \begin{cases} T, \; 0 \le |f| < (1-\alpha, /2T) \\ (T/2)\{1 - \sin[(2\pi |f| T - \pi)/(2\alpha)]\}, (1-\alpha)/2T \le |f| \le (1+\alpha)/2T \\ 0, \; |f| > (1+\alpha)/2T \end{cases}$$

$$x(t) = \frac{\sin \pi t/T}{\pi t/T} \cdot \frac{\cos \alpha \pi t/T}{1 - 4\alpha^2 t^2/T^2}$$

Let $f_0 = (1+\alpha)/2T$. At f_0^-,

$$X'(f_0^-) = (T/2) \left\{ (-\pi T/\alpha) \cos[(2\pi fT - \pi)/2\alpha] \right\} \Big|_{(1+\alpha)/2T}$$

$$= -(\pi T^2/2\alpha) \cos(\pi/2) = 0 . \tag{4.52}$$

Since $X'(f_0^+) = 0$, the first derivative is continuous. Moving to the second derivative,

$$X''(f_0^-) = -(\pi T^2/2) \left\{ (-\pi T/\alpha) \sin[(2\pi fT - \pi)/2\ ] \right\} \Big|_{(1+\alpha)/2T}$$

$$= (\pi^2 T^3/2\alpha^2) \sin(\pi/2) = \pi^2 T^3/2\alpha^2 . \tag{4.53}$$

But $X''(f_0^+) = 0$. Thus the second derivative is discontinuous. By the theorem, $x(t)$ should asymptotically (large t) fall as $1/t^3$, which in fact it does, as inspection of (4.49) shows.

4.4 PERFORMANCE OF MULTILEVEL PAM WITH RAISED COSINE PULSE SHAPING

Now that we have derived the condition for avoiding intersymbol interference and seen how a pulse characteristic meeting this condition should be shared between transmitting and receiving filters, we will return to consideration of the pulse amplitude modulation (PAM) model of Section 4.2, one of the most common applications of these ideas. We will examine error rate performance under the ideal circumstances of white Gaussian noise and no channel distortion in order to get an idea of the best one can do in such a system, and to examine the tradeoff between number of levels and bandwidth. Some of this discussion overlaps that of Section 2.2.6 in Chapter 2, but is is useful to show the relevance of the earlier results when there is a pulse train and a particular spectral shaping. The present discussion follows that in [1].

Assume, then, the system of Figure 4.10 with an ideal channel $H(f) = 1$, white Gaussian noise of two-sided spectral power density $N_0/2$ watts/Hz, and a set of L possible amplitude levels $\pm\Delta$, $\pm 3\Delta$, ..., $\pm(L-1)\Delta$. Assume further the raised cosine pulse shaping, with rolloff factor described in the last section.

As in previous examples with a white noise channel, and since the channel is ideal and the overall raised cosine pulse characteristic $X(f)$ is real and symmetric, $X(f)$ is split into identical transmitting and receiving filters $P(f)$ and $C(f)$ each equal to $|X(f)|^{1/2}$, where $X(f)$ is given by (4.48a). This corresponds to the matched filter of Chapter 2. Note that from (4.49) and Parseval's theorem, we have

$$\int_{-\infty}^{\infty} X(f)\,df = \int_{-\infty}^{\infty} P^2(f)\,df = \int_{-\infty}^{\infty} p^2(t)\,dt = 1 .$$

For this $X(f)$, the signal power at the output of the transmitter, averaged over all time and the ensemble of all possible data inputs, is

$$\begin{aligned} P_s &= \lim_{N\to\infty} E_{\{d_n\}} \left[\frac{1}{2NT} \int_{-NT}^{NT} s^2(t)\,dt \right] \\ &= \lim_{N\to\infty} (1/2NT) \left\{ E \int_{-NT}^{NT} [\sum_{n=-N}^{N} d_n p(t-nT)]^2 \, dt \right\} . \end{aligned} \tag{4.54}$$

With mutually independent data,

$$P_s = \lim_{N\to\infty} (\overline{d^2}/2NT) \sum_{n=-N}^{N} \int_{-NT}^{NT} p^2(t-nT)\,dt = \overline{d^2}/T , \tag{4.55}$$

independent of the rolloff factor α, where

$$\overline{d^2} = (\Delta^2/3)(L^2-1) \tag{4.56}$$

is the average power of the transmitted data level (see (2.42)). Thus

$$P_s = (\Delta^2/3T)(L^2-1) \ . \tag{4.57}$$

From this point the derivation is the same as that in Section 2.2.6. The probability of a symbol error is

$$P_e = 2(1-1/L)\ Q\left\{\left[\frac{6TP_s}{N_0(L^2-1)}\right]^{1/2}\right\}. \tag{4.58}$$

We recognize that TP_s is the energy in the pulse. Further, if P_N denotes the input noise power in the Nyquist bandwidth $(-\pi/T,\ \pi/T)$ in radians or $(-1/2T,\ 1/2T)$ in hertz, then

$$P_N = 1/T \text{ Hz} \cdot (N_0/2) \text{ watts/Hz} = N_0/2T \text{ watts} , \tag{4.59}$$

and

$$P_e = 2(1-1/L)\ Q\left\{\left[\frac{3}{L^2-1}\right]^{1/2}\left[\frac{P_s}{P_n}\right]^{1/2}\right\}. \tag{4.60}$$

Equivalently, using the relation for energy/bit,

$$E_b = \frac{(\text{energy/symbol})}{(\text{bits/symbol})} = \frac{P_sT}{\log_2 L} , \tag{4.61}$$

we have

$$P_e = 2(1-1/L)\ Q\left\{\left[\frac{6\log_2 L}{L^2-1}\cdot\frac{E_b}{N_0}\right]^{1/2}\right\}, \tag{4.62}$$

which reduces to (4.9) for binary-antipodal signaling when $L=2$. Note that (4.60) is equivalent to (2.43) with $E = E_b\log_2 L =$ energy/symbol interval. Curves of P_e vs. E_b/N_0 are shown in Figure 2.14. At error rates of the order of 10^{-6} or lower, doubling the number of levels, to allow transmission of one more information bits per symbol without any change in bandwidth calls for at least 7 dB increase in average transmitted power.

4.4.1 MORE LEVELS VS. MORE BANDWIDTH

As we noted at the beginning of Section 4.2 and again just above, PAM with more levels requires more power (or more energy/bit) to attain adequate spacing between levels for the same P_e. We can do a rough numerical analysis to see the penalty paid for increasing data rate without increasing bandwidth. Assume we want to double the data rate without increasing P_e. As (4.22) states, the data rate is proportional to $\log_2 L$. To double the data rate from a system with L levels, we need L^2 levels. Replacing L by L^2 in the argument of the Q function in (4.58), using an increased energy/bit E_b' (to be determined), and ignoring the algebraic factor $(1-1/L)$, we have

$$P_e = 2Q\left\{\left[\frac{6\log_2 L^2}{L^4-1} \cdot \frac{E_b'}{N_0}\right]\right\}$$

$$= 2Q\left\{\left[\frac{6\log_2 L}{L^2-1} \cdot \frac{2E_b'}{(L^2+1)N_0}\right]^{1/2}\right\}. \tag{4.63}$$

For the same P_e, the argument of (4.63) must be same as that of (4.62), i.e.,

$$E_b' = \frac{(L^2+1)E_b}{2} . \tag{4.64}$$

In contrast, when bandwidth is unlimited, (2.35) states that for binary antipodal signaling,

$$P_e = 0.5\ Q\,(\sqrt{2E_b/N_o})\ ,$$

which is independent of the data rate and the bandwidth consumed. That is, when the data rate is increased, no added signal energy per bit is needed for the same error rate. Note, however, that signal power must increase linearly with the data rate to maintain constant E_b.* Table 4.1 illustrates the penalty for conserving bandwidth.

The practical implication of this calculation is that if the bandwidth is available, higher data rates are best achieved by increasing the symbol rate rather than the number of levels. In practice, for channels that are distorted or attenuated as a function of frequency, the tradeoff between increasing the symbol rate and increasing the number of levels has to be calculated. Similarly, if the noise increases with frequency, as it would in a crosstalk-dominated environment, it would also be expected that it would be preferable to limit the bandwidth.

---

* The channel capacity (2.112) has a constant asymptotic limit (2.113) as the bandwidth becomes arbitrarily large, if average power is fixed. Here, however, where power is increasing, the asymptotic channel capacity increases indefinitely with bandwidth.

Table 4.1: Binary Antipodal Signaling, Unlimited Bandwidth

Relative data rate	1	2	4	4
Number of levels	2	2	2	2
Relative BW consumed	1	2	4	8
Energy/bit E_b'	E_b	E_b	E_b	E_b

Multilevel Signaling, Constant Bandwidth

Relative data rate	1	2	4	8
Number of levels	2	4	16	256
Relative BW consumed	1	1	1	1
Energy/bit E_b'	E_b	2.5 E_b	8.5 E_b	128.5 E_b

4.5 GENERAL ENCODING MODEL

In the immediately preceding sections, we have been dealing with signals which are independent from one signaling interval to the next. In this section we treat schemes which introduce correlation between the intervals for the purpose of shaping the spectrum of the signal. We will look at two specific examples of spectral shaping, directed respectively to elimination of DC and amelioration of intersymbol interference.

The general form of correlation encoding is depicted in Figure 4.16 [10]. The data sequence, $\{a_n\}$, is fed into a precoder which prevents *error propagation*, which is the phenomenon of a single error caused by noise inducing a whole string of subsequent errors. The output sequence, $\{b_n\}$, feeds into a line coder which introduces the correlation between signal intervals. Both a_n and b_n take on the values $0, 1, \ldots, L-1$, in general. The output sequence, $\{w_n\}$, is a sequence of integers with the desired spectral properties. As we shall see, the box labeled "Line coding" represents a considerable abstraction. As with Nyquist pulses, the operation is shared between the transmitter and the receiver.

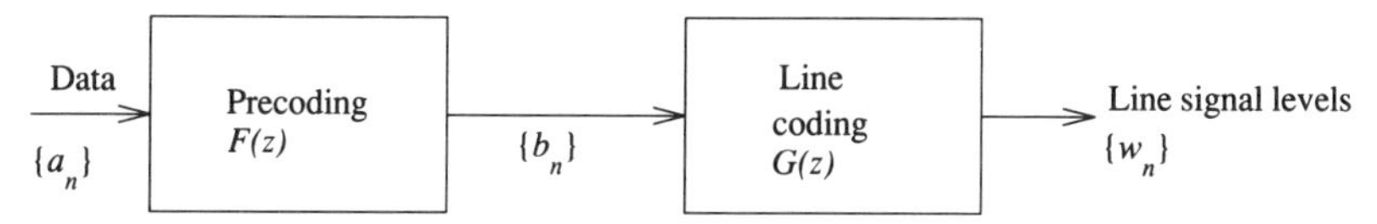

Fig. 4.16 Generation of encoded line signal levels from precoding and level coding operations. Here a_n is an element in the input data sequence, b_n is an element in the intermediate data sequence, and w_n is a symbol-interval level of the line signal.

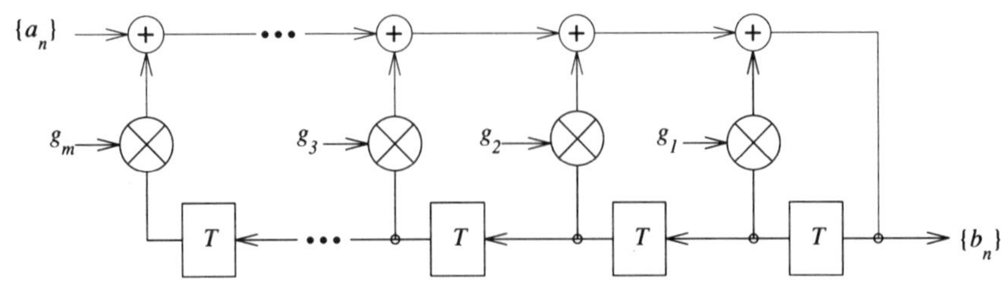

Fig. 4.17 Precoder.

The operation of the system is best explained by beginning with the line encoder which may be represented by the polynomial

$$G(z) = \sum_{i=0}^{n} g_i \, z^i \tag{4.65}$$

The sequence $\{ w_n \}$ can be written

$$w_n = \sum_{i=0}^{n} b_{n-i} \, g_i \tag{4.66}$$

where g_i, $i = 0, 1, \ldots, m$, embodies the encoding action. Without loss of generality, we may set $g_o = 1$. If w_n and b_i, $i = n-m, \ldots, n-1$, were known, the term b_n could be calculated by simple arithmetic. The difficulty is that an error in one of the b_i would propagate to succeeding symbols.

The precoder prevents propagation by performing an inverse of the line encoding. Let b_n be given by the recursive relation

$$b_n = \left[a_n - \sum_{i=1}^{n} b_{n-i} g_i \right] \mod L \,, \tag{4.67}$$

which can be written

$$a_n = \sum_{i=0}^{n} b_{n-i} \, g_i \mod L \,. \tag{4.68}$$

By taking z-transforms on both sides of (4.68)

$$B(z) = A(z)/G(z) \,, \tag{4.69}$$

where $A(z)$ and $B(z)$ are the transforms of the sequences $\{ a_n \}$ and $\{ b_n \}$, respectively. The precoder $F(z)$ of Figure 4.16 has now been defined as the inverse $1/G(z)$ of the coder. The division operation represented by (4.69) can be carried out by the linear sequential circuit shown in Figure 4.17,* where all arithmetic is modulo L.

---

* The role of linear sequential circuits, such as shown in Figure 4.17, in realizing multiplication and division of polynomials was discussed in Section 3.2.4.

Now we show the error propagation prevention mechanism. Equation (4.67) can be written

$$b_n = a_n - \sum_{i=1}^{n} b_{n-i}\, g_i + kL , \tag{4.70}$$

where k is an arbitrary integer selected to bring b_n into the range $[0,...,L-1]$, and where ordinary arithmetic is used. By substitution into (4.66), we find

$$\begin{aligned} w_n \bmod L &= [b_n + \sum_{i=1}^{n} b_{n-i}\, g_i] \bmod L \\ &= [a_n - \sum_{i=1}^{n-i} g_i + kL + \sum_{i=1}^{n-i} g_i] \bmod L = a_n . \end{aligned} \tag{4.71}$$

Thus a_n is recovered from w_n at the receiver independent of any other (possibly erroneous) earlier line signal samples.

4.5.1 AMI CODING

We now apply coding and precoding to AMI for $L=2$. As in Figure 4.18 the line encoding is with the corresponding transfer function

$$w_n = b_n - b_{n-1} \tag{4.72a}$$

$$G(z) = 1 - z \tag{4.72b}$$

with the precoding operation

$$F(z) = 1 \oplus z \oplus \cdots \oplus z^i \oplus \cdots , \tag{4.73}$$

where, as in the rest of the text, $\oplus$ denotes modulo 2 addition. In this particular case, the sequence $\{w_n\}$ is the sequence of pulse heights which are transmitted over the channel. An example of AMI coding and decoding is shown in Table 4.2. We see that alternate ones are encoded into pulses of alternating sign, thereby eliminating DC. We also see that the original data sequence $\{a_n\}$ is contained in the pulse heights. The expressions for b_n, w_n, and recovery of a_n come from formulas (4.67), (4.66), and (4.71), respectively.

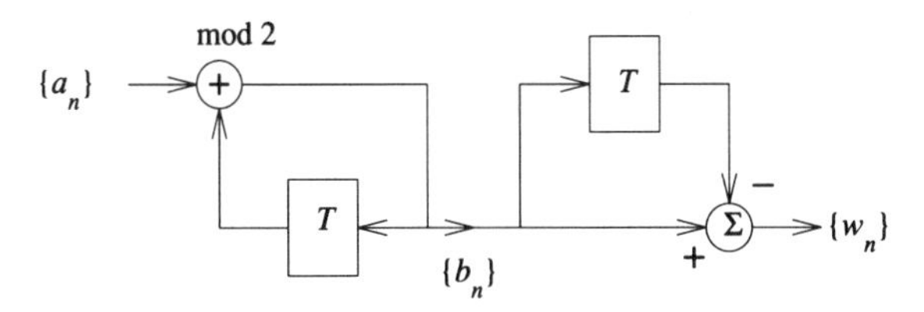

Fig. 4.18 AMI encoding.

Table 4.2: AMI Coding

n (time index)	<0	0	1	2	3	4	5	6	7	8	9	10	11	12
a_n	0	1	0	0	0	0	0	1	0	0	0	0	1	0
$b_n = a_n + b_{n-1}$ (mod 2)	0	1	1	1	1	1	1	0	0	0	0	0	1	1
$w_n = b_n - b_{n-1}$	0	1	0	0	0	0	0	−1	0	0	0	0	1	0
$a_n = w_n$ (mod 2)	0	1	0	0	0	0	0	1	0	0	0	0	1	0

4.6 CORRELATIVE LEVEL ENCODING (PARTIAL RESPONSE)

As we have seen, for Nyquist signaling the overall response of the system is such that adjacent symbol intervals are independent of one another. However, by introducing correlation in a controlled fashion, as with AMI of the last section, the overall response of the system can be shaped so as to reduce the effect of intersymbol interference [11,12]. The term *partial response* is used in connection with this technique since the response to an input symbol is spread over more than one symbol interval—the response in a single interval is partial. The crucial spectral shaping is at the edge of the Nyquist band.

The block diagram of the partial response system is shown in Figure 4.19. In this case the overall impulse response is given by

$$x(t) = p(t) * h(t) * c(t) = \sum_{n=0}^{m} g_n x_{Nyquist}(t - nT) , \tag{4.74a}$$

where g_n, $n = 0, 1, \dots, m$ are the coefficients of the encoding process (4.65) and

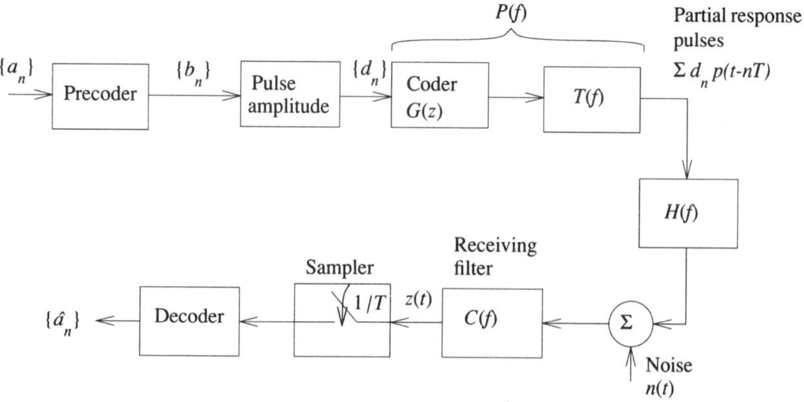

Fig. 4.19 Communication model for partial response signaling $T(f)H(f)C(f) = X_{Nyquist}(f)$.

where $x_{Nyquist}(t)$ is the minimum bandwidth Nyquist pulse of Figure 4.13. The overall transfer function is given by

$$X(f) = \begin{cases} T\sum_{n=0}^{m} g_n e^{-j2\pi n fT} = TG(e^{-j2\pi fT}), & |f| < 1/2T \\ 0, & \text{otherwise}, \end{cases} \tag{4.74b}$$

where (4.65) has been used with $\exp(j2\pi fT)$ substituted for z to express the Fourier transform. Note that the entire spectrum is contained in the band $|f| \le 1/2T$.

From (4.74b) it is clear that the pulse frequency characteristic is determined by the polynomial $G(z)$. It is also clear from Theorem 4.2 that the behavior of the pulse tail is a function of continuity at the band edges $\pm 1/2T$. The tail behavior is given by the following theorem [21]:

THEOREM 4.3

The first $k-1$ derivatives of $X(f)$ given in Theorem 4.2 are continuous if and only if $G(z)$ has $(1+z)^k$ as a factor.

PROOF

From the discussion in Section 4.6, it is evident that for arbitrary line coding of partial response signals

$$X(f) = X_{Nyquist}(f)G(e^{-j2\pi fT}) . \tag{4.75}$$

Since $G(z)$ is a polynomial in z, all of the derivatives of $G(e^{-j2\pi fT})$ with respect to f exist and are continuous. Any discontinuity of $X(f)$ is due to the discontinuity of $X_{Nyquist}(f)$ at $f = \pm 1/2T$. Thus from (4.75), we see that the first k derivatives of $X(f)$ are continuous at these points if the first $k-1$ derivatives of $G(e^{-j2\pi fT})$ are equal to zero. In order for this to obtain, it is necessary that $G(z)$ have $(1+z)^k$ as a factor.

Sufficiency: Let $G(z) = A(z)\ (1+z)^k$ where $A(z)$ is a polynomial in z not having a root at $z = -1$. We have $d^nG(z)/dz^n|_{z=-1} = 0$ for $0 \le n \le k-1$. However,

$$d^nG(e^{-j2\pi fT})/df^n|_{f=1/2T} = d^nG(z)/dz^n|_{z=-1}\ (de^{-j2\pi fT}/df) = 0 . \tag{4.76}$$

Necessity: Assume that

$$d^nG(e^{-j2\pi fT})/df^n|_{f=1/2T} = 0 , \tag{4.77}$$

which implies that

$$d^n G(z)/dz^n \big|_{z=-1} = 0 \; ; \qquad 0 \le n \le k - 1 \; . \tag{4.78}$$

Figure 4.20 shows a number of partial response pulses and their Fourier transforms. All of the pulses have more than one nonzero symbol-interval sample, and all have spectra with desirable characteristics, as we will illustrate below. The coding polynomial defines the class.

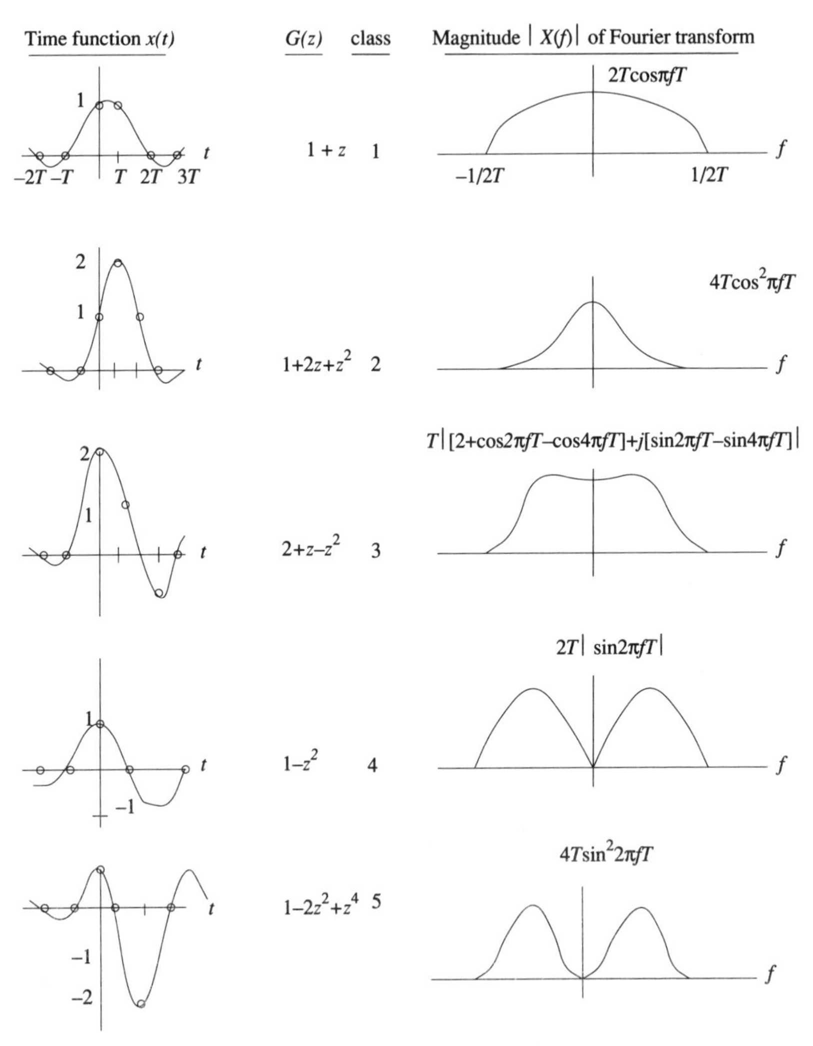

Fig. 4.20 Correlative level pulses. Energy near $f=\pm 1/2T$ Hz, and in some cases near $f=0$, is reduced by allowing controlled intersymbol interference. The class 1 pulse is sometimes drawn centered on $t=0$, implying sampling times at $t=\pm T/2, \pm 3T/2, \ldots$.

4.6.1 CLASS 1 PARTIAL RESPONSE (DUOBINARY)

For purposes of explanation, we consider the simplest example, *Class 1 partial response*, which is also called *duobinary coding* [13,14]. In this case, the overall system response, combining transmit and receive filters and the channel, is represented by the characteristic shown at the top of the right-hand column in Figure 4.20. The pulse shape $x(t)$, at the top of the left-hand column, is equivalent to the sum of a minimum bandwidth Nyquist pulse at $t=0$ and another at $t=T$, i.e.,

$$x(t) = x_{Nyquist}(t) + x_{Nyquist}(t-T), \tag{4.79a}$$

where the Nyquist characteristic is given by (4.42). Note that

$$x(nT) = \begin{cases} 1, & n = 0,1 \\ 0, & \text{otherwise.} \end{cases} \tag{4.79b}$$

The transform of $x(t)$, as shown in Figure 4.20, is given by

$$X(f) = F[x(t)] = X_{Nyquist}(f)[1 + e^{-j2\pi fT}]$$

$$= \begin{cases} 2Te^{-j2\pi fT}\cos(\pi fT), & |f| \le 1/2T \\ 0, & |f| > 1/2T \end{cases} \tag{4.79c}$$

Note that $X(f)$ has minimum bandwidth and that there is continuity at the band edge. It will be shown in the sequel that the tails of $x(t)$ decay as t^{-2} because of this continuity. (Recall that the tail decays as t^{-1} for the minimum bandwidth Nyquist pulse.)

Suppose that the transmitting and receiving filters are designed so that the overall impulse response of the system is given by $x(t)$, i.e., $x(t) = p(t)*h(t)*c(t)$ or in the frequency domain $X(f) = P(f)H(f)C(f)$ (Figure 4.19). In order to achieve the partial response characteristic, the overall response must be characterized by (4.79c) with suitable delay to insure realizability. In the sequel we shall assume that this transfer function is split evenly between the transmitter and the receiver with the channel assumed to be flat:

$$P(f) = C(f) = [X(f)]^{1/2}. \tag{4.80}$$

As in the case of PAM the transmitted pulse amplitudes are denoted by the sequence $\{d_n\}$, which takes on the values

$$d_n = 2b_n + 1 - L. \tag{4.81a}$$

in the general case, where $\{b_n\}$ is an L-ary sequence. This operation, shown in

Figure 4.19 just before the coder, is done to generate pulse levels symmetric about zero. It was, for simplicity, omitted from our earlier discussion of line signal level coding at the beginning of Section 4.5. For binary signaling this is simply

$$d_n = 2b_n - 1 \ . \tag{4.81b}$$

The transmitted sequence is then

$$s(t) = \sum_n d_n \, p(t - nT) \tag{4.82}$$

with pulse shape given by (4.80).

The output of the receive filter may be written

$$z(t) = r(t) + \nu(t) \ , \tag{4.83}$$

where $r(t)$ is the received signal component and $\nu(t)$ is the result of passing white Gaussian noise through the receive filter $c(t)$. From Figure 4.19 and (4.79a), the expected value of the output is

$$E(z(nT)) = r(nT) = d_n + d_{n-1} = 2[b_n + b_{n-1} - 1] \ . \tag{4.84}$$

By removing constant factors, we see that the decision is based on the terms $b_n + b_{n-1}$. (This is w_n in Figure 4.16.) Hence, the line coding is characterized by the polynomial

$$G(z) = 1 + z \tag{4.85}$$

and the proper precoding is, from (4.67), $b_n = a_n \oplus b_{n-1}$, which may be characterized by

$$F(z) = 1 \oplus z \oplus \cdots \oplus z^i + \cdots \ . \tag{4.86}$$

From the discussion in Section 4.5, we see that $(b_n + b_{n-1}) \bmod 2$, $r(nT) = a_n$. The truth table in Table 4.3 shows the mean values of the filter output $r(nT)$ for the various combinations of a_n and b_{n-1}. (See (4.65), (4.68), and (4.84).) As we see, each value of the three-level receiver signal output (4.84) uniquely defines a value of a_n.

Table 4.3: $r(nT) = 2\left[a_n \oplus b_{n-1} + b_{n-1} - 1\right]$

		a_n	
		0	1
b_{n-1}	0	−2	0
	1	2	0

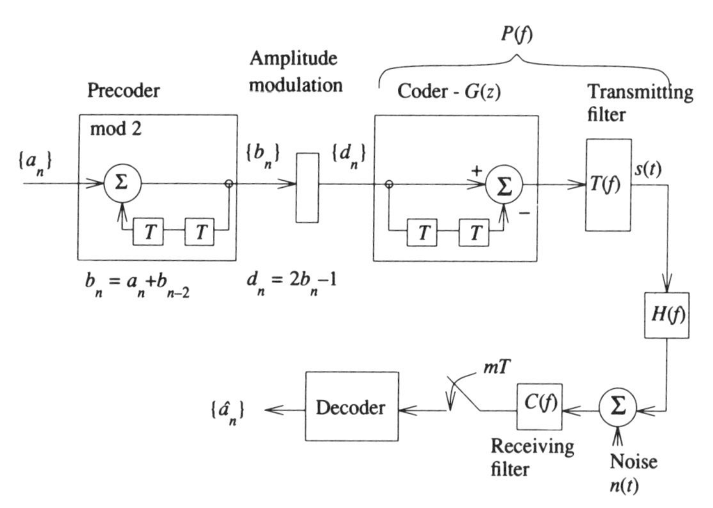

Fig. 4.21 Communication model for Class 4 partial response (modified duobinary). $P(f)\ H(f)\ C(f) = 2T|\sin 2\pi fT|$, $T(f)H(f)C(f) = X_{Nyquist}(f)$, and $s(t) = \sum_n d_n p(t-nT)$.

4.6.2 CLASS 4 PARTIAL RESPONSE (MODIFIED DUOBINARY)

We now consider the *Class 4 partial response* technique, also called *modified duobinary*. The communication model is shown in Figure 4.21. This class of signaling has the same performance as duobinary with respect to intersymbol interference with the additional virtue of having no DC. Duobinary signaling has found extensive application in single sideband data transmission (see Chapter 5) because it can be confined to the Nyquist band with relatively simple filtering.

The overall impulse response of the modified duobinary system is shown in the fourth position of the left-hand column in Figure 4.20. The impulse response of the system is

$$x(t) = x_{Nyquist}(t) - x_{Nyquist}(t-2T), \tag{4.87}$$

so that

$$x(nT) = \begin{cases} 1, & n = 0 \\ -1, & n = 2 \\ 0, & \text{otherwise} \end{cases} \tag{4.88}$$

The transform of $x(t)$ is given by

$$X(f) = F[x(t)] = X_{Nyquist}(f)\ [1 - e^{-j4\pi fT}]$$

$$= \begin{cases} 2T\ j\ e^{-j2\pi fT}\ \sin(2\pi fT), & |f| \le 1/2T \\ 0, & |f| > 1/2T \end{cases}. \tag{4.89a}$$

Table 4.4: $r(nT) = 2\left[a_n \oplus b_{n-2} - b_{n-2}\right]$

		a_n	
		0	1
b_{n-2}	0	0	2
	1	0	-2

We assume that the response is split evenly between the transmitter and the receiver (see (4.80)). Note that there is a null in the spectrum at DC. Further, there is the same continuity at the band edge as in Class 1 PR, thereby insuring pulse tails which decay as t^{-2}, as we shall see presently.

The pulse coding is very similar to what we have just described for Class 1 PR. Consequently the output of the receive filter has the expected value

$$w_n = E(z(nT)) = r(nT) = d_n - d_{n-2} = 2\left[b_n - b_{n-2}\right] \quad (4.89b)$$

implying a coding which is characterized by the polynomial

$$G(z) = 1 - z^2 = (1-z)(1+z) \ . \quad (4.89c)$$

As in AMI, the factor $(1-z)$ eliminates DC while the factor $1+z$ ameliorates intersymbol interference, as in Class 1 PR. In order to prevent error propagation, we use the precoding $b_n = a_n \oplus b_{n-2}$ which may be characterized by

$$F(z) = 1 \oplus z^2 \oplus z^4 + \cdots = 1/(1 \oplus z^2) \ . \quad (4.89d)$$

The truth table is shown in Table 4.4, as derived from (4.65), (4.68), and (4.89b).

4.6.3 PROBABILITY OF ERROR WITH SYMBOL-BY-SYMBOL DETECTION OF MODIFIED DUOBINARY SIGNALS

As we noted earlier, partial response signals give spectral shaping advantages at the expense of noise immunity when each pulse is individually detected. In this section, we derive the probability of error for symbol-by-symbol detection of a train of class 4 partial response pulses, and show the SNR penalty with respect to binary-antipodal signaling.

With signal levels 0 and ±2 at the receiver output (Table 4.4), let the boundaries of the "decide 0" signal region be ±1. (The optimum threshold between decision regions, and the sub-optimum one of Figure 4.23, are derived in Exercise 4.6.) Assume that the signal-to-noise ratio is large. It is possible that a noise sample could deflect a received sample from one part of the outer decision region all the way to the other part, but this is a very low probability event. We can assume that a noise large enough to move a received sample across the nearest decision boundary will almost always cause an error. Assume further that the original data levels are equally likely and, as usual, that the noise is white Gaussian with two-sided power density $N_0/2$ watts/Hz. Then if ν is the noise sample at the output of the receiver filter, we have, as in (4.20),

$$P_e = Pr[\nu > 1] + (1/2)Pr[\nu > 1] = (3/2)Pr[\nu > 1] \,. \tag{4.90}$$

From (4.80) and (4.89a), the random variable ν has zero mean and variance

$$\sigma^2 = N_0/2 \int_{-1/2T}^{1/2T} |C(f)|^2 df = N_0 T \int_{-1/2T}^{1/2T} |\sin 2\pi f T| df = 2N_0/\pi \,. \tag{4.91}$$

The energy per transmitted pulse is, since the pulse amplitudes are ± 1,

$$E_0 = \int_{-1/2T}^{1/2T} |P(f)|^2 df = 2T \int_{-1/2T}^{1/2T} |\sin 2\pi f T| df = 4/\pi \,. \tag{4.92}$$

Then

$$P_e = (3/2)(1/\sqrt{2\pi\sigma^2}) \int_1^\infty \exp[-z^2/2\sigma^2] dz = (3/2)\ Q[(\pi/4)\sqrt{2E_0/N_0}] \,. \tag{4.93}$$

By comparison, the argument of the $Q(\cdot)$ for binary-antipodal signaling is, from (2.35), $\sqrt{2E_0/N_0}$. The energy required by modified-duobinary signaling to achieve the same P_e as binary antipodal signaling is $(4/\pi)^2$ greater, or 2.1 dB, as shown also in Figure 4.22. A price has been paid in SNR for an advantage in spectral shaping and, not so incidentally, in a side benefit: the mutual dependence of signal level transitions (no two successive positive or negative line signal samples can occur), which can be used to detect severe degradation of a communications link.

Symbol-by-symbol detection of correlative level signals suffers from loss of SNR because past information is used only *after* detection. If one could use *all* the information in an entire transmitting sequence to make the decisions about signal levels, there might not be such a loss.

In fact, by using maximum likelihood sequence estimation, introduced in Chapter 3 for convolutional decoding, correlative level coding suffers no loss at all. In Chapter 7, the Viterbi algorithm for maximum likelihood sequence estimation is applied to received data signals. The example of a train of class 4 correlative level pulses is given in Section 7.1.6, and it is shown that the 2.1 dB loss of symbol-by-symbol detection is completely recovered.

4.6.4 HIGHER-ORDER PULSES FOR MAGNETIC RECORDING

In digital magnetic recording systems for high-density data storage, it has long been recognized that certain partial response characteristics are well

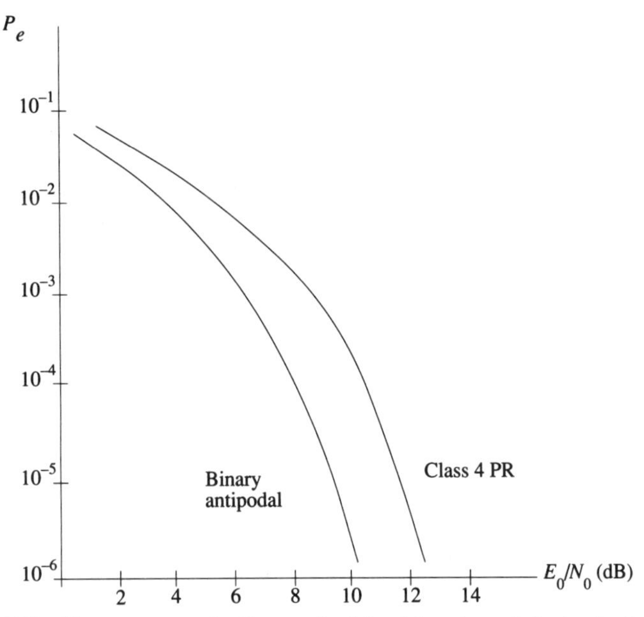

Fig. 4.22 Error rate curves for binary antipodal and class 4 correlative level pulses.

matched to the recording channel characteristics and result in higher storage density for a specified error-rate performance [15]. In particular, use of the class 4 characteristic $(1-z^2)$ and Viterbi algorithm for detection has been shown to offer improved performance [16]. But better performance still is possible with more complex partial response pulses not included in Figure 4.19. A class of higher-order partial response characteristics for magnetic recording has been proposed by Thapar and Patel [17], whose work is followed here.

The linear system shown in Figure 4.23 is characterized by a write channel transfer function (including zero-order hold)

$$H_{write}(f) = j\sin(\pi fT), |f| \leq 1/2T, \tag{4.94}$$

which corresponds to a sampled-data pulse characteristic* model of saturation recording

$$H_{write}(z) = 1 - z. \tag{4.95}$$

$H_{write}(f)$ is a high-pass characteristic which must be a component of whatever pulse spectrum is seen by the detector reading from the magnetic medium. However, the pulse spectrum should be more of a bandpass characteristic to con-

* Equation (4.95) corresponds to a time function $\delta(0) - \delta(t-T)$ with Fourier transform $2\exp[-j\pi fT] \cdot \sin(\pi fT)$, but the $\exp[-j\pi fT]$ term is only a time shift.

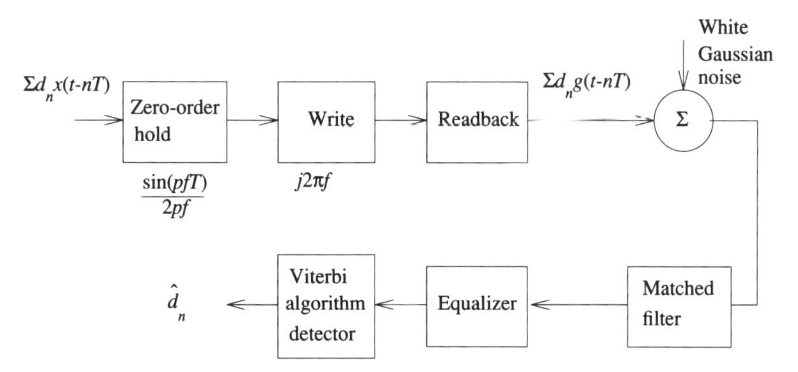

Fig. 4.23 System model for saturation recording and readback.

serve bandwidth and be consistent with other parts of the channel transfer function, namely the readback section. The class 4 partial response signal provides a spectral null at $1/2T$ Hz, producing a bandpass characteristic, but for very high recording density, the channel equalization that is necessary introduces considerable noise enhancement that reduces the SNR at the detector input.

The class of partial response signals that may more closely match the readback signal is expressed as

$$P_n(z) = (1-z)(1+z)^n, \tag{4.96}$$

where $(1+z)^n$ is a low-pass characteristic that limits bandwidth more with increasing n. The value $n=1$ produces the class 4 (duobinary) pulse; $n=2$ produces a sharper-cutoff pulse used by Forney [18].

The Fourier transform of (4.96), which can be written by simply replacing each z by $\exp(-j2\pi fT)$, is

$$H_n(f) = jT2^n \cos^{n-1}(\pi fT)\sin(2\pi fT). \tag{4.97}$$

Figure 4.24 displays $H_n(f)$ for various n. The signal energy shifts toward lower frequencies with increasing n. If, for increasing n, the symbol interval T is reduced so that roughly the same spectral region is occupied, the data rate (and thus the data storage density) increases correspondingly.

There is increased loss from intersymbol interference and reduced SNR as n is increased, so there is not an easy gain. Using the Viterbi algorithm for detection, the loss due to intersymbol interference is approximated by the ratio $d_{\min}^2/E_p$, where $d_{\min}$ is the minimum distance between soft decision sequences and E_p is the energy in the partial response pulse. Table 4.5 shows this loss for various values of n. For $n=1$ or $n=2$ there is no loss; for $n \geq 3$, the loss is 2.2 dB and it increases with n.

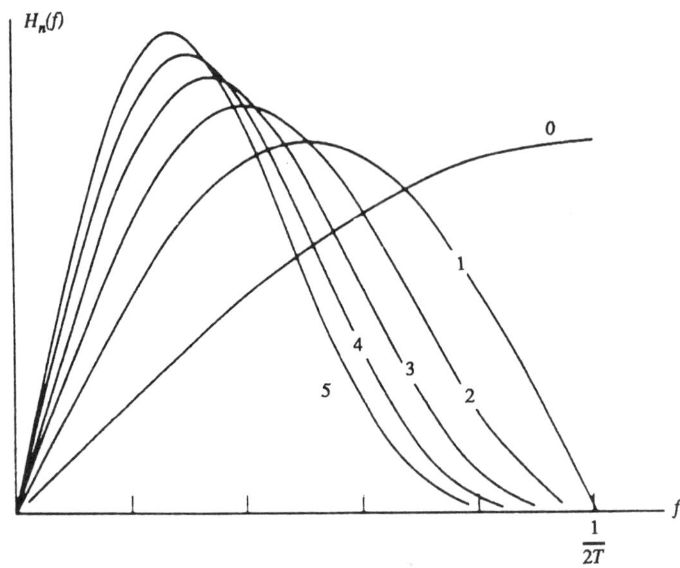

Fig. 4.24 Channel transfer functions $H_n(f)$ for fixed T and various values of n, the power of $(1+z)$ in (4.96). (From [17].)

Although it appears advisable to stick to small values of n, the loss from intersymbol interference must be traded off against the loss from noise enhancement in channel equalization that decreases with increasing n. For increasing recording density as the step response of the entire channel becomes narrower, higher values of n become optimum. For example, $n=3$ was shown in [17] to be the optimum value for a step response roughly 1/3 as wide as a step response for which $n=0$ is optimum.

Table 4.5: Parameters for Higher-Order Pulse Filters (from [17])

$$P_n(z) = (1-z)(1+z)^n$$

n	Number of levels	E_p	$d^2_{\min}$	$d^2_{\min}/4E_p$	SNR loss (dB) due to ISI
0	3	2	8	1	0
1	3	2	8	1	0
2	5	4	16	1	0
3	7	10	24	0.6	2.2
4	13	28	48	0.4	3.7
5	19	84	120	0.3	4.5

4.7 BLOCK CODES: A MULTIRATE DIGITAL FILTERING APPROACH

Partial-response signaling, described in the last section, is an example of convolutional coding (Chapter 3), since a line signal level is determined by a linear combination of past input data. We shall see a further example, for passband signals, in the section of trellis coded modulation in Chapter 5.

An alternative approach is block coding [4], where a block of input data is transformed into a different block of pulse amplitudes. In general, L symbols with amplitudes selected from an L_1-level alphabet are mapped into a block of K symbols with amplitudes selected from a K_1-level alphabet. One simple example is generation of Manchester (biphase) signals. A block of one binary input symbol is transformed into a two-pulse block at twice the rate. Another simple example, PAM, is the transformation of a block of m binary symbols into a block of one multilevel pulse taking on 2^m possible values and appearing at $1/m$ of the input rate. A less trivial example discussed in Section 4.8 is 4B3T, which transforms a block of 4 binary symbols into a block of 3 ternary symbols at 3/4 of the input rate. Since not all output sequences are used, the codings are less than 100% efficient.

Block coding is used to implement a tradeoff between bandwidth and noise immunity (level spacing) and to control spectral characteristics. There is interest in generating block codes in low-cost FIR (finite impulse response) digital filters, but linear, time-invariant digital filters have some limitations. One is that the number of output levels tends to be large unless the impulse response of the filter consists only of a few samples that are different from zero, which limits the possibilities for spectral shaping. A second is that the output rate and the input rate are the same, unlike the three examples suggested above.

A class of linear, time-varying ("multirate") digital filters for generation of block codes has been proposed by Peek and Lakeman [19]. An output sequence with a limited number of levels, desired spectral characteristics (such as nulls at band edges), and rate K/L times the input rate, where K and L are two positive integers, can be readily generated from random binary input data. The technique transforms a block of L input symbols into a block of K output symbols by a matrix transformation

$$\boldsymbol{d} = W\boldsymbol{\alpha} \,,$$

$$W = \begin{bmatrix} w_{11} & w_{12} \cdots w_{1L} \\ w_{21} & w_{22} \cdots w_{2L} \\ \vdots & \vdots \qquad \vdots \\ w_{K1} & w_{K2} \cdots w_{KL} \end{bmatrix} \tag{4.98}$$

where $\boldsymbol{\alpha}$ is an L-element block of ± 1 symbols (a column vector of length L). The transformation is realized in K FIR filters as shown in Figure 4.25.

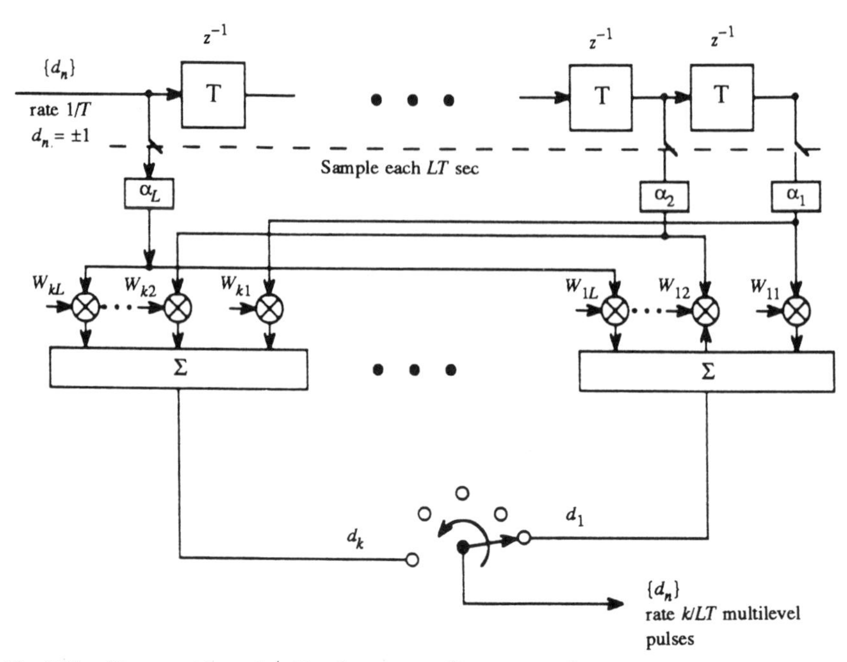

Fig. 4.25 Linear multirate signal level generator. The commutator makes a complete cycle each LT seconds, and issues K output symbols.

An example of an effective multirate coding is provided by the $(K = 6,\ L = 10)$ matrix

$$W = \begin{bmatrix} 0 & 1 & 0 & 0 & -1 & 0 & 2 & 0 & 0 & -2 \\ 0 & -1 & 0 & -1 & 0 & 0 & -2 & 0 & -2 & 0 \\ -1 & 0 & 0 & 1 & 0 & -2 & 0 & 0 & 2 & 0 \\ 0 & 0 & -1 & -1 & 0 & 0 & 0 & -2 & -2 & 0 \\ 1 & 0 & 0 & 1 & 0 & 2 & 0 & 0 & 2 & 0 \\ 0 & 0 & 1 & 0 & 1 & 0 & 0 & 2 & 0 & 2 \end{bmatrix} \tag{4.99}$$

This particular transformation provides a 6/10 reduction in rate and an increase from 2 to 7 levels. It can be shown [19] that it has a power spectral density function that is zero at zero frequency. It compares favorably with other codes [20] that have similar spectral characteristics, and performs about the same as a seven-level code with Viterbi detection [18]. The specification of multirate digital filters is not yet straightforward, but the potential exists for realizing a range of desirable line signal characteristics through simple and inexpensive digital filtering.

4.8 SIGNALING ON THE DIGITAL SUBSCRIBER ACCESS LINE

Baseband signaling has a major application in subscriber communication, via a passive cable, with a nearby access point of a digital network. For the integrated services digital network (ISDN, Chapter 1), with access points in telephone offices, ordinary twisted-pair "telephone lines" are used by subscribers for "basic" access at 160 kbps full duplex (144 kbps information rate plus 16 kbps overhead). Figure 4.26 illustrates the (baseband) transceiver at the "U" interface communicating with a similar transceiver in the serving office. At the serving office, connections are made via packet- or line-switched networks as requested by the user.

These twisted-pair access circuits, up to about 18,000 feet in length, do not contain any filters or active elements and thus are not severely bandlimited in the way telephone circuits and radio carrier systems, which include channel-bank filters, are. They do, however, suffer from a number of impairments, including a gradual falloff in frequency response, which must be taken into account in the design of a reliable 160 kbps data communication system. The capacity of the digital subscriber line was analyzed in Chapter 2, Section 2.7.2. By use of a collection of techniques described in this and later chapters, 160 kbps and higher rates are in fact possible on the great majority of subscriber lines. In fact, rates of 800 kbps full duplex and 1.5 Mbps one-way, plus 160 kbps full duplex, may be attainable on a large proportion of in-place twisted-pair subscriber lines.

The subscriber line was characterized in Chapter 1 and particularly in Figures 1.25–1.27. The major impairments, aside from the gradual falloff in frequency response, are crosstalk between physically adjacent twisted-pair lines, echoes and transmission distortion from bridged taps, and additive noise. The selection of a line code is a compromise among these impairments. To mini-

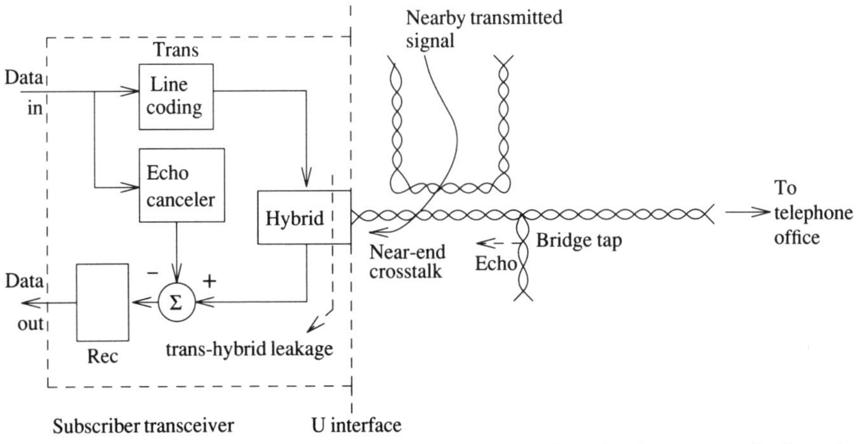

Fig. 4.26 Data communication on the twisted-pair subscriber line, showing sources of echo and crosstalk.

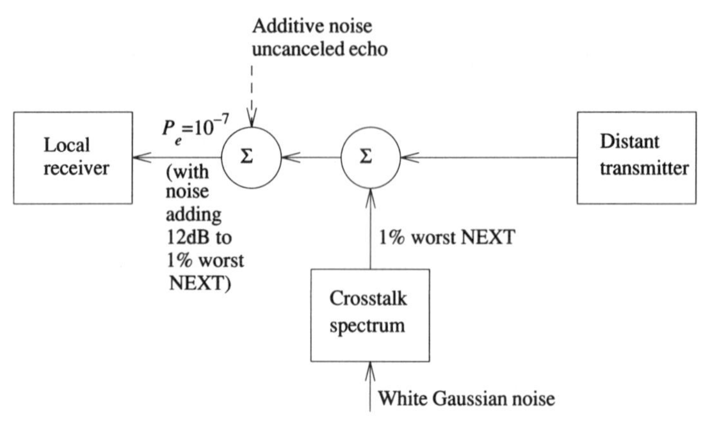

Fig. 4.27 Model for subscriber line error rate performance

mize channel distortion, a narrow (but not necessarily low frequency) signal spectrum is desirable. Near-end crosstalk (NEXT), which increases roughly as $f^{3/2}$ (see Section 1.6.2) and weights the compromise heavily, is minimized by a low-frequency, narrow-band spectrum.

In the consideration of data communication on multiple-pair cabled systems, an accepted performance criterion is that a bit error rate of 10^{-7} be achieved with 12-dB noise margin when the 1% worst NEXT is present. Noise margin is the amount of noise (and uncancelled echo) that can be tolerated without exceeding the 10^{-7} error rate. Figure 4.27 shows the model, with the 1% worst NEXT represented as the result of passing white Gaussian noise through a filter representing the crosstalk spectrum. Here NEXT is the total

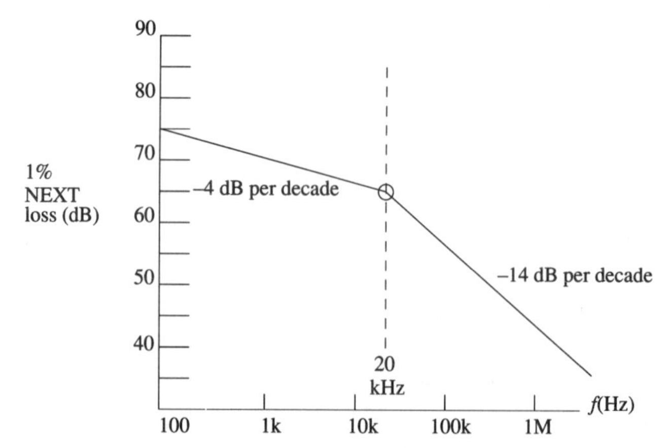

Fig. 4.28 Model of 1% worst NEXT (from Lechleider [22]). The loss in the "crosstalk channel" carrying power from all the other pairs in a cable to the test pair decreases with frequency, so that NEXT is increasing with frequency.

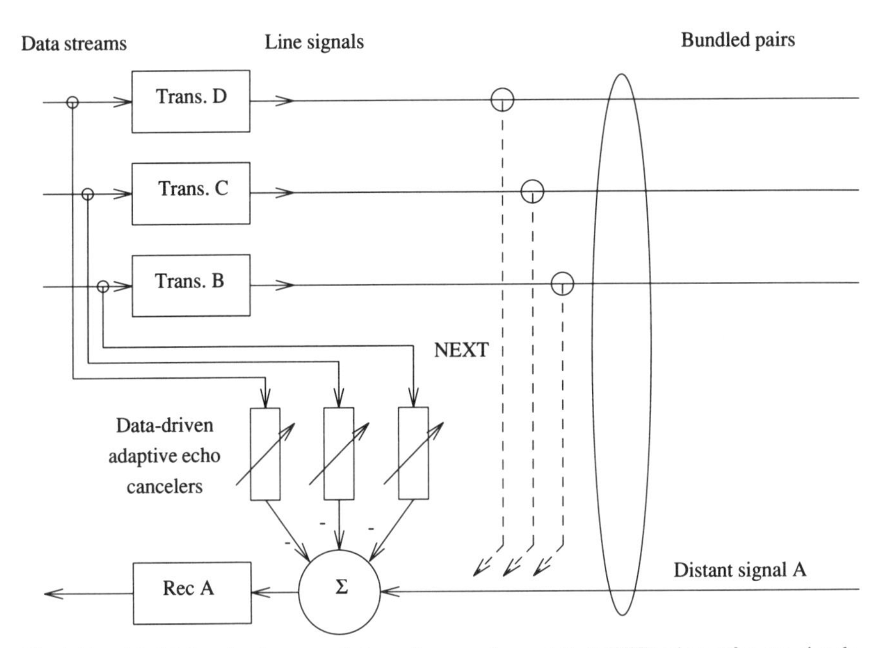

Fig. 4.29 Model for adaptive cancellation of near-end crosstalk (NEXT) when reference signals are available.

power coupled from the other 49 pairs in a 50-pair cable into the test pair, and 1% worst NEXT is the total power that is exceeded on only 1% of randomly selected test pairs. Ninety-nine percent of randomly selected test pairs will have less severe NEXT. Figure 4.28 models 1% worst NEXT loss.

In order to realize the performance against NEXT described above, the system design must consider several critical alternatives. One of the most fundamental choices is between simultaneous use of the same bandwidth vs. time multiplexing or separate spectral ranges for the two directions of transmission. If the same spectrum is simultaneously used, a relatively complex echo canceler is required to eliminate self-interference, as shown in Figure 4.29 and examined in detail in Chapter 9. The upper frequency edge is determined by the line code and pulse shape, which can be relatively simple and robust since only the bandwidth of a one-way channel is consumed. In fact, a spectrally inefficient rectangular pulse shape is specified in the ANSI ISDN Standard [23]. If time-division multiplexing, i.e., the time-compression multiplexing scheme (TCM) described in Chapter 1, or separate transmission bands are used, the total spectrum used is relatively large, more than twice as large for the same line code, greatly increasing the crosstalk level. The bandwidth can be squeezed down by multilevel signaling, but at the usual cost in noise immunity. In view of the above and the decreasing cost of processing, the echo cancellation approach appears to be favored for most future applications.

Unfortunately, the effective length of the echo impulse response, which must not extend beyond the reach of a reasonable echo canceler, increases as the proportion of signal power at low frequencies increases. This is because the passive hybrid coupler (Chapter 9) used to match two-wire to four-wire circuits performs worse at lower frequencies and echo returns at a higher level, leaving more echo to be contended with by the echo canceler. The implication is that signal energy cannot all be concentrated at low frequencies, despite the pressure from NEXT to put energy there.

The same principle of adaptive interference cancellation can, if receivers are clustered together, be used to cancel some of the NEXT (Figure 4.29). Using the inputs to all but one receiver as driving signals for a bank of adaptive noise cancelers, a replica of the total NEXT can be automatically generated and subtracted from the input to the remaining receiver [25]. This is done for each receiver. More generally, the $N(N-1)$ cancelers can be realized as a single adaptive matrix operator processing an N-dimensional input (the N receiver input with NEXT) to generate an N-dimensional output (the set of "clean" received signals). In practice, it is usually difficult, for purely administrative reasons, to locate and use the driving signals in a serving office, even though transceivers are concentrated there, and difficult to obtain them at the subscriber ends except where subscribers are fortuitously concentrated.

On top of these considerations, the achievement of a high data rate on badly distorted or noisy subscriber lines calls for a compromise between bandwidth and noise immunity, to be made in the choice of spectral shaping and in the decision whether to use more than two signaling levels.

4.8.1 LINE CODES FOR THE DIGITAL SUBSCRIBER LINE

Because of the conflicting demands of those varied impairments, and differences among telephone administrations about their relative significance, there were and remain conflicting solutions to the problem of designing an effective and economical data communication system for the digital subscriber line. Much of the controversy has concerned the choice of line code. Although 2B1Q (4-level PAM) appears to have prevailed, it is instructive to examine the alternative line codes that were considered.

The leading line code candidates, illustrated in Figure 4.30, were Manchester, modified duobinary, AMI, 4B3T, 3B2T, and 2B1Q. The Manchester, modified duobinary (partial response class 4 with precoding), and AMI (alternate mark inversion or bipolar) codes were introduced earlier and should be familiar to the reader. 2B1Q ("two bits in one symbol interval with a quartet of possible values") is simply four-level pulse-amplitude modulation. The code 4B3T is four information bits coded into three ternary symbols. The 16 information words code into only 16 of the 27 possible symbol combinations, leaving 11 others for timing and other side information or for variations in signal coding as described below. The 3B2T code is three information bits coded into two ter-

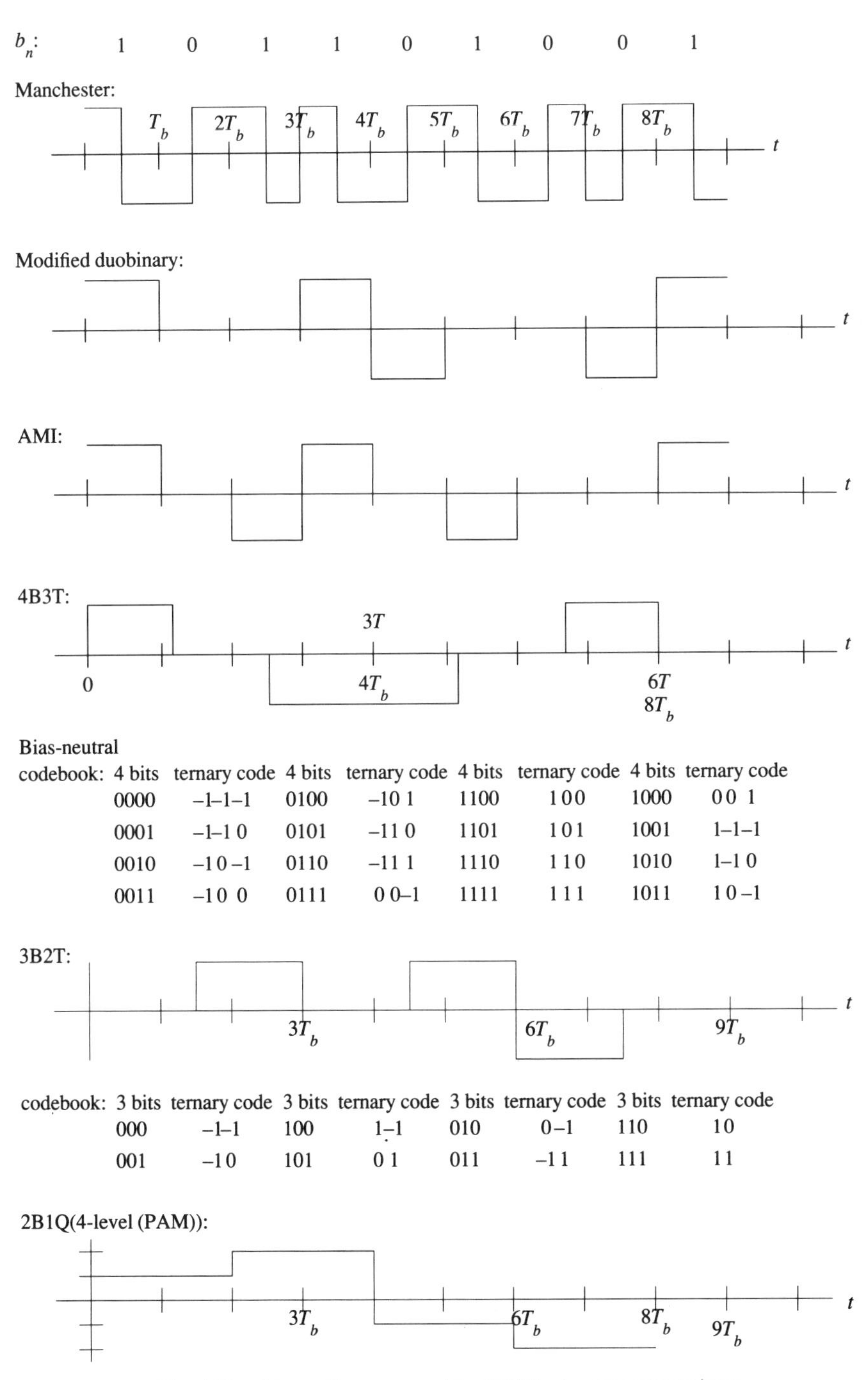

Bias-neutral codebook:

4 bits	ternary code	4 bits	ternary code	4 bits	ternary code	4 bits	ternary code
0000	−1−1−1	0100	−1 0 1	1100	1 0 0	1000	0 0 1
0001	−1−1 0	0101	−1 1 0	1101	1 0 1	1001	1−1−1
0010	−1 0 −1	0110	−1 1 1	1110	1 1 0	1010	1−1 0
0011	−1 0 0	0111	0 0−1	1111	1 1 1	1011	1 0 −1

3B2T codebook:

3 bits	ternary code	3 bits	ternary code	3 bits	ternary code	3 bits	ternary code
000	−1−1	100	1−1	010	0 −1	110	1 0
001	−1 0	101	0 1	011	−1 1	111	1 1

Fig. 4.30 Line codings proposed for ISDN basic-access transceivers.

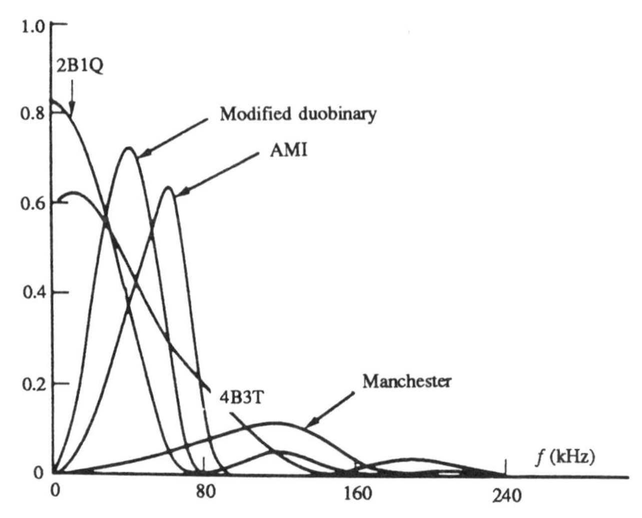

Fig. 4.31 Relative line signal power density spectra at *data* rate of 160 kbps. Symbol rate (baud) is 160,000 symbols/sec for Manchester, modified duobinary, and AMI; 120,000 symbols/sec for 4B3T; 106,667 symbols/sec for 3B2T; and 80,000 symbols/sec for 2B1Q (4-level OAM). Rectangular pulses are assumed. No normalization is implied; only the spectral shapes are significant. Most curves taken from [24]; 4B3T is simulated average from T1D1.3/86-154.

nary symbols, or eight information words coded into nine possible symbol combinations. Figure 4.31 illustrates power density spectra for a number of the line codings at a *data* rate of 160 kbps.

In the interest of an inexpensive digital implementation, squared-edge pulses are used, as we already noted. There is no adjacent channel interference problem, and the loss of energy at high frequencies is small enough to be acceptable, as is the added contribution to NEXT.

Each code has its advantages and disadvantages. Manchester is simple and has zero DC and at least one transition per symbol interval for timing recovery. Its disadvantage is consumption of a huge spectrum, and thus vulnerability to NEXT and intersymbol interference (see Figure 4.32). Modified duobinary is moderately spectrally efficient, has zero DC, and also causes minimal intersymbol interference (remember its thin tail!), but it has the disadvantage of noise margin reduction in symbol-by-symbol detection. Simulations show it is 2 to 3 dB below its more spectrally efficient competitor in immunity to NEXT and intersymbol interference (ISI, Section 4.9) on worst-case subscriber lines [22].

AMI has zero DC but no guaranteed transitions with unscrambled data input. It is a little worse than modified duobinary in spectrum and performed about 3 dB less well against NEXT and ISI in simulations, but had the advantage of requiring shorter filters in the equalizer and echo canceler.

4B3T reduces bandwidth significantly (by reducing the baud to 120,000 symbols/sec) at the cost of (unrecoverable) noise margin loss from the use of three instead of two levels. It has the added disadvantage, in general, of no guarantee of zero DC, so special steps have been defined to keep the DC level from running away.

The particular 4B3T line code [26], MMS43, considered most seriously by the T1D1 committee of the U.S. Exchange Carrier Systems Association used three code books. One codebook tended to bias the line signal positive, the second was neutral, and the third tended to bias the line signal negative. If the line signal had been running positive, the negative codebook would be switched in, and vice versa. This was an adaptive way of maintaining a zero DC line signal, rather than using a zero DC line code or scrambling customer data. In simulations, MMS43 showed 2 to 4 dB improvement over modified duobinary, for comparable implementation complexity on the worst subscriber lines. The 3B2T line code has a still lower baud of 106,667 symbols/sec, but also has the DC problem. Simulations show performance comparable to that of MMS43. The 2B1Q, or four-level PAM, line code makes the greatest baud reduction (to 80,000 symbols/sec) and showed the best performance in simulations, beating MMS43 by 2 to 3 dB on worst-case lines. These results were achieved using nonlinear echo cancellation [27,28], described in Chapter 9, and channel estimation techniques for timing recovery.

Using 2B1Q and VLSI implementation of a transceiver with decision-feedback and forward adaptive equalizers (Chapter 8) and a nonlinear echo canceler (Chapter 9), it appears possible to achieve a low error rate (10^{-7}) operating full duplex at 160 kbps on the vast majority of twisted-pair subscriber lines. The transceiver is sketched in Figure 4.33.

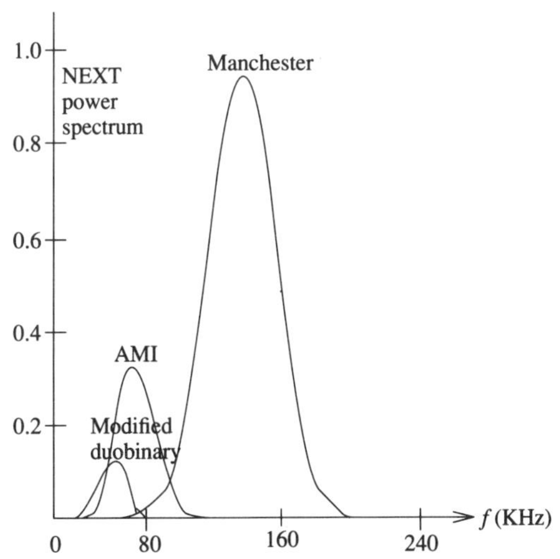

Fig. 4.32 Measured near-end crosstalk over a set of typical subscriber lines, for different line codings (from Lin and Tzeng [24]).

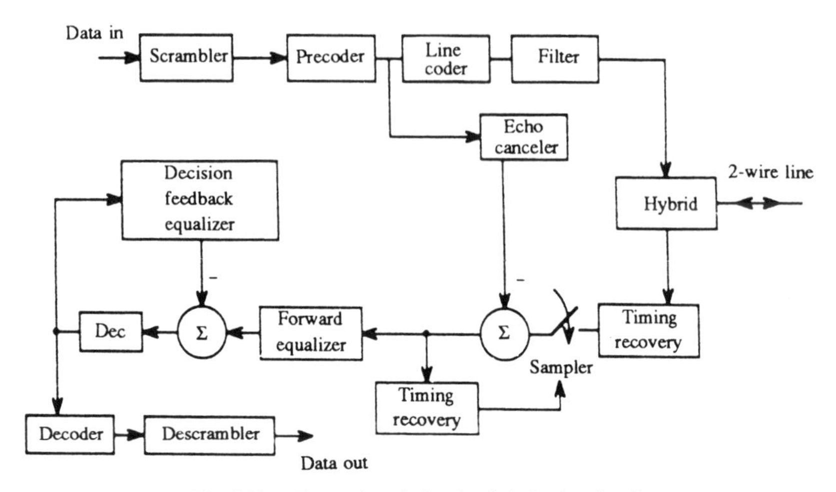

Fig. 4.33 Transceiver design for digital subscriber line.

4.8.2 SIGNALING THROUGH THE DIGITAL SUBSCRIBER LINE AT FASTER THAN 160 kbps

The ISDN line rate of 160 kbps full duplex is not the maximum possible on many subscriber twisted-pair lines. At the time of writing, development was beginning on the full-duplex high-speed digital subscriber line (HDSL) and on the asymmetric digital subscriber line (ADSL). HDSL could support access to DS1 service at 1.544 Mbps in North America. The applications could include access to data networks and telephone switches for aggregated voice/data traffic, and delivery (via ADSL) of highly compressed motion pictures from a "video on demand" library [29]. Possible approaches to these two digital subscriber line systems are illustrated in Figure 4.34.

In the HDSL implementation of Figure 4.34, two pairs each providing 800 kbps full duplex transmission are combined to realize 1.544 Mbps full duplex service, the DS1 rate. In ADSL, a single pair supports 1.544 Mbps transmission in one direction, which would usually be from a serving office to a customer location, plus either analog voice or 160 kbps full duplex for basic ISDN service.

The service requirement for HDSL is that it works in a Carrier Serving Area (CSA), which is an administrative area around a wire center. The CSA includes all 26-gauge subscriber twisted-pair lines up to 9 kft, and all 24-gauge subscriber lines up to 12 kft, with the length of bridged taps subtracted from the working length [30]. In addition, each bridged tap must be no longer than 2 kft, and the total length of bridged taps must be no more than 2.5 kft. ADSL should work over the entire nonloaded copper subscriber line plant, with the exception of some pathological lines. This corresponds to about 14.4 kft of 24

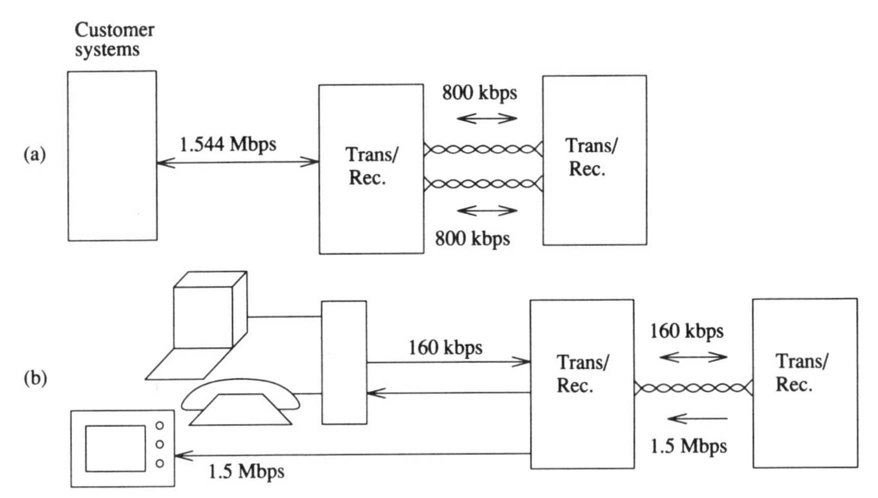

Fig. 4.34 High-speed digital subscriber line configurations; (a) Two pairs, each operating 800 bps full-duplex (HDSL). (b) One pair, operating in one direction at 1.5 Mbps and additionally full-duplex at 160 kbps (ADSL).

gauge line, and up to 18 kft for nonloaded 24-gauge line. ADSL is able to work over longer subscriber lines because it does not have the severe near-end crosstalk problem that HDSL has.

There are many alternative ways of realizing HDSL systems, but the main choice may be between baseband and passband systems [30]. The baseband system is likely to operate 800 kbps full duplex on each of two wire pairs, as illustrated in Figure 4.34, and to use 2B1Q signaling. The passband system is likely to operate at 1.6 Mbps in one direction only on each wire pair, and could use 64-point QAM (Chapter 5). An alternative passband approach would use uncoupled subchannels, e.g., "multitone" transmission (Chapter 5), which largely eliminates intersymbol interference problems. The full-duplex baseband system requires echo cancellation, and both systems are likely to use fractionally spaced forward equalization and decision feedback equalization (Chapter 8). Additional possibilities for performance-enhancing techniques include adaptive crosstalk cancellation, as shown in Figure 4.32, and trellis-coded modulation (Chapter 5). At the time of writing, there was still no unequivocal performance differentiation between the baseband and passband approaches. For ADSL, multitone appears likely to be selected for the high-rate signal.

4.9 INTERSYMBOL INTERFERENCE

Intersymbol interference is a harmful mutual interaction of pulses in a transmitted pulse train. In the usual linear system model, e.g., Figure 4.1, it is caused by distortion in $H(f)$, the channel filter, which is not compensated by the combination of the transmitter filter $P(f)$ and the receiver filter $C(f)$. When samples are taken at the sampling times $nT + t_0$, they contain unwelcome contributions from neighboring pulses as well as the desired contribution from the current pulse. That is, the overall pulse characteristic $X(f)$ is not equivalent Nyquist. The exception to the "unwelcome" statement is for partial response pulses, where correctable intersymbol interference is intentionally inserted for spectral shaping advantages.

For future reference, it is important to keep in mind that only a limited number of neighboring pulses, within the dispersion or "memory" of the channel, can contribute interference at a given sample time.

Designers of data communication equipment naturally have an interest in characterizing and measuring the effect of intersymbol interference on performance, which is the subject of the next several subsections. This is a prelude to correction of intersymbol interference, which is the function of the ideal maximum likelihood sequence estimation receiver described in Chapter 7, and of the adaptive channel equalization structures described in Chapter 8.

4.9.1 THE EYE PATTERN

The *eye pattern* is the *synchronized superposition of all possible signals q(t)* emerging from the receiving filter (Figure 4.1). Each of the signals contributing to the pattern corresponds to a particular transmitted data sequence. For a binary system, the eye pattern indeed looks like an eye, as shown in Figure 4.35. For multilevel transmission, the eye pattern looks like a group of binary eyes in a vertical column.

Eye patterns are convenient for determining at a glance the quality of the received pulse train. If the eye is tending to close from top to bottom, there are transmitted data sequences for which a small noise will cause an error when sampling the received signal. One says that the *noise margin* is small. If the eye opening is not very wide, small variations in the location of the sampling time could result in sampling at times where the noise margin is small, and thus error would be more likely.

4.9.2 PEAK DISTORTION CRITERION FOR INTERSYMBOL INTERFERENCE

When channel distortion is present, it is usually not feasible to compute the error probability, although reasonable bounds (Section 4.10.4) can sometimes be obtained. Two other criteria, *peak distortion* and *mean-squared distortion*, are easy to use and frequently preferred to error probability for comparative evaluations of data communication techniques.

Peak distortion is a worst-case criterion for determining the effect of intersymbol interference on error rate. To begin the derivation, assume that the eye opening (Figure 4.35) is unity in the absence of intersymbol interference, and denote the sampling times by $t_0 + nT$.

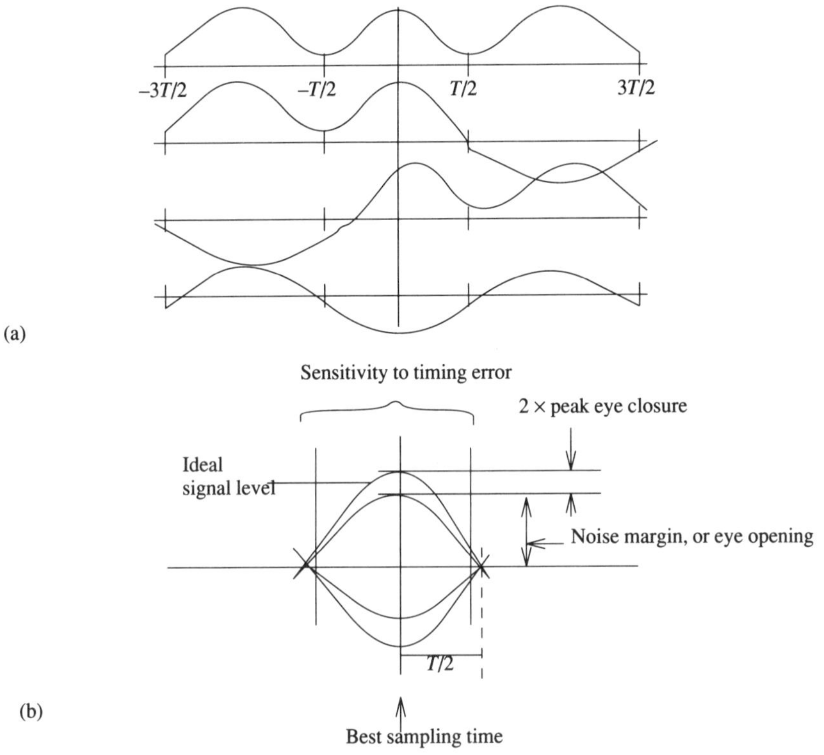

Fig. 4.35 Construction of eye pattern by superposition of receiver output waveforms resulting from all possible transmitted (binary) data sequences. (a) All possible sequences. Differences in data beyond the memory of the channel are irrelevant. (b) Result of superposition (center pulse only).

When there is intersymbol interference, the distance from the horizontal axis to the inside edge of the eye at the sampling time becomes smaller. This height, the noise margin, is one minus the peak eye closure. The peak eye closure is the maximum amount that intersymbol interference can perturb a signal level, normalized to half the ideal spacing between signal levels at the receiver output. If the end-to-end impulse response (the concatenation of $p(t)$, $h(t)$, and $c(t)$ of Figure 4.1 is $x(t)$ and the transmitted signal levels for L-level PAM are $\pm\Delta, \pm 3\Delta, ..., \pm(L-1)\Delta$, as shown in Figure 4.11, then the receiver output levels at $t = t_0$ are, in the absence of noise and intersymbol interference, $\pm x_0\Delta$, $\pm 3x_0\Delta,...,\pm(L-1)x_0\Delta$ *where* $x_0 = x(t_0)$. Our assumption of unity ideal eye opening is equivalent to $x_0\Delta = 1$, so that $x_0\Delta$ will be our normalization factor.

Assume a sequence $\{d_n\}$ of line-signal modulation levels (as in Figure 4.10). The receiving filter output sample in the absence of noise is

$$q_0 = q(t_0) = d_0 x_0 + \sum_{n \neq 0} d_n x_{-n} \quad . \tag{4.100}$$

For $n > 0$, $d_n x_{-n}$ is precursor interference from the pulse whose peak is to come nT seconds later, and for $n < 0$, $d_n x_{-n}$ is postcursor interference from pulses whose peaks occurred earlier. The *maximum value* of the summation occurs when each d_n has the largest possible magnitude $(L-1)\Delta$ and the same algebraic sign as x_{-n}. This maximum is given by

$$Max.\ interference = (L-1)\Delta \sum_{n \neq 0} |x_n| . \tag{4.101}$$

The peak eye closure, as defined above, is then

$$Peak\ eye\ closure = (1/x_0\Delta) \cdot Max.\ interference = (L-1) D_{peak} , \tag{4.102}$$

where

$$D_{peak} = (1/x_0) \sum_{n \neq 0} |x_n| \tag{4.103}$$

is defined as the *peak distortion* of the system impulse response. The eye opening, as defined earlier, is then

$$Eye\ opening = 1 - (L-1) D_{peak} . \tag{4.104}$$

This is the smallest noise margin, or noise sample amplitude, that can cause an error. It occurs only with a particular data sequence. We can thus obtain an *upper bound* on P_e by assuming this is the noise margin all of the time, whatever the data sequence. Whether or not the upper bound is tight depends on the likelihood of the worst-case data sequence. This worst case sequence may not be uncommon, e.g., a sequence of all ones or all zeros.

The worst-case probability of error is, if P_{ei} is the error probability of the ith data sequence and ν is the Gaussian noise component of the receiver output sample q_0,

$$max\ P_{ei} = Pr\{\nu > x_0\Delta[1 - (L-1) D_{peak}]\} . \tag{4.105}$$

For Gaussian noise of variance σ^2,

$$max\ P_{ei} = \int [1/\sqrt{2\pi\sigma^2}]\ \exp[-x^2/2\sigma^2]\ dx = Q\left\{\frac{x_0\Delta[1-(L-1)D_{peak}]}{\sigma}\right\} . \tag{4.106}$$

The peak distortion D_{peak}, given by (4.103), is calculated from the samples of the overall system impulse response for the general baseband system of Figure 4.10,

$$x_n = x(nT + t_0) = (1/2\pi) \int_{-\infty}^{\infty} P(f)H(f)C(f)\exp[j2\pi f(nT + t_0)]\,df . \tag{4.107}$$

Assuming that the channel characteristic $H(f)$ was not considered in the design of the received pulse, the produce $P(f)C(f)$ might, for example, be designed to yield an equivalent Nyquist channel. Note that the peak distortion is a function of the reference sampling time t_0. Normally, t_0 should be chosen to minimize D_{peak}, implying a choice for t_0 that may be quite different from the peak of the pulse response $x(t)$. Timing sensitivity problems at that t_0 where D_{peak} is minimum may make even this choice something less than optional.

4.9.3 THE RMS ERROR CRITERION

The last section described the peak distortion error criterion, which leads to an upper bound on the error probability. An alternative criterion, the RMS error criterion, gives a heavier weighting to large values of the intersymbol interference contributions $\{x_n,\ n\neq 0\}$ by considering the sum of the *squares* of these impulse response samples. It is derived as follows:

From Figure 4.10, the receiver filter output sample at time t_0 is

$$q_0 = q(t_0) = \sum_n d_n x(t_0 - nT) + \nu(t_0), \tag{4.108}$$

where

$$x_n = x(t_0 + nT).$$

As before, d_n is chosen from the level set $-(L-1)\Delta,\ldots,-\Delta,\Delta,\ldots,(L-1)\Delta$. The intersymbol interference (ISI) is

$$ISI = \sum_{n\neq 0} d_n x_{-n}. \tag{4.109}$$

If there were no ISI or noise, the output sample would ideally be $q_0 = d_0 x_0$, taking on values from the set $-x_0(L-1)\Delta,\ldots,-x_0\Delta, x_0\Delta,\ldots,x_0(L-1)\Delta$.

The *mean-square* value of the ISI is

$$E[\sum_{n\neq 0} d_n x_{-n}]^2 = E(d_n^2)\sum_{n\neq 0} x_n^2\ , \tag{4.110}$$

where we have utilized the uncorrelatedness of the $\{d_j\}$. The expectation of d_n^2 is taken over the ensemble of L equally-likely levels (Figure 4.11) and is given by, using (4.42),

$$E(d_n^2) = \Delta^2(L^2-1)/3\ . \tag{4.111}$$

Normalizing the mean-square interference by the square of the distance $x_0\Delta$ between an (ideal) output sample and its nearest decision threshold, we have

$$\begin{aligned} Mean-square\ eye\ closure &= E[\sum_{n\neq 0} d_n x_{-n}]^2/(x_0\Delta)^2 \\ &= [(L^2-l)/3x_0^2]\ \sum_{n\neq 0} x_n^2. \end{aligned} \tag{4.112}$$

We define the *mean-square distortion* (MSD) as the mean-square eye closure for a binary ($L=2$) system:

$$MSD = \frac{1}{x_0^2} \sum_{n \neq 0} x_n^2 . \qquad (4.113)$$

The effect of additive noise can also be brought in. Recall that the sum of noise and intersymbol interference components of the receiver output sample q_0 is

$$Noise + ISI = \nu + \sum_{n \neq 0} d_n x_{-n}, \qquad (4.114)$$

where ν is assumed to be a random variable of variance σ^2 and independent of the data sequence $\{d_i\}$. Thus the mean-square value of intersymbol interference *and* noise is, by the uncorrelatedness of data and noise,

$$\sigma_{tot}^2 = E[\sum_{n \neq 0} d_n x_{-n} + x_i]^2 = [\Delta^2 (L^2 - 1)/3] \sum_{n \neq 0} x_n^2 + x_i^2 . \qquad (4.115)$$

One might crudely assume that the intersymbol interference is approximated by a Gaussian random variable, allowing easy calculation of the probability of error. However, the Gaussian assumption is not very reliable in the tail of the probability distribution for noise plus ISI. Intersymbol interference and noise have quite different effects on the error rate and should not be too hastily combined. A clever approach to evaluating their combined effect is given by the Saltzberg [31] bound, described in the next section.

Mean-square distortion has a simple relationship to the equivalent Nyquist channel characteristic. Recall that

$$x(nT) = x_n = \int_{-\infty}^{\infty} X_{eq}(f) e^{j2\pi nfT} \, df , \qquad (4.116)$$

where

$$X_{eq}(f) = \begin{cases} \sum_k X(f+k/T), & |f| \leq 1/2T \\ 0, & \text{elsewhere} . \end{cases} \qquad (4.117)$$

Now

$$\int_{-1/2T}^{1/2T} |X_{eq}(f)|^2 df = T^2 \int_{-1/2T}^{1/2T} \sum_n \sum_m x(nT) x(mT) e^{2\pi j(n-m)fT} \, df$$

$$= T \sum_n x_n^2 , \qquad (4.118)$$

$$MSD = (1/x_0^2) \sum x_n^2 = (1/x_0^2) \; [(1/T) \int_{-1/2T}^{1/2T} |X_{eq}(f)|^2 df \, - \, x_0^2]$$

$$= (1/T) \int_{-1/2T}^{1/2T} |X_{eq}(f)|^2 df - 1 . \qquad (4.119)$$

Alternatively,

$$(1/Tx_0^2) \int_{-1/2T}^{1/2T} |X_{eq}(f) - x_0 T|^2 df =$$

$$(1/Tx_0^2) \int_{-1/2T}^{1/2T} [|X_{eq}(f)^2 - 2x = MSD \ , \tag{4.120}$$

where we have used the fact that

$$x_0 = \{ \int_{-1/2T}^{1/2T} X_{eq}(f) e^{j2\pi ft} \, df \} |_{t=0} \ . \tag{4.121}$$

That is, the mean-square distortion is proportional to the energy in the deviation of $X_{eq}(f)$ from the ideal characteristic $x_0 T$. We see that a knowledge of the overall channel transfer function $X(f)$ and hence of $X_{eq}(f)$ is sufficient for easy determination of the mean-square distortion.

4.9.4 THE SALTZBERG BOUND

We noted in the last section that it is not easy to calculate probability of error in the presence of intersymbol interference, even when the channel, and thus all the pulse samples, are known. Not content with peak or mean-square distortion criteria, communication analysts have pursued tight and intuitive bounds on probability of error. One of the earliest and most useful, assuming independent data, was derived by Saltzberg [31]. His approach was to subtract the larger intersymbol interference components from the signal amplitude, and add the smaller intersymbol interference components to the Gaussian noise, after which a Gaussian tail probability is computed as the error probability.

We seek an upper bound for the cumulative probability density function of the sum of intersymbol interference (ISI) and Gaussian noise. Express the receiver filter output sample $q_0 = q(t_0)$ as

$$\begin{aligned} q_0 &= d_0 x_0 + \sum_{k \neq 0} z_k + \nu \\ &= d_0 x_0 + z , \end{aligned} \tag{4.122}$$

where the $\{z_k\}$ are intersymbol interference components, defined below,

$$z = \sum_{k \neq 0} z_k + \nu \ , \tag{4.123}$$

and the noise variable ν has the probability density function

$$p_\nu(y) = [1/\sqrt{2\pi\sigma^2}\,] \exp(-y^2/2\sigma^2) . \tag{4.124}$$

It should be apparent that z is a random variable with a symmetric distribution.

In (4.122), the ISI contribution z_k from a neighboring pulse k symbol intervals away is the product of pulse sample value and data level value

$$z_k = x_{-k}\, d_k, \tag{4.125}$$

where d_k is a random variable which can take on any of the L possible data modulation values in the set $\{-(L-1), -(L-3), \ldots, (L-1)\}$. This is consistent with (4.114), but for simplicity, we have assumed that $\Delta = 1$ in the definition of data modulation levels. Thus z_k has a discrete distribution, and its probability density function can be written, assuming equally likely levels,

$$p(z_k) = (1/L) \sum_{j=1}^{L} \delta\,(z_k - \mu_j |x_k|), \tag{4.126}$$

where $\{\mu_j\}$, from $j=1$ to L, is the set of possible values of d_k. The magnitude $|x_k|$ can be used because of the symmetry of the modulation level set $\{\mu_i\}$. The parameters $\{|x_k|\}$ are, for a given channel, fixed.

How can a detection error occur? If $d_0 = -(L-1)$, then an error occurs if $z > x_0$, i.e. z carries q_0 more than halfway from $-(L-1)x_0$ to $-(L-3)x_0$. If d_0 is any of the levels from $-(L-3)$ to $L-3$, an error occurs if $|z| > x_0$, i.e. in either direction. If $d_0 = L+1$, an error occurs if $z < -x_0$. Thus

$$\begin{aligned} P_e &= [(L-2)/L]2Pr(z > x_0) + [2/L]Pr(z > x_0) \\ &= [2(L-1)/L]Pr(z > x_0). \end{aligned} \tag{4.127}$$

We will derive a *Chernoff* bound on $Pr(z > x_0)$, for which z must be the sum of independent, zero-mean random variables, which it is in this case. The Chernoff bound, which has been proved in Section 2A.8, is stated as

$$Pr(z > x) \le e^{-\lambda x}\, E(e^{\lambda z}), \tag{4.128}$$

where the parameter λ is an arbitrary positive real number.

Applying the Chernoff bound to z as defined in (4.123),

$$Pr(z > x) \le e^{-\lambda x} E(e^{\lambda \nu}) \prod_{k \ne 0} E(e^{\lambda z_k}), \quad \textit{all } \lambda > 0. \tag{4.129}$$

Now

$$E(e^{\lambda \nu}) = 1/\sqrt{2\pi\sigma^2} \int_{-\infty}^{\infty} e^{\lambda \nu}\, e^{-\nu^2/2\sigma^2}\, d\nu = e^{\lambda^2\sigma^2/2}, \tag{4.130}$$

and

$$\begin{aligned} E(e^{\lambda z_k}) &= \int_{-\infty}^{\infty} (1/L) \sum_{j=1}^{L} \delta(z_k - \mu_j |x_k|)\, e^{\lambda z_k}\, dz_k \\ &= (2/L) \sum_{j=(L/2)+1}^{L} \cosh \lambda\, \mu_j |x_k|\,. \end{aligned} \tag{4.131}$$

Thus

$$Pr(z \geq x) \leq \exp[-\lambda x + (1/2)\lambda^2\sigma^2] \prod_{k\neq 0} (2/L) \sum_{j=L/2+1}^{L} \cosh \lambda \, |x_k| \, \mu_j, \tag{4.132}$$

where σ_u^2 is the variance of the noise variable ν. Using two different upper bounds on $\cosh(x)$, Saltzberg [31] shows that for binary data ($L = 2$; $\mu_1 = -1$, $\mu_2 = +1$),

$$Pr(z > x_0) < \exp\left\{ \frac{-\left[x - \sum_{k\in K} |x_k|^2\right]}{2[\sigma^2 + \sigma_d^2] \sum_{k\notin K} |x_k|^2} \right\}. \tag{4.133}$$

The set K consists of the larger samples of the distorted pulse. The procedure for partitioning the set of pulse samples so as to provide the tightest bound is a simple iterative one. Start with all samples in the denominator sum, and iteratively delete the largest sample in that sum and add it to the numerator sum until the quantity in the brackets reaches its maximum. It is always possible to put all the samples in the denominator of the exponent (providing a looser bound), so that ISI may be approximated as a Gaussian variable.

4.9.5 OTHER BOUNDS

The Saltzberg bound is reasonable but not demonstrably optimal. The partition of interference samples into the set of larger values offsetting the signal and a set of smaller values taken as statistical augmentation of the noise power minimizes a particular bound derived from the Chernoff inequality, but other partitions might be equally reasonable. For example, one might put a certain proportion of the total interference "mass" into the first set (K), and the rest into the complementary set. The important point is that a more accurate estimate of error rate can be achieved than by lumping all of the interference mass with the Gaussian noise.

Many other bounds have been described, with improved performance under certain conditions and an overview is provided by Benedetto, Biglieri, and Castellani [32]. Lugannani [33], like Saltzberg, exploited the Chernoff inequality. Ho and Yeh [34] estimated error probability from a Hermite polynomial expansion, and Shimbo and Celebiler [35] from a Gram-Charlier expansion. Glave [36] proposed and Matthews [37] further developed a bound from the first moments of the combined intersymbol interference. Other bounds based on moments, and computational techniques for them, were described by Yao and Biglieri [38], Prabhu [39], and Cariolaro and Pupolin [40].

An entirely different class of experimental approaches is sometimes taken to estimation of error rates. These include Monte Carlo and empirical distribution modeling techniques [41] that allow estimates of rare events (such as infrequent errors) to be made from a number of observed received signal samples that is several orders of magnitude lower than that called for with direct counting estimates of the rare events. These experimental extrapolation techniques are very useful for channels that cannot be accurately characterized for analytical bounds or for which a long observation time or simulations program running time could be required for direct counting.

APPENDIX 4A POWER DENSITY FUNCTION OF A CORRELATED LINE SIGNAL

We show, in the following derivation, that the power density function is determined as the product of the energy spectrum of the pulse used in a line signal and the power spectrum of the correlated data stream modulating the pulses.

Let

$$s(t) = \sum_{n=-\infty}^{\infty} d_n p(t-nT) \tag{4A.1}$$

be a complex PAM data train. The random amplitudes $\{d_n\}$, *possibly mutually correlated*, are selected from a discrete set. The sequence $\{d_n\}$ is wide-sense stationary, with mean $\bar{d}$. The autocorrelation function of this sequence is defined as

$$r_d(m) = E\left[d_n^* d_{n+m}\right] \tag{4A.2}$$

To find the power spectrum of $s(t)$, first determine its autocorrelation function:

$$\begin{aligned}
\phi_{ss}(t+\tau_1, t) &= E\,[s^*(t)\; s(t+\tau)] \\
&= \sum_n \sum_{n'} E(d_n^* d_n')\; p^*(t-nT)p(t+\tau-n'T) \\
&= \sum_n \sum_m E(d_n^* d_n')\; p^*(t-nT)p(t+\tau-nT-mT) \\
&= \sum_m r_d(m) \sum_n p^*(t-nT)p(t+\tau-(n+m)T)\,.
\end{aligned} \tag{4A.3}$$

This last expression is periodic in t with period T, since adding any multiple of T to the value of t does not change the value of the time function; a simple translation of the variable n of the (infinite) summation restores the same expression.

The mean of $s(t)$,

$$E\,[s(t)] = \bar{d} \sum_n p(t-nT)\;, \tag{4A.4}$$

is also periodic with period T. The random process $s(t)$ is therefore a periodically stationary, or *cyclostationary*, process.

Since we are interested in the average rather than the instantaneous power spectrum, we average $\phi_{ss}(t+\tau,t)$ over one period of T to yield

$$\phi_{ss}(\tau) = \frac{1}{T} \int_{-T/2}^{T/2} \phi_{ss}(t+\tau,t)\,dt$$

$$= \frac{1}{T} \sum_m r_d(m) \sum_n \int_{-T/2}^{T/2} p^*(t-nT)\,p(t+\tau-(n+m)T)\,dt \,.$$

With the change of variable $\lambda = t-nT$, we have

$$\phi_{ss}(\tau) = \frac{1}{T} \sum_m r_d(m) \sum_n \int_{-T/2-nT}^{T/2-nT} p^*(\lambda)\,p(\lambda+\tau-mT)\,d\lambda$$

$$= \frac{1}{T} \sum_m r_d(m) \int_{-\infty}^{\infty} p^*(\lambda)\,p(\lambda+\tau-mT)\,d\lambda \,. \tag{4A.5}$$

Taking the Fourier transform,

$$S_s(f) = \int_{-\infty}^{\infty} \phi_{ss}(\tau)\,e^{-j2\pi f\tau}\,d\tau$$

$$= \frac{1}{T} \sum_m r_d(m) \int_{-\infty}^{\infty} \int_{-\infty}^{\infty} p^*(\lambda)\,p(\lambda+\tau-mT)\,e^{-j2\pi f\tau}\,d\lambda\,d\tau \,.$$

Let $\tau = \gamma-\lambda+mT$; then

$$S_s(f) = \frac{1}{T} \sum_m r_d(m)\,e^{-j2\pi mfT} \left[\int_{-\infty}^{\infty} p(\gamma)\,e^{-j2\pi f\gamma}\,d\gamma\right] \left[\int_{-\infty}^{\infty} p^*(\lambda)\,e^{-j2\pi f\lambda}\,d\lambda\right]$$

$$= \frac{1}{T} S_d(f)\,|P(f)|^2 \,, \tag{4A.6}$$

where

$$S_d(f) = \sum_m r_d(m)\,e^{-j2\pi mfT} \tag{4A.7}$$

is the power density spectrum of the amplitude level sequence.

It can be seen from (4A.6) that the power density function of a line signal can be shaped through both the pulse shape and coding of the amplitude sequence. If the coding does not introduce any correlation, then

$$r_d(m) = \begin{cases} E|d_n|^2, & m=0 \\ 0, & m \neq 0 \end{cases} ,$$

so that $S_d(f)$ is a constant

$$S_d(f) = E|d_n|^2$$

and

$$S_s(f) = \frac{1}{T} E|d_n|^2 \, |P(f)|^2$$

has a shape depending only on the pulse $p(t)$.

REFERENCES

[1] R. Lucky, J. Salz, and E. Weldon, *Principles of Data Communication*, McGraw-Hill, 1968.

[2] H. Taub and D. S. Schilling, *Principles of Communication Systems*, Second Edition, McGraw-Hill, New York, 1986.

[3] W. Stallings, *Data and Computer Communications*, Macmillan, New York, 1985.

[4] E. A. Lee and D. G. Messerschmitt, *Digital Communication*, Kluwer Academic Publishers, 1988.

[5] J. G. Proakis, *Digital Communications*, McGraw-Hill, 1983.

[6] S. Benedetto, E. Biglieri, and V. Castellani, *Digital Transmission Theory*, Prentice-Hall, New York, 1987.

[7] K. S. Shanmugan, *Digital and Analog Communication Systems*, Wiley, New York, 1979.

[8] J. A. C. Bingham, *The Theory and Practice of Modem Design*, Wiley and Sons, New York, 1988.

[9] H. Nyquist, "Certain Topics in Telegraph Transmission theory," AIEE Trans. Vol. 47, pp. 617–644, 1928.

[10] N. Q. Duc and B. M. Smith, "Line Coding for Digital Data Transmission," *Australian Telecom. Res. J.*, Vol. 11, No. 2, pp. 14–27, 1977.

[11] E. R. Kretzmer, "Generalization of a Technique for Binary Data Communication," *IEEE Trans. Communication Tech.*, Vol. COM-14, pp. 67–68, February, 1966.

[12] S. Pasupathy, "Correlative Coding: A Bandwidth Efficient Signaling

Scheme," *IEEE Communications Society Magazine*, Vol. 15, No. 4, pp. 4–11, July, 1977.

[13] A. Lender, "The Duobinary Technique for High-Speed Data Transmission," *IEEE Trans. on Commun. Electronics*, Vol. 82, pp. 214–218, May, 1963.

[14] A. Lender, "Correlative Level Coding for Binary Data Transmission," *IEEE Spectrum*, Vol. 3, pp. 104–115, February, 1966.

[15] H. Kobayashi and D. T. Tang, "Application of Partial Response Channel Coding to Magnetic Recording Systems," *IBM J. Research & Development*, Vol. 15, July, 1970.

[16] R. W. Wood and D. A. Peterson, "Viterbi Detection of Class IV Partial Response on a Magnetic Recording Channel," *IEEE Trans. on Communications*, Vol. COM-34, No. 5, May, 1986.

[17] H. K. Thapar and A. M. Patel, "A Class of Partial Response Systems for Increasing Storage Density in Magnetic Recording," *IEEE Trans. Magnetics*, Vol. MAG-23, No. 5, pp. 3666–3668, September, 1987.

[18] G. D. Forney, "Maximum-Likelihood Sequence Estimation of Digital Sequences in the Presence of Intersymbol Interference," *IEEE Trans. Information Theory*, Vol. IT-18, May, 1972.

[19] J. B. H. Peek and L. F. P. M. Lakeman, "Generating Block Line Codes with Spectrum Nulls Using Multirate Digital Filters," *Proc. IEEE Internet Conf. on Communication*, Denver, Colorado, pp. 1098–1102, June, 1991.

[20] H. Kobayashi, "A Survey of Coding Schemes for Transmission or Recording of Digital Data," *IEEE Trans. on Communications Technology*, Vol. COM-19 No. 6, pp. 1087–1100, December, 1971.

[21] P. Kabal and S. Pasupathy, "Partial-Response Signaling," *IEEE Trans. on Communications Technology*, Vol. COM-14, pp. 921–934, September, 1975.

[22] J. Lechleider, "Line Codes for Digital Subscriber Lines," *IEEE Communications Magazine*, Vol. 9, September, 1989.

[23] Integrated Services Digital Network (ISDN) — Basic Access Interface for S and T Reference Points (layer 1 Specification), ANSI T1605-1989, Amer. Nat. Standards Institute, Washington, DC.

[24] N. S. Lin and C-P. J. Tzeng, "Full-Duplex Data Over Local Loops," *IEEE Communications Magazine*, Vol. 26, No. 1, pp. 31–42, February, 1988.

[25] M. L. Honig, K. Steiglitz, and B. Gopinath, "Multi-Channel Signal Processing for Data Communications in the Presence of Crosstalk," *Proc. ICASSP '88*, April, 1988.

[26] Siemens Communications Systems, Inc., "Some Characteristics of the MMS43 Line Code," ECSA, contribution T1D1.3/85-133, Exchanges Carriers Systems Assoc., Washington, DC.

[27] O. Agazzi, D. G. Messerschmitt, and D. G. Hodges, "Nonlinear Echo Cancellation of Data Signals," *IEEE Trans. on Communications*, Vol. COM-30, No. 11, pp. 2421–2433, November, 1982.

[28] British Telecom. Res. (ab.), "2B1Q Measured with Simultaneous Echo Cancellation and Timing Adaptation," ECSA, contribution T1D1.3/86-199.

[29] M. M. Anderson, "Video Services on Copper," *Proc. IEEE ICC'91*, Denver, pp. 302–306, June, 1991.

[30] "High-Speed Digital Subscriber Lines," *J. Selected Areas in Communications*, Vol. 9, No. 6, August, 1991.

[31] B. R. Saltzberg, "Intersymbol Interference Error Bounds with Application to Ideal Bandlimited Signaling," *IEEE Trans. Information Theory*, Vol. IT-14, pp. 563-568, July, 1968.

[32] S. Benedetto, E. Biglieri, and V. Castellani, *Digital Transmission Theory,* Prentice-Hall, New York, 1987.

[33] R. Lugannani, "Intersymbol Interference and Probability of Error on Digital Systems," *IEEE Trans. Information Theory*, Vol. IT-15, pp. 682–688, November, 1969.

[34] E. Y. Ho and Y. S. Yeh, "A New Approach for Evaluation the Error Probability in the Presence of Intersymbol Interference and Additive Gaussian Noise," *Bell System Technical J.*, Vol. 49, pp. 2249–2265, November, 1970.

[35] O. Shimbo and M. Celebiler, "The Probability of Error Due to Intersymbol Interference and Gaussian Noise in Digital Communication Systems," *IEEE Trans. Communications Technology*, Vol. COM-19, pp. 113–119, April, 1971.

[36] F. E. Glave, "An Upperbound on the Probability of Error Due to Intersymbol Interference for Correlated Digital Signals," *IEEE Trans. Information Theory*, Vol. IT-18, pp. 356–362, May, 1972.

[37] J. W. Matthews, "Sharp Error Bounds for Intersymbol Interference," *IEEE Trans. Information Theory*, Vol. IT-19, pp. 440–447, 1973.

[38] K. Yao and E. Biglieri, "Multidimensional Moment Error Bounds for Digital Communication Systems," *IEEE Trans. Information Theory*, Vol. IT-26, pp. 454–464, 1980.

[39] V. K. Prabhu, "Some Considerations of Error Bounds in Digital Systems," *Bell System Technical J.*, Vol. 50, pp. 3127–3151, 1971.

[40] G. L. Cariolaro and S. Pupolin, "Moments of Correlated Digital Signals for Error Probability Evaluation," *IEEE Trans. Information Theory*, Vol. IT-21, pp. 558–568, 1975.

[41] S. B. Weinstein, "Estimation of Small Probabilities by Straight Line Extrapolation of the Trail of a Probability Distribution Function," *IEEE Trans. on Communication*, Vol. COM-19, No. 6, Part 1, December, 1971.

EXERCISES

4.1: Show that the power spectrum of a Manchester-encoded signal is given by (4.21).

4.2: Derive the receiver that minimizes probability of error for a Manchester-encoded signal in white Gaussian noise.

4.3: Show that the power spectrum of the Miller-encoded line signal is given by

$$S(f) = T\,[1-\cos(2\pi fT)][1-\cos(\pi fT)]/(\pi fT)^2 \ .$$

Hint: The Miller-encoded signal for random data can be decomposed into two uncorrelated signals for zeros and ones.

4.4: For random input data, derive the maximum likelihood detector for a Miller-encoded signal in white Gaussian noise.

4.5: For the same average power, derive (a) the decrease in modulation level separation, and (b) the equivalent loss in SNR of a four-level PAM signal in comparison with a binary antipodal signal.

4.6: For AMI signaling perturbed by white Gaussian noise, show that the optimum decision boundaries in a maximum likelihood receiver are given by "choose zero if $\cosh(2q_n/N_0) < \exp(2E_p/N_0)$" where q_n is the sampled output of the receive filter.

4.7: Consider the overall pulse shape given by the transfer function

$$X(f) = \begin{cases} 1 - (f/1000)^2/2; & |f| \leq 1000 \\ \left[(f - 2000)/1000\right]^2/2; & 1000 \leq |f| \leq 2000 \\ 0; \text{ otherwise.} \end{cases}$$

(a) Show that this pulse is Nyquist.

(b) At what rate do the tails fall?

(c) If this pulse were used for baseband signaling, what bandwidth is required?

(d) What is the rate that information can be transmitted?

4.8: Suppose that a channel with bandwidth 3000 Hz is available. Show the data rate achievable over this channel with the Nyquist signaling as a function of excess bandwidth. Assume the excess bandwidth varies between 0 and 100%.

4.9: Given a channel with transfer function

$$H(f) = (1 + a\cos(2\pi fT))/2, \quad 0 < a < 1 ,$$

show the transfer functions for the transmit and receive filters so that the signaling is Nyquist at rate $1/T$ symbols/sec with 50% excess bandwidth.

4.10: Suppose that the desired probability of error is $P_e = 10^{-5}$ and that the available bandwidth is B. Find the relationship between the rate that can be transmitted and B with excess bandwidth and E_B/N_0 as parameters. In performing this calculation ignore the factor $1 - 1/L$ in (4.58). Also, assume that L is a power of two.

4.11: Prove that for any $X(f)$ band limited to $|f| < 1/T$, the Nyquist criterion that $x(0) = 1$, $x(nT) = 0$, all $n \neq 0$, is satisfied if (for $f > 0$) the real part of $X(f)$ consists of the rectangular function plus an arbitrary function odd about $f = 1/2T$ and the imaginary part of $X(f)$ is any arbitrary even function about $f = 1/2T$.

4.12: For the class 5 partial response pulse, with Fourier transform

$$X(f) = 4T\sin^2 2\pi fT, \quad 0 < |f| \leq 1/2T ,$$

(a) Derive the pulse function $x(t)$ and plot on calibrated axes. Find the asymptotic rate of decay of $x(t)$.

(b) Using the theorem of Section 4.7, show that the theorem gives the same rate of decay as in (a).

4.13: For each of the classes of partial response signals, 2, 3, and 5,

(a) How fast do the tails fall?

(b) Find $X(f)$.

4.14: For class 4 partial response, used with L-ary signaling, $L > 2$, construct a truth table similar to Table 4.4.

4.15: Design a partial response signal for which there is no DC and for which the tails fall as t^{-4}.

4.16: For the class 2 partial response pulse, $G(z) = 1+2z+z^2$,

(a) Derive $F(z)$, the precoder in the encoding model of Figure 4.16.

(b) Construct a table similar to Table 4.2 (which was derived for AMI). Use the same data sequence $\{a_n\}$, but the appropriate formulas for b_n and y_n.

(c) Derive the peak and mean-squared intersymbol interference for a train of class 2 pulses (with $d_n = \pm 1$ modulation levels).

5

Passband Data Transmission

5.0 INTRODUCTION

Modulation is the process by which a baseband information signal is converted into a passband signal that can transit a passband channel constrained in bandwidth and possibly other ways. In order to conserve bandwidth, it is convenient to impress the information signal onto a sinusoidal carrier signal. Spread spectrum systems, on the other hand, effectively modulate information signals onto wideband carriers. We shall restrict attention here to digital modulation formats that are spectrally efficient, that is, the breadth of spectrum taken up by the modulated waveform is the same or not much more than that taken up by the baseband signal.

For a sinusoid, the only parameters that can be varied by the information signal are amplitude and phase, so that we are concerned here with amplitude or phase modulation, or some combination of the two. Frequency modulation is a type of phase modulation in which a function proportional to the integral of the information signal is added to the phase of the carrier sinusoid.

The choices made for modulation systems reflect many tradeoffs. One is between cost (implementation complexity) and performance; another is between noise immunity and spectral efficiency, defined as

$$\textit{Spectral Efficiency} = \log_2(L)/WT \text{ bps/Hz}, \tag{5.1}$$

where L = size of signal set from which selection is made each symbol interval, T = symbol interval (seconds), and W = (one-sided) bandwidth. The choice made in a particular communications situation depends on how difficult it is to obtain the desired performance through the available channel. For example, a dial-up point-of-sale terminal may use spectrally inefficient binary FSK (frequency-shift keyed) modulation and a simple modem structure in order to obtain high reliability at low cost for small data transfers at low bit rates. Only a fraction of a bps is packed into one-hertz of bandwidth. In contrast, data communication at 9600 bps through a telephone channel uses phase-quadrature and multilevel linear modulation to pack close to 4 bps into each hertz of bandwidth. Digital microwave radio transmitters may have spectral efficiencies as high as 8 bps/Hz. Linear modulation can be defined as modulation which simply translates the baseband spectrum to a higher frequency range and does not expand it.

In this chapter, we describe some of the modulation formats commonly used for data communication, with emphasis on the spectrally efficient linear, two-dimensional formats widely used in higher-speed voiceband modems and in digital microwave radio, and in the future mobile radio communication. A simplifying complex analytic notation is introduced which includes the concept of an "equivalent baseband channel." This makes the analysis of passband data communication systems similar to that of baseband systems, except that the signals and filter impulse responses are complex.

5.1 COMPLEX ANALYTIC REPRESENTATIONS [1–3]

In Figure 5.1, assume a real transmitted signal $s(t)$ with a passband spectrum (Figure 5.2), and a carrier frequency f_c within or close to this passband. The amplitude and phase representation of the information input will later be replaced by a complex data train. The general form of the passband signal is

$$s(t) = A(t) \cos[2\pi f_c t + \theta(t)]. \tag{5.2}$$

Here $A(t)$ is the amplitude of $s(t)$, and $\theta(t)$ is the phase. Varying either or both can be used to carry information. Equivalently, (5.2) can be expanded into

$$s(t) = m_i(t) \cos[2\pi f_c t] - m_q(t) \sin[2\pi f_c t], \tag{5.3}$$

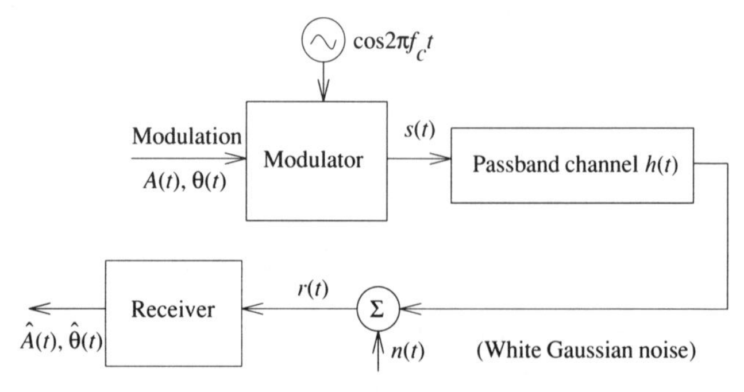

Fig. 5.1 A passband data transmission system.

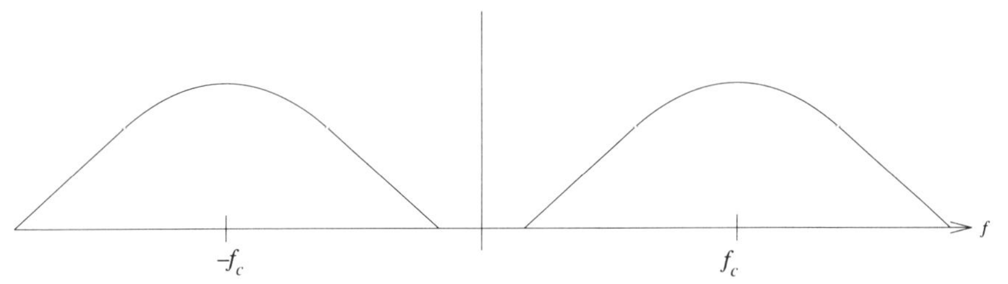

Fig. 5.2 A passband spectrum.

where

$$m_i(t) = A(t)\cos[\theta(t)], \quad m_q(t) = A(t)\sin[\theta(t)] \tag{5.4}$$

are the in-phase and quadrature modulations, respectively.

Define the complex baseband signal (using "~" to denote a complex function),

$$\tilde{m}(t) = m_i(t) + j\, m_q(t). \tag{5.5}$$

Then the complex passband signal is defined to be

$$\tilde{s}(t) = \tilde{m}(t)\, e^{j2\pi f_c t} = s(t) + j\check{s}(t), \tag{5.6}$$

where the real part of $\tilde{s}(t)$ is the actual transmitted signal $s(t)$ of (5.2), and the imaginary part of $\tilde{s}(t)$,

$$\check{s}(t) = m_i(t)\sin[2\pi f_c t] + m_q(t)\cos[2\pi f_c t], \tag{5.7}$$

is the *Hilbert transform* of $s(t)$. Thus,

$$s(t) = \text{Re}\ \tilde{s}(t) = \text{Re}\ \tilde{m}(t)e^{j2\pi f_c t} \tag{5.8}$$

The complex signal $\tilde{m}(t)$ is called the *pre-envelope* of the bandpass signal $s(t)$.

The Hilbert transform is a concept that appears often in discussions of quadrature modulation systems. It is a linear filter with transfer function (Figure 5.3)

$$\mathcal{H}(f) = -j\ \text{sgn}(f) \tag{5.9a}$$

and impulse response

$$h(t) = 1/\pi t, \quad t \neq 0. \tag{5.9b}$$

A Hilbert filter retards the phase of each spectral component by 90°, so that it turns a cosine function into a sine and a sine function into a negative cosine. Some examples of Hilbert-transformed spectra are given in Figure 5.4. It can easily be shown (Exercise 5.1), in the frequency domain, that if $m_i(t)$ and $m_q(t)$ are real functions bandlimited to $|f| < f_c$, then (5.7) is in fact the Hilbert transform of (5.3).

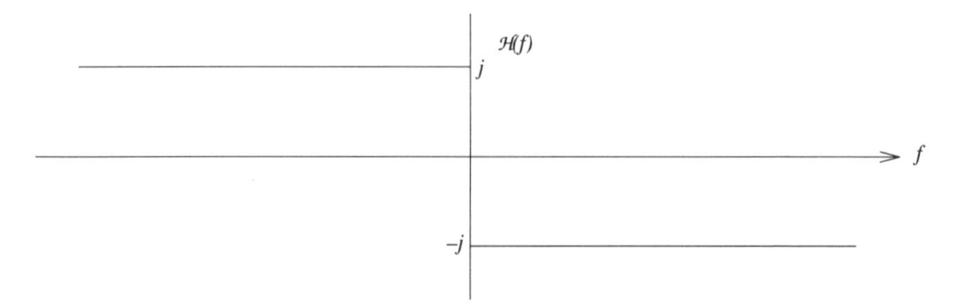

Fig. 5.3 Transfer function of ideal Hilbert filter.

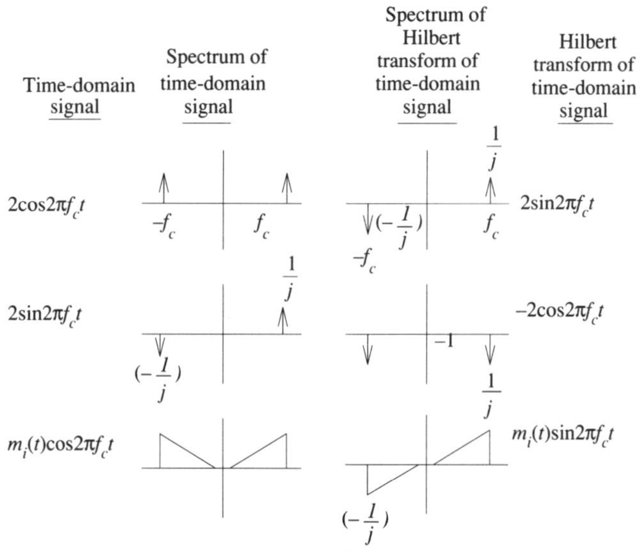

Fig. 5.4 Hilbert transforms of sinusoids and a modulated sinusoid.

A further property of the Hilbert transform is that if $\check{s}(t)$ and $\check{h}(t)$ are Hilbert transforms of $s(t)$ and $h(t)$ respectively, then

$$\check{s}(t) * \check{h}(t) = -s(t) * h(t) \ . \tag{5.10}$$

This is apparent in the transform domain, where the product of the two $[-j\text{sgn}(f)]$ functions associated with the Hilbert transforms of $s(t)$ and $h(t)$ is minus one.

A complex signal, such as $\tilde{s}(t)$ of (5.6), for which the imaginary part is equal to the Hilbert transform of the real part, is called *complex-analytic*.* Its

* A function is analytic in a region of the complex plane if it is differentiable at all points in the region, the derivative being taken in any direction.

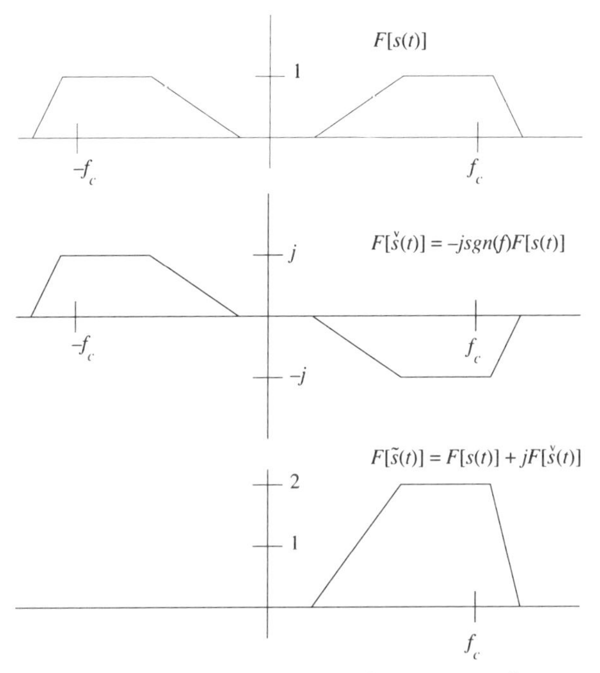

Fig. 5.5 One-sided spectrum of complex analytic signal $\tilde{s}(t) = s(t) + j\check{s}(t)$. F denotes Fourier transform.

spectrum is one-sided, as shown in Figure 5.5, and this is, in fact, an adequate definition of complex analyticity. The complex-analytic signal is a convenient and compact representation for signal analysis. All linear passband filtering operations can be done using the complex baseband signal $\tilde{m}(t)$ given by (5.5), and the complex equivalent baseband filter described in the next section. Equalizers for two-dimensional signals (modulation on both cosine and sine carriers) are similarly described as complex baseband filters.

5.1.1 THE EQUIVALENT COMPLEX BASEBAND CHANNEL

As we noted in the introduction to this chapter, the complex-variable notation for a passband system is a means to simplify notation and make analysis similar to that for a baseband system. We continue the development of this model with derivation of a complex-analytic channel $\tilde{h}(t)$ and an equivalent complex baseband channel $\tilde{h}_B(t)$. When all of these elements are defined, we will put them together in a complete complex equivalent baseband data communication system.

We begin with some manipulation of frequency spectra. Assume the real channel $h(t)$, with Fourier transform $H(f)$, through which $s(t)$ is transmitted is

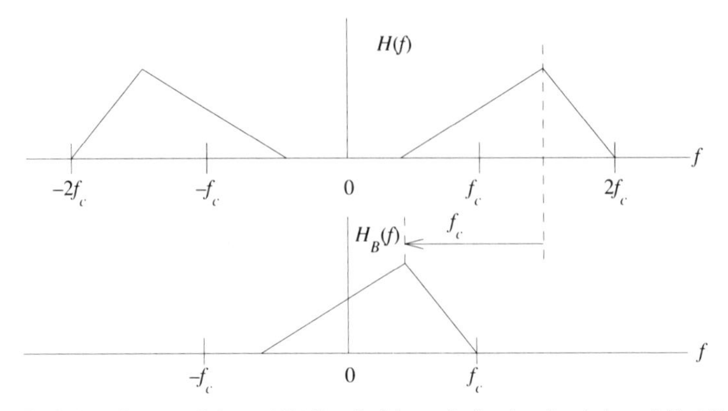

Fig. 5.6 Spectra of channel $H(f)$ and of the equivalent baseband channel $H_B(f)$.

a passband channel bandlimited to $|f| < 2f_c$, as shown in the example of Figure 5.6. *Define* a complex function $h_B(t)$ such that its Fourier transform, $H_B(f)$, is given by

$$H_B(f) = \begin{cases} 2H(f+f_c), & f + f_c > 0 \\ 0, & f + f_c < 0. \end{cases} \tag{5.11}$$

The lower curve in Figure 5.6 illustrates that $H_B(f)$ is merely the positive frequency part of $H(f)$ moved down in frequency by f_c *Hz*.

With the change of variable $f = -(\mu + f_c)$,

$$H_B(-\mu - f_c) = \begin{cases} 0, & \mu > 0 \\ 2H(-\mu), & \mu < 0. \end{cases} \tag{5.12}$$

If we replace μ with the symbol f and take the complex conjugate,

$$H_B^*(-f-f_c) = \begin{cases} 0, & f > 0 \\ 2H^*(-f), & f < 0 \end{cases} = \begin{cases} 0, & f > 0 \\ 2H(f), & f < 0. \end{cases} \tag{5.13}$$

Since for any real channel $h(t)$, $H*(-f) = H(f)$. If we replace μ with $-f$ in (5.12),

$$H_B(f - f_c) = \begin{cases} 2H(f), & f > 0 \\ 0, & f < 0. \end{cases} \tag{5.14}$$

From (5.13) and (5.14),

$$H(f) = \underbrace{H_B(f-f_c)/2}_{\text{exists only on } f>0} + \underbrace{H_B^*(-f-f_c)/2}_{\text{exists only on } f<0}. \tag{5.15}$$

We define $\tilde{h}_B(t)$ as the inverse Fourier transform of $H_B(f)$. Taking the inverse Fourier transform of (5.15), we have the sum of a function and its complex conjugate which yields twice the real part

$$\begin{aligned} h(t) &= \tilde{h}_B(t)\ e^{j2\pi f_c t}/2 + \tilde{h}_B^*(t)\ e^{-j2\pi f_c t}/2 \\ &= \mathrm{Re}[\tilde{h}_B(t)\ e^{j2\pi f_c t}]. \end{aligned} \tag{5.16}$$

The complex function

$$\tilde{h}_B(t) = h_1(t) + jh_2(t) \tag{5.17}$$

is the *equivalent baseband channel.* It is *non*-analytic because its Fourier transform $H_B(f)$ is *not* necessarily one sided, as consideration of (5.11) will show. Its components

$$h_1(t) = \mathrm{Re} \int_{-\infty}^{\infty} H_B(f) e^{j2\pi ft} df,\ h_2(t) = \mathrm{Im} \int_{-\infty}^{\infty} H_B(f) e^{j2\pi ft} df \tag{5.18}$$

are *not* a complex-analytic pair. However, the complex passband channel defined as (from (5.16))

$$\tilde{h}(t) = \tilde{h}_B(t) e^{j2\pi f_c t} = h(t) + j\check{h}(t) \tag{5.19}$$

is complex analytic because $H_B(f-fc)$, the Fourier transform of $\tilde{h}_B(t) e^{j2\pi f_c t}$, exists only on positive frequencies by definition (5.11).

EXAMPLE 1 Suppose $H(f) = \sin[\pi(|f|-f_c)/f_c]$, $-2f_c < f < 2f_c$, as shown below.

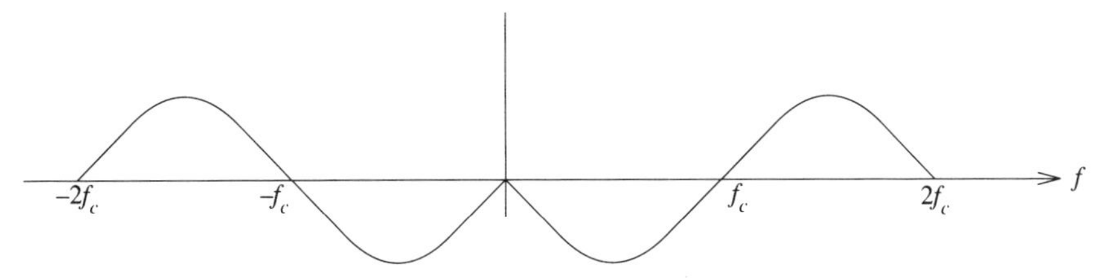

Then

$$H_B(f) = 2H(f+f_c)|_{f+f_c>0} = 2\sin(\pi f/f_c),\ -f_c < f < f_c\ .$$

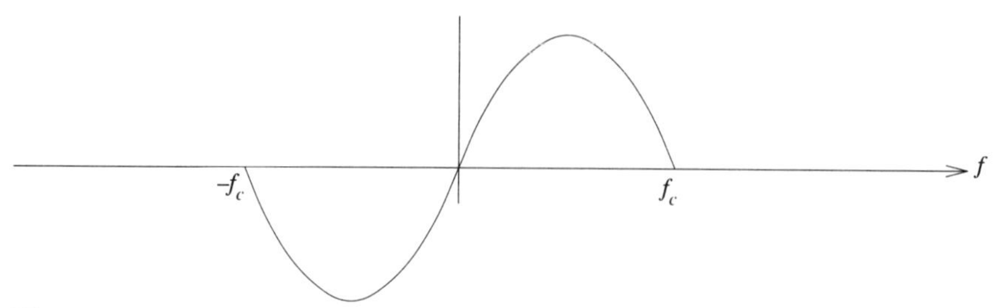

Thus

$$h_B(t) = \int_{-\infty}^{\infty} H_B(f) e^{j2\pi ft} df = \int_{-f_c}^{f_c} 2\sin(\pi f/f_c) e^{j2\pi ft} df .$$

This is nonanalytic, since it is entirely imaginary.

5.1.2 THE COMPLEX-ANALYTIC TRANSMISSION SYSTEM

Just as we have defined a complex analytic transmitted signal $\tilde{s}(t)$ in (5.6), we can define a complex analytic received signal $\tilde{r}(t)$, as shown in Figure 5.7. The expression for $\tilde{r}(t)$ as a function of the carrier frequency and an equivalent baseband received signal is defined as the complex analytic function whose real part is the actual real received signal $r(t)$. We further define $\tilde{r}_B(t)$ as the equivalent baseband received signal for which

$$\tilde{r}(t) = \tilde{r}_B(t) e^{j2\pi f_c t} \tag{5.20}$$

We will derive $\tilde{r}_B(t)$ in this section.

To see the relationship among $\tilde{m}(t)$, $\tilde{h}_B(t)$, and $\tilde{r}_B(t)$, we write, referring to Figure 5.1,

$$r(t) = s(t) * h(t) + n(t) . \tag{5.21}$$

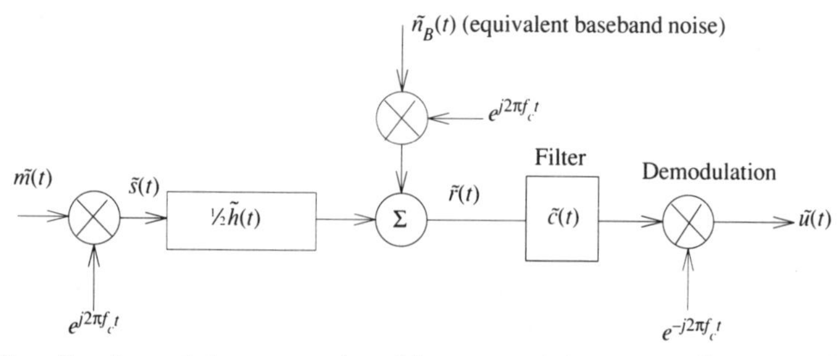

Fig. 5.7 Complex analytic representation of linear transmission system. Here $s(t) = \text{Re}\,\tilde{s}(t)$ and $r(t) = \text{Re}\,\tilde{r}(t)$ are the real transmitted and received signals respectively, and $\tilde{u}(t)$ is an estimate of the complex modulation $\tilde{m}(t)$.

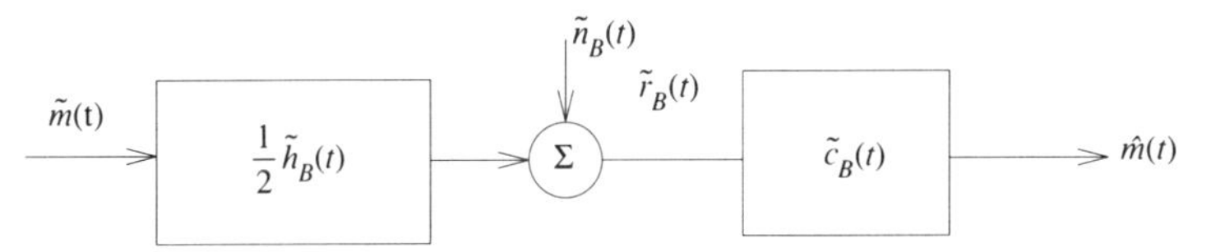

Fig. 5.8 Equivalent baseband complex transmission system.

Since, using (5.10),

$$\text{Re } [\tilde{s}(t) * \tilde{h}(t)] = 2s(t) * h(t) \tag{5.22}$$

and, from (5.6) and (5.19),

$$\tilde{s}(t) * \tilde{h}(t) = \int \tilde{s}(\tau)\tilde{h}(t-\tau)d\tau = e^{j2\pi f_c t}[\tilde{m}(t) * \tilde{h}_B(t)] \ , \tag{5.23}$$

we have

$$r(t) = \text{Re } \left\{ e^{j2\pi f_c t} \left[\frac{1}{2}\tilde{m}(t) * \tilde{h}_B(t) + \tilde{n}_B(t) \right] \right\} , \tag{5.24}$$

where $\tilde{n}_B(t)$ is an equivalent baseband noise, described in the next section. From (5.20) and (5.24),

$$\tilde{r}_B(t) = \frac{1}{2}\tilde{m}(t) * \tilde{h}_B(t) + \tilde{n}_B(t) \ . \tag{5.25}$$

This relationship defines the (complex) equivalent baseband communication system model of Figure 5.8, up to the input of the receiving filter.

The complex filtering operation in (5.25) can be realized in four cross-coupled real filters. From (5.5) and (5.17),

$$\begin{aligned} \tilde{m}(t) * \tilde{h}_B(t) &= m_i(t) * h_1(t) - m_q(t) * h_2(t) \\ &+ j\left[m_i(t) * h_2(t) + m_q(t) * h_1(t) \right] \ . \end{aligned} \tag{5.26}$$

The cross-coupled filtering structure is pictured in Figure 5.13, although in that case it is a complex passband, rather than baseband, signal that is being filtered.

With the equivalent complex baseband system fully defined, the analysis of linear passband systems can be reduced to the baseband analyses of previous chapters. Nonlinear modulation formats and nonlinear perturbations in linear systems limit the utility of equivalent baseband analysis.

5.1.3 EQUIVALENT BASEBAND NOISE

In this section, we derive the statistical properties of the equivalent complex baseband noise. This will make it possible to carry out a statistical performance analysis of the equivalent baseband transmission system. For the noise component of (5.24), we presume a bandpass noise process $n(t)$ with a spectrum limited to $|f| < 2f_c$. We represent the noise as

$$n(t) = \mathrm{Re}[\tilde{n}_B(t)e^{j2\pi f_c t}]$$
$$= n_i(t)\cos 2\pi f_c t - n_q(t)\sin 2\pi f_c t \, , \tag{5.27a}$$

where

$$\tilde{n}_B(t) = n_i(t) + jn_q(t) \tag{5.27b}$$

is the equivalent complex baseband noise. The statistics of $n_i(t), n_q(t)$, and $\tilde{n}_B(t)$ are easily derived from those of $n(t)$. Assume $n(t)$ is a stationary, zero-mean, real process. Define the autocorrelation functions (Appendix 2A)

$$\phi_{nn}(\tau) = E\Big[n(t)n(t+\tau)\Big] \tag{5.28a}$$

$$\phi_{ii}(\tau) = E\Big[n_i(t+\tau)n_i(t)\Big] \tag{5.28b}$$

$$\phi_{qq}(\tau) = E\Big[n_q(t+\tau)n_q(t)\Big] \tag{5.28c}$$

$$\phi_{iq}(\tau) = E\Big[n_i(t+\tau)n_q(t)\Big] \tag{5.28d}$$

$$\phi_{qi}(\tau) = E\Big[n_q(t+\tau)n_i(t)\Big] \, . \tag{5.28e}$$

Then (Exercises 5.2 and 5.3)

$$\phi_{ii}(\tau) = \phi_{qq}(\tau) \, , \quad \phi_{iq}(\tau) = -\phi_{qi}(\tau) \tag{5.29}$$

$$\phi_{nn}(\tau) = \phi_{ii}(\tau)\cos(2\pi f_c \tau) - \phi_{qi}(\tau)\sin(2\pi f_c \tau) \, . \tag{5.30}$$

The autocorrelation function of the complex equivalent baseband noise (5.27) is, using (5.29) and the definition for the autocorrelation function $\phi_{BB}(\tau)$ of a complex function,

$$\phi_{BB}(\tau) = E[\tilde{n}_B(t+\tau)\tilde{n}_B{*}(t)] = 2\phi_{ii}(\tau) + 2j\phi_{qi}(\tau) \, . \tag{5.31}$$

Defining the power density function $S_{BB}(f)$ as the Fourier transform of $\phi_{BB}(\tau)$, we note that if $S_{BB}(f)$ is symmetric about $f=0$, then $\phi_{BB}(\tau)$ is real and $\phi_{qi}(\tau) = 0$. Thus, for the usual case of the noise power spectrum being symmetric about the carrier, $n_i(t)$ and $n_q(t)$ are uncorrelated for all time shifts and hence for Gaussian noise they are *independent*.

From (5.30) and (5.31),

$$\phi_{nn}(\tau) = 1/2\mathrm{Re}[\phi_{BB}(\tau)e^{j2\pi f_c \tau}] \, . \tag{5.32}$$

But (Exercise 5.4)

$$F\,\{2\mathrm{Re}[\phi_{BB}(\tau)e^{j2\pi f_c \tau}]\} = S_{BB}(f-f_c) + S_{BB}(-f-f_c) \, . \tag{5.33}$$

Thus the power density spectrum of $n(t)$ is

$$S_n(f) = F[\phi_{nn}(\tau)] = 1/4\{S_{BB}(f-f_c)+S_{BB}(-f-f_c)\}\ . \qquad (5.34)$$

The auto- and cross-correlation functions have several useful symmetries in addition to (5.30). We have

$$\phi_{ii}(-\tau) = E[n_i(t)n_i(t+\tau)] = E\left[n_i(\mu)n_i(\mu-\tau)\right] = \phi_{ii}(\tau)\ . \quad (5.35)$$

From the definitions (5.28d,e), we have

$$\phi_{iq}(\tau) = E\left[n_q(t)n_i(t+\tau)\right] = E\left[n_i(\mu)n_q(\mu-\tau)\right]\ ,$$

and from (5.29)

$$\phi_{iq}(\tau) = \phi_{qi}(-\tau) = -\phi_{iq}(-\tau) \qquad (5.36)$$

Similarly,

$$\phi_{qi}(\tau) = -\phi_{qi}(-\tau)\ .$$

From (5.31)

$$\phi_{BB}^*(-\tau) = \phi_{BB}(\tau) \qquad (5.37)$$

as required for the power density function $S_{BB}(f)$ to be real. Relation (5.36) implies that $\phi_{iq}(0) = 0$, i.e., $n_i(t)$ and $n_q(t)$ are uncorrelated for a zero time shift irrespective of the shape of the noise spectrum. For Gaussian $n(t)$, samples of $n_i(t)$ and $n_q(t)$ at the same time t are independent Gaussian random variables with the same variance.

EXAMPLE 2 Suppose the noise $n(t)$ has the power density spectrum $S_n(f)$ shown:

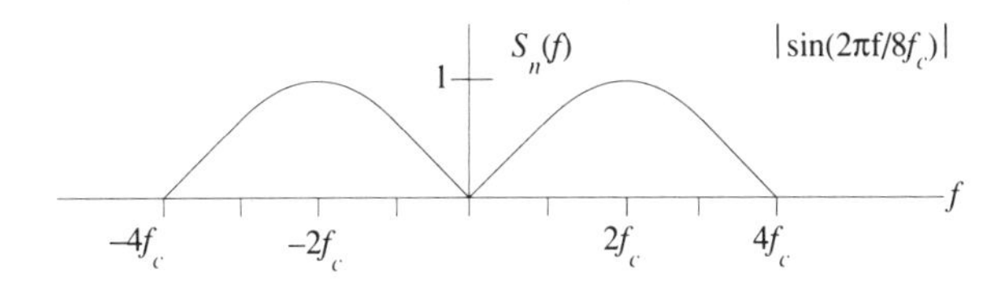

Then

$$\phi_{nn}(\tau) = \int_{-\infty}^{\infty} S_n(f)e^{j2\pi f\tau}df = \left[1+\cos 8\pi f_c\tau\right]\left\{8f_c/\pi\left[1-(8f_c\tau)^2\right]\right\}$$

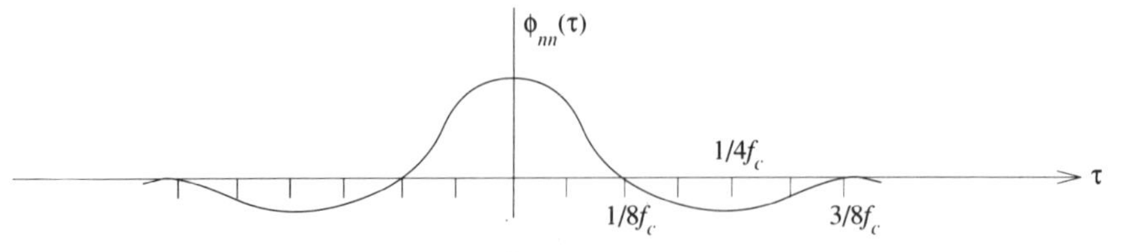

The spectral relationship (5.34) is satisfied by the equivalent baseband noise spectrum

$$S_{BB}(f) = 4\sin\left[2\pi(f+f_c)/8f_c\right], \; -f_c<f<3f_c.$$

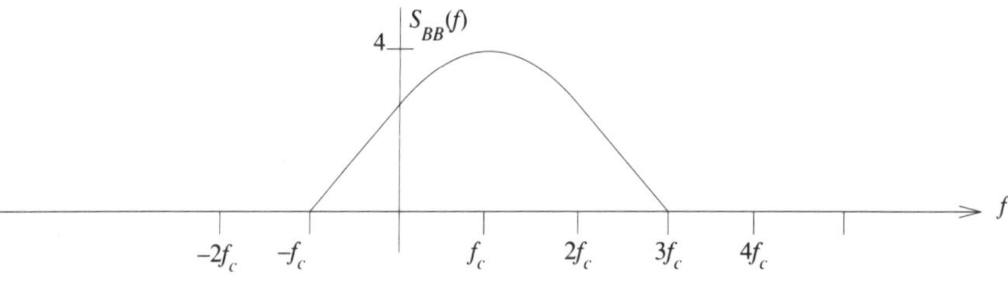

The asymmetry of $S_{BB}(f)$ implies that the autocorrelation function $\phi_{BB}(\tau)$ is complex:

$$\begin{aligned}\phi_{BB}(\tau) &= \int_{-\infty}^{\infty} S_{BB}(f)e^{j2\pi f\tau}df \\ &= \left\{8f_c/\pi\left[1-(8f_c\tau)^2\right]\right\}\left\{\left[\cos6\pi f_c\tau + \cos2\pi f_c\tau\right]\right. \\ &\quad \left. + \; j\left[\sin6\pi f_c\tau - \sin2\pi f_c\tau\right]\right\}\end{aligned}$$

Comparing this with (5.31) yields the component correlations

$$\phi_{ii}(\tau) = \left\{4f_{c/}\pi\left[1-8f_c\tau\right]^2\right\}\left\{\cos6\pi f_c\tau + \cos2\pi f_c\tau\right\}$$

$$\phi_{qi}(\tau) = \left\{4f_c/\pi\left[1-8f_c\tau\right]^2\right\}\left\{\sin6\pi f_c\tau-\sin2\pi f_c\tau\right\}$$

For white noise with two-sided spectral power density $N_0/2$, we presume that a perfect bandpass filter has limited the noise spectrum to a bandwidth $2W$ Hz about f_c, a range chosen wide enough to include all frequencies at which

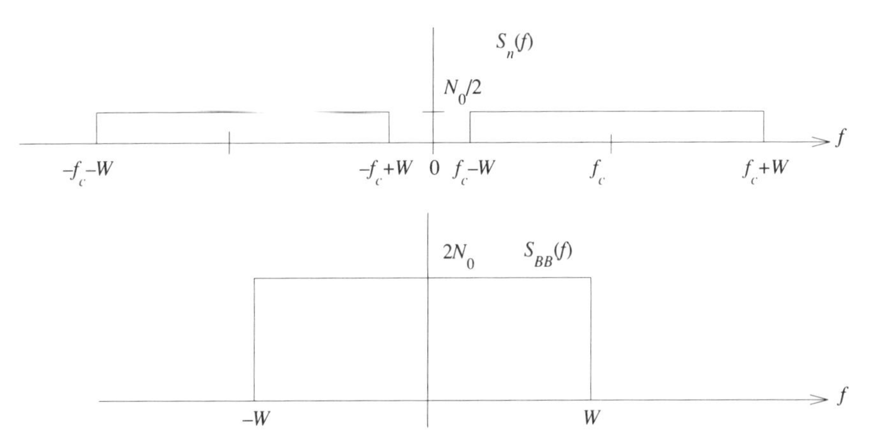

Fig. 5.9 Power density spectrum of passband white noise process $n(t)$, and of equivalent baseband noise $\tilde{n}_B(t)$.

there is transmitted signal energy. This is shown in Figure 5.9. Then from (5.34),

$$S_{BB}(f) = \begin{cases} 2N_0, & |f| \leq W \\ 0, & |f| > W \end{cases} \tag{5.38}$$

with corresponding autocorrelation function

$$\phi_{BB}(\tau) = 2N_0[\sin(2\pi W\tau)/\pi\tau]. \tag{5.39}$$

Note that $S_{BB}(f)$ is symmetric about $f=0$, fulfilling the condition stated earlier for $n_i(t)$ and $n_q(t+\tau)$ being uncorrelated for all τ. Thus the power spectral density $S_{BB}(f)$ is the sum of the power spectral densities $S_i(f)$ and $S_q(f)$ of $n_i(t)$ and $n_q(t)$ respectively, and $S_i(f)$ and $S_q(f)$ are each equal to N_0 for $|f| \leq W$, a convenience for analysis.

5.2 LINEAR MODULATION FORMATS

The complex notation we have described can be used for any of the common modulation formats used in data communications. Using a complex data train representation, the complex modulation waveform $\tilde{m}(t)$ of (5.5) is expressed as

$$\tilde{m}(t) = A(t)e^{j\theta(t)} = \sum_n \tilde{d}_n p(t-nT), \tag{5.40}$$

where the sequence of complex levels $\{\tilde{d}_n\}$ is derived from the input data sequence, and $p(t)$ may be (but is not necessarily) a complex pulse. For example, in phase modulation, the $\{\tilde{d}_n\}$ are chosen from a discrete set of complex numbers of the same magnitude and uniformly spaced phases, and $p(t)$ is a real pulse. For single sideband modulation, the $\{d_n\}$ are real and $p(t)$ is a complex analytic baseband pulse. For quadrature amplitude modulation, the $\{\tilde{d}_n\}$ are chosen from a discrete set of complex numbers placed on a uniform grid in the complex plane, and $p(t)$ is real. General AM-PM modulations use other sets of complex $\{\tilde{d}_n\}$. Our interest is focused on these linear modulations because they translate to passband the spectrum of the baseband pulse $p(t)$, thereby achieving the same high spectral efficiency as baseband signaling with spectrally efficient pulses. We will, however, examine the less spectrally efficient frequency-shift keying because of its simplicity and use in radio applications.

5.2.1 LINEAR TWO-DIMENSIONAL SIGNALS

Using the complex-analytic notation, we will derive expressions for several modulation formats. Let us assume a baseband pulse $p(t)$ (usually, but not always, real) and define the complex modulation levels

$$\tilde{d}_n = a_n + jb_n, \tag{5.41}$$

where $\{a_n\}$ and $\{b_n\}$ are two real amplitude level sequences. We have, from (5.40),

$$\tilde{m}(t) = \sum_n (a_n+jb_n)p(t-nT), \tag{5.42}$$

so that from (5.6),

$$\tilde{s}(t) = \sum_n (a_n+jb_n)p(t-nT)\exp\{j2\pi f_c t\}. \tag{5.43}$$

With $p(t)$ real,

$$s(t) = \text{Re}\,\tilde{s}(t) = \sum_n a_n p(t-nT)\cos(2\pi f_c t) - \sum_n b_n p(t-nT)\sin(2\pi f_c t). \tag{5.44}$$

The real transmitted signal can be generated in a quadrature modulator, as shown in Figure 5.10.

The complex system model of Figure 5.7 is applicable here, augmented with a detector at the output of the receiver, as shown in Figure 5.11. In the absence of intersymbol interference, the receiver filter $\tilde{c}_B(t)$ is simply a pair of identical filters — one for the real part and the other for the imaginary part of $\tilde{r}_B(t)$ — matched to $p(t)$. The baseband equivalent is shown in Figure 5.12.

It should be apparent that for individually and mutually uncorrelated signal level sequences $\{a_n\}$ and $\{b_n\}$, the spectral magnitude of $s(t)$ is identical in

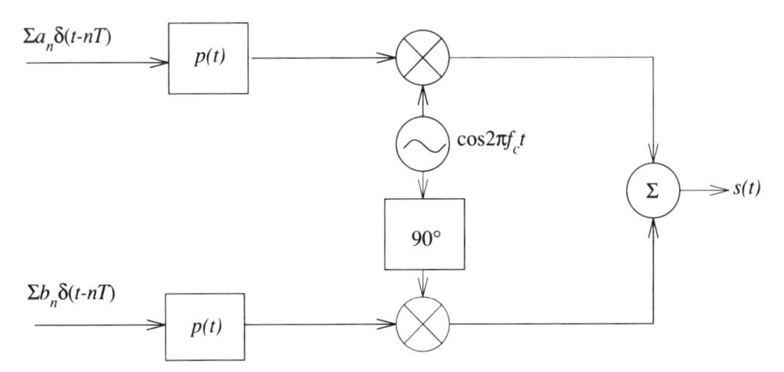

Fig. 5.10 Generation of a two-dimensional linearly modulated signal.

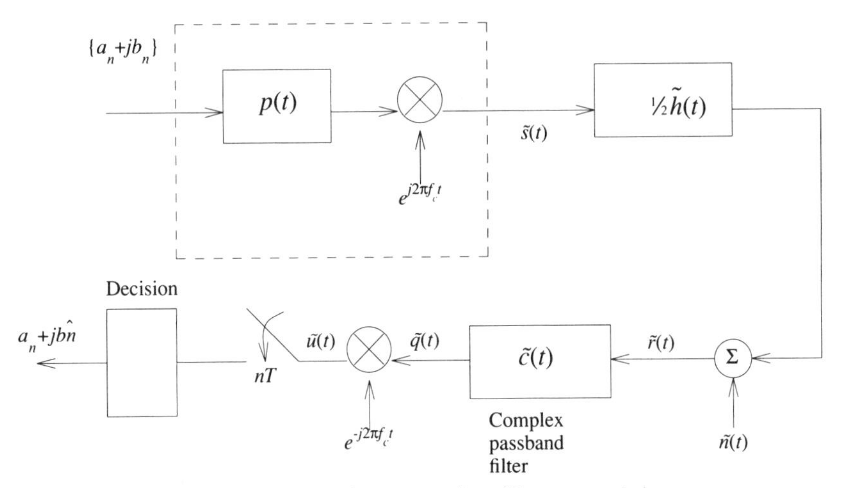

Fig. 5.11 Complex analytic representation of linear transmission system.

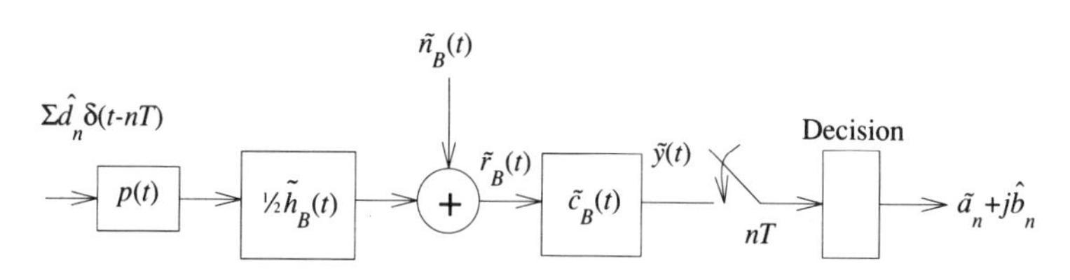

Fig. 5.12 Equivalent baseband representation.

shape to that of $p(t)$, just shifted by $\pm f_c$. That is,

$$|F[p(t-nT)\cos 2\pi f_c t]| = |Fp(t-nT) * 1/2[\delta(f-f_c) + \delta(f+f_c)]|$$
$$= 1/2\ |\exp\{-j2\pi fT\}P(f) * [\delta(f-f_c) + \delta(f+f_c)]|$$
$$= 1/2\ [|P(f-f_c)| + |P(f+f_c)|]. \tag{5.45}$$

The passband bandwidth is twice as large as the baseband bandwidth, but there are *two* passband signals, one with a cosine carrier and one with a sine carrier, so that the efficiency is the same.

In many applications, the bandwidth is limited in the passband. The same spectrally efficient techniques used in baseband signaling (see Sections 4.3 and 4.6) can be employed in passband transmission. For example, the pulse shape $p(t)$ and the receive filter $\tilde{c}(t)$ can be chosen such that the equivalent baseband channel has either a Nyquist characteristic (possibly with excess bandwidth) or a partial response characteristic.

EXAMPLE 3 A quadrature amplitude modulation (QAM) modulator, with transmitted signal given by (5.44), uses 4-level PAM data trains $\{a_n\}$ and $\{b_n\}$, raised cosine pulses with 12.5% rolloff, and has a baud of 2400 symbols/sec. The data rate of each of the baseband signals

$$\sum a_n p(t-nT), \qquad \sum b_n p(t-nT)$$

is $\log_2(4)\times 2400 = 4800$ bps, for a total data rate of 9600 bps. The spectrum occupied is (conservatively) $1.125 \times 1200 = 1350$ Hz above and below the carrier, for a total of 2700 Hz. The spectral efficiency is then 9600/2700 = 3.6 bps/Hz.

The passband QAM signal has the same total data rate with 2700 Hz bandwidth, yielding the same spectral efficiency.

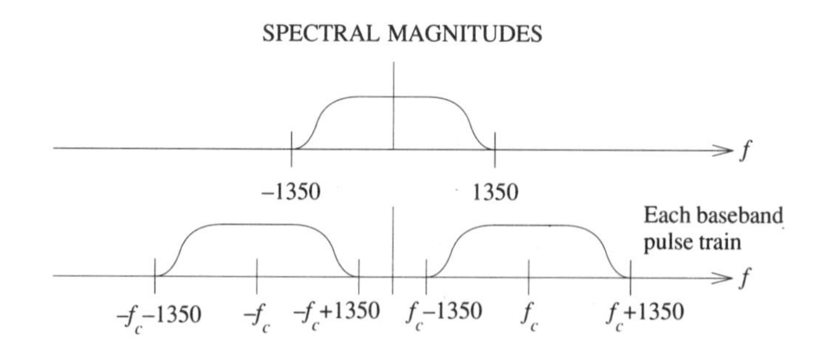

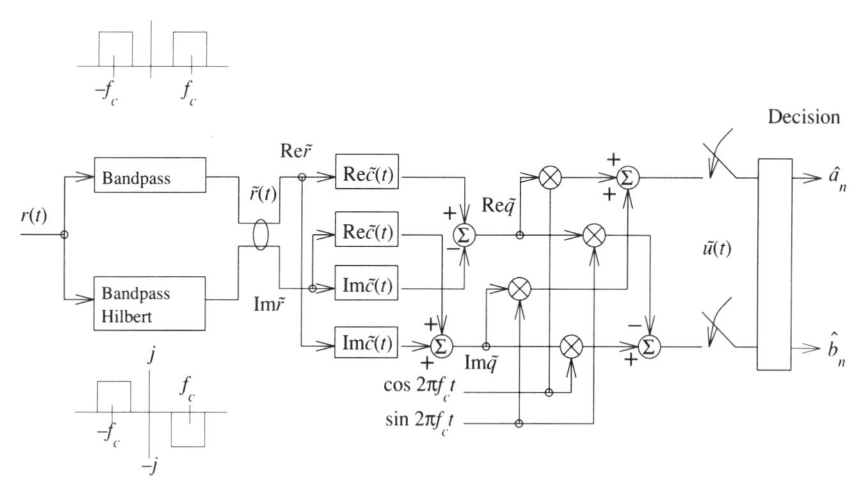

Fig. 5.13 A real receiver for linearly modulated passband signals generates a complex analytic received signal $\tilde{r}(t)$ in a "phase splitter," and then implements a complex filtering (equalization) operation prior to demodulation and detection. The phase splitter usually contributes bandlimiting filtering, as shown here.

Detection of the two-dimensional signal in a linear receiver requires creating an approximation to the complex baseband signal $\tilde{m}(t)$ and sampling at symbol intervals to recover $\tilde{d}_n = a_n + jb_n$. This is the essence of the complex receiver of Figure 5.11. The real receiver, shown in Figure 5.13, incorporates a "phase splitter" at its input to generate the complex analytic received signal $\tilde{r}(t)$. The equalizer generates an output

$$\tilde{q}(t) = \tilde{r}(t) * \tilde{c}(t) , \tag{5.46}$$

which is realized in four real filters, and the demodulator output is

$$\tilde{u}(t) = e^{-j2\pi f_c t}\tilde{q}(t) . \tag{5.47}$$

The cross-coupled structures realize the complex convolution and multiplication respectively.

Much important detail is missing from Figure 5.13 but described in later chapters. For channels with linear distortion and white Gaussian noise, the optimum configuration for the phase splitter-equalizer combination is derived in Chapter 7. When the channel characteristics can change, adaptive equalization (Chapter 8) is needed, with appropriate feedback for error information from the receiver output for use in filter adaptation. The demodulator requires its own tracking loop (Chapter 6) following phase jitter as well as recovering the proper carrier phase for coherent demodulation. Sampling time must also be derived from the incoming signal, as described in Chapter 6.

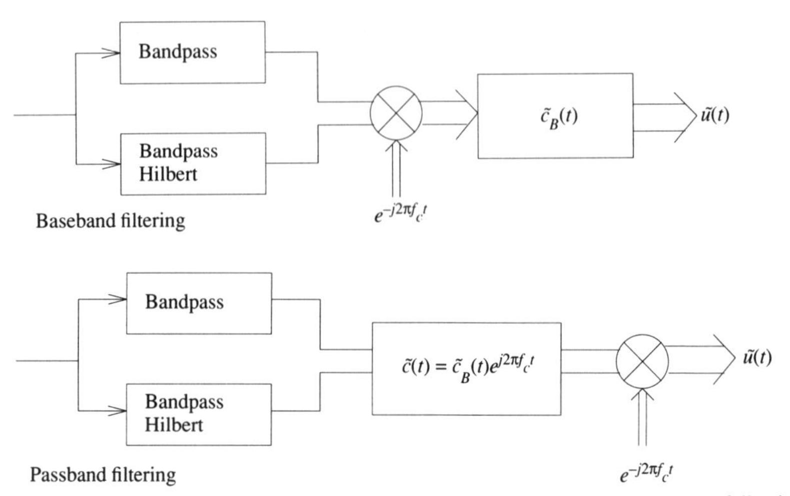

Fig. 5.14 Equivalence of receiver structures with demodulation either preceding or following filtering (equalization).

Given that these elements are provided, the data decision corresponds to finding the transmitted signal point $(a_n + jb_n)$ closest in Euclidean distance to the sampled output $\tilde{u}(nT)$. This will be made clear in Section 5.2.3.

It can be demonstrated that the order of demodulation and complex filtering in Figure 5.11 can be reversed, as illustrated in Figure 5.14. This is apparent from the equalities

$$\begin{aligned}
\tilde{r}(t)e^{-j2\pi f_c t} * \tilde{c}_B(t) &= \int_{-\infty}^{\infty} \tilde{c}_B(\tau)\, \tilde{r}(t-\tau)\, e^{-j2\pi f_c (t-\tau)}\, d\tau \\
&= e^{-j2\pi f_c t}\left[\tilde{r}(t) * \tilde{c}_B(t)e^{j2\pi f_c t}\right] \\
&= \left[\tilde{r}(t) * \tilde{c}(t)\right]\exp\left[-j2\pi f_c t\right] \qquad (5.48)
\end{aligned}$$

where the first expression describes demodulation followed by baseband filtering, and the last describes passband filtering followed by demodulation. A receiver with complex filtering preceding demodulation is desirable, since this avoids inserting the delay of the equalization filter into the control loop generating the demodulation phase, which would impair its ability to correct for channel phase jitter. This is discussed in detail in Chapter 8.

5.2.2 SINGLE SIDEBAND (SSB)

Equation (5.44) and the example given in the last section suggest a double-sideband transmitted signal $s(t)$. This is not necessarily the case. Let $\tilde{d}_n = a_n$ be

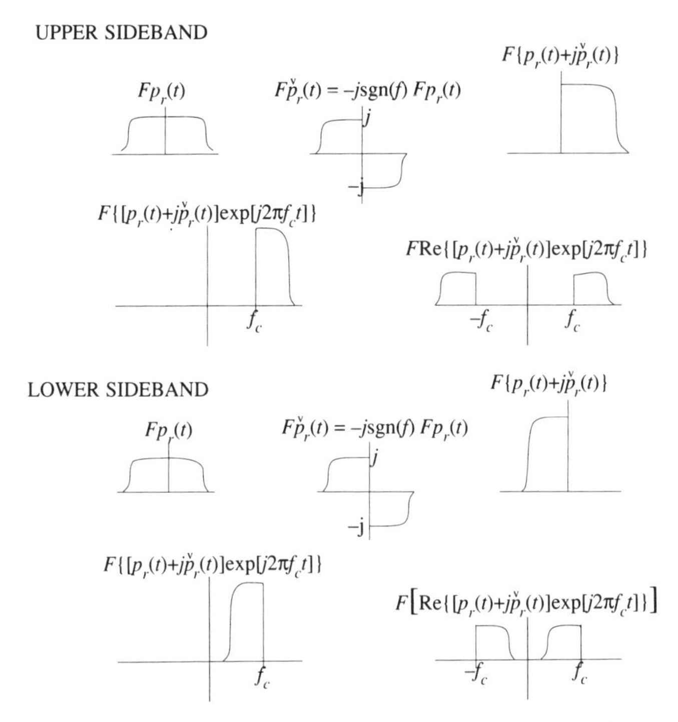

Fig. 5.15 Spectra for generation of upper- and lower-sideband signals by phasing method.

a real signal level, $p_r(t)$ a real pulse, and $\tilde{p}(t)$ a complex-analytic pulse

$$\tilde{p}(t) = p_r(t) + j\check{p}_r(t) . \tag{5.49}$$

Then,

$$\tilde{s}(t) = \sum_n a_n [p_r(t-nT) + j\check{p}_r(t-nT)] \exp\{j2\pi f_c t\}. \tag{5.50}$$

The spectrum of $p_r(t)+j\check{p}_r(t)$ is one-sided (Figure 5.15), as required for a complex analytic signal. The result is an upper-sideband signal

$$s(t) = \text{Re } \tilde{s}(t) = \sum_n a_n [p_r(t-nT)\cos 2\pi f_c t - \check{p}_r(t-nT)\sin 2\pi f_c t]. \tag{5.51}$$

With a negative sign for $\check{p}_r(t)$ in (5.50), a lower-sideband signal is produced, as shown in Figure 5.15b. Figure 5.16 shows generation of a single-sideband signal in a quadrature modulator.

Equation (5.51) and Figure 5.15 also suggest a spectral efficiency identical to that of a quadrature-modulated double-sideband signal. There is only one real

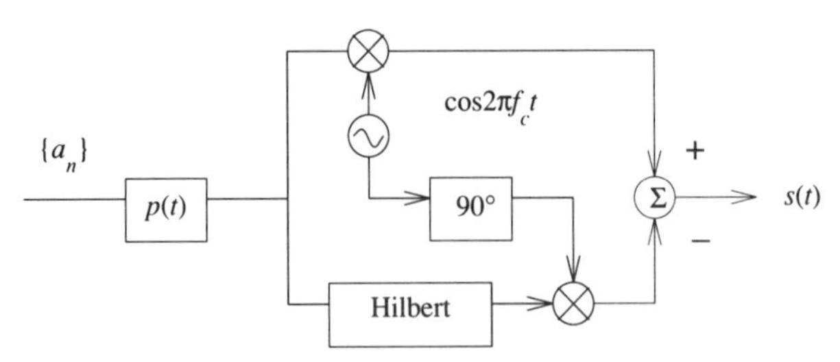

Fig. 5.16 Block diagram of upper-sideband modulator.

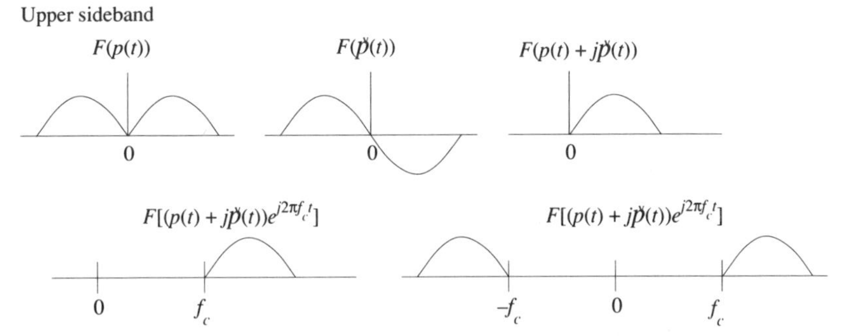

Fig. 5.17 Single sideband modulation with the modified duobinary pulse.

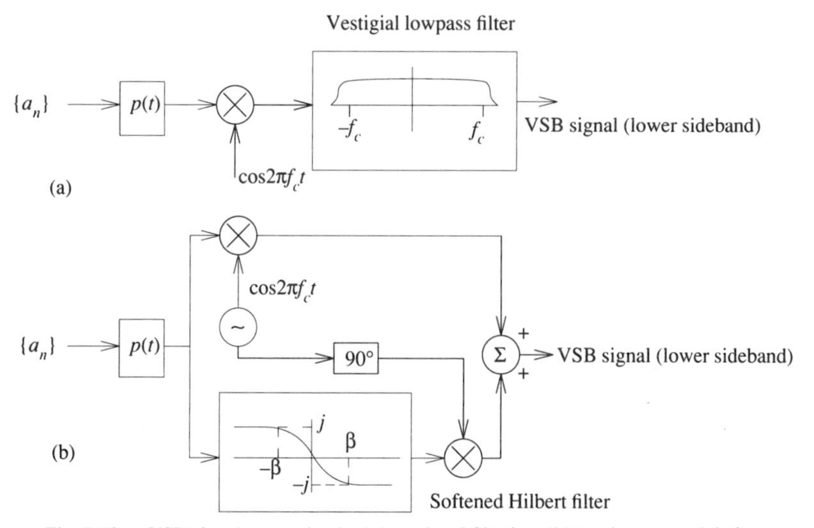

Fig. 5.18 VSB signal generation by (a) passband filtering, (b) quadrature modulation.

stream $\{a_n\}$, so that the bit rate is half that of the quadrature signal with two data streams. But the passband bandwidth is also half, so the spectral efficiency is the same.

One problem with SSB is creation of a signal with a sharp spectral edge at the carrier frequency, as illustrated in Figure 5.15. To mitigate this problem, a partial response pulse (Section 4.6) with a spectral null at $f=0$ can be used. Figure 5.17 illustrates the sequence of spectra of Figure 5.15 for a modified duobinary pulse. Of course, the noise penalty of partial response signaling must be accepted just as in baseband signaling, unless the Viterbi algorithm (maximum likelihood sequence estimation) is used for detection (Chapter 7).

5.2.3 VESTIGIAL SIDEBAND

An alternative approach to SSB filtering problems is the use of vestigial sideband (VSB) [1]. The unwanted sideband is not eliminated entirely; a vestige is passed by a softened cutoff filter (Figure 5.18a). VSB can also be produced by a quadrature modulator (Figure 5.18b) analogous to that shown in Figure 5.16 for SSB. The Hilbert filter is replaced by a soft Hilbert filter with spectrum

$$W_\beta(f) = \begin{cases} -j\,\mathrm{sgn}(f), & |f|>\beta \\ \\ \textit{smooth rolloff}, & |f|<\beta, \end{cases} \tag{5.52}$$

as illustrated in Figure 5.18b.

SSB and VSB modulations still have wide applications in analog communication, but not much any more for data communications. As shown in Chapter 6, among other disadvantages, they often require pilot tones for phase recovery that double-sideband modulation does not, and are not as amenable to trellis-coded modulation, described later in this chapter.

5.2.4 COHERENT PHASE-SHIFT KEYING (PSK)

Phase-shift keying (PSK) is perhaps the simplest form of two-dimensional linear modulation. As we noted earlier, the complex modulation levels $\{\tilde{d}_n\}$ in the baseband modulation waveform $\tilde{m}(t)$ of (5.40) are equal-length vectors with uniformly spaced angles, as illustrated in Figure 5.19. These signal *constellations*, or collections of signal points from which the $\{d_n\}$ are chosen, are concise graphical descriptions of two-dimensional modulations. For L-ary PSK, we define $\underline{\bar{d}}_n = \exp\{j\phi_n\} = \exp\{jm_n(2\pi/L)\}$, $m_n = 0,1,2,\ldots,(L-1)$. For $L=2$ (binary PSK), $\underline{\bar{d}}_n = \pm 1$.

Using this definition of $\underline{\bar{d}}_n$ in (5.40), and substituting (5.40) into (5.6) yields

$$\tilde{s}(t) = \sum_n p(t-nT)\exp\left\{j\left[\phi_n + 2\pi f_c t\right]\right\}. \tag{5.53}$$

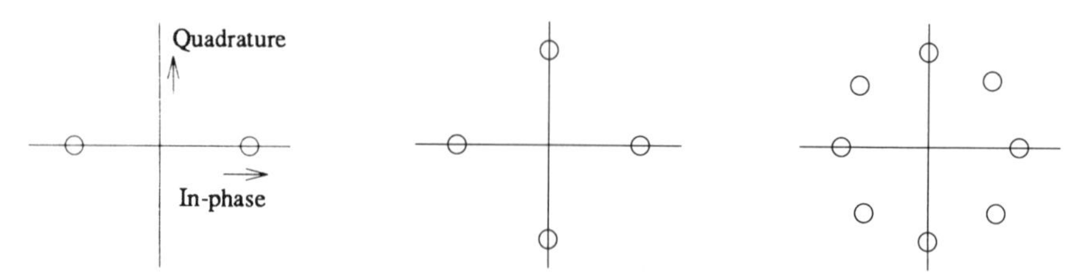

Fig. 5.19 Illustration of 2-, 4-, and 8-phase signal constellations.

The real part of $\tilde{s}(t)$, which is the actual transmitted PSK signal, has energy $T/2$ per symbol interval of length T when $p(t)$ is a rectangular pulse of unit amplitude. For coherent PSK, ϕ_n is a choice among L absolute phases specified by the data to be transmitted. For 8-phase PSK, the correspondence might be the Gray coded tribits shown next to each signal point in Figure 5.20. Gray coding in this case means that nearest-neighbor signal points are selected by data tribits that differ in only one bit position. For reasonably high signal to noise ratio, a decision error is almost always due to noise and distortion deflecting a received signal point into one of the two decisions regions adjacent to the decision region of the transmitted signal point. Gray coding assures that a symbol error, involving three bits, only results in a single bit error.

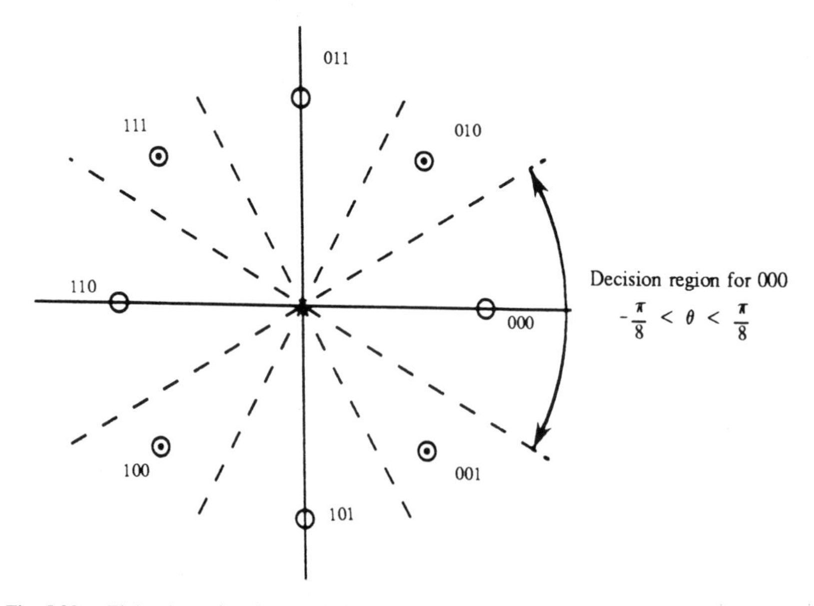

Fig. 5.20 Eight-phase signal constellation showing decision regions and Gray-coded signal points. A decision region about a signal point s_n is the set of receiver outputs (prior to the decision function) for which a decision will be made for s_n.

A PSK signal described by (5.53) can have amplitude as well as phase variations, depending on $p(t)$. With rectangular pulses, (5.53) becomes the constant-amplitude signal

$$\tilde{s}(t) = \sum_n \exp\{j\phi_n + j2\pi f_c t\} . \tag{5.54}$$

For binary PSK with rectangular pulses, the real transmitted signal in the interval $(0,T)$ is

$$s(t) = \text{Re}\ \exp\{\pm j\pi/2 + j2\pi f_c t\} = \mp \sin[2\pi f_c t],\ 0<t<T, \tag{5.55}$$

where the minus sign corresponds to the $+\pi/2$ phase. This is an example of a binary antipodal signal, for which we derived, in Chapter 2, the performance in the additive white Gaussian noise channel (for equiprobable and independent signal levels) as

$$P_e = Q\left(\left[2E/N_0\right]^{1/2}\right), \tag{2.35}$$

where $N_0/2$ is the two-sided noise spectral power density and E, the energy per symbol, is for binary signaling the energy per bit.

So far we have assumed *coherent* phase-shift keying (CPSK). An absolute phase reference is available at both transmitting and receiving ends of a communications circuit, and the received phase is measured against that reference to decide which signal point was transmitted. Even when phase coherence has been achieved, the occurrence of a phase *jump* (see Chapter 1) of any integer multiple of π/L radians can cause a series of errors. As we show in Chapter 6, the most common form of carrier recovery circuit would be insensitive to a permanent phase perturbation of this particular value, so that the receiver would not know a jump has occurred. Consequently, it is necessary to code the data into *changes* in phase, i.e., the transmitted phase is

$$\phi_n = \phi_{n-1} + \theta_n , \tag{5.56}$$

where θ_n is chosen from the set $0,\ 2\pi/L,\ \ldots,\ 2\pi(L-1)/L$. This type of coding is referred to as *differential encoding* and is commonly used in commercial modems. With a phase jump perturbing ϕ_n, the decision for that symbol interval would probably be wrong, but all subsequent decisions would be unaffected by the jump. Figure 5.21 compares this differential encoding with direct encoding.

If it is difficult to recover the phase reference, *differentially coherent* phase-shift keying (DCPSK) (next section) may be used. It employs a receiver which detects the *phase difference* $\phi_n - \phi_{n-1}$ rather than the absolute phase. Differential encoding is normally used so that this phase difference corresponds to the original data. A very simple receiver can retrieve the phase difference in DCPSK, as we will see later. There is, of course, a penalty to pay in SNR for the same error rate.

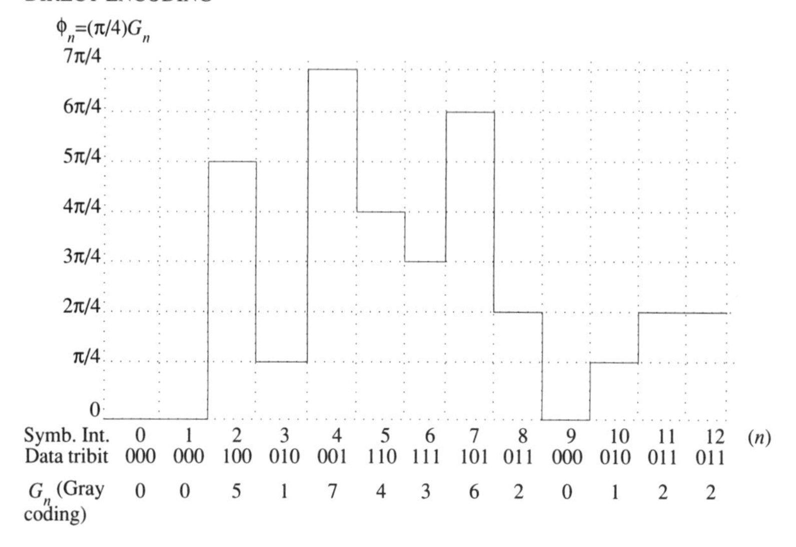

DIFFERENTIAL ENCODING

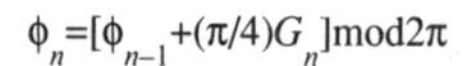

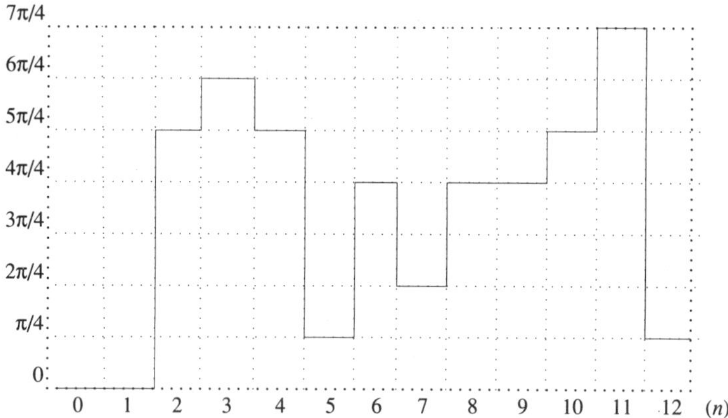

Fig. 5.21 Examples of direct and differentially encoded phase functions. The Gray coding is as shown in Figure 5.20.

Consider the phase modulation data communication system of Figure 5.22. The performance analysis for L-phase coherent PSK requires evaluation of the probability of the symbol error caused when noise makes a phase sample move outside of the decision region (Figure 5.20) corresponding to the transmitted phase. To begin this analysis, assume ϕ_0, the phase associated with the $n=0$ term in (5.53), is equal to a particular value $m(2\pi/L)$, and let us consider the

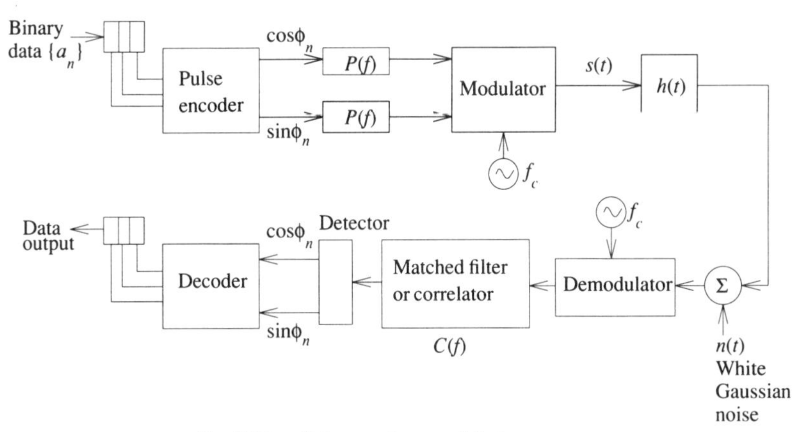

Fig. 5.22 Coherent phase modulation system.

optimum receiver for detecting ϕ_0. The real transmitted waveform is

$$s(t) = \text{Re}\left[\exp\{j2\pi f_c t\} \sum_n p(t-nT)\exp\{j\phi_n\}\right]. \tag{5.57}$$

Although in general there could be intersymbol interference from neighboring pulses ($n \neq 0$), we will assume for the moment there is none and that the channel is distortionless. The only perturbation is white Gaussian noise. This implies that each pulse can be examined individually.

Thus for purposes of analyzing the detection of ϕ_0 we can assume, with no loss of generality, the isolated pulse model

$$s_0(t) = \text{Re}\left[\exp\{j2\pi f_c t\}p(t)\exp\{jm(2\pi/L)\}\right], \tag{5.58}$$

where m is a particular value from the set $\{0, \ldots, L-1\}$.

We will analyze the performance of PSK for an ideal distortionless channel, so that the receiving filter $C(f)$ is matched to the transmitting filter $P(f)$. The complex analytic received signal (of which $r(t)$ in Figure 5.22 is the real part) is, including noise,

$$\tilde{r}(t) = \tilde{r}_B(t)\exp\{j2\pi f_c t\}$$

$$= \{p(t)\ \exp[jm\ (2\pi/L)] + \tilde{n}_B(t)\}\exp[j2\pi f_c t], \quad 0<t<T. \tag{5.59}$$

The (maximum likelihood) detector (2.21) minimizes $\int |\tilde{r}(t) - \tilde{s}_k(t)|^2 dt$ over k, where

$$\tilde{s}_k(t) = \exp\{j2\pi f_c t\}\ p(t)\exp\{jk(2\pi/L)\} \tag{5.60}$$

is a complex reference signal. This is equivalent to maximizing

$$y_k = \text{Re}\int_{-\infty}^{\infty} \tilde{r}(t)\tilde{s}_k^*(t)\ dt, \quad k \in 0,\ldots,L-1. \tag{5.61a}$$

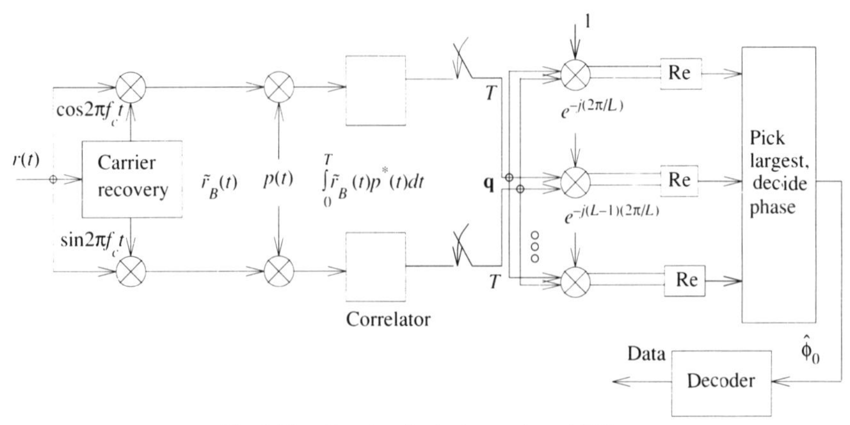

Fig. 5.23 Receiver for L-phase coherent PSK.

Substituting (5.59) and (5.60) into (5.61a),

$$y_k = \mathrm{Re} \int_{-\infty}^{\infty} \tilde{r}_B(t) p(t) \exp[-jk(2\pi/L)]\, dt$$

$$= \mathrm{Re}\left[\{ \exp[-j(2\pi k/L)] \int_{-\infty}^{\infty} \tilde{r}_B(t) p(t)\, dt \} \right], \quad k = 0,1,2,\ldots,L-1. \quad (5.61b)$$

For future convenience, we define the correlator output sample

$$q = \int_{-\infty}^{\infty} \tilde{r}_B(\tau) p(\tau)\, d\tau\,. \quad (5.61c)$$

Note that the effective interval of integration is simply the interval on which $p(t)$ is nonzero. To detect a train of symbols rather than the individual symbol examined here, a realizable correlation receiver would "integrate and dump" each symbol interval T. A receiver realizing these operations is illustrated in Figure 5.23. Note from (5.61) that only one complex integral, realizing a matched filter or correlator, has to be computed.

The correlation with L possible transmitted phase angles indicated in Figure 5.23 is equivalent to choosing the closest phase of the received signal. The probability of error given the transmitted signal point

$$\mathbf{s}_m = e^{j2\pi m/L} \quad (5.62)$$

is the probability that the received signal vector $\mathbf{q}$ at the sampler output lies *outside* the decision region illustrated in Figure 5.24. This is

$$P_e = 1 - \int_{m\text{th decision region}} p(\mathbf{q}|\mathbf{s}_m)\; dxdy\;, \quad (5.63)$$

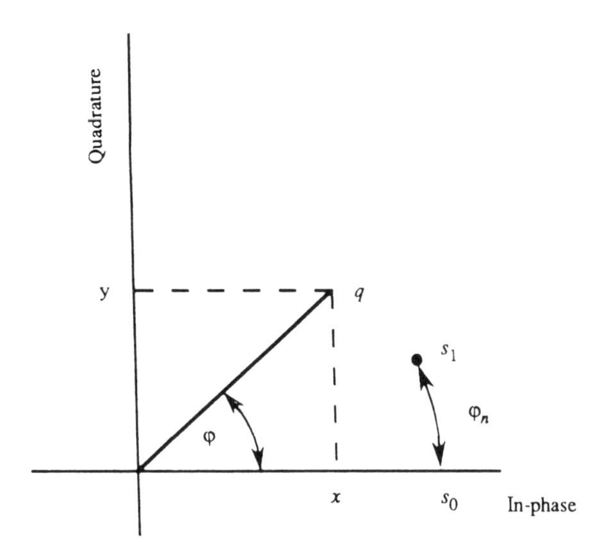

Fig. 5.24 Decision region for integration of received signal probability density function. A received signal point q is illustrated with both Cartesian (x,y) and polar (ρ,ϕ) coordinates.

where $p(\mathbf{q}\,|\,\mathbf{s}_m)$ is the probability density of the correlator output vector, given that $\mathbf{s}_m$ was transmitted. We assume that the channel is additive, white and Gaussian. Conditioned on the transmitted signal point of (5.62), the mean of $\mathbf{q}$ is, from (5.61c),

$$E(\mathbf{q}) = e^{j2\pi m/L} \int p^2(t)\,dt = E_p e^{j2\pi m/L} \tag{5.64a}$$

where E_p is the energy in the pulse. If the channel characteristic is symmetric about the carrier, the in-phase and quadrature components of the noise are independent, as described after (5.31). Each component has mean-square value

$$var(\mathbf{q}) = \frac{N_0}{2} \int p^2(t)\,dt = N_0 E_p / 2\ . \tag{5.64b}$$

Accordingly, $p(\mathbf{q}\,|\,\mathbf{s_m})$, the joint density of the in-phase and quadrature components of $\mathbf{q}$, is

$$Pr(\mathbf{q}\,|\,\mathbf{s}_m) = \frac{1}{\pi E_p N_0}\, e^{-\frac{\left[(x - E_p \cos\phi_m)^2 + (y - E_p \sin\phi_m)^2\right]}{E_p N_0}} \tag{5.65}$$

A conversion to polar coordinates, using the relationships

$$x = \rho\ \cos\ \phi,\quad y = \rho\ \sin\ \phi,\quad dxdy = \rho\, d\rho\, d\phi$$

yields

$$f(\rho,\phi|\mathbf{s}_m)d\rho d\phi = \frac{\rho}{\pi E_p N_0} e^{-\frac{\rho^2 - 2\rho E_p \cos(\phi-\phi_m) + E_p^2}{E_p N_0}} d\rho d\phi$$

$$= \frac{\rho}{\pi E_p N_0} e^{\frac{-E_p^2 \sin^2(\phi-\phi_m) - \left[\rho - E_p\cos(\phi-\phi_m)\right]^2}{E_p N_0}} d\rho d\phi \ . \qquad (5.66)$$

In order to simplify the calculations we now assume, without loss of generality, that $\phi_m = 0$. The marginal probability density function of the phase is given by

$$f(\phi) = \int_0^{\infty} p(\rho,\phi|\mathbf{s}_m)d\rho \ . \qquad (5.67a)$$

After some manipulation involving use of the approximation $Q(x) = \exp(-x^2/2)/x\sqrt{2\pi}$ (see (2.34a)), we find

$$f(\phi) \cong \frac{\cos\phi e^{-E_p \sin^2\phi/N_0}}{\sqrt{\pi N_0/E_p}} \ . \qquad (5.67b)$$

The probability of error is given by

$$P_e = 1 - \int_{-\pi/L}^{\pi/L} f(\phi)d\phi \ . \qquad (5.68a)$$

After substituting (5.67b) into (5.68a), and making the change of variable $u = \sqrt{2E_p/N_0}\sin\phi$, we find

$$P_e = 2Q\left\{[2E_p/N_0]^{1/2}\sin\pi/L\right\}$$

$$= 2Q\left\{[2E_b\log_2(L)/N_0]^{1/2}\sin\pi/L\right\} \ . \qquad (5.68b)$$

where E_b is energy per bit. The approximation is good for L as small as 4. Curves of P_e vs. E/N_0 from (5.68b) are given in Figure 5.25, as the dashed lines.

For differentially encoded data, the coherent receiver does almost as well. A data decision for the differential phase θ_n requires decisions on the absolute phases ϕ_{n-1} and ϕ_n. It is shown in [2] that for high SNR, differential encoding merely doubles the symbol error probability, a minimal degradation.

5.2.5 DIFFERENTIALLY COHERENT PSK (DCPSK)

Differentially coherent PSK is used when there are problems with carrier phase acquisition, for either technical or cost reasons. The transmitted signal is

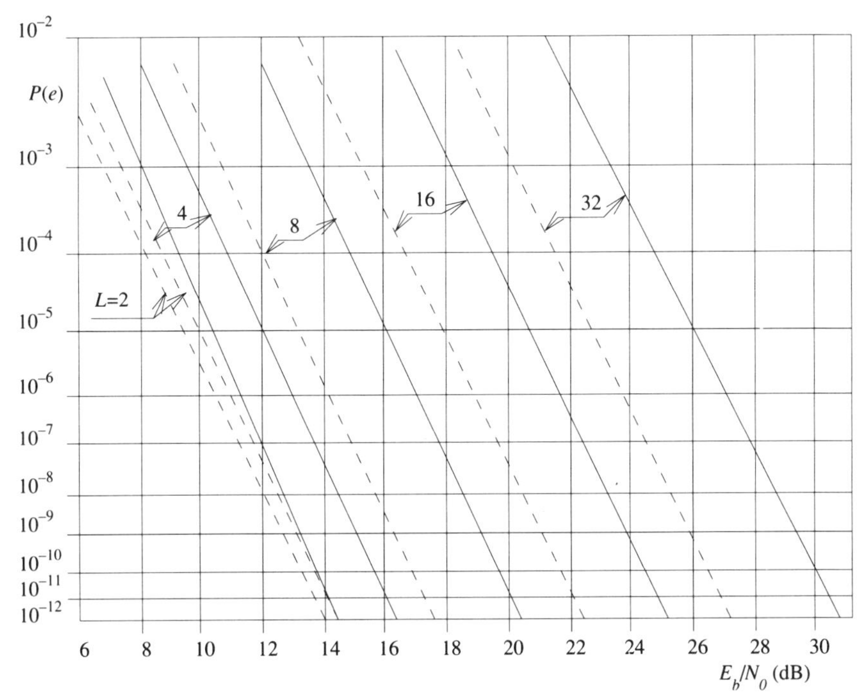

Fig. 5.25 Symbol error probability P_e for coherent and differentially coherent PSK in the white Gaussian noise channel [3].

(5.57) as before, and data are usually differentially encoded as explained in the previous section.

The only difference is in the receiver, which has the representation of Figure 5.26. This also derives from a maximum likelihood analysis [2], which is not repeated here. The receiver is effectively computing the product of the complex analytic received signal with its delayed versions yielding (neglecting noise)

$$\theta_n = \text{Angle}\left\{\tilde{r}(t)\ \tilde{r}*(t-T)\right\}\Bigg|_{(n-1)T<t<nT}$$

$$= \text{Angle}\left\{p(t)\ e^{j[\phi_n+2\pi f_c t]} \cdot p(t)e^{-j[\phi_{n-1}+2\pi f_c(t-T)]}\right\}$$

$$= \left[(\phi_n - \phi_{n-1}) + 2\pi f_c T\right] \bmod 2\pi\ . \tag{5.70}$$

If $f_c T$ is not conveniently defined as an integer, $2\pi f_c T$ can be subtracted out to yield the desired phase difference.

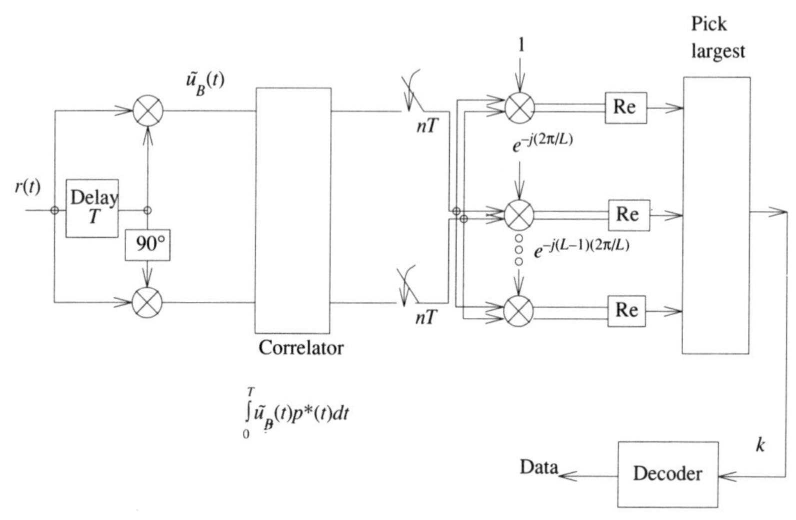

Fig. 5.26 Receiver for L-phase differentially coherent PSK.

The probability of symbol error can be analyzed as for the differentially encoded, coherently received system of the last section, and approximated as described in [3]. The result is

$$P_e \sim 2Q\left[(E_p/N_0)^{1/2}\sin(\pi/L)\right] = 2Q\left[(E_b \log_2(L)/N_0)^{1/2}\sin(\pi/L)\right] , \tag{5.71}$$

plotted as the solid curves in Figure 5.25.

Comparing this with (5.68), twice as much energy is (asymptotically) required in order to achieve the same P_e as CPSK. This is because two received noisy phase functions — and hence twice as much noise — were brought into the receiver. However, for $L=2$ (binary) the difference at high SNR is small, making DCPSK a logical choice in some applications.

5.2.6 QAM AND "OPTIMAL" TWO-DIMENSIONAL SIGNAL SETS

We have so far examined single sideband and phase modulation as examples of linear two-dimensional passband communication systems. In the case of single sideband, we had real signal points $\tilde{d}_n = a_n$ and a complex analytic pulse $\tilde{p}(t)$, as suggested by (5.50). For phase modulation, the data $\tilde{d}_n$ were complex, with equal magnitude and equally spaced on a circle in the complex plane, and $p(t)$ was a real pulse. We now turn to another popular modulation format, quadrature amplitude modulation (QAM), and to variations on it that provide more immunity against one or another channel impairment. In this modulation format, the signal points are located on $L=M^2$ equally spaced points in the com-

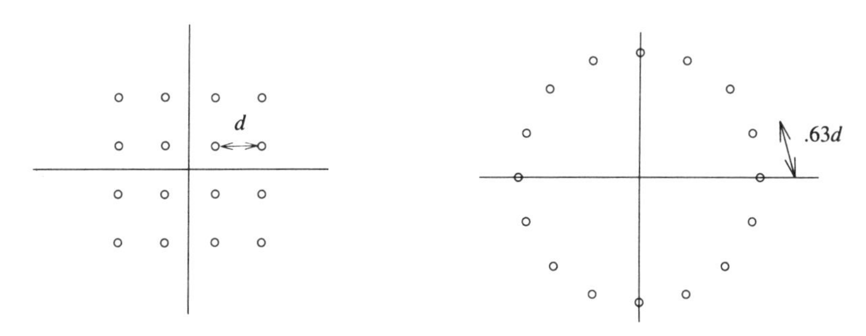

Fig. 5.27 Comparison of minimum distance between signal points in QAM and 16-phase signal constellations with the same average power 2.5 d^2.

plex plane (as in Figure 5.27a). In addition to deriving an error-rate expression for QAM valid for high SNR, we will derive signal constellations of various sizes that perform best against Gaussian noise.

We presume a real pulse $p(t)$ and begin with no restrictions on the set of L signal points $s_1,\ldots,s_L$. As already noted in connection with PSK, the set of signal points is often referred to as a *signal constellation*. Each modulation level $\tilde{d}_n = a_n + jb_n$ is selected from this set, and the transmitted signal is expressed by (5.44). The average power is usually assumed fixed, especially when comparing different constellations.

SPECTRAL EFFICIENCY VS. ERROR MARGIN

Constellations with many points are used for bandwidth efficiency, as expressed in (5.1). A 16-point constellation yields close to 4 bps/Hz, compared with the 1 bps/Hz of a two-point constellation, e.g. a simple on–off signal. But with average power constrained, signal points are closer together as their number increases, increasing the probability that noise or other perturbations will drive a signal point into the wrong decision region. As Figure 5.27 illustrates, for the same average power, a 16-phase constellation has signal points with a much smaller minimum spacing than that of 16-point QAM. We will see below that for the same average power, the minimum distance between signal points will determine the performance (against noise) of the constellation. That is the reason why constellations such as QAM, with signal points distributed more uniformly in the complex plane, are used when the need for high spectral efficiency mandates a large signal set. Figure 5.28 illustrates a 16-point QAM constellation perturbed by Gaussian noise and phase jitter.

A GOOD ESTIMATION TECHNIQUE FOR COMPARING SIGNAL CONSTELLATIONS

We will compare the performance of various constellations, but before doing that we need a way of easily determining (albeit through numerical

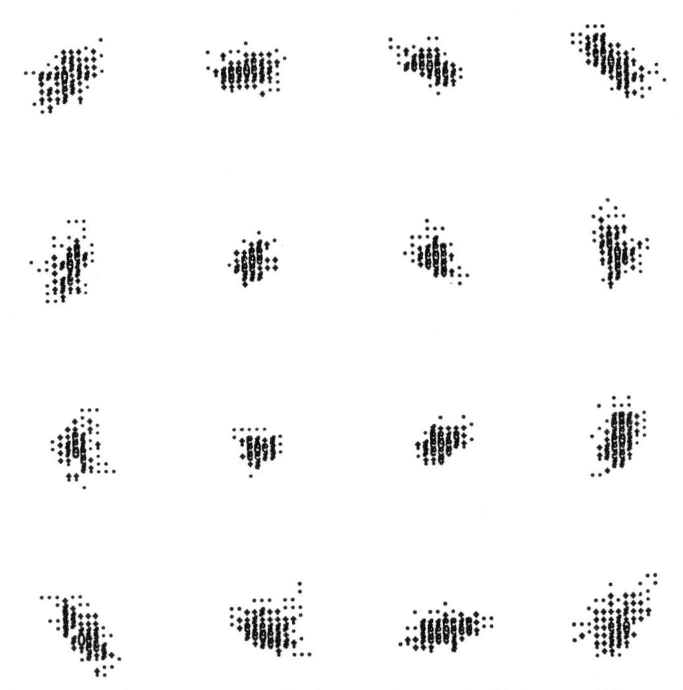

Fig. 5.28 Accumulation, over time, of noise-perturbed received 16-point QAM constellation.

analysis) what that performance is for any constellation used in a white Gaussian noise channel. We will show that if white Gaussian noise is the only perturbation, the probability of error depends only on the signal to noise ratio and the Euclidean distances between signal points. To see this, consider the equivalent baseband communication system model of Figure 5.12 with $\tilde{n}_B(t)$ a complex white Gaussian process with the power density spectrum shown in Figure 5.9. Figure 5.29 shows this system model.

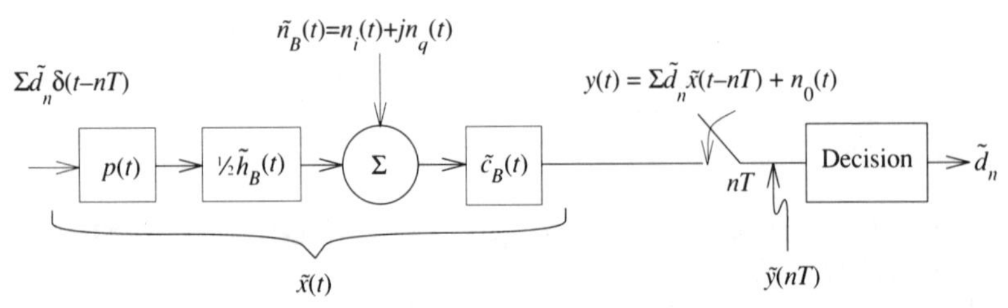

Fig. 5.29 Equivalent baseband two-dimensional signaling system with white Gaussian noise. The quadrature noise components $n_i(t)$ and $n_q(t)$ are white Gaussian and independent.

We define the overall complex linear baseband equivalent channel between (complex) modulation and the presampler receiver output as

$$\tilde{x}(t) = 1/2p(t) * \tilde{h}_B(t) * \tilde{c}_B(t). \tag{5.72}$$

The noise signal at the same point is the complex Gaussian process

$$\tilde{n}_0(t) = \tilde{n}_B(t) * \tilde{c}_B(t). \tag{5.73}$$

Thus

$$\tilde{y}(t) = \sum_n d_n \tilde{x}(t-nT) + \tilde{n}_0(t). \tag{5.74}$$

Recall, from the discussion of Section 5.1.3, that the real part $n_i(t)$ and imaginary part $n_q(t)$ of $\tilde{n}_B(t)$ are uncorrelated, and their samples, at any time t, are Gaussian random variables with the same autocorrelation function $N_0\delta(\tau)$.

At a sampling time $n = nT$, we have

$$\begin{aligned} \tilde{n}_0(nT) &= \int_{-\infty}^{\infty} \tilde{n}_B(\tau)\tilde{c}_B(nT-\tau)\,d\tau \\ &= n_{0i} + jn_{0q}, \end{aligned} \tag{5.75}$$

where

$$\begin{aligned} n_{0i} &= \int \Big[n_i(\tau) \text{ Re } \tilde{c}_B(nT-\tau) - n_q(\tau) \text{ Im } \tilde{c}_B(nT-\tau) \Big] d\tau \\ n_{0q} &= \int \Big[n_i(\tau) \text{ Im } c_B(nT-\tau) + n_q(\tau) \text{ Re } c_B(nT-\tau) \Big] d\tau . \end{aligned} \tag{5.76}$$

Since n_i and n_q are uncorrelated,

$$\begin{aligned} E\, n_{0i}^2 &= \int\int E[n_i(\tau)\; n_i(\lambda)] \text{ Re } \tilde{c}_B(nT-\tau) \text{ Re } \tilde{c}_B(nT-\lambda)\,d\tau d\lambda \\ &+ \int\int E[n_q(\tau)\; n_q(\lambda)] \text{ Im } \tilde{c}_B(nT-\tau) \text{ Im } \tilde{c}_B(nT-\lambda)\,d\tau d\lambda \\ &= N_0 \int |\tilde{c}_B(t)|^2\, dt. \end{aligned} \tag{5.77}$$

For simplicity, we assume that the integral is unity,* so that

$$E\, n_{0i}^2 = N_0/2 . \tag{5.78}$$

* For a distortionless channel, and with an ideal Nyquist characteristic split equally between transmitter and receiver, $\tilde{c}_B(t)$ is a real filter with a transfer function that is the square root of the rectangular transfer function of Figure 4.13. That is, the transfer function of $\tilde{c}_B(t)$ is equal to $\sqrt{T_1}\; -1/2T < f < 1/2T$. The integral in (5.77) is, by Parseval's theorem, equal to the integral of the magnitude squared of the transfer function, which is one.

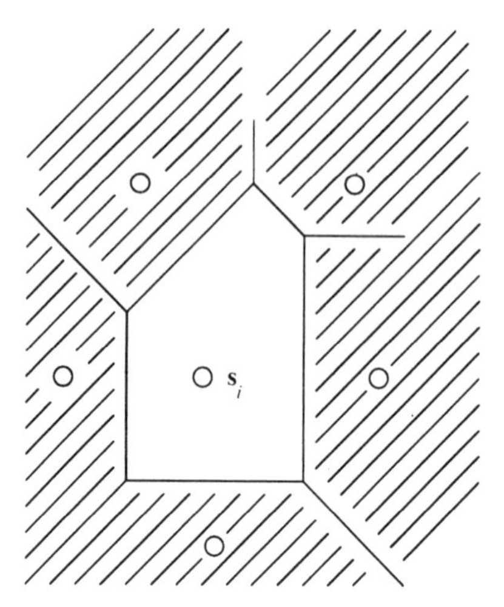

Fig. 5.30 Area of integration (shaded) for the conditional probability of symbol error if signal point $\mathbf{s}_i$ is transmitted.

Exactly the same value is obtained for $E\ n_{0q}^2$. The noise sample at the receiver output is a two-dimensional Gaussian random variable with equal probability of pointing in any direction in the complex plane. If $\tilde{x}(t)$ is an equivalent Nyquist pulse (no intersymbol interference), the signal part of $\tilde{y}(nT)$ is simply the signal point $\tilde{d}_n$.

Assume that at some sampling time nT, the signal part of $\tilde{y}(nT)$ is a particular signal point $\mathbf{s}_i$ (Figure 5.30). Then the probability density function of $\tilde{y}(nT)$ is the two-dimensional Gaussian distribution

$$p_y(\mathbf{x}|\mathbf{s}_i) = [1/2\pi N_0]\exp\{-||\mathbf{x}-\mathbf{s}_i||^2 N_0\} . \tag{5.79}$$

Note that the decision region boundaries are the perpendicular bisectors of line segments joining the signal points, as illustrated in Chapter 2, section 2.1.2, for a simple 3-point constellation. These boundaries assure that a decision is always in favor of the nearest signal point.

Evaluation of the probability of error means computing, for each signal point $\mathbf{s}_i$, the integral of $p_y(\mathbf{x}|\mathbf{s}_i)$ over the entire plane exterior to the decision region R_i for $\mathbf{s}_i$, and then averaging these conditional error probabilities over the entire constellation. Figure 5.30 shows the area of integration for the signal point $\mathbf{s}_i$. Fortunately, there is a much simpler calculation which is a good approximation for high signal-to-noise ratio [4].

To derive the asymptotic expression for symbol error probability, assume $\mathbf{s}_i$ has been transmitted and note that, if an error occurs, it is almost certainly because the receiver output sample $\mathbf{x}$ is perturbed to a decision region adjacent to R_i. The conditional probability, given that $\mathbf{s}_i$ is transmitted, that $\tilde{y}(nT)$ lies

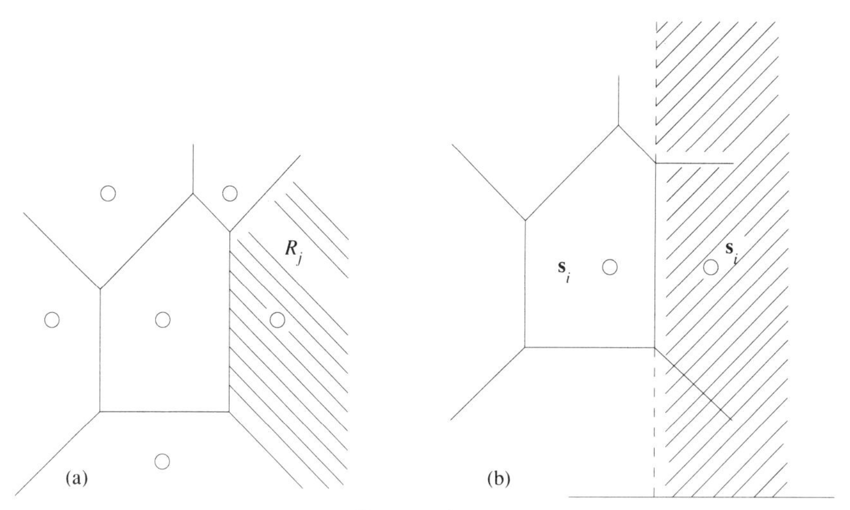

Fig. 5.31 (a) Integration area for $Pr(\tilde{y}(nT) \in R_j | s_i)$. (b) Half plane integration area yielding approximately the same value for high SNR.

inside decision region R_j, is (Figure 5.31)

$$Pr\left[\tilde{y}(nT) \in R_j | \mathbf{s}_i\right] = \int_{\mathbf{x} \in R} \left[1/2\pi N_0\right] \exp\{-||\mathbf{x}-\mathbf{s}_i||^2 / N_0\} d\mathbf{x}$$

$$\cong \int_{\mathbf{x} \in P} [1/2\pi N_0] \exp\{-||\mathbf{x}-\mathbf{s}_i||^2 / N_0\} d\mathbf{x} \ , \qquad 5.80)$$

where P_j (Figure 5.31b) is the half plane bounded by the line that is the extension of the boundary between R_i and R_j. The reason the second integral is a good approximation to the first is that for high SNR, most of the contribution to the first integral comes from a small region of R_j close to the boundary. Extending R_j in other places does not significantly change the value of the integral.

The integrand of (5.80) is circularly symmetric about $\mathbf{s}_i$, so we can arbitrarily choose coordinates so that the line bounding the half plane is parallel to one of the coordinate axes, say $\mathbf{x}_2$. Integrating on $\mathbf{x}_2$ first, that variable integrates out, leaving

$$Pr\left[\mathbf{x} \varepsilon R_j | \mathbf{s}_i\right] = \int_{x_{bdry}}^{\infty} \left[1/\sqrt{\pi N_0}\right] \exp\left\{-[x_1 - s_{i1}]^2 / N_0\right\} dx_1 , \qquad (5.81)$$

where s_{i1} is the x_1 coordinate of $\mathbf{s}_i$ and x_{bdry} is the x_1 coordinate of the half-plane boundary. But to minimize the overall symbol error probability, x_{bdry}

must lie exactly halfway between $\mathbf{s}_i$ and $\mathbf{s}_j$. Thus

$$Pr\left[\tilde{y}(nT)\varepsilon R_j|\mathbf{s}_i\right] = \int_{||\mathbf{s}_j-\mathbf{s}_i||/2}^{\infty} \left[1/\sqrt{\pi N_0}\right]\exp\left\{-x_1^2/N_0\right\}dx_1$$

$$= Q\left[\sqrt{||\mathbf{s}_j-\mathbf{s}_i||^2/2N_0}\right] . \quad (5.82a)$$

As an aside for later interest, it is not hard to show (Exercise 5.7) that for a QAM constellation (Figure 5.27a) with $L = M^2$ signal points,

$$Pr\left[\tilde{y}(nT) \in R_j/\mathbf{s}_i\right] = Q\left\{\sqrt{\frac{3P_{av}}{(M^2 - 1)N_0}}\right\}, \quad (5.82b)$$

where P_{av} is average signal power.

Applying the asymptotic expression for $Q(x)$ used in Section 5.4,

$$Pr\left[\tilde{y}(nT) \in R_j|\mathbf{s}_i\right] \cong \left[\frac{\sqrt{2N_0/\pi}}{||\mathbf{s}_i-\mathbf{s}_j||}\right]\exp\left\{-||\mathbf{s}_j-\mathbf{s}_i||^2/8N_0\right\}. \quad (5.83)$$

The overall symbol error probability is then, assuming L equiprobable signal points,

$$P_e = (1/L) \sum_i Pr\left[\tilde{y}(nT)\notin R_i|\mathbf{s}_i \; transmitted\right]$$

$$= (1/L) \sqrt{2N_0/\pi} \sum_i \sum_{j\neq i} \left[\frac{1}{||\mathbf{s}_j-\mathbf{s}_i||}\right]\exp\left\{-||\mathbf{s}_j-\mathbf{s}_i||^2/8N_0\right\}. \quad (5.84)$$

For high SNR, this expression will be dominated by those signal-point pairs that are the closest together. In fact, as the noise power N_0 becomes vanishingly small,

$$P_e = (N_{\min}N_s/L) (\sqrt{2N_0}/d_{\min}\sqrt{\pi}) \exp\left\{-d_{\min}^2/8N_0\right\}. \quad (5.85)$$

where $d_{\min} = min\ ||\mathbf{s}_j-\mathbf{s}_i||$, $i \neq j$, $N_{\min}$ = number of signal points with neighbor signal points at the minimum distance $d_{\min}$, and N_s = average number of neighbor signal points at the minimum distance for those signal points that have neighbors at the minimum distance. That is, the minimum distance between signal points is the determining quantity for the error rate.

COMPARISON OF CONSTELLATIONS

With this approximation, it is easy to compare the performances of different constellations, and even to find the optimum constellations for different numbers of signal points. Let us first compare 16-PSK with 16-QAM (Figure 5.27).

Assume average signal powers P_{PSK} and P_{QAM} respectively. For the PSK constellation, $P_{PSK}=r^2$, where r is the radius of the circle on which the points lie. Thus the Euclidean minimum separation (along the angular direction) of signal points is

$$d_{PSK} = 2r \sin(\pi/16) = 0.39r = 0.39\sqrt{P_{PSK}} . \tag{5.86}$$

In the QAM constellation of Figure 5.27, the Euclidean minimum separation d_{QAM} satisfies

$$[4(d^2/4+d^2/4) + 8(9d^2/4+d^2/4) + 4(9d^2/4+9d^2/4)]/16 = P_{QAM},$$

or

$$d_{QAM} = 0.632\sqrt{P_{QAM}} . \tag{5.87}$$

As (5.85) states, the probability of an error, i.e., the probability of a signal point being moved by noise into an adjacent decision region, is asymptotically dominated by the closest pair of signal points. Neglecting the algebraic factor before the exponential, the probabilities of the noise exceeding d_{PSK} and d_{QAM} are, respectively,

$$P_e(PSK) \sim \exp\{-d_{PSK}^2/8N_0\}$$
$$P_e(QAM) \sim \exp\{-d_{QAM}^2/8N_0\} . \tag{5.88}$$

To achieve the same error probability, d_{PSK} and d_{QAM} must be equal, i.e.,

$$0.39 \sqrt{P_{PSK}} = 0.632 \sqrt{P_{QAM}} , \tag{5.89}$$

or

$$10\log_{10}(P_{PSK}/P_{QAM} = 20\log_{10}(0.632/0.39) = 4.19 \text{ dB} . \tag{5.90}$$

This illustrates the importance of efficient "packing" of signal points in the two-dimensional signal space. As the number of signal points increases, the packing looks increasingly like a honeycomb with hexagonal structure [10]. Figure 5.32 shows constellations of several sizes optimized for Gaussian noise

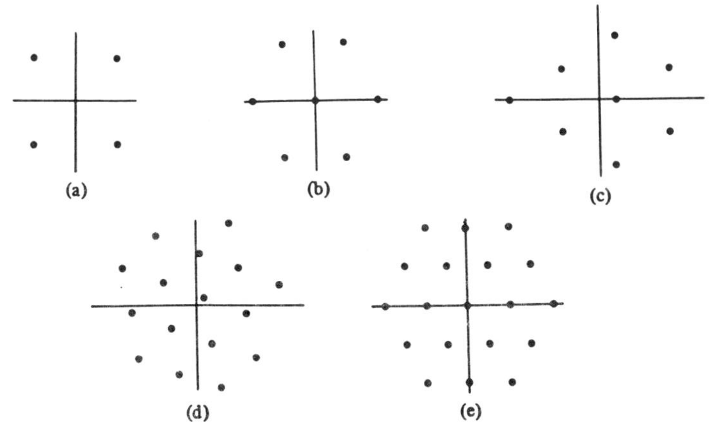

Fig. 5.32 Minimum-error signal constellations, for Gaussian noise and fixed average signal power for 4, 7, 8, 16, and 19 signal points.

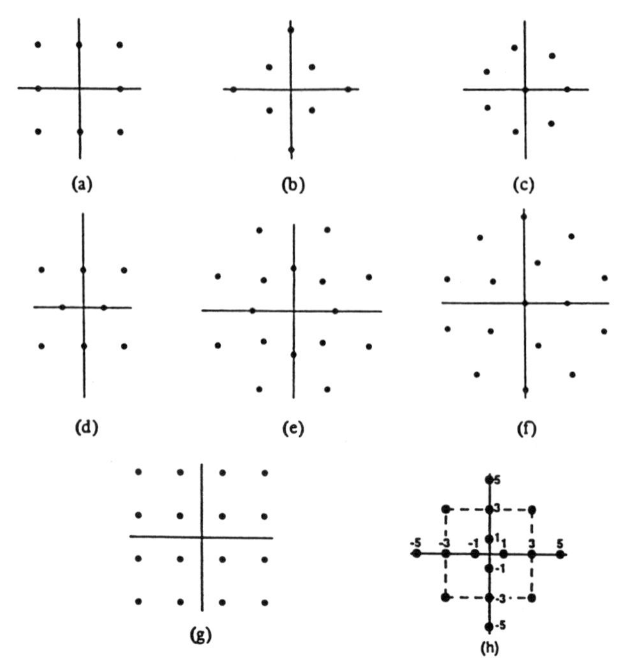

Fig. 5.33 Well-known 8- and 16-point signal constellations: (a) rectangular. (b) 4-4. (c) 1-7. (d) triangular. (e) 8-8. (f) 1-5-10. (g) QAM. (h) V.29.

and a fixed average power, as determined from a numerical gradient search procedure. The optimum 16-point constellation performs approximately 0.5 dB better than QAM. Figure 5.33 shows a number of well-known 8- and 16-point signal constellations for comparison against the optimum constellations using the asymptotic approximation. The relative performance is plotted in Figure 5.34.

As we have already noted, phase jitter (or phase noise) may also contribute to signal point displacement. This was the reasoning behind selection of the CCITT V.29 signal constellation (Figure 5.33h), as the 9.6-kbps private time standard. V.29 performs worse than the QAM constellation by about 1 dB in purely Gaussian noise but is superior in the presence of large phase jitter, of the order of 4° RMS or more, assuming Gaussian noise at a typical telephone-channel level of about –24 dB with respect to the received signal power. This is because V.29 may be viewed as four circles, each having four points. The large angular distance between points (90°) provides immunity to phase jitter (which rotates the received points).

Optimum signal constellations in the presence of Gaussian noise and phase jitter, derived in [5], are shown in Figure 5.35. A broader development is pro-

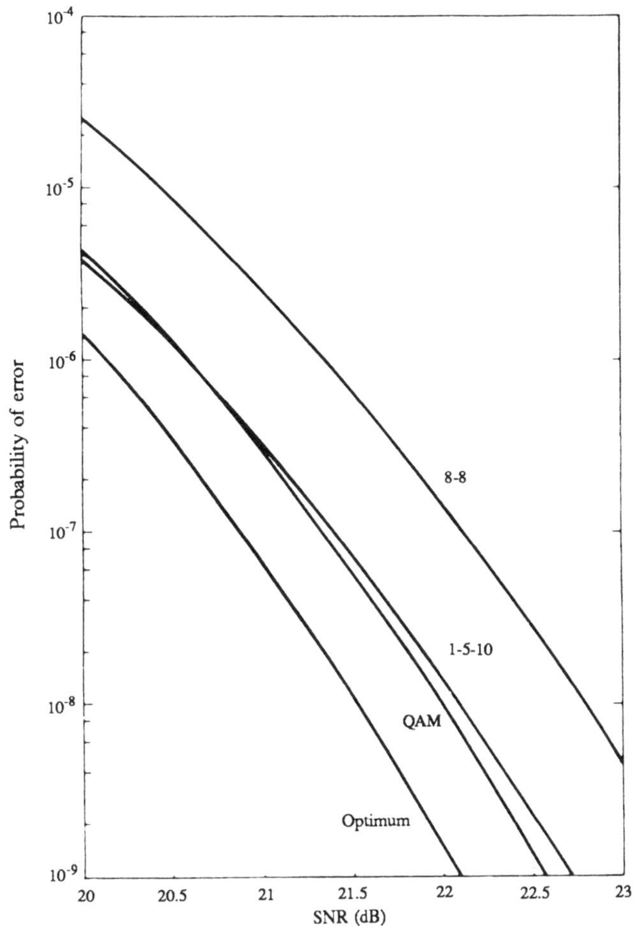

Fig. 5.34 Relative performance of 8- and 16-point signal constellations against Gaussian noise under an average power constraint.

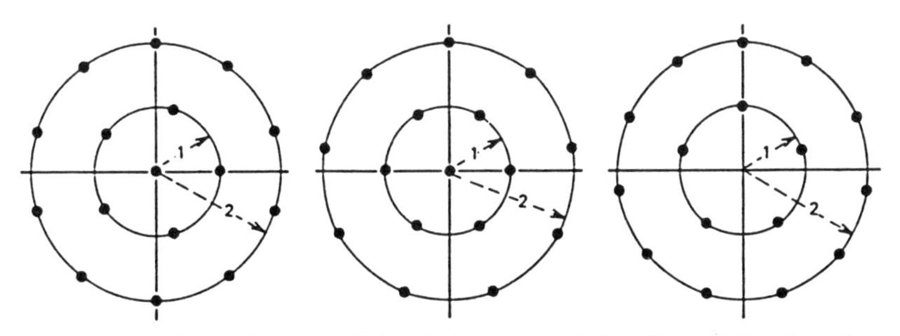

Fig. 5.35 Optimum signal constellations in the presence of phase jitter and Gaussian noise.

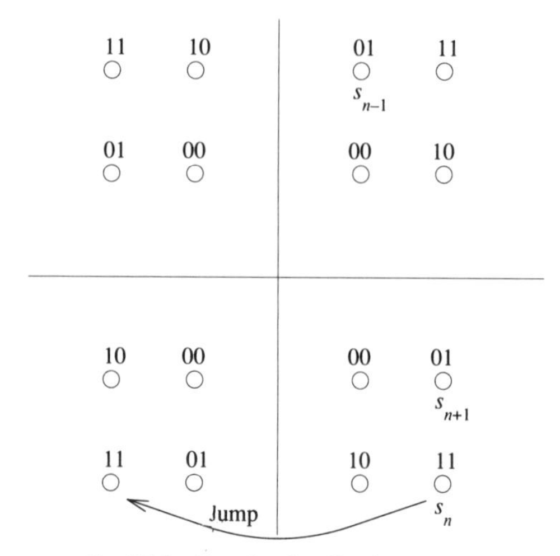

Fig. 5.36 Rotational coding for QAM.

vided in Forney et al. [6]. It is important to recognize that the carrier tracking techniques described in Chapter 6 may strongly influence the choice of a signal constellation. For example, suppose the jitter at the *input* to the receiver exceeds 4° RMS by a considerable amount. If the carrier recovery system reduces the jitter at the *output* of the receiver to < 4° RMS, then the QAM signal constellation may be preferable. At the time the V.29 signal constellation was elected as the CCITT standard, carrier recovery systems (of the type described in Chapter 6) capable of reducing the putput jitter to extremely small values were not readily available.

5.2.7 90° ROTATION INVARIANCE

The QAM signal constellation, or any constellation with 90° rotation symmetry, can be made resistant to phase jumps by a differential coding of data bits into quadrants. A sudden step in phase in the transmission channel will result only in a one-time symbol error, with no further errors. Without this differential encoding, many errors might occur during the recovery of phase synchronization over a number of symbol intervals.

The particular coding used is shown in Figure 5.36. For each set of four bits (A, B, C, D) determining a signal point in the 16-point QAM constellation, the first two bits (A, B) determine the quadrant and are differentially encoded. The quadrant for signal point $\mathbf{s}_n$ in the nth symbol interval is advanced clockwise by the number of quadrants indicated in Table 5.1 from the quadrant in which $\mathbf{s}_{n-1}$ appeared.

Assume, as a simple example, that $\mathbf{s}_{n-1}$ occupies the (0, 1) (upper left) position in the first (upper right) quadrant, that data (0, 1, 1, 1) are coded into $\mathbf{s}_n$,

Table 5.1: Clockwise Rotation (Quadrants) from $\mathbf{s}_{n-1}$ to $\mathbf{s}_n$

(A, B)	Number of quadrants
00	0
01	1
10	2
11	3

which is in the (1, 1) (lower right) position of the second (lower right) quadrant, and that a phase jump of 90° has occurred in the transmission channel. Of course, the phase jump is unlikely to be an exact multiple of 90°, but the excess carrier phase will be within tracking range, as described in Chapter 6, and can be recovered quickly. There is a symbol error from the phase jump because the receiver believes the differential encoding was for a total 180° rotation and produces an incorrect estimate (1, 0) of the differentially encoded bits (A, B). However, the position-selecting bits (C, D) will be correctly detected, since the 90° rotation from the phase jump places the lower right signal point from the second quadrant into the lower left position of the third quadrant, where it will be correctly decoded as (1, 1). In this case, the receiver's decision for (1, 0, 1, 1) differs in two bit positions from the original data (0, 1, 1, 1). A subsequent coding of (1,1,0,1) into $\mathbf{s}_{n+1}$ is correctly decoded.

It would seem desirable to also implement Gray coding in the mapping from bits to signal points. Gray coding is that mapping that minimizes the number of bits in error when a symbol error is made because of noise perturbing a signal point into a neighboring decision region. Figure 5.36 shows that the rotationally invariant constellation is not Gray coded. For if data (0, 0, 1, 0) are coded into the lower right point in the first quadrant and noise causes a decision for the upper right point in the second quadrant, the receiver will decode to 0101, differing in three bit positions rather than the maximum of one that would be possible without the constraint of rotational invariance (Exercise 5.7).

5.3 DIRECT INBAND SIGNAL GENERATION

A two-dimensional linear signal can be generated exactly as shown in Figure 5.1, by modulation of a complex baseband pulse train onto a carrier. However, this is not the only way, nor is it necessarily the most efficient way, of producing an actual line signal. An alternative way is to realize a signal generator with data input and an output which is already the modulated line signal, i.e., direct inband signal generation. The essence of this technique is reflection of the phase changes introduced by the carrier waveform back into the data driving the filter [7]. If a rational ratio of carrier frequency to pulsing rate is selected, the new set of possible data levels is finite, just as the set of original data levels is,

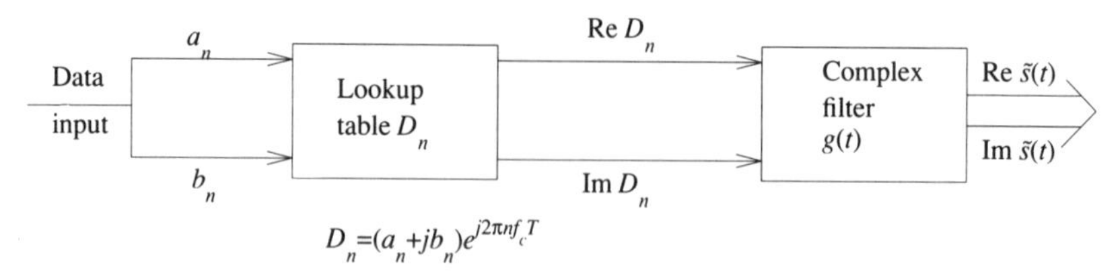

Fig. 5.37 Transmitter for direct inband signal generation.

and data values can be conveniently generated in a digital implementations by table look-up.

Consider, then, the familiar complex two-dimensional line signal

$$\tilde{s}(t) = \sum_n (a_n+jb_n)\ p(t-nT)\ e^{j2\pi f_c t} \ , \tag{5.91}$$

Rather than separate the operations of (5.91) into distinct baseband signal generation and modulation steps, write the equality

$$\begin{aligned}\tilde{s}(t) &= \sum_n (a_n+jb_n)\ p(t-nT)\,e^{j2\pi f_c t}\ e^{j2\pi nf_c t}\ e^{-j2\pi nf_c t} \\ &= \sum_n D_n g(t-nT)\ ,\end{aligned} \tag{5.92}$$

where

$$D_n = (a_n+jb_n)\,e^{j2\pi nf_cT} \tag{5.93}$$

is a phase-rotated complex data symbol and

$$g(t) = p(t)\,e^{j2\pi f_c t} \tag{5.94}$$

is a new (complex) pulse impulse response. If

$$f_cT = \frac{l}{m} \tag{5.95}$$

where l and m are integers, then the exponent in (5.93) is

$$j2\pi nf_cT = j2\pi\frac{nl}{m}\ , \quad n = 0,1,2,.... \tag{5.96}$$

The number of distinct values modulo 2π that this can take, in n increases, is m. For example, if $f_c = 1800$ Hz and $T = 1/2400$ sec, so that $l/m = 3/4$, the values of $2\pi nf_cT$ (mod 2π), $\exp(j2\pi f_cT)$, and D_n are as shown in Table 5.2.

The system diagram for the transmitter implemented as ((5.92)–(5.94)) is shown in Figure 5.37. The example of Table 5.2 is particularly simple, but in general one would use a look-up table to convert the pair (a_n, b_n) to (Re D_n, Im D_n).

One could also store time samples of the complex pulse $g(t)$ in a read-only memory, and implement (5.92) in a digital multiplexer and summer or even as a

Table 5.2
$2\pi n f_c T = 2\pi n(l/m) = 2\pi n(3/4)$

n	$2\pi n f_c T \pmod{2\pi}$	$\exp(j2\pi n f_c T)$	$D_n = (a_n + jb_n)e^{j2\pi n f_c T}$
0	0	1	$a_n + jb_n$
1	$3\pi/2$	$-j$	$b_n - ja_n$
2	π	-1	$-a_n - jb_n$
3	$\pi/2$	j	$-b_n + ja_n$
4	0	1	$a_n + jb_n$
5	$3\pi/2$	$-j$	$b_n - ja_n$
6	π	-1	$-a_n - jb_n$
.	.	.	.
.	.	.	.
.	.	.	.

look-up operation in a much larger table, with enough inputs D_n to span the memory of the pulse, as suggested in Figure 5.38. The samples of $\tilde{s}(t)$ are generated at the Nyquist interval Δ, ordinarily a fixed fraction of the symbol interval T, so that several samples $\{\tilde{s}(k\Delta)\}$ may be produced in the interval between shifting in $(a_{n-1} + jb_{n-1})$ and $(a_n + jb_n)$. An explanation of generation of

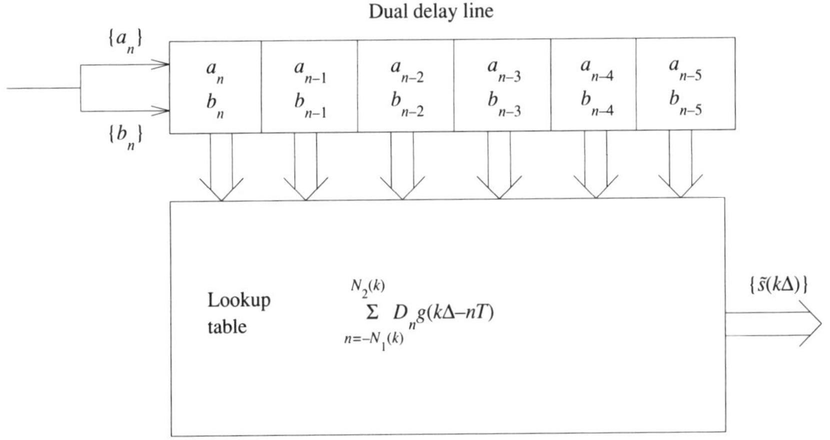

Fig. 5.38 Lookup table for complete transmitter functions. Outputs are generated at times $t_k = k\,\Delta$, where Δ is an interval small enough to satisfy the sampling theorem (Chapter 4). Note that inputs to the look-up table are supplied at symbol intervals; six data inputs are indicated for a presumed channel memory of six symbol intervals. The indices k and n are related such that all data contributing to the line signal sample $\tilde{s}(k\Delta)$ are included in the input vector, as described in Chapter 9.

line signal samples at Nyquist intervals from data at symbol intervals is given in Chapter 9, in the context of echo cancellation.

5.4 MULTITONE DATA TRANSMISSION

We assumed in previous chapters of this book, and up until now in this chapter, that data are presented to a single transmission channel and determine a line signal occupying the entire channel. They may determine signal points in two-dimensional or higher-dimensional signal space (Section 5.5), but there is one transmission channel.

In contrast to this strategy of putting all eggs in one basket, one might ask if there are circumstances in which it is more advantageous to break the data up into lower-rate substreams that traverse distinct channels.

There are three main reasons for doing this [21]. One is to reduce the noise enhancement caused by linear equalization (Chapter 7) of a distorted channel. In full-channel transmission, assigning power according to the Shannon water-pouring argument (Chapter 2) may appear to be optimum but equalization in the receiver can cause noise enhancement. By utilizing a number of narrow, frequency-division multiplexed channels instead of a single, wide channel, a water-pouring solution will achieve the same performance, relative to channel capacity, as a rectangular channel. There is no noise enhancement penalty. The second motivation for parallel data transmission is to combat transients such as impulse noise and fast fades, through use of a symbol interval on the parallel channels that is much longer than that used on the wide-channel system. The third reason is flexibility in transmission rate through using or not using particular subchannels, or sending differing numbers of bits/Hz in each subchannel. Furthermore, equalization of each subchannel becomes very simple because the channel characteristic is effectively linear.

There was early work on multitone, or parallel data transmission by Holsinger [8], Powers and Zimmerman [9], Chang [10], Darlington [11], and Saltzberg [12] describing orthogonal frequency-indexed subchannels. Weinstein and Ebert [13] described use of the DFT to generate the line signal and a time window to reduce intersymbol interference. Hirosaki [14] extended and improved the DFT system with baseband filtering and offset pulses. Kalet [15] analyzed power and bit-rate distribution over a distorted channel. Feig and colleagues [16] applied parallel data transmission to the magnetic storage channel, and Cioffi, Lechleider, Kasturia, and others [17–19] applied it to the high-speed digital subscriber loop (HDSL, Chapter 4). Spier [20] and Bingham [21] described practical considerations for voiceband modems operating at data rates up to 19.2 kbps. Ruiz, Cioffi, and Kasturia [22] combined coset coding (Section 5.8) with a DFT system to obtain high performance at moderate complexity. In general, the complexity of DFT-based parallel data transmission systems compares favorably with that of a regular, full-channel, serial data stream sys-

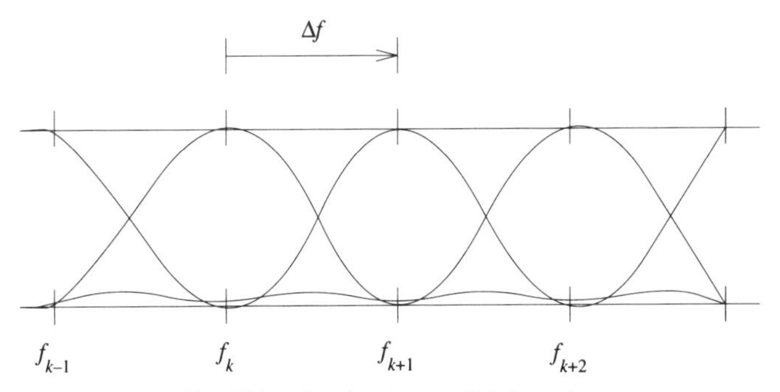

Fig. 5.39 Overlapping parallel channels.

tem. But multitone has its own vulnerabilities, particularly sensitivity to tone-type interferences, and has not made its full-channel competitors obsolete.

We will first discuss the distribution of power and bit rate across the parallel subchannels, and then describe the DFT implementation. To use multitone transmission most effectively, the total data rate R is maximized, for a given error rate P_e, total transmitted power P, and SNR, with respect to the parameters assigned to the L subchannels. The kth subchannel is modulated on subcarrier frequency f_k, and we assume, for implementation convenience, that the subchannels are of equal bandwidth. The power spectra of the individual subchannels overlap, as suggested in Figure 5.39. They do not necessarily have the identical power illustrated here. The parameters for the kth subchannel are:

$$\begin{aligned} &Power = P_k \\ &Number\ of\ bits/symbol = n_k\ , \end{aligned} \tag{5.97}$$

where

$$\sum_{k=0}^{L-1} P_k = P, \quad \sum_{k=0}^{L-1} n_k = B_s = Total\ bits/symbol\ . \tag{5.98}$$

We seek a heuristic formula for distribution of power and data rate over the subchannels. The channel transfer function magnitude, $|H(f)|$, is first quantized into a staircase function as illustrated in Figure 5.40, such that over one or several contiguous subchannels it is constant. The noise spectrum, which can include interference contributions such as NEXT (Section 4.5), is also quantized in this way, if it is not uniform.

The probability of bit error for the QAM signal in the kth subband is, assuming no interference from other subbands and using the result of Exercise (5.12),

$$P_{ek} = K\, Q\left\{ \left[\frac{3}{L_k^2 - 1} \; \frac{P_k |H_k|^2}{N_k} \right] \right\} . \tag{5.99}$$

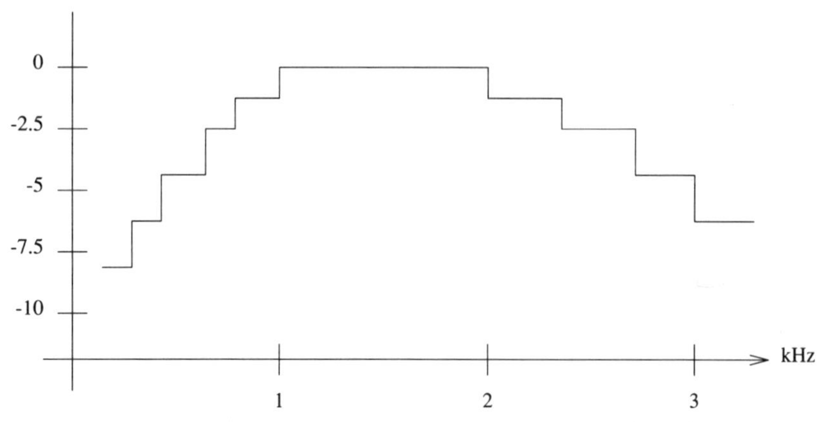

Fig. 5.40 Quantized channel transfer function magnitude $|H(f)|$.

where $L_k^2 = 2^{n_k}$, N_k is the noise power density spectrum in the kth subband, H_k is the channel transfer function in the kth subband, and K is a factor that is typically less than 4 [47]. For a specified P_{ek}, and if P_k is calculated as described below, n_k is determined and therefore the size of the subcarrier signal constellation. Kalet [34] showed that (5.99) can be inverted to yield the total number of bits per symbol at error rate P_e:

$$B_s = \sum_{k=0}^{L-1} \log_2 \left\{ 1 + \frac{3}{\left[Q^{-1}\left(P_e/4 \right) \right]^2} \frac{P_k |H_k|^2}{N_k} \right\} . \tag{5.100}$$

The selection of an optimum set of power assignments, subject to the total power constraint (5.98), is a "water-pouring" problem analogous to calculation of channel capacity (Chapter 2), but for high SNR the power assignment is approximately uniform. In the limit of an infinite number of subchannels, with the same bit error rate in each, (5.100) can be approximated by an integral, and the total bit rate is

$$R = \lim_{\Delta f \to 0} B_T/T = \int_{f_0}^{f_0+W} \log_2 \left\{ 1 + \frac{3}{\left[Q^{-1}\left(P_e/4 \right) \right]^2} \frac{P|H(f)|^2}{WN(f)} \right\} df \ , \tag{5.101}$$

where the channels are assigned to a band $\left[f_0, f_0 + W \right]$ for which the integrand is greater than 2. This is an arbitrary specification, but the smallest QAM constellation, with four points, carries 2 bits/symbol. Equation (5.101) is similar to

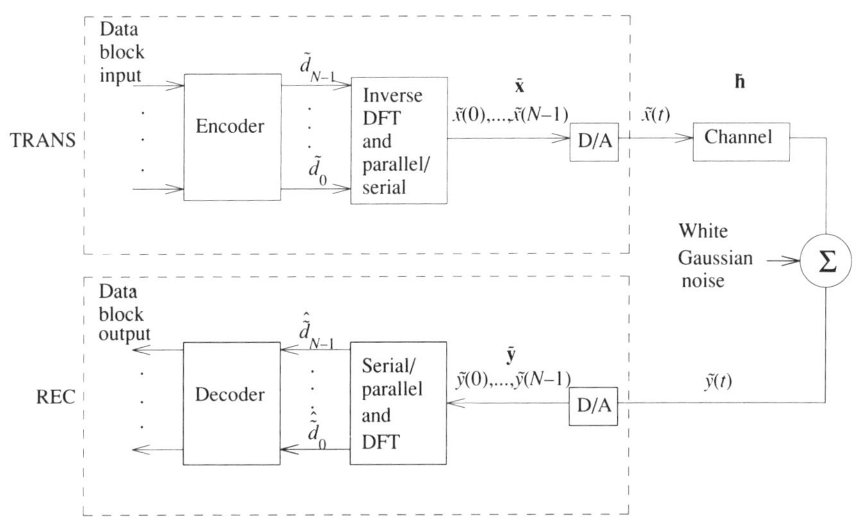

Fig. 5.41 Multitone pulsing system using the DFT.

the bit rate for a full-channel QAM system with a decision-feedback equalizer (Chapter 8), as noted by Kalet and Zervos [31].

Let us now consider the design of a multitone data pulse signal using the DFT. We follow the development of Ruiz, Cioffi, and Kasturia [22]. Assume a block of N complex data values $\tilde{d}_k$, $k = 0, ..., N - 1$, which could be the result of a coding operation, as shown in Figure 5.41. If the number of desired or usable frequency channels L is less than N, some of these inputs may be zero. For example, $\tilde{d}_0 = 0$ results in a spectral null at $f = 0$. Each complex data value $\tilde{d}_k$ is selected from a two-dimensional signal constellation C_k, which may not be the same from one value of k to another. Constellation C_k is presumed to contain W_k signal points, so that a signal-point selection carried $n_k = log_2 W_k$ bits. For example, if all the constellations are identical 4-point QAM constellations (implying also that $L = N$), then each subchannel contributing to the multitone data pulse carries 2 bits, and the pulse conveys $2N$ bits in all.

The output of the inverse DFT operation in Figure 5.41 is

$$\tilde{x}(n) = (1/\sqrt{N}) \sum_{k=0}^{N-1} \tilde{d}_k e^{j2\pi nk/N}, \ 0 \le n \le N - 1 . \tag{5.102}$$

This can be considered a sample at time nT/N of the analog pulse

$$\tilde{x}(t) = (1/\sqrt{N}) \sum_{k=0}^{N-1} \tilde{d}_k e^{j2\pi kt/T}, \ 0 \le t \le T, \tag{5.103}$$

where $f_k = k/T$ is the subcarrier frequency of the kth subchannel, and Δf of Figure 5.39 is equal to $1/T$. The spectral density illustrated there can be derived from the wide-sense cyclostationary process (which is a succession of DFT-generated pulses)

$$\tilde{s}(t) = (1/\sqrt{N}) \sum_{i=-\infty}^{\infty} \left[\sum_{k=0}^{N-1} \tilde{d}_{ki} g_k(t - iT) \right], \quad g_k(t) = e^{j2\pi kt/T}, \; 0 \le t \le T . \tag{5.104}$$

For uncorrelated, zero-mean variables $\tilde{d}_{ki}$ of variance σ_k^2, the power spectral density is [23]

$$S_s(f) = (T/N) \left\{ \sum_{k=0}^{N-1} \sigma_k^2 \operatorname{sinc}^2 \left[T(F - 2\pi k/T) \right] \right\} , \tag{5.105}$$

where $\operatorname{sinc}(x) = \sin(x)/x$.

By the properties of the DFT, the spectral overlap between frequency channels does not cause interference. Now applying the DFT to the vector $\tilde{\mathbf{x}}$ consisting of $\{\tilde{\mathbf{x}}_n, n = 0, ..., N-1\}$ will recover the original data $\{\tilde{d}_k, k = 0, ..., N-1\}$. However, there *is* interference from one received pulse to the next. It could be treated with a guard space between pulses, but *cyclic extension* [24] is a more elegant technique. Assume a channel characteristic $\mathbf{h} = h(0), ..., h(M-1)$, where $N > M$. Transmit, instead of $\tilde{\mathbf{x}}$, a pulse $\tilde{\mathbf{x}}^1$ of length $N^1 = N + M - 1$ defined as

$$\tilde{\mathbf{x}}^1 = \tilde{x}_{N-M+1}, \tilde{x}_{N-M+2}, ..., \tilde{x}_{N-1}, \tilde{x}_0, \tilde{x}_1, ..., \tilde{x}_{N-1} . \tag{5.106}$$

For a channel response consisting of M nonzero samples, the pulse $\tilde{\mathbf{x}}^1$ looks like an infinite periodic repetition of the pulse $\tilde{\mathbf{x}}$, as illustrated in Exercise 5.12. Interference between successive pulses is eliminated by ignoring the first $M - 1$ samples of the received vector $\tilde{\mathbf{y}}^1$ and accepting only the latter N samples, denoted as the vector $\tilde{\mathbf{y}}$, given by the cyclic convolution

$$\tilde{\mathbf{y}} = \tilde{\mathbf{x}} * \tilde{\mathbf{h}}, \quad \tilde{\mathbf{Y}} = \mathbf{H}\tilde{\mathbf{X}} , \tag{5.107}$$

where

$$\tilde{\mathbf{X}} = (\tilde{X}(0), \tilde{X}(1), ..., \tilde{X}(N-1)) \tag{5.108a}$$

$$\tilde{\mathbf{H}} = (\tilde{H}(0), \tilde{H}(1), ..., \tilde{H}(N-1)) \tag{5.108b}$$

are the DFTs of $\mathbf{x}$ and $\mathbf{h}$ respectively. The kth subchannel can, with knowledge of $\tilde{H}_k$, be easily equalized in the receiver [13]. However, if the original signal sets are designed according to codes matched to the channel characteristics, this is not necessary.

Cyclic extension has costs in extra pulse energy and extra uncoded bits. The SNR loss is, with appropriate average energy assumptions, $10 \log_{10} [(N + M - 1)/N]$. The alternative baseband filtering approach of Hirosaki [14] mitigates but does not eliminate the intersymbol interference.

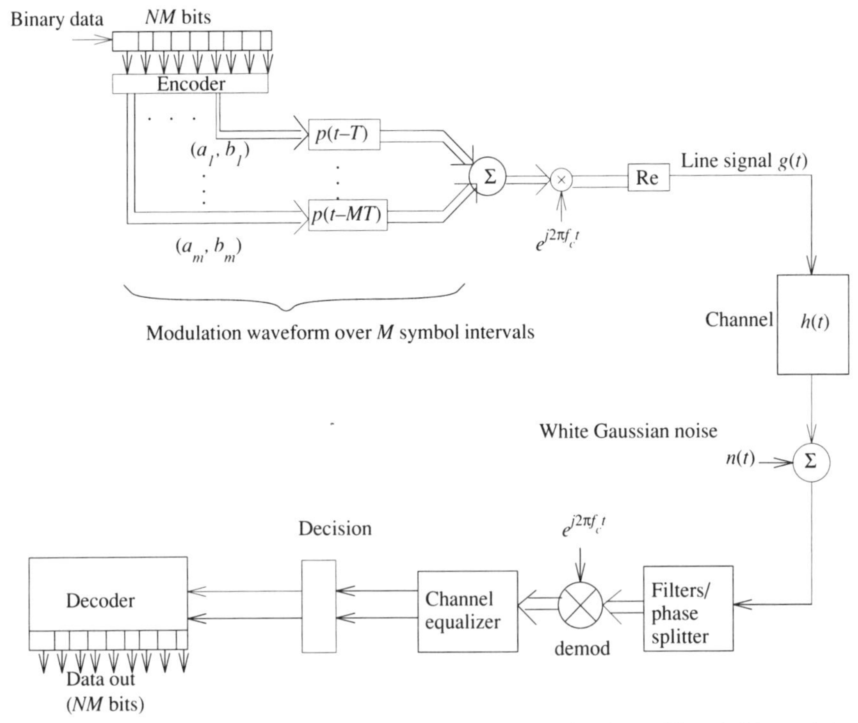

Fig. 5.42 Multi-dimensional data communication system. The signal over M symbol intervals is determined by NM bits input to the encoder.

The combination of the DFT and coset coding is shown in [22] to be particularly powerful in severely distorted channels, offering gains of 2 dB or more over the combination of a decision feedback equalizer (DFE, Chapter 8) and coset codes of equivalent complexity. These coding gains are achieved by using only a subset of the parallel channels, representing the better parts of the overall channel.

5.5 HIGHER-DIMENSIONAL SIGNALING

5.5.1 THE ADVANTAGES OF MULTIDIMENSIONAL SIGNALING

Just as two-dimensional signal constellations can offer performance improvements over use of separate one-dimensional constellations (as illustrated in Exercise 5.14), so can $2M$-dimensional signal constellations ($M>1$) offer performance improvements over two-dimensional signal constellations [25]. If N bits can be coded into 2^N points in a two-dimensional signal space, so can NM bits be coded into 2^{NM} signal points in a $2M$-dimensional space, as shown in the system diagram of Figure 5.42. For example, $N=2$ and $M=2$ implies that 4

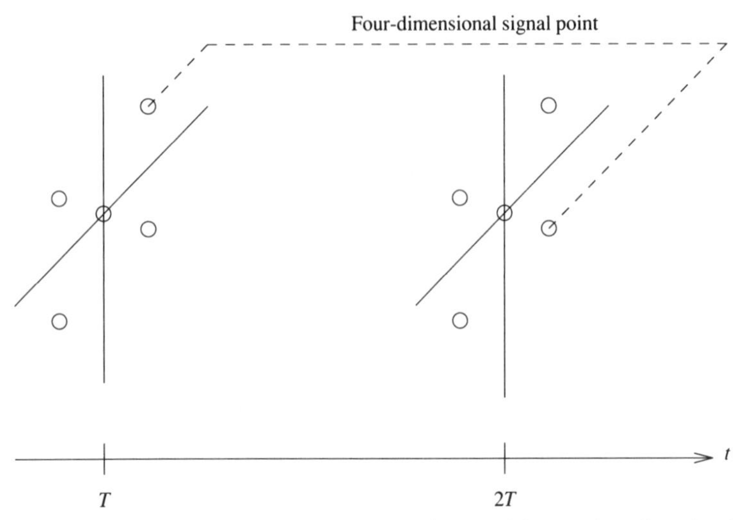

Fig. 5.43 Two successive two-dimensional signal points determining one four-dimensional signal point.

data bits code into 16 signal points in a four-dimensional space composed of two successive symbol intervals in each of which a point is used from a two-dimensional set, but *not* with the independent selection, from one symbol interval to the next, characteristic of two-dimensional signaling. In fact, the two-dimensional set used by $2M$-dimensional signaling may have more than 2^N points. The higher-dimensional constellation consists of a *subset* of all possible combinations of points appearing in the two-dimensional constellation.

It is hard to give a useful visualization of a higher-dimensional constellation, but Figure 5.43 conveys the idea of a pair of successive two-dimensional signal points determining a single four-dimensional signal point in a 16-point constellation. If all pairwise combinations of two-dimensional signal points were usable, the four-dimensional constellation would have 25 rather than 16 elements. Exercise 5.15 defines the constellation and asks the reader to compute its SNR advantage.

The objective is, of course, to ensure that the 2^{NM} points in the $2M$-dimensional signal space are further apart from one another than those in any (reasonable) 2^N-point constellation in two-dimensional space, thus achieving a smaller error probability in the presence of Gaussian noise. In the remainder of this section we will demonstrate that such improvements are possible. It is also worth mentioning that there are possibilities for conveying noninteger numbers of bits/symbol interval, unlike two-dimensional coding.

It is intuitively reasonable that the added freedom in signal-point location provided by going to higher dimensions could result in increase of the minimal separation between pairs of signal points. Of course, there are more neighbors

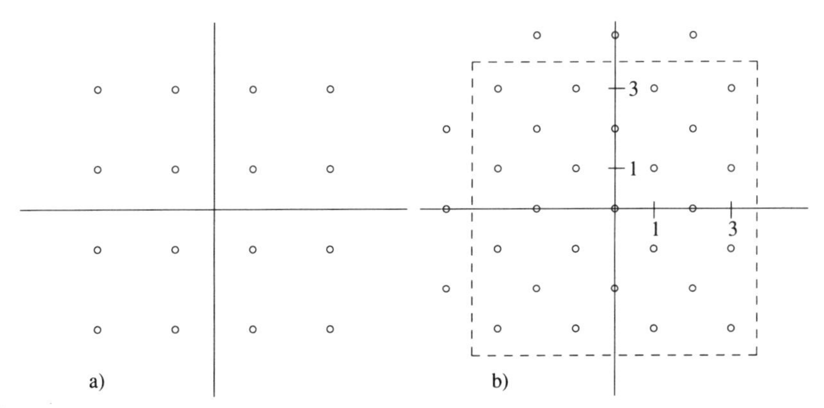

Fig. 5.44 (a) QAM two-dimensional constellation. (b) Two-dimensional projection of four-dimensional signal constellation. The signal points have coordinates that are either both even or both odd integers.

of a given signal point and thus more ways an error could be made, but this algebraic worsening of performance is more than compensated by the exponential improvement in performance demonstrated in the last section as the minimum distance between signal points increases. This improvement, at a cost in encoding delay (accumulation of incoming data to determine a multidimensional signal point) and complexity (measuring distance of received multidimensional points to signal points), does not increase indefinitely. As the dimensionality approaches infinity, the SNR gain approaches an asymptote. The maximum possible gain over two-dimensional QAM is 8 dB [25]; practical SNR gains, for M ranging from 2 to 8, are of the order of 1–4 dB.

5.5.2 A PROCEDURE FOR DERIVING AN EFFICIENT FOUR-DIMENSIONAL CONSTELLATION

Following [25], it is convenient, and close enough to optimal, to assume that signal points exist on a regular lattice that we know has desirable signal separation properties. Select the four-dimensional lattice whose coordinates (a_1, b_1, a_2, b_2) are either all even integers or all odd integers, as shown in the two-dimensional cross section of Figure 5.44b. The example given above consisted of signal points on this lattice, but only occupying 16 of the infinite number of possible four-dimensional points.

Consider selecting a four-dimensional constellation of 256 signal points, one which would offer significant improvement over 16-point QAM (Figure 5.44a). We will select the 256-point constellation as pairs of signal points drawn in two successive symbol intervals from that part of the lattice within the dashed lines. The 25 signal points within the dashed lines are closer together than points in the QAM constellation, but by selecting only a subset of

Table 5.3: Seven Basic Points for Four-Dimensional Constellation

Point	Energy	Permutations and Sign Combinations
(1111)	4	16
(2000)	4	8
(2200)	8	24
(2220)	12	32
(2222)	16	16
(3111)	12	64
(3311)	20	96

the 625 possible signal-point pairs (over two symbol intervals), the Euclidean distance between four-dimensional signal points will turn out to be larger than that of two successive independent selections of two-dimensional QAM signal points. We will illustrate, but not attempt to explain, the art of selecting the right four-dimensional subset.

First, begin with the seven "basic points" of Table 5.3. Note that the Euclidean distance between any pair of points is at least two. It is also true, although perhaps not immediately apparent, that any permutation of a basic point is at least distance two from these points or any of their permutations, and still true when some or all of the coefficients of signal points are negative rather than positive integers. By taking all permutations and assignment of + and – signs to the elements of the seven basic points, we create a total of 256 points. Note, as with our previous example, that not all of the 625 possible signal point pairs are included in the four-dimensional constellation.

The assignment of 8-bit data words to signal points is arbitrary. A Gray code would be difficult to design where there are so many closest neighbors of each 4-dimensional signal point. The 8 bits are broken into "basic point," "permutation," and "sign bit" fields. In practice, a branching tree is followed from most to least significant bit of the 8-bit data word to define the signal point.

To detect the four-dimensional signal, we must find the signal point nearest to the equalizer output vector $\mathbf{r} = (r_1, r_2, r_3, r_4)$. If there were no noise, the decision would always be the correct (transmitted) value (a_1, b_1, a_2, b_2).

An upper bound on probability of symbol error per bit transmitted for the four-dimensional constellation is derived in [25]. Probability of symbol error per bit is a convenient measure of performance for comparison of multidimensional constellation with two-dimensional constellations. The four-dimensional constellation has a 1.2 dB advantage over two-dimensional signaling at little cost in delay and complexity.

Many other higher-dimensional constellations, excelling in various ways such as minimizing peak to average power, showing invariance to 90 degrees

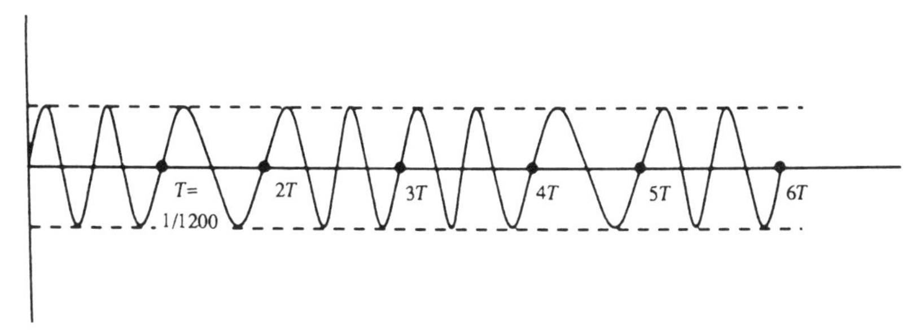

Fig. 5.45 A binary FSK waveform (f_c = 1800 Hz, f_d = 600 Hz, θ_0 = 0, $u(t)$ rectangular).

phase rotations, and yielding more gain, are known. They can be compared or coupled with trellis-coded modulation (Section 5.7), which is an alternative or combined way of squeezing SNR gain from a bandlimited channel.

5.6 FREQUENCY-SHIFT KEYING

Frequency-shift keying (FSK) conveys information by varying the carrier frequency in response to the customer data stream. FSK differs in two important respects from the linear modulation techniques we have discussed so far. First, the FSK signal is of *constant envelope,* which is important in radio applications with saturating power amplifiers, and second, it does *not* generally preserve the baseband bandwidth as the linear modulations do. For low-rate data transmission on telephone lines, and in other applications where bandwidth is not critical, it is an inexpensive, effective, and widely used modulation format. Figure 5.45 illustrates the appearance of a binary FSK waveform, and particularly the constant envelope. In Figure 5.45, the high (mark) frequency is 2400 Hz and the low (space) frequency is 1200 Hz.

For a continuous phase frequency-shift keyed (FSK) signal, the complex and real transmitted signals are, respectively,

$$\tilde{s}(t) = \sqrt{2E_b/T}\exp\{j2\pi f_c t + \theta_0 + 2\pi f_d \int_{-\infty}^{t} g(\tau)d\tau\} \tag{5.109}$$

and

$$s(t) = \sqrt{2E_b/T}\cos[2\pi f_c t + \theta_0 + 2\pi f_d \int_{-\infty}^{t} g(\tau)d\tau], \tag{5.110}$$

where

$$g(t) = \sum_{n=-\infty}^{\infty} d_n p(t-nT) \tag{5.111}$$

is a real PAM signal with data $\{d_n\}$ (possibly multilevel) and arbitrary real pulse shape $p(t)$. In (5.109) and (5.110), E_b is the energy per symbol of $s(t)$, and θ_0 is the arbitrary initial phase of the carrier, usually considered to be uniformly distributed from 0 to 2π. For binary FSK, $d_n = \pm 1$. It is also possible to generate an FSK signal by switching among unsynchronized oscillators. However, the phase discontinuities of this method produce undesirable spectral components.

The phase of $s(t)$ is

$$\theta(t) = 2\pi f_c t + 2\pi f_d \int_{-\infty}^{t} g(\tau)d\tau + \theta_0. \tag{5.112}$$

Thus the instantaneous frequency of $s(t)$ is

$$d\theta(t)/dt = 2\pi f_c + 2\pi f_d g(t), \tag{5.113}$$

where f_d is defined as the *frequency deviation* per volt of $g(t)$. The ratio

$$\Delta = 2 f_d T \tag{5.114}$$

of f_d to the Nyquist frequency $1/2T$ is called the *modulation index.* It expresses the amount the carrier frequency is changed relative to the bandwidth of a nominal (Nyquist) baseband pulse.

If the pulse shape $p(t)$ has no impulse components, the phase continuity of $s(t)$ is evident from (5.109). A step in $p(t)$, e.g. in a rectangular pulse, still results in continuous $s(t)$.

The bandwidth of a frequency-modulated signal can be considerably larger than the bandwidth of the modulation. Carson's rule [26] asserts that the bandwidth encompassing 98% of the signal energy is

$$BW_{0.98} = 2f_m + 2[(1/2\pi)\max|d\theta(t)/dt| - f_c], \tag{5.115}$$

where f_m is the highest significant frequency in the modulation signal $g(t)$. A linear modulation would have a bandwidth no greater than $2f_m$.

5.6.1 BINARY FSK

FSK is frequently used with rectangular pulses and binary data levels. If we assume that $p(t)$ is a rectangular pulse of unit height on the interval $(0,T)$, and $d_n = \pm 1$, then (5.109) can be written

$$\tilde{s}(t) = \sqrt{2E_b/T}\ \exp\{j[2\pi f_c t + \theta_0]\} \sum_n \exp\{j2\pi f_d [Q_{n-1}T + (t-nT)d_n]\} p(t-nT) \tag{5.116}$$

where

$$Q_n = \sum_{m=-\infty}^{n} d_m, \quad d_m = \pm 1, \tag{5.117}$$

is the sum of all pulse amplitude levels prior to the time $t=(n-1)T$, and $p(t-nT)$ serves as a "selection window" picking out the nth T-second interval. Figure 5.46 illustrates the instantaneous frequency and phase for a typical data sequence.

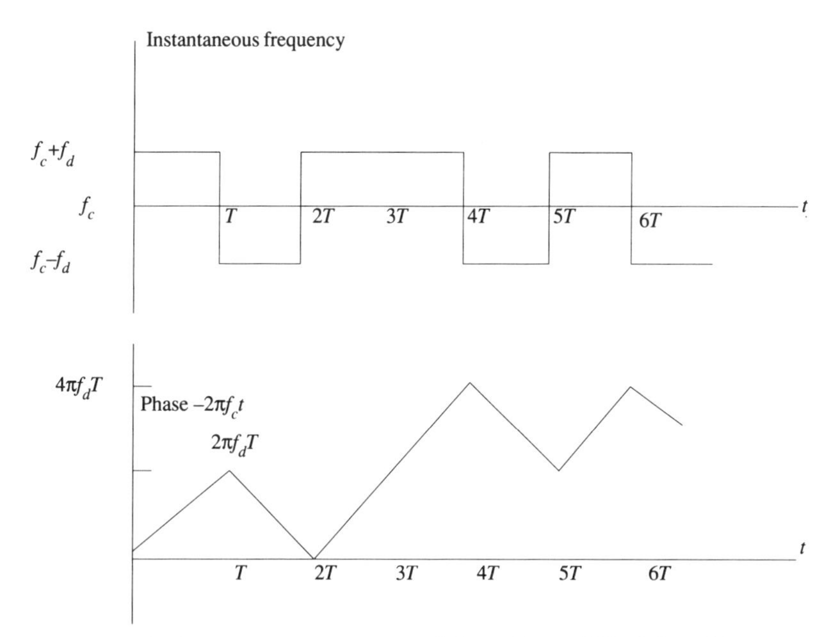

Fig. 5.46 Instantaneous frequency and phase of binary FSK of Figure 5.45.

A derivation of the power density spectrum for the multilevel FSK signal. which is somewhat involved and outside the scope of this book, is given in Lucky, Salz, and Weldon [27]. For binary modulation, the spectrum is expressed as

$$V(f-f_c) = 2E_b \left[(1/2)E(\sin\gamma/\gamma)^2 + u(t)\right], \tag{5.118}$$

where

$$u(t) = \mathrm{Re}\left\{\frac{\left[E\left(\dfrac{\sin\gamma}{\gamma}e^{-i\gamma}\right)\right]^2}{1-\cos(2\pi f_d t)\exp\left[j2\pi(f-f_c)t\right]}\right\}, \quad \cos(2\pi f_d T) < 1$$

$$u(t) = \frac{1}{2\pi}\left|E\left(\frac{\sin\gamma}{\gamma}\right)\right|^2 \sum_n \left[\delta\left[2\pi(f-f_c)T+\pi-2\pi n\right] - 1\right], \; \cos(2\pi f_d T)=1$$

and where

$$\gamma = \pi(f-f_c \pm f_d)T \tag{5.119}$$

is a random variable taking on the two values indicated with equal probability.

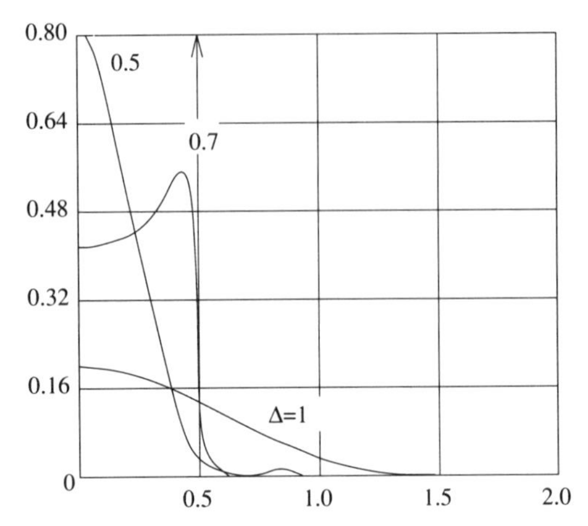

Fig. 5.47 Power density spectra for binary FSK indexed by the normalized modulation index. Here $\Delta = 2\pi f_d T/\pi = 2 f_d T$. (From Lucky, Salz, and Weldon [27], Chapter 8.)

A number of these power spectra are sketched in Figure 5.47. The general trend is that for a modulation index Δ less than 0.9, the spectra are rather smooth and occupy about the same amount of bandwidth as a double-sideband signal. For $0.9 < \Delta < 1.1$, the spectra are very sharply peaked. These strong spectral components cause crosstalk between adjacent subscriber wire pairs, so that values of Δ close to one are undesirable. Narrow-band FM corresponds to $\Delta < 1$, and wide-band FM to $\Delta > 1$.

Demodulation of FSK signals is ordinarily done using noncoherent detectors such as the limiter-discriminator shown in Figure 5.48. The front-end filter blocks out band noise (and, unfortunately, some signal components), the limiter eliminates any amplitude variation, and the differentiator produces a signal

$$v_2(t) = (d\theta(t)/dt)\sin\theta(t) = [2\pi f_c + 2\pi f_d g(t)]\sin\theta(t), \quad (5.120)$$

where $\theta(t)$ is given by (5.112), and $g(t)$, the modulation signal, by (5.111). The rectifier and post-detection filter constitute an envelope detector yielding $g(t)$. The differentiator can be approximated by a tuned bandpass filter with a linear characteristic about f_c.

Another easily implemented detector, a zero-crossing interval comparer, is shown in Figure 5.49. This circuit measures the intervals between zero cross-

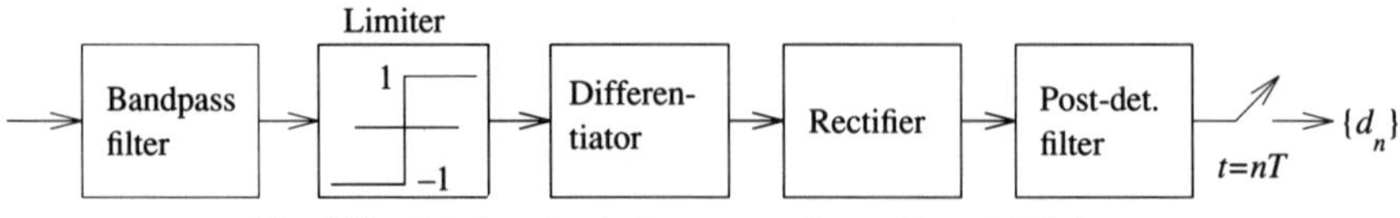

Fig. 5.48 Limiter-discriminator noncoherent binary FSK detector.

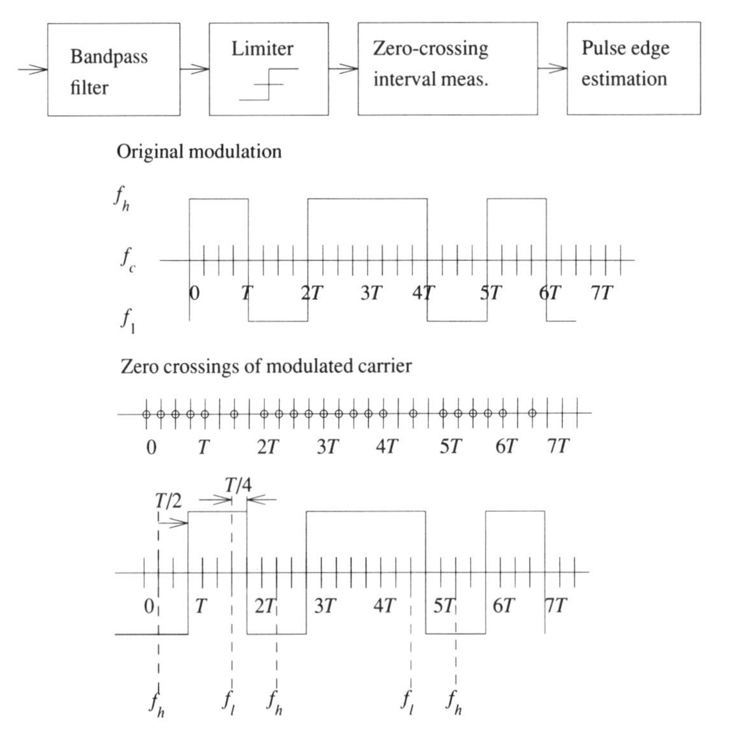

Fig. 5.49 FSK detection by zero-crossing interval measurement (example for $f_h = 2f_1$).

ings of the hard-limited received signal and makes transitions in the output waveform when it detects sufficient change in the zero-crossing interval. A simple algorithm is suggested in Figure 5.49. When a zero-crossing interval is greater than a half-period of the carrier frequency, a decision is made for the high frequency f_h, and when it is less than a half-period of the carrier frequency, a decision is made for the low frequency f_1.

Figure 5.49 indicates further details for the case when $f_h = 2f_1$. When a change occurs from f_1 to f_h, the transition in the output is delayed by $T/2$ (half a symbol interval), and when a change occurs from f_h to f_1, the transition in the output is delayed by $T/4$. This results, in the absence of noise, in the original modulation waveform being exactly reproduced at the receiver output with a delay of $3T/4$.

This algorithm is suboptimum, in the presence of noise, because it does not exploit the correlations among zero-crossing intervals in the well-defined symbol intervals characteristic of synchronous data transmission. Even for asynchronous data communication, the data comes in character bursts with well-defined symbol intervals within the burst. Algorithms for optimum location of

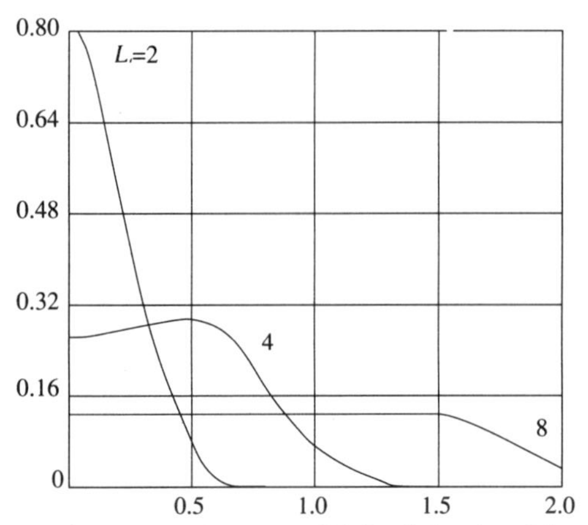

Fig. 5.50 Power density spectra, holding the modulation index $\Delta = 0.5$ constant, for different numbers L of modulation levels.

the transitions between f_1 and f_h, which are specific to the particular choices of carrier frequency and deviation ratio, have been implemented.

5.6.2 *L*-ARY FSK

If the signal levels $\{d_n\}$ in the modulation waveform $g(t)$ of (5.111) can take on L different values, the resulting signal is L-frequency FSK. As (5.109) and the usual definition for PAM signal levels suggest, these frequencies would be $-(L-1)f_d,\dots,-f_d, f_d,\dots,(L-1)f_d$. Considerable bandwidth expansion is implicit in going to higher values of L, if f_d is held constant, as shown in the curves of Figure 5.50 for $\Delta = 0.5$. If, however, Δ is decreased in proportion to L, i.e., $\Delta = 1/L$, the spectra remain narrow. The cost is in detection efficiency, since the alternative sinusoids are increasingly correlated as L increases.

5.6.3 ERROR RATE PERFORMANCE

The error-rate analysis of FSK depends heavily on the system model and parameter values, especially between narrowband ($\Delta < 1$) and wideband ($\Delta > 1$) modulation. For wideband FSK, where relatively high power efficiency is obtained at the expense of bandwidth, errors in practical receivers are due largely to the "click" phenomenon described in [27]. This is a rapid and extreme phase transition caused by a noise vector of magnitude comparable to the signal vector in the amplitude-phase plane.

For a *wideband* system with L different frequencies, it is easy to choose the modulation index such that the L alternative waveforms are orthogonal over a symbol interval (Exercise 5.13). The optimum coherent orthogonal receiver, derived in Chapter 2 as a bank of matched filters, then yields the error-rate

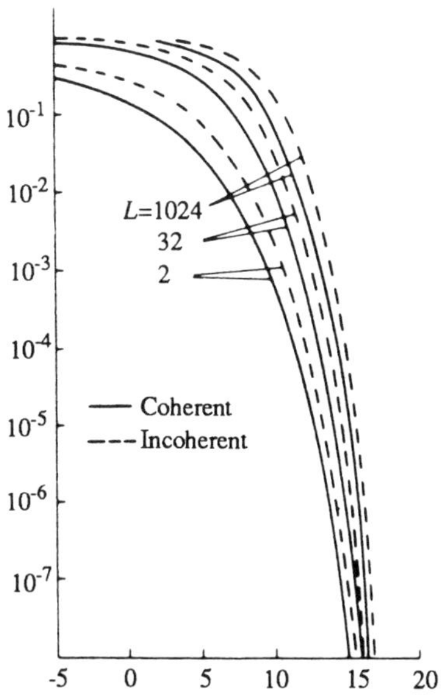

Fig. 5.51 Error probability in optimum coherent and incoherent FSK detectors (from Lucky, Salz, and Weldon [27]).

curves shown in Figure 5.51. The optimum incoherent orthogonal receiver, which does not require knowing the uniformly distributed random component of the carrier phase, performs almost as well, as also shown in Figure 5.51. In a practical discriminator receiver, which admits more noise than the optimum receivers (which effectively limit noise bandwidth), the click phenomenon significantly degrades the error rate.

5.6.4 MINIMAL-SHIFT KEYING

For binary FSK, the particular case of $\Delta = 0.5$, with the rectangular pulses characteristic of FSK, is known as *minimal-shift keying* (MSK) [28,29]. For this value of Δ, we will show that the frequency-shift keyed signal is a *linearly* modulated QPSK signal, the juncture between linear and nonlinear modulation systems.

MSK (and its generalizations in the next section) has many attractions for use in communication systems, particularly in next-generation mobile radio communications. Because only phase, not amplitude, is modulated, it has a *constant envelope,* making it suitable for the highly nonlinear amplifiers used for microwave radio communication. When data are being sent in adjacent frequency division multiplexed modulated channels, MSK has lower *crosstalk* than QPSK (quadrature phase-shift keying) if the channel separation (guardband) is more than 0.7 times the bit rate (in one channel) [33]. For these reasons, MSK was proposed for NASA's Advanced Communication Technology (ACT) satellite, to be launched in 1992. It is also being considered for mobile satellite communications, and has been used for communication with submarines.

The fact that MSK is also a linear modulation, with possibilities for coherent detection, is demonstrated as follows: From (5.116), with $\Delta=0.5$, which implies $f_d=1/4T$,

$$s(t) = \text{Re } s(t) = \sqrt{2E/T} \sum_n \cos\left\{2\pi f_c t+\theta_0+(\pi/2T)\left[Q_{n-1}T+d_n\cdot(t-nT)\right] \times p(t-nT)\right\}$$

$$= \sqrt{2E/T}\left\{m_1(t)\cos\left[2\pi f_c t+\theta_0\right] - m_2(t)\sin\left[2\pi f_c t+\theta_0\right]\right\}, \tag{5.121}$$

where

$$m_1(t) = \sum_n \cos\left\{\pi/2T\left[Q_{n-1}T+d_n\cdot(t-T)\right]\right\}p(t-nT)$$

$$m_2(t) = \sum_n \sin\left\{\pi/2T\left[Q_{n-1}T+d_n\cdot(t-nT)\right]\right\}p(t-nT)\,. \tag{5.122}$$

Assume $s(t)$ begins at $t=0$ with a prior accumulated phase of $(\pi/2)Q_{-1}=0$. Consider the interval $(2n-1)T \le t < 2nT$. Since there are an odd number of symbol intervals up to $(2n-1)T$, and the phase change in each symbol interval is $\pm\pi/2$, the accumulated phase $(\pi/2)Q_{2n-2}$ up to the interval can take on only the values $\pm\pi/2$. All other values are equivalent to these, modulo 2π.

Suppose, first, that $(\pi/2)Q_{2n-2}=\pi/2$. Then

$$m_1(t) = \cos\left\{(\pi/2)Q_{2n-2} + (\pi/2T)d_{2n-1}\cdot\left[t-(2n-1)T\right]\right\}$$

$$= -\sin\left\{(\pi/2T)d_{2n-1}\cdot\left[t-(2n-1)T\right]\right\}, \quad (2n-1)T \le t < 2nT. \tag{5.123}$$

If $d_{2n-1}=1$, this is one quarter cycle of a negative sine wave. If $d_{2n-1}=-1$, it is one quarter cycle of a positive sine wave. Both possibilities are shown in Figure 5.52. The accumulated phase in this interval is $(\pi/2)d_{2n-1}$.

Fig. 5.52 Equivalent in-phase modulation $m_1(t)$ of equation (5.121) for MSK, assuming accumulated phase of $\pi/2$ (i.e. $Q_{2n-2}=1$) up to $t=(2n-1)T$.

On the following interval $2nT \le t < (2n+1)T$, retaining our assumption that $Q_{2n-2} = 1$,

$$
\begin{aligned}
m_1(t) &= \cos[(\pi/2)Q_{2n-1} + d_{2n}\cdot(t-2nT)] \\
&= \cos[(\pi/2)Q_{n-1} + (\pi/2)d_{2n-1} + (\pi/2T)d_{2n}\cdot(t-2nT)] \\
&= \cos[\pi/2 + (\pi/2)d_{2n-1} + (\pi/2T)d_{2n}\cdot(t-2nT)] \\
&= -\sin[(\pi/2)d_{2n-1}]\cos[(\pi/2T)d_{2n}\cdot(t-2nT)]. \qquad (5.124)
\end{aligned}
$$

If $d_{2n-1}=1$, this is one-quarter of a negative cosine. If $d_{2n-1}=1$, it is one-quarter of a positive cosine. These possibilities are shown in Figure 5.52.

Thus for $Q_{2n-2}=1$,

$$
m_1(t) = -\sin\{\pi[t-(2n-1)T]/2T\}d_{2n-1}, \qquad (2n-1)T \le t < (2n+1)T, \qquad (5.125)
$$

a pulse extending over *two* symbol intervals. It can similarly be shown that for $Q_{2n-2}=-1$,

$$
m_1(t) = \sin\{\pi[t-(2n-1)T]/2T\}d_{2n-1}, \qquad (2n-1)T \le t < (2n+1)T. \qquad (5.126)
$$

A similar analysis yields $2T$-wide pulses in the quadrature modulation waveform $m_2(t)$, but displaced by T. Thus for our presumed initial accumulated phase of zero, and for $Q_{2n-2} = 1$

$$
\begin{aligned}
s(t) = &\sqrt{2E/T} \sum_n d_{2n-1}\, p_2[t-2nT+T]\cos(2\pi f_c t+\theta_0) \\
&- \sqrt{2E/T} \sum_n d_{2n}\, p_2[t-2nT]\sin(2\pi f_c t+\theta_0), \qquad (5.127)
\end{aligned}
$$

where

$$
p_2(t) = -\sin(\pi t/2T), \qquad 0 \le t \le 2T\,. \qquad (5.128)
$$

Equation (5.127) is exactly the formula for a linear two-dimensional signal. Typical in-phase and quadrature waveforms are illustrated in Figure 5.53. They

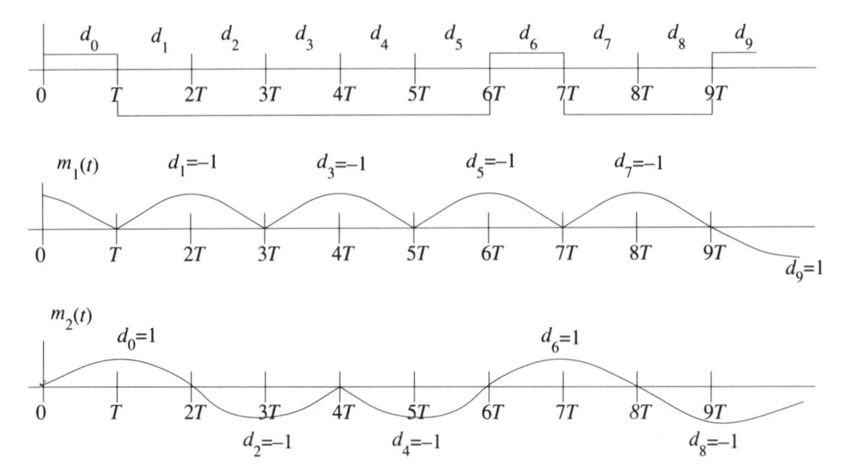

Fig. 5.53 Examples of in-phase and quadrature modulations $m_1(t)$ and $m_2(t)$ for the minimal shift-keyed signal, showing that MSK is a form of offset phase-shift keying (OPSK), itself a special case of staggered quadrature amplitude modulation (SQAM) in which a peak of the pulse train modulating the cosine carrier occurs at a null of the pulse train modulating the sine carrier, and visa versa. It has been shown [21,22] that staggering provides significant SNR enhancement in the presence of phase jitter distortion.

are a special case of offset-keyed systems in which a $T/2$ displacement between the in-phase and quadrature data trains provides a more uniform signal envelope.

Because MSK is a linear signal, its power spectrum is the same as that of the baseband pulse $p_2(t)$, which is

$$
\begin{aligned}
P(f) &= |\int \sin(\pi t/2T)\exp[j2\pi ft]\,dt|^2 \\
&= |\int \cos(\pi t/2T)\exp[j2\pi ft]\,dt|^2 \\
&= (16T^2/\pi^2)\{\cos(2\pi fT)/[1-16fT^2]\}^2 . \qquad (5.129)
\end{aligned}
$$

This power spectrum is shown in Figure 5.54 along with those for QPSK with rectangular pulses and raised cosine pulses with 12.5% rolloff. The QPSK spectra are simply those of a baseband PAM data train translated up to the carrier frequency. MSK's $2T$-wide cosine pulse has rapidly falling sidelobes, but does not have the spectral efficiency of a linear modulation with a pulse designed for spectral efficiency (and a corresponding increase in transmitter complexity).

The detection of MSK signals can be performed independently on the in-phase and quadrature data streams. As Figure 5.55 shows, assuming an ideal noiseless channel, the incoming signal $s(t)$ (5.127) is multiplied by in-phase and quadrature carrier signals, and the two product waveforms are applied to filters

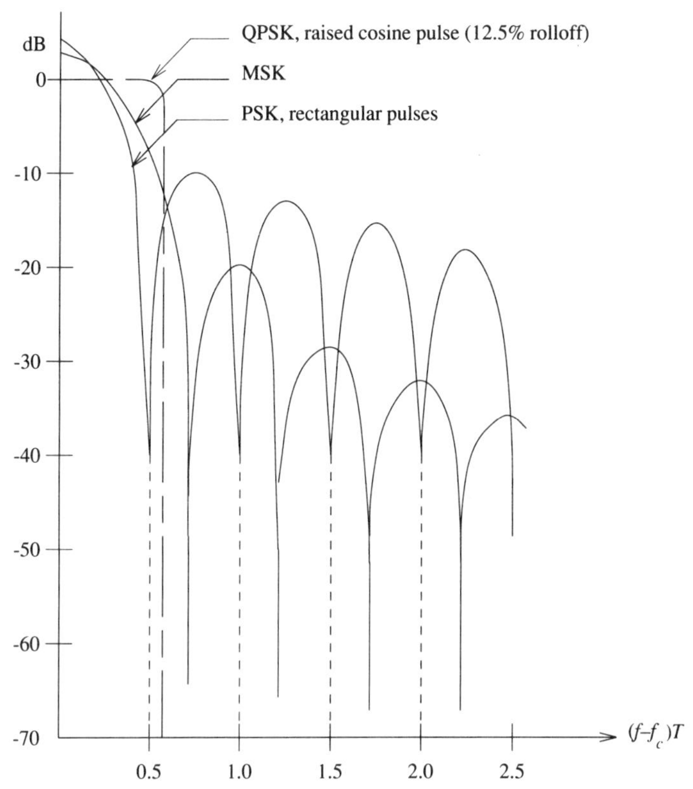

Fig. 5.54 Power density spectrum of MSK compared with spectra of PSK with rectangular and raised cosine pulses.

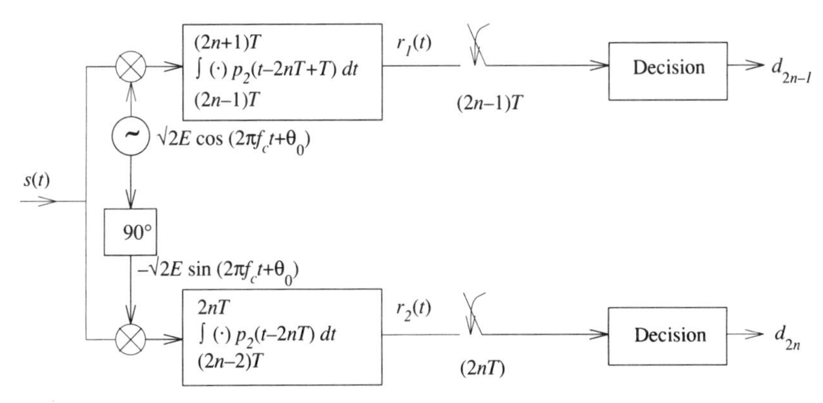

Fig. 5.55 MSK receiver, a pair of optimum binary antipodal detectors for the modulation waveforms $m_1(t)$ and $m_2(t)$.

matched to $p_2(t)$, or equivalently correlated with $p_2(t)$. This filtering in each leg, followed by sampling and detection, is the optimum receiver for binary antipodal signaling, as explained in Chapter 2.

Mathematically, in the absence of noise and assuming an integer number of periods of the cosine in $2T$, the operations in the top leg are

$$
\begin{aligned}
r_1(2nT) &= (2/T) \int_{(2n-1)T}^{(2n+1)T} s(t) \cos(2\pi f_c t+\theta_0)\ p_2(t-(2n-1)T)\,dt \\
&= (2/T) \int_{(2n-1)T}^{(2n+1)T} \{d_{2n} p_2(t-(2n-1)T)\cos(2\pi f_c t+\theta_0) \\
&\quad + m_2(t)\sin(2\pi f_c t+\theta_0)\cos(2\pi f_c t+\theta_0) p_2(t-(2n-1)T)\,dt\} \\
&= d_{2n-1} \int_{-T}^{T} \cos^2(\pi t/2T)\,dt = d_{2n-1}\ .
\end{aligned}
\tag{5.130}
$$

The error probability in the presence of white Gaussian noise is given by (2.35) for binary *antipodal* signals. Note that if the MSK signal had been detected by a symbol-interval detector for binary *orthogonal* signals, which MSK is on a symbol interval (Exercise 5.13), the probability of error would be 3 dB worse (2.36).

5.6.5 CONTINUOUS-PHASE MODULATION (CPM)

MSK is a member of a large class of continuous-phase modulation signals [29–32], characterized by phase modulation only and phase continuity but allowing various pulse shapes $p(t)$, varying number of phase shifts, L, of the data stream $\{d_n\}$, and a modulation index Δ not necessarily equal to 0.5. Some of these signals have narrower spectra or lower spectral sidelobes than MSK for the same error-rate performance. The following discussion follows Sundberg [29].

In order to realize practical maximum likelihood receivers for these signals, we will constrain the modulation index to *rational* values

$$\Delta = 2\, f_d T = 2k/p, \tag{5.131}$$

where k and p are integers with no common divisor. We will also define the normalized pulse shape

$$f(t) = (1/2T)\ p(t), \tag{5.132}$$

and its integral

$$q(t) = \int_{-\infty}^{t} f(\tau)\,d\tau, \tag{5.133}$$

where $f(t)$ is normalized so that $q(\infty) = 1/2$. Table 5.4 describes some com-

Table 5.4 Pulses $f(t)$ Commonly Used in Continuous Phase Data Signals

a) MSK

$$f(t) = \begin{cases} 1/2T, & 0 \le t \le T \\ 0 & \text{otherwise} \end{cases}$$

b) Raised cosine

$$f(t) = \begin{cases} (1/2NT)\ [1-\cos(2\pi t/NT)0], & 0 \le t \le T \\ 0 & \text{otherwise} \end{cases}$$

c) Tamed frequency modulation

$$f(t) = (1/8)[f_0(t-T) + 2f_0(t) + f_0(t+T),$$

$$f_o(t) = \sin(\pi t/T)\{1/\pi t)$$

$$- [2 - (2\pi t/T)\,cot(\pi t/T) - (\pi t/T)^2]/[24\pi t^3/T^2]\}$$

d) Spectrally raised cosine

$$f(t) = (1/NT)\{[\sin(2\pi t/NT)/(2\pi t/NT)]$$

$$[\cos(2\pi\beta t/NT)/(1-(4\beta t/NT)^2),\ 0 \le \beta \le 1]\}.$$

e) Gaussian minimum-shift keying

$$f(t) = (1/2T)\{Q[2\pi\beta_b(t-T/2)/\sqrt{1n2}]$$

$$- Q[2\pi\beta_b(t+T/2)/\sqrt{1n2}]\},\ 0 \le \beta_b T \le \infty$$

$$Q(t) = \int_{-\infty}^{t} (1/\sqrt{2\pi})\exp(-\tau^2/2)\,d\tau.$$

f) L-symbol rectangular (MSK corresponds to $N = 1$)

$$f(t) = \begin{cases} 1/2NT, & 0 \le t \le nT \\ 0 & \text{otherwise} \end{cases}$$

monly used $f(t)$ waveforms. Note that phase continuity requires only that there be no impulses in the FM modulation waveform.

The particular attraction of tamed frequency modulation and Gaussian MSK is good spectral characteristics combined with the possibility for coherent detection (modulation index = 1/2). Gaussian MSK is the leading candidate for the Europe-wide GSM (Group Speciale Mobile) digital mobile cellular communication system which is to be built in the 1990s.

We will now examine the properties of the transmitted signal for these generalized signals. From (5.110), neglecting the random phase θ_0,

$$\begin{aligned} s(t) &= \sqrt{2E/T}\ \cos\left[2\pi f_c t + 2\pi f_d \int_{-\infty}^{t} \sum_{i=-\infty}^{\infty} d_i\ p(\tau - iT)\, d\tau\right] \\ &= \sqrt{2E/T}\ \cos\left[2\pi f_c t + 2\pi\Delta \int_{-\infty}^{t} \sum_{i=-\infty}^{\infty} d_i\ f(\tau - iT)\, d\tau\right] \\ &= \sqrt{2E/T}\ \cos\left[2\pi f_c t + 2\pi\Delta \sum_{i=-\infty}^{\infty} d_i\ q(t - iT)\right]. \end{aligned} \tag{5.134}$$

Assume $f(t)$ is nonzero only over N pulse intervals, $0 \leq t \leq NT$ as illustrated in Figure 5.56 for the rectangular pulse (Table 5.4f). Then $q(t) = 1/2$, $t > NT$, and

$$s(t) = \sqrt{2E/T}\ \cos\left[2\pi f_c t + 2\pi\Delta \sum_{i=n-N+1}^{n} d_i\ q(t - iT) + \phi_n\right],$$

$$nT < t \leq (n+1)T, \tag{5.135}$$

where

$$\phi_n = \Delta\pi \sum_{i=-\infty}^{n-N} d_i . \tag{5.136}$$

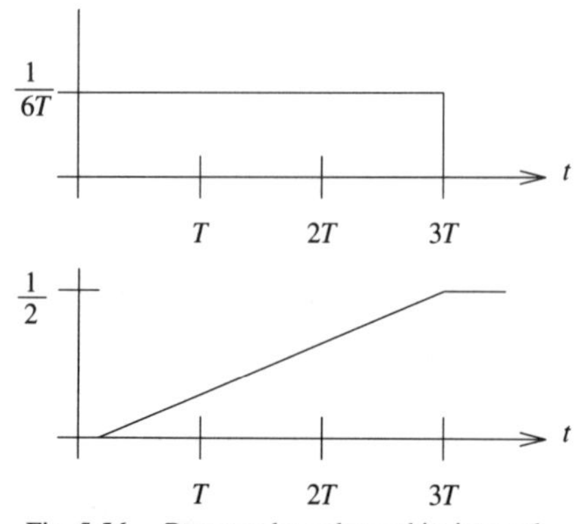

Fig. 5.56 Rectangular pulse and its integral.

For convenience, we define

$$\phi(t,\mathbf{d}) = 2\pi\Delta \sum_{i=-\infty}^{\infty} d_i q(t-iT) = 2\pi\Delta \sum_{i=n-N+1}^{n} d_i q(t-iT) + \phi_n . \quad (5.137)$$

Note that if $N=1$, (5.136) reduces to

$$\phi(t,\mathbf{d}) = 2\pi\Delta\, d_n q(t-nT) + \Delta\pi \sum_{i=-\infty}^{n-N} d_i . \quad (5.138)$$

Since k/p is a rational number, ϕ_n mod 2π has only p distinct values. For example, for $\Delta = 1/2$, it has four values $(0, \pi/2, \pi, 3\pi/2)$, and for $\Delta = 2/3$, the three values $(0, 2\pi/3, 4\pi/3)$.

The value of $s(t)$ in (5.135) is determined by a *state vector*

$$S_n = (\phi_n, d_{n-1}, d_{n-2}, d_{n-3}, \ldots, d_{n-N+1}). \quad (5.139)$$

This vector can have at most pL^{N-1} different values, so that we are considering a finite-state system describable by a trellis diagram. For a modulation, with a pulse extending over only one symbol interval, the number of different states is p. For MSK, with $\Delta = 1/2$, implying $k = 1$ and $p = 4$, there are four different possible states. For binary modulation and a pulse extending over three symbol intervals ($\Delta = 2/3$), $S_n = (\phi_n, d_n - 1, d_{n-2})$ has 12 possible values, since ϕ_n has three values and d_{n-1} and d_{n-2} have two each.

Figure 5.57 compares all phase trajectories, computed from (5.137), for MSK (rectangular pulses on one symbol interval, $\Delta = 1/2$) with those for a raised cosine pulse extending over three symbol intervals, with $\Delta = 2/3$. It is assumed, as an initial condition at $t = 0$, that $\phi_0 = 0$ and all previous data are equal to $+1$. Although all the phase trajectories are continuous, those of the signal with raised cosine pulses and $\Delta = 2/3$ are much smoother, implying better spectral properties. It should be noted that the sharp phase transitions of MSK can be softened, without changing the modulation index, by use of longer pulses. In the frequency domain, the spectra are narrowed. Pulses extending over several symbol intervals are examples of the "partial response" pulses described in Chapter 4. Just as for baseband partial response pulse trains, a maximum likelihood receiver can be realized for continuous phase modulation with rational modulation index. The power density spectra [35–38] for binary signals $(L=2)$ with various combinations of modulation index and pulse shape are shown in Figure 5.58.

5.7 TRELLIS-CODED MODULATION

The introduction to convolutional codes of Chapter 3 showed how the minimum distance between data sequences could be increased through the introduction of redundancy — sending more data bits than in the uncoded signal. When the expanded sequence must be transmitted through a noisy, distorted, and

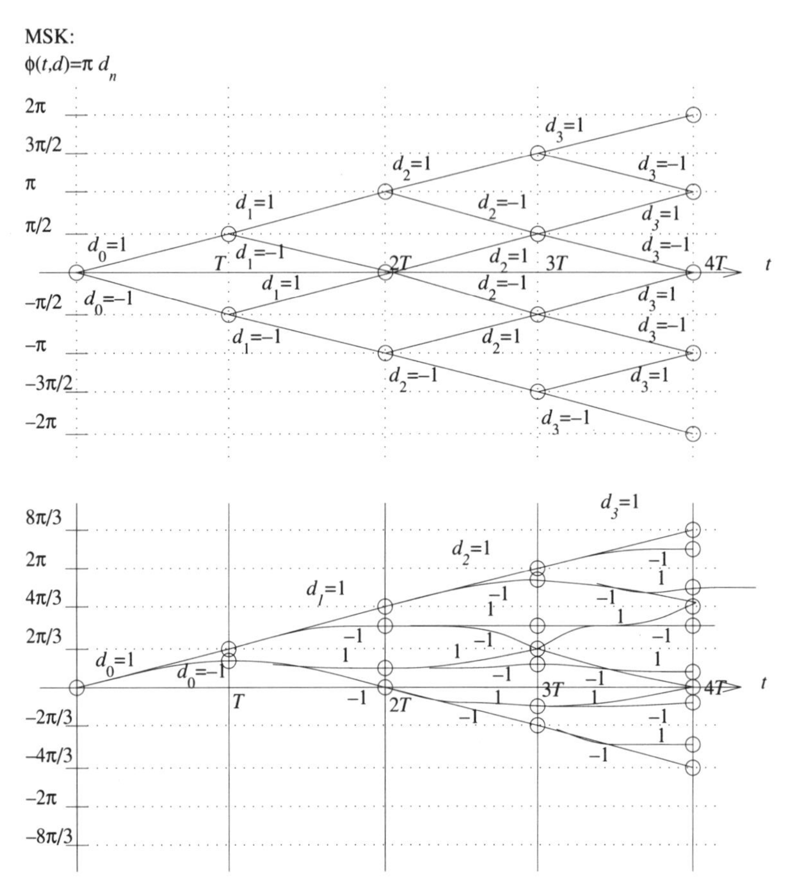

Fig. 5.57 Phase paths for (a) MSK (rectangular pulse on a single symbol interval, and $\Delta = 1/2$) and (b) another continuous phase modulation (cosine pulse extending over $N = 3$ symbol intervals, and $\Delta = 1/3$).

bandlimited channel, there is a loss of performance that must be weighed against the coding gains. This loss can be seen either in distortion caused by use of a higher symbol rate which expands the signal bandwidth into less desirable channel characteristics, or by going to an expanded signal constellation which retains the original bandwidth but reduces the noise margin between signal points because of the average power constraint. For telephone channels, the severe channel limitations make it difficult to realize any advantage from data coding. It should be noted that in channels not constrained in bandwidth, particularly those used for satellite and deep space communications, conventional (as described in Chapter 3) coding has been extremely successful [37].

In 1976, Ungerboeck [40–42] found that substantial gains could be achieved in telephone channels by convolutional coding of the signal levels

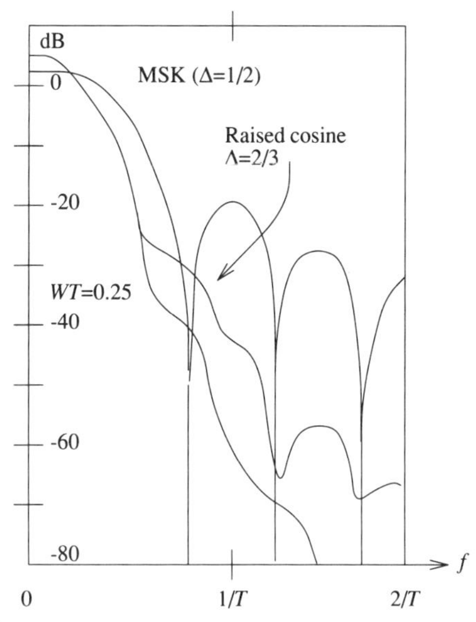

Fig. 5.58 Power density spectra for various continuous phase modulation pulses, from Sundberg [29].

rather than the source data. He unified coding and modulation to directly increase the "free distance" (minimum Euclidean distance) between coded line signal sequences. The procedure is called "channel trellis coding" because the sequence of states in the finite state machine which encodes the signal levels follows a trajectory in a trellis of possible trajectories, just as for convolutional coding of data (Chapter 3). The larger the Euclidean distance between signal sequences, the lower the error rate, which is approximated by, for large SNR,

$$P_e = N_{free} Q(d_{free}/2\sigma) \ , \qquad (5.139)$$

where N_{free} is the average number of nearest-neighbor signal sequences at distance d_{free} from a given signal sequence. A more accurate evaluation of probability of error would consider error events of slightly larger distances than the minimum distances. The transfer function approach, described in Section 3.6, uses a Markov state diagram to efficiently count the number of error events at a given distance [54].

With "codewords" consisting of modulated level sequences, a gain in minimum Euclidean separation of sequences, over the minimum distance between signal points, means that not all sequences can be codewords. Equation (5.139) can be extended, using the *transfer function* technique of Section 3.6, to incorporate the effect of signal sequences separated by more than d_{free} [52] on the error rate. Redundancy is introduced through enlargement of the signal constellation. The pulse shape is left alone and no bandwidth expansion occurs.

A sequence of transmitted signal points (coded selections from the enlarged signal constellation), arriving at the receiver perturbed by additive noise, is decoded by a soft-decision maximum-likelihood sequence decoder. Even though the minimum distance between points in the enlarged constellation is reduced from that in the original constellation (at constant average power), the coding succeeds in increasing the Euclidean distance between entire transmitted signal sequences. The net gain can approach 6 dB. The CCITT V.32 recommendation (see Table 1.5), using a signal constellation enlarged from 16 to 32 points, has been widely implemented for full-duplex data communication at rates of 9600 bps and above.

It is never necessary to more than double the number of signal points to approach optimal performance in a white Gaussian noise channel. As shown in Chapter 2 (see Figure 2.28) doubling the number of PAM levels from L to $2L$ achieves a channel capacity (with coding) of $\log_2 L$ bps at an SNR many dB below that required for good uncoded performance with L levels, and only a fraction of a dB above the information theoretic minimum SNR. Doubling again the $4L$ levels would yield negligible improvement in SNR.

5.7.1 STATE TRELLISES FOR TRELLIS-CODED MODULATION

Chapters 3 and 7 contain explanations of the concepts of generation of coded sequences in a finite-state machine, progression through a state trellis, and optimum detection by finding the best path through the trellis, as realized via the Viterbi algorithm presented in Chapter 3. For channel trellis coding, it is the transmitted signal, not the customer data, that is encoded at the transmitter and decoded at the receiver. In decoding, a decision for the most likely transmitted signal sequence is made. This decision is then translated into decisions for customer data. The encoding is by a convolutional encoder as described in Chapter 3.

As Ungerboeck noted [39], finding codes with good Euclidean distance does not follow readily from knowledge of how to find codes with good Hamming distance, and this is what delayed implementation of channel trellis coding for decades after the success of convolutional coding. Block codes, such as the multidimensional codes of Section 5.5.1, incorporate the idea of increasing the distances between transmitted level sequences, but do not allow as general a mapping as trellis-coded modulation. Ungerboeck succeeded in finding good channel codes with only a factor of two expansion in signal constellation size, and his "set partitioning" technique will be explained after an example which illustrates an effective channel trellis code.

Ungerboeck [39] provides a clear example of a four-state trellis code for eight-phase PSK. Consider the uncoded 4-PSK and coded 8-PSK signal constellations of Figure 5.59 and the convolutional encoder of Figure 5.60 (associated with the eight-point constellation). Note the factor of two signal constellation

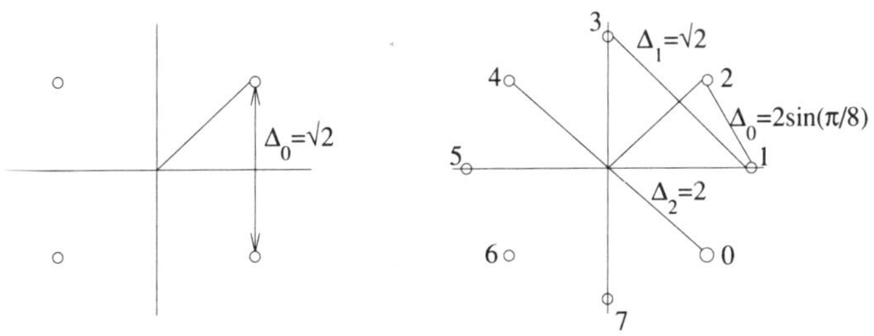

Fig. 5.59 Four-point (uncoded) and eight-point (coded) signal constellations. Each has radius one, and thus average signal power one.

expansion. Assume binary data pairs $\{x_1(n), x_2(n)\}$ are to be communicated. Using the uncoded four-phase constellation, each data pair determines one of the four signal points. In the absence of coding, decisions at the receiver are made on a symbol-by-symbol basis. The minimum free distance is simply that between adjacent signal points, or $[d_{free}]^2 = 2$.

Using the enlarged eight-point constellation and the encoding of Figure 5.60, each information dibit $[x_1(n), x_2(n)]$ plus the output of the shift register is coded into the tribit $[y_0(n), y_1(n), y_2(n)]$. The tribit determines the choice of transmitted signal as shown on the eight-point constellation.

Note that $[s_0(n), s_1(n)]$, the contents of the shift register (which operates in modulo 2 arithmetic), represents the state of the encoder, so that there are four possible states. In each symbol interval, a new dibit input causes a state transition as well as generating a tribit output. Because of the constraints of the encoder (s_1 becomes s_0), a state at time i has only two, not four, states that it can turn into at time $i+1$. Figure 5.61 shows the state trellis expressing these facts. The data written along each arm are the input dibit causing that particular transition and the resulting output tribit.

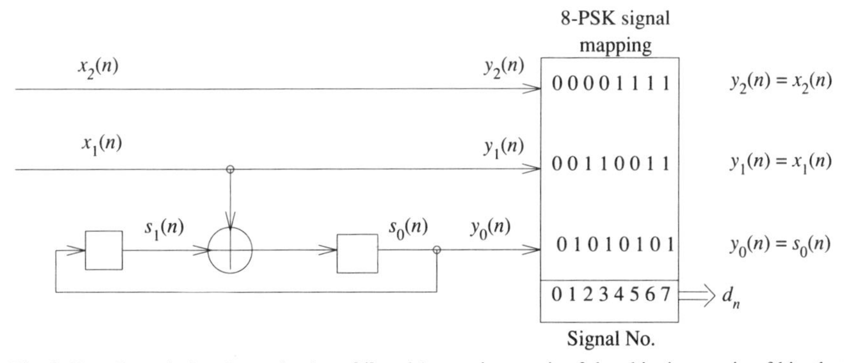

Fig. 5.60 Convolutional encoder (rate 2/3 code) mapping a pair of data bits into a trio of bits that defines a signal point, e.g., $(y_2, y_1, y_0) = (1, 0, 0)$ defines point 4 in Figure 5.59.

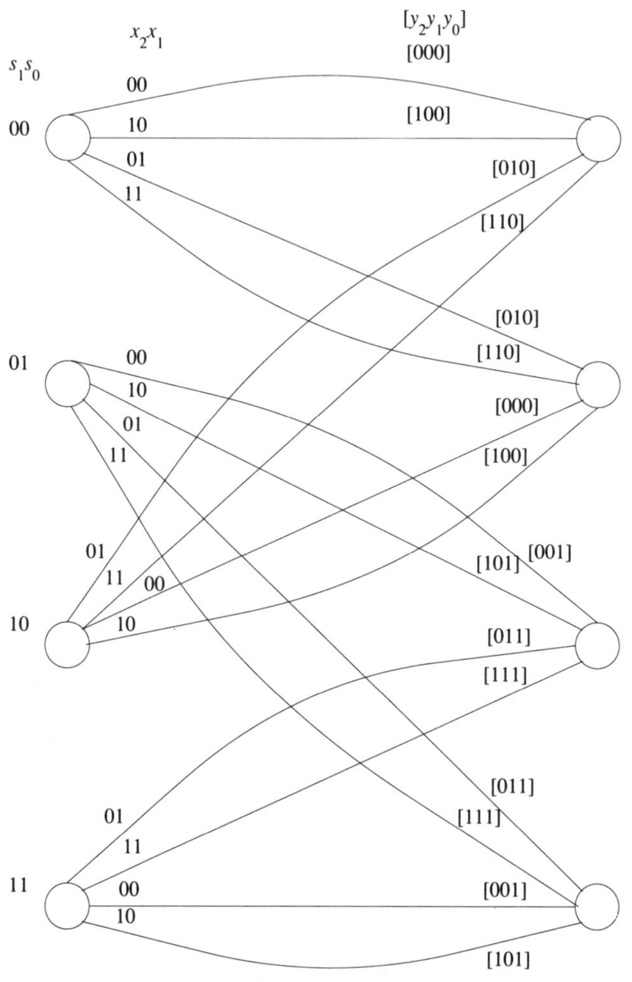

Fig. 5.61 State trellis for encoder of Figure 5.60.

It is not immediately apparent, but the 8-PSK constellation has been partitioned so that parallel branches in the trellis represent signal points at maximum separation, and branches emerging from and converging to a point have the next largest separation. Each parallel pair of arms in Figure 5.61 represents a pair of signal points with Euclidean separation of 2. At each node, the four arms entering represent signals that have Euclidean separations of at least $\sqrt{2}$, and similarly for the four arms leaving. At different nodes, the Euclidean separation of incoming arms is at least $2\sin(\pi/8)=0.3827$, the minimum signal point separation in the constellation.

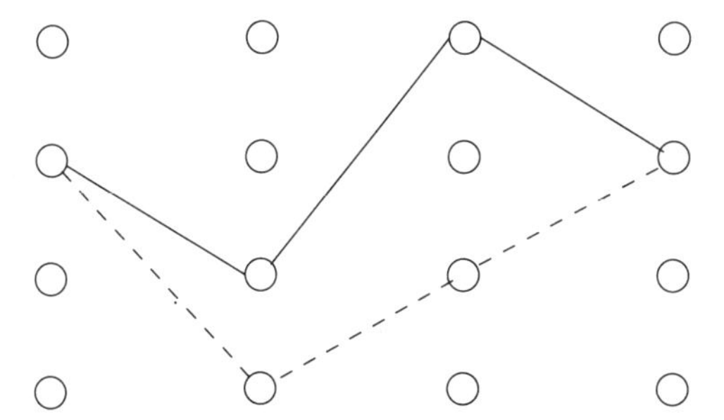

Fig. 5.62 Segments of two trajectories leaving a node separately and rejoining three symbol intervals later (for encoder of Figure 5.60).

The Viterbi decoding algorithm selects a winning trajectory through the trellis, minimizing a cost function similar to that described in Chapters 3 and 7. Suppose two candidate paths, one correct one and the other erroneous, diverge from one node and rejoin later. If they rejoin in one symbol interval, they are parallel arms with signal separation 2, i.e., $d^2 = 4$. If they go to separate nodes in the first transition, they will need a total of at least 3 symbol intervals to rejoin, as illustrated in Figure 5.62. The total squared Euclidean distance between those two path segments is

$$d^2 = (d_{leaving\ same\ node})^2 + (d_{entering\ different\ nodes})^2 + (d_{entering\ same\ node})^2$$
$$> 2 + (2 sin \pi/8)^2 + 2 > 4. \tag{5.140}$$

All other diverging trajectory path segments have even greater separation. *Thus the squared minimum free distance is determined by the parallel arms (uncoded bits) and is equal to 4.* Since the power of all PSK systems is unity, the coding gain over use of the uncoded four-phase constellation is

$$Gain = 10 \log_{10} \left[\frac{d^2_{\min}(coded\ 8-\phi)}{d^2_{\min}(uncoded\ 4-\phi)} \right] = 10 \log_{10} \frac{4}{2} = 3 \text{ dB} . \tag{5.141}$$

The four-state trellis-coded PSK curve of Figure 5.63 is computed from a simulation. At lower signal-to-noise ratio, the 3 dB gain implied by (5.139) and (5.141) is not fully realized because the probability of an error event is significant even for sequences which are not at the minimum distance from the reference sequence, so that (5.141) is no longer accurate.

As the reader will recall, the data decision sequence is determined from the state sequence of a winning trajectory. How, then, do we resolve the ambiguity to input data pairs 00 and 10 respectively? The answer is that the output data triplets 000 and 100 associated with those branches correspond to widely separated signal points, labeled "0" and "4" in Figure 5.59. This is, by design, true for every bundle of parallel branches. Once the state sequence is decided,

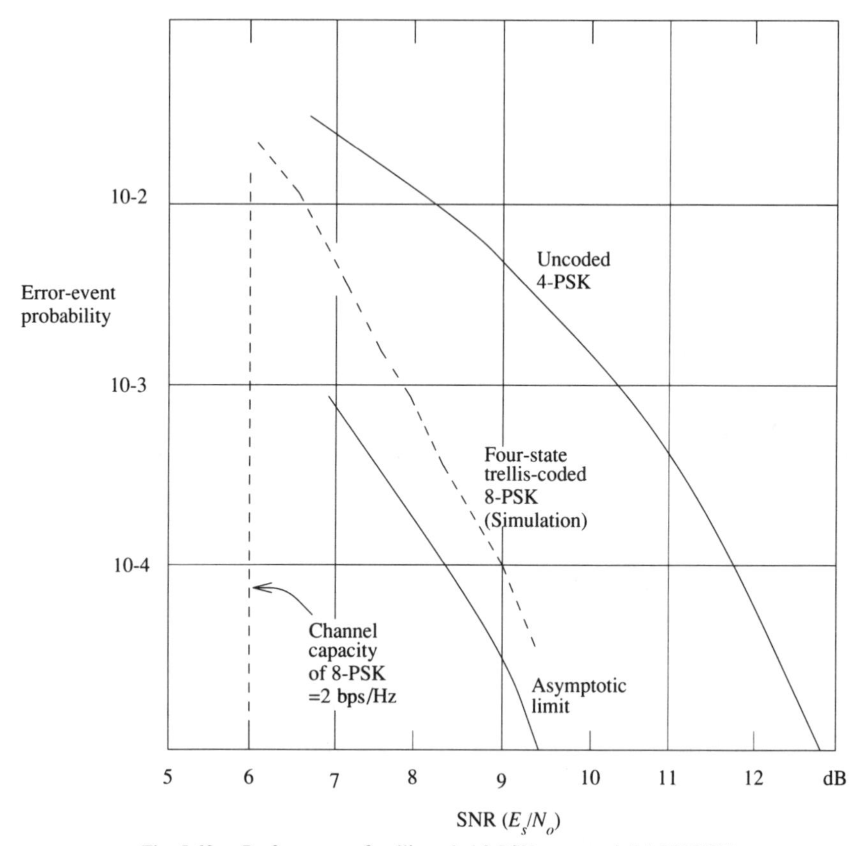

Fig. 5.63 Performance of trellis coded 8-PSK vs. uncoded 4-PSK [37].

select the parallel branch corresponding to the signal point closest to the analog received sample.

Conventional coding of the input data stream is much less powerful than trellis-coded modulation. Consider the following comparison of uncoded and conventionally coded systems with an information rate of 2 bits/symbol. Assume that the level of Gaussian noise is such that (uncoded) 4-PSK operates with a bit error rate of 10^{-5}. A convolutionally coded system using a rate 2/3 code would require 8-PSK to transmit 3 bits/symbol (to be decoded into 2 information bits/symbol). Assuming the same SNR, the 8-PSK has a bit error rate of 10^{-2}. To bring the *information* bit error rate back down to 10^{-5}, patterns of at least three bit errors would have to be corrected, which could be done by a rate 2/3 binary convolutional code with constraint length $\nu = 6$ (using the Hamming bound [27]), implying a minimum free distance of 7. A complex 64-state binary Viterbi decoder would be required. To achieve any kind of gain over the

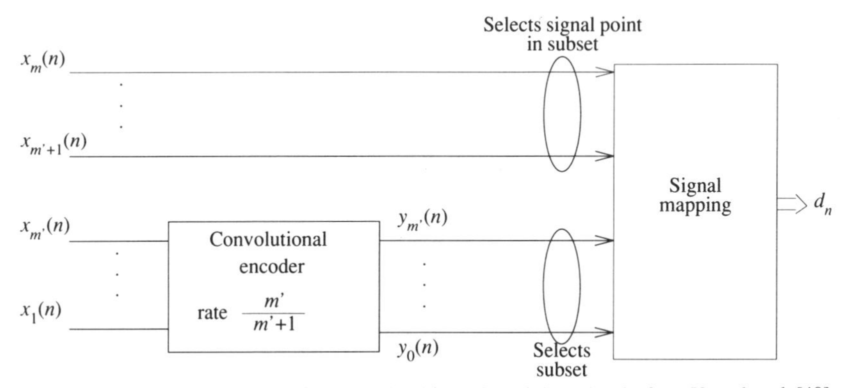

Fig. 5.64 General coding model for channel trellis coding of signal levels, from Ungerboeck [40].

uncoded 4-PSK system would be even more difficult. The lesson here is that a naive mapping of data triplets to eight phases, that does not focus on the Euclidean distance between sequences, cannot be easily corrected by conventional data coding. Ungerboeck's mapping by set partitioning does focus on Euclidean distance.

5.7.2 SET PARTITIONING

The increase of Euclidean distance between signal sequences is guaranteed through the process of *set partitioning*. First, consider the general encoding model of Figure 5.64 for the expansion of the signal constellation. With m information bits to be transmitted in a symbol interval, $m' < m$ of these bits are applied to a rate $m'/(m'+1)$ convolutional encoder, generating $m' + 1$ coded bits. These bits are used to select one of $2^{m'+1}$ *subsets* of a redundant 2^{m+1}-element signal set, i.e., one twice as large as the original signal net.

The subsets are generated by a systematic process of *set partitioning* that divides a signal constellation successively into smaller constellations with increasing smallest intraset distances Δ_i, $i = 0, 1, \ldots$. The partitioning is repeated $m' + 1$ times until $\Delta_{m'+1}$ is at least as large as the desired free distance of the trellis code being designed. The last $(m'+1)$th layer of divisions results in the group of $2^{m'+1}$ bits $y_{m'}(n), \ldots, y_0(n)$; each of the $(m'+1)$th-layer subsets is given a label consisting of an $(m'+1)$-bit word. These binary labels express how far apart signals are. If the labels of two of these bottom-layer subsets agree in the last q digits (any $q \leq m'$), but not in the next position to the left (e.g., 10111 and 11111 for $q = 3$), then the signals of the two subsets are contained within a subset at the (higher) level q in the partition tree.

Consider the same trellis-coded 8-PSK signal set introduced earlier. If the y_1 and y_0 bits are regarded as generated by a rate 1/2 convolutional encoder

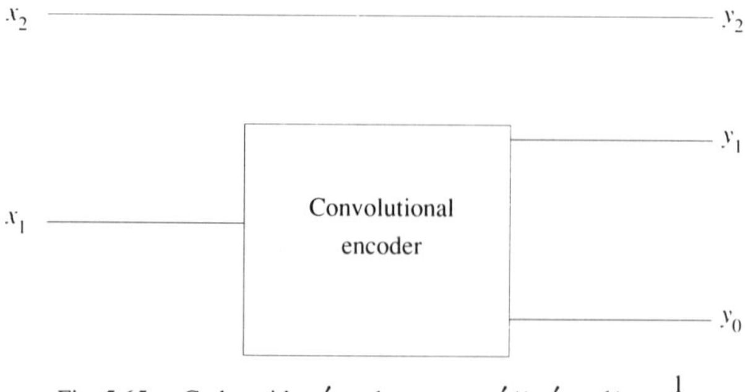

Fig. 5.65 Coder with $m' = 1$, rate $= m'/(m'+1) = \frac{1}{2}$.

driven by x_1, we have the following model corresponding to Figure 5.65: The partitioning of the 8-PSK signal set proceeds as shown in Figure 5.66. Note that if two subsets agree in the last digit but not the first digit, their combined signal set has at least distance Δ_1 between signal points.

The uncoded information bits $x_{m'+1}(n)$, ..., $x_m(n)$ are used to choose a signal point from the subset selected by $y_0(n),\dots,y_{m'}(n)$. In the example above, this is only the single bit x_2. It does not matter how these $2^{m-m'}$ signal points

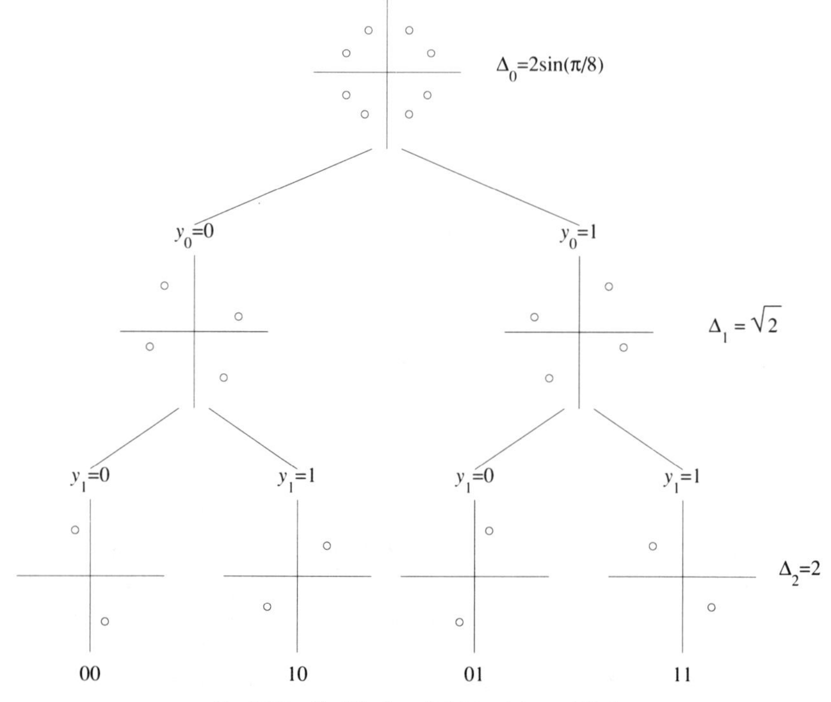

Fig. 5.66 Partitioning of eight-point constellations.

are labeled. In the code trellis, the $2^{m'-m}$ parallel transitions in each bundle correspond to these signal points within a subset.

From these considerations, the free Euclidean distance of a channel trellis code can be specified as

$$d_{free} = \min\{\Delta_{m'+1}, d_{free}(m')\}, \tag{5.142}$$

where $\Delta_{m'+1}$ is the minimum distance between parallel transitions and $d_{free}(m')$ is the minimum distance between nonparallel paths in the trellis diagram. That is, in the single-loop transitions of two paths from a shared state to another shared state, the paths are forced to come from either of two signal points in a subset, within which the minimum distance is $\Delta_{m'+1}$, the same for every subset. For multiple-hop transitions, $d_{free}(m')$ is the minimum free distance. As our initial example suggested, $d_{free}(m')$ is often less than or equal to $\Delta_{m'+1}$. Note that d_{free} depends only on the subset labels of the transitions and not on the uncoded bits. Thus, in fact, the distance properties of such a code are determined by distance properties of subsets and not on the rate of the code as determined by the number of points used from each subset of the partition.

It is not obvious how to find optimum channel trellis codes, which requires determination of $d_{free}(m')$ for candidate codes. Based on the set partitioning discussion earlier and on properties of convolutional codes, Ungerboeck showed that $d_{free}(m')$ is lower bounded by

$$d^2_{free}(m') \geq \min \sum_{|\mathbf{e}(n) \neq 0|} \Delta^2 q(\mathbf{e}(n)), \tag{5.143}$$

where $\mathbf{e}(n)$ is an error sequence between two coded signal level sequences and $q(\mathbf{e}(n))$ is the number of trailing zeros in the error sequence $\mathbf{e}(n)$ (i.e., the number of trailing positions in which the two sequences are the same). The minimization is over all nonzero error sequences (which are, by the nature of convolutional codes, themselves allowable level sequences) that deviate at some time from the all-zero error sequence and remerge with it some time later. This bound permits evaluation of a candidate code much more rapidly than would be possible if all possible distances between coded signal sequences had to be evaluated.

To search for good codes, given a sequence of minimum intra-set distances $\Delta_0 \leq \Delta_1 \leq \cdots \leq \Delta_{m'}$ and a code constraint length ν corresponding to the memory of the convolutional coder, a code-search program performs an orderly search for binary parity-check coefficients, using a set of code-rejection rules to minimize the number of times that computations of $d_{free}(m')$ have to be made. Table 5.5 gives the coding gains found for 16-PSK and "32-cross" enlarged constellations (Figure 5.67). The gains are over 8-PSK and 16-point QAM respectively.

Channel trellis-coded systems work well in the presence of Gaussian noise, but phase perturbations can be a problem. For example, trellis-coded 8-PSK degrades to the performance of uncoded 4-PSK at a carrier-phase offset of about

Table 5.5 Coding Gains in 16-PSK and 32-Cross Constellations for Various Convolutional Coding Constraint Lengths (from [Ungerboeck (40)])

Number of states 2^{ν}	Coding gain 16-PSK (dB)	Coding gain 32 cross (dB)
4	3.54	3.01
8	4.01	3.98
16	4.44	4.77
32	5.13	4.77
64	5.33	5.44
128	5.33	6.02
256	5.51	6.02

(ν = constraint length)

20°. The sensitivity to phase offset is explained by pairs of paths through the trellis whose Euclidean separation grows slowly. If phase offset rotates the received signals such that received signal points are halfway between the actual transmitted signals, the distances between the received signal level sequence and two possible transmitted signal level sequences may be the same, implying a high probability of incorrect decision. The critical value of phase rotation of 8-PSK is 22.5°.

5.7.3 ROTATIONAL-INVARIANT TRELLIS-CODED MODULATION

It is desirable that trellis-coded signals be as invariant as possible to phase rotations that leave the constellation unchanged, such as rotations of 8-PSK through multiples of 45°. If the coded signal were invariant to such rotations, standard differential encodings (Section 5.2.4), which eliminate sensitivity of such rotations, could be used. For the 32-cross expanded constellation used in

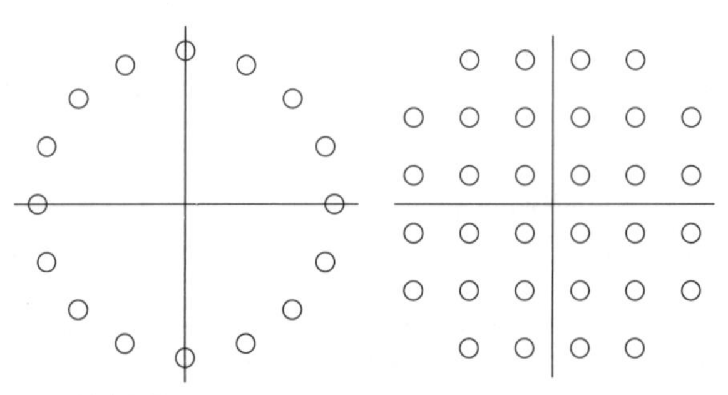

Fig. 5.67 16-PSK and 32-cross enlarged constellations.

the CCITT V.32 recommendation for 14.4 kbps data transmission over four-wire circuits, it proved impossible to find a nonsensitive linear code. This recommendation (and the V.33 recommendation for 64-QASK and 128-cross constellations) eventually adopted a *nonlinear* convolutional coding technique discovered by L. F. Wei [44] to make an 8-state code (constraint length = 3) invariant to 90° rotations, while retaining the 4 dB coding gain of the linear code. This performance holds only for the nonlinear 8-state convolutional code. For 16 states or more, no linear or nonlinear code (achieving rotational invariance) has as high a gain as the best linear code.

Following Wei [45], we will show how this special coder works and realizes 4 dB gain. The basic principle is that the convolutional coder translates a sequence of differentially coded data into a stream of transmitted symbols such that, when a phase jump of some multiple of 90° occurs and all subsequent received symbols are incremented by the amount of the jump, the resulting sequences are allowed symbol sequences for the trellis. Also, the subsequent convolutional decoding results in a data stream that, although not the same data stream that was fed into the convolutional coder at the transmitter, has the original differential coding and can therefore be translated into the correct customer data.

We begin by observing that four input bits are mapped into five bits which select a point in the 32-cross constellation of Figure 5.67. Figure 5.68 and Table 5.6 describe the differential encoder and nonlinear convolutional coder that do the mapping. To prove that the convolutional coder is nonlinear, let the state

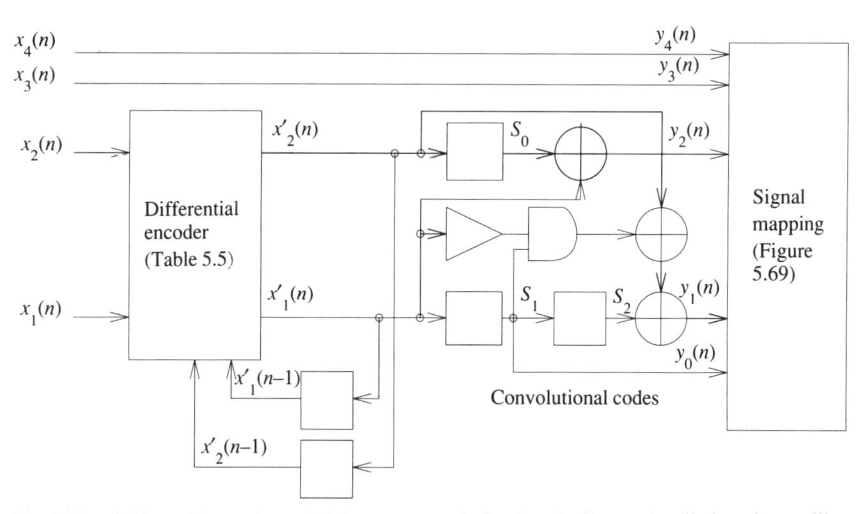

Fig. 5.68 Differential encoder and eight-state convolutional coder for rotationally invariant trellis-coded modulation [45].

Table 5.6 Differential Encoder for Figure 5.68, and Decoder Used in Receiver

Encoder			Decoder		
$x_2'(n-1)x_1'(n-1)$	$x_2(n)x_1(n)$	$x_2'(n)x_1'(n)$	$x_2'(n-1)x_1'(n-1)$	$x_2'(n)x_1'(n)$	$x_2(n)d_1(n)$
00	00	00	00	00	00
01	00	01	01	01	00
10	00	10	10	10	00
11	00	11	11	11	00
00	00	00	00	00	00
00	01	01	00	01	01
01	01	10	01	10	01
10	01	11	10	11	01
11	01	00	11	00	01
00	10	10	00	10	10
01	10	11	01	11	10
10	10	00	10	00	10
11	10	01	11	01	10
00	11	11	00	11	11
01	11	00	01	00	11
10	11	01	10	01	11
11	11	10	11	10	11

$S_2S_1S_0 = 000$, and consider three different input-output pairs:

	Input $x_2'(n)x_2'(n)$	Output $y_2(n)y_1(n)y_0(n)$
(1)	00	010
(2)	01	100
(3)	00+01=01	100

The third input is the modulo 2 sum of the first two inputs, but the third output is not the modulo 2 sum of the first two outputs. Hence, the convolutional coder is nonlinear.

The differential encoder in Figure 5.68 rotates a 2-bit input through the sequence (00,01,10,11). Current inputs of 00, 01, 10, and 11 cause respective shifts of 0, 1, 2, and 3 places. Thus a current input $x_1(n)x_2(n) = 10$ rotates a prior output $x_2'(n-1)x_1'(n-1) = 01$ by two places, to $x_2'(n)x_1'(n) = 11$.

The convolutional coder realizes the operations tabulated in Table 5.7, and shown as a trellis in Figure 5.69. (The significance of the subgroups will be explained presently.) The two uncoded bits and three output bits, designated $y_4y_3y_2y_1y_0$, select a signal constellation according to the mapping of Figure 5.70. In this figure, we see that the letters in Table 5.7, which are merely labels corresponding to $y_2y_1y_0$ values, refer to partition subsets of signal points.

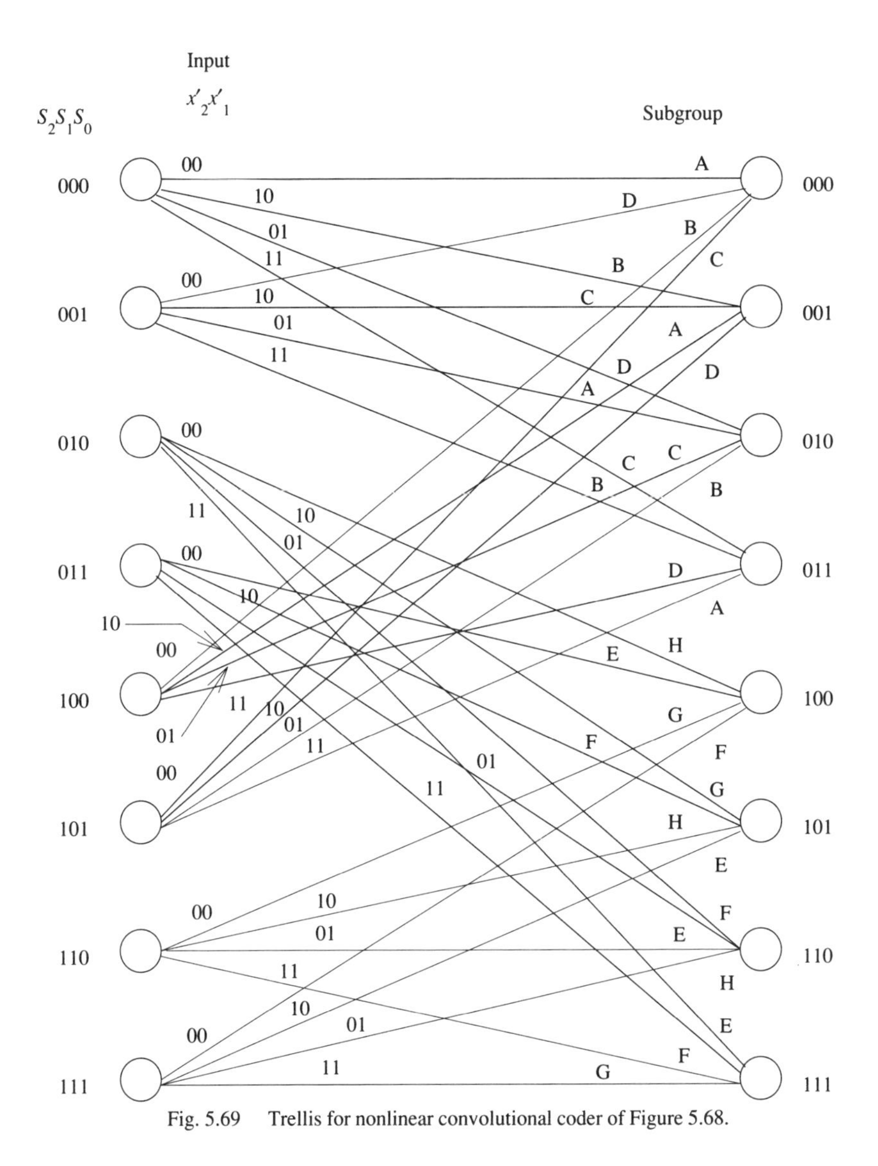

Fig. 5.69 Trellis for nonlinear convolutional coder of Figure 5.68.

The right partition subsets (A, B, C, D, E, F, G, H) are defined such that points in a given subset are (as can be seen in Figure 5.70) separated by a distance of at least four, or a squared distance $d^2 = 16$. For a specified starting state, a branch emerging from that state is actually four parallel branches corresponding to the four signal points in the subset. Other signal point separations can similarly be read from Figure 5.70. The minimal squared separation between a point in A and a point in B is eight, between a point in A and a point

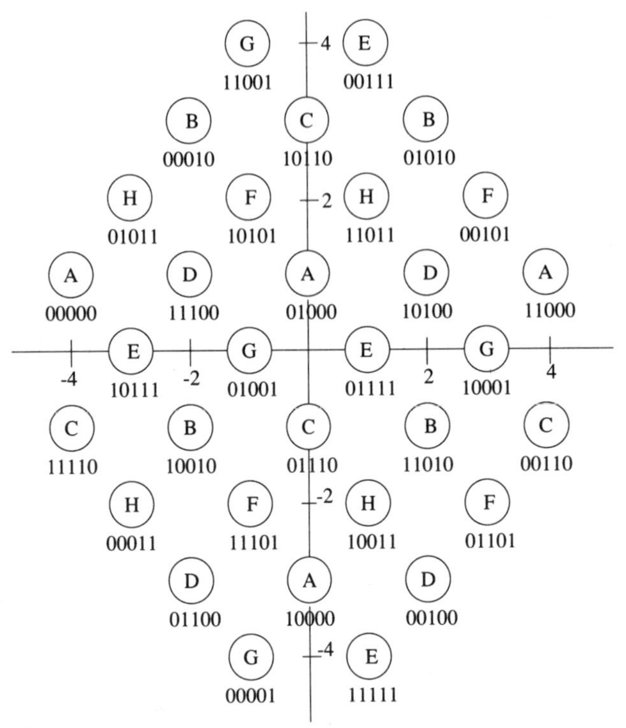

Fig. 5.70 Mapping of $y_4y_3y_2y_1y_0$ (numerical labels) into points in the 32-cross signal constellation as Figure 5.67, rotated 45°. Letters correspond to the eight partition subsets.

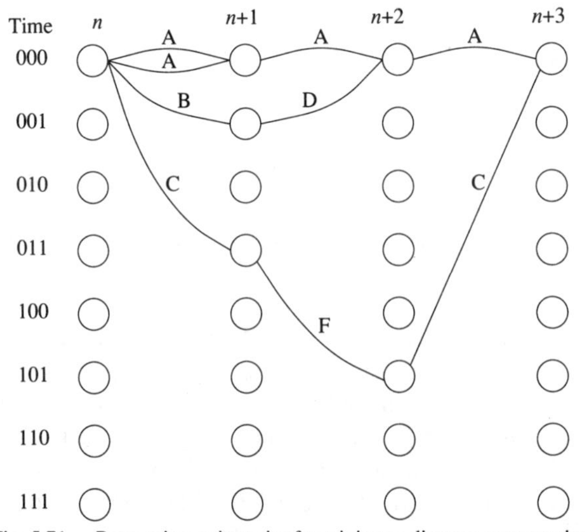

Fig. 5.71 Remerging trajectories for minimum distance computation.

Table 5.7 Convolutional Coder for Figure 5.68

Previous State $S_2S_1S_0$	Input $x_2'(n)x_1'(n)$	Output $y_2(n)y_1(n)y_0(n)$	Subgroup	New State $S_2S_1S_0$
000	00	000	A	000
	01	100	D	010
	10	010	B	001
	11	110	C	011
001	00	100	D	000
	01	000	A	010
	10	110	C	001
	11	010	B	011
010	00	011	H	100
	01	101	F	110
	10	001	G	101
	11	111	E	111
011	00	111	E	100
	01	001	G	110
	10	101	F	101
	11	011	H	111
100	00	010	B	000
	01	110	C	010
	10	000	A	001
	11	100	D	011
101	00	110	C	000
	01	010	B	010
	10	100	D	001
	11	000	A	011
110	00	001	G	100
	01	111	E	110
	10	001	H	101
	11	101	F	111
111	00	101	F	100
	01	011	H	110
	10	111	E	101
	11	001	G	111

in C is four, between a point in A and a point in D is four, and between a point in A and a point in F is two.

The gain of the nonlinear convolutional code is determined by the minimum Euclidean distance between a pair of remerging state trajectories. Fig-

Table 5.8 Squared Distances of Remerging Trajectory Pairs

Trajectory Pairs	Number of Intervals	Squared Distance
A and A	1	16
AA and BD	2	8 + 4 = 12
AAA and CFC	3	4 + 2 + 4 = 10

ure 5.71 shows three such pairs allowed by the trellis of Figure 5.69. Using our earlier observations, the squared distances between different pairs of trajectories are as indicated in Table 5.8. Consideration of longer remerge intervals yields squared distances larger than 10. Thus the minimum squared distance is 10. The coding gain is then

$$Gain = 10\log_{10}\frac{(d^2/P_{av})\,coded}{(d^2/P_{av})\,uncoded} = \log_{10}\frac{(10/10)}{(4/10)} = 4 \text{ dB}, \tag{5.144}$$

where we have noted that both the 32-cross constellation of Figure 5.70 and the uncoded 16-point QAM constellation of Figure 5.44a have average power 10, and the minimum squared distance between signal points in the QAM constellation is four.

The rotational invariance of the nonlinear convolutional coder can be proved, but will only be illustrated here. Suppose the sequence of input (customer) data is

$$\{x_2(n)x_2(n)\}\} = (\ldots,\ 00,\ 00,\ 00,\ \ldots)\ . \tag{5.145}$$

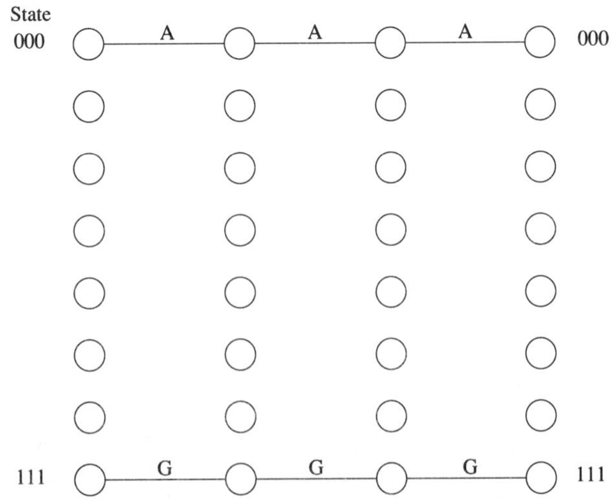

Fig. 5.72 Trajectories ...AAA... and ...GGG... corresponding to transmitted and received signals, respectively, after a 90° phase jump.

Additional customer data that do not drive the convolutional coder, but do select a signal point within a quadrant, can be ignored, since the selection is made independent of quadrant rotations. Then by Table 5.6, the output of the differential encoder is

$$\{x_2'(n)x_1'(n)\} = (..., 00, 00, 00, ...) . \qquad (5.146)$$

With this input to the convolutional coder, the trellis trajectory is the top branch in Figure 5.69, interval after interval, as shown in Figure 5.72. A signal point in subgroup A is selected each time.

Now suppose there is a 90° phase jump in the transmission channel. The subset A is rotated 90° into the signal points designated as the G subset in Figure 5.70. Observing (after the phase jump) one G signal point after another (the result of each transmitted A signal point advancing by 90°), the receiver decides (incorrectly) that the trajectory in the original coding trellis (Figure 5.69) was a succession of *lowest* signal branches, also shown in Figure 5.72. It convolutional coding decision will therefore be the sequence

$$\{x_2'(n)x_1'(n)\} = (..., 11, 11, 11, ...) \qquad (5.147)$$

that would have produced a trajectory in the lattice consisting of one G branch after another. Although this is the incorrect sequence, Table 5.6 indicates, for the differential encoder, that a sequence of inputs

$$\{x_2(n)x_1(n)\} = (..., 00, 00, 00, ...) \qquad (5.148)$$

would have produced (5.147). The receiver makes the decision (5.148) for the customer data, and we see that it is the same as the correct (5.145). The preservation of differential relationships in both the data sequence produced by the differential encoder and the transmitted symbol sequence produced by the convolutional coder leads to correct decisions about customer data despite the 90° phase jump.

Additional intuitive feel for the capability of the encoder to make correct decisions no matter what the input data sequence can be obtained by noting that each subset of Figure 5.70 rotates, in jumps of multiples of 90°, into other subsets. Table 5.9 indicates these rotations for a 90° jump. The trellis of Figure 5.69 is designed so that any allowable sequence of branches rotates into another sequence of branches such that the respective data sequences driving the convolutional coder have the same differential encoding. We already considered one example. As a second example, suppose a trajectory corresponding to convolutional coder input data trajectory ...HFHFHF... were received, corresponding to convolutional coder input data It is apparent again, as the reader can verify from Table 5.6, that the dibits of the second (and incorrect) sequence imply the same (correct) differentially coded customer data.

Table 5.9 Subset 90° Rotation Relationships

A rotates to G
B rotates to H
C rotates to E
D rotates to F
E rotates to A
F rotates to B
G rotates to C
H rotates to D

5.7.4 TRELLIS CODING BASED ON LATTICES AND COSETS

In recent years, a general theory of lattices and cosets has been developed [46–48]. A lattice is an infinitely extended regular array of points, such as all points with integer coordinates in the plane, and a coset is a regular subset, such as all points with even coordinates. The approach to set partitioning has been broadened with this methodology, leading to new codes for transmitted sequences, such as multilevel coding for coset codes. Coset techniques use abstract conceptions of set partitioning, allowing, in particular, creation of a partition subset from as many points in a coset as are needed to realize a desired data rate. This approach makes is possible to work with larger constellations and more complicated lattices.

Other efforts to systematize the design of trellis-coded modulators have focused on the relationship between ordinary convolutional codes and channel trellis codes. Calderbank [51] demonstrates an algorithmic way of finding trellis codes from binary (convolutional) codes, linking Hamming distance with Euclidean distance.

5.7.5 MULTIDIMENSIONAL TRELLIS-CODED MODULATION

There has been a further expansion of channel trellis coding into multidimensional constellations, as presented in independent early work by Forney et. al. [6] and Calderbank and Sloan [49]. Additional coding gains are possible with higher-dimensional signal constellations, which were described in Section 5.5. The four-dimensional, eight-state code described by Wei [44], with a signal constellation of 512 points, yields 4.7 dB gain over an uncoded signal and 0.7 dB gain over a trellis-coded two-dimensional signal. It conveys eight bits of information per four-dimensional symbol. By appropriate partitioning of this constellation, the minimum squared distance of the code is 16, although the minimum squared distance between signal points in a comparable two-dimensional QAM constellation is only 4.

The gain of this coded signal is easily derived, as we now do following [44]. The constellation is defined in Table 5.10. The first two coordinates

Table 5.10 512-Point Four-Dimensional Signal Constellation

Energy	Signal points with all positive signs (1/16 of all signal points)
4	(1111)
12	(3111) (1311) (1131) (1113)
20	(3311) (3131) (3113) (1331) (1133) (1313)
28	(5111) (1511) (1151) (1115)
28	(3331) (3313) (3133) (1333)
36	(5311) (1531) (1153) (3511) (1351) (1135) (5131) (3151) (1513) (1315) (3115) (5113)
36	(3333)

and the last two coordinates are each drawn from the 32-cross constellation of Figure 5.67, but only half of the 1024 possible four-dimensional points are used.

This constellation is partitioned into sixteen subsets denoted $S(\pm 1\ \pm 1\ \pm 1\ \pm 1)$. Subset $S(1111)$ is described in Table 5.11. The other fifteen subsets are generated by changing the signs of signal point components to correspond to the changes in signs in the arguments of $S(----)$. For example, since (–3,111) is an element of S(1111), (–3–111) is an element of S(1–111).

Within a given subset, the minimum squared distance between signal points is 16. This is because the subsets are defined (as is evident from $S(1111)$) so that signal points differing in only one coordinate have a distance of exactly four in that coordinate. For example, (–3–311) and (–3–3–31) differ by four in the third coordinate. Furthermore, the minimum squared distance between any two different subsets,

$$d^2(S(\mathbf{v}),\, S(\mathbf{w})) = \min_{\substack{\alpha \in S(\mathbf{v}) \\ \beta \in S(\mathbf{w})}} \{ ||\alpha - \beta||^2 \}\ , \tag{5.149}$$

is

$$d^2(S(\mathbf{v}),\, S(\mathbf{w})) = ||\mathbf{v} - \mathbf{w}||^2\ . \tag{5.150}$$

This is true because for any differing elements, $v_i \neq w_i$, of $\mathbf{v}$ and $\mathbf{w}$, the corresponding elements of α and β must also differ, and by at least two (or a squared distance of four). The squared distance is exactly four only if $\alpha = \mathbf{v}$ and $\beta = \mathbf{w}$, which have all ± 1 elements.

For a particular trellis code which we will now define, these observations guarantee that the minimum squared distance of an error event is at least 16. This convolutional code, generated in the encoder of Figure 5.73, has the required rate of eight input bits per four-dimensional symbol. The symbol is chosen from the 512 signal points by the nine output bits $(y_0, \ldots, y_8)$. This is

Table 5.11 Partition Subset $S(1111)$

Energy	Signal Points
4	(1111)
12	(–3111) (1–311) (11–3) (111–3)
20	(–3–311) (–31–31) (–311–3) (1331) (11–3–3) (1–31–3)
28	(5111) (1511) (1151) (1115)
28	(–3–3–31) (–3–31–1) (–31–3–3) (1–3–3–3)
36	(5–311) (15–31) (115–3) (–3511) (1–351) (11–35)
	(51–31) (–3151) (151–3) (1–315) (–3115) (511–3)
36	(–3–3–3–3)

fundamentally a coder with a rate of three bits per four-dimensional symbol, augmented with five uncoded bits/symbol. The trellis for this coder is sketched in Figure 5.74, and an example of the 64 parallel branches in each state transition is given in Table 5.12.

To select the symbol (signal point) from the 512 possibilities, replace each "0" in $(y_3 y_2 y_1 y_0)$ with "1", and each "0" with "–1", as illustrated in Table 5.12.

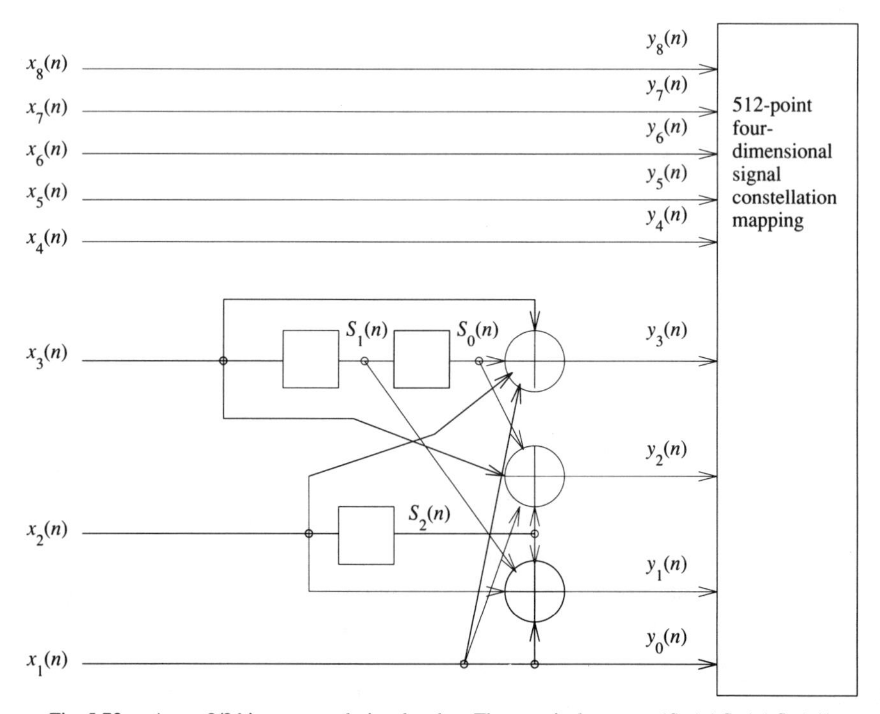

Fig. 5.73 A rate 8/9 binary convolutional coder. The state is the vector $(S_2(n) S_1(n) S_0(n))$.

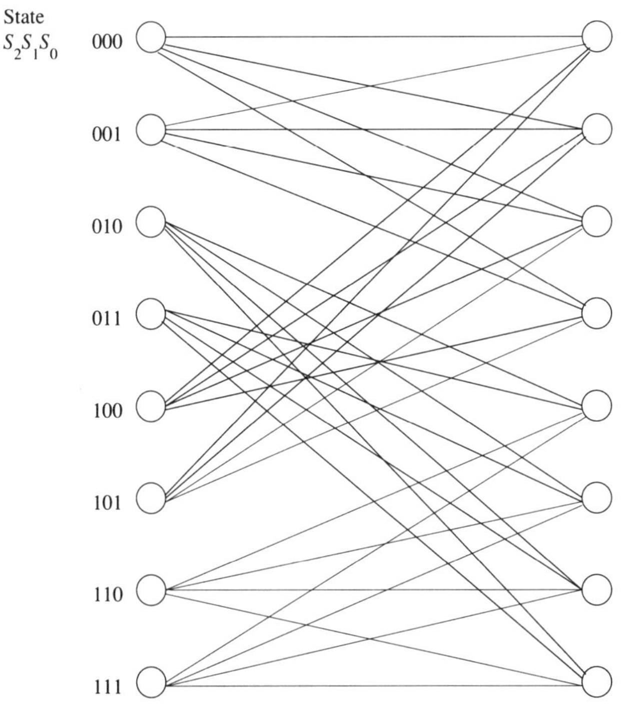

Fig. 5.74 Trellis diagram for coder of Figure 5.73. Each state transition is defined by the two input bits x_3x_2, but because of the six input bits $x_8x_7x_6x_5x_4x_1$ that do not affect the state, each branch is actually 64 parallel transitions.

This vector determines the partition subclass $S(y_3y_2y_1y_0)$, and the remaining output bits $(y_8y_7y_6y_5y_4)$ are arbitrarily assigned to symbols within that subclass, as described earlier in Figure 5.73. Consider a correct trajectory (assumed to be the sequence of topmost branches) and three possible erroneous trajectories in the lattice, illustrated in Figure 5.75. The first incorrect trajectory represents another signal in a subset and thus occupies the same branch as the correct (topmost) branch. It rejoins the correct trajectory in one symbol interval. The

Table 5.12 Illustration of 64 Parallel Branches from State (000) to State (000) (Computed from Figure 5.73)

State $S_{(n)}$	$x_3x_2x_1$	$y_ey_2y_1y_0$	Subclass	State $S(n+1)$
000	000	0000	$S(1111)$	000
000	001	1111	$S(-1-1-1-1)$	000

[These two branches exist for any particular choice of $x_8x_7x_6x_5x_4$, yielding 64 parallel branches in all.]

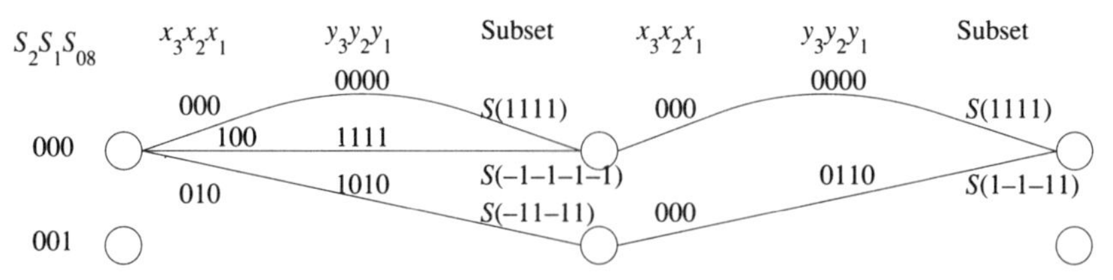

Fig. 5.75 Minimal separations between correct and two possible erroneous trajectories. The lower six states of the lattice are omitted.

correct and incorrect trajectories are parallel arms with separation $d^2 = 16$. The second incorrect trajectory is a parallel branch between the same states as the correct trajectory, but representing a signal point in a different partition subset. It also rejoins the correct one in one symbol intervals. The subset arguments between correct and incorrect branches are (1111) and (–1–1–1–1) respectively. By (5.150), the minimum squared distance between the correct and incorrect branch is equal to the squared distance of 16 between these arguments.

In the third erroneous trajectory, which rejoins the correct one two symbol intervals later, as shown by the lower example in Figure 5.75, the minimum squared distance is the sum of the minimum squared distances accumulated in each symbol interval. The minimum squared distance in each symbol interval is eight, since the arguments of the respective subsets differ in two digits and a squared distance of four is contributed from each differing digit. Thus the error trajectory again is a squared distance 16 from the correct trajectory. Longer erroneous trajectories will accumulate squared distances of at least four in each symbol interval (from subclasses differing in at least one element) and take at least four symbol intervals to rejoin the correct trajectory. Thus the minimum squared coding distance is 16.

The coding gain is

$$Gain = 10\log_{10}\frac{\left[d_{\min}^2/P\right]coded\ 512}{\left[d_{\min}^2/P\right]uncoded\ QAM} \tag{5.151}$$

The appropriate comparison is with a two-dimensional 16-point QAM signal (Figure 5.27) which conveys 4 bits per QAM symbol interval, the QAM symbol interval being half that of the four-dimensional symbol interval (for the same bandwidth). With $d=2$ spacing of signal points in the QAM constellation, $d_{\min}^2 = 4$ and the average power per symbol interval is 10. Thus $P=20$ for the two consecutive QAM symbols occurring in the same time as one four-dimensional symbol, so that is the value to be substituted into (5.151). For the 512-point four-dimensional signal constellation, the average power is equal to the average squared value of the 32 signal points in Table 5.10, or $P=27$. Thus

$$Gain = 10\log_{10}\left[\frac{16/27}{4/10}\right] = 4.717 \text{ dB} . \tag{5.152}$$

L. F. Wei [43] discusses implementation issues and a general overview is offered in the paper by Forney and Wei [50]. By coding over more than one symbol interval, a smaller expansion in the size of the signal set can yield the same coding gains as in two-dimensional (one symbol interval) coding. Moreover, 90° phase invariance can be obtained with linear codes, using differential coding on uncoded bits, as derived by Wei [46]. Trellis-coded multidimensional phase modulation, yielding asymptotic coding gains approaching 6 dB, have been described by Wei [47] and by Pietrobon, Deng, Lafanechere, Ungerboeck, and Costello [48]. This is an active research area, illustrating the gains still to be made through more sophisticated uses of coding and modulation techniques.

5.8 CONCLUSION

Passband communication will always be central to physical telecommunications, in channels as diverse as the telephone channel, the digital subscriber line, radio channels, and optical systems using wavelength division multiplexing and subcarrier modulation techniques. Although some applications appear at first to not require consideration of bandwidth conservation, the need inevitably arises. The techniques described here for effective communication in bandlimited channels will have utility into the future.

REFERENCES

[1] L. Franks, *Signal Theory*, Prentice-Hall, 1969.

[2] J. G. Proakis, *Digital Communications*, McGraw-Hill, 1983.

[3] S. Benedetto, E. Biglieri, and V. Castellani, *Digital Transmission Theory*, Prentice-Hall, 1987.

[4] G. Foschini, R. Gitlin, and S. Weinstein, "Optimization of Two-dimensional Signal Constellations in the Presence of Gaussian Noise," *IEEE Trans. on Communications,* Vol. COM-22, No. 1, January 1974.

[5] G. Foschini, R. Gitlin, and S. Weinstein, "On the Selection of Two-Dimensional Signal Constellations in the Presence of Phase Jitter and Gaussian Noise," *Bell System Technical J.,* Vol. 52, pp. 927–965, July–August 1973.

[6] D. G. Forney et al., "Efficient Modulation for Band-Limited Channels," *IEEE J., Selected Area in Communications,* SAC-2, pp. 632–647, August 1984.

[7] I. Kalet and S. Weinstein, "Inband Generation of Synchronous Linear Data Signals," *IEEE Trans. on Communications,* Vol. COM-21, No. 10, pp. 1116–1122, October 1973.

[8] J. L. Holsinger, "Digital Communication Over Fixed Time Continuous Channels with Memory," PhD Thesis, MIT, 1964.

[9] E. Powers and M. Zimmerman, "TADIM—A Digital Implementation of a Multichannel Data Modem," *Proc. IEEE ICC* '68.

[10] R. W. Chang, "High-Speed Multichannel Data Transmission with Band-limited Orthogonal Signals," *Bell System Technical J.,* Vol. 45, pp. 1775–1796, December 1966.

[11] S. Darlington, "On Digital Single-Sideband Modulators," *IEEE Trans. Circuit Theory*, Vol. OT-17, pp. 409–414, August 1970.

[12] B. R. Saltzberg, "Performance of an Efficient Parallel Data Transmission System," *IEEE Trans. on Communications,* Vol. COM-15, pp. 805-811, December 1967.

[13] S. B. Weinstein and P. M. Ebert, "Data Transmission by Frequency-Division Multiplexing Using the Discrete Fourier Transform," *IEEE Trans. on Communications,* Vol. 19, No. 5, October 1971.

[14] B. Hirosaki, "An Orthogonal Multiplexed QAM System Using the Discrete Fourier Transform," *IEEE Trans. on Communications*, Vol. COM, No. 7, pp. 982–989, July 1981.

[15] I. Kalet, "The Multitone Channel," *IEEE Trans. on Communications,* Vol. 37, No. 2, pp. 119–124, February 1989.

[16] E. Feig and A. Nadas, "Practical Aspects of DFT-Based Frequency Division Multiplexing for Data Transmission," *IEEE Trans. on Communications,* Vol. 38, No. 7, pp. 929–937, July 1990.

[17] G. P. Dudevoir, J. S. Chow, J. M. Cioffi, and S. Kasturia, "Combined Equalization and Coding for T1 Data Rates on Carrier Servicing Area Subscriber Loops," *Proc. IEEE ICC '89,* Boston, p. 5.36–5.40, June 1989.

[18] J. W. Lechleider, "The Optimum Combination of Block Codes and Receivers for Arbitrary Channels," *IEEE Trans. on Communications*, Vol. 38, No. 5, pp. 6.15–6.21, May 1990.

[19] S. Kasturia, J. T. Aslanis, and J. M. Cioffi, "Vector Coding for Partial Response Channels," *IEEE Trans. Information Theory*, Vol. 36, No. 4, pp. 741–762, July 1990.

[20] W. Spier, "Trailblazer modems: High-Powered Technology in a Low-Cost Package," *J Data & Computer Communications*, Vol. 1, No. 2, pp. 47–54, Fall.

[21] J. A. C. Bingham, "Multicarrier Modulation for Data Transmission: An Idea Whose Time Has Come," *IEEE Communications Magazine*, pp. 5–14, May 1990.

[22] A. Ruiz, J. Cioffi, and S. Kasturia, "Discrete Multiple Tone Modulation with Coset Coding for the Spectrally Shaped Channel," to be published, *IEEE Trans. on Communications*.

[23] A. Ruiz, PhD dissertation, Stanford University, June, 1989.

[24] A. Peled and A. Ruiz, "Frequency Domain Data Transmission Using Reduced Computational Complexity Algorithms," *IEEE Int. Conf. on ASSP*, Denver, pp. 964–967, April 1980.

[25] A. Gersho, and V. B. Lawrence, "Multidimensional Signal Constellations for Voiceband Data Transmission," *IEEE J. Selected Areas in Communications*, Vol. SAC-2, No. 5, pp. 687–702, September 1984.

[26] K. S. Shanmugan, *Digital and Analog Communication Systems*, Wiley, 1979.

[27] R. Lucky, J. Salz, and E. J. Weldon, Jr., *Principles of Data Communication*, McGraw-Hill, New York, 1968.

[28] H. Taub and D. L. Schilling, *Principles of Communication Systems*, McGraw-Hill, 2nd ed., 1971.

[29] C. E. Sundberg, "Continuous Phase Modulation," *IEEE Communications Magazine*, Vol. 24, No. 4, pp. 25–38, April 1986.

[30] J. Aulin and C. E. Sundberg, "Continuous Phase Modulation—Parts I and II," *IEEE Trans. on Communications*, Vol. COM-29, No. 3, pp. 196–255, March 1981.

[31] N. A. Zervos and I. Kalet, "Optimized Decision Feedback Equatization vs. Optimized Orthogonal Frequency Division Multiplexing for High-Speed Data Transmission over the Local Cable Networks," *Proc. ICC '89*, pp. 10.80–10.85, Boston, June 1989.

[32] S. A. Rhodes, "Performance of Offset QPSK Communications with Partially Coherent Detection," *Nat. Telecommun. Conference Record*, November 1973.

[33] R. D. Gitlin and E. Y. Ho, "The Performance of Staggered Quadrature Amplitude Modulation in the Presence of Phase Jitter," *IEEE Trans. of Communications*, Vol. COM-23, No. 3, pp. 348–352, March 1975.

[34] I. Kalet, "A Look at Crosstalk in Quadrature-Carrier Modulation Systems," *IEEE Trans. Communications,* Vol. 25, No. 9, pp. 884–897, September 1977.

[35] T. Aulin, and C. E. Sundberg, "Exact Asymptotic Behavior of Digital FM Spectra," *IEEE Trans. Communications,* Vol. COM-30, pp. 2438–2449, 1982.

[36] G. L. Pietrobon, S. G. Pupolin, and G. P. Tronca, "Power Spectrum of Angle Modulated Correlated Digital Signals," *IEEE Trans. Communications,* Vol. COM-30, pp. 389–395, February 1982.

[37] S. G. Wilson and R. C. Gauss, "Power Spectra of Multi-Phase Codes," *IEEE Trans. Communications,* Vol. COM-29, pp. 250–256, March 1981.

[38] G. J. Garrison, "A Power Spectral Density Analysis for Digital FM," *IEEE Trans. Communications,* Vol. COM-23, pp. 1228–1243, November 1975.

[39] A. J. Viterbi and J. K. Omura, *Principles of Digital Communications and Coding,* New York, McGraw-Hill, 1979.

[40] G. Ungerboeck, "Channel Coding with Multilevel/Phase Signals," *IEEE Trans. Information Theory,* Vol. IT-28, No. 1, pp. 55–67, January 1982.

[41] G. Ungerboeck, "Trellis-Coded Modulation with Redundant Signal Sets Part 1: Introduction," *IEEE Communications Magazine,* Vol. 25, No. 2, pp. 5–11, February 1987.

[42] G. Ungerboeck, "Trellis-Coded Modulation with Redundant Signal Sets Part 2: State of the Art," *IEEE Communications Magazine,* Vol. 25, No. 2, pp. 12–21, February 1987.

[43] A. J. Viterbi, "Error Bounds for Convolutional Codes and Asymptotically Optimum Decoding Algorithm," *IEEE Trans. Information Theory,* Vol. IT-13, No. 4, pp. 260–269, April 1967.

[44] L. F. Wei, "Rotationally Invariant Convolutional Channel Coding with Expanded Signal Space — Part I and II," *IEEE J. Selected Areas in Communications,* Vol. SAC-2, No. 5, September 1984.

[45] L F. Wei, "Rotationally Invariant Trellis-Coded Modulations with Multidimensional M-PSK," *IEEE J. Selected Areas in Communications,* Vol. SAC-7, No. 9, pp. 1281–1295, December 1989.

[46] A. R. Calderbank and N. J. A. Sloan, "New Trellis Codes Based on Lattices and Cosets," *Trans. Information Theory,* Vol. IT-33, pp. 177–195, March 1987.

[47] G. D. Forney, "Coset Codes — Part I: Introduction and Geometric Classification," *IEEE Trans. Information Theory,* Vol. 34, pp. 1123–1151, September 1988.

[48] G. D. Forney, "Coset Codes II: Binary Lattices and Related Codes," *IEEE Trans. Information Theory,* Vol. IT-34, pp. 1152–1187, September 1988.

[49] A. R. Calderbank and N. J. A. Sloan, "Four-Dimensional Modulation with an Eight-Stage Trellis Code," *AT&T Technical J.,* Vol. 64, No. 5, pp. 1005–1018, May-June 1985.

[50] G. D. Forney and L. F. Wei, "Multidimensional Signal Constellations — Part I: Introduction, Figures of Merit, and Generalized Cross Constellations," *IEEE J. Selected Areas in Communications,* Vol. 7, No. 6, pp. 877–892, August 1989.

[51] A. R. Calderbank, "Multilevel Codes and Multistage Decoding," *IEEE Trans. Communications,* Vol. 37, No. 2, pp. 222–229, March 1989.

[52] S. Pietrobon, R. Deng, A. Lafanechere, G. Ungerboeck, and D. Costello, "Trellis-Coded Multidimensional Phase Modulation," *IEEE Trans. Information Theory,* Vol. 36, No. 1, January 1990.

[53] L. F. Wei, "Trellis-Coded Modulation with Multidimensional Constellations," *IEEE Trans. on Information Theory,* Vol. IT-33, pp. 483–501, July 1987.

[54] E. Biglieri, D. Divsalar, P. J. McLane and M. K. Simon, *Introduction to Trellis Coded Modulation with Applications,* Macmillan, New York, 1991.

EXERCISES

5.1: Show that if $m_i(t)$ and $m_q(t)$ are real baseband signals bandlimited to $|f|<f_c$, then

$$\check{s}(t) = m_i(t)\sin 2\pi f_c t + m_q(t)\cos 2\pi f_c t$$

is the Hilbert transform of

$$s(t) = m_i(t)\cos 2\pi f_c t - m_q(t)\sin 2\pi f_c t\ .$$

5.2: (a) Show that $\check{n}(t)$, the Hilbert transform of a wide sense stationary noise process, is itself stationary.

(b) Show that the complex analytic passband process, $\tilde{n}_p(t) = n(t) + j\check{n}(t)$ is a wide sense stationary process whose autocorrelation function is given by

$$\tilde{\phi}_{pp}(\tau) = 2\phi_{nn}(\tau) + \check{\phi}_{nn}(\tau) ,$$

where $\phi_{nn}(\tau)$ is given by (5.28a) and $\check{\phi}_{nn}(\tau)$ is its Hilbert transform.

5.3: Show the relations of (5.29a, b) and (5.30). Start with the observation that the passband process of the previous exercise and the baseband process defined in (5.27b) are related by $\tilde{n}_B(t) = \tilde{n}_p(t)\exp(-j2\pi f_c t)$. Use the results of the previous exercise and the relations $\mathrm{Re}(Z) = (Z + Z^*)/2$ and $\mathrm{Im}(Z) = (Z - Z^*)/2j$.

5.4: Prove that

$$2F\left\{\mathrm{Re}[\tilde{p}(t) \ e^{j2\pi f_c t}]\right\} = P(f-f_c) + P(-f-f_c) ,$$

where $P(f) = F\{p(t)\}$.

5.5: What is the spectral efficiency for 8-phase signaling using the Manchester pulse shown below in each symbol interval?

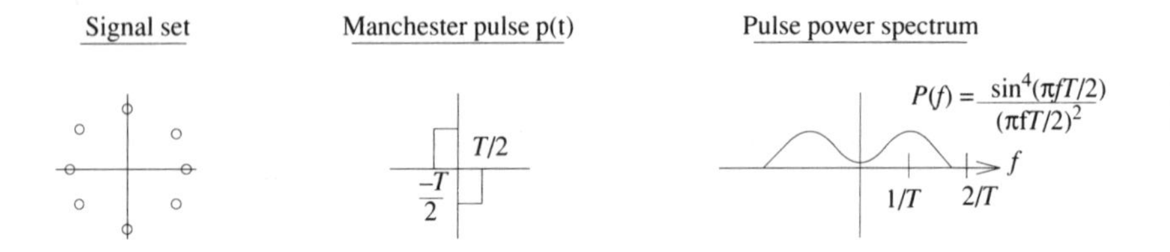

5.6: For the two signal constellations shown below, compare the average power requirement for the high signal-to-noise ratio case.

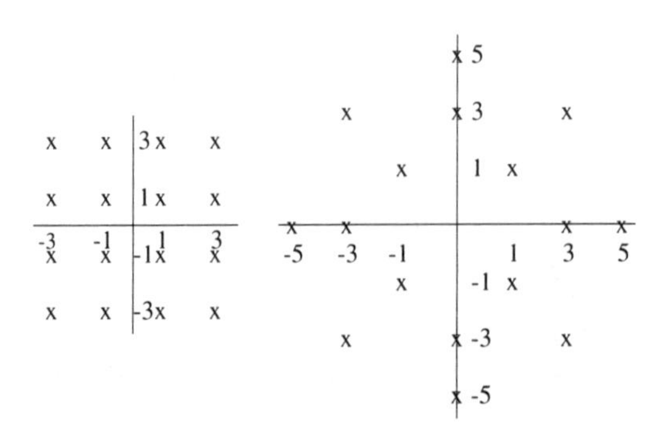

5.7: Starting from (5.82a), show that the probability of error, for a QAM constellation with M^2 signal points, is given by (5.82b).

5.8: Consider the eight-point signal constellation in Figure 5.20. Suppose that the only impairment is phase jitter which may be modeled as generated by a first-order phase-locked loop with the same center frequency as the carrier. The probability density of the phase error is given by (6.46).

(a) Write down a complete expression for the probability of error.

Now suppose that the impairment is only white Gaussian noise with independent in-phase and quadrature components.

(b) Write down an expression for the probability of error.

5.9: What are the optimum decision boundaries for the signal constellation shown on Figure 5.33 (h)? Assume that the signal-to-noise ratio is high enough that errors are made between adjacent points only.

5.10: Assume that the useful band of the voice telephone channel extends from 200 to 3200 Hz. We wish to transmit over this channel at a rate of 9600 bps using QAM. Assume that the carrier is at 1700 Hz. Also, assume that L, the number of points in the signal constellation, is a power of 2.

(a) Find the Nyquist frequency for $L = 2$, 4, 8, 16, and 32.

(b) For each of these values of L, find the maximum rolloff so that the signal stays in the band.

As will be seen in Chapter 6, decreasing the excess bandwidth of a signal increases the complexity of the timing recovery circuitry. Increasing the number of signal levels also increases cost since levels that are closer together must be distinguished. These factors are crudely reflected in the following formula for the relative cost of alternative implementations:

$$C \sim 2/\alpha + L/10,$$

where α is the rolloff.

c) Find α and L such that the cost as expressed in the above formula is minimum.

5.11: What is the Gray coding of a 16-point QAM constellation that results in the minimum number of bit errors per symbol error? Indicate 4-bit codeword for each signal point. What is the number of bit errors/symbol error for this assignment?

5.12: Assume a block of $N=7$ and a channel memory of $2T/N$ (i.e., $M=3$), where T is the pulsing (symbol) interval in the parallel data transmission system of Section 5.4. What are the 7 outputs of the cyclic convolution of the transmitted vector $\tilde{\mathbf{x}} = (\tilde{x}_0, \ldots, \tilde{x}_6)$ and channel impulse response $\tilde{\mathbf{h}} = (\tilde{h}_0, \tilde{h}_1, \tilde{h}_2)$?

5.13: Show the alternative binary FSK waveforms on the interval $[0,T]$,

$$s_{\pm}(t) = \cos[2\pi f_c t + 2\pi f_d t + \theta_0], \quad f_c = n/2T > f_d,$$

are orthogonal only for $\Delta = 2f_d T = k/2$, k an integer ≥ 1. Minimum-shift corresponds to $k=1$.

5.14: Derive the power disadvantage (in dB) from coding 2 bits/symbol interval into (one-dimensional) PAM with that from coding 2 bits/symbol into two-dimensional (four signal points) QAM with the same probability of symbol error. Assume, as a valid approximation for high SNR in a Gaussian noise channel equal minimum distance between signal points is equivalent to equal symbol error probability.

5.15: To illustrate the potential power gain of multidimensional signaling, consider the following comparison of a 2D with a 4D system, assuming a symbol interval $T=1$ second and equally likely signal points.

4 signal points:

1	1
1	−1
−1	1
−1	−1

Bits/symbol interval=2

16 signal points in 2 symbols

0	0	0	0	1	−1	−1	−1
1	1	1	1	−1	1	1	1
1	1	1	−1	−1	1	1	−1
1	1	−1	−1	−1	1	−1	1
1	1	−1	1	−1	−1	−1	−1
1	−1	1	1	−1	−1	1	1
1	−1	1	−1	−1	−1	1	−1
1	−1	−1	1	−1	−1	−1	1

Bits/symbol interval=2

(1) What is the minimum (Euclidean) distance between signal-point pairs for the 2D constellation and for the 4D constellation?

(2) What is the average power for each constellation?

(3) What is the power gain (in dB) of the 4D constellation?

5.16: Derive the explicit lookup table matrix for generation of the complex level output D_n (for inband signal generation) from the complex level input $(a_n + jb_n)$, where:

$$a_n = \pm 1, \qquad b_n = \pm 1$$

$$f_c = 2000 \text{ Hz}$$

$$T = 1/2400 \text{ sec}.$$

6

Synchronization: Carrier and Timing Recovery

6.0 INTRODUCTION

The theme of this chapter might well be "...timing is everything." In the course of our discussion in Chapter 4, we saw that the detection of a baseband digital data sequence presumed proper timing at the receiver. (See Section 4.10, particularly.) The same requirement for timing is present in the detection of passband signals; however, as we saw in Section 5.2, carrier phase coherency is also necessary. The roles of each of these synchronization subsystems are shown in Figure 6.1a. The carrier tracking system provides an estimate of the received carrier phase $\hat{\theta}$, while the timing recovery system provides an estimate of the proper sampling epoch to the receiver sampling system $\hat{\tau}$. The effect of a poorly designed carrier loop will be to increase the dispersion of the received symbols about their nominal values, bringing the received points considerably closer to the decision boundaries and decreasing the margin against an error (caused say by a noise burst); of course, large phase perturbations can cause errors *without* any noise. In Figure 6.1b we show how the transmitted symbol s_1 is rotated by phase jitter to the point u, and then further distorted by noise to the point z; note that the received point is within the decision region associated with s_2 so that an error will be made. Similarly, timing phase errors will cause the receiver to sample away from the maximum eye opening, and reduce the margin for error. It is the purpose of this chapter to present the various aspects of synchronization.

We will begin by assuming that these phases are constant (but unknown). This will facilitate the derivation of maximum-likelihood (ML) phase estimates. In order to treat the more common situation of time-varying phases, we will need to depart from ML estimation and introduce *data-directed* estimates that exploit the digital nature of the information-bearing signal.

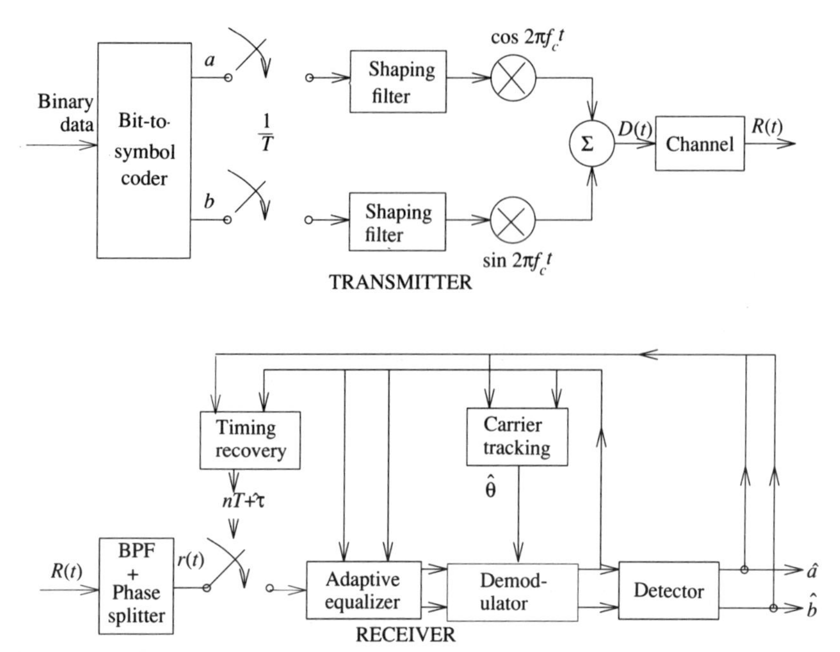

Fig. 6.1a An in-phase and quadrature data transmission system. The feedback loops are shown for adjusting the equalizer, carrier, and timing control loops.

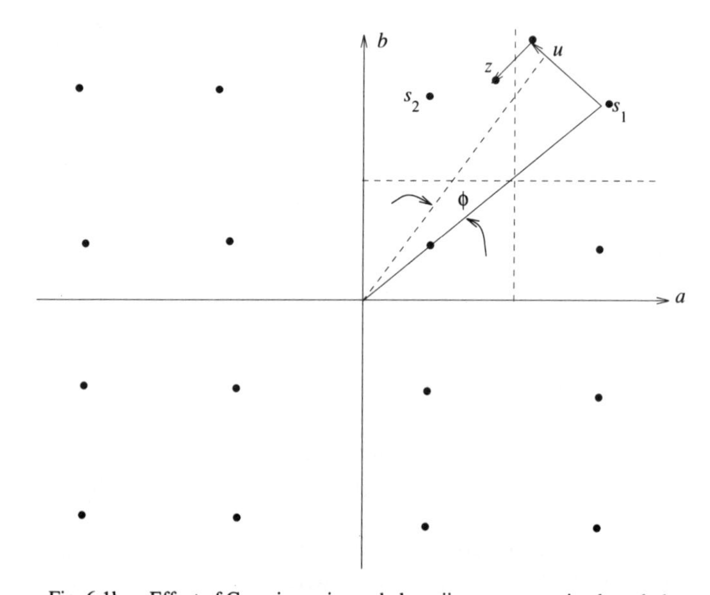

Fig. 6.1b Effect of Gaussian noise and phase jitter on transmitted symbol.

There are many common threads in the recovery of carrier and of timing. In both cases, we distinguish two fundamental modes: *acquisition and tracking*. In the acquisition or startup mode the task is to rapidly and accurately estimate the parameters pertinent to synchronization. Often this is carried out with the aid of special predetermined transmitted sequences. In the tracking or steady-state mode after acquisition, estimates are derived from the modulated data signal. The use of separate pilot-tones to convey carrier and timing phase has the disadvantages of using up power and introducing harmonic distortion when channel nonlinearities are present. In the tracking mode, the synchronization system is required to strike a balance between narrow bandwidth to suppress fluctuations, or noise, and wide bandwidth to follow variations in the parameter being estimated. A complicating factor is modulation by the data signal, and one measure of the effectiveness of a synchronization system is its performance relative to that attainable in the absence of modulation.

A common component of synchronization systems, carrier phase or timing recovery in either acquisition or tracking modes, is the *phase locked loop* (PLL). The PLL provides extremely narrow band filtering, so that the output phase estimate of a carrier of a timing signal locks to the input phase. As we shall see, the PLL arises naturally as an approximation to the maximum likelihood estimate of the phase. The ML estimate of phase and the PLL are treated in Sections 6.1 and 6.2, respectively.

Among synchronization systems, a distinction can be made between *open loop* and *closed loop* systems. In the context of synchronization, a closed loop system uses a quantity available at the system output to update the current output. Open loop systems process the received signal to generate the desired estimate without explicit use of the system output. The PLL is an example of a closed loop system.

A particular form of closed loop system is *data directed*. As the name implies, in this technique explicit use is made of prior data decisions in order to estimate the current phase (which is then used to make the current decision). Experience has shown that, for the conditions under which systems usually operate, occasional errors have a negligible effect on system performance and superior estimates are obtained. Non-data-directed and data-directed carrier recovery are treated in Sections 6.3 and 6.4, respectively.

Timing recovery is treated in Section 6.5. As in the case of the PLL, it can be shown that the standard approaches, either open or closed loop, can be related to the ML estimate. Joint carrier phase and timing recovery are the subject of Section 6.6.

As mentioned above, carrier and timing are derived from data modulated signals. In order for these operations to be carried out independently of the particular data sequence being transmitted, a device called the *scrambler* is deployed. The scrambler insures that the line signal contains enough transitions for synchronization. We take particular interest in the most common implementation, the *self-synchronizing* scrambler. This material is contained in Section 6.7.

We begin our discussion by recalling that in previous chapters on baseband and passband transmission, it was assumed that timing and carrier phase (and frequency) were available without error. Up until now, for the QAM signal defined in Chapter 5,

$$y(t) = \sum_n a_n h(t-nT+\tau) \cos\left[2\pi f_c t+\theta_c\right] - \sum_n b_n h(t-nT+\tau) \sin\left[2\pi f_c t+\theta_c\right] , \tag{6.1}$$

where a_n and b_n are the in-phase and quadrature data symbols, and $h(t)$ is the impulse response presented to the receiver, we have assumed that the timing phase (τ) and frequency ($1/T$) and the carrier phase, θ_c, and frequency, f_c, were known to the receiver. In this chapter we discuss several techniques for acquiring and tracking these parameters, as well as assessing the penalty, or system sensitivity, to the inevitable residual errors: for example, suppose we obtain a carrier phase estimate that is not precisely equal to the actual phase. We would like to know the error rate of the QAM system with a residual phase error. Such information will help us to decide, within the class of double sideband modulation, if we want to choose a signal constellation that is more immune to phase jitter than the square-QAM signal set (defined in Chapter 5), but which is perhaps more sensitive to Gaussian noise. Or, if we choose the square-QAM modulation we must recognize that we need a high-quality PLL to reduce the phase error to a relatively small value. Furthermore, the different sensitivities to timing and carrier phase errors of double sideband QAM and single sideband systems (SSB) will be described, and may, in certain circumstances, provide a rationale for deciding between these two classes of systems. While not explicitly treated in this text, the sensitivity of high-performance trellis coded modulated (TCM) systems to carrier phase errors is a performance issue that must not be overlooked; also, the fact that TCM systems produce decisions with a delay, can necessitate modifications of data-directed, carrier recovery systems.

6.1 OPTIMUM (MAXIMUM LIKELIHOOD) CARRIER PHASE ESTIMATION

A key component in carrier phase and timing phase acquisition and tracking is the phase-locked loop. We shall be discussing several realizations of this device; however, in all cases its basic function is to estimate the phase of a signal immersed in noise. In order to motivate the form of the PLL, we derive the optimum estimator of the phase of a constant-amplitude carrier immersed in white Gaussian noise. In a practical system where the carrier is modulated by data, this simple situation can be often approached by a limiting and filtering. For this model the received signal may be written as

$$y(t;\theta) = A \cos(2\pi f_c t + \theta_c) + n(t); \; 0 \le t \le KT, \tag{6.2}$$

where the observation interval, KT, is a large number of symbol intervals. We shall assume that the (optimum) estimate is obtained by maximizing the likelihood of the received signal, $y(t)$, with respect to θ. It can be shown (see Exercise 6.3) that a Bayesian estimate based on a quadratic cost function and with θ_c uniformly distributed in $(-\pi,\pi)$ leads to the same criterion (see Section 2.2). For the moment, we will assume that θ_c is a constant (but unknown) phase angle.

Recalling the expression for the likelihood of a signal in Gaussian noise (see Chapter 2), we write the desired likelihood function for the unknown phase as

$$L(\hat{\theta};\theta) = L[y(t) \mid \theta = \hat{\theta}] = e^{\frac{1}{2N_0}\int_0^{T_0}\left[y(t) - A\cos(2\pi f_c t + \hat{\theta})\right]^2 dt}, \tag{6.3a}$$

where T_0 is the *observation interval*, and the unknown phase, θ, is embodied in $y(t)$ [we use the shorthand expression $y(t)$ in place of $y(t;\theta)$]. The observation interval is an important parameter in the design of an estimator because it characterizes the amount of allowable delay, and it is inversely proportional to the loop bandwidth for the parameter tracking schemes we discuss. The maximum likelihood (ML) estimate, $\hat{\theta}_{ML}$, is obtained by minimizing the exponent in (6.3a), or by setting the derivative of

$$l(\hat{\theta}) \equiv \int_0^{T_0}\left[y(t) - A\cos(2\pi f_c t + \hat{\theta})\right]^2 dt, \tag{6.3b}$$

with respect to $\hat{\theta}$, equal to zero; performing the differentiation, we get the integral relationship

$$\dot{l}(\hat{\theta}) = \int_0^{T_0}\left[y(t)\sin(2\pi f_c t + \hat{\theta}_{ML}) + \sin(4\pi f_c t + 2\hat{\theta}_{ML})\right] dt = 0, \tag{6.4}$$

where the dot denotes differentiation with respect to $\hat{\theta}$. If the integration interval is arbitrarily large, then the second term under the integral will approach zero, and we find that the maximum likelihood estimate for the carrier phase, $\hat{\theta} = \hat{\theta}_{ML}$, satisfies the implicit equation

$$\dot{l}_K(\hat{\theta}) = \int_0^{T_0} y(t)\sin(2\pi f_c t + \hat{\theta}_{ML})\, dt = 0, \tag{6.5}$$

where the subscript K denotes an observation interval $T_0 = KT$ seconds. In Figure 6.2a we show a closed-loop mechanization of $\hat{\theta}_{ML}$, where the box labelled VCO, which will be described in more detail in Section 6.2, generates a sinusoid whose phase is proportional to the error signal. In Section 6.2, it will be shown that Figure 6.2a is representative of the *phase-locked loop* (PLL) class

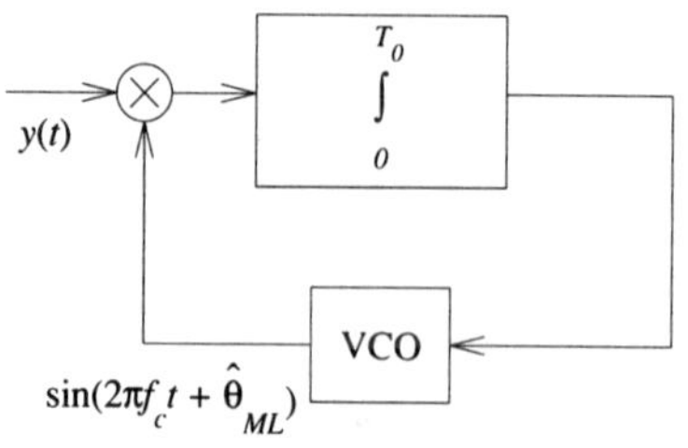

Fig. 6.2a Closed-loop mechanization of system for determining $\hat{\theta}_{ML}$.

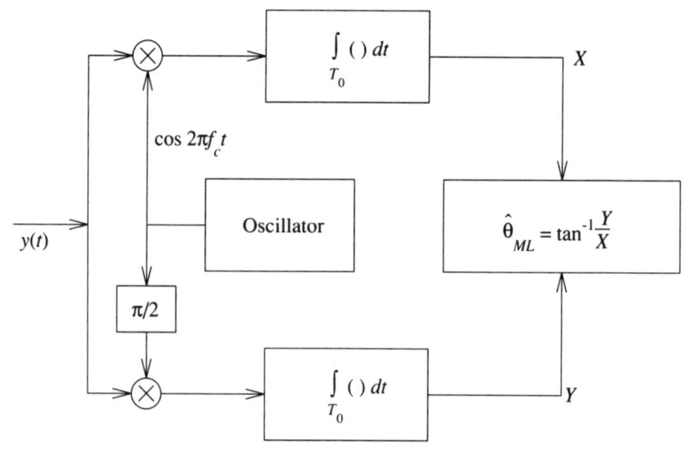

Fig. 6.2b An open-loop system for determining $\hat{\theta}_{ML}$ of an unmodulated carrier.

of tracking systems.

$$\hat{\theta}_{ML} = -\tan^{-1} \frac{\int_0^{T_0} y(t) \sin 2\pi f_c t \, dt}{\int_0^{T_0} y(t) \cos 2\pi f_c t \, dt} . \tag{6.6}$$

Note that as the noise vanishes, $\hat{\theta}_{ML} \to \tan^{-1} \tan\theta = \theta$. In Figure 6.2b we show an open-loop implementation of (6.6). To implement the optimum estimator, this configuration must be abandoned when the carrier phase varies with time.

Before we move on to more practical receivers, we can obtain an estimate of the variance of the ML phase estimate by expanding $y(t;\theta)$ in a first-order Taylor series about $\theta = \hat{\theta}_{ML}$. Expanding $y(t;\theta)$ in this manner we have

$$y(t;\theta) = y(t;\hat{\theta}_{ML}) + (\hat{\theta}_{ML} - \theta) \frac{\partial y(t;\hat{\theta}_{ML})}{\partial \hat{\theta}_{ML}} , \tag{6.7}$$

and multiplying each side of (6.7) by $\sin(2\pi f_c t + \hat{\theta}_{ML})$ and integrating over the observation interval, we have from (6.5) that

$$0 = \int_0^{T_0} y(t;\theta) \sin(2\pi f_c t + \hat{\theta}_{ML})\, dt = \int_0^{T_0} y(t;\hat{\theta}_{ML}) \sin(2\pi f_c t + \hat{\theta}_{ML})\, dt$$

$$+ (\hat{\theta}_{ML} - \theta) \int_0^{T_0} A \sin^2(2\pi f_c t + \hat{\theta}_{ML})\, dt \,. \tag{6.8}$$

Substituting (6.2) into (6.8) and retaining the nonzero terms gives

$$\hat{\theta}_{ML} - \theta = \frac{\int_0^{T_0} n(t) \sin(2\pi f_c t + \theta)\, dt}{A \int_0^{T_0} \sin^2(2\pi f_c t + \theta)\, dt} , \tag{6.9}$$

and letting the spectral density of $n(t)$ be N_0, the variance of the estimation error is given by

$$E(\hat{\theta}_{ML} - \theta)^2 = \frac{N_0}{A^2 T_0} \,. \tag{6.10}$$

Note that the variance of the error is *inversely* proportional to both the observation interval (T_0), and the signal-to-noise ratio (A^2/N_0), which is an intuitively pleasing result. We shall see a similar result in connection with PLLs.

A useful performance measure for ML estimates of any parameter is the Cramér–Rao lower bound on the variance of the estimate [1]. For unbiased estimates, i.e., $\hat{\theta} = \theta$, the Cramér–Rao lower bound is

$$Var[\hat{\theta}(\mathbf{Y}) - \theta] = E[\hat{\theta}(\mathbf{Y}) - \theta]^2 \geq \left[E\left\{ \left[\frac{\partial \ln p(\mathbf{Y}|\theta)}{\partial \theta} \right]^2 \right\} \right]^{-1} , \tag{6.11}$$

where $\mathbf{Y}$ is the observation vector. It is shown in [1] that if the lower bound is attained, then the estimate is dubbed an efficient estimate and it must be a maximum likelihood estimate. If we evaluate (6.11) for $\hat{\theta}_{ML}$ defined by (6.6), we obtain the right-hand side of (6.10); thus $\hat{\theta}_{ML}$ is an efficient estimate.

So far, we have concentrated on parameter estimation, but often the phase will vary with time (and may be characterized as a random process). A general treatment of estimation of random processes, referred to as *maximum a posteriori* (MAP) estimation, is provided in [1]; to estimate (or track) random processes a closed-loop estimation is preferred. The phase-locked loop, which is introduced in Section 6.2, is an approximation to the optimum MAP carrier

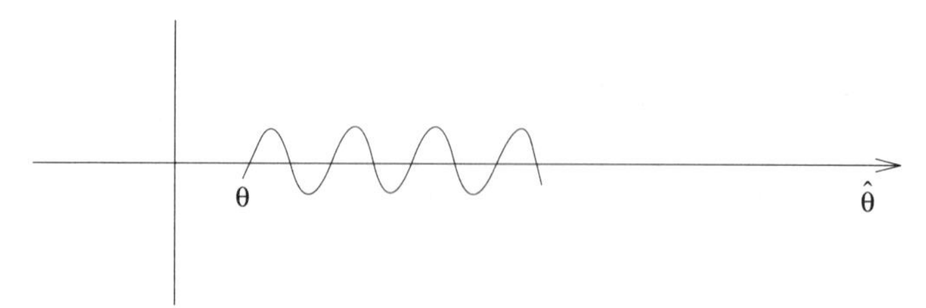

Fig. 6.3 The error function $\sin(\hat{\theta}-\theta)$.

phase estimate [2]. By way of motivating a closed-loop detector, we need an estimate that satisfies (6.5) and which will have the same variance as $\hat{\theta}_{ML}$. Averaging the expression for the contribution to $\dot{l}_K(\hat{\theta})$ in (6.5) over a symbol period, gives

$$E_\theta \int_{kT}^{(k+1)T} \left[A \cos(2\pi f_c t + \theta) + n(t) \right] \sin(2\pi f_c t + \hat{\theta})\, dt \sim \frac{AT}{2} \sin(\theta - \hat{\theta}) \tag{6.12}$$

which is a function of the phase error. As shown in Figure 6.3, this function has a root at the desired phase angle, θ, as well as roots at $\theta + m\pi$. Suppose we construct an algorithm that updates the phase estimate according to the recursive rule

$$\hat{\theta}(t + \Delta) = \hat{\theta}(\Delta) - \beta \dot{l}_\Delta(\hat{\theta}) , \tag{6.13}$$

where from (6.5) we have

$$\dot{l}_\Delta(\hat{\theta}) = \int_t^{t+\Delta} y(t) \sin(2\pi f_c t + \hat{\theta})\, dt , \tag{6.14}$$

and β is a constant of proportionality. Thus the phase estimate evolves according to a difference equation whose increment is $\beta \dot{l}_\Delta(\theta)$. This is certainly a closed-loop system, but how are (6.13) and (6.14) related to $\hat{\theta}_{ML}$? We first observe that if the system described by (6.13) settles down, then it must stop at a value $\hat{\theta}$ such that

$$m(\hat{\theta};\theta) \equiv E\left[\dot{l}_\Delta(\hat{\theta}) \right] = 0 , \tag{6.15}$$

where the expectation is with respect to $\hat{\theta}$ and where $m(\hat{\theta};\theta)$ is called a *regression* function. To see this, consider the expected value of both sides of (6.13). If the recursion stopped at any other value of $\hat{\theta}$, then there would be a non-zero driving term in this equation and we would have a contradiction, since the algorithm would *not* have terminated. We have already evaluated (6.15) in deriving (6.12), and so we know that when (6.13) settles down it must be the case that $\hat{\theta} = \hat{\theta}_{ML} \to \theta$. This provides the rationale for choosing the recursive algorithm

(6.13); if we let $\Delta \to 0$, we have a differential equation for the evolution of $\hat{\theta}$

$$\frac{d\hat{\theta}}{dt} = -\beta \dot{l}(\hat{\theta}) = -\beta\, y(t) \sin\,(2\pi f_c t + \hat{\theta}) \ , \tag{6.16}$$

where β is a constant. If we choose the increment, Δ, to be a symbol interval, T, then we have a time-discrete estimator governed by the recursive difference equation

$$\hat{\theta}_{n+1} = \hat{\theta}_n - \beta \int_{nT}^{(n+1)T} y(t) \sin\,(2\pi f_c t + \hat{\theta}_n)\, dt \ , \tag{6.17}$$

where $\hat{\theta}_n \equiv \hat{\theta}(nT)$, and β is a constant chosen small enough so that the recursion converges.

The rationale leading to (6.17) has broad implications, and will form part of the foundation for the extremely important *least mean square* (LMS) class of adaptive algorithms introduced in Chapter 8. The above approach provides the link between open-loop ML estimates and the closed-loop systems needed to track time-varying phase angles. The general approach is to define a regression function of the estimate and unknown parameter, $m(\hat{\theta};\theta)$, where

$$m(\hat{\theta};\theta) = E\left[\frac{dL(\hat{\theta};\theta)}{d\hat{\theta}}\right] , \tag{6.18}$$

and if $m(\hat{\theta};\theta)$ has a unique root at $\hat{\theta} = \theta$, then the closed-loop recursion

$$\hat{\theta}_{n+1} = \hat{\theta}_n - \beta \frac{dL(\hat{\theta};\theta)}{d\theta}\bigg|_{\theta=\hat{\theta}_n} , \quad n = 1, 2, 3, \ldots \tag{6.19}$$

will converge to the desired phase angle, i.e., $\hat{\theta}_n \to \theta$, provided β is chosen small enough. The quantity β is called the *step size* of the algorithm and the loop bandwidth is proportional to the step size. A functional implementation of the closed-loop system (6.19) is shown in Figure 6.4. While the above discussion has taken place in the context of carrier phase estimation, the approach leading to the closed-loop estimates (6.18) and (6.19) holds for the estimation of arbitrary parameters. We will use this framework for the development of systems that provide timing acquisition later in this chapter.

We now make the connection between the generic closed-loop implementation of Figure 6.4 and the classic phase-locked loop, which is an excellent closed-loop approximation of the MAP carrier phase estimate. The essential idea is to demonstrate that the value of the estimate for which the loop of Figure 6.4 achieves lock is the unknown phase, and that the variance of the error, $(\hat{\theta} - \theta)$, is the same as that given by (6.10). Implementation considerations will then determine the specific realization of the PLL.

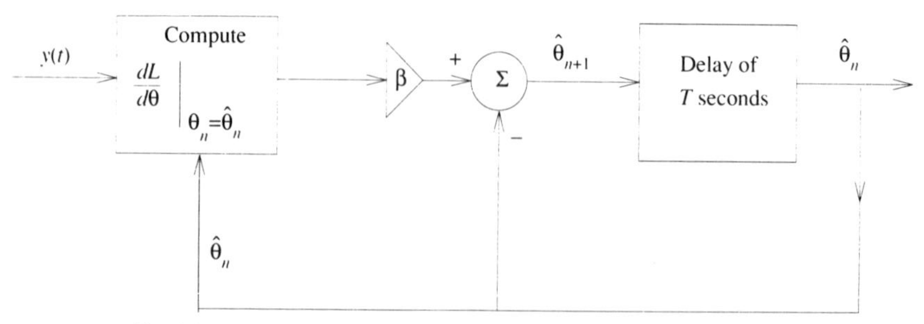

Fig. 6.4 Implementation of the closed-loop system governed by (6.19).

6.2 THE PHASE-LOCKED LOOP (PLL)

The phase-locked loop shown in Figure 6.5 is a closed-loop realization of the maximum likelihood phase estimator (or more generally of the maximum *a posteriori* estimate when $\theta(t)$ is a random process) discussed in the previous section [2]. The PLL strives to produce a phase estimate, $\hat{\theta}(t)$ equal to the input phase, $\theta(t)$, such that the error signal, $e(t)$, becomes vanishingly small. The key element of the PLL is the voltage-controlled oscillator (VCO), whose output frequency $(d\hat{\theta}/dt)$ varies in a linear fashion according to the amplitude of the input (error signal) voltage. The output of the VCO can be written

$$z(t) = K_1 \sin \hat{\theta}(t) , \tag{6.20}$$

where the instantaneous radian frequency of $z(t)$ is given by

$$\frac{d\hat{\theta}}{dt} = K_2 e(t) + 2\pi f_0 , \tag{6.21}$$

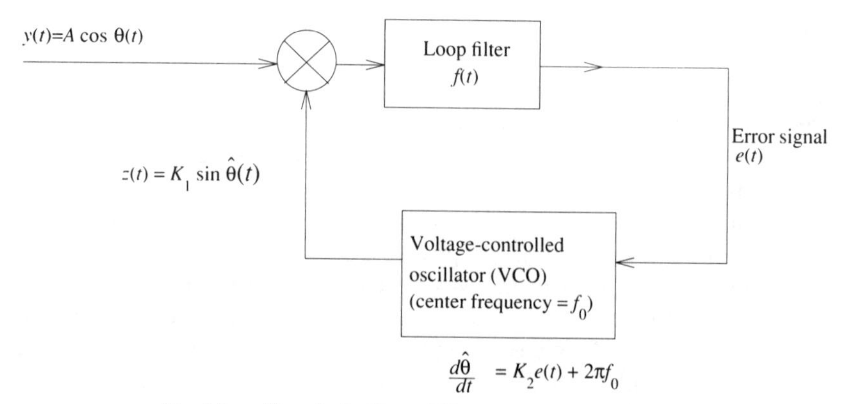

Fig. 6.5 Phase-locked loop (PLL), continuous time version.

where it is assumed that the receiver has a nominal knowledge of the transmitter frequency, and where K_2 is the sensitivity of the VCO to the input voltage, $e(t)$, and f_0 is its center frequency. The low-pass filter, $f(t)$, will filter out the second harmonic contained in the product $y(t) \cdot z(t)$. Ideally, the PLL bandwidth is narrow enough so that it is not significantly affected by the information-bearing signal.

To get more insight into the operation of the PLL we shall consider the input to the PLL to be the noise-free signal

$$y(t) = A \cos\,[2\pi f_0 t + \theta_1(t)] \,. \tag{6.22}$$

The low-pass filter in the loop eliminates the second harmonic component. Referring to Figure 6.5, the error signal is given by

$$e(t) = \frac{AK_1}{2} \int_0^t f(t-u) \sin\,[\theta(u) - \hat{\theta}(u)]\, du \,. \tag{6.23}$$

The following equation describes the operation of the PLL,

$$\frac{d\hat{\theta}}{dt} = 2\pi f_0 + \frac{K_2}{2} \int_0^t f(t-u) AK_1 \sin\,[\theta(u) - \hat{\theta}(u)]\, du \,. \tag{6.24}$$

Defining the phase error, $\phi(t) \triangleq \theta(t) - \hat{\theta}(t)$, $\hat{\theta}_2(t) = \hat{\theta}(t) - 2\pi f_0 t$, and the loop gain $K = K_1 K_2/2$, gives the dynamic equation for the evolution of the PLL:

$$\frac{d\phi(t)}{dt} = \frac{d\theta(t)}{dt} - 2\pi f_0 - AK \int_0^t f(t-u) \sin \phi\,(u)\, du \,. \tag{6.25}$$

In [2] it is shown that the PLL, as represented by (6.25), is an excellent approximation to the optimum estimator of the time-varying phase $\theta(t)$. Equation (6.25) gives rise to an equivalent baseband formulation of the PLL as shown in Figure 6.6, where $\theta_1(t) = \theta(t) - 2\pi f_0 t$.

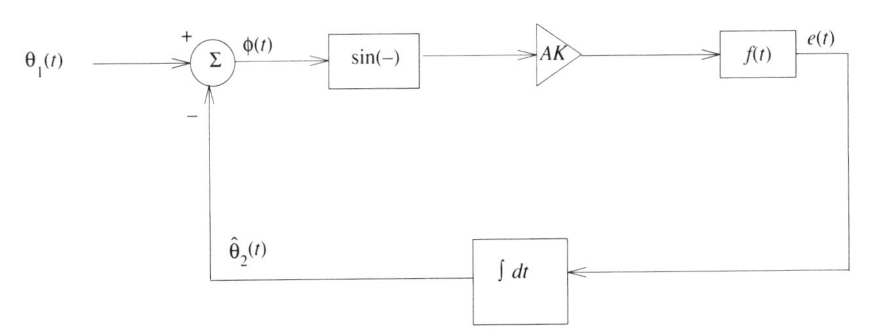

Fig. 6.6 Equivalent baseband model of the phase-locked loop.

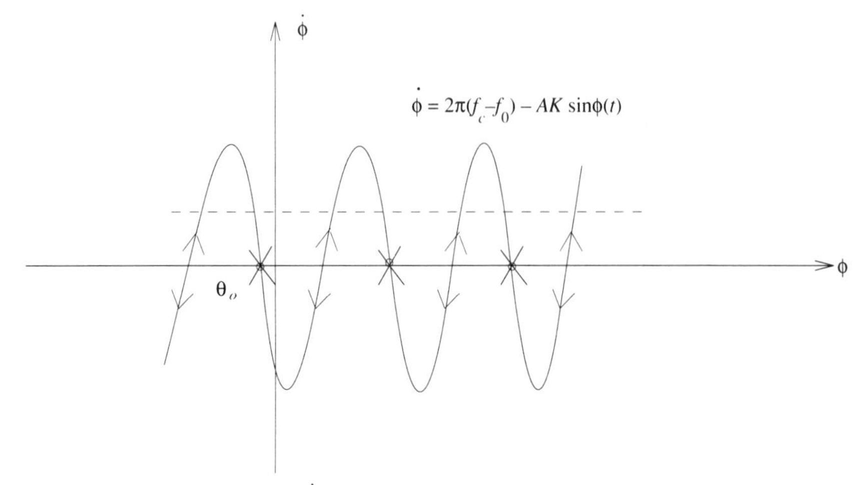

Fig. 6.7 A phase plane plot of $\dot{\phi}$ and $\phi(t)$ for a first-order PPL. With $AK \geq 2\pi(f_c - f_0)$, the loop has the stable operating points $\theta_0 \pm n2\pi$, where $\theta = 0$.

We now consider the operation of the *first-order* PLL. In this case the filter is not present in the loop, and $f(t) = \delta(t)$. The action of the VCO serves to filter out high-order frequency terms. If the instantaneous phase of the input is the carrier plus a constant phase offset,

$$\theta(t) = 2\pi f_c t + \theta_c , \tag{6.26}$$

then substituting into (6.25) we have

$$\begin{aligned} \frac{d\phi(t)}{dt} &= \frac{d\theta}{dt} - 2\pi f_0 - AK \sin \phi(t) \\ &= 2\pi(f_c - f_0) - AK \sin \phi(t) . \end{aligned} \tag{6.27}$$

A phase-plane plot (i.e., a plot of ϕ vs. $\dot{\phi}$) of (6.27) is shown in Figure 6.7. If $AK > 2\pi(f_c - f_0)$, as is assumed in the figure, the loop has stable points of operation at $(\theta_0 \pm n2\pi)$, since the equation $\dot{\phi} = 0$ will have a stable solution at $\phi = \theta_0 = \sin^{-1}[2\pi(f_c - f_0)/AK]$. Note that the derivative is positive (negative) for values of ϕ slightly less (greater) than $\theta_0 \pm 2\pi n$, $n = 0, 1, \ldots$. Thus if ϕ reaches a steady state value at one of these points, after a small perturbation the loop will be restored to the equilibrium position. In contrast, if the steady state is at the (unstable equilibrium) points $\theta_0 \pm (2n-1)\pi$, $n = 0, 1, \ldots$, a small perturbation would lead to a new operating point.

The relationship $[2\pi(f_c - f_0)/AK]$ is significant, in that it gives the *hold-in range*

$$2\pi(f_c - f_0) < AK \tag{6.28}$$

a first-order PLL; i.e., the range of frequency deviation of the incoming carrier from the center frequency of the VCO for which it is possible to *track* phase. If the incoming carrier drifts to larger deviations, the PLL will not be able to hold to a zero-phase error. Of course, the hold-in range can be increased by increasing the loop gain; however, as we shall see, this increases the deleterious effects of noise.

For the first-order PLL, (6.28) also gives the *lock-in range*; i.e., the range of frequency variation for which the incoming phase will be *acquired.* For deviations outside this range, the PLL will not be able to acquire the incoming carrier. For higher-order loops (i.e., where the loop filter is not just an all-pass filter) there is a hysteresis effect owing to the fact that the hold-in range and the lock-in range are not equal, the former being larger [2]. For these higher-order loops, the range of frequencies between the two is the *pull-in range* where the PLL acquires the phase of the incoming signal after slipping several cycles. Our analysis of the nonlinear behavior of the PLL is limited in scope to consideration of small phase errors for a first-order system. Before we turn to a discussion of a linearized model of the PLL, it should be pointed out that, as the loop error becomes large, the loop loses lock. A more detailed discussion of the nonlinear behavior of the pull-in and hold-in ranges, as well as acquisition time, can be found in the classic books by Viterbi [2] and Lindsey [3] and the recent reference [16].

6.2.1 LINEAR MODEL OF THE PHASE-LOCKED LOOP

Much insight into the behavior of the phase-locked loop can be obtained by studying the linearized loop model. This linearization is valid for small loop errors. In the analysis of the PLL the nonlinear term is $\sin\phi(t)$. This arises through the control of the oscillator frequency in the VCO. Although we have been concerned with a sinusoidal nonlinearity there are a number of other implementation approaches which give rise to other kinds of nonlinearities, such as a sawtooth (see Chapter 13 in [4]).

A fruitful approach to the analysis of the PLL is the small signal analysis in which it is assumed that the phase error is small enough so that we can make the approximation $\sin\phi(t) \cong \phi(t)$. The loop is said to be in *phase lock*, and this would be the situation that we expect to encounter in the tracking mode. In this case (6.25) becomes

$$\frac{d\phi(t)}{dt} = \frac{d\theta_1(t)}{dt} - AK\int_0^t f(t-u)\,\phi(u)\,du\,. \tag{6.29}$$

The equation is linear, and taking Laplace transforms we get

$$s\,\Phi(s) = s\,\theta_1(s) - AK\,F(s)\,\Phi(s), \tag{6.30}$$

where $\Phi(s) = L[\phi(t)]$, $\theta_1(s) = L[\theta_1(t)]$ and $F(s) = L[f(t)]$, and $L[\cdot]$

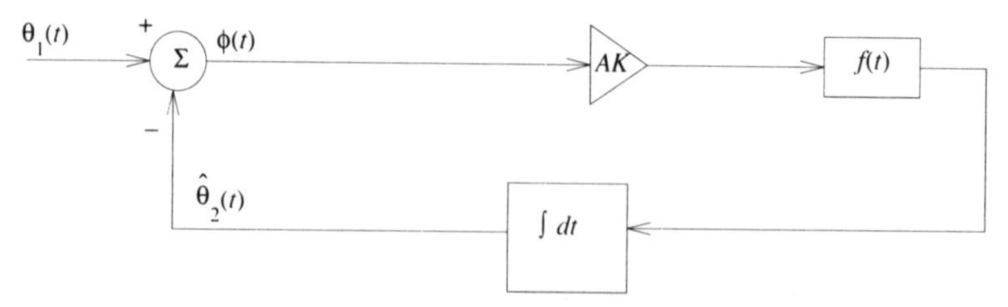

Fig. 6.8 Equivalent baseband PLL: small error signal linear approximation.

denotes the Laplace transform operation. Rearranging the terms in (6.30) gives

$$\Phi(s) = \theta_1(s)/(1 + AK\, F(s)/s) \tag{6.31}$$

or

$$H(s) = \frac{\hat{\theta}_2(s)}{\theta_1(s)} = \frac{AK\, F(s)/s}{1 + AK\, F(s)/s}\ , \tag{6.32}$$

which is the *closed loop transfer function* between the input signal, $\theta_1(t)$, and the PLL estimate, $\hat{\theta_2}(t)$. A linear model of the loop is as shown in Figure 6.8. Note that $H(s)$ has the canonical form of a feedback system. The inverse transform of $h(t) \triangleq L^{-1}(H(s))$ is the closed loop impulse response. For stability of the system, it is required that the zeros of $1 + AK\, F(s)/s$ be in the left half-plane.

The specific behavior of the PLL discussed in the foregoing is a function of the linear filter, $f(t)$. The form of $F(s)$ determines the input signals that can be tracked. We begin by considering a *first*-order loop given by $F(s) = 1$ with input $\theta(t) = 2\pi f t + \theta_0$. Taking Laplace transforms we have

$$\theta_1(s) = \frac{2\pi(f - f_0)}{s^2} + \frac{\theta_0}{s}\ . \tag{6.33}$$

From (6.32) and (6.33) the transform of the phase error is given by

$$\Phi(s) = \frac{2\pi(f - f_0)}{s(s + AK)} + \frac{\theta_0}{s + AK}\ . \tag{6.34}$$

From the final value theorem, we see that

$$\lim_{t\to\infty} \phi(t) = \lim_{s\to o} s\, \Phi(s) = \frac{2\pi(f - f_0)}{AK}\ . \tag{6.35}$$

Thus the first-order PLL can track an initial phase shift, but produces a steady-state phase error in the presence of a frequency offset. However by increasing the gain factor, AK, the error may be reduced. The error that is reflected in the output phase is a constant phase offset of $2\pi(f - f_0)/AK$ radians, however, the loop tracks the frequency difference perfectly.

Table 6.1 Phase-Locked Loop Tracking Capabilities [2]

$\theta(t)$	$F(s)$	$H(s)$	$\lim_{t\to\infty} \phi(t)$	Loop noise BW (B_L)
$2\pi ft+\theta_0$	1	$AK/(s+AK)$	$2\pi(f-f_0)/AK$	$AK/4$
$2\pi ft+\theta_0$	$1+a/s$	$AK(s+a)/(s^2+AKs+AKa)$	0	$\frac{AK+a}{4}$
$2\pi ft+\theta_0$	$\frac{s+a}{s+\varepsilon}$	$AK(s+a)/(s^2+AK+\varepsilon)s+AKa$	$\frac{\varepsilon}{a}\frac{2\pi(f-f_0)}{AK}$	$\frac{AK(AK+a)}{4(AK+\varepsilon)}$
$\frac{1}{2}Rt^2+2\pi ft$ $+\theta_0$	$1+\frac{a}{s}$	$AK(s+a)/(s^2+AKs+AKa)$	$\frac{R}{aAK}$	$\frac{AK+a}{4}$
$\frac{1}{2}Rt^2+2\pi ft$ $+\theta_o$	$1+\frac{a}{s}+\frac{b}{s^2}$	$\frac{AK(s^2+as+b)}{s^3+AKs^2+AAKs+bAK}$	0	$AK\frac{(aAK+a^2-b)}{4(aAK-b)}$

Now consider a *second*-order loop for which $F(s) = 1 + a/s$. In this case the loop transfer function is

$$H(s) = AK(s+a)/(s^2+AK\ s+a\ AK)\ . \tag{6.36}$$

For the input given in (6.33) we have that the transform of the phase error is given by

$$\Phi\ (s) = \frac{2\pi(f-f_0)\ +\ \theta_0 s}{s^2\ +\ AKs\ +\ aAK}\ . \tag{6.37}$$

Again applying the final value theorem, we see that in this case there is no tracking error. This analysis applies in a straightforward way to higher-order loops and inputs with the recognition that higher-order systems can track a wider range of input, but are harder to stabilize [2]. Table 6.1 summarizes the performance of various order loops in response to selected inputs. We will elaborate on the last column shortly. The term R in the table models a Doppler shift in the input frequency. Notice the role of the loop gain, K, in reducing the tracking error.

6.2.2 EFFECT OF NOISE ON PLL OPERATION

So far, we see that even when there is a loop error, the output error can be made arbitrarily small by increasing the loop gain K. We now consider the effect of the PLL system, including the loop gain, on the noise at the loop output. For a first-order loop we consider additive noise at the receiver input that is confined to a narrow band symmetric around the carrier frequency. In Section 5.1 it is shown that this noise may be expressed in terms of in-phase and quadra-

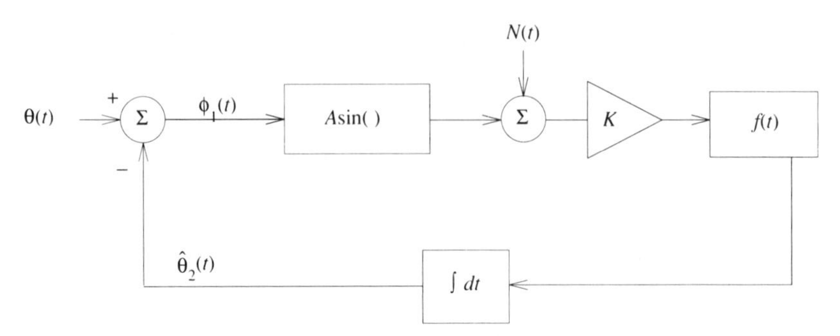

Fig. 6.9 Baseband equivalent PLL: noise analysis.

ture components as

$$n(t) = n_1(t)\cos 2\pi f_0 t + n_2(t)\sin 2\pi f_0 t\ , \tag{6.38}$$

where $n_1(t)$ and $n_2(t)$ are independent Gaussian processes with mean zero and identical spectral densities. The spectral densities of $n_1(t)$ and $n_2(t)$ are the baseband equivalent spectral density of $n(t)$, and after multiplication and low-pass filtering we arrive at the baseband circuit for this system, shown in Figure 6.9. In the presence of noise, the loop equation, (6.25), becomes

$$\frac{d\phi(t)}{dt} = \frac{d\theta_1(t)}{dt} - K\int_0^t \left[A\sin\phi(u) + N(n)\right] f(t-u)\,du\ , \tag{6.39}$$

where

$$N(t) = -n_1(t)\sin\left[\hat{\theta}(t) - 2\pi f_0 t\right] + n_2(t)\cos\left[\hat{\theta}(t) - 2\pi f_0 t\right]\ .$$

It has been argued [2] that due to the narrow bandwidth of the loop output (including the noise) we may treat $N(t)$ as baseband white Gaussian noise with the same power spectral density as the input noise, namely $N_0/2$.

Once again we assume that the phase error is small so that $\sin\phi(t) \cong \phi(t)$. The superposition property of linear systems can be used to determine the output noise spectrum. With no input signal we find that the output noise power spectral density is given by

$$S_\phi(2\pi f) = \left|\frac{KF(2\pi f)/j2\pi f}{1 + AKF(2\pi f)/j2\pi f}\right|^2 S_n(2\pi f)\ , \tag{6.40}$$

where $S_n(2\pi f)$ is the input noise spectral density. In terms of $H(j2\pi f)$, the closed loop transfer function, defined by (6.32), the error spectrum with a white noise input is given by

$$S_\phi(2\pi f) = \frac{N_0}{2A^2}\ |\ H(j2\pi f)\ |^2. \tag{6.41}$$

The mean squared phase error, σ_ϕ^2, is

$$\sigma_\phi^2 = \frac{N_0}{A^2} \int_0^\infty | H(j2\pi f) |^2 \, df \triangleq \frac{N_0}{A^2} B_L = \frac{B_L}{\rho} , \qquad (6.42)$$

where ρ is the signal-to-noise ratio (A^2/N_0), and the equivalent loop-noise bandwidth is defined by

$$B_L = \int_0^\infty | H(2\pi f) |^2 df . \qquad (6.43)$$

The loop-noise bandwidth is the bandwidth of an ideal low-pass filter whose output variance is σ_ϕ^2 when the input is a white process with one-sided density N_0/A^2. Note the similarity between (6.42) and (6.10), since the observation interval, T_0 is inversely proportional to the loop bandwidth, B_L. For the first-order loop, (6.43) can be evaluated and the loop bandwidth shown to be

$$B_L = AK/4 , \qquad (6.44)$$

and consequently the mean-square output phase error due to noise is

$$\sigma_\phi^2(noise) = \frac{1}{4} \frac{N_0}{A} K . \qquad (6.45)$$

The output noise power will be minimized if B_L is kept small. However, since the transfer function between the phase error, $\phi(t)$, and the input, $\theta_1(t)$, is $1 - H(j2\pi f)$, the signal-dependent contribution to the error power is

$$\sigma_\phi^2(signal) = \int_{-\infty}^{\infty} |1 - H(j2\pi f)|^2 S_{\theta_1}(2\pi f) df . \qquad (6.46)$$

This expression, which represents the loop tracking error, will depend on the power spectrum of the input, $S_{\theta_2}(2\pi f)$, but will be inversely proportional to K. This demonstrates the conflicting requirements between minimizing the output noise power (make K small) and minimizing the tracking error (make K large), and is illustrative of the general conflict that we have mentioned previously: a narrow loop (hence minimal output noise) may mean that the loop will not have enough bandwidth to lock onto the incoming signal. The last column of Table 6.1 contains the loop noise BW for various loop filters. Observe that the feedback constant in the PLL, K, can be decreased to synthesize a filter of arbitrarily narrow bandwidth from a fixed bandwidth loop filter. Second-order PLLs offer a better compromise between output noise and tracking of the input signal; however, the design of a PLL can always be improved by *a priori* knowledge of the input process. In Chapter 10, we describe an *adaptive* PLL which adjusts its structure in response to the characteristics of the input phase process; if a signal is not present, the loop-adaptive PLL will minimize the output power (which is all noise).

Before we leave the PLL, we summarize the exact analysis [2] of the phase error at the output of a first-order loop. Using diffusion theory and the resulting nonlinear, partial-differential, Fokker–Planck equation, it has been shown that in response to a sinusoid of unknown phase imbedded in Gaussian noise, where the center frequency of the VCO and the carrier frequency are *equal*, that the phase error, ϕ, has the Viterbi–Tikhonov distribution

$$Q(\phi) = \frac{\exp(\alpha \cos \phi)}{2\pi I_0(\alpha)} .$$

In the above equation, α is the signal-to-noise ratio

$$\alpha = \frac{A^2}{N_0 B_L} \tag{6.47}$$

and $I_0(\alpha)$ is the zeroth-order, modified Bessel function. For large signal-to-noise ratios, we have $I_0(\alpha) \cong \exp(\alpha)/(2\pi/\alpha)^{1/2}$, so that the distribution of the phase error becomes

$$Q(\phi) = \frac{e^{-\alpha\phi^2/2}}{(2\pi/\alpha)^{1/2}} , \tag{6.48}$$

which is Gaussian distribution with variance $N_0 B_L / A^2$. A sketch of this probability density is shown in Figure 6.10. Note that when the signal-to-noise ratio is small, the density is uniform over $(-\pi, \pi)$.

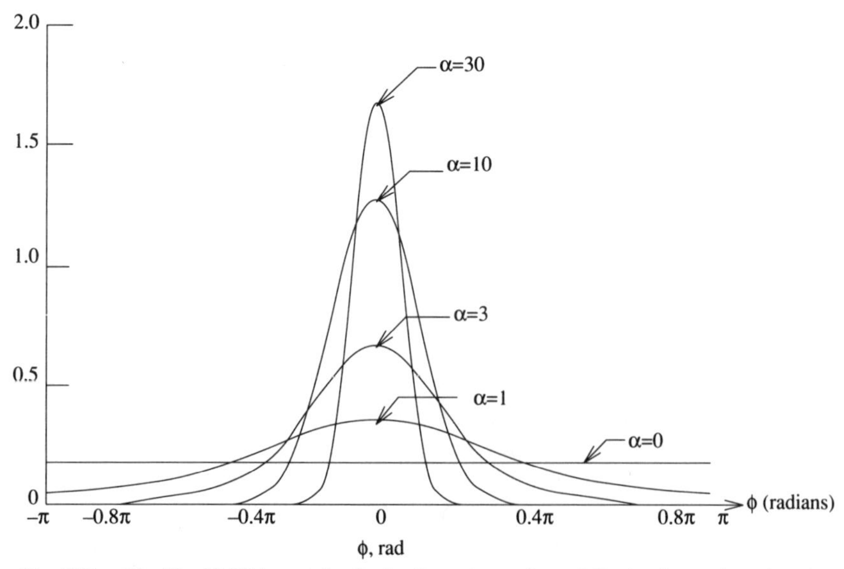

Fig. 6.10 The Viterbi–Tikhonov distribution for various values of the signal-to-noise ratio (α).

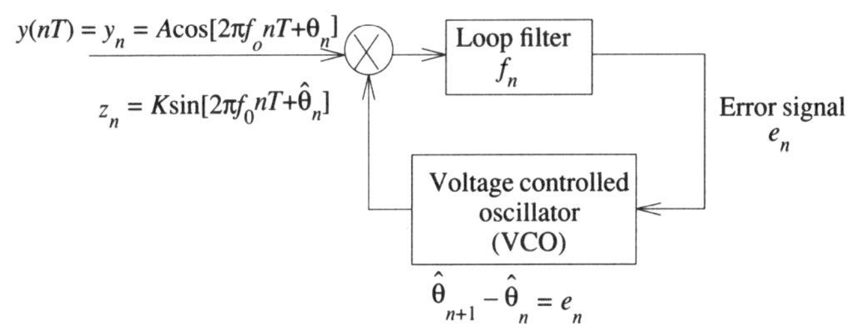

Fig. 6.11 Phase-locked loop (PLL): discrete-time version.

Note that the signal-to-noise ratio (6.47) is proportional to the input SNR, A^2/N_0 and is inversely proportional to the loop bandwidth, B_L. It is a salient property of a phase-locked loop that the output SNR remains the same and is not affected, as the input varies over the loop *hold-in range*. In contrast to a conventional filter, the PLL acts like a narrowband filter, whose center frequency *tracks* the frequency of the input sinusoid. It is also important to remember that the first-order PLL will only lock if (6.29) is satisfied. Any attempt to *increase* the output SNR by decreasing the gain, K, will *decrease* the range over which the loop will lock. It is incumbent upon the designer to determine the amount of uncertainty in the input frequency and to design the loop filter accordingly. The adaptive PLL presented in Chapter 10 will provide the designer with additional flexibility, when it is expected that the input phase will vary over a broad frequency range.

6.2.3 DISCRETE-TIME PLLs

The discrete-time PLL has a somewhat different behavior from the continuous-time PLL we have just discussed, in that the loop gain in the discrete-time case has to be within a certain range for the PLL to be stable. A discrete-time PLL is shown in Figure 6.11, where the discrete time VCO has the property that

$$\hat{\theta}_{n+1} - \hat{\theta}_n = e_n , \tag{6.49}$$

where $\hat{\theta}_n$ and $\hat{\theta}_{n+1}$ are successive time samples of the phase estimate. Assuming that the loop filter, f_n, rejects second harmonics, the filter output is given by

$$e_n = AK \sum_m f_m \sin\phi_{n-m} \approx AK \sum_m f_m \phi_{n-m} , \tag{6.50}$$

where the phase error, ϕ_n, is defined by

$$\phi_n = \theta_n - \hat{\theta}_n , \tag{6.51}$$

and we have assumed small error signal (i.e., linear) operation. Combining (6.49)–(6.51) we have the difference equation governing loop operation

$$\hat{\theta}_{n+1} - \hat{\theta}_n = AK \sum_m f_m \, [\theta_{n-m} - \hat{\theta}_{n-m}] \, . \tag{6.52}$$

Taking z-transforms of the above equation we have

$$(z - 1)\, \hat{\theta}(z) = AK\, F(z)\, [\theta(z) - \hat{\theta}(z)] \,, \tag{6.53}$$

with the obvious correspondence in notation between time sequences and transforms. From (6.53) we compute the transfer function between the input phase and the phase estimate as

$$\frac{\hat{\theta}(z)}{\theta(z)} = \frac{AK\, F(z)}{(z - 1) + AK\, F(z)} \,. \tag{6.54}$$

Consider a first-order loop where $F(z) = 1$, then the loop transfer function becomes

$$\frac{\hat{\theta}(z)}{\theta(z)} = \frac{AK}{z - 1 + AK} \,, \tag{6.55}$$

which has a pole at $z_p = 1 - AK$. For the system to be stable, $|z_p| < 1$, or $0 < AK < 2$. The upper limit on the loop gain needed for stability is in contrast to the continuous-time, first-order PLL, where there is no constraint on the loop gain.

6.3 CARRIER RECOVERY: NON-DATA-AIDED SYSTEMS

In our study of the phase-locked loop it was assumed that the signal was a simple unmodulated sinusoid. In this section we shall consider the effect on PLL operation when there is information modulating the carrier. The question is: how can sinusoids containing carrier (and timing) phase information be generated from information-bearing bandpass signals? As previously discussed, we do not explicitly consider out-of-band systems transmitting separate pilot tones. Our analysis of the maximum likelihood phase estimate, and the associated PLL carrier tracking systems, which assumed an unmodulated carrier (this tone may be viewed as an out-of-band pilot tone), does, however, provide a benchmark for the performance of pilot-tone-based systems. The synchronization systems that we discuss strive to achieve the performance of the out-of-band unmodulated carrier tracked by a PLL. In this section we discuss synchronization systems that do *not* make use of data decisions to form the phase estimates; hence these systems are referred to as *non-data-aided carrier recovery systems.*

To aid in our understanding of carrier phase recovery, and timing recovery as well, we briefly discuss the concept of *cyclostationarity* [5]. A process is said to be cyclostationary if moments of all orders are periodic in time. This may be viewed as a generalization of stationary processes whose moments are constant with time. In further analogy with stationary processes, a process, $y(t)$, is said to be *wide-sense cyclostationary* if the mean, $E[y(t)]$, and the autocorrelation, $E[y(t)\,y(t+\tau)]$, are both periodic functions of t. As we shall see presently, the modulated waveforms that we are considering are cyclostationary processes if the phase is treated as an unknown, but random, parameter.

We begin the analysis of modulated signals with the double-sideband signal expressed in complex analytic notation as (see Section 5.1)

$$y(t) = \text{Re}\left[A(t)\exp(j2\pi f_c t + \theta)\right] \tag{6.56}$$

where, for the moment, we assume that $A(t)$ is a stationary *real* random process representing the baseband modulating signal. The mean and autocorrelation of $y(t)$ can be shown respectively, to be

$$E[y(t)] = \bar{A}\cdot\exp[j2\pi f_c t + \theta] \tag{6.57a}$$

$$\begin{aligned} R_{yy}(t,t+r) &= E[y(t)y(t+\tau)] \\ &= \frac{1}{2}\text{Re}[R_{aa}(\tau)\exp(j2\pi f_c\tau)] \\ &+ \frac{1}{2}\text{Re}[R_{aa}(\tau)\exp(j4\pi f_c t + j2\pi f_c t + j2\theta)] \end{aligned} \tag{6.57b}$$

where $\bar{A}$ is the mean and $R_{aa}(\tau)$ is the autocorrelation function of A(t), $[R_{aa}(\tau) \triangleq E[A(t)A(t+\tau)]$. Note that $y(t)$ is wide-sense cyclostationary since the mean and the autocorrelation are periodic (with period $1/f_c$) in t. Since the periodic functions in (6.57) are multiplied by smooth functions, such as $R_{aa}(\tau)$, spectral lines cannot, in general, be obtained directly from $y(t)$. For example, $E\,[A(t)]$ generally equals zero, in which case (6.57a) does not provide any information about the carrier frequency or phase. However, suppose the data pattern has a DC value, i.e., $E\,A(t) = A$ (a constant), then $E\,[y(t)] = A\exp[j2\pi f_c t + \theta]$, and $y(t)$ can be fed directly into a linear filter or a PLL to extract the carrier phase and frequency. The transmitter power would have to be increased to provide a DC value in the transmitted sequence; since this is an undesirable and unnecessary waste of power, almost without exception systems are designed with $E\,A(t) = 0$ and the received signal cannot be *linearly* processed to generate a carrier recovery system. Thus, to derive carrier synchronization, the received signal must be further processed in a nonlinear system to derive spectral lines.

A standard technique for generating the spectral lines needed to extract carrier phase information is to feed the modulated waveform into a square-law or other nonlinear device followed by a phase-locked loop with center frequency at twice the carrier frequency (see Figure 6.12). The output of the squarer contains a sinusoidal component at twice the carrier frequency. As we have seen in the last section, the PLL acts as a narrow-band filter centered around $2f_c$, and it serves to smooth the waveform around this frequency to produce a time-averaged version of its input. In contrast to the case of an unmodulated carrier, where a pure tone would be produced at the squarer output (and the PLL would extract the tone from the background noise), the PLL needs to have a narrower bandwidth to extract a tone from the signal-dependent "jitter" present about $2f_c$.

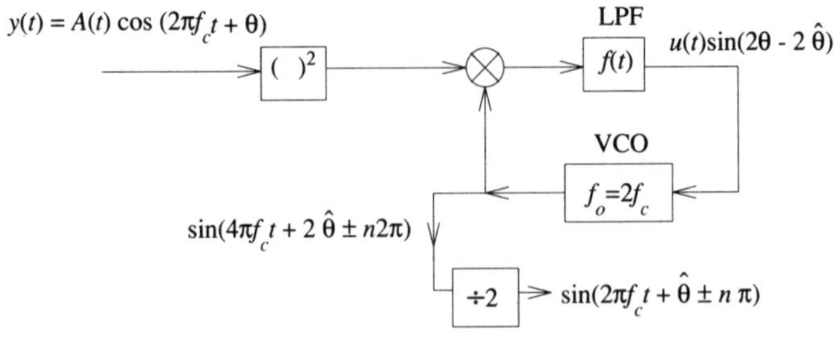

Fig. 6.12 Squaring loop carrier recover system. (Note the $\pm\pi$ ambiguity in the system output.) The signal $u(t)$ is the low-pass filtered version of $A^2(t)$.

From (6.2), we have

$$y^2(t) = \frac{A^2(t)}{2} \; [1 + \cos(4\pi f_c t + 2\theta)]$$

whose average value is

$$E\,[y^2(t)] = \frac{1}{2} \, R_{aa}\,(0)\;[1 + \cos(4\pi f_c t + 2\theta)] \; . \tag{6.58a}$$

If the PLL has unity gain in the neighborhood of $2f_c$, then the mean value of the output has frequency $2f_c$, phase 2θ, and amplitude $1/2\; R_{aa}\;(0) = 1/2\; E\;(A^2(t))$. Assuming that the PLL is operating with a small error, the loop output is $\sin\,(4\pi f_c t + 2\hat{\theta})$ where

$$2\frac{d\hat{\theta}}{dt} = Ku(t)\,\sin\,(2\theta - 2\hat{\theta}) \approx Ku(t)\;[\theta(t) - \hat{\theta}(t)] \; , \tag{6.58b}$$

and where $u(t)$ is the low-pass filtered version of $A^2(t)$. If the amplitude of $u(t)$ fluctuates slowly, a reference carrier can be generated by clipping the LPF output to remove amplitude fluctuations. This result is fed into a frequency divider which halves frequency and phase to produce $\cos\,(2\pi f t + \hat{\theta} \pm n\pi)$. Note that the phase angle is the same as that of the original signal, but that there is a 180° phase ambiguity since $\cos\,(4\pi f_c t + 2\theta) = \cos\,(4\pi f_c t + 2\theta + 2\pi)$, and dividing the argument by 2 produces the phase ambiguity of π radians, which can be handled by *differential encoding* whereby information is conveyed by the difference between encoded phase angles (see Chapter 5). The squarer/PLL system may be characterized as a non-data-directed, closed-loop system, which can provide acquisition and/or tracking.

An alternative form for tracking the phase of a double-sideband suppressed carrier signal is the Costas loop [3], shown in Figure 6.13. The advantage of the Costas loop is that a frequency divider is not required. It can be shown that the Costas loop is equivalent to a square-law device followed by a PLL. The

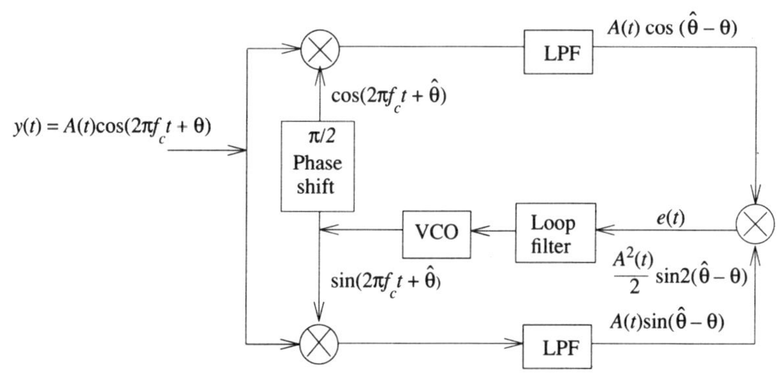

Fig. 6.13 A Costas loop for tracking the phase of a double-sideband suppressed-carrier signal. The Costas loop is equivalent to the squarer/PLL system of Figure 6.12.

equivalence is established by showing that the inputs to the loop filters are the same in both configurations, i.e.,

$$e(t) = \frac{KA^2(t)}{2} \sin 2(\hat{\theta} - \theta), \tag{6.59}$$

provided the output of the VCO is taken as $1/2(K^2)$ in the squarer/PLL configuration and as K in the Costas loop. In the absence of noise, the relationship between the signals in the Costas loop can be used to establish (6.59). The Costas loop and the squaring loop, though they differ in the location of the loop nonlinearity, produce equivalent estimates, since the loop error signals are the same.

6.3.1 CARRIER RECOVERY FOR QAM SYSTEMS: NON-DATA-DIRECTED SYSTEMS

Until now we have only considered double sideband suppressed carrier systems (DSBSC): We now turn our attention to the consideration of QAM signals, i.e., signals of the form $y(t) = \mathrm{Re}[A(t)\exp(j\,2\pi f_c t + j\theta)]$, where $A(t) = a(t) + jb(t)$. Let us assume that $a(t)$ and $b(t)$ are independent zero-mean stationary processes. The autocorrelation function of $y(t)$ is

$$R_{yy}(t+\tau,t) = \frac{1}{2}\,\mathrm{Re}[\{R_{aa}(\tau) + R_{bb}(\tau)\}\exp(j2\pi f_c\tau)] + \frac{1}{2}\,\mathrm{Re}[\{R_{aa}(\tau) - R_{bb}(\tau)\}\exp(j4\pi f_c t + j2\pi f_c\tau + j2\theta)], \tag{6.60}$$

and since the QAM signal is of the suppressed carrier class, (6.60) does not contain a tone. If we pass the modulated waveform through the squarer-PLL circuit

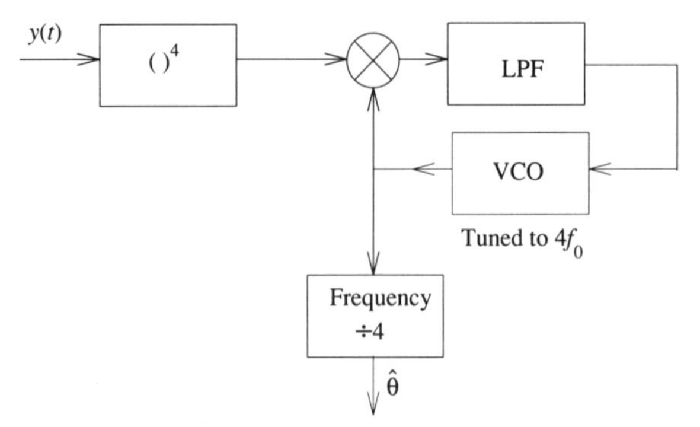

Fig. 6.14 A fourth-power synchronizer for carrier recovery in 4-PSK and QAM systems.

of Figure 6.12, the average signal has the value

$$E[u(t)] = \frac{1}{2}\,[R_{aa}(0) - R_{bb}(0)]\ \mathrm{Re}[\exp(j4\pi f_c t + 2j\theta)]\ . \quad (6.61)$$

This system will produce a tone at $2\pi f_c$, unless $a(t)$ and $b(t)$ have the same power, in which case $E[u(t)] = 0$. There are many popular modulated signals for which $R_{aa}(0) = R_{bb}(0)$, including 4-phase PSK and rectangular QAM. For this class of signals the squarer/PLL *cannot* be used to generate a tone; however, we can generate a tone by processing the received signal with a higher-order nonlinearity. If we replace the squarer by a *fourth-power* device, as shown in Figure 6.14, we can show that

$$\begin{aligned} y^4(t) &= \frac{1}{8}\ \mathrm{Re}[A^4(t)\ \exp(j8\pi f_c t + j4\theta)] + \frac{3}{8}|A(t)|^4 \\ &+ \frac{1}{2}\ \mathrm{Re}[|A(t)|^2\ A^2(t)\ \exp(j4\pi f_c t + j2\theta)]\ . \end{aligned} \quad (6.62a)$$

Now if the bandpass filter is tuned to $4f_0$, it is a tedious, but straightforward calculation to show that

$$E[u(t)] = \frac{1}{4}\ \mathrm{Re}\left[\left\{\overline{a^4} - 3\left[\overline{a^2}\right]^2\right\}\exp(j8\pi f_c t + j4\theta)\right]. \quad (6.62b)$$

where $R_{aa}(0) = R_{bb}(0) = \overline{a^2}$. Again, a divide-by-4 frequency divider can be used to recover the frequency and phase.

6.3.2 OUTPUT PHASE JITTER FOR THE SQUARING LOOP

For the case of an unmodulated carrier, the output jitter is given by (6.42) or (6.45) for the first-order PLL. With a squaring loop, the jitter due to noise will be increased as we will now demonstrate for a double sideband signal. Referring to Figure 6.12, the signal at the output of the squarer is given by

$$y^2(t) = \left[A(t) + n_c(t)\right]^2 \cos^2(2\pi f_0 t + \theta) + n_s^2(t) \sin^2(2\pi f_0 t + \theta)$$
$$- 2\left[A(t) + n_c(t)\right] n_s(t) \cos(2\pi f_0 t + \theta) \sin(2\pi f_0 t + \theta) . \quad (6.63)$$

Since the multiplier is followed by a low-pass filter, we need only focus on the low-frequency signal at the LPF input; this signal is given by

$$y^2(t) \times \sin(4\pi f_c t + 2\hat{\theta})\Big|_{LPF} = \left[A^2(t)+2A(t)\, n_c(t)+n_c^2(t)+n_s^2(t)\right] \sin 2(\hat{\theta}-\theta)$$
$$- 2\left[A(t)\, n_s(t)+n_c(t)\, n_s(t)\right] \cos 2(\hat{\theta}-\theta) . \quad (6.64)$$

Note that the term $A^2(t)\ \sin\ 2(\hat{\theta} - \theta)$ is the desired signal and the remainder is (signal-dependent) noise. After some tedious algebra, and using the fact that for Gaussian random variables

$$E\, n_c^4 = 3\sigma^4 , \quad (6.65)$$

we find that the variance of the output noise is given by

$$\sigma_\phi^2 = \left(\frac{1}{\rho} + \frac{1}{4\rho^2}\right) B_L , \quad (6.66)$$

where the signal-to-noise ratio ρ is given by A^2/N_0. Comparing (6.66) with (6.42) for the unmodulated carrier, we see that due to the presence of data modulation, the output noise is larger by the factor $B_L/4\rho^2$. Lindsey and Simon [6] have called this additional phase jitter the *squaring loss*. At high signal-to-noise ratios, the performance is that attainable with an unmodulated carrier.

Recall that for 4-PSK and rectangular QAM systems, the squarer could not be used to recover the carrier, and that a fourth power system was needed. It can be shown [6] that the fourth-power synchronizer has a *quadrupling loss*, but at high signal-to-noise ratios the loop achieves the performance attainable with an unmodulated carrier.

6.4 CARRIER RECOVERY: DATA-AIDED SYSTEMS

In the non-data-aided carrier recovery systems presented above, the presence of a modulated signal forced us to abandon the simple PLL approach to carrier recovery in favor of the squarer/PLL with its higher output-phase errors. In fact, the approaches described above could be used for analog systems (since the digital nature of the signal was not exploited). If the receiver has access to the transmitted data, or a reliable replica (i.e., the output decisions when the error rate $\approx 10^{-4}$ or better) we will demonstrate that the carrier synchronizer does not need a squarer/PLL to estimate the carrier phase. Furthermore, such

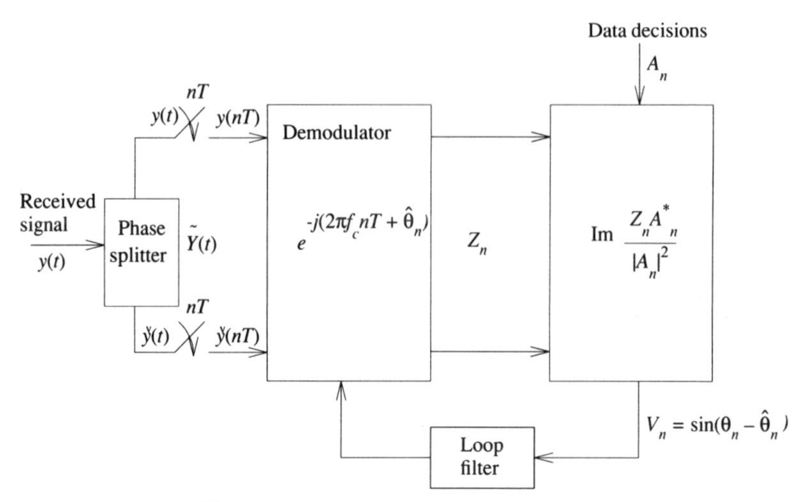

Fig. 6.15 A data-directed carrier recovery PLL.

systems provide a more accurate estimate than the squarer/PLL approach described above, since a squaring loss is not incurred. Certainly, the transmitted data is not available to the receiver during steady-state operation. However, the detected data decisions will be correct a very high percentage of the time, so these decisions may be used in place of the actual data. During start-up, when the error rate will be high, a replica of the transmitted startup sequence can be stored at the receiver and used for carrier acquisition. To begin our discussion, consider the data-directed carrier recovery system shown in Figure 6.15 with the received QAM signal

$$y(t) = \text{Re } A(t)\ e^{j(2\pi f_c t + \theta_n)} . \tag{6.67}$$

If we assume that there is no intersymbol interference and the data decisions are available at the receiver, then the waveform $A(t)$ may be readily reconstructed and used to "wipe-off" the modulation. The loop we describe was originally proposed by Falconer in his classic paper [7]. Passing the received signal $y(t)$ through a 90° phase splitter produces the components of the analytic signal

$$\tilde{Y}(t) = y(t) + j\check{y}(t) = A(t)\ e^{j(2\pi f_c t + \theta_n)}, \tag{6.68}$$

where $\check{y}(t)$ is the Hilbert transform of $y(t)$. Demodulating by $e^{-j(2\pi f_c t + \theta_n)}$ and sampling at the decision instants, nT, gives the sampled demodulated output

$$Z_n = A_n e^{-j(\hat{\theta}_n - \theta_n)} . \tag{6.69}$$

with the phase error $\hat{\theta}_n - \theta_n$. We use the known data symbols to form the loop error signal

$$V_n = \text{Im}\ \frac{Z(nT)A_n^*}{|A_n|^2} = \text{Im}\ e^{-j(\hat{\theta}_n - \theta_n)}$$
$$= -\sin\,(\hat{\theta}_n - \theta_n)\ , \tag{6.70}$$

which for small phase errors is proportional to the desired phase error. The signal V_n is fed into a loop filter to provide the estimate. Following our discussion of discrete-time PLLs, a simple first-order filter would be

$$\hat{\theta}_{n+1} = \hat{\theta}_n + KV_n\ , \tag{6.71a}$$

which produces the first-order (discrete-time) PLL

$$\hat{\theta}_{n+1} = \hat{\theta}_n - K\sin\,(\hat{\theta}_n - \theta_n)\ , \tag{6.71b}$$

where K is the loop gain. It can be shown that the data-directed estimate, (6.71), can be obtained as the MAP phase estimate assuming a time-discrete Gaussian model for θ_n and that the data decisions, A_n, are available at the receiver. Recall from our discussion of the transfer function of discrete-time PLLs, that if $0 < K < 2$, the linearized version of (6.71b) will converge to the desired phase θ_n. However, if the decisions are produced by a trellis-coded demodulator, a decision delay of L symbols produces a correction term $\hat{\theta}_{n-L} - \theta_{n-L}$. To stabilize such loops, the value of K will have to be reduced by a factor of L. This may substantially alter the loop dynamics.

There are several nice properties of this closed-loop, data-directed PLL, including the fact that the nature of the signal constellation does *not* affect the structure of the tracking loop. Recall that the squarer/PLL had to extract a tone from the modulated error term, $A^2(t)\ \sin\,[4\pi f_c t + 2(\theta - \hat{\theta})]$, where average value is given by (6.58a). The primary benefit of the data-directed PLL is the improved performance that can be achieved by making use of the (presumed correct) data decisions. At low SNR, the designer can of course use the nonlinear/PLL carrier recovery system discussed above, but at moderate-to-high SNR levels, when data decisions reliably replicate the transmitted data, the data-directed loop has become the preferred carrier recovery system.

Because of the popularity and practical significance of the data-directed PLL, we delve deeper into the performance of this system. First, we consider the steady-state (i.e., small error) operation of (6.71b), i.e.,

$$\hat{\theta}_{n+1} = \hat{\theta}_n - K\,(\hat{\theta}_n - \theta_n) \tag{6.72a}$$

$$= (1 - K)\,\hat{\theta}_n + K\theta_n\ . \tag{6.72b}$$

Taking the discrete-time Fourier transform of both sides of (6.72b), the transfer function of the data-directed PLL is given by

$$\frac{\hat{\theta}(f)}{\theta(f)} = F(f) = \frac{K}{\exp(j2\pi fT) - 1 + K}, \tag{6.73}$$

where the obvious notational correspondence between transforms and time-domain signals has been made. This transfer function for a first-order filter can be used to compute the response of the loop to phase jitter at a frequency of f Hz. Such jitter is typical on telephone channels, often consisting of one or more sinusoids, which typically are at harmonics of the power line frequencies, and we will assume sinusoidal jitter of magnitude J, i.e., $\theta_n = \mathrm{Re}[J \exp(j2\pi fnT)]$.

Now let us consider the effect of additive noise on the linearized phase-tracking algorithm. Assume the complex, equalized, demodulated output can be written

$$Z_n = A_n \exp[-j(\hat{\theta}_n - \theta_n)] + W_n, \tag{6.74}$$

where $W_n = w_n + j\check{w}_n$ is a complex Gaussian random variable with zero-mean. Although in general, successive output noise samples will be correlated after receiver filtering, we assume that are uncorrelated to simplify the results. The effect of this simplification should be minor if the phase-tracking bandwidth is much smaller than the data bandwidth, or if the frequency response of the channel and of the equalizer are both nearly flat. Thus, with $\langle\ \rangle$ denoting expectation and N_0 the variance of W_n, we have $\langle w_n w_m\rangle = \langle \check{w}_n \check{w}_m\rangle = (N_0/2)\delta_{nm}$, and $\langle w_n \check{w}_m\rangle = 0$. Then (6.72b) the loop for updating $\hat{\theta}_n$ can be written, after using the standard linearizing approximation for small phase errors, as

$$\hat{\theta}_{n+1} = (1-K)\hat{\theta}_n + K\theta_n + K\,\mathrm{Im}\left[\frac{W_n}{A_n}\right]. \tag{6,75}$$

The random variable $\mathrm{Im}\,(W_n/A_n)$ is not Gaussian unless $|A_n|$ is constant (pure phase modulation). However, assuming the information symbols and noise are independent, they are zero-mean and statistically independent with variance $(N_0/2)\langle 1/|A_n|^2\rangle$

From (6.73) $\hat{\theta}(f) - \theta(f) = [F(jf) - 1]\,\theta(f)$, and we can use the superposition principle for linear systems, to write the error in the output of the phase-locked loop as

$$\hat{\theta}_n - \theta_n = \mathrm{Re}\,\{J[F(f) - 1]\exp(j2\pi fnT)\} + v_n, \tag{6.76}$$

where the first term on the right-hand side is the loop's response to the input jitter and the second term is the filtered noise. The filtered noise sequence $\{v_n\}$ satisfies

$$v_{n+1} = (1-K)v_n + K\,\mathrm{Im}(W_n/A_n), \tag{6.77}$$

and therefore has zero-mean and steady-state variance

$$\lim_{n\to\infty} \langle v_n^2\rangle = \frac{KN_0}{2(2-K)}\,\frac{1}{\langle|A|^2\rangle}. \tag{6.78}$$

From (6.76) the mean-square error in the phase estimate for single-tone jitter at radian frequency f rad/sec is thus

$$\sigma_\phi^2 = <(\hat{\theta}_n - \theta_n)^2> = \frac{|J|^2}{2} |F(f) - 1|^2 + <v_n^2> , \quad (6.79a)$$

which from (6.73) and (6.78) is

$$<(\hat{\theta}_n - \theta_n)^2> = \frac{|J|^2}{2} \frac{4 \sin^2 (2\pi fT/2)}{K^2 + 4(1-K) \sin^2 (2\pi fT/2)} + \frac{KN_0}{2(2-K)} \frac{1}{<|A|^2>} . \quad (6.79b)$$

Note the strong similarity between the second term in (6.79b), which is the output jitter due to noise, and (6.45), the output jitter for a PLL with an unmodulated input. With $K << 2$, the expressions are identical; this demonstrates that the data-directed PLL can, under these circumstances, achieve the *same* performance as a system without any data-induced jitter. The residual RMS phase jitter, given by the square root of the above expression, is plotted as a function of the loop gain, K, for signal-to-noise ratios, $<|A|^2>/N_0$, of 30 dB and 22 dB in Figures 6.16a and 6.16b, respectively. In each case, a 16-point QAM constellation is assumed. For a given loop bandwidth (which is proportional to K) the higher the phase jitter frequency, f, relative to the symbol rate $1/T$, the greater the residual RMS jitter. This is expected since, when the jitter changes more rapidly, the first-order transfer function $F(j2\pi f)$ will produce a smaller output (and hence a larger phase error). The curves show the case of no jitter (in which case the residual jitter results from noise entering the discrete-time phase-locked loop) and also the cases of 14-degree peak-to-peak jitter with $fT = 1/48$ and with $fT = 1/20$. The choice of bandwidth of the decision-directed phase-tracking loop, determined by K, should be governed by the highest expected phase jitter frequency. If the spectrum of the phase jitter is known, a higher-order phase-locked loop may permit more effective phase tracking by reducing the noise in the loop output.

For given values of RMS residual phase jitter, the error probability can be approximated as in Chapter 5, where we averaged the output phase error over the Viterbi–Tikhonov distribution. For example, we find from Figure 6.16b that the residual RMS phase jitter is about 2.5 degrees in the 16-point QAM systems, for $K = 0.3$, when the channel has a signal-to-noise ratio of 22 dB and 14 degrees peak-to-peak channel phase jitter with frequency 1/48 that of the symbol rate. In Figure 6.17a-b, we show the sensitivity of the 16-point QAM and the V.29 systems described in Chapter 5 to residual phase jitter with a Viterbi–Tikhonov distribution. From Figure 6.17a we find that the resulting error probability is about 4×10^{-7}. With the same value of K, in the absence of phase jitter the system has an error probability of about 5×10^{-8}; with $K = 0$, and no phase jitter, the error probability is about 10^{-8}. Note how quickly the QAM constellation degrades as a function of residual jitter, while the V.29 constellation can

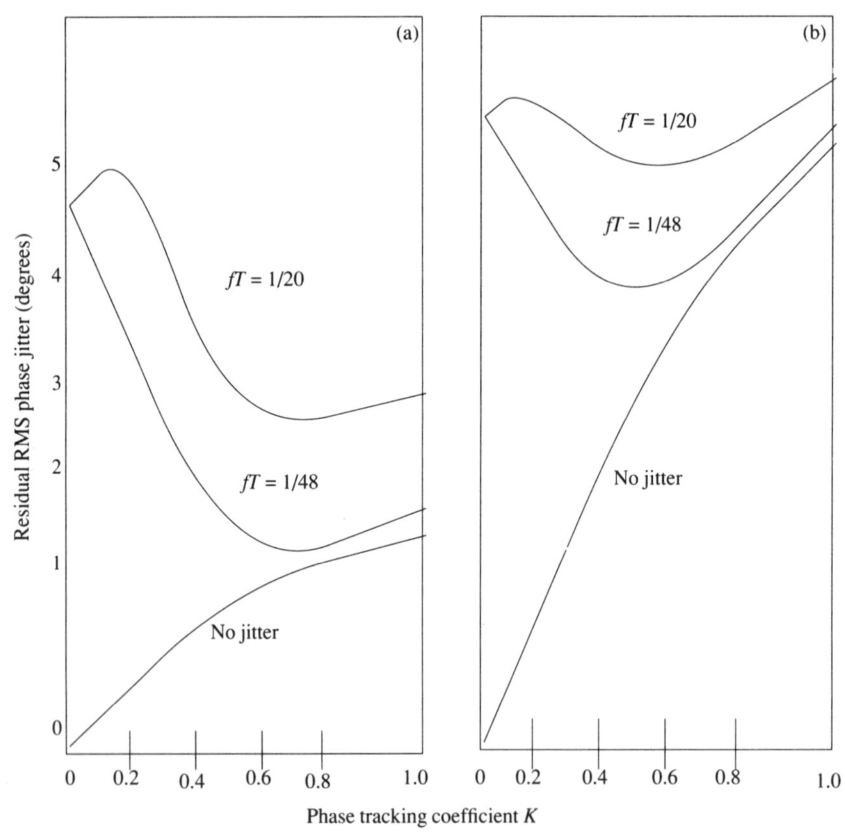

Fig. 6.16 (a) Residual RMS phase jitter for a channel with 30-dB SNR and 14 degrees peak-to-peak phase jitter. (b) Residual RMS phase jitter for a channel with 22-dB SNR and 14 degrees peak-to-peak phase jitter.

tolerate substantially more phase jitter (recall that its signal points are separated by 90°), but has a ~1 dB penalty in Gaussian noise. The signal constellation selected for a particular application should be based on the amount of phase jitter expected and the performance of the receiver PLL (i.e., the amount by which the PLL reduces the input jitter to residual phase jitter).

This concludes our discussion of carrier-tracking loops; we again emphasize that the data-directed PLL described above can achieve the same performance as a PLL presented with an unmodulated carrier. The data-directed PLL suffers no loss, in contrast to the squarer/PLL, in removing the degradation caused by the modulation and because of this fact has become the standard carrier recovery method in high-performance QAM systems.

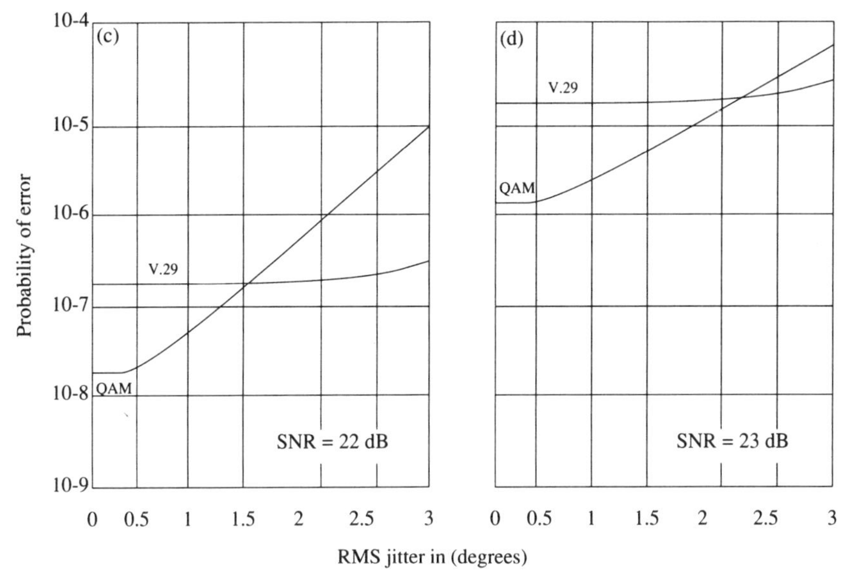

Fig. 6.17 Error rate vs. jitter with a phase-locked loop: (a) average power constraint; (b) peak power constraint.

6.5 TIMING RECOVERY

The proper recovery and tracking of timing frequency and phase is essential for the reliable detection of modulated data signals. For satisfactory performance in multilevel pulse-amplitude modulated (PAM) systems, the timing error must be held to an increasingly smaller fraction of a symbol interval as the number of levels increase. Recall the eye diagram described in Chapter 4, which shows the sensitivity of PAM systems to timing phase errors. As we have remarked at the beginning of this chapter, we focus on timing recovery techniques that are in-band, i.e., separate tones for the recovery of the clock signals are not provided. For this in-band, or self-clocked mode of operation the received signal is fed into a timing circuit and processed (typically through a nonlinearity followed by a PLL) to produce a timing wave where the frequency of zero crossings is the desired timing frequency and whose phase can be related to the timing phase that is appropriate for the receiver. As with carrier phase and frequency recovery, the received signal will generally be processed nonlinearly to generate spectral lines and then fed into a PLL which will lock onto the desired tone. Again we consider two classes of self-clocked timing recovery: non-data-aided and data-aided. We begin with a discussion of maximum likelihood (ML) timing recovery.

6.5.1 MAXIMUM LIKELIHOOD TIMING RECOVERY SYSTEM

We begin our discussion of timing recovery with the baseband PAM signal

$$x(t) = \sum_{k=-\infty}^{\infty} a_k \, p(t - kT - \tau) \,, \tag{6.80a}$$

where T is the signaling interval and τ is a time shift, possibly random, introduced by the channel. If $p(t)$ is a Nyquist pulse, sampling at the "right time," i.e., $t = nT + \tau$ reproduces the symbol. If $p(t)$ is a partial response pulse, correct sampling gives a controlled amount of intersymbol interference. However if sampling is at an instant sufficiently far away from the desired epoch, the intersymbol interference can obliterate the information (see the eye diagram). Our discussion of timing recovery is somewhat different in one central aspect than carrier phase recovery. For the latter system, the desired carrier phase is unambiguous. If the system pulse is highly distorted, then the desired timing epoch may not even exist, in the sense that there is no epoch for which reliable performance is achieved. In this case equalization of the pulse is necessary before a suitable epoch can be identified.

The form of the optimum timing recovery circuit is quite similar to that of the carrier phase recovery, a nonlinear device usually a squarer, followed by a phase tracker. In order to motivate this configuration, consider the maximum likelihood estimator detector of timing phase. The received signal plus noise can be written

$$y(t) = \sum_{m=1}^{N} a_m p(t - mT - \tau) + n(t) \,, \tag{6.80b}$$

where we will focus on estimating the epoch, τ. Here we assume that $p(\cdot)$ is relatively undistorted so that there is a sampling phase, τ, that produces reliable decisions. We will assume that the free-running receiver timing frequency is close enough to T^{-1}, so that the difference can be absorbed in the tracking of τ.

We begin with the simplest case of binary data and no intersymbol interference. We also assume white Gaussian noise with double-sided density $N_0/2$. Following the same sequence of steps that lead to (2.18)–(2.21), the likelihood of $y(t)$ conditioned on τ, and averaged over the transmitted sequence a_n, $n = 1, 2, \ldots, N$ is

$$L[y(t)|\tau] = E_{a_1, a_2, \ldots, a_N} \exp\left[-\frac{1}{2N_0}\int_0^{NT} [y(t) - \sum_{n=1}^{N} a_n p(t - nT - \tau)]^2 \, dt\right] , \tag{6.81}$$

where the *observation interval*, NT, represents the amount of delay in the estimation process and will be inversely proportional to the loop bandwidth in a closed-loop implementation of the ML estimate. Expanding the squared term in the exponent, we observe that

i) $\int_0^{NT} y^2(t)\, dt$ is independent of τ and $a_1, a_2, \ldots, a_N$.

ii) If the observation interval, NT, is large relative to the symbol interval, T, then $\int_0^{NT} \left[\sum_n a_n p(t - nT - \tau)\right]^2 dt$ will also be independent of τ and $a_1, a_2, \ldots, a_N$.

Eliminating those terms that do not depend on τ, we see that to determine the maximum likelihood estimate, $\hat{\tau}_{ML}$, we need to perform the average

$$L[y(t)|\hat{\tau}] = E_{a_1, \ldots, a_N}\left[\exp \frac{-1}{2N_0} \sum_{n=1}^{N} a_n \int_0^{NT} y(t)\, p(t - nT - \hat{\tau})\, dt\right]$$

$$= \prod_{n=1}^{N} E_{a_n} \exp\left[\frac{-a_n}{2N_0} \cdot q_n(\hat{\tau})\right], \tag{6.82}$$

where we have used the independence of the data symbols, and where $q_n(\hat{\tau})$ is defined by

$$q_n(\hat{\tau}) = \int_0^{NT} y(t)\, p(t - nT - \hat{\tau})\, dt, \tag{6.83}$$

which is recognized as the output, at times $nT + \hat{\tau}$, of a filter *matched* to the transmitted pulse, $p(t)$. Taking the average of (6.82) with respect to the data symbol, taking logarithms, and neglecting constants we have to *maximize*

$$l(\hat{\tau}) \equiv L[y(t)|\hat{\tau}] \sim \sum_{n=0}^{N} \ln \cosh \frac{q_n(\hat{\tau})}{2N_0}, \tag{6.84}$$

in order to obtain $\hat{\tau}_{ML}$.

A functional block diagram of the above operations is shown in Figure 6.18. Of course, like the ML carrier phase estimate of Figure 6.3, this approach is not practical since the receiver has to wait the entire observation period to determine $\hat{\tau}$. Note that, for moderate to small values of x (i.e., moderate SNR), since $\ln \cosh x \sim x^2/2$, the simplified likelihood expression,

$$l(\hat{\tau}) \sim \sum_{n=0}^{N} q_n^2(\hat{\tau}), \tag{6.85}$$

may be maximized. Note that (6.85) contains a squaring operation. An expression similar to (6.85) is obtained directly if we assume that the data symbols

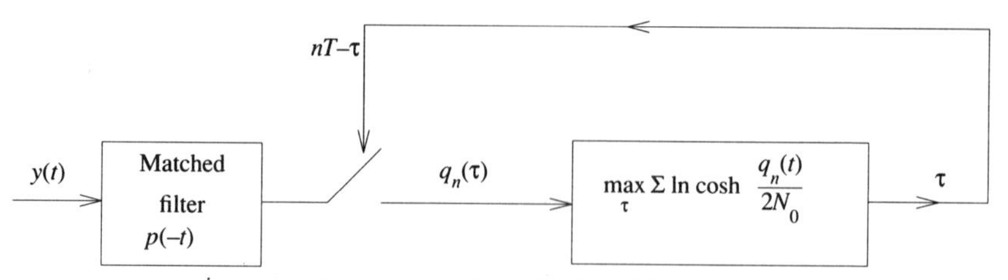

Fig. 6.18 Open-loop maximum likelihood timing recovery.

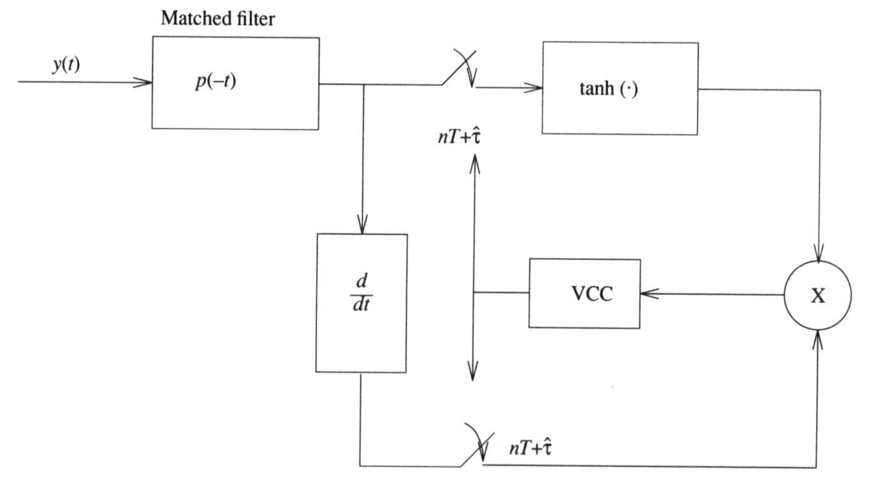

Fig. 6.19a Closed-loop approximation to maximum likelihood timing recovery.

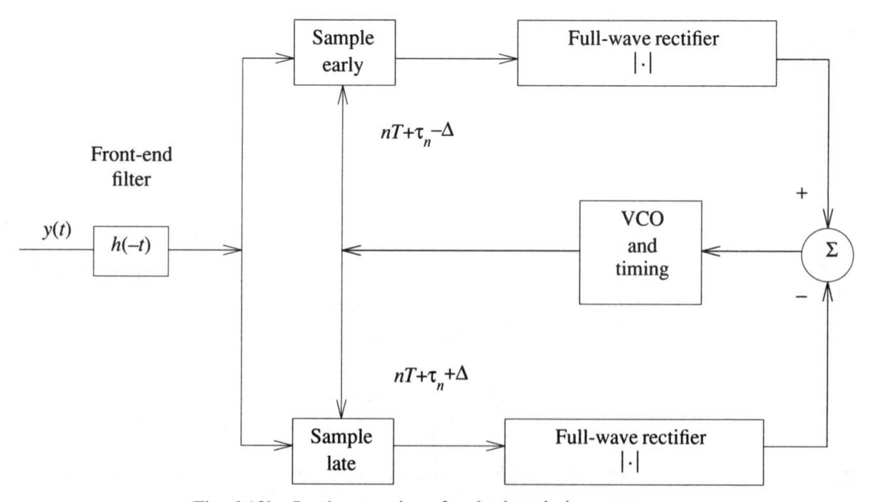

Fig. 6.19b Implementation of early–late timing recovery.

have a Gaussian distribution, rather than a binary one. Thus when there is not much knowledge of the data symbols (low SNR or a Gaussian distribution), the ML estimate involves maximizing a function that processes the *squared* output samples of a matched filter. To move toward a realizable timing loop, we follow an approach similar to that used in Section 6.1, where we introduced a regression function and considered the contribution to the ML estimate during each symbol period. We define the regression function

$$m(\hat{\tau};\, \tau) = E\, \frac{dl(\tau)}{d\tau}\,, \tag{6.86}$$

where the regression function will have a root at $\hat{\tau} = \tau$, the desired epoch. The closed-loop recursion becomes

$$\hat{\tau}_{n+1} = \hat{\tau}_n - \beta \left. \frac{dl(\tau)}{d\tau} \right|_{\tau=\hat{\tau}_n}, \qquad n = 1, 2, 3, \ldots \,. \tag{6.87}$$

Using (6.84), we have the recursion

$$\hat{\tau}_{n+1} = \hat{\tau}_n - \beta \dot{q}_n(\hat{\tau}_n) \tanh\left(q_n(\hat{\tau}_n)/2N_0\right)\,, \tag{6.88}$$

where the "dot" denotes differentiation with respect to time. This system is shown in Figure 6.19a, where VCC is a voltage-controlled clock that implements the recursion (6.88). As the SNR increases, the tanh(·) nonlinearity can be replaced by sgn(·) since the tanh(·) function approaches a sgn(·) function, or a limiter, whose output will be data decisions. If we had assumed decision-directed operation, i.e., the a_n's are known at the receiver, then we would also arrive at (6.88). Note that if the output decisions are correct, then the loop (6.88) will stop adaptation when $E\, a_n \dot{q}_n(\tau) = 0$, and $\hat{\tau}_{ML}$ is at the maximum of the pulse correlation function, $\int p(t)p(t - \tau)\,dt$. Figure 6.19b shows an *early–late* gate [8] realization of the system, where the unrealizable operation of differentiation is replaced by sampling of the waveform just before and after the receiver sampling phase.

6.5.2 SQUARER-BASED TIMING RECOVERS

Motivated by the squarer-based expression (6.85) for the likelihood as a function of the timing phase, we now discuss the most common form of timing recovery circuit. The squarer, or envelope-based timing recovery system can be either open or closed loop depending on whether the nonlinear operation occurs before or after the sampler. We consider the circuit shown in Figure 6.20, where the PLL or bandpass filter is tuned to the timing frequency $1/T$ Hz. Note that the emphasis here is on obtaining a timing wave of the correct frequency. Typically, the envelope of a PAM waveform will be cyclostationary, so that a timing tone can be derived.

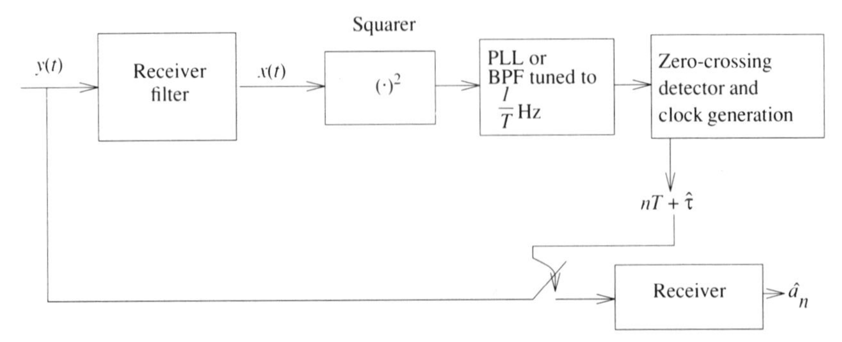

Fig. 6.20 Squaring timing recovery circuit.

Let us assume that a_n in (6.80a) is a zero-mean, stationary sequence with independent elements. Observe that the square of $x^2(t)$ has a periodic mean value,

$$E[x^2(t)] = \overline{a^2} \sum_k p^2(t - kT - \tau) = E[x^2(t+kT)] , \tag{6.89}$$

with a period equal to the transmitted symbol period. By proper filtering, this signal can be used for timing recovery. For a QAM signal, we form $E[x(t)\ x^*(t)]$, where $x^*(t)$ is the complex conjugate of $x(t)$, which produces an expression proportional to (6.89). To determine the strength of the fundamental component of (6.89), which is the desired timing tone, we use the Poisson sum formula

$$\sum_{k=-\infty}^{\infty} h(t - kT) = \frac{1}{T} \sum_{l=-\infty}^{\infty} H(l/T)\ e^{j\,2\pi\, l\, t/T} , \tag{6.90}$$

where $h(t)$ and $H(f)$ are Fourier transform pairs. Using (6.89) and (6.90), we express the mean of the squared input signal by the Fourier series

$$E[x^2(t)] = \frac{\overline{a^2}}{T} \mathrm{Re} \sum_l D_l \exp \frac{j2\pi l(t - \tau)}{T} , \tag{6.91}$$

where

$$D_l = \int_{-\infty}^{\infty} P\ (l/T - f)\ P(f)\ df . \tag{6.92}$$

The coefficient of the desired tone, D_1, is sketched in Figure 6.21, and is proportional to the area under the shaded portion of the curves. Notice that this is a function of the excess bandwidth of the signal. If $P(f)$ is a minimum-bandwidth Nyquist transfer function, then $D_1 = 0$, since there is no overlap between $P(1/T - f)$ and $P(f)$. In this case timing recovery using squaring is *not* possible. Of course, minimum bandwidth systems are generally not used (although severe channel distortion could reduce the received bandwidth to the minimum, or less). For typical, but efficient, data communications systems we

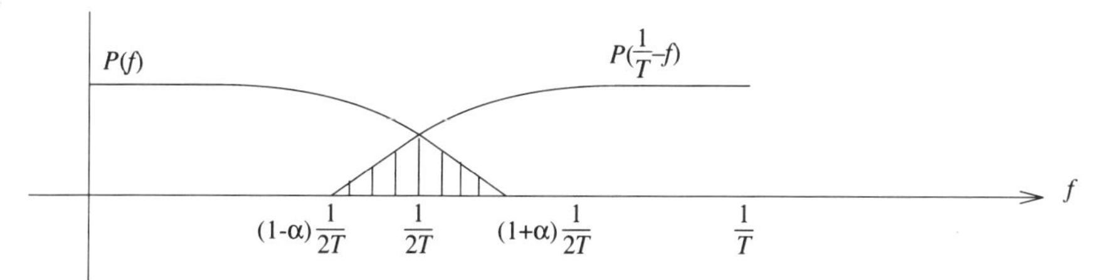

Fig. 6.21 Computation of the coefficient of the timing–frequency tone in a squaring system. $P(f)$ is the baseband pulse shape and α is the roll-off. The power in the tone is proportional to the area under the shaded portion of the curve.

have $|P(f)| = 0$ for $|f| > 1/T$ (see Figure 6.21). Thus there are only three nonzero terms in (6.92), $l = 0, \pm 1$ for any value of f. Passing the output of the squarer through a narrowband filter with unit gain at $f = 1/T$, the recovered waveform is

$$E[w(t)] = \frac{\overline{a^2}}{T} \operatorname{Re}\left[D_1 \exp\left[\frac{j\,2\pi\,(t-\tau)}{T}\right]\right] \tag{6.93}$$

$$= \frac{\overline{a^2}}{T} D_1 \cos\left[\frac{2\pi}{T}(t-\tau)\right] .$$

The zero crossings of the reference waveform are the desired time instants displaced by $T/4$ seconds, and a simple time shift produces the desired epoch. While the mean-squared signal has been shown to contain a tone at the symbol rate, we must remember that the receiver is presented with $x^2(t)$, not $Ex^2(t)$, and the squared signal can be written as

$$x^2(t) = \left[\sum a_n\, p(t - nT - \tau)\right]^2$$

$$= \sum_n a_n^2\, p^2(t - nT - \tau) + \sum_{n \neq m}\sum a_n\, a_m\, p\,(t - nT - \tau)\; p\,(t - mT - \tau) \tag{6.94}$$

Since $a_n^2 = 1$, the first term is periodic and produces the symbol-rate tone, while the second term is the jitter (or signal dependent noise) that perturbs the zero crossings of the periodic signal. Referring to Figure 6.21 we know that only the signal power in the roll-off region contributes to the tone, the remaining power contributes to the jitter. This observation suggests *prefiltering* the input to the squarer, to a narrow region about $1/2T$ Hz, so as to reduce the jitter in the timing recovery circuit. Such a system is shown in Figure 6.22.

When there is very little excess bandwidth, two alternatives are available: use of a higher-order nonlinearity or (for systems that use fractionally spaced equalizers) monitoring the dynamics of the equalizer tap weights [14–15] Mazo

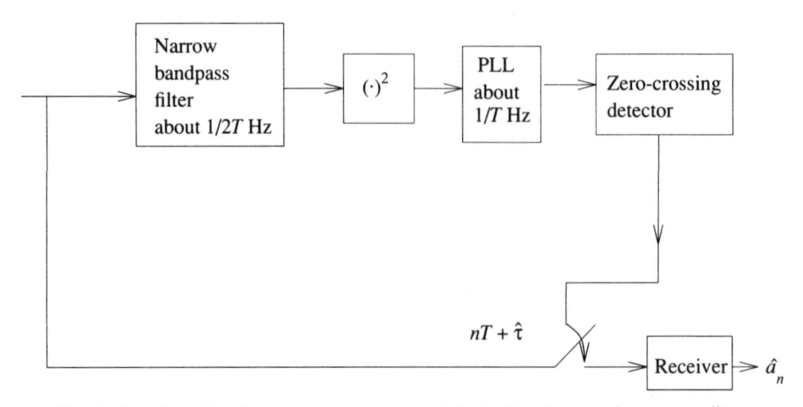

Fig. 6.22 Squaring timing recovery circuit with pre-filtering to reduce output jitter.

[9] has shown that a fourth-power circuit will produce a tone, even the absence of excess bandwidth, and for small excess bandwidth will have a larger tone-to-jitter ratio than the squaring loop.

Note that for the QAM signal

$$y(t) = \sum_n a_n h(t-nT+\tau) \cos\left[2\pi f_c t+\theta_c\right] - \sum_n b_n h(t-nT+\tau) \sin\left[2\pi f_c t+\theta_c\right] , \tag{6.1}$$

the squared envelope of the signal, $y^2(t) + \check{y}^2(t)$, where $\check{y}(t)$ is the Hilbert transform of $y(t)$, has an average value

$$E\{y^2(t) + [\check{y}(t)]^2\} = \left[\overline{a^2} + \overline{b^2}\right] \sum_n h^2(t - nT - \tau) . \tag{6.95}$$

This expression is very similar to (6.89) and will provide the same quality timing tone as the baseband system described above. It is important to observe that (6.95) is *independent* of the carrier phase θ_c. This means that for QAM signals, timing can be extracted without regard to the carrier phase. Often, QAM receivers, as in Figure 6.23, have a front-end filter that produces the Hilbert transform needed to generate (6.95). This signal is processed by a PLL to extract the timing frequency and epoch (similar to the subsystem shown in Figure 6.22); a separate signal path is taken to provide an estimate of the carrier phase. If the carrier recovery system is of the data-aided type, then the data decisions are fed back to aid in the updating of the carrier phase estimate. The ability to achieve independence of timing and carrier recovery in QAM systems is an important feature of double-sided modulation. As we show in Section 6.6,

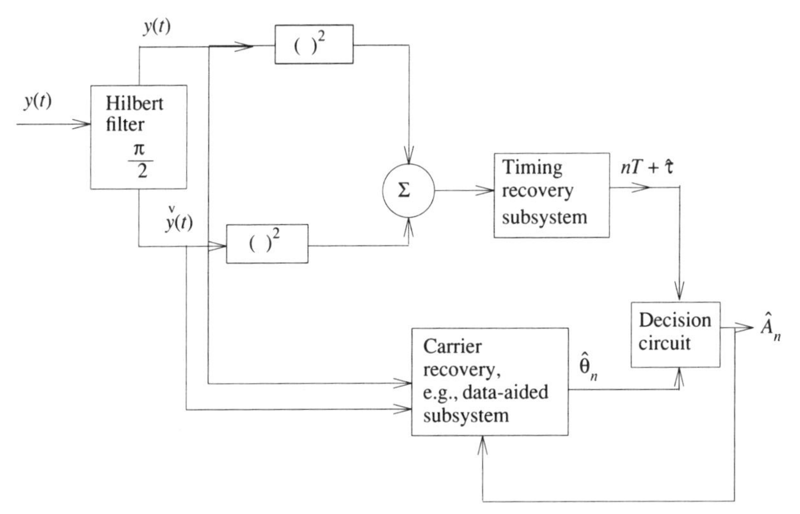

Fig. 6.23 A QAM receiver that decouples timing and carrier recovery.

asymmetrical modulation methods such as VSB and SSB do not readily lend themselves to independent timing and carrier recovery systems.

6.5.3 SYMBOL RATE ALL DIGITAL TIMING RECOVERY [10]

Until now the timing recovery algorithms that we have discussed have either been open loop and analog (the squarer) or closed loop and involving substantial amounts of analog signal processing (such as the loop shown in Figure 6.19a that uses the derivative of the input signal). In this section we discuss an all-digital timing recovery method that uses a minimum of signal processing. With the current trend toward fully digital receivers we are confronted with a challenge. The signal in such a receiver is sampled and A/D converted at the input. It is available only at discrete time intervals for further processing. Sampling could be performed at a rate high enough to allow a complete reconstruction of the signal, and analog timing recovery schemes (such as the squarer) could then be "digitized" and would still perform in a functionally equivalent way. However, at high rates such an approach is a complex and expensive solution which leaves much to be desired (e.g., the high A/D conversion rate that is needed with such a scheme). Furthermore, in many echo-cancellation-based systems (see Chapter 9), it is desirable to sample in synchronism with the symbol rate and many such systems actually use only one sample per baud interval for signal processing. Such a low sampling rate is justified because the final decisions at the output are also based on samples taken at the symbol rate, and the behavior of the data signal between the sampling instants is immaterial. Another difference between analog and digital processing is depicted in Figure

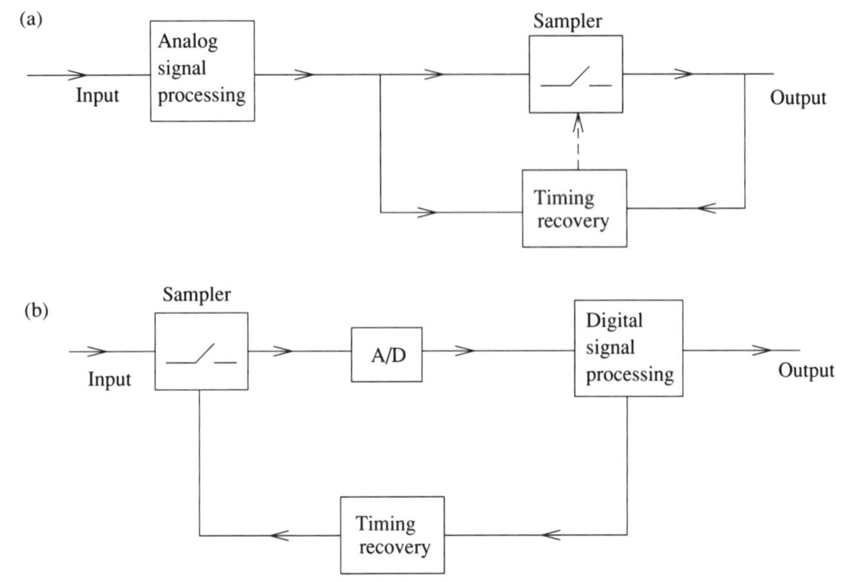

Fig. 6.24 Block diagram of timing recovery loop for baseband data receiver. (a) analog signal processing. (b) digital signal processing.

6.24 for the simple case of a synchronous baseband data receiver. In the analog version, the signal is sampled after processing and conditioning, whereas in the digital receiver the unconditioned and probably distorted signal is sampled. Timing recovery from the digitized signal must always be achieved with a feedback loop, in contrast to analog processing, where nonfeedback schemes are possible.

The key to all-digital timing recovery is to construct monotonic regression functions of the desired epoch which can be computed solely from baud-rate samples and whose root is close to the minimum of a reasonably chosen performance measure. Consider the regression function

$$m(\tau) = h_1 - h_{-1} = h(\tau + T) - h(\tau - T) , \tag{6.96}$$

where $h(\cdot)$ is the system impulse response at the input to the receiver. Finding the root of this function is equivalent to forcing the two pulse samples, on either side of the main sample, to be equal. In Figure 6.25 we show $m(\tau)$ for a cosine roll-off pulse with various roll-offs (α). Note the linearity of the function around the origin, and the relative constancy of the slope as a function of roll-off. These two properties guarantee that a zero-seeking algorithm can be designed to achieve the condition $m(\tau) = 0$.

It should be emphasized that, in unequalized systems, selection of (6.96) as the function which will provide the sampling epoch, is only appropriate when

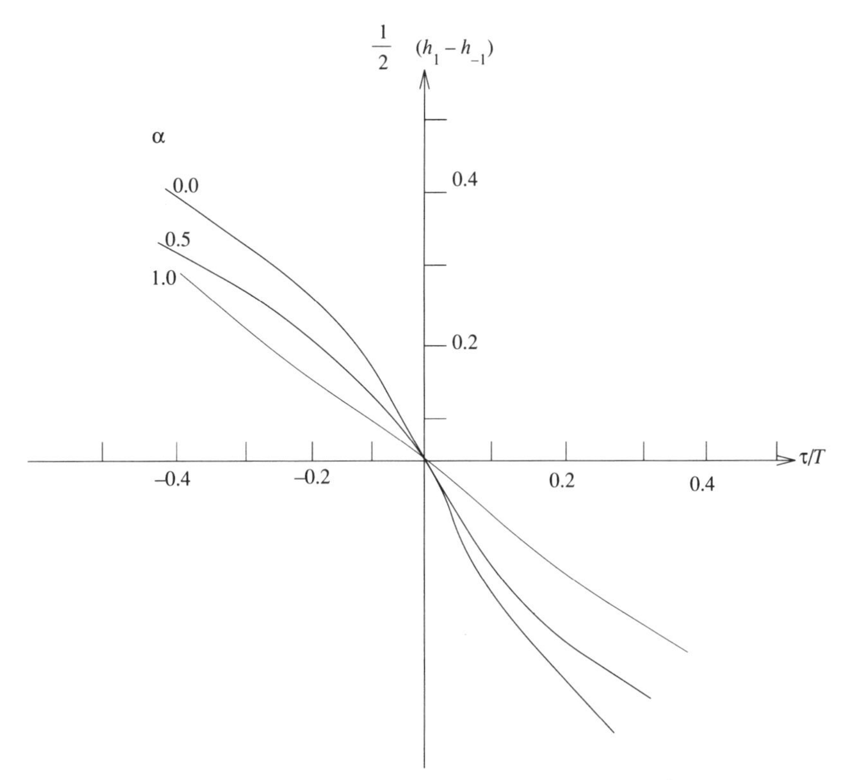

Fig. 6.25 Phase detector characteristics of a loop which forces $h_1 = h_{-1}$ (Nyquist channel with cosine roll-off α).

the class of channels has been sufficiently well characterized so that an epoch satisfying (6.96) produces a satisfactory error rate. With this caveat we proceed to iteratively find the root of $m(\tau)$; to do so we need to find a function whose average is $m(\tau)$. Consider the linear combination

$$z_n(\tau) = x_n\, a_{n-1} - x_{n-1}\, a_n \,, \tag{6.97}$$

where $\{x_n\}$ are the samples of the received signal. It is straightforward to show that the average of $z_n(\tau)$ with respect to the data symbols and noise produces the desired function

$$E\Big[z_n(\tau)\Big] = m(\tau) = h_1 - h_{-1} \,. \tag{6.98}$$

To find the value of τ such that $m(\tau) = 0$, the timing phase would be adjusted according to

$$\hat{\tau}_{n+1} = \hat{\tau}_n - K\, z_n(\hat{\tau}_n) \,. \tag{6.99}$$

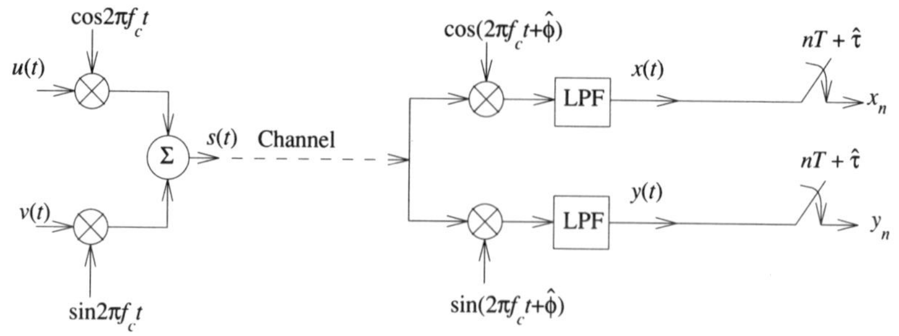

Fig. 6.26 Quadrature transmission system.

There are many regression functions, $m(\tau)$ and signal functions $z_n(\tau)$ that can be invented, and the designer's selection would be influenced by the error rate achieved, complexity, jitter about the regression root, and the robustness of the timing epoch to different channel distortion.

6.6 JOINT CARRIER AND TIMING RECOVERY

In the preceding sections we described individual carrier and timing recovery techniques. In Figure 6.23, we showed that for QAM systems it was possible to provide independent carrier and timing recovery systems. Here we will discuss the feasibility, desirability, and performance of *joint* estimation of θ and τ. In principle, joint estimates should not be worse than the individual estimates. We will discuss the important differences between systems that use symmetric modulation methods, such as QAM, where the carrier frequency is in the center of the passband spectrum and systems that use asymmetrical modulation, such as SSB, where the carrier frequency is at the edge of the spectrum. We deviate from our philosophy of maximum-likelihood (decision-directed) estimation (see Exercise 6.19) and use the mean-square error as our performance measure. This metric allows us to readily evaluate system performance.

We begin our discussion by assuming a transmitted signal

$$s(t) = u(t) \cos 2\pi f_c t + v(t) \sin 2\pi f_c t , \tag{6.100}$$

where $u(t)$ and $v(t)$ are baseband data signals which depend on the particular system under consideration. At the receiver, we will assume a quadrature demodulation circuit (see Figure 6.26), which is functionally equivalent to using a Hilbert filter (e.g., Fig. 6.15). The receiver has a phase error ϕ and a timing error τ. The received signal is multiplied with $\cos(2\pi f_c t + \phi)$ and $\sin(2\pi f_c t + \phi)$, and after suppression of the double frequency terms, the signals

$$x(t) = u(t) \cos\phi - v(t) \sin\phi + n_1(t) \tag{6.101}$$

$$y(t) = u(t) \sin\phi + v(t) \cos\phi + n_2(t) \tag{6.102}$$

are obtained, where $n_1(t)$ and $n_2(t)$ are independent Gaussian noise components introduced by the channel. By assuming an ideal channel we have separated timing and carrier recovery from equalization. This is partly justified because the equalizer usually changes slowly in the steady state mode, so that transient phase changes of the channel are first absorbed by the faster-reacting adaptive timing and carrier recovery loops. In addition there are some channels (like radio links) which do not show the large amount of linear distortion.

Before we proceed, we will say a few words about the baseband signals $u(t)$ and $v(t)$. The following table summarizes the modulation methods of interest:

$$\text{DSB:} \quad u(t) = \sum a_n h(t-nT), \quad v(t) = 0$$

$$\text{QAM:} \quad \begin{cases} u(t) = \sum a_n h(t-nT) \\ v(t) = \sum b_n (t-nT) \end{cases}$$

$$\text{SSB:} \quad \begin{cases} u(t) = \sum a_n h(t-nT) \\ v(t) = \check{u}(t) = \sum a_n \check{h}(t-nT); \text{ where } \check{U}(f) = -j \text{ sgn}(2\pi f)\cdot U(f) \end{cases}$$

$$\text{VSB:} \quad \begin{cases} u(t) = \sum a_n h(t-nT) \\ v(t) = u_\beta(t) = \sum a_n h_\beta(t-nT); \end{cases}$$

$$\text{where } U_\beta(f)/U(f) = \begin{cases} -j \sin(2\pi fT/2\beta), \text{ if } |f| \le \beta/2T \\ -j \text{ sgn}(2\pi f), \text{ if } |f| \ge \beta/2T \end{cases}$$

$$\text{PSK:} \quad \begin{cases} u(t) = \sum \cos(\theta_n) h(t-nT) \\ v(t) = \sum \sin(\theta_n) h(t-nT) \end{cases} \tag{6.103}$$

As a performance measure, we define the mean-square error (MSE) between $x(t)$ and $u(t)$ at the sampling times,

$$\varepsilon^2 = E\{[x(\tau+nT) - u(nT)]^2\} = E\{[y(\tau+nT) - v(nT)]^2\} \tag{6.104}$$

as both our performance measure and regression function (see (6.86)). As with regression functions, to derive updating algorithms we take the gradients with respect to τ and ϕ, which are given by

$$\begin{aligned} \frac{\partial \varepsilon^2}{\partial \tau} &= 2E\left\{e(\tau+nT)\ \frac{\partial x(\tau+nT)}{\partial \tau}\right\} \\ \frac{\partial \varepsilon^2}{\partial \phi} &= -2E\left\{e(\tau+nT)\ y(\tau+nT)\right\}, \end{aligned} \tag{6.105}$$

and the update equations are

$$\hat{\tau}_{n+1} = \hat{\tau}_n - \alpha \frac{\partial \varepsilon^2}{\partial \tau} \bigg|_{\tau=\hat{\tau}_n}$$

$$\hat{\phi}_{n+1} = \hat{\phi}_n - \beta \frac{\partial \varepsilon^2}{\partial \phi} \bigg|_{\phi=\hat{\phi}_n} , \tag{6.106}$$

where $e(\tau + nT)$ is the error signal between the actual and the desired data output at the sampling instants. The above notation applies directly to DSB, VSB, and SSB systems. It is also valid for the cases of PSK and QAM, since optimization with respect to one component, say in-phase, of the mean-square error will also minimize the MSE in the other channel. A very significant advantage can however be gained in QAM-type systems if the cost function (6.104) is modified so as to contain the sum of the mean-square errors in the in-phase and quadrature channels. We will discuss such a symmetrical system in a moment.

The cost function (6.104) can be expressed in the form

$$\varepsilon^2 = E\{[u(\tau+nT)\cos\phi - v(\tau+nT)\sin\phi + n_1(\tau+nT) - u(nT)]^2\}. \tag{6.107}$$

Since the noise is uncorrelated with the data signals, the mean-square error can be expressed in terms of the individual correlation functions,

$$\varepsilon^2 = R_{uu}(\tau,\tau)\cos^2\phi + R_{vv}(\tau,\tau)\sin^2\phi + R_{uu}(0,0)$$
$$- R_{uv}(\tau,\tau)\sin 2\phi - 2R_{uu}(\tau,0)\cos\phi + 2R_{vu}(\tau,0)\sin\phi + \sigma^2 . \tag{6.108}$$

where σ^2 is the noise variance. As we have noted earlier in this chapter, the random processes $u(t)$ and $v(t)$ are not stationary in general, but are cyclostationary processes. Their correlation functions are defined by

$$E\{u(t+\tau)u(t)\} \triangleq R_{uu}(t+\tau,t) \tag{6.109}$$

and are periodic in t with period T. When the bandwidth is limited to the Nyquist frequency, the process $u(t)$ becomes weakly stationary and the correlation functions depend on the time difference only.

Using (6.108), the gradients can be expressed as

$$\frac{\partial \varepsilon^2}{\partial \tau} = \frac{\partial}{\partial \tau} R_{uu}(\tau,\tau)\cos^2\phi + \frac{\partial}{\partial \tau} R_{vv}(\tau,\tau)\sin^2\phi - \frac{\partial}{\partial \tau} R_{vu}(\tau,\tau)\sin 2\phi$$
$$- 2\frac{\partial}{\partial \tau} R_{uu}(\tau,0)\cos\phi + 2\frac{\partial}{\partial \tau} R_{vu}(\tau,0)\sin\phi \tag{6.110}$$

$$\frac{\partial \varepsilon^2}{\partial \phi} = [R_{vv}(\tau,\tau) - R_{uu}(\tau,\tau)]\sin 2\phi - 2R_{uv}(\tau,\tau)\cos 2\phi$$
$$+ 2R_{uu}(\tau,0)\sin\phi + 2R_{vu}(\tau,0)\cos\phi . \tag{6.111}$$

We will now discuss (6.108)–(6.111) with respect to the modulation schemes of interest.

All cross-correlation terms appearing in the above expressions are zero in symmetrical modulation schemes. This is easily verified for QAM (where u and v are independent), and it is trivial in the case of DSB ($v(t) = 0$). It is not so for SSB and VSB where the same symbols are used in the u and v signals.

Assume now that the carrier phase error ϕ has settled to $\phi = 0$, but that the timing has still some offset. The gradient with respect to the carrier then becomes

$$\left.\frac{\partial \varepsilon^2}{\partial \phi}\right|_{\phi=0} = -2R_{uv}(\tau,\tau) + 2R_{vu}(\tau,0) \ . \tag{6.112}$$

Since the cross terms are zero, this expression is zero for symmetrical modulation systems (as one would expect, because ϕ has settled and need not be updated any more), and the effect of an offset in either the timing or carrier loop will *not* affect the operation of the *other* loop. In nonsymmetrical modulation systems, however, (6.112) is not zero, which means that the carrier phase is shifted away from its zero position to an angle which yields less mean-square error for that particular timing offset. Consequently, an error in one loop will affect operation of the other loop. This different behavior for the two types of systems will affect the performance and capabilities of any joint $\phi-\tau$ loops.

6.6.1 SYSTEMS WITH SYMMETRICAL MODULATION

For QAM and we have

$$R_{uu}(\tau,\tau) = R_{vv}(\tau,\tau) \tag{6.113}$$

and thus we can express the mean-square error as

$$\varepsilon^2 = R_{uu}(\tau,\tau) + R_{uu}(0,0) - 2R_{uu}(\tau,0)\cos\phi + \sigma^2 \tag{6.114}$$

and the gradients become

$$\frac{\partial \varepsilon^2}{\partial \tau} = \frac{\partial}{\partial \tau} R_{uu}(\tau,\tau) - 2\frac{\partial}{\partial \tau} R_{uu}(\tau,0)\cos\phi \tag{6.115}$$

$$\frac{\partial \varepsilon^2}{\partial \phi} = 2R_{uu}(\tau,0)\sin\phi \ . \tag{6.116}$$

Note that for any given τ, the mean-square error is always minimized by choosing $\phi = 0$. From our discussions of maximum likelihood timing estimation, we note that the derivative of both $R_{uu}(\tau,\tau)$ and $R_{uu}(\tau,0)$ is zero at $\tau = 0$, the best sampling time is always at $\tau = 0$ independent of the carrier offset ϕ. *The effects of timing and carrier errors are thus completely decoupled for systems with symmetric modulation.*

As an example, we will discuss a QAM system with zero excess bandwidth. This can be the limiting case of PAM Nyquist signaling or a partial response system, for example duobinary. Since these signals are stationary, the update equations, (6.106), are simplified to

$$\hat{\tau}_{n+1} = \hat{\tau}_n + 2\alpha\, R'_{uu}(\hat{\tau}_n)\cos\phi_n \tag{6.117}$$

$$\hat{\phi}_{n+1} = \hat{\phi}_n - 2\beta\, R_{uu}(\hat{\tau}_n)\sin\phi_n\ , \tag{6.118}$$

where $R'_{uu}(\tau) = \partial R_{uu}(\tau)/\partial\tau$. The above equations can be linearized for small errors, ϕ_n and τ_n, in the region of interest; any attempt to solve the difference equations in their transcendental forms would overlook the fact that the symbol estimation is not reliable for large offsets. Linearization around the equilibrium point yields

$$\tau_{n+1} \approx \tau_n[1 + 2\alpha\, R''_{uu}(0)] \tag{6.119}$$

$$\phi_{n+1} \approx \phi_n[1 - 2\beta\, R_{uu}(0)] \tag{6.120}$$

and "one-step convergence" can be achieved if the step sizes of the two loops are chosen according to

$$\alpha = -\frac{1}{2R''_{uu}(0)}\ , \qquad \beta = \frac{1}{2R_{uu}(0)}\ . \tag{6.121}$$

It should be noted that these optimum step sizes are inversely proportional to the signal power. An accurate automatic gain control is therefore necessary in channels with an unknown or time-varying attenuation (fading). Since the stability limits are given by

$$-\alpha\, R''_{uu}(0) < 1, \qquad \beta\, R_{uu}(0) < 1\ , \tag{6.122}$$

a transient gain increase of more than 3 dB, which would not be immediately compensated by a slow-working AGC, could drive a critically adjusted loop into instability. This can however easily be prevented by inhibiting the loop during such transients.

Finally, we give the values of $R''_{uu}(0)$ for the cases of QAM Nyquist signaling and QAM-duobinary partial response signaling (the power $R_{uu}(0)$ is normalized to unity):

$$\text{Nyquist PAM:}\quad R''_{uu}(0) = \frac{\pi^2}{3T^2} \tag{6.123}$$

$$\text{Duobinary PR:}\quad R''_{uu}(0) = -\frac{\pi^2}{3T^2}\left[1 - \frac{2}{\pi^2}\right]. \tag{6.124}$$

It should be kept in mind that the evaluation of the correction terms in (6.117) and (6.118) generally involve time averaging over a large number of symbols. The length of averaging that must be done in practice depends on the variance of the expressions whose expected value is given in (6.105) and (6.106). It can be shown that this variance can be dramatically reduced, and in some cases even be completely eliminated, by modifying the cost function accordingly to the symmetrical structure of QAM systems, so that the actual training time can be significantly reduced. The improvement is realized by modifying the cost function so as to minimize the sum of the mean-square errors in both demodulated outputs, rather than considering the mean-square error of only one channel. This results in an algebraic cancellation of cross terms rather than the ensemble average cancellation of cross terms [when we use the asymmetric mean-square error (6.104)]. These results can be extended to nonsymmetric modulation systems, VSB and SSB; however there is a fundamental problem since the signals $u(t)$ and $v(t)$ are no longer independent of each other and the cross-correlations are not necessarily zero. For the symmetric systems, carrier phase offset results in an eye pattern closure without influencing the optimum sampling time. For such systems, a symmetrically structured algorithm provides substantial advantages: if the correction terms were available, convergence may take place in one step; practically no averaging is required to calculate the gradient (the variance of its estimate is determined by noise only when the timing has settled). The loop can thus react extremely fast (for start-up or jitter tracking). For SSB and VSB systems, a carrier phase offset results simultaneously in a closure and a shift of the eye pattern (owing to the timing change). For minimum error, the timing phase must be correspondingly shifted. Due to the strong interaction and signal correlation in the two loops, a joint algorithm for timing and carrier recovery can be expected to converge very slowly. It is possible to devise techniques which decorrelate the loop signals so that faster convergence can be achieved. However, the problems inherent in recovering carrier and timing for VSB and SSB systems are severe enough that insertion of special tones, called *pilot tones* [11], at the band edges or at the carrier frequency is often required for successful demodulation. Pilot tones are undesirable because: (1) stringent filtering is needed to extract the tones, (2) the power in the tones is at the expense of signal power, and (3) channel nonlinearities may produce in-band spectral lines that degrade performance. It is for these reasons that symmetrical modulation systems, such as QAM, have come to predominate in bandwidth-efficient communication systems, where there are stringent requirements on the magnitude of timing and carrier phase jitter.

6.7 PERIODIC INPUTS AND SCRAMBLERS

In our discussion of the carrier and timing recovery techniques presented in this chapter, it was tacitly assumed that the customer data symbols were truly random. A fair percentage of the time, business machines, personal computers, workstations and other terminal equipment will actually be transmitting an idle code or framing symbols while the (human) user is thinking. Rather than ceasing transmission when there are no data to be transmitted, idle codes are transmitted to maintain the required power level on the network facilities and to maintain the receiver AGC at a nominal level. In this section we look at the effect of idle codes on synchronization systems. Since periodic idle codes produce discrete lines in the received spectrum, we determine the conditions under which synchronization systems can satisfactorily operate in the presence of periodic idle codes. In most receivers a dense input line spectrum is needed to maintain satisfactory operation, and a scrambler will be used to randomize customer data which has a small period. It is critical that the equalizer be presented with a closely spaced line spectrum; for example, if the input period were two symbol intervals, it is clear that the equalizer can only compensate for distortion at two frequencies in the Nyquist band. Consequently, at the instant when the data returns from the periodic to the random mode, the equalizer frequency response will be far from the optimum (for random data) value, and the distortion at the equalizer output could be much larger than the channel distortion. A scrambler is typically used to provide an equalizer input that has a close-to-random (i.e., flat) line spectrum. These devices come in pairs: the scrambler, which is placed at the transmitter, and the descrambler, which inverts the scrambling operation, is placed at the receiver. The scrambler and descrambler are, respectively, linear feedback and feedforward shift register systems that generally produce output sequences that have a period of $2^N - 1$ symbols, where N is the number of stages in the shift register.*

6.7.1 TIMING RECOVERY WITH PERIODIC INPUT SEQUENCES

Data communications synchronization systems should be transparent to the user, i.e., periodic patterns of transmitted sequences should have no effect on the operation of the system. We first examine the effect of the user transmitting periodic sequences in time intervals where there are no customer data to transmit. Suppose we have a sequence of complex numbers $\{C_k\}$ which have a period of N symbol intervals, i.e.,

$$C_k = C_{k+N} \,, \quad k = 0, \pm 1, \pm 2, \ldots, N-1 \,. \tag{6.125}$$

We define the discrete Fourier transform of this sequence as

$$C(k\Omega) = \sum_{n=0}^{N-1} c_n e^{-jkn\Omega T} \,, \quad k = 0,\ 1,\ \ldots,\ N-1 \tag{6.126}$$

where $\Omega = 2\pi/NT$. Notice that the symbol period, T, is not critical since $\Omega T = 2\pi/N$. The inverse transform is given by

$$c_n = \frac{1}{N} \sum_{k=0}^{N-1} C(k\Omega)\ e^{jkn\Omega T} \,. \tag{6.127}$$

* These are maximal length shift register sequences described in Section 3.2 in connection with cyclic codes.

Consider the periodic QAM signal

$$s(t) = \text{Re} \sum_{n=-\infty}^{\infty} C_n p(t - nT) e^{j2\pi f_c t} \tag{6.128}$$

applied to the square-law timing recovery system. The output of the square law device is

$$r(t) = s^2(t) = \frac{1}{2} \left| \sum_{n=-\infty}^{\infty} C_n p(t-nT) \right|^2 + \frac{1}{2} \text{Re} \left\{ \left[\sum_{n=-\infty}^{\infty} C_n p(t-nT) \right]^2 e^{j4\pi f_c t} \right\} . \tag{6.129}$$

The second term on the right-hand side of (6.129) is dominated by high-frequency components which can be filtered out; consequently, we focus on the first term which is the envelope of $s(t)$. Since a product in the time domain is equivalent to convolution in the frequency domain, the Fourier transform of the first term in (6.129) is given by

$$\begin{aligned} R(f) &\triangleq F\left[\left| \sum_{n=-\infty}^{\infty} C_n p(t - nT) \right|^2 \right] \\ &= \left[\sum_{n=-\infty}^{\infty} C_n e^{-j2\pi f n T} P(f) \right] \circledast \left[\sum_{m=\infty}^{\infty} C^*{}_m e^{-j2\pi f m T} P(f) \right] , \end{aligned} \tag{6.130}$$

where $\circledast$ denotes convolution in the frequency domain and $P(f)$ is the Fourier transform of $p(t)$. Performing the convolution yields

$$\begin{aligned} R(f) = \sum_{k=0}^{N-1} \sum_{l=0}^{N-1} & C(k/NT) C(l/NT) \sum_n \sum_m P(k/NT + n/T) P(l/NT + m/T) \\ & \circledast \delta(f - [k-l]/NT - [m+n]/T) . \end{aligned} \tag{6.131}$$

This shows that there are tones spaced at intervals $1/NT$ throughout the spectrum. The desired tone is at $1/T$ and all the other sidelobe tones may be regarded as interference. The power in the desired tone is

$$R[1/T] = \sum_{k=0}^{N-1} |C(k/NT)|^2 \, P(k/NT) P(1/T - k/NT) , \tag{6.132}$$

while the power in an interfering tone is

$$\begin{aligned} & R(1/T + j/NT) \\ & = \sum_{k=0}^{N-1} C(k/NT) \delta((k-j)/NT) P(k/NT) P((k-j)/NT - 1/T), \quad j = 1,2,\ldots . \end{aligned} \tag{6.133}$$

The signal $r(t)$ is fed into a PLL which acts like a narrow-band filter around $f = 1/T$. Tones outside the band $(1/T - B/2,\ 1/T + B/2)$, where B is the bandwidth of the loop, are rejected. As we have seen earlier in this chapter, the strength of the tone to be tracked is a function of the pulse rolloff through the term $P(k/NT)P(1/T - k/NT)$. Again, we observe that the squaring approach fails if there is zero excess bandwidth. In order to reduce the magnitude of the sidelobes, the signal $s(t)$ in the timing loop may be passed through a filter centered at $1/2T$ which precedes the square-law device. From the foregoing it is clear that the frequency content of the line signal is strongly a function of the modulating data. Suppose for example that the all-zero sequence is transmitted $C(k\Omega) = \delta_{k0}$ and $N = 1$. We have

$$S(f) = \sum_{n=-\infty}^{\infty} G(n/T)\ \delta(f - f_0 - n/T)\ , \tag{6.134}$$

and suppose that the pulse has an excess bandwidth of less than 100% so that

$$S(f) = G(0)\ \delta(f - f_0)\ , \tag{6.135}$$

which is a single tone at the carrier frequency. Clearly this tone cannot be used to recover timing. An example spectrum is as shown in Figure 6.27, where $f_0 = 1800$ Hz, $1/T = 1200$ and the pulse has 100% excess bandwidth. Thus if a customer transmits all zeroes for a period of time, no timing information is present in the received signal. The PLL will drift since it is driven only by noise. When transmission of random data is resumed, timing must be recovered all over again.

6.7.2 SCRAMBLING SYSTEMS

The universally favored approach to increasing the period of sequences is scrambling. The simplest form of scrambling is to add a long pseudo-noise (PN) sequence to the data sequence at the transmitter and to subtract it (via modulo 2

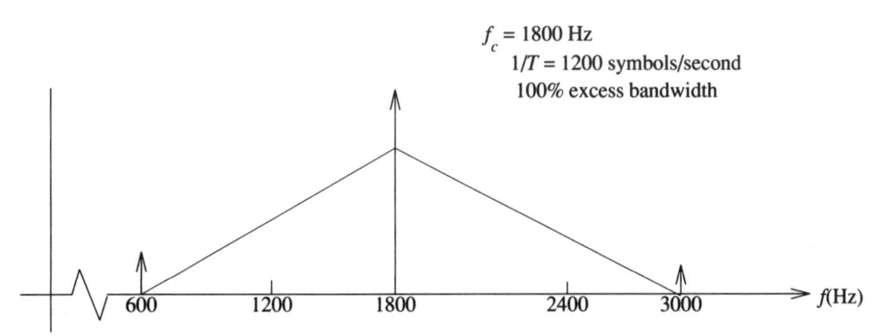

Fig. 6.27 Spectrum at output of square law timing recovery circuit when customer data is the all-zero sequence.

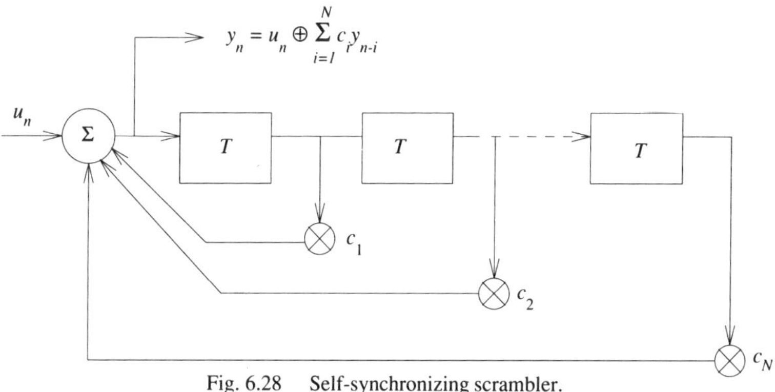

Fig. 6.28 Self-synchronizing scrambler.

addition) at the receiver. In the case of all-zero transmission, the transmitted sequence will have a period equal to the length of the PN sequence, and there is a rich spectrum for timing recovery, i.e., many harmonics of $1/NT$. The problem with this straightforward approach is its vulnerability to a dropped or added bit. In this event, the wrong sequence is subtracted at the receiver and disastrous error propagation ensues.

An approach for which the effects of synchronization loss are relatively short lived is the so-called *self-synchronizing scrambler* [12]. The scrambler and descrambler are the linear sequential circuits shown in Figures 6.28 and 6.29 respectively, and consist of feedback and feedforward shift registers. Data symbols are fed into the scrambler every T seconds and added modulo 2 to past outputs to produce the current output. The inputs to the delay elements shown in Figure 6.28 are delayed by T seconds. The output of the scrambler is then encoded for transmission over the channel. After decoding at the receiver, the resulting sequence is put through a descrambler, shown in Figure 6.29, where the original sequence is recovered. The inverse scrambler is self-synchronizing, and it will eventually cleanse itself of a transmission error once the error has propagated through the shift register. The number of errors in the descrambler

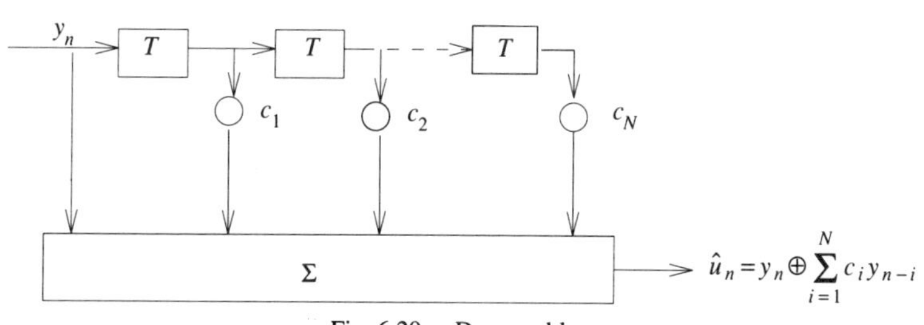

Fig. 6.29 Descrambler.

output sequence is the number of channel errors multiplied by the number of nonzero tap weights in the shift register.

To demonstrate that the scrambler and descrambler are indeed inverse operations refer to Figure 6.28 to see that the scrambler output is

$$y_n = u_n \oplus \sum_{i=1}^{N} c_i \, y_{n-i} \, , \tag{6.136a}$$

where u_n is the input data sequence, y_n the scrambler output sequence, and c_i are the shift register tap weights. From Figure 6.29, the inverse scrambler output is

$$\hat{u}_n = v_n \oplus \sum_{i=1}^{N} c_i \, v_{n-i} \, , \tag{6.136b}$$

where v_n is the input. If there are no channel errors, then the scrambler output, y_n, is the descrambler input v_n, and the descrambler output becomes

$$\hat{u}_n = y_n \oplus \sum_{i=1}^{N} c_i y_{n-i} \, , \tag{6.137a}$$

and substituting (6.136a) into (6.137a) gives the desired result,

$$\hat{u}_n = u_n \oplus \sum c_i y_{n-i} \oplus \sum_{i=1}^{N} c_i y_{n-i} = u_n \, , \tag{6.137b}$$

that the inverse scrambler output is indeed the customer data. While the feedforward and feedback operations could be interchanged at the transmitter and receiver, it is essential that the feedback system not be used at the receiver. This precludes *error propagation*.

6.7.3 THE SCRAMBLER AS A LINEAR SEQUENTIAL CIRCUIT

In order to analyze the operation of the self-synchronizing scrambler, we return to the consideration of linear sequential circuits which were introduced in Section 3.2.2. In general, the output of a linear sequential circuit consists of the sum of the *free response* and the *forced response*. The free response is due solely to initial conditions within the circuit with no input. If the circuit is in the quiescent state, i.e., it has zero initial conditions, there is no output without an input. The forced response is the output when an input is applied to a circuit in the quiescent state. As in the case of conventional linear circuits, the forced response can be found by convolving the discrete impulse response with the input sequence. Thus, if h_n, $n = 0,1,\ldots$, is the impulse response of a circuit and u_n is its input at time nT, $n = 0,1,\ldots$, then the output at time nT is

$$y_n = \sum_{k=0}^{n} h_k u_{n-k}, \tag{6.138}$$

where $\sum$ indicates summation modulo 2. If we take the z-transform of both sides of (6.138), we find

$$Y(z) = U(z)H(z), \tag{6.139}$$

where $U(z)$ and $H(z)$ are the z-transforms of u_n and h_n, $n = 0,1,...$, respectively.

Consider the N-stage scrambler shown in Figure 6.28 with feedback coefficients $c_1,...,c_N$. The output at time nT, y_n is given by

$$\begin{aligned} y_n &= c_1 s_{1n} \oplus c_2 s_{2n} \oplus \cdots \oplus c_N s_{Nn} \oplus u_n \\ s_{1n} &= y_{n-1} \\ s_{in} &= s_{i-1,n-1} \qquad i \geqq 2, \end{aligned} \tag{6.140}$$

where u_n is the input at time nT and s_{kl} is the output of the kth delay element at time l. The output of the scrambler may be rewritten as

$$y_n = \sum_{i=1}^{N} c_i y_{n-i} \oplus u_n .$$

At the descrambler we form

$$z_n = y_n \oplus \sum_{i=1}^{N} c_i y_{n-i} ,$$

which recovers the input when there are no channel errors.

Now let the input be the impulse

$$u_n = \begin{cases} 1 & n = 0 \\ 0 & n > 0 \end{cases}$$

and

$$s_{i0} = 0, \quad \text{for all } i.$$

The output sequence can be written

$$y_n = \begin{cases} 1 & n = 0 \\ \sum_{i=1}^{n} c_i y_{n-i} & n \leqq N \\ \sum_{i=1}^{N} c_i y_{n-i} & n > N. \end{cases} \tag{6.141}$$

If we take z-transforms of both sizes, we find after some manipulation that the transform of the impulse response, i.e., the transfer function, is given by

$$H(z) \triangleq \sum_{k=0}^{\infty} y_k z^k = 1 \bigg/ \left[1 - \sum_{i=1}^{\infty} c_i z^i\right] . \tag{6.142}$$

The z-transform of the forced response of the scrambler can be found from (6.139) and (6.142).

The free response of the scrambler can also be found from (6.140) when $u_n = 0$, for all n. We begin by assuming a particular initial state vector. Assume that the output of all of the delay elements but one are zero. Let the nonzero output be that of the ith delay element and denote the output by s_{i0}. It can be shown that y_n^i, the output of the scrambler due solely to state s_{i0}, is

$$y_n^i = \begin{cases} c_i s_{i0} & n = 0 \\ \sum_{j=1}^{n} c_j y_{n-j}^i + c_{i+n} s_{i0} & 0 < n \leqq N - i \\ \sum_{j=1}^{N} c_j y_{n-j}^i & n > N - i. \end{cases} \tag{6.143}$$

If we take the z-transform of both sides of Eq. (6.143), we find that the z-transform of the response to initial condition s_{i0} is

$$Y^i(d) = \left[s_{i0} \sum_{k=0}^{N-1} c_{i+k} z^k \right] \Big/ \left[1 - \sum_{j=1}^{N} c_j z^j \right]. \tag{6.144}$$

Now, to find the response to any initial condition $s_{10}, s_{20}, \ldots, s_{m0}$, we sum over i. Thus, the z-transform of the free response is

$$Y_{free}(x) = S(z) \Big/ \left[1 - \sum_{j=1}^{N} c_j z^j \right] = S(z)H(z), \tag{6.145}$$

where $S(z) \triangleq \sum_{i=1}^{N} s_{i0} \sum_{k=1}^{N-1} c_{i+k} z^k$.

A fact that is crucial to our analysis in the sequel is that the polynomial $S(z)$ spans the space of polynomials of degree $N - 1$ in GF(2).* By choosing the initial conditions s_{i0}, $i = 1,2,\ldots,N$, $S(z)$ can be any polynomial of degree $N - 1$. To show this, suppose we have the polynomial $T(z) = t_0 + t_1 d + \ldots + t_{N-1} z^{N-1}$. Equating $T(z)$ and $S(z)$ term by term, we have N equations in N unknowns, $s_{10}, s_{20}, \ldots, s_{N0}$. The equations can be represented in the form

$$\begin{bmatrix} c_N & 0 & \cdots & 0 \\ c_{N-1} & c_N & \cdots & 0 \\ \vdots & \vdots & & \\ c_1 & c_2 & \cdots & c_N \end{bmatrix} \begin{bmatrix} s_{10} \\ s_{20} \\ \vdots \\ s_{N0} \end{bmatrix} = \begin{bmatrix} t_{N-1} \\ t_{N-2} \\ \vdots \\ t_0 \end{bmatrix}. \tag{6.146}$$

* For definitions of this and other algebraic concepts, see Section (3.2).

The $N \times N$ lower triangular matrix in (6.146) is nonsingular (since $c_N \neq 0$); therefore the N simultaneous equations have a unique solution.

From the foregoing, we see that the total response of a scrambler to an input, with transform $U(z)$ is

$$Y(z) = [U(z) + S(z)]/\phi(z), \tag{6.147}$$

where $\phi(z) \triangleq 1 - \sum_{i=1}^{N} c_i z^i$ is the transform of the feedback coefficients. The above equation completely describes the behavior of the scrambler to any input for any given initial state. In order for the scrambler to work properly, it is required that $\phi(z)$ be primitive polynomial, meaning that it is irreducible and has exponent $2^N - 1$.*

Because of their importance in timing recovery, we focus upon periodic input sequences. The z-transform of a periodic sequence is of the form $R(z)/(1 - z^{\lambda})$, where λ is the period of the sequence and $R(z)$ is a polynomial, of degree less than λ, in z over $GF(2)$. To illustrate, suppose we have a series of elements in $GF(2)$, 1011, 1011, ... Using the relationship for a geometric progression we find that the z-transform of this series is $(1 + z^2 + z^3)/(1 - z^4)$. In general, it can be shown that the z-transform of a periodic time series is of the form $P(z)/Q(z)$, where $P(z)$ and $Q(z)$ are polynomials over a Galois field. Now suppose that the polynomials $P(z)$ and $Q(z)$ are relatively prime, i.e., they contain no common factors. We can write

$$P(z)/Q(z) = P(z)Q'(z)/(1-z^l) , \tag{6.148}$$

where l is the exponent of $Q(z)$. From (6.148) it is evident that the period is l. If $P(z)$ and $Q(z)$ are not relatively prime, the common factors are removed to obtain a ratio of polynomials that are relatively prime. Finally, consider the product of the ratios $P(z)/Q(z)$ and $R(z)/S(z)$ where all common factors are removed. The period is the least common multiple, (lcm), of the exponents of $Q(z)$ and $S(z)$.

Suppose that a periodic sequence is fed into a scrambler, i.e.,

$$U(z) = P(z)/Q(z),$$

where $P(z)$ and $Q(z)$ are relatively prime. From (6.147) the output is

$$Y(z) = [S(z)Q(z) + P(z)]/[\phi(z)Q(z)]. \tag{6.149}$$

We consider first the case where $\phi(z)$ and $Q(z)$ are relatively prime. If the

---

* Recall from Section 3.2.2 that an irreducible polynomial has no factors and that the exponent of a polynomial is the minimum value of l for which the polynomial divides $1 - z^l$ evenly (no remainder).

numerator and denominator of (6.149) are relatively prime, then the output is periodic with period L, where $L = \text{lcm}\,(l, 2^m - 1)$. However, we will show that given $P(z)$ and $Q(z)$, there is a set of initial conditions for which

$$S(z)Q(z) + P(z) = T(z)\phi(z), \tag{6.150a}$$

where $T(z)$ has degree $l-1$. When (6.150a) holds, the output period is l. Thus, assuming that all initial states are equiprobable, with probability 2^{-N} the initial state will be such that the scrambler "locks up" and the output period equals the input period. (As we have previously mentioned, this is a very undesirable situation if the input period is small.) To support (6.150a) we cite the following theorem. There exist (unique) polynomials $T'(z)$ and $S'(z)$ such that

$$S'(z)Q(z) + T'(z)\phi(z) = 1 \tag{6.150b}$$

only if $Q(z)$ and $\phi(z)$ are nonzero relatively prime polynomials over $GF(2)$. Now multiply both sides of the above equation by $-P(z)$ and let $S(z) = -P(z)S'(z)$ and $T(z) = P(z)T'(z)$. We make use of the fact that $S(z)$ spans the space of polynomials of degree $N - 1$ to guarantee that every $S'(z)$ there corresponds a $S(z)$. We now summarize the above.

Let the scrambler be defined by the primitive polynomial $\phi(z) = 1 - \sum_{i=1}^{N} c_i z^i$ and also suppose that the transform of the input to the scrambler is $P(z)/Q(z)$, where $P(z)$ and $Q(z)$ are relatively prime. It is also assumed that $\phi(z)$ and $Q(z)$ are relatively prime. For a particular set of initial conditions, the output period of the scrambler is the input period, l, where l is the exponent of $Q(z)$. For all other initial conditions the output period is the least common multiple of l and $2^N - 1$. We will now show that the requirement that $\phi(z)$ and $Q(z)$ be relatively prime in satisfied whenever the exponent of $Q(z)$ is not a multiple of $2^N - 1$. The proof is by contradiction. Suppose $\phi(z)$ and $Q(z)$ are not relatively prime, then it is possible to write*

$$Q(z) = R(z)\phi^j(z), \qquad j = 1, 2, \ldots, \tag{6.151}$$

where $R(z)$ is a polynomial, with exponent r, which is relatively prime to $\phi(z)$. The exponent of $Q(z)$ is $\text{lcm}[r, p^k(p^{N-1})]$, where k is such that $p^{k-1} < j \leq p^k$. Clearly, the exponent of $Q(z)$ is a multiple of $p^N - 1$, thus proving the desired result. Thus, when the input period is less than p^{N-1}, which is the usual case, the $\phi(z)$ and $Q(z)$ are relatively prime. It is interesting to note that even if the input to the scrambler has an exponent which is a multiple of $2^N - 1$, it may be that $Q(z)$ and $\phi(z)$ are still relatively prime. For example, $Q(z)$ can be the reciprocal polynomial to $\phi(z)$.

---

* Since $\phi(z)$ is irreducible, we could not write $Q(z)$ as a factor of $\phi(z)$.

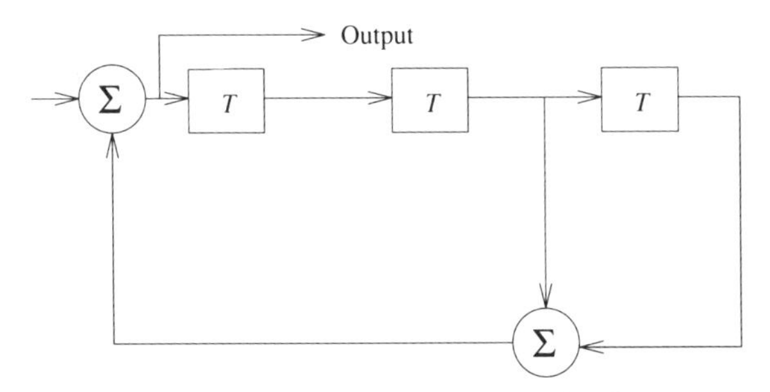

Fig. 6.30 Three-bit scrambler, $C(z) = z^3 + z^2 + 1$.

Consider now the situation when $Q(z)$ and $\phi(z)$ are not relatively prime. As above, we can then factor $Q(z)$ in the form $Q(z) = \phi^j(z)R(z), j \geq 1$, where $R(z)$ is either 1 or a polynomial relatively prime to $\phi(z)$. From (6.149) the z-transform of the output is

$$Y(z) = [S(z)\phi^j(z)R(z) + P(z)]/[\phi^{j+1}(z)R(z)]. \tag{6.152}$$

Since by assumption $P(z)$ is relatively prime with $Q(z) + \phi^j(z)R(z)$, the numerator and denominator of (6.152) are relatively prime. Since $\phi^{j+1}(z)$ and $R(z)$ are relatively prime, the output period is then the least common multiple of $2^k(2^N - 1)$ and r, with k given by $2^{k-1} < j + 1 \leq 2^k$. Note that this result holds independently of initial conditions.

In summary, we see that the output period is contingent on whether or not $\phi(z)$ and $Q(z)$ are relatively prime. It was shown that if the exponent of $Q(z)$ is not a multiple of $2^N - 1$, then $\phi(z)$ and $Q(z)$ are relatively prime. However, when $Q(z)$ and $\phi(z)$ are not relatively prime the output period must be determined from (6.152).

We may illustrate these concepts by means of examples. Consider the scrambler shown in Figure 6.30. Recall that $z^3 + z^2 + 1$ is a primitive polynomial with exponent 7 (see Section 3.2). Let the input sequence be 1010 ... with transform

$$U(z) = 1/(1 - z^2) .$$

From (6.152) the output sequence is

$$Y(z) = \left[1 + (1 - z)^2 \, S(z)\right] \Big/ (1 - z^2)\,(1 + z^2 + z^3) ,$$

where $S(z)$ represents the initial state of the scrambler. For all but one of the initial states, the output period is lcm(2,7) = 14. For the initial state $s_1 = 0$, $s_2 = 0$, $s_3 = 1$, for example, the output sequence is 00100111101100, 00100111101100 ... . In contrast, for initial state $s_1 = 1$, $s_2 = 0$, $s_3 = 1$, the output sequence is 101010 ... . Using (6.145) and (6.150a), we see that in this latter case the state polynomial is

$$S(z) = z^2 + z + 1 .$$

Clearly, the lockup condition, where the output period equals the input period, would give unsatisfactory performance with respect to timing recovery. If the initial conditions are random, the probability of the unique initial condition which will cause lockup is 2^{-N}. For a 23-bit scrambler the probability of lockup is approximately 10^{-16}, an event that is rare in comparison with the other events that can disrupt transmission. A way to avoid lockup is to detect lockup and to cause a change in the scrambled sequence. For example, suppose that a sequence of zeros greater than the length of the scrambler is detected at the output of a scrambler. A single one is added to the scrambler input. This can be reversed at the descrambler output since the receiver is counting zeros as well.

The single drawback to the self-synchronizing scrambler is error multiplication. An error made in the channel is multipled by the number of non-zero taps in the scrambler. Thus in designing a scrambler, primitive polynomials* which have few terms are used. For example, a 23-bit scrambler with characteristic polynomial $Q(z) = z^{23} + z + 1$ is used giving a three-fold multiplication of errors. Error multiplication does not seem to be so great a problem as to preclude the use of scramblers. Errors can be corrected before the inverse scrambler, and moreover in ARQ systems, detecting a few more errors in a block does not require significantly more redundant bits. Error-detection based systems may serve to inhibit the length of a scrambler since it would not be desirable to spread errors over more than one data block.

REFERENCES

[1] H. L. Van Trees, *Detection, Estimation, and Modulation Theory,* Wiley, 1968.

[2] A. J. Viterbi, *Principles of Coherent Communications,* McGraw-Hill, 1966.

[3] W. C. Lindsey, *Synchronization Systems in Communications,* Prentice-Hall, 1972.

[4] E. A. Lee and D. G. Messerschmitt, *Digital Communication,* Kluwer Academic Publishers, 1988.

[5] L. E. Franks, "Carrier and Bit Synchronization in Data Communications — A Tutorial Review," *IEEE Trans. on Communications,* Vol. Com.-28, No. 8, pp. 1107-1121, August 1980.

* A table of primitive polynomials may be found in Reference [13].

[6] W. C. Lindsey and M. K. Simon, *Telecommunication Systems Engineering*, Prentice-Hall, 1973.

[7] D. D. Falconer, "Jointly Adaptive Equalization and Carrier Recovery in Two-Dimensional Digital Communication Systems," *Bell System Tech. Journal,* Vol. 55, No. 3, March 1976.

[8] R. D. Gitlin and J. Salz, "Timing Recovery in PAM Systems," *Bell System Tech. Journal,* Vol. 50, pp. 1645-1669, May/June 1971.

[9] J. E. Mazo, "Jitter Comparison of Tones Generated by Squaring and Fourth-Power Circuits," *Bell System Tech. Journal,* Vol. 57, pp. 1489-1498, May/June 1978.

[10] K. H. Mueller and M. Müller, "Timing Recovery in Digital Synchronous Data Receivers," *IEEE Trans. on Communications,* Vol. 24, No. 5, pp. 516-531, May 1976.

[11] R. W. Lucky, J. Salz, and E. J. Weldon, Jr., *Principles of Data Communication,* McGraw-Hill, 1968.

[12] R. D. Gitlin and J. F. Hayes, "Timing Recovery and Scramblers in Data Transmission," *Bell System Tech. Journal,* Vol. 54, No. 2, March 1975.

[13] W. W. Peterson and E. J. Weldon, *Error Correcting Codes*, Second Edition MIT Press, Cambridge, Mass., 1972.

[14] R. D. Gitlin and H. C. Meadors, Jr., "Center Tap Tracking Algorithms for Timing Recovers," *AT&T Tech. Journal,* Vol. 66, Issue 6, pp. 63-78, Nov/Dec. 1987.

[15] G. Ungerboeck, "Fractional Tap-Spacing Equalizer and Consequences for Clock Recovery in Data Modems," *IEEE Transactions on Communications,* COM-24, No. 8, pp. 856-864, August 1976.

[16] H. Meyr and G. Ascheid, *Synchronization in Digital Communications,* John Wiley, 1990.

EXERCISES

6.1: Are the timing and carrier recovery systems of Figure 6.1 open or closed loop? Are they independent of each other?

6.2: (a) What is the period of the dotting signal, $\sum_n (-1)^n h(t-nT)$?

(b) Determine the Fourier coefficient for the nth harmonic of the dotting signal.

(c) Determine the spectrum for the double dotting signal 1, 1, -1, -1, 1, 1, -1, -1, ...

6.3: Show that the maximum likelihood estimate of the carrier phase [given by (6.6)] is equivalent to the maximum *a posteriori* estimate.

6.4: Suppose that the filter in a phase-locked loop is the passive circuit shown on Figure 6E.1. Write down the second order differential equation describing the system.

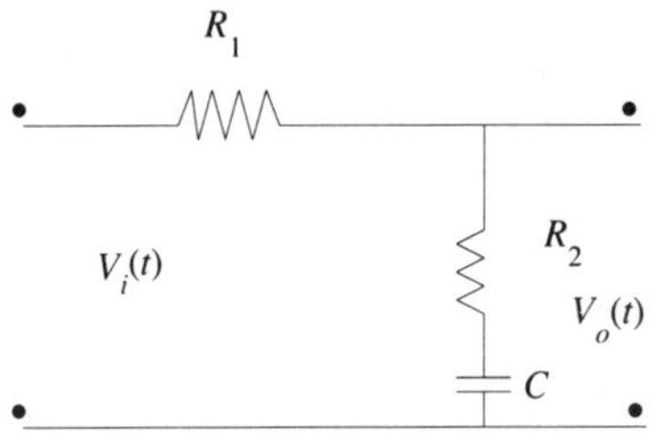

Fig. E6.1 Passive loop filter.

6.5: (a) For the PLL loop described in Exercise 6.4, find the closed-loop transfer function for the case of small phase error.

(b) What is the steady state phase error for the case of frequency and phase offset?

(c) How would you design the loop filter in order to reduce the steady-state phase error?

6.6: Find the loop equivalent noise bandwidth for the PLL described in Exercise 6.5.

6.7: Repeat Exercise 6.5 for the PLL with the active filter shown in Figure E6.2.

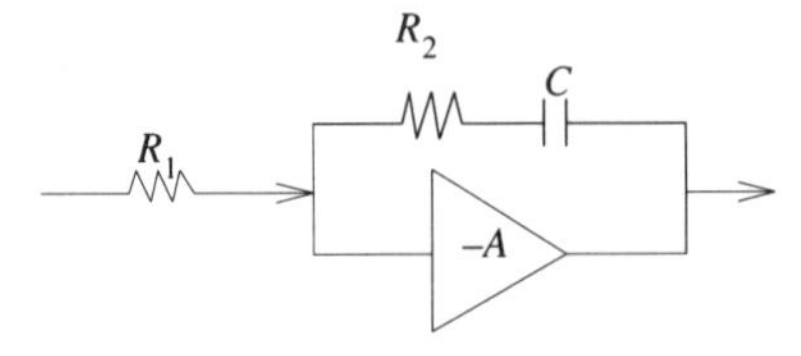

Fig. E6.2 Active loop filter.

6.8: Repeat Exercise 6.6 for the PLL with the active filter shown in Figure E6.2.

6.9: Suppose that the input to a first-order PLL has a frequency offset $\Delta = f_c - f_0$. Suppose also that the loop has phase jitter due to white Gaussian noise.

(a) Under the small signal approximation, write down the probability density of the steady-state phase error.

(b) Suppose that the phase-locked loop operates in a PSK system (see Chapter 5) such that an error of ϕ_0 degrees causes a bit error. Assuming that phase error is the only source of bit error, find the probability of error.

(c) Find the value of loop gain which minimizes the probability of bit error [approximation (2.34b) may be useful here].

6.10: Derive equations (6.57a) and (6.57b).

6.11: Derive equation (6.60).

6.12: Derive equation (6.62b).

6.13: Why is the early-late timing recovery scheme of Figure 6.19b a good approximation to the system shown in Figure 6.19a with tanh(·) replaced by sgn(·)?

6.14: For a raised cosine pulse, with rolloff α, by how much is the timing jitter reduced if a pre-filtering circuit is used?

6.15: For a Nyquist pulse with a linear rolloff of α, find the amplitude of the waveform produced by the timing recovery system of Figure 6.20.

6.16: Suppose that a system uses a linear rolloff Nyquist pulse with $\alpha = T/4$. If the input sequence has period three with elements $0, j, -j$, sketch the spectrum of the line signal.

6.17: Show that when the modulating pulse $p(t)$ has minimum bandwidth (i.e., zero rolloff), that the signal $\sum_n a_n p(t - nT)$ has a stationary correlation function.

6.18: Derive the joint data-directed timing and carrier phase estimates for a QAM signal.

6.19: Prove the right-hand side of equation (6.104); i.e., that $E[x(\tau+nT) - u(nT)]^2 = E[y(\tau+nT) - v(nT)]^2$.

6.20: Consider the three-bit scrambler with polynomial $C(z) = z^3 + z + 1$.

(a) What is the output period when the input is the repeated sequence 1011 if there is no lockup?

(b) Find the z-transform of the output sequence for the input sequence of part (a).

(c) Find the initial state for which the scrambler locks up.

6.21: We have a two bit scrambler with defining polynomial $C(z) = z^2 + z + 1$.

(a) What is the output period when the input is the repeated sequence 1011 if there is no lockup?

(b) Find the z-transform of the output sequence for the input sequence of part (a).

(c) Find the initial state for which the scrambler locks up.

(d) Calculate the output periods for various initial conditions when the input sequence is 100100 ... .

6.22: Find the 10-bit scrambler which minimizes error propagation. (You will need to consult one of the references for this.)

7

Optimum Data Transmission

7.0 INTRODUCTION

In Chapter 4, we examined signal design for baseband pulse transmission, and described the compromises among the bandwidth of the transmitted signal, noise immunity, and mitigation of intersymbol interference inherent in any design. Peak and mean-square intersymbol interference were defined, and it was shown how linear channel distortion can degrade performance. This chapter describes system structures that are optimum, either with respect to maximizing the probability that a sequence of symbols is correctly received or minimizing the output mean-square error; these receivers are primarily designed to correct the degradation caused by noisy channels and linear distortion. It is not possible, except in certain singular cases, to achieve the performance of an impairment-free system, or that of a system which attains the matched filter bound (which is equivalent to the transmission of isolated pulses). A further limitation discussed in Chapter 2 was the assumption of pulse-by-pulse (i.e. symbol-by-symbol) detection, which is optimum only in the absence of intersymbol interference. In the case of partial response signaling, we described precoding operations that eliminated the intentional intersymbol interference, but with a penalty in noise immunity. This penalty can be avoided if pulse-by-pulse detection is replaced by a process (the maximum likelihood receiver) that uses the entire received sequence for detection. This technique, the Viterbi algorithm, was applied to the decoding of convolutional codes in Chapter 3 and to the decoding of trellis codes in Chapter 5. In this chapter, the same technique will be applied to the detection of a sequence of amplitude-modulated pulses.

The goal of the optimum, or maximum likelihood receiver, is to produce data decisions at the receiver output which utilize the entire received signal to produce decisions that are most likely to be right; it is generally more practical to use alternative optimization criteria because of limitations on receiver complexity. We shall derive several suboptimum system structures that give the best possible performance under different constraints. Commercial modems do, in fact, implement algorithms close to those described in this chapter.

Our "optimal" structures depend on the exact characteristics of the transmission channel, which are not ordinarily known either to the system designer or to the user. How, then can such structures be realized in practice? The solution is in *automatic* and *adaptive* mechanisms which can learn the channel characteristics; the systems are described in the chapter following this one. For the time being, we shall be concerned only with the derivation of the desired optimal structures, and shall assume that means exist for learning the values of their channel-dependent parameters.

The system models used here, although widely recognized to be very useful, are somewhat narrowly defined. Primarily, they relate to pulse-amplitude modulated (PAM) transmission through channels, which exhibit only linear distortion and additive noise. Real (i.e., baseband) impulse responses are used here for simplicity, but the structures derived are applicable to passband systems with linear quadrature amplitude modulation (QAM), using the complex baseband equivalent channels described in Chapter 5.

Nonlinear distortion, impulse noise, transient, and other interruptive phenomena, as well as time-varying carrier phase perturbations are all excluded from these models. Some of these additional impairments can be partly compensated by adjunct structures, such as the phase trackers described in Chapter 6. It is important, however, to recognize the sensitivity of some of the optimum receivers to impairments such as phase jitter. Often there is a choice of either explicitly compensating for an impairment or designing a system which is fairly insensitive to the impairment under discussion (e.g., choosing a modulation method, such as trellis coding, that is relatively insensitive to phase jitter or nonlinearities, rather than designing a high-performance phase-locked loop). The operational environment (i.e., which impairments dominate), implementation complexity and cost, and available design choices are all inputs to the design decision.

In this chapter, the performance measures being optimized are chosen for relevance and mathematical tractability. Specifically, in Section 7.1 we shall define the optimum (maximum likelihood) system as that which provides, at the receiver output, the entire data symbol sequence most likely to have been produced at the transmitter on the basis of observations made at the receiver input. This maximum likelihood system turns out to be a *nonlinear* processor, whose essential subsystem is most efficiently realized through an application of dynamic programming procedure known as the *Viterbi algorithm* [1] (VA). This concept, which has been introduced in Chapter 3 in connection with the decoding of convolutional codes and applied to the decoding of trellis codes in Chapter 5, will be extended in this chapter to accommodate the deconvolution of PAM signals passed through a distorting linear channel. The computational complexities associated with implementation of the VA will be described, along with some engineering compromises that produce realizable systems.

In Section 7.2 we derive the *optimum linear receiver* which is defined to be that linear filter which exhibits the least mean-square error (MSE) between its output at symbol intervals and the transmitted data symbol values. This structure is generalized, in Section 7.3, to include use of prior (and assumed to be correct) decisions in what is known as a *decision feedback* equalizer. The overall mean-square error is used as the performance criterion, rather than probability of symbol error, because it leads to tractable linear equations for the optimum linear receiver parameters, and it will also be the measure of choice for adapting these structures (adaptation of the receiver parameters is treated in Chapter 8). As indicated by the Saltzberg bound (see Chapter 4), there is usually a close correlation between the mean-square error and the probability of a symbol error.

7.1 MAXIMUM LIKELIHOOD SEQUENCE ESTIMATION (MLSE): THE VITERBI ALGORITHM

Suppose we wish to design a receiver, not necessarily linear, which makes an accurate decision about the entire transmitted data sequence on the basis of observing the entire received signal. The method adopted is to select the decision sequence which is most likely, on the basis of the observation, to have been the transmitted sequence. It is presumed that all compensation for channel distortion is to be done at the receiver, working with whatever signal shape arrives from the combination of transmitter and channel, and allowing an arbitrary processing delay. It would, of course, be better to design a system which optimizes both transmitter and receiver, since there is, at least for isolated pulse transmission, an optimum division of linear filtering between transmitter and receiver [4] (see Chapter 4), but this would require a feedback channel, from receiver to transmitter, which is not always available. Instead, a compromise transmitter filter is often used, with all adjustable or adaptive elements at the receiver.

The maximum likelihood sequence receiver using the Viterbi algorithm was derived in Chapter 3, for hard and soft decision decoding of convolutional codes. In this chapter we derive the maximum likelihood sequence estimation (MLSE) receiver for a signal set consisting of PAM waveforms and show that it can be realized via the VA. Each of the possible received noiseless signals corresponds to the transmission of a particular data sequence. In selecting as our performance criteria the maximization of the probability that the sequence of decisions is correct we should realize the following limitations of this measure:

(1) As the sequence gets longer, the probability of a sequence error will generally approach unity,

(2) Sequences with an arbitrarily large number of errors are weighed the same as a sequence with one error,

(3) The probability of a symbol or bit error is not simply related to the sequence error.

The relatively simple and elegant implementation (for channels with short-duration impulse responses), the mathematical tractability in determining the performance of the MLSE receiver, as well as the comparable performance with the optimum symbol detector (at high SNR), has made the MLSE an attractive system to investigate.

An appreciation of the complexity associated with a "brute force" approach to maximum likelihood detection can be achieved by realizing that if L is the symbol alphabet size, i.e., the number of different pulse levels, and N is the total number of symbols in the entire session, there are L^N different possible transmitted signals. At first, this would suggest an unreasonably complex receiver, since L^N likelihood integrals (see Section 2.2) would have to be computed and compared with one another. For $L=2$ (binary data) and a message of $N=1000$ symbols, a straightforward approach would mean 2^{1000} computations, an impossibly large number (which grows exponentially with time). The salient feature of the Viterbi algorithm is that the finite channel memory allows application of the same recursive principles used in decoding convolutional codes, to reduce the complexity to the order of L^M operations per symbol, where M is the channel dispersion, or "memory," in symbol intervals. So the complexity of the VA is *exponential* in the number of states, but *linear* in the number of transmitted symbols, or time.

As demonstrated in Chapter 3, the Viterbi algorithm is an application of *dynamic programming*, and is based on a clever reorganization of the observed received process into a discrete-time process with a finite number of possible values at each sample time. This process maps the received signal into one produced *by a one-step, vector* Markov channel — the present value is dependent only on the most recent past value of the vector (called the state) — and is assumed to be embedded in additive white Gaussian noise, whose mean and autocorrelation are known.

We will first define the MLSE problem more precisely, and then show how the Viterbi algorithm can produce the MLSE estimate. Figure 7.1 is the model for a baseband data communication system. A sequence of levels $\{a_n\}$ corresponding to input data or derived from it by precoding, is applied to the transmitter at symbol intervals of T seconds. The transmitted PAM signal is

$$s(t) = \sum_n a_n \, g(t-nT) \; , \tag{7.1}$$

where $g(t)$ may or may not be an equivalent Nyquist pulse. After passage through the channel, $h(t)$, the combination of signal and noise observed at the

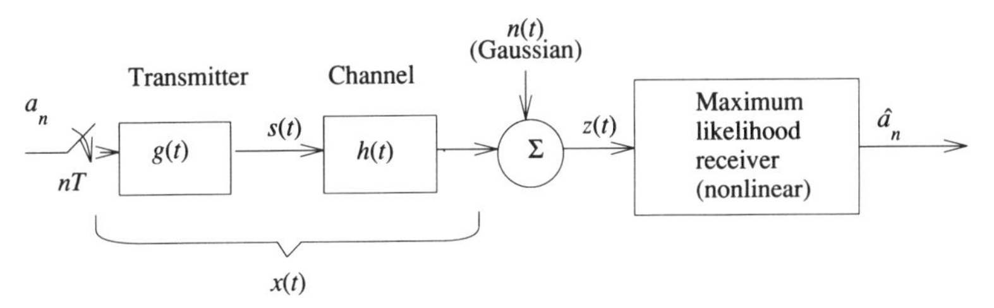

Fig. 7.1 Baseband data transmission through a linear additive noise channel with a maximum likelihood (nonlinear) receiver.

receiver input is

$$z(t) = \sum_m a_m \, x(t-mT) + n(t) \, , \tag{7.2}$$

where the overall pulse response, $x(t)$, is given by $x(t) = g(t)*h(t)$, and $n(t)$ is presumed to be a Gaussian process, but not necessarily white. The analysis of system performance is easier if $n(t)$ is white (as explained in Section 2.2, an invertible, "whitening filter" is often included in the receiver to compensate for the spectral coloring of the noise introduced in the front end of the receiver), but the derivation of the Viterbi algorithm does not require this assumption.

In addition to the Gaussian nature of the noise, we make the following assumptions:

(1) The dispersion, or *memory*, of the channel is limited to a finite time, MT seconds, as illustrated in Figure 7.2.

(2) The transmitter, rather than producing $s(t)$ for all time, is turned on at some arbitrary time N_1T and off at a later time N_2T, where $N_2-N_1>>M$.

Since the memory of the channel can allow arrival of signal energy at the receiver as late as $t = N_2T + MT$, we observe the received signal $z(t)$ on the time interval $N_1T \leq t \leq (N_2+M)T$ to learn as much as possible about the

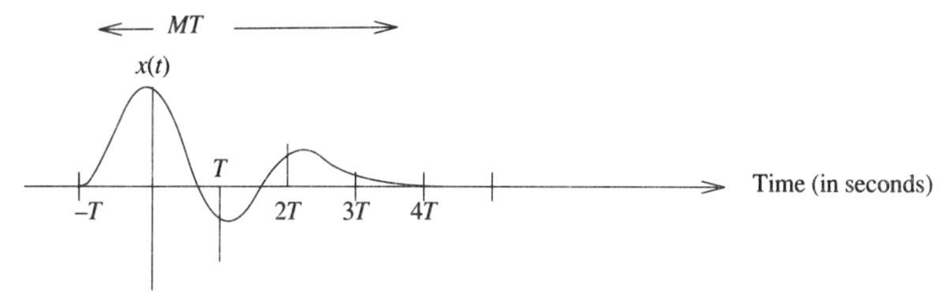

Fig. 7.2 Finite channel dispersion, or *memory*. In this case, the maximum separation between significant samples, at T second intervals, of the channel impulse response is from $-T$ to $3T$, so that $M = 4$.

transmitted sequence $\{a_n\}$, $N_1 \le n \le N_2$. The receiver produces a decision sequence $\hat{a}_{N_1}, \hat{a}_{N_1+1}, \ldots, \hat{a}_{N_2}$ which maximizes the *a posteriori* probability of a particular data sequence, given observation of the received waveform. The probability of interest is given by

$$Pr\left[a_{N_1}, \ldots, a_{N_2} \;\middle|\; z(t), \quad N_1 T \le t \le (N_2+M)T\right]. \tag{7.3}$$

Note that this performance measure treats all error events equally, i.e., the weighting given to a sequence that has ten errors is the same as a sequence having one error.

The process $z(t)$ can, if it is essentially bandlimited, be closely approximated by a Karhunen–Loève expansion, as described in Appendix 2B. The expansion coefficients for white noise are uncorrelated, and in the Gaussian noise case are independent Gaussian random variables $\{\zeta_k\}$. With this formalism in mind, by using Bayes's rule, the likelihood of the received sequence is

$$Pr\left[a_{N_1},\ldots,a_{N_2} \mid z(t): N_1 T \le t \le (N_2+M)T\right] = Pr\left(z(t) \mid a_{N_1},\ldots,a_{N_2}\right) \times \frac{Pr(a_{N_1},\ldots,a_{N_2})}{P(z(t))}. \tag{7.4}$$

Since all the symbols are assumed equiprobable, $Pr(a_{N_1},\ldots,a_{N_2})$ is independent of the data sequence, as is the denominator probability density, and the maximization of (7.4) with respect to the data sequence is equivalent to maximization of the likelihood function

$$Pr\left[z(t) \mid a_{N_1}, \ldots, a_{N_2}\right] = \frac{1}{(2\pi N_o)^{K/2}} \exp\left\{-\frac{1}{2N_o}\int\left[z(t) - \sum_n a_n x(t-nT)\right]^2 dt\right\}. \tag{7.5}$$

The maximum likelihood sequence estimate of $a_{N_1}, \ldots, a_{N_2}$ is that data sequence maximizing (7.5), or equivalently, *minimizing* the log-likelihood function

$$\ell\,(a_{N_1}, \ldots, a_{N_2}) = \int\left[z(t) - \sum_n a_n x(t-nT)\right]^2 dt. \tag{7.6}$$

Observe that (7.6) is a measure of the squared distance between the received signal, $z(t)$, and the possible transmitted signals. In other words, the optimum MLSE detector finds that transmitted signal which is closest to the received signal, where the distance is measured as the square root of the integral of the squared difference between the two signals.

7.1.1 MINIMIZATION OF THE LOG-LIKELIHOOD FUNCTION

Expanding (7.6), the log-likehood function to be minimized, we need only minimize

$$u_{N_2} = u(a_{N_1}, \ldots, a_{N_2}) = -2 \sum_{k=N_1}^{N_2} a_k y_k + \sum_{k=N_1}^{N_2} \sum_{l=N_1}^{N_2} a_k a_l r_{k-l}, \quad (7.7)$$

where we have defined the *matched filter* output, y_k, at $t = kT$ by

$$y_k = \int_{N_1T}^{(N_2+M)T} z(t)x(t-kT)\,dt = \sum_n a_n r_{k-n} + \eta_k. \quad (7.8)$$

The terms in (7.8) are defined as follows. The noise at the matched filter output is given by

$$\eta_k = \int_{N_1T}^{(N_2+M)T} n(t)x(t-kT)\,dt, \quad (7.9)$$

which is a zero-mean Gaussian variable, whose correlation function is given by the channel correlation function

$$r_{k-l} = \int_{N_1T}^{(N_2+M)T} x(t-kt)x(t-lT)\,dt, \quad N_1 \le k,l \le N_2 . \quad (7.10)$$

The $\{y_n\}$, which are the signal-dependent observables , are samples, at T-second intervals, of the output of a filter *matched* to the overall channel impulse response $x(t)$. The term r_{k-l} depends only on the time difference $(k-l)T$ because of the symmetry of the integrand. Note that r_0 is the energy of the pulse at the matched filter output. Furthermore, the finite support MT of $x(t)$ implies that $r_{k-l} = 0$, $|k-l|>M$. Note that processing of y_n requires knowledge of the channel to form the matched filter output and to compute the channel correlation function. The Gaussian noise variables $\{\eta_k\}$ are correlated, but can be made uncorrelated by a whitening filter.

EXAMPLE 1 VITERBI ALGORITHM FOR THE NYQUIST CHANNEL
Suppose $x(t)$ corresponds to square-root shaping, at the transmitter, of a Nyquist pulse. Thus $x(t)$ is such that $|X(f)|^2$ is a Nyquist spectrum. Note that the complementary square root shaping will be provided by the matched filter at the receiver, so that

$$r_k = \int_{-\infty}^{\infty} x(t)x(t - kT)\,dt = \int_{-\infty}^{\infty} |X(f)|^2 e^{-j2\pi fkT} df .$$

Since $|X(f)|^2$ is the transform of a Nyquist pulse, we have that

$$r_k = E_x \, \delta_k ,$$

where $E_x = \int_{-\infty}^{\infty} x^2(t)\,dt$ is the pulse energy. Since $r_k \neq 0$ only for $k=0$, the

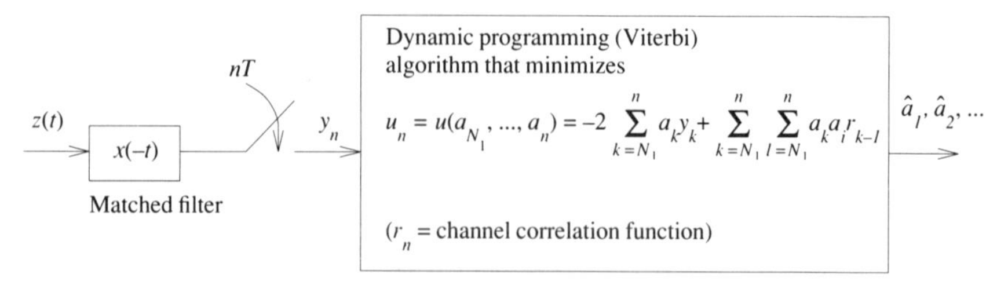

Fig. 7.3 The maximum likelihood receiver processes the matched filter output, y_n, and uses the Viterbi algorithm to produce output decisions, $\hat{a}_1, \hat{a}_2, ...$, which minimize the log likelihood u_n.

log-likehood (7.7) becomes

$$u_n = 2\sum_{k=1}^{n} a_k y_k + E_x \sum_{k=1}^{n} a_k^2 = E_x \sum_{k=1}^{n} \left[\left(a_k - \frac{y_k}{E_x} \right)^2 - \frac{y_k^2}{E_x^2} \right],$$

which can be optimized in *symbol-by-symbol* manner by choosing a_k to be the quantized (or sliced) signal closest to the scaled received sample, y_k/E_x. Thus for a purely Nyquist channel, optimum reception can be done on a symbol-by-symbol basis and is simply a matched filter followed by a slicer. This result is to be expected, since for a Nyquist channel the received samples are independent; thus nothing is to be gained by processing a block of symbols.

The Viterbi algorithm,* with its iterative procedure for minimizing (7.7), is found by separating out the elements of (7.7), involving the most recently transmitted symbol, a_n:

$$u_n = u_{n-1} - 2a_n y_n + 2a_n \sum_{k=n-M}^{n-1} a_k r_{n-k} + a_n^2 r_0 \ . \tag{7.11}$$

Note that the summation is only over the M nonzero elements of the original (N, n) index range. The maximum likelihood receiver, as shown in Figure 7.3, processes the synchronously sampled output of the matched filter so as to recursively minimize the log-likelihood, (7.11), via the Viterbi algorithm.

7.1.2 THE STATE VECTOR

As in Section 3.5.2 we define the *state vector*, **s**, of the observed process $\{y_n\}$. It is, at time *nT*, the vector of all past data levels (including a_n) which will, together with the next data level a_{n+1}, determine the signal portion of the

---

* The reader may recall that the Viterbi algorithm is used in Section 3.5.2 to detect a sequence eminating from a convolutional encoder. The present derivation of the Viterbi algorithm for a sequence with intersymbol interference parallels this earlier derivation.

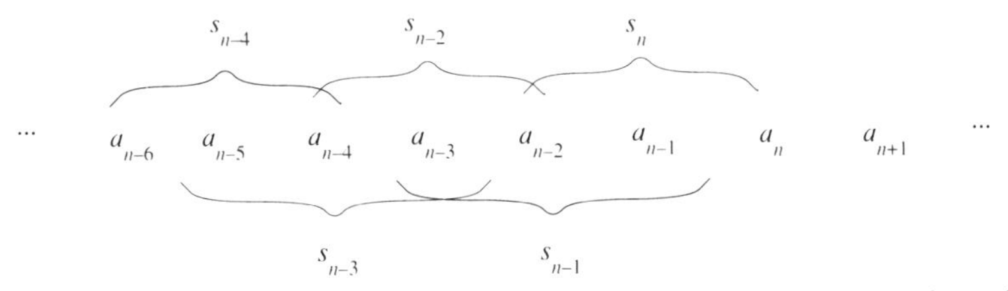

Fig. 7.4 Definition of state vectors for a channel with a memory of $M = 3$ symbol intervals.

observed value y_{n+1}. With a channel memory of M symbols, and prime denoting transpose, the state vector has the M elements $\boldsymbol{s}'_n = \{a_{n-M+1}, \ldots, a_n\}$, as indicated in Figure 7.4 for $L=2$ and $M=3$.* These states are often represented as points on a vertical scale, a convenient concept for later discussion. Note that if levels are drawn from an alphabet of L values, $\boldsymbol{s}_n$ has L^M possible values, and the function that we are striving to minimize can be written, from (7.11), as

$$u_n = D(\boldsymbol{s}_1, \boldsymbol{s}_2, \ldots \boldsymbol{s}_n) = u_{n-1} + d(\boldsymbol{s}_{n-1}, \boldsymbol{s}_n; y_n) , \tag{7.12}$$

and where

$$d(\boldsymbol{s}_{n-1}, \boldsymbol{s}_n; y_n) = -2a_n y_n + 2a_n \sum_{k=n-M}^{n-1} a_k r_{n-k} + a_n^2 r_0 \tag{7.13}$$

is called the transition metric (or incremental distance) from $\boldsymbol{s}_{n-1}$ to $\boldsymbol{s}_n$.

The sequence of state vectors $\boldsymbol{s}_{N_1}, \ldots, \boldsymbol{s}_n$ contains no more nor less information than the data (level) sequence $a_{N_1-M+1}, \ldots, a_n$, so that estimation of the state sequence (or trajectory) is completely equivalent to estimation of the level sequence. We will use the notation $u(\boldsymbol{s}_n)$ interchangeably with u_n to designate the function to be minimized at time nT.

By recasting the problem as the estimation of a state sequence, we have the equivalent problem of estimating a one-step, vector Markov process. It is one step because $\boldsymbol{s}_n$ depends only on $\boldsymbol{s}_{n-1}$ and not on earlier states (which are built up from symbols older than the channel memory and the current data symbol, a_n). For this type of random process, the optimum state trajectory (the one that minimizes u_n) can be built up from linkages of least-cost (i.e., least-distance) transitions. Taking advantage of this construction, we will determine the best overall state trajectory as best of the L^M *sequence* of states with least cost (i.e. minimum u) among the $(L^M)^n$ candidate state sequences at time nT that lead up to the L^M possible values of the final state.

The basis for this reasoning will be shown below, but the intuitive appeal of this approach can easily be shown by contradiction. From (7.12) we see that u_n is a cumulative function of the states up to time $n-1$, the current state, and the signal sample values. Suppose the segment of candidate optimal states terminat-

---

* In the convolutional encoder the memory was the number of stages.

ing at s_n did *not* minimize u_n, then replacing that segment by the sequence of states which minimizes u_n, will obviously result in a lower value of u_n and will not have any adverse affect on values of u_{n+1}, u_{n+2}, ..., etc. Thus the additive nature of the criterion u_n, suggests that the receiver construct, for each state, candidate trajectories, which are themselves least-cost trajectories leading to the final state. This is the very same *principle of optimality* used in the derivation following (3.43) for the decoding of convolutional codes.

Applying the above argument to (7.11), consider a sequence of states which terminates with a *particular choice* for $\boldsymbol{s}_n$, which we will imprecisely continue to denote as $\boldsymbol{s}_n$. With the above notation, the cost associated with the best state trajectory leading up to the given (or conditioned) state is defined as $D(S_n)$, where

$$D(s_n) \triangleq \min_{\boldsymbol{s}_{n-M+1}, \ldots, \boldsymbol{s}_{n-1} | \boldsymbol{s}_n} D(\boldsymbol{s}_{n-M+1}, \ldots, \boldsymbol{s}_n) \tag{7.14}$$

If a_n is the final symbol transmitted, then the overall minimum value of u_n and the best state sequence is determined from the L^M candidates by finding $\boldsymbol{s}_n$ which minimizes $U(\boldsymbol{s}_n)$. Continuing with our minimization we have

$$D(\boldsymbol{s}_n) = \min_{\boldsymbol{s}_{n-M+1}, \ldots, \boldsymbol{s}_{n-2} | \boldsymbol{s}_{n-1}, \boldsymbol{s}_n} \left[D(\boldsymbol{s}_{n-M+1}, \ldots, \boldsymbol{s}_n) \right] = \min_{\boldsymbol{s}_{n-1} | \boldsymbol{s}_n} \left\{ \min_{\boldsymbol{s}_{n-M+1}, \ldots, \boldsymbol{s}_{n-2} | \boldsymbol{s}_{n-1}, \boldsymbol{s}_n} \left[D(\boldsymbol{s}_{n-M+1}, \ldots, \boldsymbol{s}_{n-1}) + d(\boldsymbol{s}_{n-1}, \boldsymbol{s}_n; y_n) \right] \right\} \tag{7.15}$$

Conditioning on the previous, $\boldsymbol{s}_{n-1}$, and current, $\boldsymbol{s}_n$, states and using the signal sample, y_n, in the inner minimization of (7.15) permits d to be computed from (7.13). Also, the conditioning on $\boldsymbol{s}_n$ is irrelevant to performing the minimization over $\boldsymbol{s}_{N_1+M-1}, \ldots, \boldsymbol{s}_{n-2}$. Thus (7.15) can be simplified to

$$D_{\min}(\boldsymbol{s}_n) = \min_{\boldsymbol{s}_{n-1} | \boldsymbol{s}_n} \left\{ \min_{\boldsymbol{s}_{n-M+1}, \ldots, \boldsymbol{s}_{n-2} | \boldsymbol{s}_{n-1}} \left[D(\boldsymbol{s}_{n-M+1}, \ldots, \boldsymbol{s}_{n-1}) \right] + d(\boldsymbol{s}_{n-1}, \boldsymbol{s}_n; y_n) \right\} \tag{7.16}$$

The minimization can be put in the desired recursive form by recognizing that the inner term in the brackets may be written as $D(\boldsymbol{s}_{n-1})$. Combining (7.14) and (7.16) we have the dynamic programming recursion

$$D(\boldsymbol{s}_n) = \min_{\boldsymbol{s}_{n-1} | \boldsymbol{s}_n} \left\{ D(\boldsymbol{s}_{n-1}) + d(\boldsymbol{s}_{n-1} \boldsymbol{s}_n; y_n) \right\} . \tag{7.17}$$

7.1.3 THE VITERBI ALGORITHM

In performing the minimization required in (7.17), the substantial overlap of the contents of $\boldsymbol{s}_{n-1}$ and $\boldsymbol{s}_n$ means that one cannot move from an arbitrary value of $\boldsymbol{s}_{n-1}$ to a given value of $\boldsymbol{s}_n$. There are, in fact, exactly L values of $\boldsymbol{s}_{n-1}$ which can be predecessor states to a particular choice of $\boldsymbol{s}_n$. Thus to evaluate (7.17), we range across those states $\boldsymbol{s}_{n-1}$ that lead to $\boldsymbol{s}_n$; the presence of the signal sample, y_n, will make the selection of the minimum value of $\boldsymbol{s}_{n-1}$ vary with the value of the received signal sample.

The minimization, conditioned on a choice of value for the latest states, $\boldsymbol{s}_n$, is equivalent to determination of that value of $\boldsymbol{s}_{n-1}$ for which the sum of the transition metric $d(\boldsymbol{s}_{n-1}, \boldsymbol{s}_n; y_n)$ and the least-cost path up to $\boldsymbol{s}_{n-1}$ is a minimum. Determination of the least-cost path up to $\boldsymbol{s}_{n-1}$ is, according to (7.17), the same as the original minimization (7.11), except moved one symbol interval earlier in time.

A three-step iterative rule for determining the optimum state trajectory leading up to a given current state (at any time index n) can be expressed as follows:

(1) For each value of $\boldsymbol{s}_n$, determine the L values of the transition metric $d(\boldsymbol{s}_{n-1}, \boldsymbol{s}_n; y_n)$ corresponding to the L values of $\boldsymbol{s}_{n-1}$ which could lead to $\boldsymbol{s}_n$.

(2) For each possible value of the L allowed predecessor state $\boldsymbol{s}_{n-1}$, add the metric $D(\boldsymbol{s}_{n-1})$ of the optimal state trajectory up to that value of $\boldsymbol{s}_{n-1}$ to the transition metric $d(\boldsymbol{s}_{n-1}, \boldsymbol{s}_n; y_n)$ from that value of $\boldsymbol{s}_{n-1}$ to the given current state $\boldsymbol{s}_n$.

(3) Designate as a candidate optimal path that sequence of states leading to $\boldsymbol{s}_n$ corresponding to the smallest of the L metrics evaluated in step (2). This smallest metric is designated $D(\boldsymbol{s}_n)$ and saved. The optimal path to each state is also stored.

These three steps comprise the Viterbi algorithm. If they are followed at each value of n through $n = N_2$, they will determine the best state trajectory to each possible value of the final state $\boldsymbol{s}_{N_2}$. A minimization over those L^M survivors will yield the globally optimum state trajectory, and hence the most likely sequence of transmitted levels. However, as we will show later, *merges* will imply that all data decisions do not have to wait until this final step.

Since there are L^M possible states at each observation time, the ensemble of all possible states at all observation times can be graphically represented as a trellis structure (Figure 7.5), in which movement along the horizontal direction represent observation times and the vertical points represent the different possible states. A sequence of states is a trajectory through this lattice.

As previously noted, a given current state cannot arise from any arbitrary predecessor state, since states overlap in content. For example, with $L = M = 2$, the state (1,-1), which at time nT represents ($a_{n-1} = -1$, $a_n = 1$), can arise only from predecessor states ($a_{n-2} = 1$, $a_{n-1} = -1$) and ($a_{n-2} = -1$, $a_{n-1} = -1$), and not from predecessor states ($a_{n-2} = -1$, $a_{n-1} = 1$) and

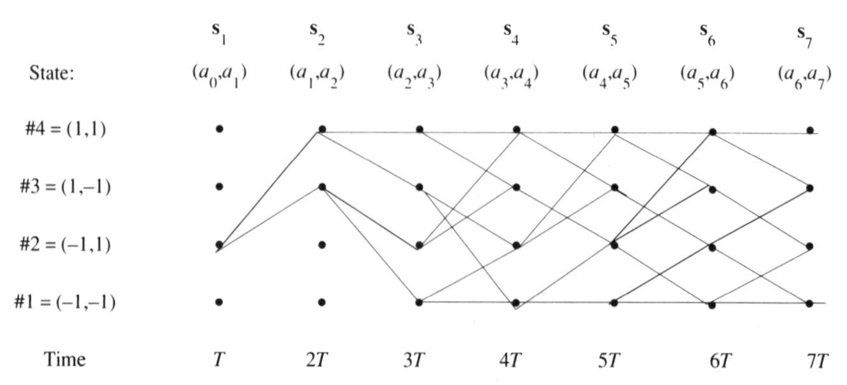

Fig. 7.5 Trellis structure for state transitions in the Viterbi algorithm, illustrated for $L=2$ levels and channel memory $M=2$. A "tree" of possible transition paths is illustrated for the example of initial data levels $a_0=-1$ and $a_1=1$. Note that a state at time nT can only have arisen from that subset agreeing in a_{n-1}. For example, $S_5=(1,-1)$ can follow only $S_4=(1,1)$ or $S_4=(-1,1)$.

$(a_{n-2} = 1, a_{n-1}=1)$. The number of allowable predecessor states is, as noted above, equal to L. The number of transition metrics which must be computed at each step is L (possible predecessors for each state) $\times$ L^M (the number of states) $= L^{M+1}$.

7.1.4 OTHER APPLICATIONS OF THE VITERBI ALGORITHM [8]

The development leading to (7.17) made the following assumptions related to the finite memory of the system:

i) The function to be optimized is a function of a finite-state vector

ii) The evolution of the state vector is Markovian (i.e., the new value of the state depends only on the immediately preceding state and the current received signal sample)

iii) The function under discussion can be split into a portion which depends on prior states and a portion that depends on the current signal sample (the sufficient statistic) and the transition from the most recent to the current state.

Thus for any finite-state Markov system, where it is desired to optimize a function that satisfies (i)–(iii), the Viterbi algorithm (VA) can be used to find the optimum sequence of states. The principal attraction of the VA is that complexity grows exponentially with the number of states, not exponentially with time. The potential drawbacks are that (1) the system may not be known *a priori* (e.g., the channel impulse response is generally unknown) and adaptive techniques may be needed to learn the system parameters or (2) the number of states may be so large that a reduced-state system is the only implementation option.

The Viterbi algorithm has been applied to problems as diverse as the maximum likelihood reception of MSK modulation [8], fiber optic signals [9], speech and text recognition systems [10], and joint source and channel coding [11].

EXAMPLE 2 In order to illustrate the Viterbi algorithm, we go through a step-by-step example. Suppose that the system has the following parameters. $L=2$ (data values ± 1), $M=2$, $r_0 = 1$, $r_{\pm 1} = .5$ and $r_{\pm 2} = -.25$. For purposes of illustration, we detect a sequence which is only seven symbols long. The successive outputs of the matched filter are $y_1 = 1.5$, $y_2 = 1.0$, $y_3 = 0.5$, $y_4 = 1.0$, $y_5 = -1.5$, $y_6 = -3.0$, $y_7 = 0.5$. The steps required to detect this sequence are as follows.

(1) For each of the $L^M = 4$ initial states calculate the "initial distance," $U(\boldsymbol{s}_2)$, to state $\boldsymbol{s}_2$ from (7.7) with $N_1 = 1$ and $N_2 = 2$. These values, along with those obtained in the succeeding steps of the algorithm, are shown in Figure 7.6. For example, for $a_1 = -1$ and $a_2 = +1$, the initial distance is +2.

(2) The minimum distance to the next state, $\boldsymbol{s}_3$, is found by application of (7.17). For example, let $\boldsymbol{s}_3 = (-1, +1)$. The predecessor states are (-1,-1) and (+1,-1) with initial distances 12 and 4, respectively. The incremental distances $d(\boldsymbol{s}_2, \boldsymbol{s}_3, y_3)$ are calculated from (7.13). These are -.5 and -1.5, respectively. Thus, the minimum distance to (-1,+1) is 2.5 through (+1,-1), as indicated by the solid line.

(3) Repeat step (2) for each of the succeeding states $\boldsymbol{s}_4, \boldsymbol{s}_5, \boldsymbol{s}_6, \boldsymbol{s}_7$.

(4) Optimize over the final state $\boldsymbol{s}_7$. From Figure 7.6, we see that the optimum path ends at (–1,+1). The optimum path goes through the succession of states (+1,+1), (+1,–1), (–1,+1), (+1,–1), (–1,–1) and, finally (–1,+1). The detected sequence is then $a_1 = +1$, $a_2 = +1$, $a_3 = -1$, $a_4 = +1$, $a_5 = -1$, $a_6 = -1$, and $a_7 = +1$.

7.1.5 MERGES

As in Chapter 3 the merge phenomenon allows us to make decisions before the whole sequence has been received. Figure 7.6 provides an example. At state $\boldsymbol{s}_6$, we notice that all of the optimum paths to the four possible values pass through the (–1,+1) value of state $\boldsymbol{s}_4$. This means that no matter what the future path is, we are certain that the optimum passes through this point. Accordingly, we decide $a_1 = +1$, $a_2 = +1$, $a_3 = -1$, and $a_4 = +1$.

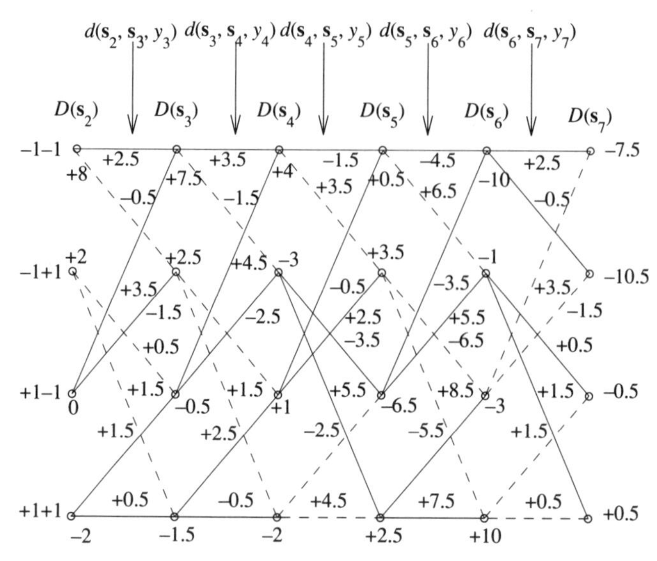

Fig. 7.6 Lattice (trellis) diagram used for Viterbi decoding.

In this particular example, the merge took place in minimum time, $M = 2$ steps; however, in general merging is a random phenomenon. It could happen that no decision is possible until the end of the sequence. However, in cases of practical interest, merges generally occur persistently, and the probability of running for a period of time longer than the channel memory without observing a merge is small. In order to provide decisions at a constant rate to the end user, the usual practice is to provide storage registers of a length which is somewhat longer than the expected interval between merges. Usually, this is three or four times the length of the channel memory. If a merge has not occurred at this time, a merge is forced by choosing the most probable of the state sequences stored in the registers.

7.1.6 PERFORMANCE OF THE VITERBI ALGORITHM: MINIMUM DISTANCE

In Chapter 4 we showed that, for uncoded systems with a Nyquist pulse response, the probability of error depends upon the ratio of the squared minimum distance between transmitted *symbols* to the background noise power. In Chapters 3 and 5 we showed that the performance of block, convolutional, and trellis-coded modulation, again for an ideal channel, depends upon the *minimum distance* between allowed signal *sequences*. Here we address the performance of the Viterbi algorithm, initially for uncoded modulation, but for a system with arbitrary channel characteristics. While implementation of the Viterbi algorithm is often too complex for practical systems, having a measure of the performance of the optimum receiver provides a useful bound on the performance of sub-optimum, but realizable, receiver structures. We will take several approaches to estimating performance. The first approach uses the fundamental M-ary signal detection framework presented in Chapter 2. In this approach we bound the probability that the ML sequence detector makes an error. However, there is a problem with this approach: as the length of the data sequence becomes large, this probability approaches one. In Section 7.1.7 we calculate the probability of a symbol error.

We can derive a bound on the performance of the Viterbi detector by rewriting (7.2) as the M-ary decision problem

$$z(t) = \sum_{n=1}^{N} a_n{}^{(m)} x(t-nT) + n(t) = s_m(t) + n(t), \qquad \begin{cases} m = 1,2,\dots,L^N \\ 0 \le t \le (N+M)T, \end{cases} \tag{7.18}$$

where the number of possible signals is $M = L^N$ over a time interval of $(N+M)T$ seconds in duration. From (2.9c) we know that the probability of the maximum likelihood receiver making an error, when the signal $s_j(t)$ is transmitted, is given by

$$P_{ej} = Pr\left[\bigcup_{\substack{i \ne j \\ i=1}}^{M} \left\{ \ell_i[z(t)|s_j(t)] > \ell_j[z(t)|s_j(t)] \right\}\right], \tag{7.19}$$

where $\ell_i[z(t)|s_j(t)]$ is the likelihood associated with the signal $s_i(t)$ given that $s_j(t)$ is the transmitted sequence. Applying the union bound (see Section 2.1.3) to (7.19) gives

$$\begin{aligned} P_{ej} &\le \sum_{\substack{i \ne j \\ i=1}}^{M} Pr\left\{ \ell_i[z(t)|s_j(t)] > \ell_j[z(t)|s_j(t)] \right\} \\ &= \sum_{\substack{i \ne j \\ i=1}}^{M} Pr\left\{ \int_0^{NT} z(t)[s_j(t)-s_i(t)]\,dt \le \frac{1}{2}\int_0^{NT} [s_j^2(t)-s_i^2(t)]\,dt \right\}. \end{aligned} \tag{7.20}$$

We observe that the random variable appearing in (7.20) is a zero-mean Gaussian random variable with variance

$$E\int_0^{NT}\int_0^{NT} n(t)n(\tau)\epsilon_{ij}(t)\epsilon_{ij}(\tau)\,dt\,d\tau = \sigma^2 \int_0^{NT} \epsilon_{ij}^2(t)\,dt , \tag{7.21}$$

and we define the error signal, $\epsilon_{ij}(t)$, and the error sequence, $\epsilon_n^{(ij)}$, by

$$\epsilon_{ij}(t) = s_j(t)-s_i(t) = \sum_n \left[a_n^{(i)} - a_n^{(j)}\right] x(t-nT) = \sum_n \epsilon_n^{(ij)} x(t-nT) \tag{7.22}$$

and

$$\epsilon_n^{(ij)} = a_n^{(i)} - a_n^{(j)} . \tag{7.23}$$

With this notation the probability appearing inside the summation of (7.20) is given by

$$Q\left(d_{ij}/2\sigma\right) , \tag{7.24}$$

which is in terms of the distance, d_{ij} between allowable, but not identical, signal sequence pairs.* The squared minimum distance is

$$d_{ij}^2 = \int_0^{NT} \in_{ij}^2(t)\,dt = \int_0^{NT} \left[\sum_n \in_n^{(ij)} x(t-nT) \right]^2 dt , \tag{7.25}$$

The requirement that $s_i(t) \neq s_j(t)$ is reflected in the condition that not all $\in_n$ may be zero. Combining (7.20)–(7.25), we have

$$P_{ej} \leq \sum_{\substack{i \neq j \\ i=1}}^{M} Q\left(d_{ij}/2\sigma\right) , \tag{7.26}$$

which represents an upper bound on the probability that an error will be made in detecting the *entire* sequence. Surely, for any system, as time evolves, individual symbol errors will be made, and many terms in (7.26) will be nonzero and thus the probability of an error occurring in the detected sequence approaches one. The merit in the above formulation is to demonstrate that it is the distance between signal *sequences* that determines the error rate. For a given transmitted signal, $s_j(t)$, most of the distances, d_{ij}, will be quite large, and consequently their contribution to (7.26) will be quite small. As with the discussion of the Viterbi algorithm in Chapter 3, (7.26) will often be dominated by those terms for which d_{ij} is a *minimum* and for high signal-to-noise ratios we have

$$P_e \lesssim K_1\, Q\left(d_{\min}/2\sigma\right) , \tag{7.27}$$

where

$$d_{\min}^2 = \min_{i \neq j} d_{ij}^2 = \min_{\substack{\{\in_n\} \\ \text{all } \in_n \neq 0}} \int_0^{NT} [\sum_{n=1}^{NT} \in_n x(t-nT)]^2\,dt , \tag{7.28}$$

and K_1 is a constant that accounts for those signals that are at a minimum distance from the transmitted signal. (Note that as $NT \to \infty$, $K \to 1$ due to the large number of possible error events at $d_{\min}$). The constraint that $s_i(t) \neq s_j(t)$ is embodied by the condition that all $\in_n$ may not be zero in the search to minimize the above expression. Note that if a_n takes on the levels ±1, ±3, then $\in_n$ will search over 0, ±2, ±4, ±6, and that the evaluation of (7.28) is generally a complex task, often requiring the use of the Viterbi algorithm itself to perform the minimization. We will continue our discussion of performance later in the chapter when we introduce the subject of error events.

---

* The term $Q(\cdot)$ in (7.25) was defined in (2.32a).

Under certain conditions on the channel pulse response, $x(t)$, the minimizing sequence will be $\epsilon_n = 0$ $(n \neq 0)$, then

$$d_{\min}^2 = \epsilon_0^2 \int_0^{NT} x^2(t)\,dt \quad , \tag{7.29}$$

which is recognized as the same exponent as the matched filter bound described in Chapter 4. In other words, for certain pulses types it is possible that the Viterbi receiver will utilize all the energy in the transmitted pulse—and provide performance equivalent to that of a system where pulses are transmitted in isolation.

EXAMPLE 3 Consider binary signaling on the partial response channel shaping

$$|X(f)|^2 = \begin{cases} 1 + \dfrac{1}{2}\cos 2\pi f T\,, & |f| \leq \dfrac{1}{2}T \\ 0, & |f| > \dfrac{1}{2}T \end{cases} .$$

Assume that the shaping is split between the transmitter and receiver, i.e. the transmit filter is $\sqrt{X(f)}$. The receiver matched filter will have the same $\sqrt{X(f)}$ characteristic. From (7.28) we have

$$d_{\min}^2 = \min_{\text{all } \epsilon_n \neq 0} \int_0^{MT} \left[\sum_n \epsilon_n x(t-nT) \right]^2 dt = \min \sum_n \sum_m \epsilon_n \epsilon_m \int_{-\infty}^{\infty} |X(f)|^2 e^{-j2\pi f(n-m)T} df \ ,$$

and for the above channel, $r_k = \delta_k + \frac{1}{2}\delta_{k-1} + \frac{1}{2}\delta_{k+1}$, so

$$d_{\min}^2 = \min_{\epsilon_n} \sum_{n=1}^{N} \left(\epsilon_n^2 + \epsilon_n \epsilon_{n-1} \right) = \frac{1}{2}\epsilon_N^2 + \frac{1}{2}\sum_{n=1}^{N} \left(\epsilon_n + \epsilon_{n-1} \right)^2 ,$$

where the minimization is done over the values $\epsilon_n = 0, \pm 2$. Since

$$\sum_n \epsilon_n^2 = \sum_n \epsilon_{n-1}^2$$

it follows that

$$\sum_n \left(\epsilon_n^2 + \epsilon_n \epsilon_{n-1} \right) = \frac{1}{2} \sum_n \left(\epsilon_n + \epsilon_{n-1} \right)^2 .$$

Since this is a sum of positive terms, it is clear that the choice of $\epsilon_n = 0$, $n \neq 0$ achieves the minimum, and results in $d_{\min}^2 = 2$. Substituting the value $d_{\min} = 2$ into (7.27) gives

$$P_e \sim Q\left[\sqrt{2/\sigma}\right] , \tag{7.30}$$

which is the same expression as (2.35) for binary antipodal Nyquist (no intersymbol interference) signaling. Thus, for the above partial response channel, the Viterbi decoder uses all the pulse energy to achieve performance equivalent to the matched filter bound. Note that in contrast to the example presented for the Nyquist channel, the processing cannot be done on a symbol-by-symbol basis, and the Viterbi decoder is needed to achieve the matched filter bound.

EXAMPLE 4 The Viterbi algorithm attains the matched filter bound if the eye is open. See Exercise 7.7.

7.1.7 ERROR EVENTS

We have shown that the Viterbi receiver may be visualized as a trellis which evolves with time. Once merges occur, the trellis has a single maximum likelihood sequence of states emanating backwards in time from the most recent merge. Of course, even with the maximum likelihood detector, errors will occur; *error events* are the diverging of the transmitted symbol sequence and the detected maximum likelihood sequence. (The memory in the system allows only certain paths to be realized.) Wherever these two sequences diverge, an

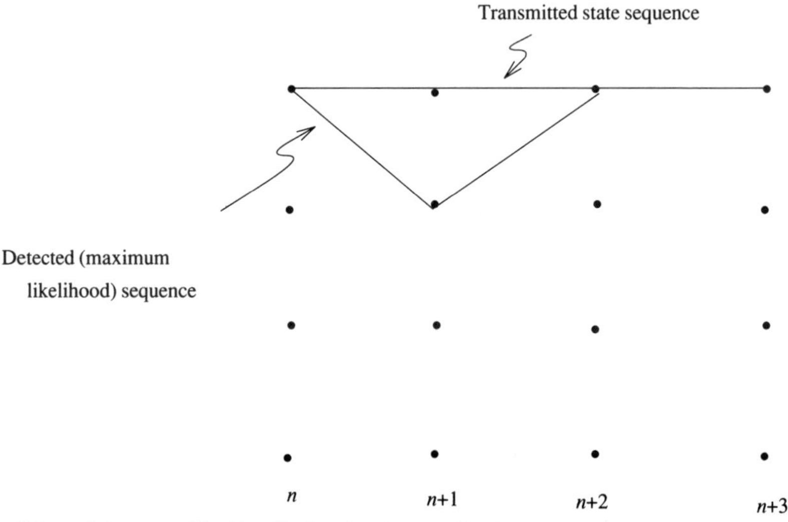

Fig. 7.7a A four-state Viterbi trellis showing the transmitted sequence and an erroneously detected sequence. The error event shown differs from the transmitted sequence in only one position.

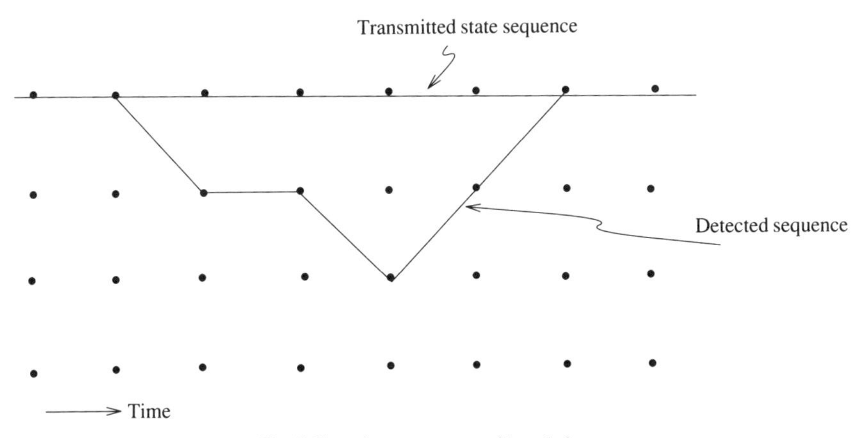

Fig. 7.7b An error event of length four.

error event is said to occur. Error events can be of varying length, so errors occur in bursts; in Figure 7.7a we show a four-state trellis with a transmitted sequence and the detected sequence which differ in one position; this is an error event of length one (here, only $\in_0 \neq 0$, and all other $\in_n = 0$). In Figure 7.7b we show an error event of length four. With such an error event, errors occur in bursts, not individually. The erroneously detected maximum likelihood sequence corresponds to a legitimate sequence of symbols and as such is constrained in the trajectory it follows. In general, the characterization of the length of error events is very channel dependent. The error event framework may be used to obtain very meaningful upper and lower bounds on the symbol error rate. In so doing, we follow the original approach of Forney [1]. For an error event, extending from time n_0 to $n_0 + l$, to occur three subevents must occur:

(1) At time index n_0, the detected state, $\hat{\mathbf{s}}_{n_0}$, equals the transmitted state, $\mathbf{s}_{n_0}$.

(2) The detected state sequence is consistent with the allowed trajectories through the trellis.

(3) Over the interval $n_0 \leq n < n_0 + l$, the accumulated likelihood of the detected trajectory exceeds that along the transmitted trajectory.

7.1.8 SYMBOL ERROR RATE

In order to calculate the probability of a symbol error at time $t = LT$, we observe that error events may be regarded as stationary events; i.e., the probability of the Lth symbol being in error does not depend upon the actual time index. Errors at LT can be caused by error events starting at any index $-\infty < m < L$, so the probability of a symbol error at $t = LT$ may be expressed as

$$Pr\ (error\ at\ LT) = \sum_i \sum_{m=-\infty}^{L} Pr\ (error\ at\ LT | E_i^m)\ Pr\ (E_i^m), \quad (7.31)$$

where $Pr(E_i^m)$ is the probability that the *ith* error event began at $t = mT$. We note that

$$Pr\ (error\ at\ LT|\ E_i^m) = \begin{cases} 0 \text{ if } E_i^m \text{ does not have an error at } LT \\ 1 \text{ if } E_i^m \text{ has an error at } LT \end{cases} \quad (7.32)$$

and that by stationarity $Pr\ (E_i^m)$ is independent of the starting time m. So using the union bound

$$Pr\ (error\ at\ LT) \le \sum_i Pr(E_i) \sum_{m=-\infty}^{L} Pr\ \left[error\ at\ LT | E_i^m\right]$$

$$= \sum_i W_i\ Pr\ (E_i), \quad (7.33)$$

where by using (7.32) we see that the inner summation counts the number of errors in E_i as it slides through the range $(-\infty, L)$ (i.e., $W_i =$ the number of errors [the weight] of the ith error event). Note that (7.33) is independent of the time LT. The probability of an error event is the probability that an erroneous signal sequence, S_j, has a higher likelihood than the transmitted signal which is given by

$$Pr\ (E_i) = \sum_j Pr\ (E_i | S_j\ transmitted)\ Pr\ (S_j) \le \sum_j Q \left[\frac{d_{ij}}{2\sigma}\right] Pr\ (S_j), \quad (7.34)$$

where we made use of the previously derived result (7.24).

Using the above, we have that

$$Pr\ (error) \le \sum_i W_i \sum_j Q \left[\frac{d_{ij}}{2\sigma}\right] Pr\ (S_i). \quad (7.35)$$

For high SNR, (7.35) can be further approximated since the minimum distance error event dominates all the other terms, giving

$$Pr\ (error) \lesssim K\ Q \left[\frac{d_{\min}}{2\sigma}\right]. \quad (7.36)$$

Forney was also able to derive [13] a *lower* bound on the symbol error rate that involves the same quantities appearing in (7.36), i.e.,

$$P_e \ge K_1 Q \left[\frac{d_{\min}}{2\sigma}\right]. \quad (7.37)$$

In summary, the performance of the maximum likelihood Viterbi sequence detector has been shown to depend principally on the *minimum distance* between allowed transmitted sequences. For small amounts of intersymbol interference, it is often the case that the minimum distance error event is a single symbol error and the symbol error probability is identical to the matched filter bound. Finally, the fact that we are using a *linear* form of modulation implies that the minimum distance is independent of the choice of transmitted data sequence.

7.2 WHITENED MATCHED FILTER RECEIVER

We now present an alternative approach to the implementation of the maximum likelihood Viterbi receiver. This approach, known as the *whitened matched filter* [1], is useful in its own right both in performance evaluation of the receiver and in suggesting suboptimum receiver structures. We begin with the observation that the samples, y_k, at the output of the matched filter form a set of *sufficient statistics*; that is, they totally summarize, for the maximum likelihood receiver, the information contained in the received signal. With this observation, the maximum likelihood problem can be reformulated as optimum-sequence detection based on the sampled data model

$$y_k = \sum_n a_n r_{k-n} + \eta_k \tag{7.10}$$

at the output of the matched filter. Recall that the noise η_k, which has been colored by passing through the matched filter, is a zero mean Gaussian process whose autocorrelation function is given by the pulse-correlation function r_k. For convenience we will assume a very long observation interval, and the Fourier transform of r_k, which is also the power spectrum of the noise η_k, is then given by

$$R_\eta(f) \equiv T \sum_k r_k e^{-j2\pi fkT} = \sum_k \mid X(f+\frac{k}{T})|^2, \quad |f| \leq \frac{1}{2}T. \tag{7.38a}$$

Since the noise spectrum, which is also the transfer function of the pulse at the output of the matched filter, is a positive real function, it can be factored into

$$R_\eta(f) = F(f)F(-f), \tag{7.38b}$$

which implies that the noise correlation function, r_k, can be written as the convolution

$$r_k = f_k * f_{-k} , \tag{7.39}$$

where f_k is the causal (i.e., one-sided in time) inverse transform of $F(f)$.

Following the sampler by an *invertible* linear filter does not affect the optimality of the detector, since if it did, the detector could simply incorporate the inverse filter at its front end. (This observation can be used to find the optimum detector for channels with colored noise. See Exercise 7.3.) With this in mind we construct the *whitened matched filter* (using the noise-whitening filter $F^{-1}(f)$) shown in Figure 7.8, whose output, v_k, is given by

$$v_k = \sum_l a_l f_{k-l} + w_k \quad , \quad k=0,1,2,... \tag{7.40}$$

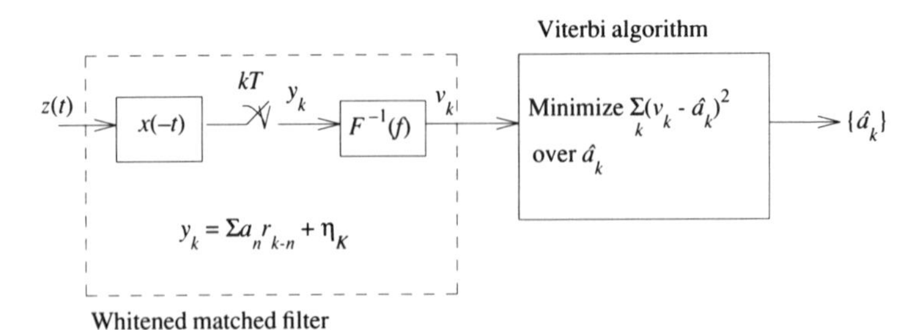

Fig. 7.8 Whitened matched filter form of Viterbi receiver.

where the noise, w_k, has been whitened by the filter $F^{-1}(f)$, and the output system pulse is the one-sided (or causal) pulse f_k defined by (7.38) and (7.39). With the above construction, we now have a discrete-time characterization of the original continuous-time, maximum-likelihood, detection problem. Following the development in Chapter 2, the optimum receiver for the discrete-time channel model given by (7.40) selects the data sequence $\{\hat{a}_k\}$ which is at the minimum Euclidean distance from the sequence $\{v_k\}$, i.e., the detector selects $\{\hat{a}_k\}$ that

$$\min_{\hat{a}_k} \sum_k (v_k - \hat{a}_k)^2 . \tag{7.41}$$

Clearly, performance will again be determined by the minimum distance error event; i.e.,

$$P_e \le K\, Q\left[\frac{d_{\min}}{2\sigma}\right], \tag{7.42}$$

where $d_{\min}$ is, as before, defined in terms of the error sequence, $\in_n$, as

$$d_{\min}^2 = \min_{\substack{\in_n \\ \text{all } \in_n \neq 0}} \sum_k \left[\sum_n \in_n f_{k-n}\right]^2 = \min_{\substack{\in_n \\ \text{all } \in_n \neq 0}} \sum_n \sum_m \in_n \in_m r_{n-m}, \tag{7.43}$$

where, as above,

$$r_k = \sum_n f_n f_{k-n} = f_k * f_k . \tag{7.44}$$

The whitened matched filter receiver provides the motivation for several of the suboptimal, but highly practical, receivers considered later in this chapter.

7.3 SUBOPTIMUM MLSE STRUCTURES

7.3.1 MEMORY TRUNCATION AND STATE DROPPING

The data storage requirement for implementation of the Viterbi algorithm will be unreasonably large unless the channel "memory" is quite small. For high-speed data transmission using quadrature four-level signals on telephone channels, in which the channel memory can easily be 10 symbol intervals, the number of possible state vectors is $L^M = 4^{10} = 1{,}048{,}576$. With retention of an M-element state-path trajectory for each possible current state, in which each element requires $\log_2 L^M = 20$ bits storage capacity, the total storage capacity required, not even mentioning the computational burden, is a substantial $L^M \cdot M \cdot \log_2 L^M = 2.1 \times 10^8$ bits. In order to reduce memory and computational requirements to reasonable levels, two suboptimal approaches have been suggested. One approach [7], is to reduce the channel memory (dispersion) by partial linear equalization prior to maximum likelihood sequence estimation, and the other is to "drop" those current states [6] and [14] (and state trajectories leading up to them), which are relatively improbable. Both approaches have been found to perform somewhat better than a linear equalizer alone under certain severe channel conditions, although neither can claim to be optimal in the same sense as the full MLSE receiver.

7.3.2 MOTIVATION FOR LINEAR RECEIVER

In the next section we will describe the optimum linear and related receivers. These receivers have a *raison d'être* in their own right, by virtue of the fact that such structures are implementable in a cost-conscious manner, can be made to readily adapt to the *a priori* unknown or changing channel environment, and provide exceptional performance for all but the most severe channel conditions. In this section we will provide a rationalization and transition from the nonlinear structure that implements the maximum likelihood sequence detector to detectors that process the received samples in a linear manner.

Recall from (7.8) that the maximum likelihood detector minimizes

$$u_N = -2\sum_{k=1}^{N} \hat{a}_k y_k + \sum_{k,l=1}^{N} \sum \hat{a}_k \hat{a}_l r_{k-l} = -2\mathbf{y}_N' \hat{\mathbf{a}}_N + \hat{\mathbf{a}}_N' R \hat{\mathbf{a}}_N \,, \quad (7.45)$$

by finding the optimum sequence of *discrete-valued* data symbols. In the right-hand side of (7.45) we have used quadratic form notation where the prime denotes vector transpose, $\mathbf{y}_N$ is the vector of matched filter output samples, and R is a Toeplitz matrix whose elements, $r_{ij} = r_{i-j}$, are the channel correlation coefficients. Suppose, in the interest of finding a less complex and linear receiver, the requirement that $\{\hat{a}_n\}$ be discrete valued is relaxed. With $\{\hat{a}_n\}$ thought of as continuous variables, the nature of the problem has changed from a detection to an estimation problem; the estimator will be followed by a simple quantizer that produces the final decisions. Within this framework, minimization of u_N may be accomplished by direct differentiation. Differentiation of (7.45) with respect to $\hat{\mathbf{a}}_N$ gives a linear set of equations with solution

$$\hat{\mathbf{a}}_N = R^{-1} \mathbf{y}_N \ . \quad (7.46)$$

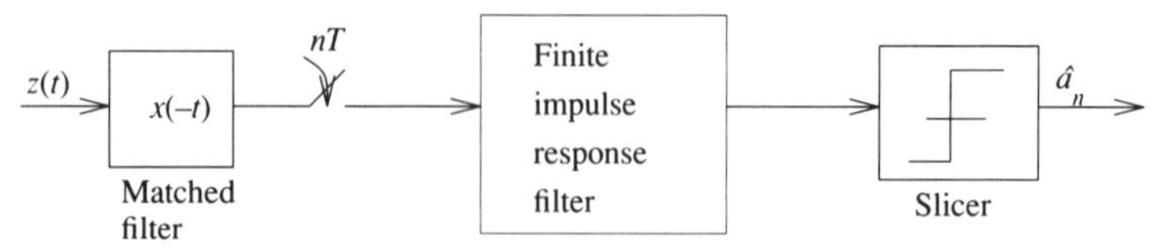

Fig. 7.9 Linear receiver derived from maximum likelihood criteria assuming data symbols are continuous variables.

The above is a linear operation on the matched filter output samples, and an individual estimate $\hat{a}_n$ is given by the finite impulse response (FIR) filtering operation

$$\hat{a}_j = \sum_{l=1}^{M} r_{j-l}^{(-1)} y_l \ , \tag{7.47}$$

where $r_{j-l}^{(-1)}$ denotes the elements of R^{-1}. In general, R^{-1} would not be known *a priori* and the adaptive techniques to be described in Chapter 8 could be used to estimate R and/or R^{-1}. Figure 7.9 shows the linear receiver described by (7.47); in the next section it will be shown that the optimum (in the mean-square error sense) linear receiver is comprised of a matched filter followed by a FIR filter.

The above derivation shows that a linear receiver is the best means of estimating, in a maximum likelihood sense, continuous-valued pulse amplitude modulated data symbols embedded in additive Gaussian noise. In the next section we start with a linear receiver and optimize the structure relative to the mean-square performance criterion.

7.4 THE OPTIMUM LINEAR RECEIVER (EQUALIZER)

We have seen that a data communication system of finite complexity but not constrained to linear processing can (at least theoretically) realize a maximum likelihood estimate of the transmitted data sequence. That is, it can produce the sequence most likely to have been transmitted on the basis of an observation of the noisy received signal. Since this is the best one could hope to do, any other structure will be suboptimal.

This does not mean that the MLSE receiver is preferable to all other structures. We have noted that substantial (even though finite) complexity and compromises (related to adaptation to unknown channel environments, forcing of merges, etc.) are often associated with practical implementation of the Viterbi algorithm. The best reason, however, for using relatively simple linear receiver structures is that they usually offer entirely adequate performance. With experience and tradition added to these considerations, it is not surprising that linear (or quasi-linear) receiver structures are found in almost all commercial modems. In this section, we will derive the optimum linear receiver and evaluate its performance. In the next chapter we will show how this and related receivers can be made to adapt to unknown channel conditions.

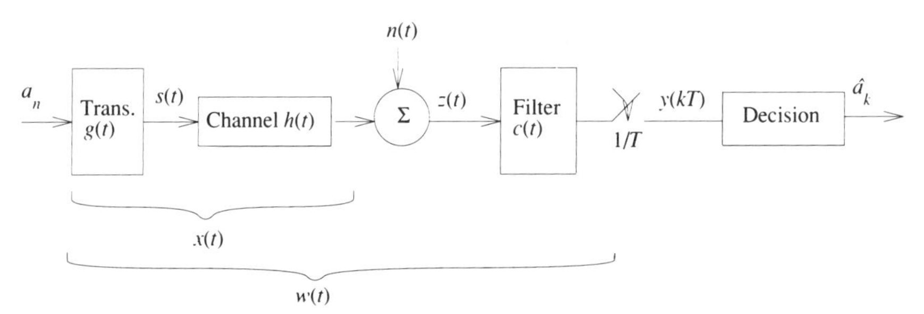

Fig. 7.10 Baseband data transmission system with linear receiver.

We shall assume, as indicated in Figure 7.10, a baseband PAM data communication system in which the transmitted pulse shape has already been selected. The criteria for pulse shaping on a noisy but undistorted channel have been derived in Chapter 4, and are usually adhered to, albeit loosely, in practical engineering work. The transmitted waveform is

$$s(t) = \sum_n a_n g(t-nT) , \tag{7.48}$$

where the $\{a_n\}$ are independent, zero-mean data values, and the received waveform, after the signal has passed through a channel $h(t)$, is

$$z(t) = \sum_m a_m x(t-mT) + n(t) . \tag{7.49}$$

Here, as in the last section, the pulse presented to the receiver is

$$x(t) = g(t) * h(t) , \tag{7.50}$$

and $n(t)$ is white Gaussian noise with two-sided power density $N_0/2$ watts/Hz. We wish to determine the receiver filter, $c(t)$, such that the sampled receiver output sequence $\{ y(kT) \}$, where

$$y(t) = z(t) * c(t) , \tag{7.51}$$

differs as little as possible from the data sequence $\{ a_k \}$. Ideally, we want the *decision*,

$$\hat{a}_k \triangleq \text{data level closest to } y(kT) ,$$

to be identical to a_k, and a rational performance criterion would be the proportion of correct decisions. Unfortunately, even with the constraint of a linear receiver it is very difficult to analyze or optimize performance according to this criterion. For these reasons, we pursue the minimization of the *mean-square error* (MSE)

$$MSE = \langle [y(kT)-a_k]^2 \rangle , \tag{7.52}$$

where the expectation indicated by the angle brackets is the ensemble average with respect to the data symbols and the additive noise. It should also be noted that, in general, the channel characteristics are unknown to the receiver. In order to achieve a meaningful receiver, techniques must be available for the *adaptation* of the receiver to the precise channel environment. As we will show in Chapter 8, the mean-square error forms the basis for the most useful adaptation techniques found in practical systems. Since the Saltzberg bound described in Chapter 4 only provides an upper bound on the error rate in terms of the MSE, the reader should be aware that it is possible to decrease the MSE without a corresponding decrease in error rate. Nevertheless, as per the Saltzberg bound, a lowered MSE usually indicates a decreasing error rate, and our objective is to *minimize the MSE by proper selection of the receiver filter* $c(t)$.

7.4.1 FORMULATION OF THE MEAN-SQUARE ERROR

Expanding (7.52), we write the MSE as

$$MSE = \langle y^2(kT) - 2a_k y(kT) + a_k^2 \rangle . \tag{7.53}$$

Defining the overall pulse shape from transmitter input to receiver output as

$$w(t) = x(t) * c(t) , \tag{7.54}$$

we see that the sampled receiver output is

$$y(kT) = \sum_m a_m w(kT - mT) + v(kT) , \tag{7.55}$$

where $v(t) = n(t) * c(t)$ is the noise component of the receiver output. With $P = \langle a_n^2 \rangle$, the mean-square error is given by

$$MSE = \langle \left[\sum_m a_m w_{k-m} + v_k \right]^2 - 2a_k \sum_m a_m w_{k-m} + a_m^2 \rangle , \tag{7.56}$$

so

$$MSE/P = \int \int \left[A(t,\tau) + \sigma^2 \delta(t-\tau) \right] c(t) c(\tau) dt d\tau - 2 \int x(-t) c(t) dt + 1 , \tag{7.57}$$

where

$$\langle n(t) n(\tau) \rangle = \frac{N_0}{2} \delta(t-\tau)$$

$$\sigma^2 \triangleq N_0/P$$

$$A(t,\tau) \triangleq \sum_m x(mT-t) x(mT-\tau) . \tag{7.58}$$

7.4.2 MINIMIZATION OF THE MSE

The minimum MSE with respect to the function $c(t)$ is found by equating the first variation of MSE (with respect to $c(t)$) to zero. Setting the variation of the MSE, with respect to $c(t)$, to zero gives

$$\int [A(t-\tau) + \sigma^2 \delta(t-\tau)] c(\tau) d\tau = x(-t) . \tag{7.59}$$

From (7.54), the sample of the equalized pulse at time nT is

$$w_n = \int x(nT-t)\,c(t)\,dt, \tag{7.60}$$

so that the left-hand side of (7.59) can be rewritten (recalling definition (7.58)) as

$$\int [\sum_n x(nT-t)x(nT-\tau) + \sigma^2\delta(t-\tau)]\,c(\tau)\,d\tau = \sum_n w_n x(nT-t) + \sigma^2 c(t)\,. \tag{7.61}$$

Thus (7.59) becomes

$$\sum_n w_n x(nT-t) + \sigma^2 c(t) = x(-t)\,. \tag{7.62}$$

Solving for $c(t)$,

$$c(t) = \frac{x(-t)}{\sigma^2} - \sum_n \frac{w_n}{\sigma^2} x(nT-t) = \sum_n c_n x(nT-t)\,, \tag{7.63}$$

where the coefficients, c_n, are defined as

$$c_0 = (1-w_0)/\sigma^2 \tag{7.64a}$$

$$c_n = -\,w_n/\sigma^2,\ n\neq 0\,. \tag{7.64b}$$

Note that the above does not specify $c(t)$, since $c(t)$ is implicit in the $\{w_n\}$ via (7.60). We will determine $c(t)$, or equivalently the coefficient $\{c_n\}$, as a function of the channel impulse response and noise power. However, (7.63) suggests an important physical configuration of the optimal fixed receiver filter.

7.4.3 INTERPRETATION OF THE OPTIMUM LINEAR RECEIVER

Recall that a *matched filter*, having impulse response $x(t)$, is the front end of a receiver designed for optimum detection of the signal $\sum_n a_n x(t-nT)$ in additive white Gaussian noise. Taking the Fourier transform of (7.63) gives

$$C(f) = \sum c_n e^{-j2\pi fnT} X^*(f)\,, \tag{7.65}$$

which is the cascade of the matched filter, $x(-t)$, and a tapped delay line. In other words, the optimum filter $c(t)$, of (7.63) is the weighted sum of outputs of the matched filter *delayed* by different amounts nT. The structure which produces a weighted sum of time-delayed versions of a signal is a *transversal filter*, and is pictured along with the matched filter which precedes it, in Figure 7.11. Ericson [20] has shown that for every "reasonable" performance criterion, the optimum receiving filter can be realized as a matched filter followed by a tapped delay line. A timing recovery circuit, of the type discussed in Chapter 6, is also shown in the figure. Note that since the equalizer tap spacing is the same as the sampling rate, the sampler could be moved to the input of the equalizer (which is shown as an analog filter), and the resulting arrangement, would have an

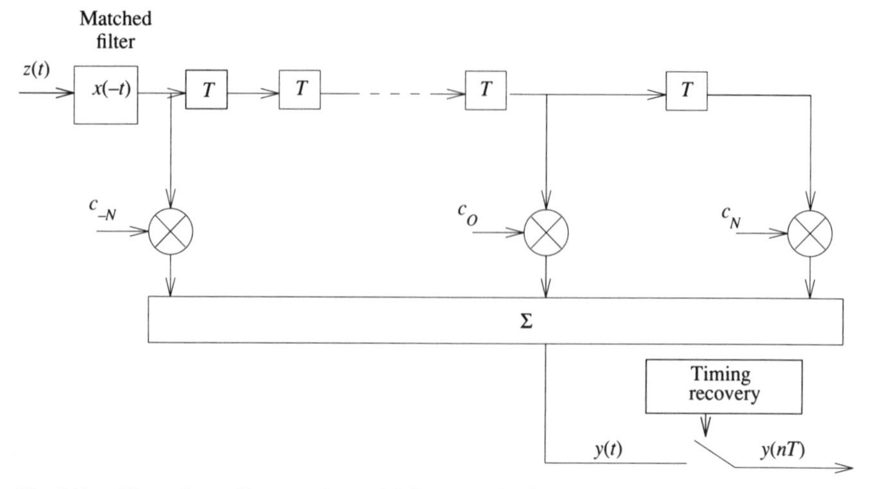

Fig. 7.11 The optimum linear receiver, $c(t)$, is a cascade of a matched filter and tapped delay line (transversal filter) equalizer.

analog-to-digital converter (A/D) followed by a digitally implemented equalizer. The transversal filter alone is often referred to as a *synchronous equalizer*.

In Chapter 8, we will describe how the equalizer coefficients can be made to automatically adapt, without human assistance, to minimize the mean-square error; thus it is not necessary to know the best equalizer coefficients, *a priori*. Although the receiver of Figure 7.11 works reasonably well if the matched filter is approximated by the average channel response, considerably better performance is possible on severely delay-distorted channels, by making the matched filter also adaptive. An effective way to do this is to combine the matched filter

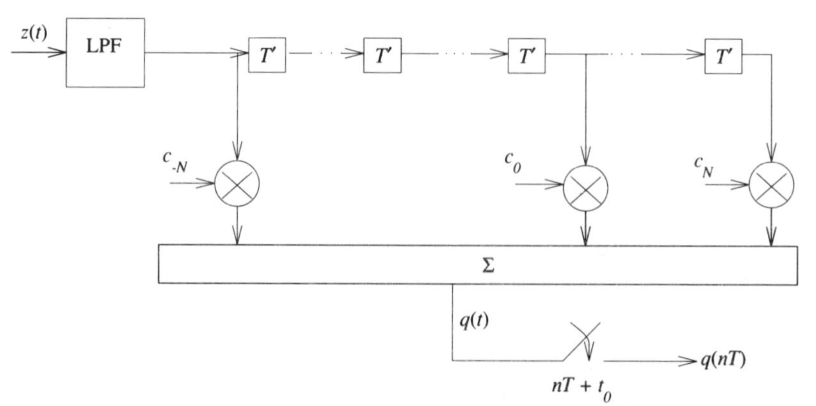

Fig. 7.12 A fractionally-spaced equalizer combining the operations of matched filter and synchronous equalizer.

and the synchronous equalizer into *one* adaptive transversal filter as shown in Figure 7.12, whose output is denoted by $q(t)$ with taps at spacings T' which are smaller than T. Typically, $T' = T/2$, and this structure is known as a *fractionally-spaced equalizer* (FSE). The low-pass filter (LPF) is not a critical component and provides the conventional front-end antialiasing function needed to limit the noise at the input to the sampled data system.

It is useful to isolate the sampling epoch shown in Figure 7.12. This is done by recognizing that sampling the equalizer output $q(t)$ at the epoch t_0 is equivalent to preceding a fixed-epoch sampler by a pure delay of t_0 seconds. The delay has a transfer function $e^{-j2\pi f t_0}$, which may be considered as part of the channel's phase characteristic.

7.4.4 SPECTRAL PLANS FOR SYNCHRONOUS AND FRACTIONALLY-SPACED EQUALIZERS

Before proceeding with the detailed determination of the optimum filter shapes for the fractionally-spaced and synchronous equalizer, it is important to point out the differences in the spectral characteristics of each of these systems. For the synchronous equalizer, typical spectra for the pulse presented to the equalizer, $X(f)$, and the (periodic) transform of the tap coefficients, $C_T(f)$, are shown in Figure 7.13. Note that the period of $C_T(f)$ is $1/T$. For the synchronous equalizer, the signal at the input to the equalizer is given by (neglecting noise)

$$r(t) = \sum_m a_m x(t - mT) \quad , \tag{7.66}$$

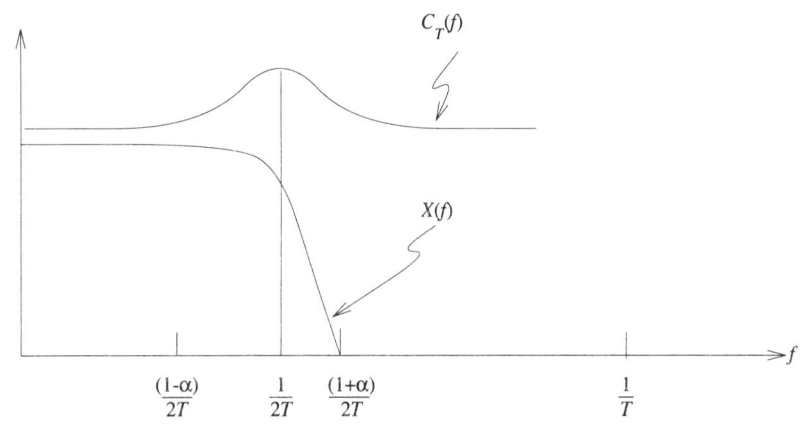

Fig. 7.13 Typical spectra for a synchronous equalizer. $X(f)$ is the spectrum of the pulse presented to the equalizer and $C_T(f)$ is the transform of the tap coefficients. The excess bandwidth is specified by the parameter α.

and the output samples are

$$y(nT+t_0) = \sum_m a_m w(nT-mT + t_0) \ , \tag{7.67}$$

where w_n is the output pulse. From the Nyquist criterion (see Chapter 4) we know that the output pulse has a periodic transform given by

$$W_T(f) = \sum_l c_l e^{-j2\pi f l T} \sum_k X\left[f+k/T\right] \exp\left[-j(f+k/T)t_0\right] = C_T(f) \cdot X_T(f) \ . \tag{7.68}$$

In (7.68), $X_T(f)$ is the folded, or aliased, spectrum of the input pulse (including the affect of the timing phase). Note that since $C_T(f) = C_T(f+k/T)$, the synchronously spaced equalizer can only act to modify the folded spectrum $X_T(f)$, as opposed to directly modifying $X(f)e^{-j2\pi f t_0}$. In other words, because its period is $1/T$, the synchronous equalizer cannot exercise independent control over the channel response on both sides of the rolloff region about $f=1/2T$. If, because of a severe phase characteristic and/or a poor choice of t_0, a null is created in the rolloff portion of the folded spectrum $X_T(f)$, then all the synchronous equalizer can do to compensate for this null is to synthesize a rather large gain in the affected region; this leads to severe performance degradation because of the noise enhancement at these frequencies.

Consider, on the other hand, the *fractionally-spaced* equalizer, with taps spaced $T' < T/(1 + \alpha)$ seconds apart (with $\alpha \geq 0$), that samples the input every T' seconds. Note that the input sampling frequency, $1/T'$, is at least twice the highest frequency component of the baseband signal. As shown in Figure

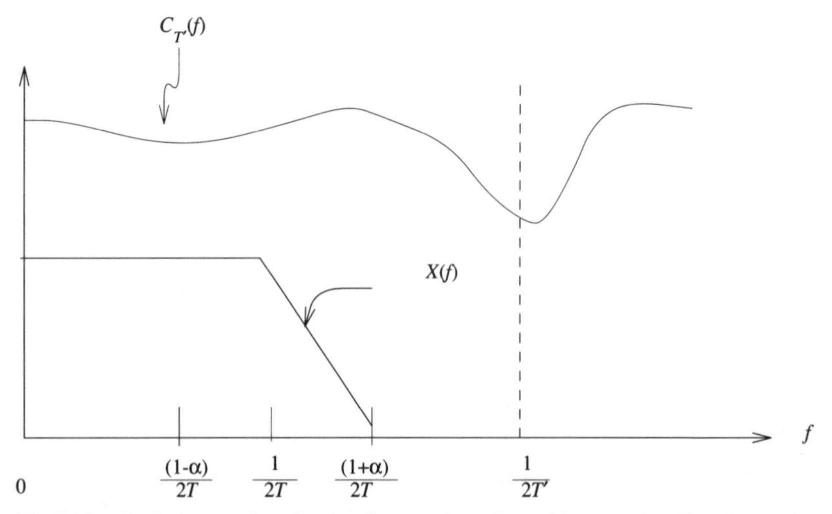

Fig. 7.14 Typical spectra for a fractionally-spaced equalizer with tap spacing $T' < T/(1+\alpha)$.

7.14, the equalizer has the (periodic) transfer function

$$C_{T'}(f) = \sum_i c_i e^{-j2\pi f i T'} .$$

Note that if $1/2T' \geq (1+\alpha)/2T$, then the first repetition interval of the transfer function $C_{T'}(f)$ includes the rolloff portion of the spectrum, as shown in Figure 7.14. We assume that, for digital implementation purposes, T' is an appropriate rational fraction of T. For an FSE receiver, the equalizer input is sampled at the rate T', but the equalizer output is still sampled at the rate T, since data decisions are made at symbol intervals. The equalized spectrum, just *prior* to the output sampler, is periodic (with period $1/T'$) and is given by

$$W_{T'}(f) = C_{T'}(f) \sum_k X\left(f + k/T'\right) \exp\left[j\left(f + k/T'\right) t_0\right], \quad (7.69)$$

and for systems where $1/T' \geq (1 + \alpha)/T$, only the $k = 0$ term survives, i.e.,

$$W_{T'}(f) = C_{T'}(f) X(f) e^{j2\pi f t_0}, \qquad |f| \leq \frac{1}{2T'} . \quad (7.70)$$

The salient aspect of (7.70) is that $C_{T'}(f)$ acts on $X(f)e^{j2\pi f t_0}$ *before* aliasing, with respect to the output sampling rate, is performed. Thus $C_{T'}(f)$ can compensate for any timing phase — by synthesizing a transfer characteristic of the form $e^{-j2\pi f t_0}$. Clearly, such compensation is highly desirable since it is done without any noise enhancement and avoids the extreme sensitivity to timing phase associated with the conventional equalizer. In order to compensate for channel phase distortion of the form $e^{-j\theta(f)}$, the FSE synthesizes the conjugate transfer function $e^{j\theta(f)}$. Thus, a significant property of the FSE is that it can compensate for *any* delay distortion *without* noise enhancement. For the synchronous equalizer, a "bad" timing phase is one which produces nulls in the folded spectrum of $X(f)$. In the FSE, a "good" timing phase is generated, regardless of the input sampling epoch, such that the FSE does the minimum amount of amplitude enhancement. After sampling the equalizer output at the rate $1/T$, the output spectrum is periodic with period $1/T$ and is given by

$$W_T(f) = \sum_l W_{T'}(f+\ell/T) = \sum_l C_{T'}(f+\ell/T) X(f+\ell/T) \exp\left[-j(f+\ell/T)t_0\right] . \quad (7.71)$$

Note that (7.71) differs from (7.68) in that it is the (aliased) sum of equalized components, rather than equalization of a sum of aliased components. It is evident that a FSE is capable of much more than compensating for a poor choice of timing phase. With a properly chosen tap spacing ($T' \leq [1/(1 + \alpha)]T$), the FSE has the compensation capability of an analog filter. As such, *the FSE can realize the optimum linear receiver.*

7.4.5 DETERMINATION OF THE OPTIMUM TAP WEIGHTS FOR THE FRACTIONALLY-SPACED EQUALIZER

We now continue our discussion of the determination of the optimum tap weights, the coefficients c_n of expression (7.63).

Substituting (7.58) and (7.63) into (7.59) gives

$$\int \sum_m x(mT-t)x(mT-\tau) \sum_n c_n x(nT-\tau)\,d\tau + \sigma^2 \sum_n c_n x(nT-t) = x(-t) . \tag{7.72}$$

Recall the channel correlation function can be written as

$$r_{nm} = \int x(t-nT)x(t-mT)\,dt = \int x(nT-\tau)x(mT-\tau)\,d\tau , \tag{7.73}$$

So that (7.72) becomes

$$\sum_n \sum_m r_{m-n} c_n x(mT-t) + \sigma^2 \sum_n c_n x(nT-t) = x(-t) . \tag{7.74}$$

The solution for $\{c_n\}$ is most readily found in the frequency domain. Taking the Fourier transform of (7.74), with respect to t, gives

$$\sum_m \sum_n r_{m-n} c_n X(-f) e^{-j2\pi fmT} + \sigma^2 \sum_n c_n X(-f) e^{-j2\pi fnT} = X(-f) . \tag{7.75}$$

At every frequency for which $X(-f) \neq 0$, this expression can be divided by $X(-f)$ to give

$$\sum_n \sum_m r_{m-n} c_n e^{-j2\pi fmT} + \sigma^2 \sum_n c_n e^{-j2\pi fnT} = 1 . \tag{7.76}$$

Let us now define the periodic transfer functions (period $1/T$) of the tap weights and the channel correlation function respectively by

$$C_T(f) = \sum_n c_n e^{-j2\pi fnT} \tag{7.77a}$$

$$Q_T(f) = \sum_m r_m e^{-j2\pi fnT} = \frac{1}{T} \sum_n \left| X(f + n/T) \right|^2 , \tag{7.77b}$$

where the proof of the last equality is given in Exercise 7.11.

With these definitions and a change of variable, we have that the first term in (7.76) can be expressed as

$$\sum_n \sum_m r_{m-n} c_n e^{-j2\pi fmT} = \sum_l \sum_n r_l c_n e^{-j2\pi f(n+l)T} = Q_T(f) C_T(f) \tag{7.78}$$

and (7.76) becomes

$$Q_T(f) C_T(f) + \sigma^2 C_T(f) = 1 . \tag{7.79}$$

Thus the Fourier transform of the set of tap weights is

$$C_T(f) = \frac{1}{\frac{1}{T} \sum_n |X(f + n/T)|^2 + \sigma^2} . \tag{7.80}$$

The tap weights can be obtained numerically by an inverse Fourier transform of (7.80). It should be emphasized that (7.80) is derived on the assumption that the transversal filter of Figure 7.11 is preceded by a filter exactly matched to the preceding channel $x(t)$.

The overall optimum receiver transfer function (matched filter plus transversal filter), as specified by (7.65), is

$$C(f) = \frac{X^*(f)}{\frac{1}{T} \sum_n \left| X(f + n/T) \right|^2 + \sigma^2} . \tag{7.81}$$

Note that the receiver strikes a balance between *inverting* the folded magnitude-squared channel response, $\sum |X(f + n/T)|^2$, over those frequency regions where the noise is small

$$\left[\sigma^2 << 1/T \sum_n |X(f + n/T)|^2 \right]$$

and *matching* the channel over the frequency regions where the noise is large

$$\left[\sigma^2 >> 1/T \sum_n |X(f + n/T)|^2 \right] .$$

If the receiver simply inverted the channel, the receiver would unduly enhance the noise over these frequency regions where the noise power greatly exceeded the signal power; for a flat noise spectrum, this *noise enhancement* would occur over frequency regions where the channel had a relative null. Thus, the optimum linear receiver strikes a balance between noise enhancement and channel inversion.

A very important property of the optimum linear receiver is the lack of dependence of the receiver on the phase characteristics of the channel, including the timing phase; not surprisingly, the overall mean-square error will also be independent of these characteristics. Combining the above equations, we have that the overall channel, from transmitter input to receiver output, has the transfer function

$$W(f) = X(f)C(f) = \frac{|X(f)|^2}{\frac{1}{T} \sum_n |X(f + n/T)|^2 + \sigma^2} . \tag{7.82}$$

Note that the overall transfer function, $W(f)$, is *independent* of the phase characteristics of the channel. There are several points to be noted:

- Since the sampling phase of the receiver using a fractionally spaced equalizer may be absorbed as part of the channel, the performance of a system that implements the optimum linear receiver will be independent of the sampling epoch (timing phase).
- More generally, the performance of a system with an optimum linear receiver will be independent of the channel phase characteristics. Incorporation of the conjugate channel phase characteristics in the optimum linear receiver ensures this independence, and since the noise power cannot be affected by pure phase compensation, compensation for the channel phase characteristic is achieved without enhancing the noise.

Since the optimum linear receiver consists of an arbitrarily long tapped delay line, the degree to which practical receivers, which emulate the structures described above, achieve independence from the choice of sampling phase and channel phase characteristic will depend on the number of taps employed and the specific channel characteristics at hand. The formulas and results given above assume that the channel characteristics are known at the receiver, and an arbitrarily large number of taps are available for implementation. In Chapter 8 we address the subject of adapting the finite-tap linear receiver to the characteristics of representative channels encountered in practice.

7.4.6 MINIMIZED MEAN-SQUARE ERROR

It is of interest to note the relationship between the overall channel $W(f)$ and the mean-square error at the receiver output. Substituting the result (7.59) (which is obtained by equating to zero the first variation of the MSE with respect to the receiver impulse response $c(t)$) into expression (7.57) for the ratio of MSE to signal power, we find that the minimum mean-squared error is

$$MSE_{\min}/P = 1 - \int x(-t)c(t)\,dt = 1 - \int X(f)C(f)\,df\ . \tag{7.83}$$

If the integral over the infinite interval is broken into a sum of integrals over intervals of size $1/T$ Hz, we have that

$$MSE_{\min}/P = \int_{-\frac{1}{2T}}^{\frac{1}{2T}} \frac{\sigma^2 T}{\frac{1}{T}\sum_n |X(f + n/T)|^2 + \sigma^2}\,df\ . \tag{7.84}$$

With regard to (7.84), note that as expected, the minimum mean-squared error does *not* depend either or the sampling epoch or the channel delay characteristics. Also, note that as the noise becomes vanishingly small (i.e., $\sigma^2 \rightarrow 0$), the equalizer inverts the channel and the $MSE_{\min}$ vanishes.

7.4.7 OPTIMIZATION OF THE SYNCHRONOUS EQUALIZER

In this section we contrast the performance of the optimum linear receiver with that obtainable using the synchronous equalizer structure. The essential differences between these receiver structures are the fixed pre-filter and symbol-rate tap spacing in the synchronous equalizer. In some applications, the simplicity gained in a synchronous equalizer by a lower input sampling rate may more than compensate for the performance loss relative to the optimum linear receiver. We calculate performance by using notation similar to that used for the optimum linear receiver, where we let the overall *sampled* pulse shape from transmitter input to receiver output by

$$w_n = x_n * c_n \tag{7.85}$$

with the sampled receiver output given by

$$y(nT) = \sum_n a_m w(nT - mT) + v(nT), \tag{7.86}$$

and where $v(nT) = n(t) * c(t)|_{t=nT}$ is the noise component at the receiver output. Note that since the input to the equalizer is a sequence of samples, the system processes *sampled* (i.e., aliased) pulse shapes rather than continuous time waveforms. As before, the mean-squared error is given by

$$MSE = \langle y^2(nT) - 2a_n y(nT) + a_n^2 \rangle , \tag{7.87}$$

and it can be shown that

$$MSE/P = \sum_n \sum_m R_{n-m} c_n c_m - 2 \sum_n x_{-n} c_n + 1 , \tag{7.88}$$

where

$$R_m = \sum x_n x_{n-m} . \tag{7.89}$$

Taking Fourier transforms ($T \sum_m \cdot e^{j2\pi fmT}$) of both sides of (7.88) gives

$$MSE/P = T^2 R(f)|C(f)|^2 - 2TC(f)X_{eq}^*(f) + 1 , \tag{7.90}$$

where $R(f)$ is the Fourier transform of R_m, and $C(f)$ and $X_{eq}(f)$ are the Fourier transforms of $\{c_n\}$ and $\{x_n\}$ respectively. Taking the variation of (7.90) with respect to $C(f)$ gives

$$C_{opt}(f) = \frac{X_{eq}^*(f)}{TR(f)} = \frac{X_{eq}^*(f)}{|X_{eq}(f)|^2 + 1} = \frac{X_{eq}^*(f)}{|\sum_n X(f + \frac{n}{T})|^2 + 1} . \tag{7.91}$$

Note the similarity between the transform of the optimum FSE, (7.80), and (7.91). While the equations are similar in form, they differ in that $X_{eq}(f)$ has explicit dependence on the channel delay characteristics and the receiver epoch.

7.5 DECISION FEEDBACK EQUALIZATION

7.5.1 MOTIVATION AND STRUCTURE

In this section we describe a structure, known as the *decision feedback equalizer* (DFE), which makes use of data decisions that have already been made, to provide improved performance (provided the decisions are correct). Since the fractionally-spaced equalizer effectively compensates for delay distortion, the anticipated benefit of the DFE is to improve performance in the presence of moderate to severe amplitude distortion. To motivate the DFE structure, consider the impulse response, after some linear preprocessing, shown in Figure 7.15. In terms of this pulse response, and the additive noise sample, ν_n, the received sample, z_n, is given by

$$z_n = a_n x_0 + \sum_{m<0} a_{n-m} x_m + \sum_{m>0} a_{n-m} x_m + \eta_n , \tag{7.92a}$$

where the first term is the desired data symbol, the second term involves pre-cursor channel samples, and the third term involves post-cursor samples (as well as previously detected data).

If we assume that *prior decisions are correct*, then improved reception of the current data symbol can be achieved by subtracting the effect of the preceding decisions from the current sample. This processing of received samples by a DFE is depicted in Figure 7.16, where the input to the slicer is given by

$$z_n - \sum_{m>0} \hat{a}_{n-m} x_m = a_n x_0 + \sum_{m<0} a_{n-m} x_m + \sum_{m>0} (a_{n-m} - \hat{a}_{n-m}) x_m + \eta_n \tag{7.92b}$$

$$= a_n x_0 + \sum_{m<0} a_{n-m} x_m + \eta_n \quad (\text{assuming } \hat{a}_m = a_m) . \tag{7.92c}$$

Note that the term $\sum_{m>0} \hat{a}_{n-m} x_m$ can be realized via a tapped delay line,with inputs $\hat{a}_n$ and weights $\{x_m\}$. In arriving at (7.92c) from (7.92b), we have used our assumption that the data decisions that have been previously made are correct. With this assumption, the probability of the *current* decision being in error is determined by the amount of precursor intersymbol interference and the

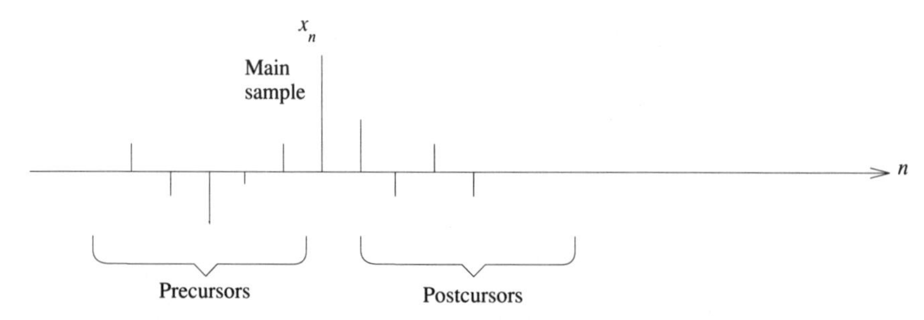

Fig. 7.15 A typical impulse response, x_n, showing precursors and postcursor samples.

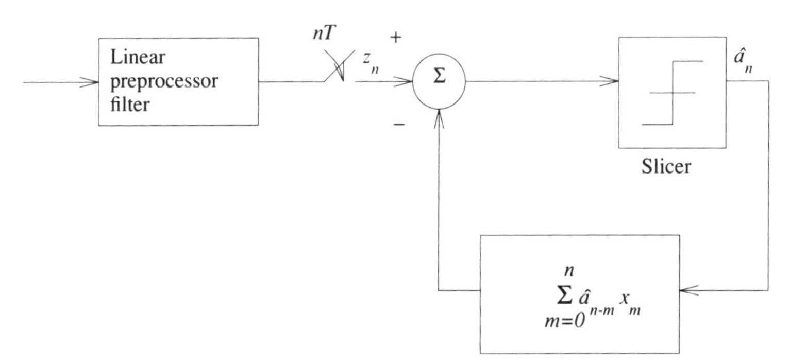

Fig. 7.16 The decision feedback principle.

noise power. Applying the Saltzberg bound to the DFE structure of Fig. 7.16, the probability of the current data symbol being in error is bounded by

$$P_e \leq \exp\left[-x_0^2/(2\sigma^2 + \sum_{m<0} x_m^2)\right] . \tag{7.93}$$

7.5.2 OPTIMUM DECISION FEEDBACK EQUALIZATION

As we have previously observed, the optimum linear receiver can compensate perfectly for delay distortion (i.e., by inverting the channel delay distortion, without increasing the noise at the receiver output), but strikes a balance between inverting the channel's amplitude distortion and enhancing the output noise. A decision feedback equalizer has the potential to compensate for amplitude distortion without providing noise enhancement. Since the signals fed back through the tapped delay line in Figure 7.16 are data decisions, they do not contain an explicit noise* component, and consequently the DFE does not provide any noise enhancement. The objective of this section is to describe the optimum DFE structure (i.e., with arbitrary complexity). We follow the format and discussion of Salz [15] in his seminal paper on decision feedback equalization.

We begin with the baseband model shown in Figure 7.17, where $\{a_n\}$ are independent, odd-integer-valued, $(\pm 1, \pm 3, \ldots, \pm(L-1))$, random data symbols, $h(t)$ is the channel pulse response, $n(t)$ is additive noise with spectral density $N_0/2$, $w(t)$ is the pulse response of the linear feedforward equalizer, $\{b_n\}$ are the coefficient of the feedback section of the DFE, and $\{\hat{a}_k\}$ are the data decisions. Note that v_k is the analog output of the DFE that is presented to the slicer to produce the decisions $\{\hat{a}_k\}$.

---

* Of course, the data decision will be influenced by the input noise at the receiver. If correct decisions are assumed, then the noise has no affect on the linear combination of these decisions.

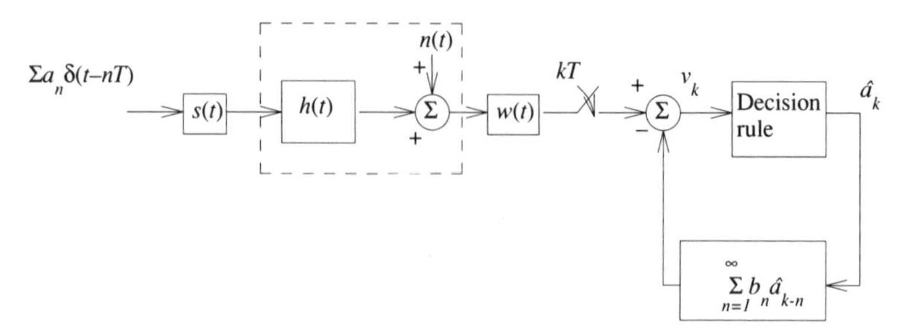

Fig. 7.17 Model for infinitely long decision feedback equalizer.

The performance measure to be minimized with respect to $w(t)$ and $\{b_n\}$, for a given transmitted power, is the mean-squared error

$$MSE_1 = E\left[(v_k - \hat{a}_k)^2\right]. \tag{7.94}$$

While (7.94) represents the function we would like to optimize, the data decision $\hat{a}_k$ depends on $w(t)$ and $\{b_n\}$ in a very nonlinear way, so we will modify (7.94) to the familiar mean-square error assuming that prior decisions are correct, i.e.,

$$E = E\left[(v_k - a_k)^2\right]. \tag{7.95}$$

The quantity, E will serve as a lower bound on the actual mean-square error MSE_1. We let $r(t)$ denote the overall impulse response up to the subtractor, i.e., We let $r(t)$ denote the overall impulse response up to

$$r(t) = s(t) * h(t) * w(t) , \tag{7.96}$$

and with this notation, and assuming prior decisions are correct, we have that the input to the quantizer, v_k, is given by

$$v_k = \sum_{n=\infty}^{\infty} r_n a_{k-n} - \sum_{n=1}^{\infty} b_n a_{k-n} + n(t)*w(t)\Big|_{t=kT} \tag{7.97}$$

and the mean-square error is

$$E = E\left\{ \sum_{n=-\infty}^{-1} r_n a_{k-n} + \sum_{n=1}^{\infty} (r_n - b_n) a_{k-n} + n(t)*w(t)\big|_{t=kT} + (r_0 - 1) a_k \right\}^2 \tag{7.98}$$

Expanding (7.98) we have that

$$E = \sigma_a^2 \sum_{n=-\infty}^{-1} r_n^2 + \sigma_a^2 (r_0 - 1)^2 + \sigma_a^2 \sum_{n=1}^{\infty} (r_n - b_n)^2 + \sigma^2 \tag{7.99}$$

where

$$\sigma_a^2 = E\{a_n^2\} = \frac{L^2 - 1}{3} \tag{7.100}$$

$$\sigma^2 = \frac{N_0}{2} \int_{-\infty}^{\infty} w^2(t)\,dt \, . \tag{7.101}$$

It is evident that E is minimized by choosing $b_n = r_n$ $(n=1,2,...)$. Thus the feedback coefficients are the causal samples of the overall impulse response, and are implicitly dependent upon the linear portion of the receiver, $w(t)$. With this observation, E becomes

$$E = \sigma_a^2 \left[\sum_{n=-\infty}^{o} r_n^2 - 2r_0 + 1 + \frac{\sigma^2}{\sigma_a^2} \right] , \tag{7.102a}$$

or

$$E = \sigma_a^2 \left[\sum_{n=-\infty}^{-1} r_n^2 + (r_0 - 1)^2 + \frac{\sigma^2}{\sigma_a^2} \right] , \tag{7.102b}$$

which indicates that the mean-square error can be minimized by choosing $w(t)$ such that both the pulse precursors and the output noise are minimized, while keeping the center pulse sample close to unity.

The shape of $w(t)$ can be determined by writing (7.102) as

$$\frac{E}{\sigma_a^2} = 1 + \sum_{n=-\infty}^{0} \left[\int_{-\infty}^{\infty} w(\xi) p(nT-\xi)\,d\xi \right]^2 - 2\int_{-\infty}^{\infty} w(\xi) p(-\xi)\,d\xi + \frac{N_0}{2\sigma_a^2} \int_{-\infty}^{\infty} w^2(\xi)\,d\xi \, , \tag{7.103}$$

where $p(t) = s(t) * h(t)$ is the pulse response presented to the receiver. Taking the first variation with respect to $w(\cdot)$, results in the following integral equation for $w(t)$

$$p(-t) = N_0' w(t) + \int_{-\infty}^{\infty} w(\xi) \left[\sum_{n=-\infty}^{0} p(nT-\xi) p(nT-t) \right] d\xi \, , \tag{7.104}$$

where $N_0' = N_0 / 2\sigma_a^2$. If we let

$$U_n = \int_{-\infty}^{\infty} w(\xi) p(nT-\xi)\,d\xi \, , \tag{7.105}$$

then combining (7.104) and (7.105) we have that

$$w(t) = \sum_{n=-\infty}^{0} g_n p(nT-t) \, , \tag{7.106}$$

where

$$g_n = \begin{cases} \dfrac{1}{N_0'} \left\{ 1 - U_0 \right\}, & n = 0 \\[2ex] \dfrac{-U_n}{N_0'}, & n \le -1 \end{cases} \, . \tag{7.107}$$

Equation (7.106) may be interpreted as a matched filter, $p(-t)$, followed by a one-sided (anticausal) tapped delay line with weights equal to g_n. Note that this structure is quite analogous to the optimum linear receiver, with the significant difference that the one-sided tapped delay line replaces the two-sided delay line of the linear receiver. Before proceeding with a more detailed determination of the coefficients $\{g_n\}$, it is important to note that the matched filter and anti-causal TDL structure is similar to the whitened matched filter front end of the Viterbi receiver of Figure 7.8, where the output pulse, f_k, is one-sided but causal. If our criterion was to choose the feedforward filter to force zeroes in the pre-cursor samples of the overall pulse response [13], then the front ends of the DFE and of the whitened-matched-filter Viterbi receiver would be identical. In our discussion we have chosen the feedforward filter to minimize the mean-square precursor intersymbol interference (ISI) and filtered noise. Qualitatively, the feedforward filter in a decision feedback equalizer tries to minimize precursor ISI, while leaving the feedback loop to deal with the causal ISI. This makes eminent sense, since the feedback filter would like to be presented with an input pulse response with a maximum main pulse sample (to provide the maximum margin against the additive noise), and only postcursors (which can be compensated for by the feedback coefficients).

Returning to our analysis of the optimum DFE, we can solve for $\{U_n\}$ by multiplying both sides of (7.104) by $p(kT-t)$, $k \le 0$, then integrating from minus to plus infinity. Performing these operations gives the linear system of Wiener–Hopf equations

$$R_k = N_0' U_k + \sum_{n=-\infty}^{0} R_{n-k} U_n \qquad k=0,-1,\dots, \tag{7.108}$$

where the (even) pulse correlation function, R_k, is defined by

$$R_k = R_{-k} = \int_{-\infty}^{\infty} p(-t)p(kT-t)\,dt. \tag{7.109}$$

The above set of equations can be solved for $\{U_k\}_{-\infty}^{0}$ by the Wiener–Hopf techniques described in Appendix 7A. As we show there, the solution, in terms of the discrete Fourier transform of the one-sided sequence $\{U_k\}_{-\infty}^{0}$, is given by

$$U(\theta) = \sum_{n=-\infty}^{0} U_n e^{in\theta} \tag{7.110}$$

$$= 1 - \frac{N_0'}{F^-(\theta)\gamma_0}, \tag{7.111}$$

where the power spectral density $F(\theta)$ is factored according to

$$F(\theta) = F^+(\theta)F^-(\theta) = R(\theta) + N_0' = \sum_{n=-\infty}^{\infty} f_n e^{in\theta}, \tag{7.112}$$

with the one-sided transforms defined by

$$F^+(\theta) \equiv \sum_{n=0}^{\infty} \gamma_n e^{in\theta}$$

$$F^-(\theta) \equiv \sum_{-\infty}^{0} \gamma_n e^{in\theta} \quad . \tag{7.113}$$

Furthermore, because $F(\theta)$ is the Fourier transform of a covariance function, we have that

$$F^-(\theta) = F^+(-\theta) = [F^+(\theta)]^* \ . \tag{7.114}$$

Note that γ_0 appearing in (7.111) is the DC term in the one-sided transform $F^+(\theta)$. Recall that the overall filter $w(t)$ is the cascade of a filter matched to the impulse response presented to the receiver, $p(-t)$, and a one-sided (anticausal) tapped delay line with weights equal to g_n. These latter weights are chosen so that the feed-forward filter, $w(t)$, minimizes the sum of the precursor ISI and noise power.

We now turn our attention to the minimized mean-square error. We let $w_0(t)$ denote the impulse response of the optimized receiving filter. Substituting $w_0(t)$ into (7.104) and multiplying both sides of the equation by $w_0(t)$ and integrating from minus to plus infinity gives

$$\int_{-\infty}^{\infty} p(-t)\, w_0(t)\, dt = N_0' \int_{-\infty}^{\infty} w_0^2(t)\, dt + \sum_{n=-\infty}^{0} \left[\int_{-\infty}^{\infty} p(nT-t)\, w_0(t)\, dt \right]^2 . \tag{7.115}$$

Substituting (7.115) into (7.103) we have that the minimized mean-square error is

$$\frac{E_{\min}}{\sigma_a^2} = 1 - \int_{-\infty}^{\infty} w(t) p(-t)\, dt = 1 - U_0 = N_0' g_0 \ . \tag{7.116}$$

To proceed further we note from (7.111) that since $F^+(\theta)$ and $F^-(\theta)$ both have the same DC term, γ_0, we have

$$U_0 = 1 - \frac{N_0'}{\gamma_0^2} \tag{7.117}$$

and consequently

$$E_{\min} = \sigma_a^2 \frac{N_0'}{\gamma_0^2} \ . \tag{7.118}$$

In [12], Salz has shown how γ_0^2 can be calculated and that the minimized mean-squared error can be written as

$$E_{\min} = \sigma_a^2 \exp\left\{ -T \int_{-1/2T}^{1/2T} \ln\left[Y(f) + 1 \right] df \right\} , \tag{7.119}$$

where

$$Y(f) = \frac{1}{N_0' T} \sum_{n=-\infty}^{\infty} |P(f - n/T)|^2 .$$

It is instructive to compare the minimum mean-square error using an infinitely long DFE, (7.119) with that obtained using a linear equalizer (7.84). Using the following version of Jensen's inequality

$$\exp\left[\int f(x)\,dx\right] \le \int \exp[f(x)]\,dx , \tag{7.120}$$

we can write

$$E_{\min}(DFE) = \sigma_a^2 \exp\left\{-T \int_{-1/2T}^{1/2T} \ln[Y(f)+1]\,df\right\} \le \sigma_a^2 T \int_{-1/2T}^{1/2T} \exp\ \ln[Y(f)+1]\ \ df$$

$$= \sigma_a^2 \int_{-1/2T}^{1/2T} \frac{T}{\frac{1}{T}\sum_{n=-\infty}^{\infty} |P(f - n/T)|^2}\,df = E_{\min}(\text{optimum linear receiver}).$$

This establishes the not too surprising fact that a DFE (assuming correct decisions) will have a smaller mean-square error than the linear receiver. Note that the DFE and linear receiver will have the same mean-square error only when $Y(f)$ is a constant. When $Y(f)$ is a constant, the pulse presented to the receiver only contains delay distortion (which can be compensated for by the conjugate transfer function of the matched filter); thus, when the pulse has amplitude distortion, the DFE will outperform the linear receiver.

7.5.3 PERFORMANCE COMPARISON BETWEEN LINEAR AND DECISION FEEDBACK RECEIVERS

To illustrate the different levels of performance attainable with the linear and decision feedback receivers, we consider the channel shown in Figure 7.18. In Figure 7.19 we plot the mean-square errors for linear and decision feedback receivers for binary transmission. Note the sharp knee at the Nyquist rate of 6 kbps for the linear receiver, and the more graceful deterioration of the DFE's performance. Thus, as expected, a channel with severe amplitude distortion, or a sharp amplitude cutoff, will benefit substantially from a DFE. In Chapter 8 we will present a more thorough comparison of the linear and decision feedback receivers within the constraints of practical implementation complexity.

7.5.4 ERROR PROPAGATION IN DECISION FEEDBACK EQUALIZERS

The above performance of the DFE was based on the assumption that the feedback decisions were all correct. Suppose that a decision error is made; then, the erroneous decision reduces the margin against noise for future decisions;

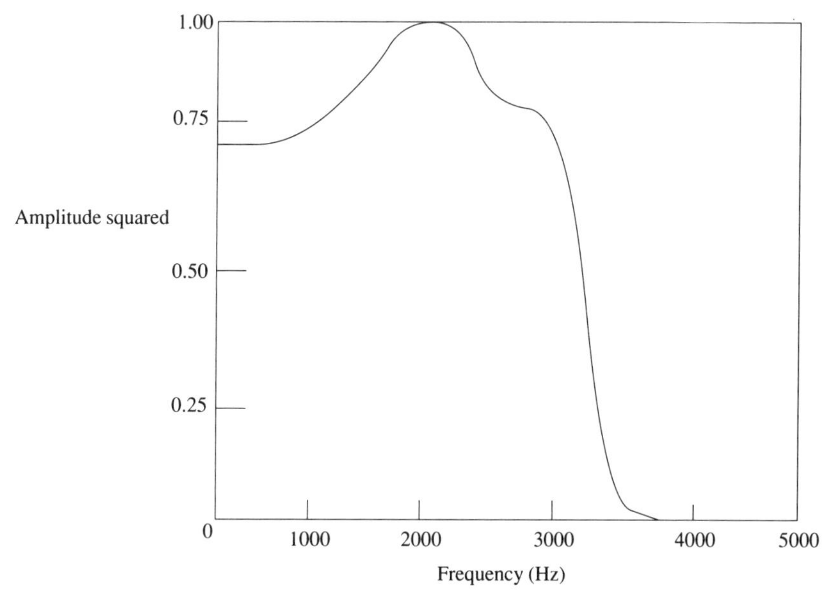

Fig. 7.18 Amplitude-squared characteristics of typical voiceband channel.

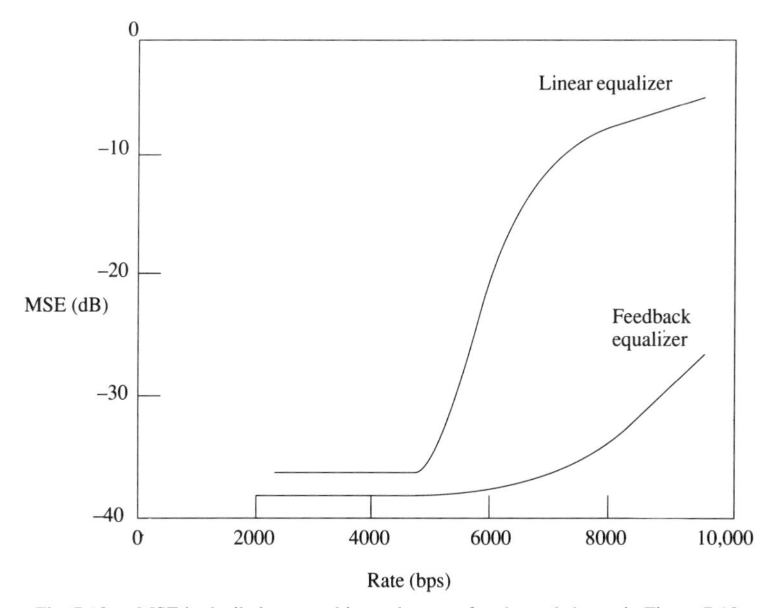

Fig. 7.19 MSE in decibels versus binary data rate for channel shown in Figure 7.18.

thus the probability of the next symbol being in error is increased. Since the DFE is a feedback system, there is a concern as to whether or not errors can *propagate* indefinitely; this raises the question as to whether or not the actual error probability of the DFE is superior to that of the linear equalizer on any or all channels. Duttweiler, Mazo, and Messerschmitt [14] have demonstrated for a *finite*-length feedback section with J taps, that if the worst-case intersymbol interference when the J feedback taps are set to zero is less than the original signal voltage (i.e., the eye is open in the absence of feedback taps), then the error probability is multiplied by at most a factor of 2^J relative to the error probability in the absence of decision errors at high signal-to-noise ratios. We can demonstrate the intuitive reasonableness of this bound by observing that:

(1) All errors will be cleansed from the feedback delay line when J consecutive correct decisions are made.

(2) When errors are being made, the probability of error is no worse than 1/2.

(3) If we let K denote the number of symbols it takes to make J correct decisions (K is the duration of the error propagation), then, since a single error produces on average $K/2$ errors, the average error rate is $(K/2) \cdot P_0$ (where P_0 is the probability of error given that the past J decisions are correct).

Thus the error rate reduces to determining the duration of the error burst, K, For a fair coin, the average number of coin tosses, K to get J consecutive heads (no errors) is easily computed to be $2\ (2^J - 1)$. Thus the average error rate is approximately 2^J times the probability of making the first error. So, if $J = 3$, the error rate is increased by less than an order of magnitude by error propagation.

7.5.5 MODULO-ARITHMETIC-BASED TRANSMITTER EQUALIZATION THAT ELIMINATES ERROR PROPAGATION

If the number of feedback taps is large, then the error propagation may severely affect the overall error rate. Since the data symbols are available at the transmitter, it is reasonable to explore the placement of an equalizer at the transmitter. Let us also suppose that we have a receiver feed-forward filter (as in Figure 7.17) that produces a causal impulse response. A linear filter placed at the transmitter to invert the channel would be unstable if the channel had zeroes in its transfer function, which when inverted, leads to unstable poles. Tomlinson [15] has proposed a nonlinear filter that can invert a causal impulse response, but which does *not* have the potential for error propagation or for unstable operation. The arrangement is shown in Figure 7.20, where a channel with z-transform $H(z)$ is equalized (or precoded) at the transmitter using modulo-N arithmetic. In

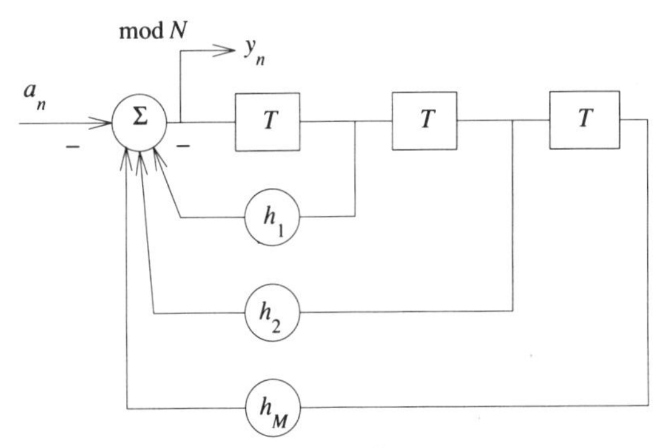

a) Modulo-N inverse filter: $[H(z)-1]^{-1}$ mod N

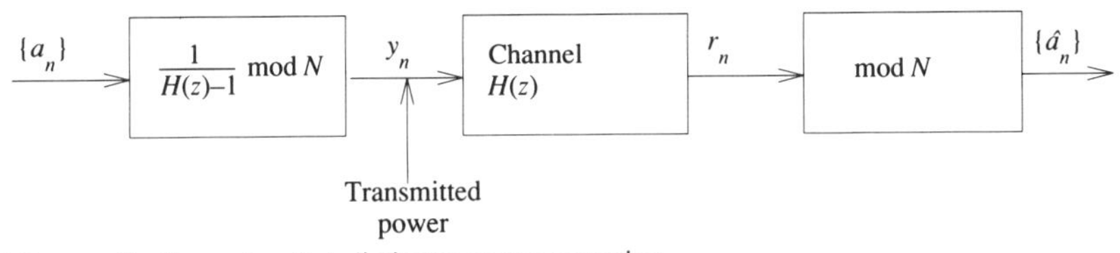

b) Pre-equalization system that eliminates error propagation

Fig. 7.20 (Tomlinson) modulo arithmetic transmitter equalization.

Figure 7.20a the coefficients $\{h_i\}_{i=1}^{M}$ are the terms in the transform $H(z)$, except that h_0 is missing. The adder is a modulo-N adder which operates as a normal adder except that

(a) If the result of the summation is greater than $N/2$, N is subtracted an integral number of times until the result is less than $N/2$.

(b) If the result of the summation is less than $-N/2$, N is added an integral number of times until the result is greater than $-N/2$.

Thus the output of the modulo-N adder has always a magnitude between $-N/2$ and $N/2$, and, as a consequence, the filter is *always stable*.

The parameter N is chosen large enough to accommodate the transmitted symbols, and equalization is accomplished as follows. referring to Figure 7.20b, in terms of z-transforms, the input to the equalizer is $A(z)$ and the output is $Y(z)$. The output of the modulo-N adder is

$$\begin{aligned} Y(z) &= A(z) - Y(z)\,[\,h_1 z^{-1} + h_2 z^{-2} + \ldots + h_M z^{-M}\,] - bN \\ &= A(z) - Y(z)\,[\,H(z) - 1\,] - bN, \end{aligned} \tag{7.121}$$

where b is an integer reflecting the modular operation. The channel output, r_n, has the z-transform

$$R(z) = H(z)\ Y(z) = \left[1 + (h_1 z^{-1} + ... + h_M z^{-M}) \right] Y(z)$$
$$= Y(z) + (h_1 z^{-1} + ... + h_M z^{-M})\ Y(z) \ , \tag{7.122}$$

and substituting (7.121) for the first term on the right-hand side of (7.122) gives

$$R(z) = A(z) - Y(z) \left[h_1 z^{-1} + h_2 z^{-2} + ... + h_M z^{-M} \right] - bN \tag{7.123}$$
$$+ Y(z) \left[h_1 z^{-1} + h_2 z^{-2} + ... + h_M z^{-M} \right]$$
$$= A(z) - bN.$$

Passing r_n through the modulo-N operation at the receiver gives

$$\hat{a}_n = a_n \ . \tag{7.124}$$

Thus a pre-equalizer at the transmitter with a transfer function of $1/[H(z) - 1]$ mod N can equalize a causal pulse.

The Tomlinson filter has the advantage that error propagation is eliminated and the noise at the receiver is not colored. Another virtue of placing a DFE capability at the transmitter is to eliminate the difficulties associated with combining the decoding of trellis-coded modulation (which provides delayed decisions) and conventional DFE [21]. The power of a DFE cannot be fully exploited when its input is the (delayed) outputs of a TCM decoder, since those weights up to and including the delay are rendered useless. One potential drawback of Tomlinson filtering is the added transmitted power appearing at the output of the modulo-N inverse filter. Mazo and Salz [16] have shown that the increase in transmitted power is rather modest. Another difficulty, the need for the transmitter to know the channel characteristics, can often be overcome by a duplex communications link and use of the adaptive techniques described in the next chapter.

7.6 CHAPTER SUMMARY

In this chapter we first developed the optimum maximum likelihood sequence receiver, which can be implemented as a matched filter followed by a nonlinear system that uses the principles of dynamic programming to realize the Viterbi algorithm receiver. This receiver finds the sequence of data symbols that are closest, in an integrated squared error sense, to the received signal. Performance of the receiver depends on the minimum distance between allowed signal sequences, and for certain channels this distance is such that all the energy in the received pulse is available for detection — giving a performance equivalent to that attainable using parallel, independent channels to detect a single isolated pulse by matched filter reception. Our attention then turned to the optimum *linear* receiver, an equalizer that minimizes the output mean-square error and whose structure can be realized by either a matched filter followed by a synchronous tapped delay line or equivalently by a tapped delay line whose taps are finely spaced (typically at half a symbol period). The tap spacing of this latter fractionally spaced equalizer (FSE) is such that the system can equalize the total channel characteristic. This provides a system that can compensate for arbitrary channel delay characteristics, without enhancing the noise at the receiver output. Finally, the decision feedback equalizer (DFE) was described as a means of using previously detected decisions to provide additional performance gain in the presence of severe amplitude distortion.

APPENDIX 7A THE WIENER–HOPF DECISION FEEDBACK EQUATION

In this appendix we solve the following set of equations for $\{U_k\}_{k=-\infty}^{0}$:

$$
\begin{aligned}
R_k &= N_0' U_k + \sum_{n=-\infty}^{0} R_{n-k} U_n, \qquad k = 0, -1, -2, \ldots \\
&= \sum_{n=-\infty}^{0} F_{n-k} U_n ,
\end{aligned}
\tag{7.108}
$$

where R_k is a correlation function defined by

$$
R_k = R_{-k} = \int_{-\infty}^{\infty} p(-t) p(kT-t)\, dt \tag{7.109}
$$

and $F_k = R_k + N_0' \delta_k$ (where $\delta_k = 1, k=0$; $\delta_k = 0, k \neq 0$). It is well known that a correlation function can be factored into the convolution of two (one-sided) sequences $\{f_k^-\}_{-\infty}^{0}$ and $\{f_k^+\}_0^{\infty}$, such that

$$
F_k = \sum_{j=0}^{\infty} f_j^+ f_{k-j}^- \quad \text{for all } k . \tag{7A.1}
$$

If we define the sequence $\{X_n\}_{-\infty}^{\infty}$ via the equation

$$
R_k = \sum_{j=0}^{\infty} f_j^+ X_{k-j} \qquad \text{all } k , \tag{7A.2}
$$

then substituting (7A.1) and (7A.2) into (7.108) gives

$$
\sum_{j=0}^{\infty} f_j^+ \left\{ X_{k-j} - \sum_{n=-\infty}^{0} U_n f_{n-k-j}^- \right\} = 0 , \quad k = 0, -1, -2, \ldots \tag{7A.3}
$$

Note that a solution for $\{U_n\}_{n=-\infty}^{0}$ of the equation

$$
X_k = \sum_{n=-\infty}^{0} U_n f_{n-k}^- , \qquad k \leq 0 \tag{7A.4}
$$

is also a solution of (7A.3). We now turn our attention to the solution of (7A.4) by defining the two-sided Fourier transform of the sequence $\{X_n\}_{-\infty}^{\infty}$ by

$$X(\theta) = \sum_{n=-\infty}^{\infty} X_n e^{in\theta} . \tag{7A.5}$$

Taking the transform of both sides of (7A.2), we obtain

$$R(\theta) = F^+(\theta) X(\theta) . \tag{7A.6}$$

Additionally, the one-sided, $\sum_{n=-\infty}^{0}$, transform of (7A.4) is

$$X^-(\theta) = U^-(\theta) F^-(\theta) , \tag{7A.7}$$

where $X^-(\theta)$ is derived from (7A.6) by noting that

$$X^-(\theta) = \left[\frac{R(\theta)}{F^+(\theta)} \right]_- , \tag{7A.8}$$

where $[\cdot]_-$ stands for "projection to negative integers only." The projection is obtained by expanding (7A.8) in a two-sided Fourier series and retaining the part of the series containing the DC term and the negative coefficients. In terms of the above notation, the transform $U^-(\theta)$ of the desired quantity $\{U_n^-\}_{n=-\infty}^{0}$ is given by

$$U^-(\theta) = \frac{1}{F^-(\theta)} \left[\frac{R(\theta)}{F^+(\theta)} \right]_- . \tag{7A.9}$$

We can proceed further by noting that since $F(\theta) = F^+(\theta) F^-(\theta)$ and $F(\theta) = R(\theta) + N_0'$, we have that

$$\left[\frac{R(\theta)}{F^+(\theta)} \right]_- = \left[\frac{F(\theta) + N_0'}{F^+(\theta)} \right]_- = \left[\frac{F^+(\theta) F^-(\theta) + N_0'}{F^+(\theta)} \right]_- = F^-(\theta) + \frac{N_0'}{\gamma_0} \tag{7A.10}$$

where γ_0 is the DC coefficient of $F^+(\theta)$. Combining (7A.9) and (7A.10) we have

$$U^-(\theta) = 1 - \frac{N_0'}{F^-(\theta)\gamma_0} . \tag{7A.11}$$

Finally, we demonstrate how $F^{\pm}(\theta)$ can be calculated from $F(\theta)$. Since $F(\theta)$ is a power spectral density, we know that $F(\theta) > 0$ for $0 \le \theta \le 2\pi$, thus $\ln F(\theta)$ may be expanded in a two-sided Fourier series

$$\ln F(\theta) = \sum_{n=-\infty}^{0} \gamma_n e^{in\theta} + \sum_{n=0}^{\infty} \gamma_n e^{in\theta} \tag{7A.12}$$

and exponentiation of (7A.12) gives

$$F^{+}(\theta) = \exp\left\{\sum_{n=0}^{\infty} \gamma_n e^{in\theta}\right\}$$

$$F^{-}(\theta) = \exp\left\{\sum_{n=-\infty}^{0} \gamma_n e^{in\theta}\right\}. \tag{7A.13}$$

REFERENCES

[1] G. D. Forney, Jr., "Maximum Likelihood Sequence Estimation of Digital Sequences in the Presence of Intersymbol Interference," *IEEE Trans. on Information Theory*, Vol. IT-18, pp. 363–378, May 1972.

[2] G. Ungerboeck, "Adaptive Maximum Likelihood Receiver for Carrier Modulated Data Transmission Systems," *IEEE Trans. on Comm.*, Vol. COM-22, No. 5, pp. 624–636, May 1974.

[3] J. F. Hayes, "The Viterbi Algorithm Applied to Digital Data Transmission," *IEEE Communications Society Magazine*, Vol. 13, No. 2, pp. 5–16, March 1975, reprinted in *IEEE Communications Society's Tutorials in Modern Communications*, Computer Science Press, 1983.

[4] R. W. Lucky, J. Salz, and E. J. Weldon, Jr., *Principles of Data Communications*, McGraw-Hill, 1968.

[5] J. G. Proakis, *Digital Communications*, McGraw-Hill, 1983.

[6] F. L. Vermeulen and M. E. Hellman, "Reduced State Viterbi Decoder for Channels with Intersymbol Interference," *1974 International Conference on Communications* Conf. Record, pp. 37B1–37B4.

[7] D. D. Falconer and F. R. Magee, Jr., "Adaptive Channel Memory Truncation for Maximum Likelihood Sequence Estimation," *Bell System Tech. J.*, Vol. 52, No. 9, pp. 1541–1562, Nov. 1973.

[8] G. D. Forney, Jr., "The Viterbi Algorithm," *IEEE Proc.*, Vol. 61, No. 3, pp. 268–278, March 1973.

[9] G. J. Foschini, R. D. Gitlin, and J. Salz, "Optimum Direct Detection for Digital Fiber Optic Communication Systems," *Bell System Tech. J.*, Vol. 54, No. 8, pp. 1389–1430, Oct. 1975.

[10] L. R. Rabiner and B. H. Juang, "An Introduction to Hidden Markov Models," *IEEE ASSP Magazine*, pp. 4–16, January 1986.

[11] E. Ayanoglu and R. M. Gray, "The Design of Joint Source and Channel Trellis Waveform Encoders," *IEEE Trans on Information Theory*, Vol. IT-33, No. 6, pp. 855–865, November 1987.

[12] J. F. Hayes, T. M. Cover and J. B. Riera, "Optimum Sequence Detection and Optimum Symbol-by-Symbol Detection: Similar Algorithms," *IEEE Trans. on Communications*, Vol. COM-30, No. 1, pp. 152–157, Jan. 1982.

[13] G. D. Forney, Jr., "Lower Bounds on Error Probability in the Presence of Large Intersymbol Interference," *IEEE Trans. on Communications*, Vol. COM-20, pp. 76–77, Feb. 1972.

[14] G. J. Foschini, "A Reduced State Variant of Maximum Likelihood Sequence Detection Attaining Optimum Performance for High Signal-to-Noise Ratios," *IEEE Trans. on Information Theory*, Vol. IT-23, No. 5, pp. 605–609, Sept. 1977.

[15] J. Salz, "Optimum Mean Square Decision Feedback Equalization," *Bell System Tech. J.*, Vol. 53, No. 8, pp. 1341–1373, Oct. 1973.

[16] R. Price, "Nonlinearly Feedback Equalized PAM Versus Capacity, for Noisy Filter Channels," in *Conf. Rec., Int. Conf. Commun.*, Philadelphia, PA, June 1972.

[17] D. L. Duttweiler, J. E. Mazo, and D. G. Messerschmitt, "An Upper Bound on the Error Probability in Decision Feedback Equalization," *IEEE Trans. on Information Theory*, Vol. IT-20, No. 4, pp. 490–497, July 1974.

[18] M. Tomlinson, "New Automatic Equalizer Employing Modulo Arithmetic," *Electron, Lett.*, Vol. 7, pp. 138–139, March 1971.

[19] J. E. Mazo and J. Salz, "On The Transmitted Power in Generalized Partial Response," *IEEE Trans. on Communications*, Vol. COM-24, No. 3, pp. 348–352, March 1976.

[20] T. Ericson, "Structure of Optimum Receiving Filters in Data Transmission Systems," *IEEE Trans. on Information Theory*, Vol. IT-17, pp. 352–353.

[21] G. D. Forney, Jr. and M. V. Eyuboglu, "Combined Equalization and Coding Using Precoding," *IEEE Communications Magazine*, Vol. 30, No. 12, pp. 25–35, December 1991.

EXERCISES

7.1: For a 16-point, 2400-baud QAM system, how many multiplications per second does a Viterbi algorithm receiver have to perform if the channel has a memory of 10 symbols?

7.2: (a) For the parameters of the Viterbi algorithm example presented in Section 7.1.4, determine the optimum decisions for the following received samples: $y_1 = 0.5$, $y_2 = 1.0$, $y_3 = 1.5$, $y_4 = 3.0$, $y_5 = 0.5, y_6 = 1.5, y_7 = 2.0$.

(b) For the same received samples, find the optimum decision sequence if the channel memory is unity and the data symbol levels are 0, ±1.

7.3: Consider the model of (7.2)

$$z(t) = \sum_m a_n x(t - mT) + n(t),$$

but where $n(t)$ has a non–white power spectral density, $N(f)$. Derive the structure of the maximum likelihood sequence detector.

Hint: Recall that invertible, linear operations can precede the sequence detector.

7.4: Consider 7.2, when $N(f) = \beta|X(f)|^2$. This approximates a channel where the noise is due to crosstalk of a signal with the same spectrum as the transmitted signal. Sketch the maximum likelihood sequence detector. What factors determine the error rate?

7.5: Consider *continuous-phase* FSK defined by

$$s(t) = cos[\omega(u_k)t + \theta_k], \qquad kT \le t \le (k + 1)T$$

where $\omega(u_k)$ is the radian frequency selected by u_k and θ_k is selected such that

$$\omega(u_{k-1})(k-1) + \theta_{k-1} \equiv \omega(u_k)kT + \theta_k \quad \text{modulo } 2\pi .$$

This condition defines continuous phase FSK and introduces memory into the modulation process. Suppose

$$\begin{aligned} &\omega(0)T \equiv 0 \\ &\omega(1)T \equiv \pi \qquad \text{mod} 2\pi . \end{aligned}$$

Then if $\theta_0 = 0, \theta_1 = 0$ or π depending upon whether u_1 equals zero or one, and $\theta_k = 0$ or π depending upon whether an even or odd number of ones has been transmitted,

(a) Show that $y_k \equiv s(kT)$ may be described in terms of a two-state process, x_k, having values 0 or π.

Hint: Write
$y_k = \cos[\omega(u_k)t + x_k] = cosx_k \cos\omega(u_k)t, kT \le t < (k + 1)T$

(b) Draw a trellis diagram for the evolution of the states.

(c) Draw a block diagram showing the state evolution.

7.6: Show for the pulse $h_k = \delta_k - \delta_{k-1}$ that the minimum distance error event has a single error.

7.7: Show that the Viterbi algorithm attains the matched filter bound if the eye is open.

7.8: Suppose that the channel consists of a cubic nonlinearity following the linear filter, but before the additive noise.

(a) What is the structure of the maximum likelihood detector?

(b) Under what conditions can the Viterbi algorithm be used to realize the receiver?

(c) Give an approximate bound on the error rate.

7.9: Derive equation (7.37).

7.10: What is the maximum likelihood detector for the channel of (7.2) followed by any invertible nonlinearity?

7.11: Derive equation (7.77b)

8

Automatic and Adaptive Equalization

8.0 INTRODUCTION

Up to this point in the text, we have made two key assumptions in discussing the structures described in Chapter 7: we have assumed arbitrary receiver complexity and we have also assumed that the channel characteristics are known at the receiver. For the maximum likelihood sequence estimation receiver, implemented via the Viterbi algorithm, the number of states was allowed to grow without bound and the observables used to compute the transition metrics are outputs of a (presumed known) filter matched to the channel. In the optimum linear receiver, the matched filter appears again along with a tapped delay line equalizer of arbitrary length. In the optimum linear receiver which does not use a matched filter, the tap weights are dependent on the channel covariance matrix. In practice, the channel characteristics are generally not known. If a dialed telephone line is used, the channel is different on each call. Even for private or leased channels, the characteristics may be known only within certain limits. For many channels, such as fading radio systems, phase perturbations and other time-varying channel variations are present, requiring constant tracking to avoid deterioration of performance. The optimum receivers we have described in the preceding chapters would be of academic interest only if it were not possible to *adapt* the parameters appearing in their structures to accurately model the actual channel or a function of the channel, such as its inverse.

Motivated by reducing the optimum linear receiver and related structures to realizable systems, this chapter concentrates on *structures* and *adaptation techniques* of the element which has had the largest impact on data communications, the *equalizer*. Except for extremely distorted channels, a properly designed adaptive equalizer can transform an unknown, severely dispersive channel into a near-ideal one, with data transmission approaching the matched filter performance bound. Based upon the designer's *a priori* knowledge of the range of expected channel characteristics, a structure (generally a tapped delay line) is chosen to model the channel (or its inverse) and an adaptive algorithm is designed to adjust the parameters in the structure (e.g., the tap weights in a tapped delay line filter) so as to minimize a performance measure such as the mean-square error. We describe several important equalizer structures and algorithms that have emerged over the past twenty-five years, including the tapped delay line filter adjusted to minimize the mean-square error. Self-adjusting linear filters, principally adaptive equalizers and echo cancellers (see Chapter 9), are at the heart of higher-speed telephone line modems, and have new importance in digital microwave radio, twisted pair (ISDN), undersea, and optical communication systems, and have strongly influenced the broad discipline of linear adaptive systems.

Probability of error is the most appropriate performance criterion for overall system evaluation and equalizer adaptation, but, as we have noted before, the mathematical analysis becomes extremely difficult. The early equalizers pioneered by R. W. Lucky [1] were designed to provide implementation simplicity and to minimize peak distortion, or equivalently maximize the eye opening, since at that time, this was felt to be an important indicator of the error rate. This criterion led directly to the "zero forcing" equalizer, which seeks an overall impulse response with zeros at all sample points within the time span of the equalizer except, of course, the non-zero center sample. However, this approach ignores the effect that such a system response will have on the noise at the receiver output, and moreover, when the eye is initially closed, the parameter adjustment algorithm is not guaranteed to maximize the eye opening, even when a training sequence is provided. *Zero-forcing* algorithms, which were of great interest in voice-grade modems in the 1960s because they are simple to implement in digital circuits, may in fact experience a renaissance as designers consider equalization of gigabit per second optical systems (see Chapter 10). In this same time period, Widrow and Hoff [2] proposed the *least mean-square* (LMS) error algorithm for general purpose adaptive filtering, when a training sequence is available. This powerful, yet simple algorithm, which minimizes the output mean-square error, has seen broad application in many areas of communications where adaptive systems are required, including antenna systems, predictive speech systems, and neural networks (which was one of Widrow's original applications for the LMS algorithm). The paramount commercial application of the LMS algorithm has been to adaptive equalization and echo cancellation in data communications systems. The LMS algorithm is useful both for the initial training of adaptive equalizers, and its utility is substantially enhanced by exploiting Lucky's significant observation [3] that once reliable decisions are available at the receiver output, these decisions could replace the training sequence so that adaptation may be continuous in order to track slow changes in the channel.

The mean-square error (MSE) has now been recognized as the most convenient and meaningful criterion from an implementation (ease of) analysis, and performance viewpoints. As shown by the Saltzberg bound (see Chapter 4), the error rate can be shown to decrease monotonically with the MSE. Our discussion will focus on the steady-state and transient performance that can be obtained with linear, synchronously spaced (i.e., at a symbol interval) and fractionally spaced [4] tapped delay line equalizers and on an analysis of the LMS algorithm for adjusting the equalizer during its convergence to the optimum settings. We will also discuss an alternative linear structure, the lattice filter [5], that can be shown to have desirable implementation properties, as well as several nonlinear structures, such as the decision feedback equalizer [6] and the linear canceler [7,8], that improve performance, beyond that attainable with a linear structure, in the presence of severe amplitude distortion. We will also describe adjustment algorithms that are an alternative to the LMS techniques for minimizing the mean-square error. These algorithms, which include the Kalman [9,10] and Fast Kalman [11] classes of estimation techniques, strive to provide more rapid convergence of the weights of the tapped delay line equalizer to their optimum settings. The Fast Kalman approach should really be referred to as *computationally fast,* since it provides the same rate of convergence as the Kalman algorithm, but with an implementation complexity approaching that of the LMS algorithm.

Our principal interest in this chapter is in determining and demonstrating how closely we can approximate, *in practice*, the performance of the optimum structures described in the previous chapter. Specifically, we will describe adaptation techniques that permit the rapid and reliable convergence of the equalizer coefficients to their desired settings.

There is an important distinction to be made between automatic and adaptive equalization. *Automatic equalization* refers to the initial adjustment of the equalizer tap weights with the aid of a *training sequence* available at the receiver. An *adaptive* or decision-directed *equalizer* is adjusted without the aid of such a reference sequence. In practice, especially for higher-speed ($\geq$2 bps/Hz) data transmission, an equalizer frequently will first train in an automatic mode and then enter an adaptive mode. Low- to medium-speed systems can often train in the decision-directed mode.

8.1 SCOPE OF EQUALIZATION APPLICATIONS

Before proceeding with the detailed discussion of the adaptive structures and algorithms, it is worthwhile to reflect on the scope of the application of adaptive equalization technology. The dominant area of application has been to the voice-grade telephone channel, because the narrow channel bandwidth (~3 kHz) has permitted relatively sophisticated signal processing to be implemented in a cost-effective manner, and the continuing quest for ever-higher data rates has made adaptive equalization a necessity for systems providing more than 2 bps/Hz. As of this writing, equalization, along with the other foundation technologies of echo cancellation and trellis-coded modulation, have propelled voice-band modems to a rate of 19.2 kbps, which represents an efficiency of 8 bps/Hz that closely approaches the Shannon channel capacity of the medium. For other channels, where the bandwidth is substantially greater than the telephone channel, implementation of adaptive equalizers is more costly, and some of the reduced complexity algorithms (such as zero forcing) initially employed in voice-grade modems may be appropriate. For example, the intersymbol interference in 90 Mbps radio channels operating in the 4/6 and 10/12 GHz bands, which is due to multipath, can be effectively equalized [12]; however, the rapidly fading characteristics of microwave radio channels puts a premium on adaptive techniques that can quickly acquire the equalizer parameters appropriate for the fading channel. As was mentioned earlier, adaptive equalization is a key technology for the implementation of 144 kbps basic-rate ISDN service; in Chapter 10 we describe potential applications to Gbps fiber optic systems.

8.2 BASEBAND EQUIVALENT SYSTEM

We begin with a discussion of the equalization of baseband systems. Later in this chapter we discuss systems that combine adaptive equalization and the data-directed phase-locked loop described in Chapter 6. With this in mind, we review the discussion of Chapter 5 on how a baseband-equivalent model may be derived from a passband system.

The transmitted signal, $s(t)$, in an in-phase and quadrature, or quadrature amplitude modulated (QAM), data transmission system of the type described in Chapters 5 and 6 may be represented as the *real* part of the analytic signal.

$$\tilde{s}(t) = s(t) + j\check{s}(t) = \sum_n \tilde{d}_n p(t-nT)e^{j2\pi f_c t}, \tag{8.1}$$

where $\tilde{d}_n$ denotes the complex discrete-multilevel data sequence, $a_n + jb_n$, $p(t)$ is the (generally real) baseband transmitter pulse shaping, $1/T$ is the symbol rate, α is the pulse rolloff, f_c is the carrier frequency, and $\check{s}(t)$ is the Hilbert transform of $s(t)$. For (8.1) to be an analytic signal, it is required that $f_c > (1 + \alpha)/2T$. As explained in Chapter 5, complex notation is a compact means of denoting either passband or in-phase and quadrature signals, as well as system pulse responses. As shown in Figure 8.1, $s(t)$ is transmitted through the passband (about f_c) channel $h(t)$, with impulse response

$$h(t) = h_1(t)\cos 2\pi f_c t - h_2(t)\sin 2\pi f_c t = \mathrm{Re}\{(h_1(t) + jh_2(t))e^{j2\pi f_c t}\}$$
$$= \mathrm{Re}\{\tilde{h}_B(t)e^{j2\pi f_c t}\}, \tag{8.2}$$

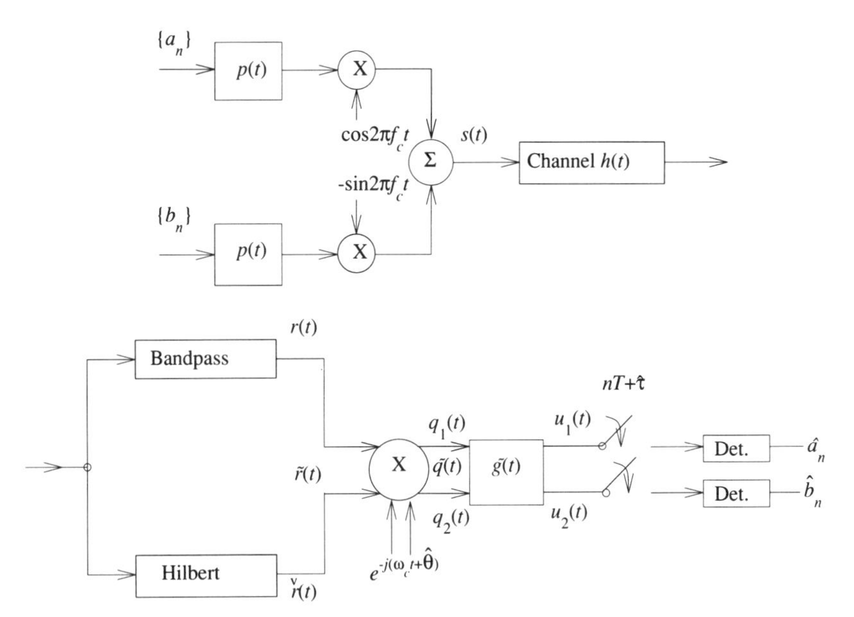

Fig. 8.1 QAM data transmission system. Variables with overtildes are complex, i.e., they have in-phase and quadrature components.

where the complex baseband-equivalent channel is defined by

$$\tilde{h}_B(t) = h_1(t) + jh_2(t). \tag{8.3}$$

Since $\tilde{h}_B(t)$ is a passband channel shifted down to DC, its spectrum will not, in general, be symmetric about zero frequency [this accounts for the nonzero quadrature pulse, $h_2(t)$]. The received analytic signal has the representation

$$\tilde{r}(t) = r(t) + j\check{r}(t) = \sum_n \tilde{d}_n \tilde{x}_B(t-nT)\, e^{j(2\pi f_c t+\theta)} + \tilde{v}(t)\, e^{j2\pi f_c t}, \tag{8.4}$$

where $r(t)$ and $\check{r}(t)$ are the in-phase and quadrature components of the received signal (and are a Hilbert transform pair), $\tilde{x}_B(t)$ is the baseband-equivalent received pulse which is given by the convolution of $\tilde{h}_B(t)$ with $p(t)$, θ is the phase shift of the channel, and $\tilde{v}(t)$ is the complex noise signal.

If the received analytic signal is demodulated by a phase-coherent receiver, then the demodulated signal can be written as

$$\tilde{q}(t) = \tilde{r}(t)\, e^{-j(2\pi f_c t+\hat{\theta})} = \sum_n \tilde{d}_n \tilde{x}_B(t-nT) + \tilde{v}(t), \tag{8.5}$$

where $\hat{\theta}$ is the phase of the demodulator. Next, the demodulated signal, $\tilde{q}(t)$, is processed by a the receiving filter or equalizer

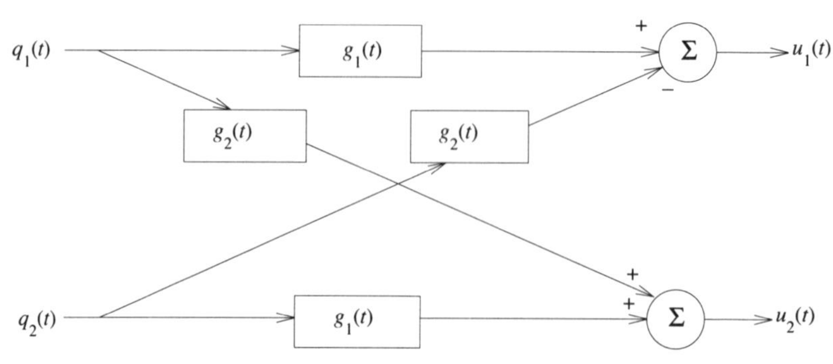

Fig. 8.2 Complex filtering (equalization) operation.

$$\tilde{g}(t) = g_1(t) + jg_2(t). \tag{8.6}$$

In representing $\tilde{g}(t)$ by (8.6) we have used the notation $g_1(t)$ and $g_2(t)$, rather than $g(t)$ and $\check{g}(t)$, to emphasize that the receiving filter does *not* correspond to an analytic pulse (since $g_1(t)$ and $g_2(t)$ are not constrained to have any special relationship). The equalized signal at the filter output is given by

$$\begin{aligned}\tilde{u}(t) &= \tilde{q}(t) * \tilde{g}(t) = \left[q_1(t)+jq_2(t)\right] * \left[g_1(t)+jg_2(t)\right] \\ &= \left[q_1(t)*g_1(t)-q_2(t)*g_2(t)\right] + j\left[q_2(t)*g_1(t)+q_1(t)*g_2(t)\right] \\ &= u_1(t)+ju_2(t).\end{aligned} \tag{8.7}$$

As shown in Figure 8.1, the in-phase and quadrature output signals $u_1(t)$ and $u_2(t)$ are synchronously sampled at $t = nT + \hat{\tau}$ and quantized to provide the data decision $\hat{a}_n$ and $\hat{b}_n$.

With this model, note that the filtering operation

$$\tilde{u}(t) = \tilde{q}(t) * \tilde{g}(t) \tag{8.8}$$

corresponds to the four convolutions described above and depicted in Figure 8.2. From the above equations, we observe that QAM passband transmission can be conveniently and compactly described in terms of complex notation.* For the sake of simplicity, in introducing the subject of adaptive equalization we let $\tilde{d}_n = a_n$ and $g_2(t) = 0$, i.e., we first consider a baseband model. This model is only equivalent to the actual passband model if $h_2(t) = 0$; this implies that the channel has even amplitude and odd phase symmetry about the carrier frequency. While this is generally not the case in practice, the simplifying assumption of a baseband system is useful for the introduction of the equalization

* The reader may recall that complex analytic notation was introduced in Section 5.1.

material presented in this chapter. We will return to complex notation later in this chapter, when we consider integrated systems that perform equalization and carrier tracking using a common error signal. Readers who feel comfortable with complex (passband) notation can interpret all signals and filter coefficients in our discussion as complex quantities, as per the relationships described above.

8.3 MINIMIZATION OF THE MEAN-SQUARE ERROR BY THE GRADIENT ALGORITHM

To begin our discussion of adaptive equalization, we recall the optimum linear receiver structure which was shown in Chapter 7 to be a cascade of a matched filter and a tapped delay line. Let us first assume that since the channel is unknown, we will replace the matched filter by a fixed noise-rejecting filter and concentrate our efforts on adapting the tapped delay line filter. With this in mind, consider the simplified baseband data transmission system shown in Figure 8.3. In this model, $\{a_n\}$ are the successive data symbols in the PAM data signal, T is the symbol interval, $p(t)$ and $h(t)$ are the impulse responses of the transmitter and channel filters respectively. Although the independent, identically distributed random variables a_n are generally called "data symbols," they may result from a coding of an input data stream into a sequence of pulse levels. For binary transmission, a_n is usually assumed to take on the values ± 1, and for multilevel transmission the symbol values are generally the equally spaced levels ± 1, ± 3, ± 5,... . The additive white Gaussian noise, with two-sided spectral power density $N_0/2$, is denoted by $N(t)$. The receiver consists of a fixed front-end, noise-rejecting filter $f(t)$, and an adjustable equalizer with impulse response $c(t)$ and output $y(t)$. The receiver sampling phase is t_0 sec, and $\hat{a}_n$ denotes the quantized sampler output at time $nT + t_0$, i.e., the decision.

As defined in the last chapter, the mean-square error, MSE, is the average of the squared difference between the equalizer output and the transmitted symbol, i.e.,

$$MSE \equiv E[y(nT+t_0) - a_n]^2 , \tag{8.9}$$

where E denotes the ensemble average over the data sequence and the additive noise. Assume that the equalizer is realized as the tapped delay line filter shown

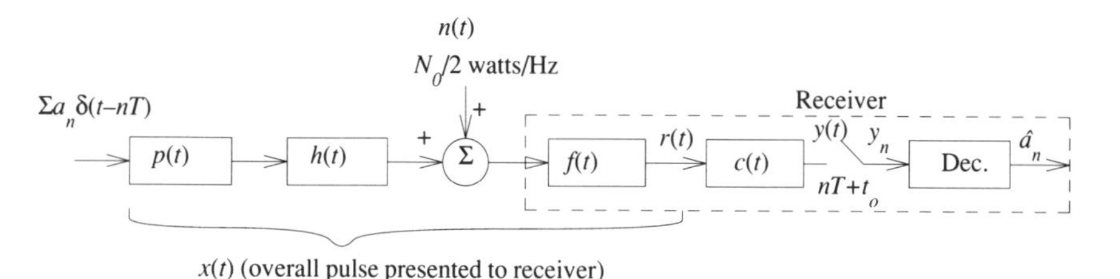

Fig. 8.3 Baseband PAM data transmission system.

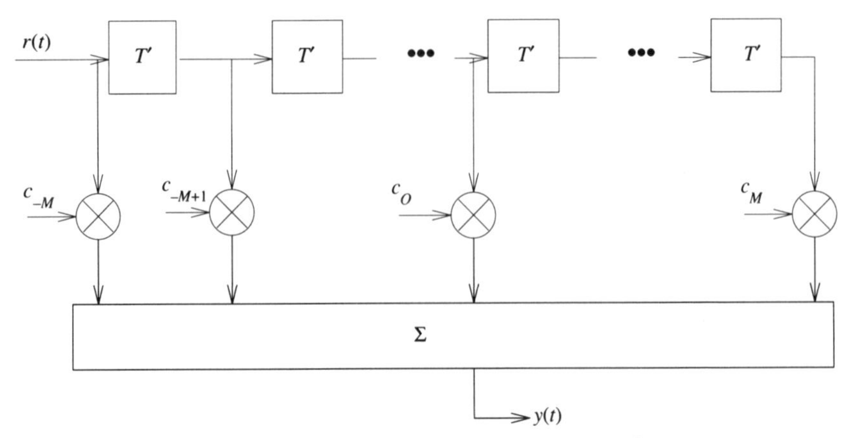

Fig. 8.4 Tapped delay line equalizer, with delays $T' \leq T$.

in Figure 8.4 with tap weights each spaced T' sec apart. Recall from Chapter 7 that $T' = T$ for a synchronous equalizer and $T' < T$ for a fractionally spaced equalizer (FSE). The equalizer output is given by

$$y(t) = \sum_{m=-M}^{M} c_m r(t-mT'), \tag{8.10a}$$

and the output at the sampling instants is given by

$$y_n \equiv y(nT+t_0) = \sum_{m=-M}^{M} c_m r(nT-mT'+t_0) , \tag{8.10b}$$

where we have assumed a $2M+1$ tap equalizer with tap weights $c_{-M},\ldots,c_0,\ldots,c_M$ and a zero-delay reference tap located in the physical center of the equalizer. The reference tap is chosen such that, if the channel presented to the equalizer is already Nyquist, then all taps would be zero except the reference tap. Selection of a reference tap location is a convenient mechanism for developing the theory of adaptive equalization, but could be misconstrued as implying a noncausal structure. In practice, an equalizer will have exactly the same structure shown in Figure 8.4, but will have an output that is delayed by half the time span of the equalizer.

It is not necessary for the equalizer to have an equal number of taps before and after the "center" tap even though we assume so here. The tap weights are to be adjusted to achieve the minimum MSE at the equalizer output. Note that in a digital implementation it is not possible to sample the received signal at any desired intervals T'. The receiver would, instead, be realized by sampling the received signal at intervals T' that are an integer fraction of a symbol interval and then perform decimation to effectively sample the equalizer output at symbol intervals. As described in Chapter 7, equalizers that have their tap weights spaced an appropriate fraction of a symbol interval apart are called fractionally spaced equalizers (FSE). In the previous chapter, some of the significant performance advantages of FSEs have been described.

If we define the transposed tap voltage (row) vector at time nT as

$$\mathbf{r}'_n = \Big[r(nT+MT'+t_0),\dots,r(nT+t_0),\dots,r(nT-MT'+t_0)\Big] \tag{8.11}$$

and the transposed tap weight (row) vector at time nT as

$$\mathbf{c}'_n = \Big[c_{-M},\dots,c_0,\dots,c_M\Big],$$

where the prime denotes transpose, then the sampled equalizer output is given by the inner product

$$y_n = y(nT+t_0) = \mathbf{c}'\mathbf{r}_n, \tag{8.12}$$

and the mean-square error can be expressed as

$$MSE = E(e_n^2) = E(y_n-a_n)^2 = E[y_n^2-2a_ny_n+a_n^2]$$

$$= E\left[\sum_i\sum_j c_ic_jr(nT-iT'+t_0)\,r(nT-jT'+t_0) - 2a_n\sum_i c_ir(nT-iT'+t_0) + a_n^2\right], \tag{8.13a}$$

where e_n is the error signal, y_n-a_n, that will figure so prominently in our discussion of adaptive systems. The MSE can also be written in the following compact vector format, which will be extremely useful for our subsequent discussion on adaptive algorithms:

$$MSE = E(e_n^2) = E(\mathbf{r}'_n\mathbf{c}_n - a_n)^2 . \tag{8.13b}$$

If we assume that the data symbols and noise samples, $N(nT)$, are mutually and individually independent sequences, we observe that

$$E\{r(nT-iT'+t_0)\,r(nT-jT'+t_0)\} = E\Big\{[\sum_l a_lx(nT-iT'+t_0-lT) + N(nT-iT'+t_0)]$$

$$[\sum_m a_mx(nT-jT'+t_0-mT)+N(nT-jT'+t_0)]\Big\}$$

$$=\sum_l\Big[E(a_e^2x(nT-iT'+t_0-lT)x(nT-jT'+t_0-lT)\Big]+E\Big[N(nT-iT'+t_0)N(nT-jT'+t_0)\Big]$$

$$= P\left[\sum_m x(mT-iT'+t_0)x(mT-jT'+t_0) + \frac{\sigma^2}{P}\delta_{ij}\right],$$

where

$$P \triangleq Ea_n^2 \ , \tag{8.14a}$$

and σ^2 is the variance of a sample of the noise emerging from the receiver input filter $f(t)$ (see Figure 8.3). Again, we remind the reader that throughout the text we have assumed that $Ea_n = 0$. We will define

$$r_{ij} \triangleq \sum_m x(mT - iT' + t_0)x(mT - jT' + t_0) + \frac{\sigma^2}{P}\delta_{ij} \tag{8.14b}$$

and the matrix R with elements r_{ij} as the *channel covariance matrix.* We also use the notation

$$Ea_n \sum_i c_i r(nT - iT' + t_0) = P \sum_i c_i x(-iT' + t_0) = P\ \mathbf{c}_n'\mathbf{x} \ , \tag{8.14c}$$

where

$$\mathbf{x}' = \left[x(MT' + t_0), \ldots, x(t_0), \ldots, x(-MT' + t_0)\right] \tag{8.14d}$$

is the row vector of samples, at intervals T', of the channel impulse response presented to the equalizer. That is, $x(t)$ is the convolution of $p(t)$, $h(t)$, and $f(t)$. It is worth noting, for future reference, that if $T' = T$ then R has identical terms for a fixed value of $i - j$, i.e., R is a *Toeplitz* matrix. The utility of R being Toeplitz will become evident later in this chapter.

With the above notation, the MSE given by (8.13) can be written as

$$MSE = P\left[\mathbf{c}'\overline{\mathbf{r}_n\mathbf{r}_n'}\mathbf{c} - 2\mathbf{c}'\overline{a_n\mathbf{r}_n} + \overline{a_n^2}\right] = P[\mathbf{c}_n'R\mathbf{c}_n - 2\mathbf{c}_n'\mathbf{x} + 1] , \tag{8.15a}$$

where the overbar denotes expectation and the channel correlation matrix is defined by

$$R = \overline{\mathbf{r}_n\mathbf{r}_n'} \ , \tag{8.15b}$$

and the pulse response is represented by the vector

$$\mathbf{x} = \overline{a_n\mathbf{r}_n} \ . \tag{8.16}$$

Expression (8.15b) is recognized to be a *quadratic form* having elliptical contours of equal MSE in tap-weight space, as illustrated in Figure 8.5, for an equalizer with only two tap weights, c_1 and c_2. The dot in the center of the contours corresponds to the value of the tap weight vector yielding minimum MSE. A three-dimensional plot of MSE as a function of c_1 and c_2, depicted in Figure 8.6, looks like a bowl whose lowest point corresponds to the dot shown in Figure 8.5. The correlation matrix, R, will be instrumental in determining the properties of the MSE, as a function of the tap weights. It will be useful to note that R is a positive semidefinite matrix; a positive semidefinite matrix has the property that for any nonzero vector $\mathbf{z}$, $\mathbf{z}'R\mathbf{z} \geq 0$. The positive semidefinite property of R follows from (8.15a) and (8.15b) by observing that since

$$R = \overline{\mathbf{r}_n\mathbf{r}_n'},$$

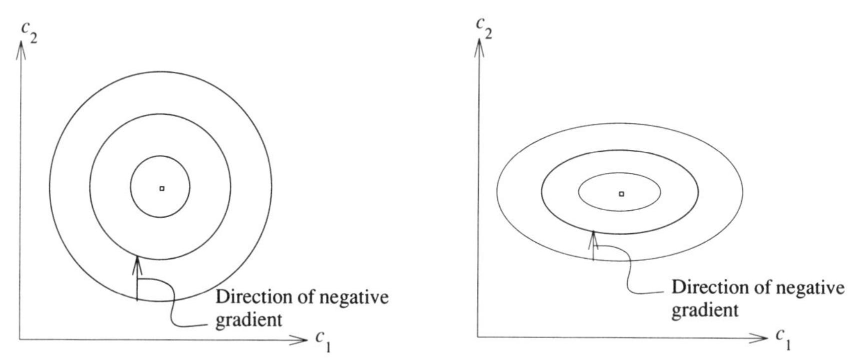

Fig. 8.5 Contours of equal mean-square error in a two-dimensional tap weight space showing the effect of eigenvalue spread. Left shows a small eigenvalue spread and right shows a larger eigenvalue spread.

it must be that for any nonrandom vector **z**

$$\mathbf{z}'R\mathbf{z} = \mathbf{z}'\overline{\mathbf{r}_n\mathbf{r}_n'\mathbf{z}} = \overline{(\mathbf{z}'\mathbf{r}_n)^2} \geq 0 . \tag{8.17}$$

The MSE is, in fact, a *convex function* of the tap weights as suggested by Figure 8.6. This means that there are no relative minima, and thus, the MSE has a single global minimum. A search for the minimum that slides down the inside of the bowl-shaped function will be guaranteed to reach the minimum. Convexity is proved in Appendix 8A by showing that the MSE satisfies the convexity

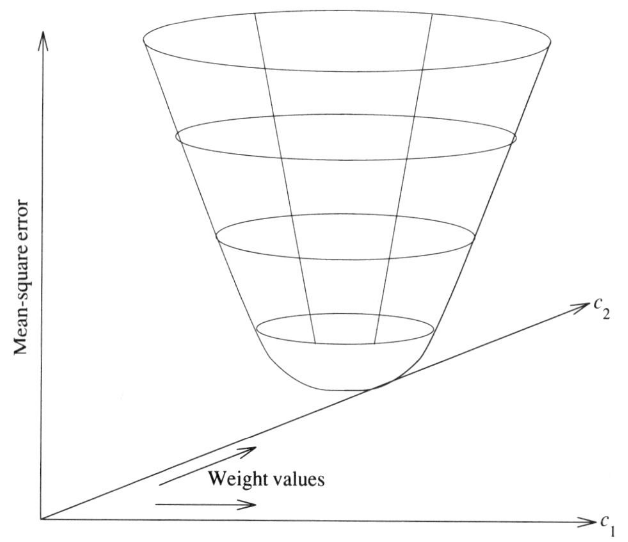

Fig. 8.6 Mean-square error (MSE) as a (convex) function of the tap weight.

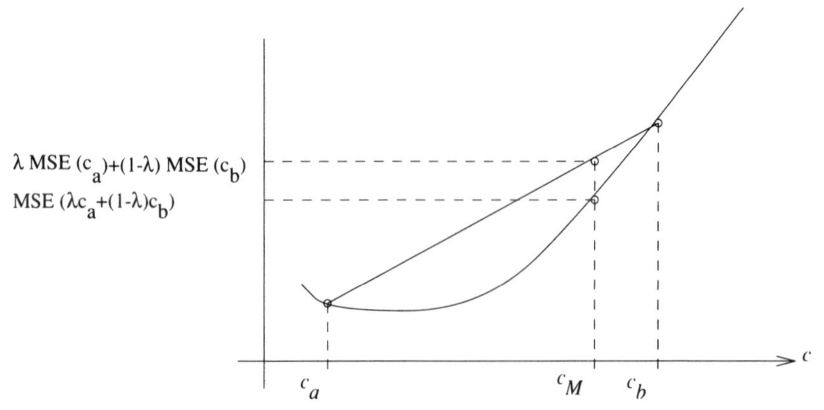

Fig. 8.7 Definition of convexity, illustrated for one-dimensional argument.

definition

$$MSE\ [\lambda \mathbf{c}_a + (1-\lambda)\mathbf{c}_b] \le \lambda MSE[\mathbf{c}_a] + (1-\lambda) MSE[\mathbf{c}_b], \tag{8.18}$$

where $\mathbf{c}_a$ and $\mathbf{c}_b$ are any two values of the tap-weight vector, and $0 \le \lambda \le 1$. This definition means that the MSE for some $\mathbf{c}_M$ between $\mathbf{c}_a$ and $\mathbf{c}_b$ always lies below a straight line joining the values of MSE at $\mathbf{c}_a$ and $\mathbf{c}_b$, as illustrated in Figure 8.7.

8.3.1 HOW MANY TAPS ARE NEEDED?

So far, we have established that the mean-square error is a unimodal, convex function of the tap weights which can be written in the compact vector and matrix notation of (8.14)–(8.17). Our primary objective is to provide techniques for determining the optimum tap weights of a finite-length equalizer, when the channel is *not* known to the receiver. For finite-length equalizers, the Fourier transform techniques of Chapter 7 are not applicable. There is much to be learned by studying the idealized circumstance where the channel is presumed known. From an investigation of this situation, bounds can be obtained on the performance attainable for the real-world (unknown channel) situation. If the channel characteristics are assumed known, then (8.15a) can be differentiated to determine the optimum tap weights. Suppressing the time index on the tap weight vector and differentiating (8.15b) with respect to the tap weights gives the linear equation

$$R\mathbf{c} - \mathbf{x} = 0, \tag{8.19a}$$

which has to be solved for the optimum set of tap weights. If R is positive definite (see Exercise 8.3, where we show that R will be positive definite if the impulse pulse response presented to the equalizer does not have any nulls in its discrete Fourier transform), then the optimum set of tap weights is obtained by

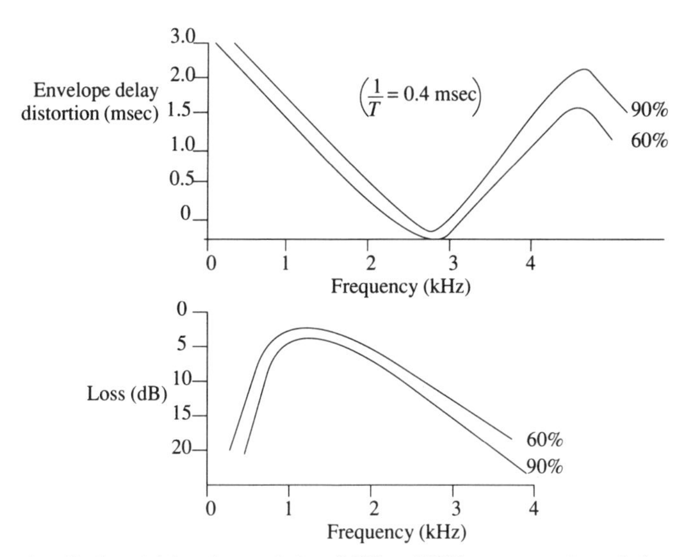

Fig. 8.8 Amplitude and delay characteristics of 60% and 90% worst-case channels from the 1982 Bell System survey.

inverting the correlation matrix, R, to give

$$\mathbf{c}_{opt} = R^{-1}\mathbf{x}. \tag{8.19b}$$

The minimized MSE is computed by substituting the optimum tap-weight vector, $\mathbf{c}_{opt}$, into (8.15a) giving

$$E(\mathbf{c}_{opt}) = P\left[1 - \mathbf{x}'R^{-1}\mathbf{x}\right]. \tag{8.20}$$

For a finite number of taps, (8.19) and (8.20) do not provide much insight as to the dependence of the minimum MSE on the properties of the channel. The optimum tap weight vector, $\mathbf{c}_{opt}$, and the minimized MSE, $E_{opt}(\mathbf{c}_{opt})$, can be computed given the channel pulse response, $\{x_n\}_{-M}^{M}$ and the channel correlation matrix R. The typical dependence of the MSE as a function of the number of equalizer taps will be discussed by the way of example. In Figure 8.8 we show the amplitude and delay characteristics for the 60th and 90th percentile worst-case channels from the 1982 Bell System survey [13] of voice-grade leased channels. (Only 10% of the channels are worse than the 90% channel.) Figure 8.9 shows the mean-square error obtained in a real-time experiment, as a function of equalizer length (plotted as the number of symbols), for synchronous and fractionally spaced equalization of a 4800 bps (2400 symbols/sec) transmission over the 60 percent and 90 percent worst-case channels of Figure 8.8. Observe the difference in steady-state performance between the FSE and the synchronous equalizer. This is not surprising, since the delay distortion is quite severe (with $T = 0.4$ ms, the delay spread at the band edge is ~5 symbols for the

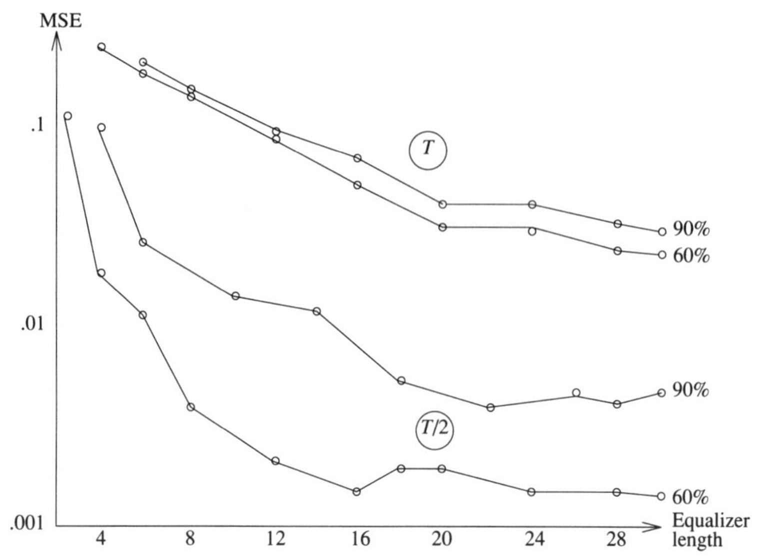

Fig. 8.9 Mean-square error versus equalizer length for synchronous and fractionally spaced equalization of 4800 bps (2400 symbols/sec) transmission over the channels of Figure 8.8.

90 percent channel). Note that at a length of 16 symbols, the FSE equalizer has reached its asymptotic value, while the synchronous equalizer has just about reached its (smaller) asymptotic value at a span of 32. Since these two points correspond to the same number of tap weights for the respective equalizers, it is clear that in this case the FSE is the unequivocal choice. In general, it is not possible to answer the question: for the same number of taps, which system will perform better: a FSE or a synchronous equalizer? Prudent engineering judgment would require a characterization of the range of channels that are expected to be encountered in practice and then doing a detailed set of evaluations producing a series of curves like those in Figure 8.9. Insight into the required number of taps can be obtained by considering the channel characteristics shown in Figure 8.8. Consider the delay curves, which measure the relative delay experienced by the different frequency components. Relative to the reference frequency of 1800 Hz, energy at 3000 Hz will be delayed by ~3.0 msec. If the symbol rate is 2400 symbols per second, then there is 0.4 msec between transmitted symbols, or the *delay spread* of the channel is 3.0/0.4 or 7.5 symbol intervals. Such a dispersion of pulse energy would require an equalizer spanning at least 7.5 symbols, which is somewhat less than the equalizer span required (as per Figure 8.9). If the symbol rate were reduced to 600 symbols per second, then the delay spread would be ~2 symbols, and an equalizer may not be required. Further insight into the required equalizer span may be obtained if

either the amplitude or delay characteristics have sinusoidal ripples. Such ripples might be caused by the filters encountered by the signal, since many filter design techniques only meet the desired filter characteristics at specific values of frequency and have a sinusoidal ripple between those points. For this situation, the theory of *paired echoes* may be applied [3]. Consider a channel with a filter amplitude characteristic that has a ripple of period T_0, i.e.,

$$X(f) = 1 + a \cos 2\pi f T_0 , \tag{8.21a}$$

superimposed on a channel with an otherwise ideal pulse response, $g(t)$, with a spectrum $G(f)$. The corresponding impulse response is then given by

$$\begin{aligned} x(t) &= \frac{1}{2}\int_0^\infty G(f) \left[1 + a \cos 2\pi f T_0\right] \cos 2\pi f t \, df \\ &= g(t) + \frac{a}{2}\, g(T - T_0) + \frac{a}{2}\, g(t + T_0) \,. \end{aligned} \tag{8.21b}$$

As the name "paired echo" suggests, the presence of sinusoidal amplitude ripple in the channel produces two echoes; one echo is delayed by T_0 seconds and the other echo is advanced by the same amount. This time spread of $2T_0$ when normalized by the symbol period will give a measure of the equalizer span required. Similar results can be derived for sinusoidal ripple in the delay characteristics (see Exercise 8.4). The above approximations can in some circumstances serve to provide a useful estimate of the equalizer span required or to ascertain if equalization is required at all. Exact results can always be obtained for a specific channel by numerical solution of (8.19) for a given number of tap weights.

8.3.2 STEADY-STATE PERFORMANCE OF FRACTIONALLY-SPACED EQUALIZERS

To further illustrate the advantages of fractionally-spaced equalization over synchronous equalization we describe a number of computer simulation runs which were made for different equalizer configurations and for channel distortion of varying severity. The system tested was a 9.6 kbps QAM system, having a symbol rate of 2400 symbols/sec and an excess bandwidth of 12 percent, with the four-level transmitted-symbol alphabet $\{\pm 1, \pm 3\}$ and a signal-to-noise ratio of 28 dB. For each run the steady-state mean-square error was measured after a sufficiently long period of adaptation, and the FSE was of the $T/2$ type. Amplitude and delay-distortion characteristics are illustrated in Figure 8.10 for the three linear channels which were simulated. The "Good" channel has low distortion, and is well within the limits of standard voice-grade telephone channel conditioning, i.e., the "Basic" conditioning [13] illustrated in Figure 8.11. The "Bad-Phase" and "Bad-Slope" channels have, respectively, severe phase distortion and severe amplitude distortion, placing these channels just outside the defining boundaries of basic-conditioned channels. In Figures 8.12–8.14 we compare the performance of a 24-tap synchronous T equalizer and a 48-tap $T/2$

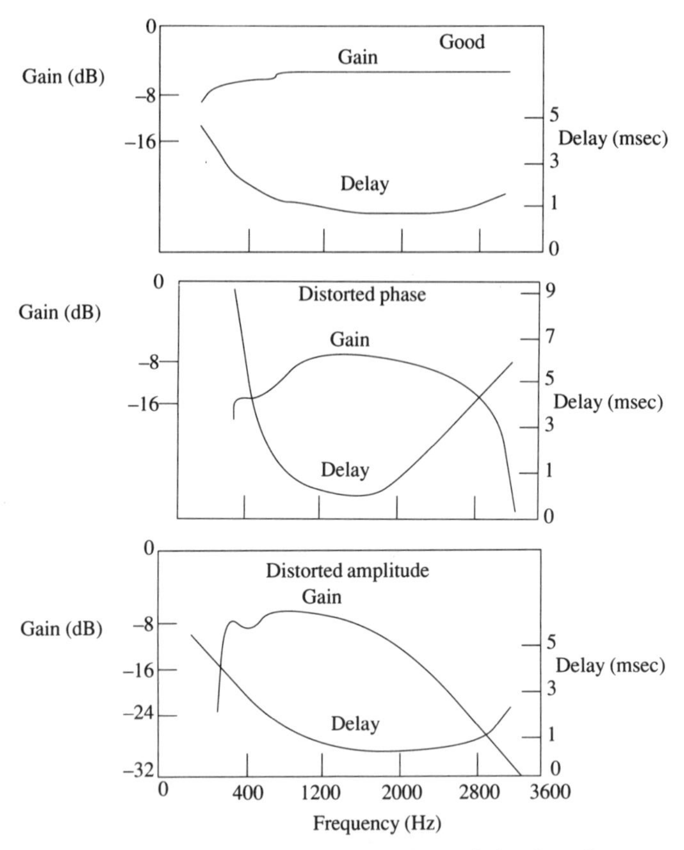

Fig. 8.10 Characteristics of simulated transmission channels.

equalizer on the three test channels. Performance is examined for five timing epochs within a symbol interval, and is measured by the output signal-to-noise ratio, defined as

$$SNR_{out} = 10\log\frac{P_{BB}}{MSE},$$

where P_{BB} is the received baseband average signal power (a constant), and MSE is the measured output mean-square error. The received signal is normalized so that the ratio of the signal power at the output of the receiving filters to the power of the additive noise, at the same point in the system, is 28 dB. Thus if the equalizer could "undo" the channel distortion without enhancing the noise, then the output SNR would be 28 dB. A further degradation, which is not taken into account in these simulations, is the loss of power through the channel. It is apparent that the performance of the fractionally-spaced equalizer is almost independent of the timing epoch, in sharp contrast to that of the synchronous

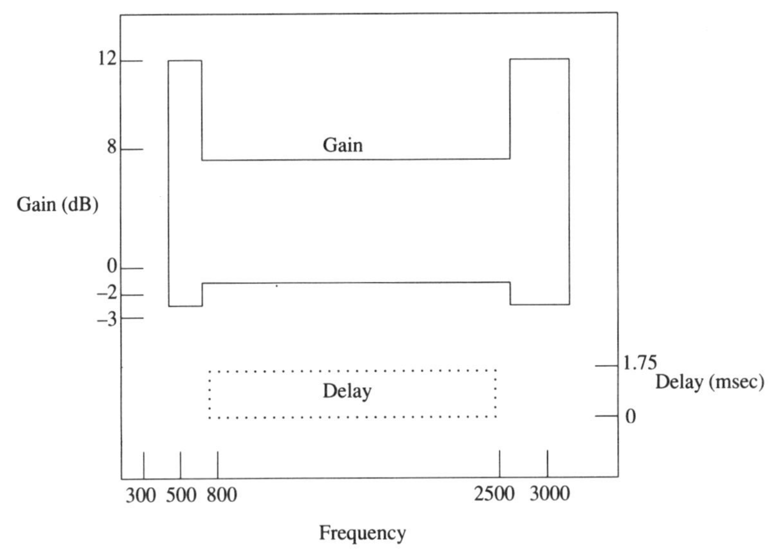

Fig. 8.11 Defining boundaries of "Basic"-conditioned channels.

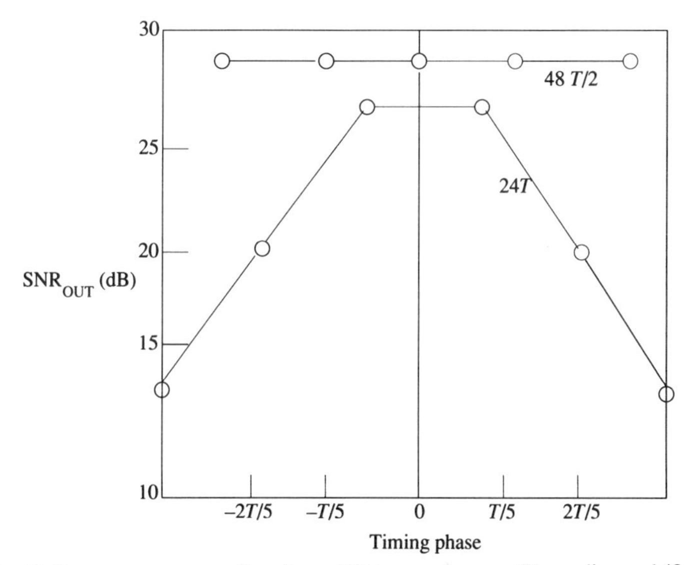

Fig. 8.12 Performance versus sampling phase of 24-tap synchronous (T) equalizer and 48-tap ($T/2$) equalizer on the "Good" channel of Figure 8.10. Results from a computer simulation of a 9600 bps (2400 symbols/sec) QAM modem.

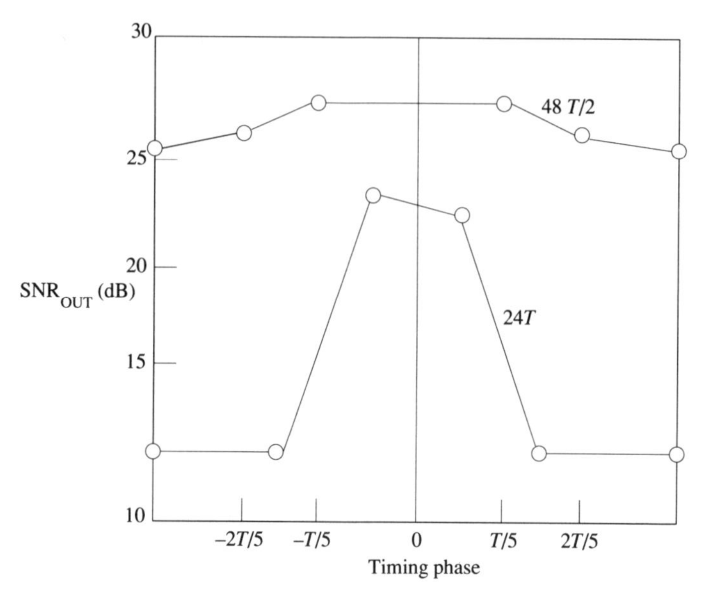

Fig. 8.13 Performance on "Bad-Phase" channel of Figure 8.10 of a fractionally spaced equalizer ($T/2$) and a synchronous (T) equalizer.

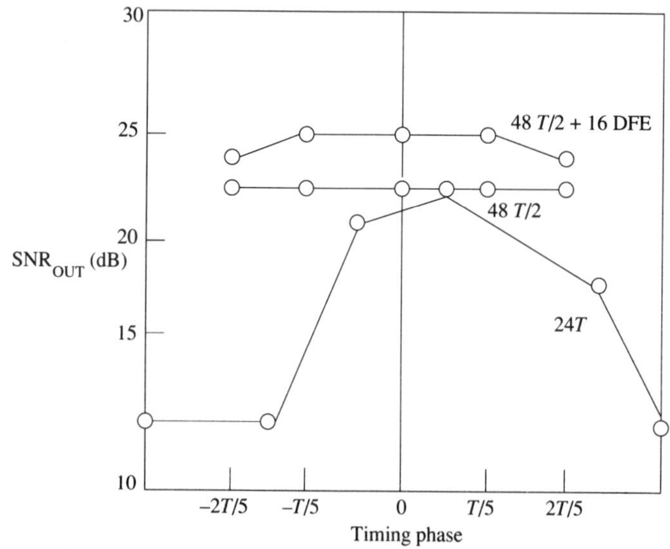

Fig. 8.14 Performance on "Bad-Slope" channel. The top curve is for a receiver which incorporates both a 48-tap FSE and a 16-tap decision-feedback equalizer (DFE).

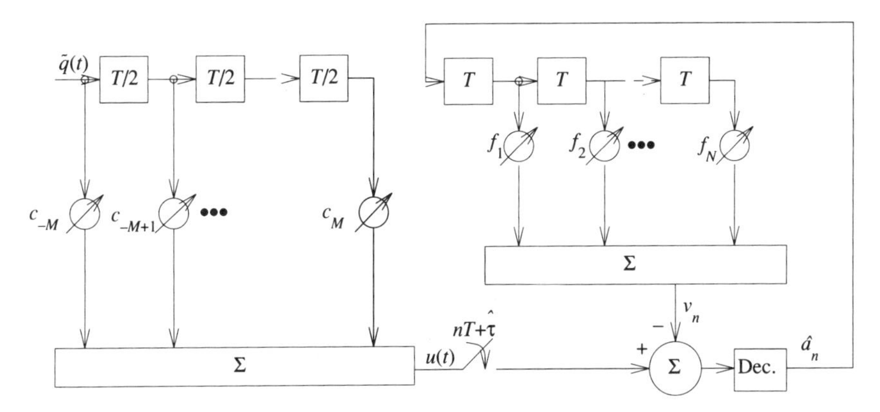

Fig. 8.15 QAM data receiver combining a fractionally-spaced equalizer with tap weights $\{c_i\}$ and a decision feedback equalizer with tap weights $\{f_i\}$.

equalizer. This is consistent with our analysis in Chapter 7 for infinite length equalizers which showed that for the fractionally-spaced equalizer the minimum mean-square error is independent of the sampling epoch. It is also significant that the performance of the fractionally-spaced equalizer on the "Bad-Phase" channel is significantly better than that achieved by the synchronous equalizer even for the best sampling phase. The capability of the FSE for phase equalization, before folding the spectrum about the Nyquist frequency, is seen to be an important advantage on channels with severe phase distortion. With the addition of a decision-feedback equalizer (DFE), with feedback taps, $\{f_i\}_{i=1}^{N}$, shown in Figure 8.15, compensation for severe amplitude distortion is also improved, as illustrated in Figure 8.14. The simulation of an FSE with $3T/4$ tap spacing [still greater than the minimum tap spacing of $T/(1+\alpha)$, where $\alpha = 0.12$ is the percentage of excess bandwidth] resulted in performance comparable to that of the $T/2$ equalizer. A $3T/4$ equalizer needs only 2/3 as many taps as a $T/2$ equalizer to span a given channel dispersion; this choice of tap spacing can not only reduce implementation complexity but, as we shall see later in this text, also improve steady-state performance when digital resolution is a consideration.

8.3.3 ADAPTIVE EQUALIZATION

We now return to our discussion of techniques which insure the adaptation of the equalizer tap weights, so that the system will settle at the weights which minimize the mean-square error. As the number of taps becomes large, for a synchronous equalizer we see that the left-hand side of (8.19a) approaches the convolution

$$\sum_{m=-\infty}^{\infty} R_{n-m}\, c_m = x_{-m}, \qquad -\infty \le m \le \infty. \tag{8.22}$$

By taking Fourier transforms of both sides of (8.22) we may determine that the spectrum of the equalizer coefficients is given by

$$C(f) = \frac{X^*(f)}{R(f)} = \frac{X^*(f)}{\left|\sum_k X(f + k/T)\right|^2 + \frac{\sigma^2}{P}}, \qquad (8.23)$$

which is identical to (7.91).

While the above expressions can be used to calculate the optimum tap weights when the channel characteristics are known, iterative techniques are appropriate to search for the minimum of the mean-square error when the channel is unknown. We begin our discussion with the method of steepest descent, which is an iterative algorithm for adjusting the tap weight vector when the channel parameters are available. The algorithm begins with an arbitrary initial value for the tap vector and has increments in the direction of the negative gradient of the MSE surface with respect to the tap weights. The generic form of the steepest-descent algorithm is

$$\mathbf{c}_{n+1} = \mathbf{c}_n - \beta \frac{\partial E}{\partial \mathbf{c}_n}, \quad n = 0,1,2,..., \qquad (8.24a)$$

where $\mathbf{c}_n$ and $\mathbf{c}_{n+1}$ are the tap weights at the nth and $(n+1)$th iterations respectively, β is a proportionality constant, and $\partial E/\partial \mathbf{c}_n$ is the gradient of the MSE with respect to the tap weights evaluated at $\mathbf{c}_n$.

As shown in Figure 8.16 for a one-tap equalizer, the negative of the gradient always points toward the minimum value c_{opt}. If the constant of proportionality is such that the tap weights moves to the right of c_{opt}, then the change in sign of the gradient provides the proper restoring force towardsthe minimum point. From (8.15b) we can directly calculate that the gradient of MSE with respect to $\mathbf{c}$ is proportional to

$$\nabla E_{\mathbf{c}} \sim \boldsymbol{Rc} - \boldsymbol{x} \ , \qquad (8.24b)$$

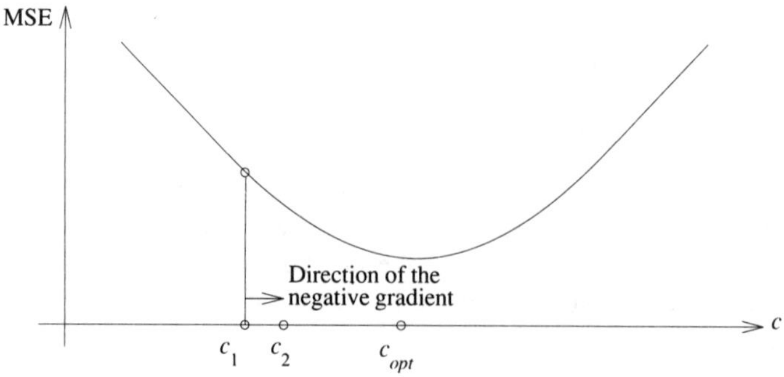

Fig. 8.16 Minimization of the MSE via the steepest descent algorithm. The tap weight c_2 at the second iteration is the initial tap weight plus a term in the direction of the negative gradient. The optimum setting is denoted by c_{opt}.

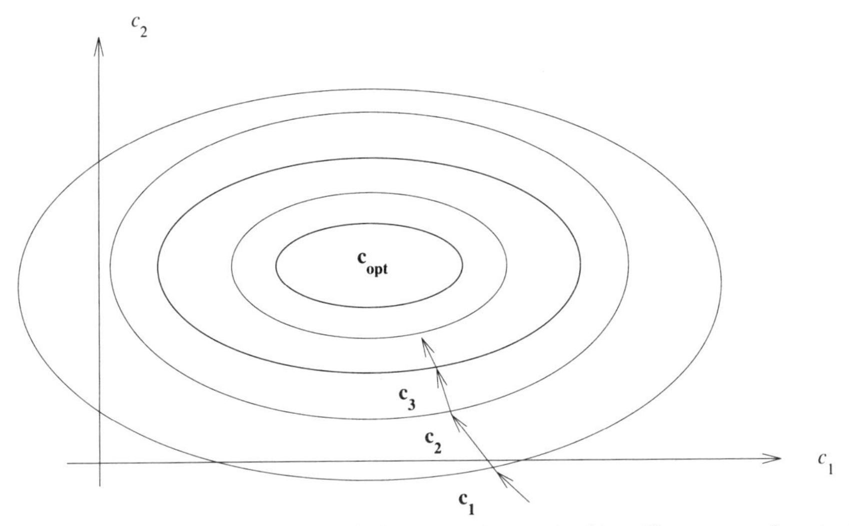

Fig. 8.17 Minimization of the MSE via the steepest-descent algorithm. The contours of constant MSE are shown along with a tap weight trajectory that follows the steepest-descent algorithm.

and from (8.13) or (8.24b) we derive the explicit expression for each component of the gradient

$$\frac{1}{2P}\frac{\partial MSE}{\partial c_m} = \frac{\partial}{\partial c_m}\left[\sum_{i=-M}^{M}\sum_{j=-M}^{M} r_{ij}c_i c_i - 2\sum_{i=-m}^{M} c_i x(-iT'+t_0) + 1\right]$$

$$= \sum_{i\,=\,m}^{M} r_{mi}\, c_i - x(-mT'+t_0) = [R\mathbf{c}]_m - [\mathbf{x}]_m, \qquad (8.24c)$$

where $[\cdot]_m$ denotes the mth component of the vector in the brackets. The intuitive appeal of the steepest-descent algorithm is the fact that the optimum tap weight setting, which is given by (8.19b), is precisely the unique point at which the gradient is zero. From (8.24a) observe that when the gradient is zero, the steepest-descent algorithm will stop making adjustments, so this condition defines the tap weight values where the steepest-descent algorithm will settle. In Figure 8.17 we show the evolution of the tap weights according to (8.24), where the superscript denotes time evolution. Note how the tap weights move toward the optimum setting in a deliberate fashion. Even though (8.24) presumes knowledge of the channel, such iterative techniques may still be computationally efficient under these circumstances. For example, direct matrix inversion of (8.19a) by a Gauss elimination technique to solve for $\mathbf{c}_{opt}$, will require on the order of $(2M+1)^3$ multiplications. Since the steepest-descent algorithm requires $2M+1$ multiplications per iteration, if the algorithm converges in L steps, and if $(2M+1)\ L < (2M+1)^3$ (which will generally be the case), then the iterative approach will be computationally more efficient than direct matrix inversion.

The steepest-descent algorithm only makes use of *currently available* information; this is a strength in that the algorithm, and the algorithms derived from it when the gradient is not available (e.g., the least-mean squares (LMS) algorithm), are simple to implement. On the other hand, as we shall see later, the transient performance of the algorithm may be improved by making use of the additional information contained in the past signal samples. If we substitute the expression for the gradient, given by (8.24b) into (8.24a), we have the *recursive steepest-descent algorithm*

$$\mathbf{c}_{n+1} = \mathbf{c}_n - \frac{\beta}{2P}\frac{\partial MSE}{\partial \mathbf{c}_n} = \mathbf{c}_n - \beta(R\mathbf{c}_n - \mathbf{x}), \quad n = 0,1,2,\ldots, \tag{8.25a}$$

where n is the iteration index (corresponding to discrete time—if adjustments are made at the symbol rate) and β is a positive constant called the *step size.* The algorithm takes small steps in the direction opposite to the gradient of the MSE, which is the direction known to provide the maximum rate of decrease of the MSE.

To demonstrate the convergence of this algorithm, also called the *known gradient* algorithm (since we have assumed knowledge of the channel characteristics, the gradient is available), we examine convergence of the tap weight error vector $\varepsilon_n = \mathbf{c}_n - \mathbf{c}_{opt}$. This error vector, which measures the difference between the current tap weight vector and the optimum setting, obeys the recursion

$$\begin{aligned}\varepsilon_{n+1} &= \mathbf{c}_{n+1} - \mathbf{c}_{opt} = \varepsilon_n - \beta(R\mathbf{c}_n - R\mathbf{c}_{opt}) \\ &= (I - \beta R)\varepsilon_n, \quad n = 0,1,2,\ldots,\end{aligned} \tag{8.25b}$$

where we have used the relation $\mathbf{x} = R\mathbf{c}_{opt}$ appearing in (8.19a). Thus the tap weight error is governed by a first-order autonomous difference equation, and after n iterations, the tap weight error vector is given by

$$\varepsilon_{n+1} = (I - \beta R)^{n+1}\varepsilon_0 . \tag{8.25c}$$

To proceed with the analysis, we use the fact [14] that the correlation matrix, R, can be diagonalized and put in the form

$$R = P\Lambda P', \tag{8.26a}$$

where Λ is a diagonal matrix whose elements are the eigenvalues of R,

$$\Lambda = \begin{bmatrix}\lambda_{-M} & & 0 \\ & \ddots & \\ 0 & & \lambda_M\end{bmatrix}, \tag{8.26b}$$

and P is the matrix of whose columns are the corresponding eigenvectors of R,

$$P = [\mathbf{e}_{-M}, \mathbf{e}_{-M+1}, \dots, \mathbf{e}_M] . \tag{8.26c}$$

The diagonalization of R follows directly from the eigenvalue equation, $R\mathbf{e}_i = \lambda_i \mathbf{e}_i$ for each eigenvalue–eigenvector pair, which may be rewritten in matrix form as $RP = P\Lambda$; it should also be noted that since the eigenvectors are orthonormal, the matrix of eigenvectors satisfies the relation defining orthogonal matrices, $PP' = I$ (where I is the identity matrix, so that $P^{-1} = P'$). Thus $RP = P\Lambda$ or

$$R = P\Lambda P' . \tag{8.26d}$$

The $2M+1$ dimensional elliptical contours of equal mean-square error, shown in Figure 8.5, have axes whose lengths are proportional to the eigenvalues $\{\lambda_i\}_{-M}^{M}$. Using the above diagonalization, we can write the following equations for the evolution of the tap weight error:

$$\begin{aligned} \varepsilon_{n+1} &= (I-\beta R)^{n+1}\varepsilon_0 = (I-\beta P\Lambda P')^{n+1}\varepsilon_0 \\ &= [P(I-\beta\Lambda)P']^{n+1}\varepsilon_0 = P(I-\beta\Lambda)^{n+1}P'\varepsilon_0 . \end{aligned} \tag{8.27}$$

In arriving at (8.27) we made use of the fact that $PP' = I$, so that $P(I-\beta\Lambda)P' \;\; P(I-\beta\Lambda)P' = P(I-\beta\Lambda)^2 P'$. Consequently, the tap-weight error will converge to zero, provided that each term in the diagonal matrix $(I-\beta\Lambda)^{n+1}$ approaches zero, or equivalently

$$|1-\beta\lambda_i| < 1, \quad i = -M, \dots, M . \tag{8.28}$$

This condition is hardest to satisfy for the largest eigenvalue, $\lambda_{\max}$. If $1 - \beta\lambda_{\max}$ is positive, then (8.28) requires that $\beta\lambda_{\max} > 0$, and if $1-\beta\lambda_{\max}$ is negative, then $\beta\lambda_{\max} < 2$. Thus if

$$0 < \beta < 2 / \lambda_{\max} , \tag{8.29}$$

the tap weights will converge to the optimum setting.

Since $\lambda_{\max}$ is less than the sum of all the eigenvalues of R, and since this sum is equal to the trace of the matrix, we have the upper bound $\lambda_{\max} \le (2M+1)\overline{r_n^2} = \text{trace } R$.

CONVERGENCE RATE OF THE STEEPEST-DESCENT ALGORITHM

It is instructive to consider the *rate of convergence* of the norm of the tap error vector. If we let

$$Z_n = \varepsilon_n' \varepsilon_n \tag{8.30}$$

denote the power in the tap error vector, we can obtain the following interative equation for Z_n directly from (8.25b),

$$Z_{n+1} = \varepsilon_{n+1}'\varepsilon_{n+1} = \varepsilon_n'(I-\beta R')(I-\beta R)\varepsilon_n = \varepsilon_n'\left[I-2\beta R+\beta^2 R^2\right]\varepsilon_n , \tag{8.31}$$

where we have made use of the symmetry of R. To proceed further, we make use of the following eigenvalue bounds on quadratic forms [14]: for any vector, $\mathbf{y}$, the quadratic form $\mathbf{y}'R\mathbf{y}$ is upper and lower bounded in terms of the maximum (λ_{max}) and minimum eigenvalues (λ_{min}) of R by

$$\lambda_{min} \mathbf{y}'\mathbf{y} \leq \mathbf{y}'R\mathbf{y} \leq \lambda_{max} \mathbf{y}'\mathbf{y} \,. \tag{8.32}$$

Applying (8.32) to (8.31) gives the following upper bound on the error vector norm

$$Z_{n+1} \leq \left(1 - 2\beta\lambda_{min} + \beta^2\lambda_{max}^2\right) Z_n \,, \tag{8.33}$$

and Z_n will approach zero provided that

$$\beta \leq 2\frac{\lambda_{min}}{\lambda_{max}^2} = \frac{2}{\lambda_{max}} \frac{\lambda_{min}}{\lambda_{max}} \,, \tag{8.34}$$

and since $\lambda_{min} / \lambda_{max} \leq 1$, (8.34) is a more stringent requirement than (8.29). Letting

$$\gamma = 1 - 2\beta\lambda_{min} + \beta^2\lambda_{max}^2 \,, \tag{8.35}$$

the norm of the error vector approaches zero *exponentially*, with a time constant γ, according to the following recursion,

$$Z_{n+1} \leq \gamma^n Z_0 \,, \tag{8.36}$$

The actual convergence rate will depend on eigenvalues λ_{min} and λ_{max} and the choice of the step size β. Convergence will be fastest (or more precisely, the bound will decay the quickest) if γ is minimized with respect to β. Setting the derivative of (8.35) with respect to β to zero, we determine that the optimum value of β is

$$\beta_{opt} = \frac{\lambda_{min}}{\lambda_{max}^2} \,, \tag{8.37}$$

or half of the maximum permissible value of β [according to (8.34)]. With $\beta = \beta_{opt}$, we have that

$$Z_{n+1} \leq \left[1 - \left(\frac{\lambda_{min}}{\lambda_{max}}\right)^2\right]^n Z_0 \,. \tag{8.38}$$

Note that the rate of convergence depends on the ratio of the maximum-to-minimum eigenvalues of R, and if all the eigenvalues of R, are the same, then $\lambda_{min} = \lambda_{max}$ and convergence is achieved in one step! It is important to note that the last statement is true for the steepest-descent algorithm, which assumes knowledge of the exact gradient. If the gradient is available, and R has identical eigenvalues, then the contours of equal MSE are circles, and certainly the minimum can be reached in one step, since the gradient points directly to the optimum tap setting.

The steepest-descent algorithm can be modified to take advantage of the above observation by premultiplying the gradient by the inverse of the correlation matrix, R. Such an algorithm is called *Newton's method* and is of the form

$$\mathbf{c}_{n+1} = \mathbf{c}_n - \beta R^{-1}(R\mathbf{c}_n - \mathbf{x}) \ , \tag{8.39}$$

which is to be compared with (8.25a). For Newton's algorithm the tap error vector evolves according to

$$\varepsilon_{n+1} = (1-\beta)\varepsilon_n \ , \tag{8.40}$$

and with $\beta = 1$ we have one-step convergence. We again emphasize that the above behavior is for the known-gradient algorithm, and will be contrasted with the convergence characteristics when the channel (and hence the gradient) is unknown.

We continue our discussion of the steepest-descent algorithm by examining the behavior of the mean-square error, when the tap weights are adjusted via the steepest-descent algorithm. From (8.13b) we can write the MSE as

$$\begin{aligned} E_n &= E_{opt}(\mathbf{c}=c_{opt}) + \varepsilon_n' R \varepsilon_n \\ &= E_{opt} + \varepsilon_n' R \varepsilon_n \ , \end{aligned} \tag{8.41}$$

where E_{opt} is the minimum value of the MSE, i.e., when $\mathbf{c} = \mathbf{c}_{opt}$.

The dynamic behavior of the mean-square error is characterized by the quadratic term

$$q_n \equiv \varepsilon_n' R \varepsilon_n , \tag{8.42}$$

whose evolution is bounded by

$$q_{n+1} < (1 - 2\lambda_{\min}\beta + \lambda_{\max}^2 \beta^2) q_n , \tag{8.43}$$

which is identical to (8.33). Hence, both the norm of the tap error and the mean-square error for the steepest-descent algorithm decay exponentially with the same time constant and with the same requirements on the step size, β, for convergence.

8.4 THE LEAST-MEAN-SQUARE (LMS) ESTIMATED-GRADIENT ALGORITHM

Now we turn our attention to the real-world problem: providing a means for adjusting the tap weights to their optimum setting when the channel characteristics are not known. One approach would be to frequently sound the channel to learn or estimate the pulse samples and to then compute the actual gradient for use in the steepest-descent algorithm. Such an approach would of course be slow and complex in that different types of circuitry would be necessary for each stage of operation. The least-mean-square (LMS) error *adaptive* filtering algorithm proposed by Widrow and Hoff [2] has the following desirable characteristics for application to the equalization of data communication systems:

(1) It can be used to find *the* optimum tap weights of a tapped delay line (TDL) adaptive equalizer

(2) It lends itself to simple, digital implementation

(3) It can be used in training (*automatic*) and tracking (*adaptive* or *decision-directed*) modes

(4) It can be applied to more complex equalization structures that use a tapped delay line, or related structures, as a building block (as well as a host of other systems such as echo cancelers)

(5) It exhibits robust performance in the presence of implementation imperfections or simplifications, or even some limited system failures.

8.4.1 THE LMS ALGORITHM FOR TAPPED DELAY LINE EQUALIZERS

The term adaptive is used to describe the LMS algorithm, because knowledge of the channel characteristics is not needed for the algorithm to converge — hence the LMS algorithm will adapt to the particular channel in use. Widrow proposed that the gradient needed in (8.25a) be replaced by a particular unbiased *estimate*, and that the quality of the estimate be continuously improved by small corrections. The LMS algorithm is motivated by considering the gradient or the MSE [as given by (8.13)]. Differentiating (8.13b) with respect to $\mathbf{c}$, and interchanging the linear operations of averaging and differentiation gives

$$\frac{\partial E_{opt}}{\partial \mathbf{c}} = \frac{\partial}{\partial \mathbf{c}}[Ee_n^2] = E\left[\frac{\partial}{\partial \mathbf{c}} e_n^2\right] = 2E\left[e_n \frac{\partial e_n}{\partial \mathbf{c}}\right] = 2E[e_n \mathbf{r}_n] \tag{8.44}$$

where the *error signal*

$$e_n = y(nT+t_0) - a_n$$

is the difference between the equalizer output, $y(nT+t_0)$, and the *desired* or *reference* level, a_n. In (8.44) $\mathbf{r}_n$ is the vector of signal samples in the tapped delay line. The significant observation made by Widrow and Hoff was that the unbiased estimate $\partial/\partial\mathbf{c}\ e_n^2$ could be used in place of $\partial E/\partial\mathbf{c}$ in a modified steepest descent type of algorithm. Thus, the gradient of the unaveraged or *instantaneous* squared error, which is given by

$$\frac{\partial}{\partial \mathbf{c}} e_n^2 = 2e_n \mathbf{r}_n = 2(y_n - a_n)\mathbf{r}_n, \tag{8.45}$$

is used to provide the incremental term in the LMS algorithm. The LMS algorithm has the property that the tap vector increment is the product of the equalizer output error signal and the vector of samples resident in the tapped delay line.

If the increments are weighted by a sufficiently small constant, called the step size, then the algorithm will average (8.45), so that the correction term will be given by the correlation between the error signal and the signal samples appearing at the input of the adaptive filter. Using (8.45), the correction term for the mth tap weight, c_m, is given by

$$\frac{\partial}{\partial c_m}(y_n - a_n)^2 = 2(y_n - a_n)\frac{\partial y_n}{\partial c_m} = 2(y_n - a_n) r(nT + t_0 - mT') = 2e_n r_{n-m}, \tag{8.46}$$

which is the correlation of the error signal with the signal sample at the mth tap. Note that the error signal is common to each correction term. Using this estimate of the gradient of the MSE yields the celebrated *least-mean squares* (LMS) algorithm

$$\mathbf{c}_{n+1} = \mathbf{c}_n - \beta[y(nT + t_0) - a_n]\mathbf{r}_n \tag{8.47}$$

$$= \mathbf{c}_n - \beta e_n \mathbf{r}_n. \tag{8.48}$$

A realization of a structure for implementing the LMS algorithm is shown in Figure 8.18. In order to compute (8.45), the data symbols have to be known, or very reliable estimates have to be available. This is generally achieved by initially using a known sequence to train the equalizer. This mode of operation, with a *training sequence*, is called *automatic equalization*. Automatic equalization must, of course, cease when actual data transmission begins. This leads to the concept of *adaptive equalization*, invented by Lucky [1], where the data

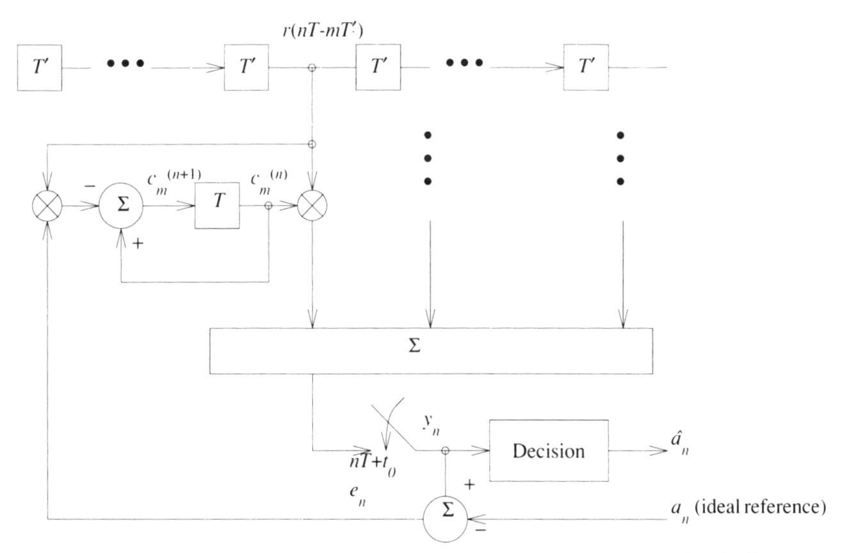

Fig. 8.18 Automatic equalization where the tap weights are adjusted using an ideal reference (or correct data decisions) for training.

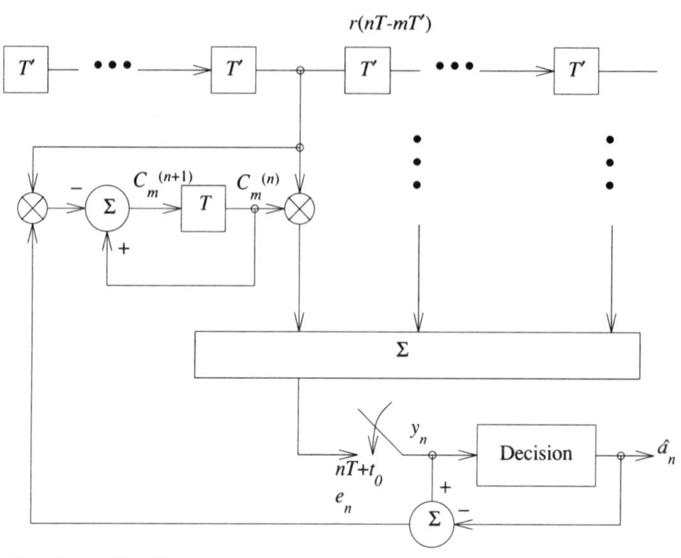

Fig. 8.19 Equalizer with adjustment of tap weights using data decisions (adaptive equalization) to form the error signal.

value, a_n, is replaced by the decision $\hat{a}_n$. Thus, *adaptive LMS equalization* is described by the *nonlinear algorithm*

$$\mathbf{c}_{n+1} = \mathbf{c}_n - \beta[y(nT+t_0) - \hat{a}_n]\mathbf{r}_n , \tag{8.49}$$

with an implementation illustrated in Figure 8.19. In Figure 8.19, note that the error is labeled $\hat{e}_n$, to indicate that the quantized equalizer output is being used to form the reference signal $\hat{a}_n$. Note that if errors are made infrequently (i.e., the probability that $\hat{a}_n \neq a_n$ is very small), the choice of a relatively small step size β will insure that (8.49) performs the same as (8.48). We will return to the difficult subject of evaluating the performance of adaptive equalization in the presence of decision errors, but for now we will proceed to analyze the LMS algorithm under the assumption that either a training sequence is available or decision errors are made at an error rate low enough so as not to affect the behavior of the algorithm. The first property we wish to demonstrate is that, if the LMS algorithm converges, it must converge in the mean, to the optimum setting. Averaging (8.48) and assuming the algorithm has converged, on the average, to the tap weight vector $\mathbf{c}_\infty$, then the LMS relation becomes

$$\mathbf{c}_\infty = \mathbf{c}_\infty - \beta\overline{e_\infty \mathbf{r}_\infty} , \tag{8.50a}$$

where e_∞ and $\mathbf{r}_\infty$ are, respectively, the value of the error signal and the vector of received samples at convergence. Examining (8.50a) it is clear that, if the algorithm has converged, then the average correction term must be zero, or

$$\overline{e_\infty \mathbf{r}_\infty} = 0 \ . \qquad (8.50b)$$

Evaluating (8.50b) we have from (8.15b) and (8.16) that

$$\overline{e_\infty r_\infty} = \overline{\mathbf{r}_\infty (\mathbf{r}'_\infty \mathbf{c}_\infty - a_\infty)} = R\mathbf{c}_\infty - \mathbf{x} = 0 \ , \qquad (8.50c)$$

and we see that $\mathbf{c}_\infty$ is indeed $\mathbf{c}_{opt}$. Thus, if the LMS algorithm converges, the steady-state tap vector is the optimum setting.

8.4.2 LMS ADAPTATION OF DECISION FEEDBACK EQUALIZERS

A decision feedback equalizer (DFE), which is depicted in Figure 8.15, has been shown to provide improved performance in the presence of moderate-to-severe slope distortion. The LMS algorithm can be used to jointly adapt both the feedforward coefficients (the c_n's) and the feedback coefficients (the f_n's) based on a *common* error signal

$$\begin{aligned} e_n &= v_n - a_n \\ &= \mathbf{c}' \ \mathbf{q}_n - \mathbf{f}' \ \hat{\mathbf{a}}_n - a_n \ , \end{aligned} \qquad (8.51a)$$

where $\mathbf{q}_n$ is the vector of samples in the forward delay line and $\hat{\mathbf{a}}_n$ is the vector of decisions in the feedback delay line at the nth sampling instant. By defining the augmented vector, $\mathbf{w}$, containing all the tap weights

$$\mathbf{w} = \begin{bmatrix} \mathbf{c} \\ \\ -\mathbf{f} \end{bmatrix}, \qquad (8.51b)$$

and the vector $\mathbf{s}$, consisting of the signal samples in the two delay lines

$$\mathbf{s} = \begin{bmatrix} \mathbf{q}_n \\ \\ \hat{\mathbf{a}}_n \end{bmatrix}, \qquad (8.51c)$$

the common error signal can be written as

$$e_n = \mathbf{w}'\mathbf{s} - a_n . \qquad (8.51d)$$

If we compute the MSE by squaring and averaging (8.51d), we immediately see that the MSE is a convex function of $\mathbf{w}$ (see Exercise 8.7). Thus there is a unique value of $\mathbf{w}' = (\mathbf{c}', -\mathbf{f}')$ that minimizes the MSE, and the LMS algorithm can be used to adaptively update the feedforward and feedback taps using the *common* error signal, e_n. Using the LMS framework, we update $\mathbf{c}$ and $\mathbf{f}$ by minimizing the instantaneous squared error with respect to each of the quantities. This leads to the following LMS updating equations for a DFE:

$$\mathbf{c}_{n+1} = \mathbf{c}_n - \alpha e_n \mathbf{q}_n$$
$$\mathbf{f}_{n+1} = \mathbf{f}_n + \beta e_n \hat{\mathbf{a}}_n \ . \tag{8.52}$$

The LMS adaptation of DFE tap weights can be extended to the passband structures discussed later in this chapter. Similarly, the fast-converging (Kalman) algorithms, which we will discuss in the next section, can be extended to DFEs in a straightforward manner.

8.4.3 CONVERGENCE RATE AND RESIDUAL ERROR OF THE LMS ALGORITHM

In this section we will determine the rate of convergence of the MSE to its steady-state value, when the taps are adjusted via the LMS algorithm. Since the LMS algorithm is continually being driven by a correction term, it should be noted that even when the mean of the tap vector has converged, the taps will fluctuate about the optimum value. This noise-like fluctuation will cause the residual steady-state MSE to be larger than that given by (8.20). We will calculate the magnitude of this *misadjustment* as a function of the one variable parameter in the LMS algorithm, the step size β. As with the tracking loops discussed in Chapter 6, selection of the step size is not as simple as it may first appear. We will show that a large value of β is desirable for rapid convergence, while a small value of β is desirable to minimize the misadjustment. If we expect the channel environment to be (slowly) time varying, it will be shown that even when "steady state" has been reached, it is desirable to keep the step size at a moderate level so that changes in the channel characteristics can be tracked. Of course, one strategy is to gear shift from an initial large step size to a substantially smaller one as the algorithm approaches convergence. We begin our discussion by assuming the automatic adjustment of the equalizer tap weights according to the LMS algorithm. It is straightforward to show (see Exercise 8.8) that the average tap error is governed by (8.25b). In other words, the *average tap error* is the same whether or not the gradient is known. As we shall show below, the MSE evolves in a fundamentally different manner for the LMS implementation than for the known gradient (steepest-descent) case. Later in the chapter we will point out the distinctions in behavior of the LMS algorithm for synchronous and fractionally spaced equalizers. So far, our treatment is general enough to accommodate both structures. To study the evolution of the MSE we substitute the tap-error vector at the nth iteration,

$$\boldsymbol{\varepsilon}_n = \mathbf{c}_n - \mathbf{c}_{opt} \ ,$$

into (8.25b), and find that

$$E_n \,/\, P = E_{\min} \,/\, P + q_n \ , \tag{8.53}$$

where

$$q_n = E[\varepsilon_n' R \varepsilon_n] \tag{8.54}$$

is the *excess mean-square error* at the nth iteration. Convergence of the equalizer can be examined in terms of the convergence of q_n to its ultimate steady-state value.

To examine this convergence, we subtract $\mathbf{c}_{opt}$ from (8.48) to obtain an iterative equation for the tap-error vector

$$\begin{aligned} \varepsilon_{n+1} &= \mathbf{c}_{n+1} - \mathbf{c}_{opt} = \mathbf{c}_n - \mathbf{c}_{opt} - \beta e_n \mathbf{r}_n \\ &= [I - \beta \mathbf{r}_n \mathbf{r}_n'] \varepsilon_n - \beta \mathbf{r}_n e_{n_{opt}} , \end{aligned} \tag{8.55}$$

where

$$e_n(opt) \triangleq \mathbf{c}_{opt}' \mathbf{r}_n - a_n \tag{8.56}$$

is the instantaneous error when the taps are at their optimum settings. Recalling the diagonalization of the channel correlation matrix

$$R = P' \Lambda P, \tag{8.57}$$

where Λ is the $(2M+1)\times(2M+1)$ diagonal matrix of eigenvalues of R, and P is the orthogonal matrix whose columns are the eigenvectors of R, we define the rotated tap-error vector

$$\mathbf{y}_n = P \varepsilon_n . \tag{8.58}$$

The excess mean-square error (8.54) can be expressed as

$$q_n = E\,[\varepsilon_n' R \varepsilon_n] = E\,[\mathbf{y}_n' \Lambda \mathbf{y}_n] = \sum_{i=-M}^{M} \lambda_i \; E[y_{ni}^2] , \tag{8.59}$$

where y_{ni} is the ith component of y_n. From (8.55) and (8.58) we write

$$\mathbf{y}_{n+1} = [I - \beta P \mathbf{r}_n \mathbf{r}_n' P'] \mathbf{y}_n - \beta e_n(opt) P \mathbf{r}_n , \tag{8.60}$$

as the iterative equation satisfied by the rotated tap-error vector. Combining (8.59) and (8.60) gives the following expression for the excess MSE,

$$\begin{aligned} q_{n+1} = E \Big\{ & [\mathbf{y}_n'(I - \beta P \mathbf{r}_n \mathbf{r}_n' P') - \beta e_n(opt) \mathbf{r}_n' P'] \Lambda [(I - \beta P \mathbf{r}_n \mathbf{r}_n' P') \mathbf{y}_n \\ & - \beta e_n(opt) P \mathbf{r}_n] \Big\} . \end{aligned} \tag{8.61}$$

The above unwieldy expression can be simplified into a single first-order recursion by making several approximations that are generally valid in practice. Perhaps the most needed assumption we make is the *independence assumption*, where we assume that the sequence of signal vectors appearing in the equalizer delay line are independent of each other. Clearly, the vectors that differ by one index (a symbol interval) are highly dependent since they have common elements, except for the most recent sample. The assumption allows us to decouple the ε_n and $\mathbf{r}_n$ dependent terms appearing in (8.61). Since iteration of (8.61) shows that ε_n depends on all the prior signal samples $\{\mathbf{r}_m\}_0^{n-1}$, the independence assumption is needed for us to perform independent averaging of the terms appearing in (8.61). Fortunately, experimental results agree with the behavior predicted by the analysis based upon the independence assumption. Mazo [15] has developed an elegant convergence analysis without the independence assumption; he has shown that for very small values of step size β, the equations derived from the independence assumption provide an accurate model of the evolution of the mean-square error. In practice, the results predicted by using the independence assumption hold over a wide range of step size values.

First, we examine one of the cross terms implicit in (8.61) by assuming that the minimum error $e_n(opt)$ is statistically independent of all tap-voltage vectors, $\mathbf{r}_n$; since its average is zero we can eliminate the term

$$E[e_{n_{opt}} \mathbf{r}_n' P \Lambda (I - \beta P \mathbf{r}_n \mathbf{r}_n' P') \mathbf{y}_n] = 0 . \tag{8.62}$$

The above assumption is motivated by (8.50b), which states that the minimum error and the vector $\mathbf{r}_n$ are uncorrelated. Again, experimental results support the logic in extending the lack of correlation between $e_n(opt)$ and $\mathbf{r}_n$ to independence. Using this assumption, we can write one of the squared terms as

$$\begin{aligned} E[e_n^2(opt) \mathbf{r}_n' P' \Lambda P \mathbf{r}_n] &= E[e_n^2(opt)] E[\mathbf{r}_n' R \mathbf{r}_n] \\ &= E_{\min}\ E[\mathbf{r}_n' R \mathbf{r}_n] \le E_{\min} \cdot \lambda_{\max} (2M+1)\ E(r_n^2) \ , \end{aligned} \tag{8.63}$$

where $E(r_n^2)$ is the power of the tap-voltage samples, and we have applied the eigenvalue bound

$$\mathbf{r}_n R \mathbf{r}_n \le \lambda_{\max} \mathbf{r}_n' \mathbf{r}_n = \lambda_{\max} \sum_{m=-M}^{M} [r_m^{(n)}]^2 \tag{8.64}$$

to the term $E\mathbf{r}_n' R \mathbf{r}_n$.

Continuing our examination of (8.61), using the independence assumption and the eigenvalue bound, we bound one of the cross terms as follows:

$$\begin{aligned} E[\mathbf{y}_n' P \mathbf{r}_n \mathbf{r}_n' P' \Lambda P \mathbf{r}_n \mathbf{r}_n' P' \mathbf{y}_n] &\le \lambda_{\max}\ E[\mathbf{y}_n' P \mathbf{r}_n \mathbf{r}_n' P' P \mathbf{r}_n \mathbf{r}_n' P' \mathbf{y}_n] \\ &= \lambda_{\max} (2M+1) E(r_n^2) E[\mathbf{y}_n' \Lambda \mathbf{y}_n] \\ &= \lambda_{\max} (2M+1) E(r_n^2) q_n \ . \end{aligned} \tag{8.65}$$

The term which contributes a negative sign,

$$\langle\ \mathbf{y}_n' P \mathbf{r}_n \mathbf{r}_n' P' \Lambda\ \mathbf{y}_n\ \rangle = \langle\ \mathbf{y}_n'\ \Lambda^2\ \mathbf{y}_n\ \rangle = \sum_{i=-M}^{M} \lambda_i^2\ \langle\ \mathbf{y}_{ni}^2\ \rangle \ , \tag{8.66a}$$

has a very significant influence on both convergence and steady-state behavior and must be treated more delicately. What is needed is a good lower bound on (8.66a); however, the most direct lower bound,

$$\langle\ \mathbf{y}_n'\ \Lambda^2\ \mathbf{y}_n\ \rangle \ge \lambda_{\min}\ \langle\ \mathbf{y}_n'\ \Lambda\ \mathbf{y}_n\ \rangle \ ,$$

which involves the minimum eigenvalue, λ_m, is (in general) too loose, since just one small eigenvalue will drastically reduce the magnitude of this term.

There is, unfortunately, no tighter lower bound, since if the only significant component $\langle y_{ni}^2 \rangle$ is associated with the smallest eigenvalue, as is possible when there are no restrictions on these components, then the above bound can be achieved. In practice, this is an extremely unlikely event (the mean-square tap errors, $\langle y_{ni}^2 \rangle$, are pretty much equal in value), and the bound is unduly pessimistic. We choose to approximate rather than (lower) bound this term. If, the $\langle y_{ni}^2 \rangle$ are relatively uniform for all i, then a reasonable approximation is

$$\langle \mathbf{y}_n' \ \Lambda^2 \ \mathbf{y}_n \rangle \approx \bar{\lambda} \langle \mathbf{y}_n' \ \Lambda \ \mathbf{y}_n \rangle = \bar{\lambda} q_n \ , \tag{8.66b}$$

where $\bar{\lambda}$ is defined as the average eigenvalue

$$\bar{\lambda} = \frac{1}{2M+1} \sum_{i=-M}^{M} \lambda_i \,. \tag{8.66c}$$

A more appropriate approximation when there are many very small eigenvalues is the average of the set of *significant* eigenvalues, e.g. those containing 95% of the eigenvalue mass. As we shall show in Appendix 8B, fractionally spaced equalizers, because of their spectral nulls, can have many small eigenvalues.

Using these bounds and approximations in (8.61) finally yields a first-order recursion for the excess MSE,

$$q_{n+1} < [1 - 2\beta\bar{\lambda} + \lambda_{\max}\beta^2(2M+1)E(r_n^2)]q_n + \lambda_{\max}(2M+1)E(r_n^2)\beta^2 E_{\min} \,. \tag{8.67}$$

To apply the above equation to systems which use a fractionally spaced equalizer (FSE), some of the terms appearing in (8.67) must be appropriately interpreted. In systems which use a FSE, the received signal is sampled at the rate $1/T'$, where $1/T'$ is greater than twice the highest-frequency component of the baseband signal. Note that if the time span of an FSE is kept constant, the number of tap weights is in inverse proportion to T'. The channel correlation matrix, R, which is Toeplitz for a synchronous equalizer, is no longer Toeplitz for a FSE. It is shown in the Appendix 8B that, for $T' = T/2$ and an infinitely long FSE, exactly half the eigenvalues are zero and the other half tend to follow a uniform sampling of the aliased magnitude-square channel characteristic. The ith eigenvector corresponding to the nonzero eigenvalues is given approximately as a sinusoid of frequency $f_i = i(2N+1)/T$, $i = 0, 1, \ldots, N$. The eigenvectors corresponding to the zero eigenvalues have most of their spectral energy concentrated near $1/T$ Hz. In this light, we re-examine the above bounds to see if they are still reasonably tight for a suitably long FSE. Recall that (8.64) was obtained by using the bound

$$\sum_{-M}^{M} \lambda_i s_i^2 = \mathbf{s}' \ \Lambda \ \mathbf{s} \le \lambda_M \sum_{-M}^{M} s_i^2 \,.$$

Since half the eigenvalues will be quite small, we have as a tight bound that

$$\sum_{-M}^{M} \lambda_i s_i^2 \approx \sum_{-M/2}^{M/2} \lambda_i s_i^2 \leq \lambda_M \sum_{-M/2}^{M/2} s_i^2 ,$$

where the indices greater than $M/2$ will be associated with the zero eigenvalues. We can, however, recover the full summation by noting that $\mathbf{s}_i$, a component of $\mathbf{s} = P\mathbf{r}$, is given by the convolution of the input samples and the ith eigenvector. For $|i| > M/2$, this convolution is equivalent to passing the received bandlimited signal through a narrow-band filter centered at $1/T$ Hz, and is thus close to zero. We can conclude that

$$\sum_{-M}^{M} \lambda_i s_i^2 \leq \lambda_M \sum_{-M/2}^{M/2} s_i^2 \cong \lambda_M \mathbf{s}'\mathbf{s} = \lambda_M \mathbf{r}'\mathbf{r},$$

and the bounds remain valid. We now reconsider the term

$$\begin{aligned} \langle \mathbf{y}' \Lambda^2 \mathbf{y} \rangle &= \sum_{-M}^{M} \lambda_i^2 \langle y_i^2 \rangle \approx \sum_{-M/2}^{M/2} \lambda_i^2 \langle y_i^2 \rangle \approx \bar{\lambda} \sum_{-M/2}^{M/2} \lambda_i \langle y_i^2 \rangle \\ &\approx \bar{\lambda} \sum_{-M}^{M} \lambda_i \langle y_i^2 \rangle = \bar{\lambda} \langle \mathbf{y}' \Lambda \mathbf{y} \rangle = \bar{\lambda} q, \end{aligned} \tag{8.68}$$

where $\bar{\lambda}$ is an average eigenvalue over the set of *significant* eigenvalues of the channel covariance matrix. In obtaining (8.68), we have again assumed that the $\langle y_i^2 \rangle$ are fairly uniform (in contrast to the s_i^2, which depend critically on the index i), and we interpret $\bar{\lambda}$ as the average of the "nonzero" eigenvalues of the channel correlation matrix.

In practice, it is not difficult to estimate $\bar{\lambda}$ for a FSE, as the eigenvalues, λ_i, tend to approach zero quite rapidly. A reasonable criterion is the average eigenvalue over the partial set of eigenvalues containing all but a small fraction (perhaps 5 percent) of the eigenvalue mass. With this in mind, we can apply (8.67) to both synchronous and fractionally spaced equalizers. We now return to our discussion of the rate of convergence of the MSE and the residual MSE.

8.4.4 BOUNDS ON THE STEP SIZE FOR CONVERGENCE

We first investigate the conditions under which the excess MSE will decrease with time. Now in order to for the mean-square error to decay it is clear from (8.67) that

$$|1 - 2\beta\bar{\lambda} + \lambda_M \beta^2 (2M+1)\langle r_n^2 \rangle| < 1$$

or

$$\beta \leq \beta_{\max} = \frac{2\bar{\lambda}}{\lambda_M} \frac{1}{(2M+1)} \frac{1}{\langle r_n^2 \rangle} . \tag{8.69}$$

Even with all the bounds and approximations which have been made in reaching (8.67), a significant difference is readily apparent in the maximum allowable step size for the known-gradient algorithm and the estimated-gradient (LMS) algorithm. We have shown that, for the known-gradient algorithm to converge, it is required that $0 \leq \beta \leq 2/\lambda_M$. The maximum step size for the LMS algorithm is considerably smaller than that for the known-gradient (steepest descent) algorithm, as can be deduced by considering (8.34) and (8.69). Since the trace of $R = (2M+1)\langle r_n^2 \rangle = \sum_{-M}^{M} \lambda_i$, we have from (8.69) that

$$\beta_{\max} = \frac{2}{\lambda_M} \frac{\bar{\lambda}}{(2M+1)} \frac{1}{\langle r_n^2 \rangle} = \frac{2}{\lambda_M} \frac{\bar{\lambda}}{\sum_{i=-M}^{M} \lambda_i} < \frac{2}{\lambda_M(2M+1)} . \tag{8.70}$$

Thus, the maximum permissible step size for the LMS (estimated-gradient) algorithm is reduced, by a factor equal to the number of tap weights, from the maximum step size permitted in the steepest-descent (known-gradient) algorithm. This reduction in step size is, in part, a consequence of the need to average out the random fluctuations of each tap which are due to noise and the random data pattern.

By differentiating the right-hand side of (8.67) with respect to β we obtain the step size, β_n^*, which provides the maximum rate of convergence [relative to the bound (8.67)]:

$$\beta_n^* = \frac{\bar{\lambda} q_n}{\lambda_M(2M+1)\langle r_n^2 \rangle} \cdot \frac{1}{[q_n + E_{opt}]} . \tag{8.71a}$$

Note that β_n^* is a function of time, n, and of the generally unknown (to the receiver) quantities q_n and E_{opt}. During the initial stages of convergence, $q_n >> E_{opt}$, so that (8.71a) becomes the constant value

$$\beta_0^* = \frac{\bar{\lambda}}{\lambda_M(2M+1)\langle r_n^2 \rangle} = \frac{1}{2}\beta_{\max} , \tag{8.71b}$$

and q_n converges exponentially towards a steady-state value. Thus a useful rule is: *the initial step size should be half the maximum permissible step size.* As we have shown in Appendix 8B, since the eigenvalues sample the channel spectra, the nulls in a distorted channel will produce a large spread in $\bar{\lambda}/\lambda_{\max}$. This suggests a reduction in the step size for most rapid convergence with highly distorted channels.

As convergence nears completion, the steady-state step size, β, resulting in a specified mean-square error, $E_{opt} + q_\infty$, is found by equating the two sides of (8.67) giving,

$$\beta_\infty = \frac{2\bar{\lambda}}{\lambda_M} \cdot \frac{1}{(2M+1)\langle r_n^2 \rangle} \cdot \frac{q_\infty}{q_\infty + E_{opt}} = \frac{q_\infty}{q_\infty + E_{opt}} \beta_{\max} . \tag{8.72}$$

One might think that we would desire that $q_\infty = 0$, which implies that the steady-state step size, β_∞, approach zero. In practice, the large initial step size is generally changed in discrete steps, which is often referred to as gear-shifting, to a final value, β_∞, which is generally kept at some small-to-moderate value since:

(1) The channel may change, in which case it is required that the LMS algorithm retain a tracking capability

(2) As we shall show in Chapter 10, with a digital implementation a *larger* error may be produced when $\beta_\infty \to 0$ than when β_∞ is a small but finite value.

8.4.5 RESIDUAL MEAN-SQUARE ERROR

From (8.53) we know that

$$E_\infty = E_{opt} + P \cdot q_\infty \tag{8.73}$$

and from (8.67) we can solve for the excess mean-square error at convergence, q_∞, and the steady-state MSE

$$q_\infty = E_{opt}/(\beta_{\max}/\beta_\infty - 1) \ , \tag{8.74a}$$

$$E_\infty = E_{opt}/(1 - \beta_\infty/\beta_{\max}) \ . \tag{8.74b}$$

If we assume that $\beta_{\max} / \beta_\infty >> 1$, then

$$q_\infty \sim \frac{\beta_\infty}{\beta_{\max}} E_{opt} \tag{8.75}$$

is a measure of the residual MSE. In practice, the ratio $\beta_\infty/\beta_{\max}$ can be made quite small, so that the residual fluctuation about the MSE, sometimes called the *misadjustment*, can be severely attenuated. As we show in Chapter 10, sufficient precision is required in a digital implementation of the LMS algorithm to reduce β_∞, and have q_∞ be correspondingly reduced according to (8.75).

8.4.6 SPEED OF CONVERGENCE OF THE LMS ALGORITHM

In this section we contrast the speed of convergence of the LMS algorithm with that of the steepest-descent algorithm. Recall that the steepest-descent algorithm could be made to converge in one step if all the eigenvalues of R were the same. Thus the speed of convergence of the steepest-descent algorithm is governed by the eigenvalue spread. We will show that this is not the only factor that determines the speed of convergence of the LMS algorithm. If we select $\beta = \beta_0^*$, according to (8.71b), then the initial convergence of the LMS algorithm is exponential with the time constant, γ, i.e.,

$$q_{n+1} < \gamma q_n \ , \tag{8.76a}$$

where

$$\gamma = 1 - \frac{\bar{\lambda}}{\lambda_{\max}} \cdot \frac{\bar{\lambda}}{(2M+1) \langle r^2 \rangle} \ , \tag{8.76b}$$

and iterating this bound gives

$$q_{n+1} < \gamma^n q_0 \, . \tag{8.77}$$

Recalling the definition of $\bar{\lambda}$, we have that

$$\gamma = 1 - \frac{\bar{\lambda}}{\lambda_{\max}} \frac{\sum \lambda_i}{(2M+1)^2 \langle r^2 \rangle} \, , \tag{8.78}$$

and since the trace of R can be written as either the sum of the diagonal terms or the sum of the eigenvalues, we have that

$$\text{trace } R = \sum_{-M}^{M} \lambda_i = (2M+1) \langle r^2 \rangle \, , \tag{8.79}$$

which leads to the compact expression for the time constant

$$\gamma = 1 - \frac{\bar{\lambda}}{\lambda_{\max}(2M+1)} = 1 - \frac{1}{\rho(2M+1)} \, . \tag{8.80}$$

In (8.80), ρ is defined as the eigenvalue ratio, $\lambda_{\max}/\bar{\lambda}$, and increasing the length of the equalizer has the same effect as if the eigenvalue ratio were increased; similarly, increasing the eigenvalue ratio is equivalent to lengthening the equalizer. Note that (8.80), which is the time constant for the LMS algorithm, has a similar but fundamentally different form than that of (8.38), which describes the convergence of the steepest-descent algorithm. *While the time constant of the LMS algorithm does indeed depend on the eigenvalue ratio, the number of adaptive weights has a comparable effect on the rate of convergence of the LMS algorithm.*

It should also be noted again that in contrast to the steepest-descent algorithm, even if the eigenvalues were all equal, the LMS algorithm would have the nonzero time constant

$$\gamma = 1 - \frac{1}{2M+1} \, , \tag{8.81a}$$

which implies that convergence of the LMS algorithm *cannot* be achieved in one step. With γ given by (8.80), we have

$$q_{n+1} \le \left[1 - \frac{1}{(2M+1)\rho} \right] q_n + \lambda_{\max}(2M+1)\overline{r^2}\beta^2 E_{opt} \, , \tag{8.81b}$$

and at start-up the second term is generally much smaller than the first term; so the initial delay of the residual MSE is dominated by

$$q_{n+1} \lesssim \left[1 - \frac{1}{(2M+1)\rho} \right] q_n \, , \tag{8.81c}$$

and the MSE evolution is given by

$$E_{n+1} \leq \left[1 - \frac{1}{(2M+1)\rho}\right]^n E_0 + E_{opt} \left\{\frac{1}{1 - \beta_\infty/\beta_{\max}}\right\}. \quad (8.81d)$$

From (8.81c) we see that it takes approximately $2.3\rho(2M+1)$ iterations for q_n to decrease by an order of magnitude. This illustrates the sensitivity of the convergence time to the number of taps and the eigenvalue ratio.

It is important to realize that two factors can slow convergence of the LMS algorithm: a large eigenvalue ratio $\bar{\lambda}/\lambda_{\max}$ (which is related to the degree of distortion in the folded power spectrum of the channel) and a large number of adjustable tap weights. While it might appears that increasing the number of tap weights can do no harm (since the E_{opt} will decrease as a function of $2M+1$), the discussion in this section has shown that the rate of convergence can be adversely impacted by an unnecessarily large number of adaptive elements. We will show later in the text (Chapter 10) that increasing the number of taps will amplify the adverse affects that limited precision has on the steady-state MSE achievable by an adaptive equalizer. It is a general design principle that the number of adjustable tap weights should be kept to the minimum needed to achieve the level of equalization required.

8.5 FAST CONVERGENCE VIA THE KALMAN (RECURSIVE LEAST-SQUARES) ALGORITHM

In the last section we saw that the rate of convergence of the LMS algorithm was determined by the eigenvalue ratio and the number of adaptive weights. In this section we describe a powerful theoretical approach that leads to an adaptive algorithm (called the Kalman or recursive least-squares algorithm) that dramatically reduces the time it takes for an adaptive tapped delay line filter to acquire its optimum settings. It will be shown that the desired (optimum) tap weights are identical for the Kalman and LMS algorithms.

Fast-convergence adaptive algorithms are desirable for two reasons:

(1) Such algorithms could be used during system start-up, where the time needed for equalizer adaptation is time that is not available for the transmission of customer data. If the equalizer training time is an appreciable fraction of the time duration of customer data frames, then a slowly converging algorithm can significantly degrade the user throughput.

(2) If the channel characteristics are changing at an appreciable rate (e.g., a fading channel) then a fast-converging algorithm may provide an useful alternative to the tracking capability of the LMS algorithm.

These algorithms are intended to provide the system with a maximum amount of "up time."

The striking property of the Kalman algorithms is that they typically provide convergence to the optimum tap setting in a number of iterations of the algorithm equal to the number of tap weights, independent of the channel characteristics. This is to be contrasted with the exponential convergence of the LMS algorithm, whose time constant is highly dependent on the eigenvalue spread and the number of adaptive elements. For an equalizer with $(2M+1)$ weights, it will be shown that the Kalman algorithm requires on the order of $(2M+1)^2$ multiplications per iteration, while the LMS algorithm requires only $(2M+1)$ multiplications per adjustment. Thus the signal processing complexity to implement a Kalman algorithm may be prohibitive in some applications (depending on the symbol rate and the number of weights).

The Kalman algorithms described in this section depart from the LMS philosophy in a fundamental way: the performance criterion is changed from the *statistical* error measure used in the LMS algorithm to a *deterministic*, sum-of-squared-errors criterion based on the received signal samples. The convergence of the LMS algorithm suffers from the use of a *statistical* performance criterion (the MSE), while the Kalman signal processing is based on minimizing the *actual* sum of squared errors determined by the received signal samples. This deterministic least-squares criterion is minimized at every iteration, while the LMS algorithm is only striving to converge to the Wiener solution.

We will show that the dramatically improved performance of the Kalman algorithm (i.e., convergence in $(2M+1)$ steps) is due to two factors:

(1) The algorithm orthogonalizes the signal vectors presented to the adaptive filter (i.e., the processed signal vectors have a unity eigenvalue ratio). This has the effect of negating the impact of eigenvalue spread on the convergence of the algorithm, and the algorithm uses a Gram–Schmitt orthogonalization procedure to synthesize the optimum weight vector from these linearly independent received signal vectors. With high probability, the first $(2M+1)$ received signal vectors are linearly independent, and the Kalman algorithm is structured to synthesize the optimum tap vector in terms of these signal vectors. In contrast, the LMS algorithm moves to the next tap setting without exploiting the linear independence of the input vectors.

(2) By using a nonstatistical performance criterion, there is no longer a need to use a small step size, which is inversely proportional to the number of tap weights, to "average out" the random data fluctuations. This will have the effect of providing performance that is much less sensitive to the number of tap weights than the LMS algorithm.

The fast-converging algorithm, which we refer to as the Kalman algorithm, is also known as the *recursive least-squares* (RLS) technique, since, as we shall show, the criterion recursively finds tap vectors that provide a least-squares fit to the received signal samples. The term Kalman algorithm is used because of its resemblance to the vector, state-space stochastic filtering techniques pioneered by Kalman [16] and because Godard, in his original work on fast-converging algorithms, [9] used a special case of the Kalman filtering model to derive an algorithm for rapid adjustment of equalizer tap weights. The use of the term Kalman algorithm in the context of rapid equalizer convergence is often a source of confusion. The classic Kalman filter [16] solves the same estimation or prediction problem as the Wiener filter, but in the time domain, as opposed to the frequency domain Wiener approach. The Kalman filter is a *fixed*, nonadaptive recursive structure that uses knowledge of the system structure and statistics to perform estimation. In the context of adaptive equalization, we could indeed consider the Kalman structure as a valid equalizer structure (instead of the transversal filter), but such a recursive structure is hard to adapt and offers little, if any, performance improvement over the transversal filter. In the discussion of equalizer tap weight convergence via the Kalman algorithm, we apply the original Kalman estimation framework to devise a technique for the rapid adjustment of the weights of a transversal filter (TDL) equalizer; i.e., the equalizer structure is still the familiar tapped delay line.

The Kalman (or RLS) algorithm, then, will be applied to provide rapid convergence of the tap weights of a TDL equalizer. Later on in the chapter we will consider another equalizer structure, the *lattice* equalizer. For the lattice both LMS (or gradient) as well as RLS algorithms can be derived. To further complicate matters, a class of algorithms called *Fast Kalman* algorithms [11] will also be briefly discussed. These algorithms, which are really *computationally fast* Kalman algorithms, have the *same* speed of convergence properties as the Kalman algorithm, but provide a reduction in the number of multiplications per iteration to the order of the number of adaptive weights (i.e., similar to the LMS algorithm). Thus these algorithms are computationally efficient realizations of the Kalman algorithm. To balance the excitement of the fast-converging/low-complexity promise offered by the fast Kalman algorithms, researchers have reported that the fast Kalman algorithms exhibit extreme sensitivity to the inevitable imprecisions associated with digital implementation which sometimes cause unstable operation.

8.5.1 PERFORMANCE MEASURE FOR THE KALMAN ALGORITHM

A performance measure that is based only on the received signal samples and the training sequence can be constructed as the *cumulative squared (or least-squares) error measure*

$$E_n = \sum_{i=1}^{n} \alpha^{n-i} e_i^2, \tag{8.82}$$

where $0 < \alpha \leq 1$ is a weighting factor that is used to weight recent signal samples more heavily. A value of $\alpha = 1$ indicates that all the errors are weighted equally (which may be fine for initialization, but is not suitable for the tracking mode), while a value on the order of 0.9 may be suitable for time-varying channel environments. The weighting, or forgetting factor α strikes a balance between noise averaging and tracking, just as the step size does in the LMS algorithm. In (8.82) e_i is, as before, the equalizer output error at the ith decision instant,

$$e_i = y_i - a_i = \mathbf{c}'\mathbf{r}_i - a_i \,. \tag{8.83}$$

Note that the value of the criterion (8.82) changes with each received sample, in contrast to the (statistical) MSE which is a constant. The objective is to find a recursive algorithm for determining the value of $\mathbf{c}_n$ that minimizes E_n. Note that $\mathbf{c}_n$ is viewed as a fixed but unknown tap vector that minimizes E_n at *each* sampling instant.

Substituting (8.83) into (8.82) gives

$$E_n = \sum_{i=1}^{n} \alpha^{n-1}(\mathbf{c}_n'\mathbf{r}_i - a_i)^2 = \sum_{i=1}^{n} \alpha^{n-i}(\mathbf{c}_n'\mathbf{r}_i - a_i)(\mathbf{r}_i'\mathbf{c}_n - a_i) \tag{8.84a}$$

$$= \mathbf{c}_n' \left[\sum_{i=1}^{n} \alpha^{n-i}\mathbf{r}_i\mathbf{r}_i' \right] \mathbf{c}_n - 2\mathbf{c}_n' \sum_{i=1}^{n} \alpha^{n-i} a_i \mathbf{r}_i + \sum_{i=1}^{n} \alpha^{n-i} a_i^2 \,. \tag{8.84b}$$

The value $\mathbf{c}_n$ that minimizes the quadratic form E_n is obtained by differentiating (8.84b), with respect to $\mathbf{c}_n$, giving the set of linear equations

$$R_n\mathbf{c}_n = \hat{\mathbf{x}}_n \tag{8.85a}$$

or

$$\mathbf{c}_n = R_n^{-1}\hat{\mathbf{x}}_n \,, \tag{8.85b}$$

for the optimum set of weights. In (8.85) the matrix and vector are defined by

$$R_n = \sum_{i=1}^{n} \alpha^{n-i}\mathbf{r}_i\mathbf{r}_i' \tag{8.86a}$$

$$\hat{\mathbf{x}}_n = \sum_{i=1}^{n} \alpha^{n-i} a_i \mathbf{r}_i \,. \tag{8.86b}$$

It is clear that R_n and $\hat{\mathbf{x}}_n$ converge (as $n \to \infty$) respectively to the channel correlation matrix, R, and the channel pulse vector, $\mathbf{x}$. Because of the sample-mean appearance of $\hat{R}_n$ and $\hat{\mathbf{x}}_n$, one would expect, certainly with $\alpha = 1$, that $\mathbf{c}_n \to \mathbf{c}_{opt}$. as $n \to \infty$. Our intention is *not* to solve (8.85a) afresh each time a new sample $\mathbf{r}_n$ is received, but to develop an algorithm that computes $\mathbf{c}_n$ in a *recursive* fashion as new signal vectors arrive. We begin by noting that both R_n and $\hat{x}_n$ can be written recursively as

$$R_n = \alpha R_{n-1} + \mathbf{r}_n\mathbf{r}_n' \tag{8.87a}$$

$$\hat{\mathbf{x}}_n = \alpha\hat{\mathbf{x}}_{n-1} + a_n\mathbf{r}_n \,. \tag{8.87b}$$

The key to obtaining a recursive estimate for the tap weights is to use the matrix inversion lemma (which is proved in Appendix 8C), which says that if R_n

satisfies (8.87a), then its inverse, R_n^{-1}, satisfies the recursion

$$R_n^{-1} = \frac{1}{\alpha}\left\{R_{n-1}^{-1} - \frac{R_{n-1}^{-1}\mathbf{r}_n\mathbf{r}_n' R_{n-1}^{-1}}{\alpha + \mathbf{r}_n' R_{n-1}^{-1}\mathbf{r}_n}\right\}. \tag{8.88}$$

If we let D_n denote R_n^{-1}, we introduce the *Kalman gain vector*

$$\mathbf{K}_n = \frac{1}{\alpha + \mu_n} D_{n-1}\mathbf{r}_n , \tag{8.89}$$

where the scalar μ_n is defined by

$$\mu_n \equiv \mathbf{r}_n' D_{n-1}\mathbf{r}_n . \tag{8.90}$$

Note that since D_{n-1} approaches R^{-1}, the quantity $\mathbf{r}_n' D_{n-1}^{1/2} D_{n-1}^{1/2}\mathbf{r}_n$ approaches a sequence with a flat spectrum (i.e., one with a unity eigenvalue ratio). Using the above definitions in (8.88) gives the following recursion for the inversion correlation matrix,

$$D_n = \frac{1}{\alpha}\left\{D_{n-1} - \mathbf{K}_n\mathbf{r}_n' D_{n-1}\right\}. \tag{8.91}$$

Postmultiplying both sides of (8.91) by $\mathbf{r}_n$, we get

$$D_n\mathbf{r}_n = \frac{1}{\alpha}\left\{D_{n-1}\mathbf{r}_n - \mathbf{K}_n\mathbf{r}_n' D_{n-1}\mathbf{r}_n\right\}, \tag{8.92}$$

and using (8.89), we have the relation

$$D_n\mathbf{r}_n = \frac{1}{\alpha}\left\{[\alpha+\mu_n]\mathbf{K}_n - \mu_n\mathbf{K}_n\right\} = \mathbf{K}_n , \tag{8.93}$$

which states that the Kalman vector is the orthogonalized input sequence. Since

$$\mathbf{c}_n = R_n^{-1}\hat{\mathbf{x}}_n = D_n\hat{\mathbf{x}}_n , \tag{8.94}$$

and using (8.87b) and (8.91) we have, after some manipulation, the sought-after recursion for the tap weight vector at each iteration

$$\mathbf{c}_n = \frac{1}{\alpha}\left\{[D_{n-1} - \mathbf{K}_n\mathbf{r}_n' D_{n-1}]\cdot[\alpha\hat{\mathbf{x}}_{n-1} + a_n\mathbf{r}_n]\right\}$$

$$= D_{n-1}\hat{\mathbf{x}}_{n-1} + \frac{1}{\alpha}a_n D_{n-1}\mathbf{r}_n - \mathbf{K}_n\mathbf{r}_n' D_{n-1}\hat{\mathbf{x}}_{n-1} - \frac{a_n}{\alpha}\mathbf{K}_n\mathbf{r}_n' D_{n-1}\mathbf{r}_n .$$

After some manipulation, using (8.90), (8.93), and (8.94) gives

$$\mathbf{c}_n = \mathbf{c}_{n-1} - \mathbf{K}_n[\mathbf{c}_{n-1}'\mathbf{r}_n - a_n] = \mathbf{c}_{n-1} - e_n\mathbf{K}_n , \tag{8.95}$$

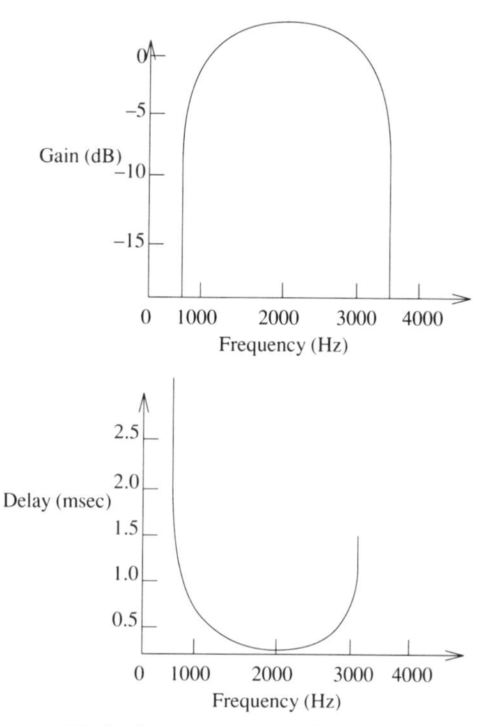

Fig. 8.20 Channel characteristic for fast-converging equalization experiment: an eigenvalue ratio of 18.6, a 31-tap equalizer, and a SNR of 31 dB were used.

where we note that $\mathbf{c}'_{n-1}\mathbf{r}_n$ is the output of the equalizer at the nth data decision. Thus for a given $\mathbf{c}_{n-1}$ and D_n updating is accomplished by using the newly received signal $\mathbf{r}_n$ to compute the Kalman gain vector, (8.89), and update the matrix inverse, (8.91), and to update the coefficients via (8.95).

The impressive convergence properties of the Kalman algorithm will be illustrated for the channel shown in Figure 8.20, which has an eigenvalue ratio of 18.6. In Figure 8.21 we compare, via simulation, the convergence of the LMS and the Kalman algorithms. Recall that it takes roughly $2.3(2M+1)\rho$ iterations for the LMS algorithm to decrease the MSE by an order of magnitude. With $(2M+1)\rho = \sim 600$, we see that the simulation is consistent with our analysis. Note the rapid and remarkable convergence of the Kalman algorithm in $2(2M+1)$ iterations (this includes 31 symbols to fill the equalizer delay line).

Some comments regarding the Kalman algorithm are in order:

(1) If we combine (8.93) and (8.95), we note that the increment to the algorithm is in the direction $D_n\mathbf{r}_n$, and since $D_n \rightarrow R^{-1}$, we note that the algorithm tends to produce orthogonal increments.

(2) The number of multiplications is proportional to $(2M+1)^2$, and the majority of these operations occur in the updating of D_n.

(3) The Kalman algorithm converges to the neighborhood of the minimum value of the MSE in ~$2(2M+1)$ iterations, independently of the channel eigenvalue ratio.

8.5.2 CONVERGENCE OF THE KALMAN ALGORITHM

In this section we provide analytic support for the rapid convergence of the excess MSE when the Kalman algorithm is used to adjust the equalizer tap weights. Since the criterion, E_n, which the algorithm minimizes is non-statistical, we have to introduce some averaging to display the mean-square error. Our approach is to use the evolution of the tap weights, defined by the Kalman equations described in the last section, to calculate the MSE, which is given by the familiar relation

$$E_n = E\varepsilon_n' R\varepsilon_n + E_{opt} = q_n + E_{opt} . \tag{8.53}$$

Since the Kalman algorithm produces a tap vector that satisfies

$$\mathbf{c}_n = R_n^{-1}\mathbf{x}_n ,$$

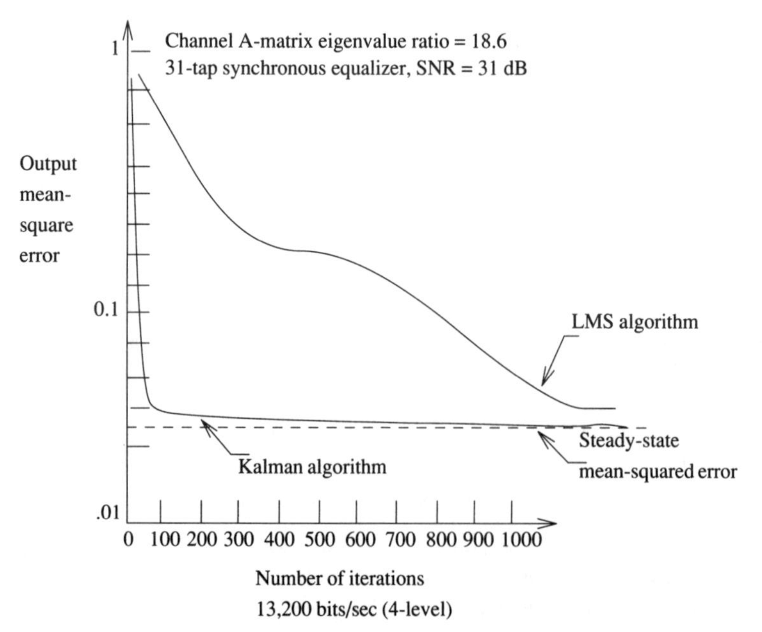

Fig. 8.21 Simulation comparison of convergence of LMS and Kalman algorithms.

we model the convergence to the steady state solution by

$$\mathbf{c}_n = \mathbf{c}_{opt} + \varepsilon_n , \tag{8.96}$$

where ε_n is the random tap-error vector each of whose components has an identical variance, $E_{opt}/n\overline{r^2}$. This modeling reflects the fact that sample-mean estimates such as R_n and $\hat{\mathbf{x}}_n$ converge with a variance proportional to the background noise-to-signal ratio (E_{opt} may be viewed as the background noise as convergence is approached [thus $E_{opt}/\overline{r^2}$ is the steady-state noise-to-signal ratio]) and inversely proportional to the number of iterations. With this model, we have that

$$\varepsilon_n = \mathbf{c}_n - \mathbf{c}_{opt} , \tag{8.97}$$

and the residual MSE can be evaluated as

$$q_n = \overline{\varepsilon_n' R \varepsilon_n} = \overline{\zeta_n' R \zeta_n} = \sum_{i,j=-M}^{M}\sum R_{i-j} \overline{\zeta_i^{(n)} \zeta_i^{(n)}} , \tag{8.98}$$

where $\zeta_i^{(n)}$ is the ith component of the tap error vector at the nth iteration. It is reasonable to assume that the ith and jth components of the tap error are uncorrelated, so that

$$\overline{\zeta_i^{(n)} \zeta_j^{(n)}} = \frac{E_{opt}/\overline{r^2}}{n} \cdot \begin{cases} 1, & i=j \\ 0, & i \neq j . \end{cases} \tag{8.99}$$

The residual MSE can now be evaluated as

$$q_n = \sum_{i=-M}^{M} \frac{E_{opt}}{\overline{r^2} \cdot n} R_0 = (2M+1) \frac{E_{opt}}{n \cdot \overline{r^2}} \overline{r^2} , \tag{8.100}$$

which combined with (8.53) gives

$$E_n = E_{opt} \left\{ 1 + \frac{(2M+1)}{n} \right\} , \tag{8.101}$$

which describes the convergence of the MSE to its optimum value. Note that the MSE can be reduced to within 3 dB of the minimum value in $(2M+1)$ iterations. This is consistent with the experiment reported in the last section. Also note that (8.101) is independent of any channel conditions (such as the eigenvalue ratio), and that at convergence there is no fluctuation about the minimum attainable error (the weighting factor is unity). Note that after $(2M+1)$ iterations, the number of multiplications is $\sim(2M+1)^3$, which is the number required by a direct matrix inversion assuming the channel were known. Thus the Kalman algorithm can achieve the same steady-state error, with the same number of multiplications required in a direct solution for the optimum weights. Of course, the "direct" solution is generally not attainable since the channel is not known.

Finally, the approximate formula for the convergence of the Kalman algorithm, as given by (8.101) should be contrasted with (8.81d), which describes the convergence of the LMS algorithm. A more detailed analysis of the *initial* convergence of the Kalman algorithm [10] shows that, with n being the number of iterations, the initial convergence of the transient term is inversely proportional to $n!$.

8.6 FAST KALMAN ALGORITHMS: KALMAN ALGORITHMS WITH REDUCED COMPLEXITY

As we pointed out, the Kalman algorithm requires about $(2M+1)^2$ multiplication per iteration (or symbol interval). The fast Kalman algorithms provide the same speed of convergence as the Kalman algorithm, but with reduced complexity. In fact, the Fast Kalman algorithms achieve a complexity on the order of the LMS algorithm, i.e., $(2M+1)$ multiplications per iteration. A detailed discussion of the fast algorithms, of which there are many variants (differing in the number of multiplications) would take us far afield (Alexander [17] provides a general discussion of various Fast Kalman algorithms). The Fast Kalman algorithms take advantage of the "shifting" property of the correlation matrix R_n. The "shifting" property recognizes that the vectors $\mathbf{r}_n$ and $\mathbf{r}_{n-1}$ differ by only a shift and changes in the first and last components. Falconer and Ljung [11] have shown that the recursion (8.91) can be carried out in about $10(2M+1)$ multiplications, and Cioffi and Kailath [18] have further reduced the number of multiplications to $7(2M+1)$. With this level of complexity it should be relatively easily to implement Fast Kalman algorithms.

The dramatic improvements in complexity reduction promised by the Fast Kalman algorithms, along with their extremely rapid convergence, is unfortunately tempered by their sensitivity to roundoff, or other errors, inherent in a digital implementation. Many authors [17] have provided evidence which shows that roundoff errors introduced in the Fast Kalman algorithm cause it to go unstable when the forgetting factor is less than unity. Cioffi and Kailath [18] give a detailed discussion of the various techniques that provide modest improvements in stabilizing the Fast Kalman algorithms. At the time of this writing, Slock and Kailath [19] and Hariharan and Clark [35] have published a promising approach to stabilizing the Fast Kalman algorithm by introducing redundancy at a slight increase in computational complexity.

8.7 LATTICE FILTERS: ANOTHER STRUCTURE FOR FAST-CONVERGING EQUALIZATION

Our discussion of fast-converging algorithms has, so far, assumed a tapped delay line structure. If we recall Newton's algorithm for the known gradient algorithm

$$\mathbf{c}_{n+1} = \mathbf{c}_n - \beta R^{-1}(R\mathbf{c}_n - \mathbf{x}) , \tag{8.39}$$

and the Kalman (RLS) algorithm

$$\mathbf{c}_{n+1} = \mathbf{c}_n - \mathbf{K}_{n-1}\, e_n = \mathbf{c}_n - e_n\, D_n \mathbf{r}_n \,, \tag{8.95}$$

we see that one common ingredient they have is the pre-processing of either the tap vector (in 8.39) or the signal vector (in 8.95) to provide orthogonal inputs (e.g., $D_n \mathbf{r}_n$) for the algorithm to iterate. Knowing that the Kalman algorithm provides rapid convergence, has motivated researchers to consider the potential of a lattice filter [5,17], which was originally developed for orthogonalization of speech samples, to pre-process the signal samples. The intent is to have the lattice filter produce uncorrelated signal samples at each of its output stages. One stage of a lattice filter is shown in Figure 8.22a; the lattice filter is generally composed of a concatenation of several such 2×2 stages. Each stage would replace the delay element of a transversal filter. The following relations describe the operation of the lattice stage,

$$\begin{aligned} b_n^{(1)} &= f_n^{(1)} \\ f_n^{(m+1)} &= f_n^{(m)} - k'_m\, b_{n-1}^{(m)} \\ b_n^{(m+1)} &= k''_m f_n^{(m)} + b_{n-1}^{(m)} \,, \qquad \begin{array}{l} m = 1,2, \ldots M \\ n = 1,2, \ldots \end{array} \end{aligned} \tag{8.102}$$

where n denotes the iteration index (time) and m denotes the index of the lattice section, $f_n^{(m)}$ and $b_n^{(m)}$ are known as the forward and backward error residuals, and k'_m and k''_m are the reflection, or partial correlation coefficients. It can be shown that if $E|f_n^{(m)}|^2$ and $E|b_n^{(m)}|^2$ are minimized for all values of m, through the choice of k'_m and k''_m, then for every instant, n, the $b_n^{(m)}$ are uncorrelated as a function of m [17], i.e.,

$$E\, b_n^{(m)} b_n^{(l)} = 0, \text{ if } m \neq l \,. \tag{8.103}$$

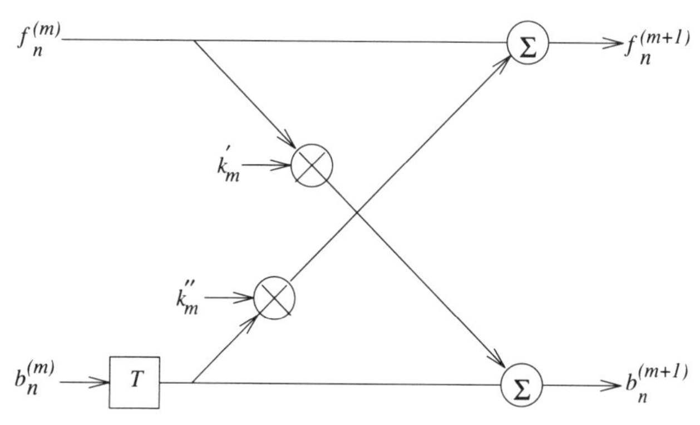

Fig. 8.22a A stage of lattice filter.

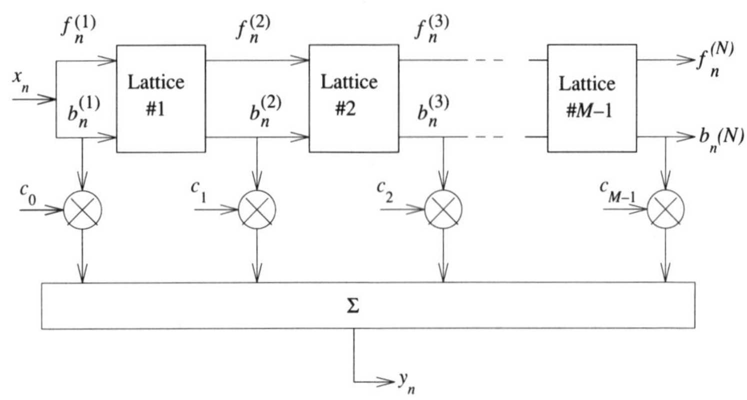

Fig. 8.22b A lattice equalizer.

If the signal samples, r_n, are applied to the lattice-based equalizer shown in Figure 8.22b, then the samples $[b_n^{(m)}]$ will be uncorrelated as m varies for fixed n. These uncorrelated signals are then fed to the *tapped delay line* section with coefficients $\{c_i\}_{i=1}^{M}$. If the reflection coefficients are updated as per (8.102) and the TDL weights are updated according to LMS type algorithms, the system is referred to as a *gradient lattice* algorithm. If the lattice and TDL tap weights are updated via least-squares algorithms, then the structure is called the *least-squares lattice*. As before, the gradient lattice applies the *statistical* LMS philosophy to the lattice structure, while the least-squares systems recursively minimize a *deterministic* cumulative squared error criterion to determine the tap weights. The least-squares lattice and the TDL Kalman algorithms have identical mathematical descriptions, so, in the absence of implementation differences, their performance will be the same. It should be noted that since the reflection coefficients are independent of the number of stages in the filter, the lattice filter can add or delete stages without the need to recalculate these coefficients. However, since the associated TDL coefficients need to be recalculated whenever a stage is added or deleted, this property of the lattice filter (which is quite useful in the original speech orthogonalization application) is not a great advantage in the equalization application.

Satorius and Pack [20] have compared the convergence properties of lattice equalizers, for mean-square or least-squares adjustment algorithms with the mean-square TDL equalizer (the Kalman TDL system has performance identical to that of the least-squares lattice). The convergence curves are shown in Figure 8.23. Note that both the least-squares and gradient lattices provide performance that is independent of the eigenvalue ratio, while the convergence time of the LMS algorithm is seen to be quite sensitive to this quantity. Note that the LS-lattice converges quite a bit faster than the gradient lattice (which is always faster than the LMS algorithm).

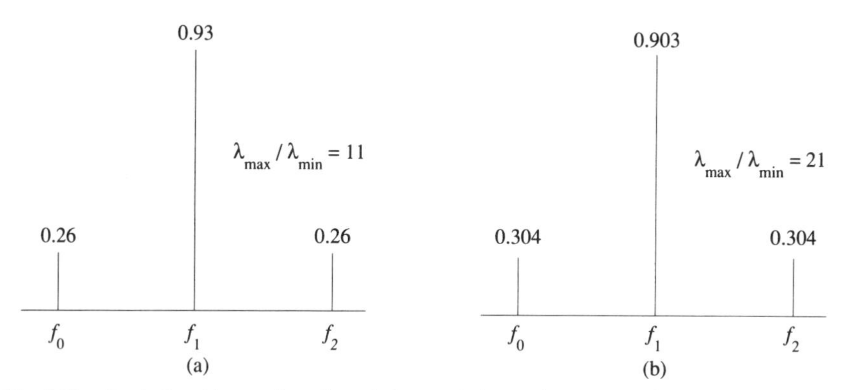

Fig. 8.23a Equivalent discrete-time channel characteristics used to generate the convergence results in Figures 8.23b and 8.23c.

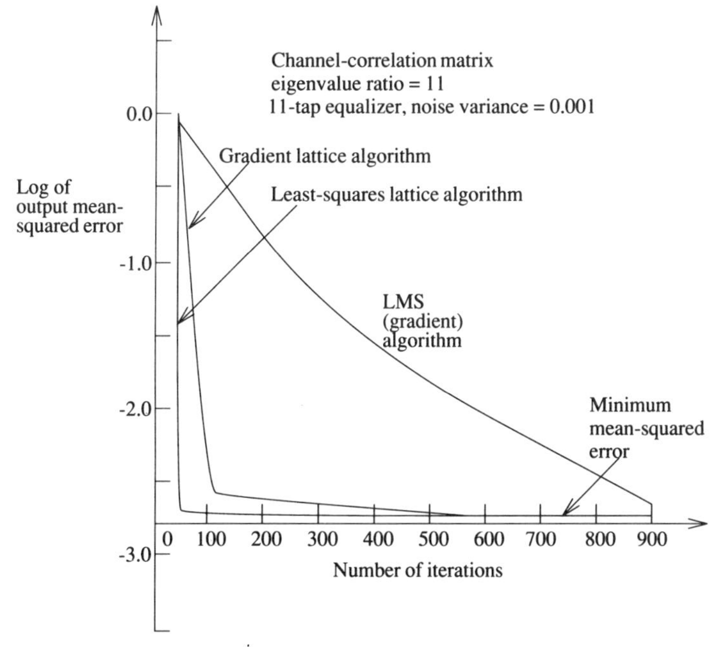

Fig. 8.23b Comparison of the convergence rate of lattice equalizers and the LMS algorithm for a channel with an eigenvalue ratio of 11.

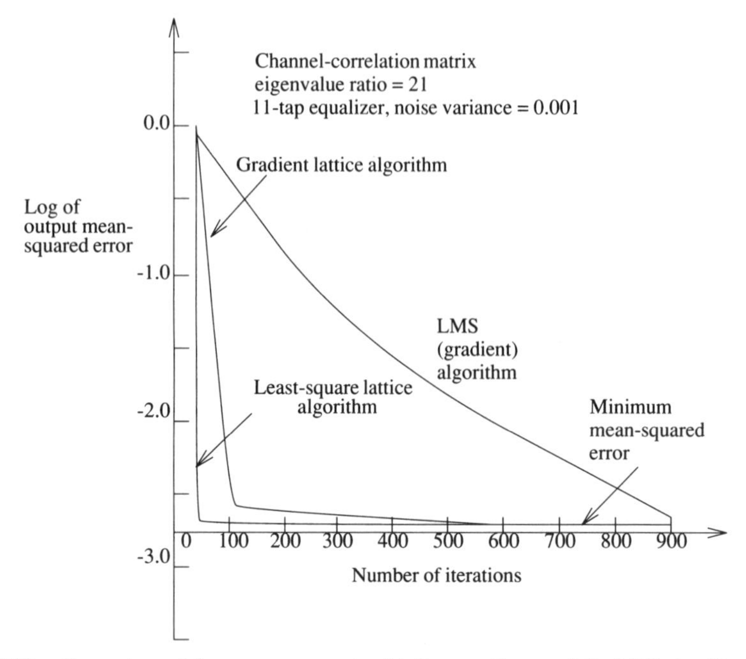

Fig. 8.23c Comparison of the convergence rate of lattice equalizers and the LMS algorithm for a channel with an eigenvalue ratio of 21.

8.8 TRACKING PROPERTIES OF THE LMS AND THE RECURSIVE LEAST-SQUARES ALGORITHMS

All adaptive systems experience a degradation in tracking nonstationary channels, as a result of the conflict inherent in choosing a step size (in the LMS algorithm) or forgetting factor (in the RLS algorithm) that strikes a balance between a small steady-state error (narrow bandwidth) and rapid tracking (large bandwidth). Eleftheriou and Falconer [21] have compared the performance of the LMS and RLS algorithms in the presence of a time-varying channels. Under some simple, but reasonable, assumptions of Markovian channel evolution, they have shown that when the step size in the LMS algorithm and the forgetting factor in the RLS algorithm are properly chosen, the two adaptive filters have the *same* tracking behavior. The details are provided in Appendix 8D. These results are important in that they are yet another example of the power and robustness of the LMS algorithm.

8.9 COMPLEXITY COMPARISON

We conclude our discussion on fast-converging algorithms with a summary of the complexity of the various options. It is hard to give a precise account of

Table 8.1: Complexity Comparison of LMS and Kalman Algorithm

Number of Operations per Iteration		
Algorithm	T Equalizer	$T/2$ Equalizer
LMS (TDL)	$2N$*	$4N$
Kalman (TDL)	$2N^2+5N$	$8N^2+10N$
Fast Kalman (TDL)	$7N+14$	$24N+45$
RLS Lattice	$15N-11$	$46N$

* N is the number of tap weight being adjusted.

complexity of the various RLS algorithms because of their many realizations. In Table 8.1 [22], we list the number of multiplies-plus-add operations required per iteration for synchronous and fractionally spaced equalizers. Note that the Fast Kalman is the most efficient type of RLS algorithm, and comparable in complexity to the LMS algorithm. Advocates of the RLS lattice algorithm say that the added complexity is worth the reported better numerical stability properties.

8.10 CYCLIC EQUALIZATION

There is another aspect of equalizer convergence, that we wish to explore in some detail. During an initial training period when the eye pattern may be closed, automatic equalization with an ideal reference, as illustrated in Figure 8.18, may be necessary for rapid convergence or indeed to guarantee convergence at all. There is, however, a difficulty, which is illustrated in Figure 8.24, in synchronization of the ideal reference data stream with the tap voltage samples in the equalizer. If the channel delay is not accurately known, the reference

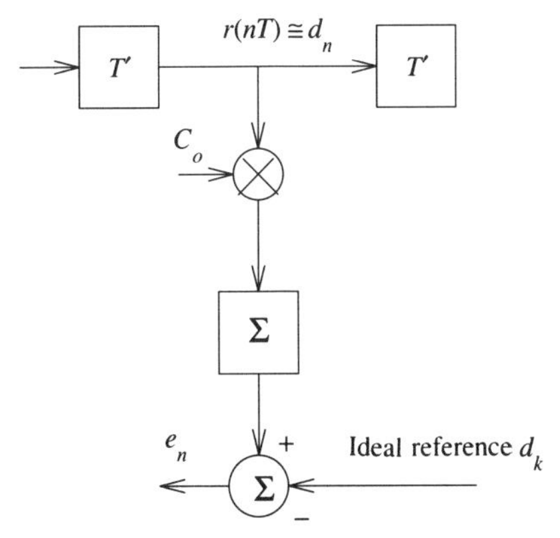

Fig. 8.24 Synchronizing a reference data train and the center tap voltage using a periodic stored reference. The problem is to align the indices k and n.

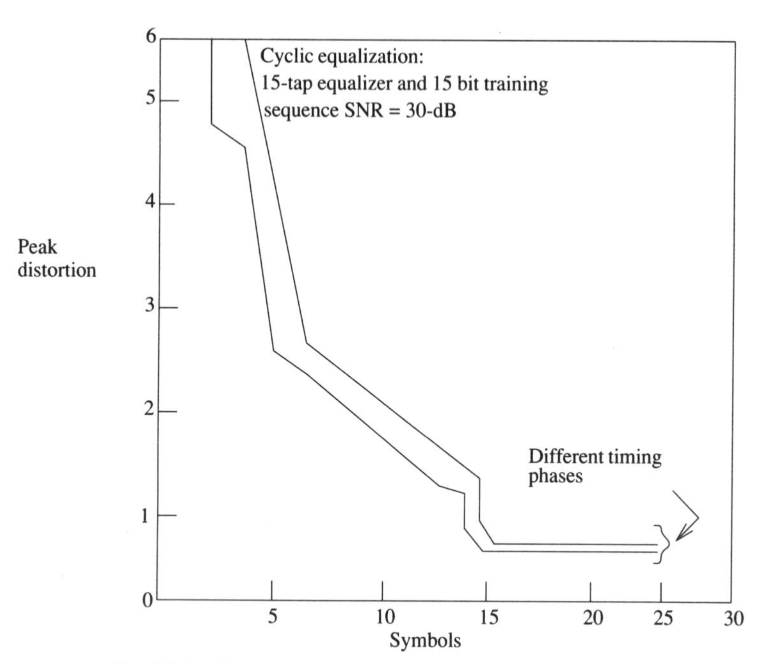

Fig. 8.25 Start-up behavior with cyclic equalization (from [23]).

level at time kT may not relate to any of the elements of the tap voltage vector. That is, the reference level may correspond to a part of the incoming signal that either has not arrived yet or has already passed through the equalizer. Ideally, of course, the reference level corresponds to a voltage at or near the center tap of the equalizer.

A solution to this problem, devised by Mueller and Spaulding [23], is to transmit a repetitive (which is stored at the receiver) reference data sequence of length equal to the equalizer, so that the current reference data level d_k is highly correlated with one or more tap voltages somewhere in the equalizer. Once the equalizer has converged, via the LMS algorithm, a cyclic shift of the tap weights will place the larger tap weights near the center of the equalizer. With proper selection of a reference data sequence, the algorithm can be shown to converge in a number of symbol intervals equal to the length in symbol intervals of the equalizer, as illustrated in Figure 8.25. In contrast to the convergence of the Kalman algorithm, cyclic equalization generally does *not* find the optimum tap setting in $2M+1$ iterations. After $2M+1$ iterations, the taps will be such that they equalize the channel only at the discrete frequencies contained in the input line spectra.

An appropriate reference sequence, for an equalizer with length equal to 2^n-1 symbol intervals, is a shift-register-generated pseudorandom (PR)

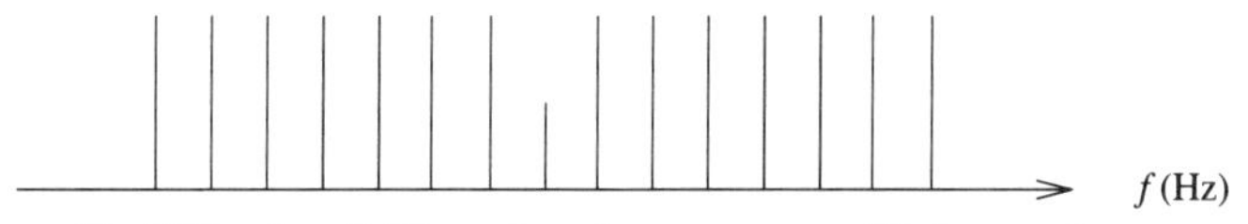

Fig. 8.26 Length 15 pseudorandom sequence and its line spectrum.

sequence of length $2^n - 1$. A PR sequence has a line spectrum (Figure 8.26) with equal amplitudes at uniform frequency intervals, except for a smaller line at $f = 0$. As noted above, use of a PR sequence in cyclic equalization will produce an end-to-end signal spectrum which satisfies the Nyquist criterion at $2^n - 1$ frequencies.

Equalization is possible in the time generally equivalent to the length of the equalizer (and the repetitive data sequence) because the tap weights are the solution of a set of simultaneous period linear equations generated over that period. An example of a three-tap synchronous equalizer is given in Figure 8.27, where the input samples are cycled back through the equalizer and the ideal reference is stored at the receiver. Note that the tap voltages have the same period, $3T$, as the cyclic data sequence. The three simultaneous equations for the equalizer output samples at sample times t_0, $t_0 + T$, and $t_0 + 2T$ are

$$y_1 = y(t_0) = r_1 c_1 + r_2 c_2 + r_3 c_3$$
$$y_2 = y(t_0 + T) = r_3 c_1 + r_1 c_2 + r_2 c_3$$
$$y_3 = y(t_0 + 2T) = r_2 c_1 + r_3 c_2 + r_1 c_3$$

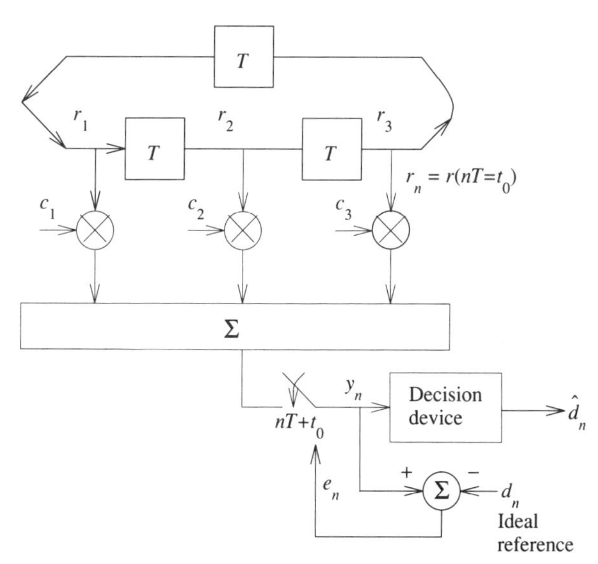

Fig. 8.27 Cyclic equalization for 3-tap synchronous equalizer and repetitive data sequence of length 3. The ideal reference is used to form the error sequence.

or

$$y_n = \sum_{m=1}^{3} c_m \, r_{n-m}, \qquad n = 1, 2, 3 .$$

The above convolution is a *circular* convolution, in that the input and output are periodic with a period of three symbols. This means that the channel correlation matrix will be *circulant* (each row is a circular shift of another row) in addition to being Toeplitz. In [23] it is shown that when the channel is Nyquist, with timing offset from the optimum epoch, then perfect equalization is obtained after a number of updates equal to the number of equalizer weights. More generally, it has been observed in practice with non-Nyquist channels that the well-controlled correlation properties of the periodic training sequences tend to reduce the equalizer convergence time by at least a factor of two, when compared to the settling time in the presence of random data. In addition, approximate values of the optimum weights can be achieved in a number of iterations equal to the equalizer length. Refer to the example shown in Figure 8.25 for a 15-tap synchronous equalizer. The space between the curves is for various sampling phases within a symbol interval. With the eye initially closed, 15 iterations are sufficient to open the eye (so that the system can switch to decision-directed operation as well as provide synchronization).

8.11 ZERO-FORCING EQUALIZATION

The original work of R. W. Lucky [3] was directed toward zero-forcing automatic and adaptive equalization. While zero-forcing has been superceded by the LMS approach, for those applications where the speed requirements are modest enough so that the LMS algorithm may be implemented, there may be future applications, say in Gbps high-speed fiber optic systems, where the simplicity of zero-forcing may make it the system of choice (see Chapter 10). For a tapped delay line equalizer, the goal of zero-forcing equalization is to produce an equalized pulse response that is Nyquist. Referring to Figure 8.3, a zero-forcing system will strive to produce a Nyquist equalized pulse, x_n,

$$x_n = \sum_m c_m \, h_{n-m} \tag{8.104a}$$

in terms of the equalizer coefficients, c_n, and the overall pulse presented to the equalizer, h_n. For a finite length equalizer, (8.104a) is an underdetermined system, since the equation(s)

$$\sum_{m=-M}^{M} c_m \, h_{n-m} = x_n = \begin{cases} 1, & n = 0 \\ 0, & n \neq 0 \end{cases} \tag{8.104b}$$

have $2M+1$ unknowns. To get a system of equations with a unique solution,

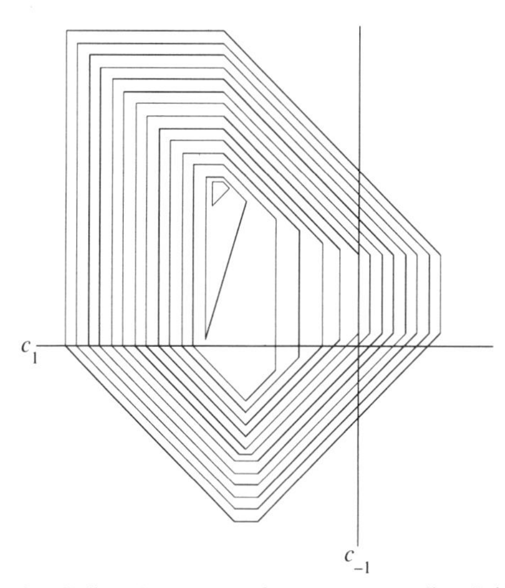

Fig. 8.28 Equal peak distortion contours for a two-tap equalizer (with an open eye).

Lucky [1] proposed solving the *finite* set of equations,

$$\sum_{m=-M}^{M} c_m\, h_{n-m} = \begin{cases} 1, & n = 0 \\ 0, & -M \le n \le M, \quad n \ne 0, \end{cases} \tag{8.105a}$$

which forces the center $\pm M$ samples of the equalized pulse to zero (while keeping the center sample at unity). Lucky showed that if the unequalized *eye* is open (see Chapter 4), then the solution of (8.105a) minimizes the *peak distortion* defined in Chapter 4, and the contours of peak distortion are the piecewise linear regions shown in Figure 8.28. On the other hand, if the eye is closed, then forcing the center $2M$ samples to zero (excluding the main sample) is no longer guaranteed to minimize the peak distortion. In this case, zero-forcing may produce large equalized pulse samples, x_n, for $|n|>M$. Another drawback of zero-forcing is that the effect of the equalizer on the output noise is not accounted for; consequently, if there is a null in the channel, the zero-forcing equalizer will most certainly amplify the noise appearing in the frequency band near the null.

The drawbacks of zero-forcing (needing an open eye to minimize peak distortion and possible noise enhancement) may be outweighed by the need to implement a *multiplication-free* equalization algorithm. Microwave radio and fiber optic systems, which have extremely high data rates (implying expensive and bulky signal processing), frequently have low-noise environments that make zero-forcing an attractive system because of its low complexity.

To generate an algorithm that will only stop when the zero-forcing condition,

$$x_n = \begin{cases} 1, & n = 0 \\ 0, & -M \le n \le M, \quad n \ne 0 \end{cases}, \tag{8.105b}$$

is achieved we observe that the correlation of the error sample with delayed data decisions is

$$\overline{a_{n-m} e_n} = \begin{cases} h_o - \overline{a^2}, & n = m \\ h_m, & n \ne m, \end{cases} \quad -M \le m \le M \tag{8.106}$$

where the error signal, e_n, has been defined in (8.13). If the data power, $\overline{a^2}$, is normalized to unity, then (8.106) gives a measure of how close h_m is to unity for the main sample and how far from zero the remaining samples are. From our discussion of adaptive algorithms, we realize that the *zero-forcing algorithm*

$$\mathbf{c}_{n+1} = \mathbf{c}_n - \beta\, e_n\, \mathbf{a}_n \tag{8.107}$$

will only stop when condition (8.106) is satisfied. In (8.107), the vector $\mathbf{a}_n$ is comprised of the $2M+1$ current data symbols. As shown in Figure 8.29, the ith

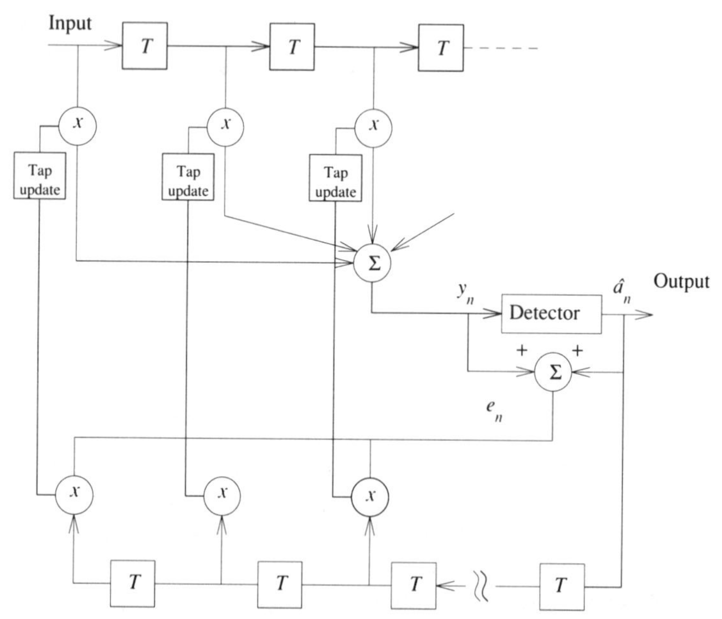

Fig. 8.29 An adaptive zero-forcing equalizer. Note that the error is correlated with the data decisions.

equalizer tap is updated by correlating the error signal with the ith data symbol offset from the reference. Note that zero-forcing is multiplication free since the common error signal can be shifted and added in response to the discrete-valued data symbols. Also observe that the correction term is in the direction of the data vector $\mathbf{a}_n$; recall that the LMS algorithm makes a correction in the direction of the signal vector at the input to the delay line, $\mathbf{r}_n$. As expected, the tap weights produced by the two algorithms will be different (unless we have an approximately noiseless, distortionless channel so that $\mathbf{r}_n = \mathbf{a}_n$).

8.12 PASSBAND EQUALIZATION

In Chapter 5 we showed two canonical structures for implementing carrier recovery baseband and passband equalization. These structures are shown in Figure 5.14, which is repeated as Figure 8.30. From one viewpoint, the two structures are equivalent if demodulation is performed with a *fixed* phase angle since equalization and demodulation are interchangeable. However, if the demodulator uses a data-directed phase-locked loop of the type described in Chapter 6, then there is a preferred arrangement for the equalizer and the PLL. The salient factor is the different value of delay in the PLL control loop depending upon which configuration is chosen. This is critical since in general, the phase jitter varies much more rapidly than the channel impulse response: thus, the PLL needs to be a relatively quick-reacting system, compared to the equalizer. A PLL output has two components: $f_c nT$ (which is the predictable phase update) and $\hat{\theta}_n$ (which is the estimate of the jitter including any difference between the receiver and transmitter carrier frequencies). The routine part of the

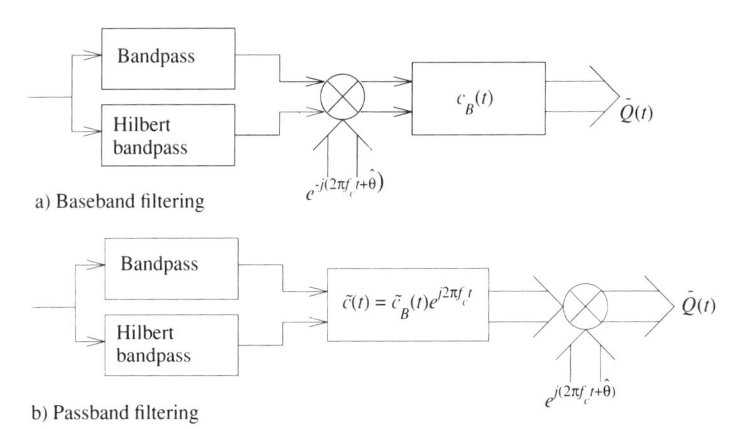

Fig. 8.30 Equivalent receiver structures with demodulation either (a) preceding or (b) following equalization.

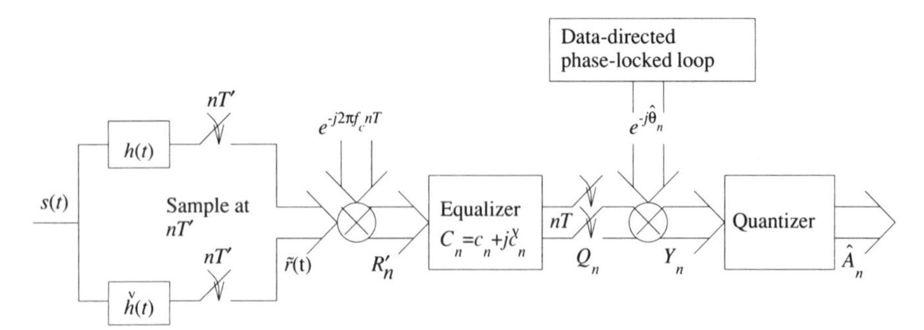

Fig. 8.31 An implementation equivalent to Figure 8.30b.

demodulator that removes the accumulating phase $f_c nT$ can, of course, precede or come after the equalizer, since it is the delay in producing $\hat{\theta}_n$, the time-varying jitter, which is the focus of this discussion. An implementation equivalent to Figure 8.30b is shown in Figure 8.31 where the removal of the $f_c nT$ phase contribution precedes the equalizer. Note that equalization is after "demodulation" by $\exp[-j2\pi f_c nT]$ but before the relevant demodulation by $\exp[-j\hat{\theta}_n]$. This arrangement is referred to as *passband equalization.* Suppose the channel is distortionless and the PLL precedes the equalizer in the *baseband* equalizer arrangement. Under these circumstances, the equalizer will have a unity center tap and all the other taps will be zero; from the viewpoint of the PLL updating loop, the equalizer appears as a *delay* of half the span of the equalizer (measured in symbol periods). As we will show in Chapter 10, a data-directed PLL operating in the presence of a delay of *MT* seconds, will have to *reduce* its step-size (or bandwidth) by a factor of *M*. Consequently, the tracking ability of the loop will be substantially impaired and for a system with 2400 updates per second, the frequencies of jitter that can be tracked may be < 5 Hz (jitter is typically in the 20 to 180 Hz range). In the *passband* equalizer arrangement, the equalizer precedes carrier tracking and the delay of the equalizer is *not* in the PLL control loop. Thus, passband equalization is the preferred arrangement when the receiver must track rapidly varying jitter.

We now consider the updating of the passband equalizer tap weights in conjunction with a data-directed PLL for QAM systems. Using notation similar to (8.4), in terms of the transmitted QAM data symbols, $A_n = a_n + j\check{a}_n$, we write the received signal at the output of the Hilbert filter shown in Figure 8.30.

$$\tilde{r}(t) = \sum_n A_n \, \tilde{x}_B(t-nT) \, e^{j(2\pi f_c t+\theta)} + \tilde{\gamma}(t) \, e^{j2\pi f_c t} \tag{8.4}$$

The signal, $\tilde{r}(t)$, can be synchronously or fractionally sampled, and the passband equalizer with a structure like that shown in Figure 8.2 produces, at the rate $1/T$, the equalized output signal

$$Q_n = q_n + j\check{q}_n = \sum_{k=-M}^{M} C^*_k R_{n-k} , \qquad (8.108a)$$

where

$$q_n = \sum_{k=-M}^{M} \left[c_k r_{n-k} + \check{c}_k \check{r}_{n-k} \right] \qquad (8.108b)$$

$$\check{q}_n = \sum_{k=-M}^{M} \left[c_k \check{r}_{n-k} - \check{c}_k r_{n-k} \right] \qquad (8.108c)$$

and the following complex and in-phase and quadrature signal identification have been made,

$$R_n = r_n + j\check{r}_n \qquad (8.108d)$$

$$C_n = c_n + j\check{c}_n . \qquad (8.108e)$$

Again, we remind the reader that the system variables are described by expressions involving complex quantities or by pairs of equations for the corresponding two real system variables. While $(q_n, \check{q}_n)$ and $(r_n, \check{r}_n)$ are both Hilbert transform pairs, the tap weights c_n and $\check{c}_n$ are in general not Hilbert pairs. The equalizer output is demodulated using the phase estimate $\hat{\theta}_n$ to form the demodulated sample

$$Y_n = y_n + j\check{y}_n = Q_n \exp\left[-j(2\pi f_c nT + \hat{\theta}_n) \right] , \qquad (8.108f)$$

which is quantized to produce the data decision, $\hat{A}_n$. In order to adjust the complex (in-phase and quadrature) equalizer tap weights, we introduce the *complex error* defined by

$$\varepsilon_n = Y_n - A_n = e_n + j\check{e}_n . \qquad (8.109)$$

To jointly optimize the equalizer tap coefficients and the demodulator phase, we minimize the average of the sum of the squared in-phase and quadrature errors, i.e., the mean-square complex error,

$$\langle |\varepsilon_n|^2 \rangle = \langle e_n^2 \rangle + \langle (\check{e}_n)^2 \rangle . \qquad (8.110)$$

As shown in Exercise 8.10, for most QAM signal constellations, we could minimize the squared in-phase error (or the squared quadrature error) by itself and obtain the same steady-state solution. A detailed examination comparing the averaging of $|\varepsilon_n|^2$ and e_n^2 reveals that, while the average is indeed the same, certain cross terms cancel algebraically in computing $\overline{|\varepsilon_n|^2}$ while they cancel as a result of statistical averaging in the computation of $\overline{e_n^2}$. This means that an adaptive algorithm minimizing $\overline{e_n^2}$, while simpler to implement, will take longer to converge since there will be larger fluctuations about the desired correction term.

8.13 JOINT OPTIMIZATION OF EQUALIZER TAP COEFFICIENTS AND DEMODULATION PHASE

This section is based on the fundamental work of D. D. Falconer [24] who first proposed a procedure for the efficient joint optimization of passband equalizers and demodulators. To elucidate the relationships governing the optimal tap vector $\mathbf{C}_n$ and demodulator phase $\hat{\theta}_n$ (both of which may be functions of time), we assume that successive data symbols are uncorrelated; i.e.,

$$\langle A_l A_m \rangle = 0 \qquad \text{all} \quad l \neq m \tag{8.111}$$
$$\langle A_l A^*_m \rangle = \langle |A|^2 \rangle \delta_{lm},$$

where δ_{lm} is the Kronecker delta function. Note that the cross correlation of the data symbols with the sampled phase-splitter outputs results in

$$\langle A^*_n R_n \rangle = \mathbf{X}_n \exp[j(2\pi f_c nT + \theta_n)] \langle |A|^2 \rangle , \tag{8.112}$$

where

$$\mathbf{X} \equiv \begin{bmatrix} X_{-M} \exp(-j2\pi f_c MT) \\ \cdot \\ \cdot \\ \cdot \\ X_M \exp(j2\pi f_c MT) \end{bmatrix} \tag{8.113}$$

is the complex impulse-response vector of the combination of the transmitter pulse filter and the channel, truncated to $2M + 1$ samples. The complex channel correlation matrix or A matrix is defined to be

$$A \equiv \frac{\langle R_n R^*_n \rangle}{\langle |A|^2 \rangle} . \tag{8.114}$$

This is a symmetric, complex (Hermitian) matrix ($A^* = A$) whose $(l - m)$th element is

$$A_{lm} = \sum_n \mathbf{X}_n \mathbf{X}^*_{n+m-l} \exp[j2\pi f_c (l - m)T] + \rho_{l-m} , \tag{8.115}$$

where $\{\rho_{l-m}\}$ is the noise autocorrelation and the asterisk denotes the transposed conjugate vector. As with the A matrix for real systems, it is positive semidefinite. (For any vector $\mathbf{u}$, $\mathbf{u}^* Au = \langle |\mathbf{u}^* R_n|^2 \rangle \geq 0$.)

We can rewrite the complex mean-square error ε_n in terms of A and $\mathbf{X}$, which are fundamental characteristics of the channel, as

$$\varepsilon_n = \{\mathbf{C}_n - A^{-1}\mathbf{X} \exp[-j(\theta_n - \hat{\theta}_n)]\}^* \cdot A\{\mathbf{C}_n - A^{-1}\mathbf{X} \exp[-j(\hat{\theta}_n - \theta_n)]\} + 1 - \mathbf{X}^* A^{-1} \mathbf{X} . \tag{8.116}$$

Because the matrix A is positive semidefinite, ε_n has the unique minimum

$$\varepsilon_{\min} = 1 - \mathbf{X}^* A^{-1} \mathbf{X} , \tag{8.117}$$

which is achieved when $\mathbf{C}_n$ and $\hat{\theta}_n$ satisfy

$$\mathbf{C}_n = \mathbf{C}_{n\,opt}(\hat{\theta}_n) \equiv A^{-1}\mathbf{X}\exp[-j(\hat{\theta}_n - \theta_n)] \,. \qquad (8.118)$$

Observe that the solution (8.118) is *not* unique; there is an infinitude of combinations $(\mathbf{C}_n, \hat{\theta}_n - \theta_n)$ that yield the minimum. However, for any specific choice of $\hat{\theta}_n$ (including zero), there is a unique optimum choice of C_n. Indeed, this is a manifestation of the *tap-rotation* property of the passband equalizer which was pointed out by Gitlin, Ho, and Mazo [25]. In particular, when there is no attempt to estimate θ_n ($\hat{\theta}_n = 0$), then any amount of frequency offset Δ ($\theta_n = 2\pi n\Delta T$) causes $\mathbf{C}_{n\,opt}$ to "rotate" with frequency Δ. However, a typical adaptive equalizer whose tap coefficients may not be permitted to change by more than about 1 percent from one symbol interval to the next (in order to provide a satisfactory steady-state mean-square error) will not be able simultaneously to equalize the channel effectively and to rotate $2\pi\Delta$ radians per symbol interval even for moderate amounts of frequency offset. Consequently, the equalizer taps could not be expected to track typical phase jitter components accurately.

In this section we focus on the joint operation of the adaptive equalizer and the data-directed PLL introduced in Chapter 6 (which removes the major burden of tracking from the slowly adapting equalizer). Assuming this separate phase-angle-tracking algorithm is successful so that the phase error $(\hat{\theta}_n - \theta_n)$ remains constant, we observe, that the mean-square error, using (8.118), is given by

$$\varepsilon_n \equiv \frac{1}{\langle |A|^2 \rangle} \langle |\mathbf{C}^*{}_n \mathbf{R}_n - A_n \exp[j(2\pi f_c nT + \hat{\theta}_n)]|^2 \rangle \,. \qquad (8.119)$$

From (8.119) observe that if the equalizer's reference signal, for the purpose of adapting its tap coefficients, is $\{A_n \exp[j(2\pi f_c nT + \hat{\theta}_n)]\}$, the reference signal rotates *in synchronism* with the frequency-offset and phase-jittered carrier of the received signal, and hence the equalizer tap coefficients do not have to rotate if $\hat{\theta}_n - \theta_n$ remains constant.

If the gradients of ε_n, with respect to the real tap coefficient vectors c_n and $\check{c}_n$, are denoted, respectively, by $\nabla\mathbf{c}_n\varepsilon_n$ and $\nabla\check{\mathbf{c}}_n\varepsilon_n$, and if we define the gradient with respect to $\mathbf{C}_n$ to be

$$\nabla_{\mathbf{C}_n}\varepsilon_n \equiv \nabla_{c_n}\varepsilon_n + j\nabla_{\check{c}_n}\varepsilon_n \,,$$

then the gradient of the right-hand side of (8.116) can be written

$$\nabla_{\mathbf{C}_n}\varepsilon_n = 2\{A\mathbf{C}_n - \mathbf{X}\exp[-j(\hat{\theta}_n - \theta_n)]\} \,. \qquad (8.120)$$

Since A is positive semidefinite, $\nabla\mathbf{c}_n\varepsilon_n = 0$ is a necessary and sufficient condition for ε_n to attain its minimum value. If the receiver knew $\mathbf{X}\exp(j\theta_n)$, and A, and could calculate this gradient during each symbol interval, then in the nth symbol interval it could use the steepest-descent (known gradient) algorithm to update its estimate of $\mathbf{C}_n$ as follows:

$$\mathbf{C}_{n+1} = \mathbf{C}_n - \frac{\beta}{2} \nabla_{\mathbf{C}_n} \varepsilon_n . \tag{8.121}$$

In this equation, $\mathbf{C}_n$ is the estimate of the correct tap coefficient vector in the nth symbol interval and $\beta/2$ is a positive constant. For the moment, we defer consideration of a LMS estimated-gradient algorithm that does not require knowledge of A and $\mathbf{X}$.

In Chapter 6 we have discussed a data-directed algorithm for providing the sequence $\{\theta_n\}$. The reasonable assumption that the carrier phase error varies slowly with time allows us to treat θ_n as a quasi-static parameter that can be estimated, in each symbol interval, from present and past received signal samples, $\mathbf{R}_n$, and data symbols, A_n. Following the data-directed PLL development in Chapter 6, the derivative of ε_n with respect to $\hat{\theta}_n$ is, for a fixed value of C_n, given by

$$\nabla_{\hat{\theta}_n} \varepsilon_n = -2 \text{ Im } \{\mathbf{C}^*{}_n \mathbf{X} \exp[-j(\hat{\theta}_n - \theta_n)]\} . \tag{8.122}$$

The estimate $\hat{\theta}_n$ is thus updated as

$$\hat{\theta}_{n+1} = \hat{\theta}_n - \frac{\alpha}{2} \nabla_{\hat{\theta}_n} \varepsilon_n , \tag{8.123}$$

where $\alpha/2$ is a constant. In general, α should be large relative to the equalizer's constant β, to ensure that the estimate $\hat{\theta}_n$ can closely track a varying angle θ_n, thereby obviating the need for the passband equalizer taps to follow it closely. Use of substantially different step sizes in the two loops will effectively decouple the systems, and let the PLL track jitter and the equalizer produce a satisfactorily small mean-square error.

Suppose the angle θ_n is not time-varying ($\theta_n = \theta$). Then the stationary points of the algorithms (8.121) and (8.123) are the solutions of the equations

$$\nabla_{\mathbf{c}} \varepsilon_n = 0$$

or

$$A\mathbf{C} = \mathbf{X} \exp[-j(\hat{\theta} - \theta)] , \tag{8.124}$$

and

$$\nabla_{\hat{\theta}} \varepsilon_n = 0$$

or

$$\text{Im } \{\mathbf{C}^* \mathbf{X} \exp[-j(\hat{\theta} - \theta)]\} = 0 . \tag{8.125}$$

It is easy to show from the Hermitian property of A that, if (8.124) is true, then (8.125) is true. Furthermore, A is positive semidefinite and, thus, expression (8.116) for the mean-square error shows that the infinite set of stationary points, defined by (8.124), are the only global minima.

The following question arises: Starting with fixed initial values, $\mathbf{C}_0$ and $\hat{\theta}_0$, and assuming $\theta_n = \theta$ for all n, do the gradient algorithms (8.121) and (8.123) jointly converge to a stationary point? Note that by defining

$$\mathbf{Z}_n = \begin{bmatrix} \mathbf{C}_n \\ \hat{\theta}_n \end{bmatrix}, \qquad P = \begin{bmatrix} \frac{\beta}{2} & 0 \\ 0 & \frac{\alpha}{2} \end{bmatrix}$$

and

$$\nabla_{\mathbf{Z}_n} \varepsilon_n = \begin{bmatrix} \nabla_{C_n} \varepsilon_n \\ \nabla_{\hat{\theta}_n} \varepsilon_n \end{bmatrix},$$

we can combine (8.121) and (8.123) by writing

$$\mathbf{Z}_{n+1} = \mathbf{Z}_n - P \nabla_{\mathbf{Z}_n} \varepsilon_n . \tag{8.126}$$

As we know, if α and β are chosen small enough, the sequence $\{Z_n\}$ converges, in the mean-square sense, to the point where $\nabla_{\mathbf{Z}} \ \varepsilon_n = 0$.

As with the baseband equalization, the receiver does not know, *a priori,* the channel correlation matrix A and the impulse response vector $\mathbf{X}$. We know that the receiver can use the LMS algorithm to approximate the gradient search algorithm by using the gradients with respect to $\mathbf{C}$ and $\hat{\theta}_n$ of the actual, magnitude-squared complex error

$$|E_n|^2 = |\mathbf{C}^*{}_n \mathbf{R}_n - \hat{A}_n \exp[j(2\pi f_c nT + \hat{\theta}_n)]|^2 . \tag{8.127}$$

The A_n used in this calculation is initially an ideal reference known to the receiver, and, during normal operation, it is the receiver's output decision $\hat{A}_n$ in the nth interval. Thus, a decision-directed stochastic approximation algorithm corresponding to (8.121) and (8.123) is

$$\mathbf{C}_{n+1} = \mathbf{C}_n - \frac{\beta}{\langle |A|^2 \rangle} \{\mathbf{R}_n \mathbf{R}^*{}_n \mathbf{C}_n - \hat{A}^*{}_n \mathbf{R}_n \exp[-j(2\pi f_c nT + \hat{\theta}_n)]\}$$

$$= \mathbf{C}_n - \frac{\beta}{\langle |A|^2 \rangle} \mathbf{R}_n (Q^*{}_n - \hat{Q}^*{}_n) , \tag{8.128a}$$

where $\hat{Q}_n = \hat{A}_n \exp[j(2\pi f_c nT + \hat{\theta}_n)]$ is the "rotated" reference for the equalizer in the nth interval, using the receiver's decision $\hat{A}_n$, and

$$\hat{\theta}_{n+1} = \hat{\theta}_n + \frac{\alpha}{|A_n|^2} \operatorname{Im} \{\mathbf{C}^*{}_n \mathbf{R}_n A^*{}_n \exp[-j(2\pi f_c nT + \hat{\theta}_n)]\}$$

$$= \hat{\theta}_n + \frac{\alpha}{|A_n|^2} \operatorname{Im} (Q_n \hat{Q}^*{}_n) , \tag{8.128b}$$

which can also be written as

$$\hat{\theta}_{n+1} = \hat{\theta}_n + \alpha / |A_n|^2 \text{ Im } \{Y_n \hat{A}^*_n\} .$$

Since the equalizer removes most of the intersymbol interference, (8.128) describes a discrete time first-order, phase-locked loop. The jointly adaptive passband equalizer and data-aided carrier recovery system is shown in Figure 8.32.

Note that there is a phase ambiguity in the receiver's decisions inherent in suppressed-carrier systems with symmetric signal constellations, using decision-directed phase tracking. For example, the QAM signal constellation is quadrantially symmetric, and, therefore, constant 90-degree errors in the phase of the receiver's decisions $\{\hat{A}_n\}$ are undetectable. This source of ambiguity is customarily removed by differentially encoding the transmitted data onto the points of the signal constellation, so that phase *differences* between successive decisions $\{\hat{A}_n\}$, rather than absolute phase values, convey information. Falconer [26] has shown that if the equalizer step size is small relative to the phase-locked loop step size, and if the equalizer taps are initialized from zero, the deterministic gradient algorithm is completely unaffected by the PLL. Under these circumstances, the equalizer and the PLL are essentially decoupled. Falconer also shows that in response to frequency offset, the PLL reduces the rate of rotation of the equalizer taps to $\beta/(\alpha + \beta)$ of the original frequency offset. Consequently, the degradation in system mean-square error due to frequency offset is typically quite small.

The two-dimensional adaptive receiver structure described above can be extended in a straightforward manner to systems employing decision feedback equalization (see Exercise 8.12).

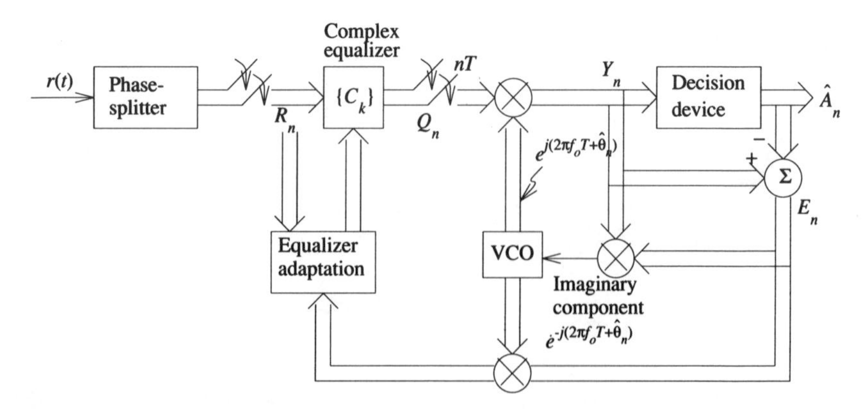

Fig. 8.32 Combined adaptive passband equalizer and data-aided carrier recovery.

8.14 ADAPTIVE CANCELLATION OF INTERSYMBOL INTERFERENCE [7,8]

It has been demonstrated, in our discussion of decision feedback equalization, that the assumption of correct decisions can lead to improved performance by canceling (or subtracting out) the effect of the postcursor ISI. In this section, we extend the DFE concept to the cancellation of precursor ISI to provide further performance improvement in the presence of severe slope distortion. The linear canceler is shown in Figure 8.33, where the linear equalizer provides moderately reliable, tentative decisions $\hat{a}_n$ that will increase the reliability of the final decisions, $\tilde{a}_n$. Assume for the moment that the filter W is a pure delay of dT seconds, then the sampled receiver input is

$$r(nT) = a_n x_0 + \sum_{m>n} a_m x_{n-m} + \sum_{m<n} a_m x_{n-m} + \upsilon_n \,, \tag{8.129}$$

where the first term is proportional to the desired data symbol, the second and third terms correspond to intersymbol interference (ISI) introduced by "future" and "past" data symbols, and the last term is the received noise sample. A DFE can be used to cancel the effect of the post-cursor ISI. Let us focus on canceling the precursor ISI; with a delay dT in the upper leg, the tentative decisions will be just the data symbols needed to form an estimate of the precursive ISI, $\sum_{m>n} \hat{a}_m x_{n-m}$. Referring to Figure 8.33, if the final decisions, $\tilde{a}_n$, are used in a DFE mode to form an estimate of the ISI, then the two estimates can be combined to produce

$$I_n = \sum_{m>n} \hat{a}_n x_{n-m} + \sum_{m<n} \tilde{a}_n x_{n-m} \,, \tag{8.130}$$

which is an estimate of the ISI. The tentative decisions can also be used to cancel the "causal" ISI, but is seems preferable to use the most reliable final (DFE) decisions for that purpose. If the tentative decisions have an error rate of 10^{-4}, then an isolated error will still allow ISI cancellation since there will be many correct decisions in the future delay line of Figure 8.33. Thus an output error rate of 10^{-7} is possible. Note that the canceler processes data symbols so that the noise cannot be enhanced by the canceler; the fractionally spaced filter, W, will perfectly compensate for the presence of channel delay distortion. If $\hat{a}_n$ and $\tilde{a}_n$ are correct and the filter W is a matched filter, as it will be under some general conditions, then subtracting the ISI estimate from the matched filter output produces a signal

$$r(nT) - I_n = a_n x_0 + \upsilon_n \tag{8.131}$$

which is equivalent to the reception of a single isolated pulse. Under these circumstances the matched-filter bound described in Chapter 4 would be achieved. Of course no receiver can do better than the maximum likelihood sequence detector (Viterbi algorithm), whose performance is determined by $d_{\min}^2$ (as per our discussion in Chapter 7).

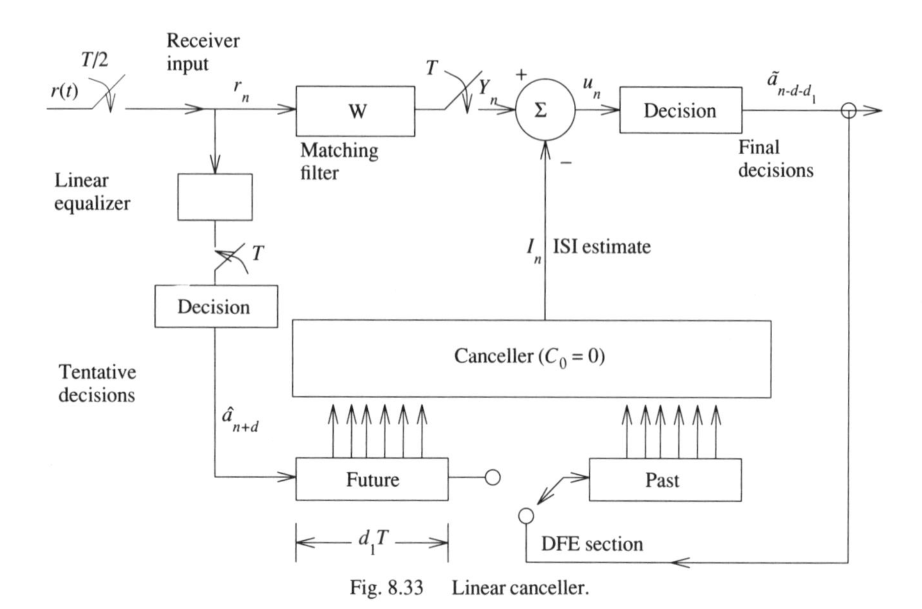

Fig. 8.33 Linear canceller.

It can be shown, under fairly general conditions, that the filter W is indeed a matched filter (see Exercise 8.13). Recall that a matched filter maximizes the signal-to-noise ratio at the sampling instant, and is the optimum receiver in the absence of ISI; since the canceler filter, C, strives to remove the residual ISI appearing at the output of the matched filter, W should become a matched filter if cancellation is successful. Recall that the matched filter also removes the channel phase characteristics without a noise penalty — thus enabling the canceler to emulate the fractionally spaced equalizer's ability to perfectly compensate for delay distortion. Before providing some simulated linear canceler performance results, we describe canceler adaptation. The linear equalizer is adapted independently of W and C by using the error signal associated with its own output.

The matching filter and the canceler are adapted based on the final error signal

$$\tilde{e}_n = u_n - a_n = y_n - I_n - a_n$$

$$= \mathbf{w}' \, \mathbf{r}_n - \mathbf{c}'^{1} \, \hat{\mathbf{a}}_n - \mathbf{c}'^{2} \, \tilde{\mathbf{a}}_n - a_n \tag{8.132}$$

where $\mathbf{c}_1$ and $\mathbf{c}_2$ are the tap weights associated with the left- and right-hand sections of the canceler, respectively (as noted before, $\mathbf{c}_2$ could be set to zero if the delay is increased to allow the delay line to span the total range of precursor and postcursor ISI). Note that the center tap of the canceler, c_0, must be set to zero, to prevent the canceler from making use of the current data symbol. Applying

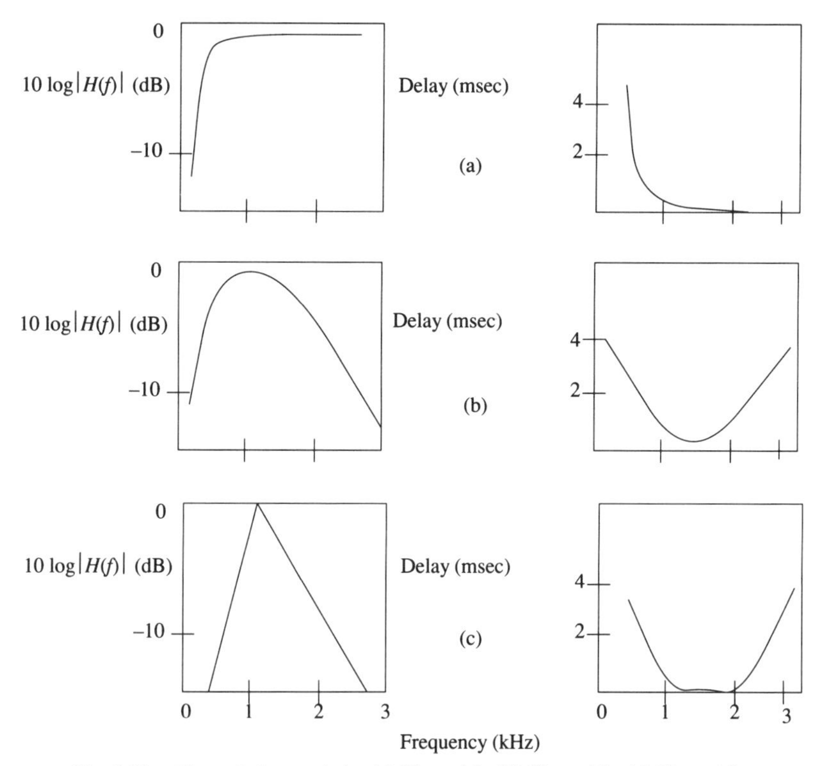

Fig. 8.34 Channel characteristics (a) Channel 1. (b) Channel 2. (c) Channel 3.

the LMS philosophy, to minimize $(\tilde{e}_n)^2$, directly generates the tap-update relations (see Exercise 8.14).

The performance of the linear canceler will be evaluated over the three channels shown in Figure 8.34. Channel 1 has a flat amplitude response over the entire voiceband frequency with little delay distortion, except near the lower band edge. Channel 2 has moderate amplitude and delay distortion, which just meets the private-line conditioning, and Channel 3 has several amplitude distortion. Pseudo-random digital data at levels ±1 and ±3 are used to modulate the 16-point quadrature amplitude modulation (QAM) transmitter at a rate of 9.6 kbps and the resulting signal is sent over one of the channels. Gaussian noise is introduced into the channel. The input signal-to-channel noise ratio (SNR_i), which will be taken to be either 30, 24, or 20 dB, represents an upper bound on the attainable output SNR.

In the simulations, the canceller filters $\{C_k\}$ and $\{W_k\}$ are allowed to start adapting after 1000 iterations to ensure that the linear equalizer (LE) has con-

Table 8.2: Output SNR for Linear Equalizer, Decision Feedback Equalizer, and Canceller

	Channel 2			Channel 3		
Input SNR (dB)	20	24	30	20	24	30
Linear equalizer	18.0	22.1	26.8	15.4	20.4	22.5
Decision feedback equalizer	19.1	23.0	27.9	17.3	22.0	27.1
Canceller	19.7	23.2	28.5	18.4	23.7	29.2

verged sufficiently and is able to provide good tentative decisions for the canceller. The performance of the linear canceller (LC) is measured by the output signal-to-noise ratio (SNR_0). This is defined as the ratio of the average data symbol power to the output noise power, which is taken to be the time average of the squared error at the LC output. In the simulations, both LC filters span 31 data symbols so that the matching filter has 62 $T/2$ taps, while the canceler has 31 taps. The LE has 64 $T/2$ taps to sufficiently span the channel-impulse response. An initial step size of 0.0005 is chosen for updating all the tap coefficients and it is reduced to 0.00005, when the taps have nearly converged, to get a small final mean-square error.

Table 8.2 summarizes the results of several simulation runs, for the LE, DFE, and LC, over Channels 2 and 3 and with various SNR_i. Results for Channel 1 are not presented because all three schemes do not suffer any significant degradation. These results are based on the use of an ideal carrier demodulating phase. Note that for Channel 2, the LC is able to attain signal-to-noise ratios which are very close to SNR_i, for all three cases. The gain is about 2 dB over the LE and about 1 dB over the DFE. If we can use the rule of thumb that every 1 dB gain corresponds to an order of magnitude reduction in error rate, then the LC would have an error rate two orders of magnitude less than the LE and one order less than the DFE. The results for Channel 3, again, show the improvement over both LE and DFE. We see that the LE performance degrades significantly when the channel has considerable slope distortion, and on the average, it suffers losses of about 4 dB from the input SNR. The LC shows improvements of approximately 3 dB over the LE and 1 dB over the DFE in all three cases.

While the above results show substantial improvement in the output SNR with a linear canceler, the real merit of the scheme is improvement in the error rate. A skeptic might claim that, since the analysis is based on correct tentative decisions, why go any further — use the tentative decisions as the final ones. While the above simulations were short enough so that the linear equalizer made no errors, it is certainly clear that the linear canceler provides a substantial margin against noise. Furthermore, in an extension of the canceler concept to the mitigation of nonlinear distortion [27], the linear equalizer made numerous errors which were removed by the nonlinear canceler.

8.15 BLIND EQUALIZATION

There are several applications in digital data communications when start-up and retraining of an adaptive equalizer has to be accomplished *without* the aid of a training sequence. Hence, the system has to be trained "blind." Of course, if the eye is open, decision-directed operation should work fine. We are interested in those circumstances where the eye is closed, and the conventional decision-directed operation will fail. Mazo [28] has shown that decision-directed operation can result in a multimodal error surface, and that reliable startup can be difficult to achieve. With this background, consider a multipoint network (as described in Figure 1.13). A blind retrain is needed if a channel from the master to one of the tributary stations goes down at any time following the initial training period, and it is desired to retrain only the corresponding tributary. Of course, if the portion of the communication line common to all tributaries fails, then it would be best to retrain all of the tributaries in the conventional manner. While periodically storing the equalizer taps will provide a solution, if the line drops momentarily so as not to change its transfer function in any appreciable way, we are interested in those circumstances when there is no alternative to blind equalization. Apart from the multipoint system, blind equalization is often required due to severe fading in digital microwave links where a reverse channel (to call for a retrain) is not available and in transmission monitoring, where a training sequence is not supplied for the benefit of the monitoring receiver. What is needed then in blind equalization applications is a bootstrapping algorithm that brings the receiver's adaptive equalizer to the point where it can reliably switch to a decision-directed algorithm. It is recognized that, in exchange for not requiring data decisions, blind equalization algorithms may require one to two more orders of magnitudes of time to converge. There are two basic algorithms for blind equalization: the constant modulus algorithm (CMA) [29] and the reduced constellation algorithm (RCA) [30,31]. Both the CMA and the RCA algorithms measure a property of the highly distorted signal, where the property (e.g., average power in the constellation) is sensitive to changes in the equalizer tap weights; restoring the property to its desired value is equivalent to restoring the equalizer to a set of tap values for which the eye is open. The parameters governing blind equalization are compared, in Table 8.3, with ideal reference and decision-directed operation.

8.15.1 CONSTANT MODULUS ALGORITHM (CMA)

Referring to Figure 8.35, where q_n denotes the complex adaptive equalizer output during the nth symbol period, CMA equalizes the QAM signal constellation by finding the set of tap weights that solve the following optimization problem:

$$\underset{\mathbf{c}}{\text{Minimize}}\; E\left[\left(|q(n)|^2 - R^2\right)^2\right], \tag{8.133}$$

Table 8.3: Summary of Adaptive Equalization Algorithms

Type	Usage	Speed of Convergence	Step Size	Most Suitable P_e Range	Desired Response
Ideal Reference	Start-up	Fast	Large	$10^{-2} < P_e < .5$	Training sequence $d_k = a_k$
Decision Directed	Steady state	Medium	Small	$P_e < 10^{-2}$	Quantized equalizer output $d_k = \hat{a}_k$
Blind	Start-up or restart	Slow	Medium	$10^{-2} < P_e < .5$	Signs or phase of equalizer output

where R is a positive real constant which can be interpreted as the radius shown in Figure 8.35. The motivation for the above cost function is that the difference between the modulus of the signal, $|q|^2$, and a reference level, R^2, characterizes the amount of intersymbol interface (ISI) plus noise at the equalizer output independently of the carrier phase and signal constellation. This is evident by noting that (8.133) is unchanged by substituting $q(n)\, e^{j\phi}$ for q_n, where ϕ is an arbitrary phase shift. The demodulated equalizer output could also have been substituted for $q(n)$. A change in signal constellation will not affect the ultimate convergence of an adaptive algorithm based on (8.133) provided that the constellation *kurtosis*, κ, defined as

$$\kappa \triangleq \frac{E[|a(n)|^4]}{\left(E[|a(n)|^2]\right)^2}$$

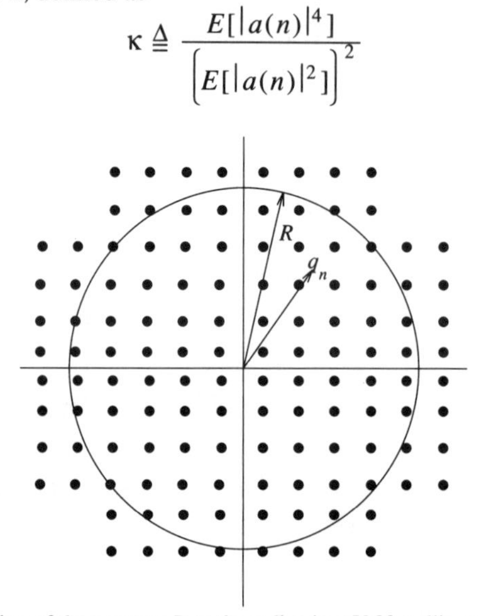

Fig. 8.35 Interpretation of the constant R as the radius in a V.33 trellis-coded 128-QAM constant modulus algorithm. The received signal sample is denoted by q_n.

is less than 2 and some other basic symmetry conditions are satisfied. The constant R is chosen subject to the constraint that the average gradient of the cost function, with respect to the tap vector, is zero when the channel is perfectly equalized, i.e., the gradient is unbiased. Under this constraint, R^2 can be shown to be given by the following formula:

$$R^2 = \frac{E[|a(n)|^4]}{E[|a(n)^2]} \quad . \tag{8.134}$$

For 128-QAM, $R^2 = 110$. If R^2 was chosen to be other than this value, CMA would converge to a 128-QAM constellation with signal point levels linearly scaled to satisfy (8.134). A less-than-optimum value of R^2 may cause instabilities when switching to the decision-directed mode.

By taking α times the negative gradient of the cost function with respect to the tap vector, substituting instantaneous values for time-averaged ones, and then adding the resulting expression to the current tap vector, it is straightforward to derive the CMA tap update recursion as

$$\mathbf{c}(n+1) = \mathbf{c}(n) - \alpha q(n) \left[|q(n)|^2 - R^2 \right] \mathbf{r}(n) \tag{8.135}$$

where α is the positive real step size which controls convergence time and steady state excess *MSE*, and $\mathbf{r}(n)$ is the vector appearing in the equalizer delay line during the nth output symbol period. When the CMA tap update recursion converges, typically we would switch to the familiar decision-directed mode, by means of the tap update recursion

$$\mathbf{c}(n+1) = \mathbf{c}(n) - \alpha \left[y(n) - \hat{a}(n) \right] e^{j\hat{\theta}(n)} \mathbf{r}(n) \quad . \tag{8.136}$$

Apart from the intuitive reasoning that the modulus of a QAM signal is a reasonable indicator of the degree of distortion present in the equalized signal, Godard [30] has shown that the set of equalizer tap weights that minimize (8.133) is close to the set of weights that minimize $E[|q_n|^2 - |a_n|^2]$, a quantity which depends on the data symbols a_n.

8.15.2 REDUCED CONSTELLATION ALGORITHM

Consider a conventional 16-point QAM system, where the receiver detects the 4 bits per symbol. In the blind-equalization environment, the channel has degraded so that the output cannot be resolved to 4 bits. Recognizing this situation, the reduced constellation algorithm (RCA) attempts to resolve the output to 2 bits, as in Figure 8.36. For the purpose of blind equalization, the reduced constellation will be one of the four points $\hat{b} = (\pm R, \pm R)$, and the error signal, $\tilde{e}_n$, will be the difference between the received point, q_n, and the closest reduced-constellation reference point $\hat{b} = (\pm R, \pm R)$.

The reduced constellation algorithm (RCA) equalizes the QAM signal constellation by finding those tap weights that solve the following optimization

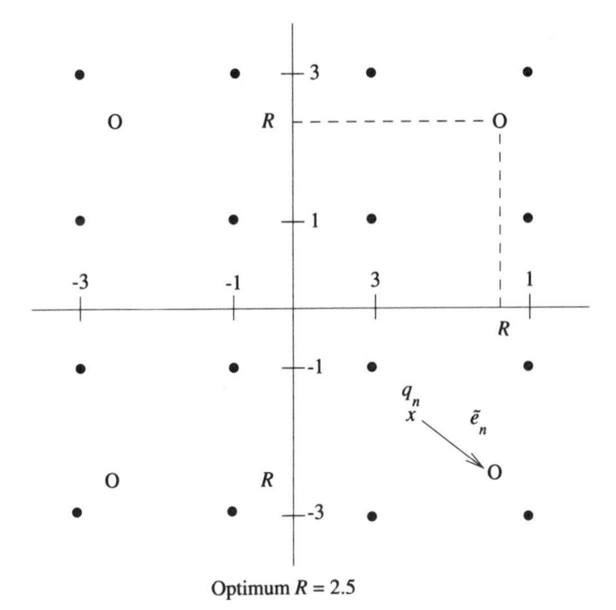

Fig. 8.36 16-point QAM used for blind equalization experiment.

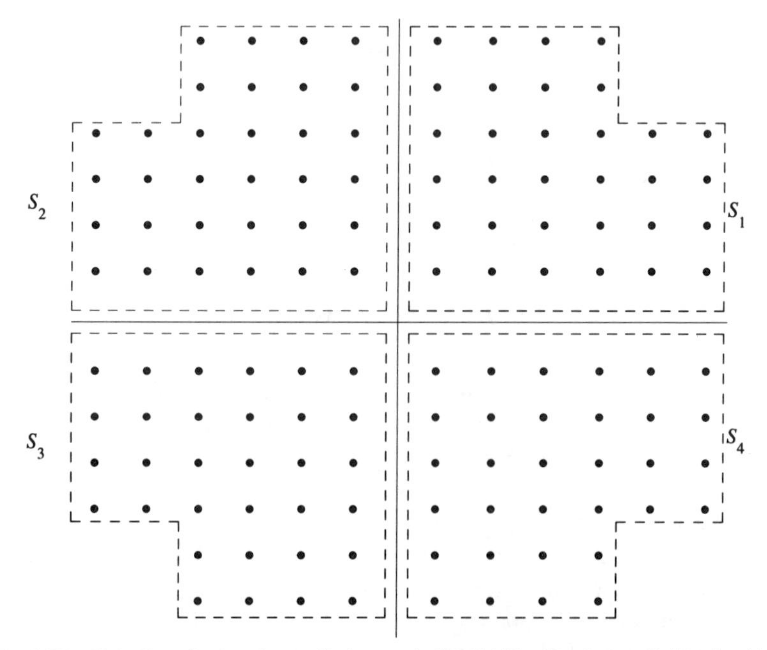

Fig. 8.37 Selection of reduced constellation sets in 128-QAM reduced constellation algorithm.

problem:

$$\underset{\mathbf{c}}{\text{Minimize}} \quad E\left[\left|q(n) - \hat{b}(n)\right|^2\right] \tag{8.137}$$

where $\hat{b}(n)$ is the reduced constellation point closest to $q(n)$. Note the similarity of this optimization to the conventional decision-directed problem. The motivation for the above cost function is that it characterizes the amount of ISI plus noise at the equalizer output, and is more likely to converge from a "cold start" than the decision-directed algorithm. However, this cost function is neither independent of carrier phase nor signal constellation.

The reduced constellation is formed in the following manner:

1. Heuristically select L sets of points $\{S_i\}_{i=1}^{L}$ from the full constellation. The selection is generally done in such a way as to balance rapid and reliable convergence with ease of implementation. A typical selection for 128-QAM is shown in Figure 8.37.

2. Form the reduced constellation using *one* point per set, such that the average gradient of the cost function with respect to the tap vector is zero when the channel is perfectly equalized. Letting the subscript l denote the set S_l, the subscript k the constellation point a_k within the lth set, complex conjugate, and $\in$ "is a member of," the resulting formula for the reduced constellation is

$$b_l = \frac{\sum_{k \in S_l} |a_k|^2}{\left[\sum_{k \in S_l} a_k\right]^*} \quad , \quad l = 1,2,\ldots,L \quad . \tag{8.138}$$

The 128-QAM reduced constellation, as partitioned in Figure 8.37, would be the four points $b_l = \pm 7.45 \pm j7.45$, and for 16-QAM, $b_l = \pm 2.5 \pm 2.5$. As with CMA, if the b_l $(l = 1,2,3,4)$ were chose to be other than those values based on the signal point levels $\{\pm 1, \pm 3, \pm 5, \pm 7, \pm 9, \pm 11\}$, RCA would converge to a 128-QAM constellation with signal point levels linearly scaled to satisfy (8.138). Again, this may cause instabilities when switching to decision-directed mode.

By taking α times the negative gradient of the cost function with respect to the tap vector, substituting instantaneous quantities for time-averaged ones, and adding the resulting expression to the current tap vector, the RCA tap update recursion becomes

$$\mathbf{c}(n+1) = \mathbf{c}(n) - \alpha \left[q(n) - \hat{b}(n)\right] e^{j\hat{\theta}(n)} \, \mathbf{x}(n) \quad . \tag{8.139}$$

As with CMA, when this algorithm converges, we switch to the decision-directed recursion. Typically convergence takes more than an order of magnitude, more time than LMS convergence with a training sequence. The RCA

Table 8.4: Adaptive Equalization Structures

Type	Structure	Channel Energy Used[a]	Comments
Linear synchronous	Tapped delay line (TDL)	h_0^2	• Standard till mid-70s
Linear fractionally spaced	Tapped delay line	h_0^2	• Current standard • Optimum linear receiver • Performance independent of delay dispersion
Decision feedback (DFE)	Nonlinear feedback TDL	$\sum_{n \geq 0}^{N} h_n^2$	• Improves performance with slope distortion • Can be combined with linear equalizer • Possibility of error propagation • Used for ISDN
Linear canceller	Three TDL's	$\sum_{n=-M}^{N} h_n^2$	• Uses "M" tentative "future" decisions and "N" past DFE decisions to cancel ISI • Can be realized using a feed forward structure
Maximum likelihood (Viterbi algorithm)	Dynamic programming	$d_{\min}^2 \leq \sum_n h_n^2$	• Optimum sequence detector • Not inherently adaptive • Generally complex to implement • Useful as a performance bound

$$d_{\min}^2 = \min_{\substack{e\varepsilon_n \\ \varepsilon_0 \neq 0}} \int [\sum_n \varepsilon_n h(t-nT)]^2 \, dt$$

a. $\{h_n\}$ are the samples of the pulse response at the receiver input.

algorithm opens up the eye so that decision-directed operation may resume after a number of iterations approximately one to two orders of magnitude more than that required when an ideal reference is available.

8.16 CHAPTER SUMMARY

This chapter has provided the theoretical framework and practical insight into the structures, adaptation algorithms, and (steady-state and transient) performance of adaptive equalizers. Several structures have been considered, beginning with the linear, synchronous, and fractionally-spaced tapped delay line

Table 8.5: Adaptive Equalization Algorithms[a]

Type	Origin	Rate of Convergence (Number of Iterations)	Complexity (Multiplications/ Iteration)	Comments
Zero-forcing	Lucky	~100N	N (Adds/ Subtracts)	• Minimizes peak distortion • No multiplications • Useful when distortion is small
Least-mean square (LMS)	Weiner filter (Widrow)	~10N	~N	• Robust ubiquitous • Minimizes mean-square error • Can be used for tracking
Kalman (recursive least squares)	Kalman filter	~N	~N^2	• Minimizes sum of squared errors (deterministic) • Same steady-state error as LMS • Tracking performance ~ LMS
(Computationally) Fast Kalman	Kalman	~N	~N	• Minimizes sum of squared errors • Algorithms tend to be unstable in limited precision environments

a. These algorithms can also be applied to other structures such as lattice filters. N = number of adaptive elements.

equalizer and concluding with several nonlinear equalizers (*nonlinear* in processing their input signals but forming a *linear* combination of the adjustable weights, so that the output mean-square error has a unique optimum setting). The attributes of these structures are summarized in Table 8.4. We have discussed equalizer adaptation, emphasizing the ubiquitous least-mean square (LMS) algorithm; attention was also directed to the fast-converging family of Kalman algorithms. Joint adaptation of a data-directed phase-locked loop and an equalizer was presented. The features of the algorithms we have considered are summarized in Table 8.5.

While adaptive equalization is a relatively mature technology, the press to ever-higher speeds of operation (e.g., in fiber optic systems) seems certain to provide the stimulus for continued progress in this field.

APPENDIX 8A CONVEXITY OF THE MEAN-SQUARE ERROR

To demonstrate convexity we must show the mean-square error (MSE) satisfies the condition:

$$MSE[\lambda \mathbf{c}_a + (1-\lambda)\mathbf{c}_b] \leq MSE[\mathbf{c}_a] + (1-\lambda)\ MSE[\mathbf{c}_b]\ , \tag{8.18}$$

where the MSE is defined by

$$MSE = \mathbf{c}'\ R\ \mathbf{c} - 2\ \mathbf{c}'\ \mathbf{x} + 1\ . \tag{8.15}$$

For the MSE to be convex, using (8.15a) and (8.18) we must prove that for any value, $\mathbf{c}_M = \lambda\ \mathbf{c}_a + (1-\lambda)\mathbf{c}_b$, that lies between $\mathbf{c}_a$ and $\mathbf{c}_b$, the following inequality is satisfied:

$$\left[\lambda \mathbf{c}_a + (1-\lambda)\mathbf{c}_b\right]' R \left[\lambda \mathbf{c}_a + (1-\lambda)\mathbf{c}_b\right] - 2[\lambda \mathbf{c}_a + (1-\lambda)\mathbf{c}_b]'\mathbf{x}$$

$$+ 1 \leq \lambda \left[\mathbf{c}_a' R \mathbf{c}_a - 2\mathbf{c}_a'\mathbf{x} + 1\right] + (1-\lambda)\left[\mathbf{c}_b' R \mathbf{c}_b - 2\mathbf{c}_b'\mathbf{x} + 1\right]. \tag{8.A.1}$$

Expanding (8.A.1) gives

$$\lambda^2 \mathbf{c}_a' R \mathbf{c}_a + \lambda(1-\lambda)\mathbf{c}_b' R \mathbf{c}_b + \lambda(1-\lambda)\mathbf{c}_a' R \mathbf{c}_b +$$

$$(1-\lambda)^2 \mathbf{c}_b' R \mathbf{c}_b - 2\lambda \mathbf{c}_a'\mathbf{x} - 2(1-\lambda)\mathbf{c}_b'\mathbf{x} + 1$$

$$\overset{?}{\leq} \lambda \left[\mathbf{c}_a' R \mathbf{c}_a - 2\mathbf{c}_a'\mathbf{x} + 1\right] + (1-\lambda)\left[\mathbf{c}_b' R \mathbf{c}_b - 2\mathbf{c}_b'\mathbf{x} + 1\right].$$

Replacing $\lambda^2 \mathbf{c}_b' R \mathbf{c}_a$ with $\lambda \mathbf{c}_a' R \mathbf{c}_a + (\lambda^2 - \lambda)\mathbf{c}_a' R \mathbf{c}_a$, and similarly for $(1-\lambda)^2 \mathbf{c}_b' R \mathbf{c}_b$, we have

$$\lambda \mathbf{c}_a' R \mathbf{c}_a + (\lambda^2 - \lambda)\mathbf{c}_a' R \mathbf{c}_a + \lambda(1-\lambda)\mathbf{c}_a' R \mathbf{c}_b + \lambda(1-\lambda)\mathbf{c}_a' R \mathbf{c}_b +$$

$$(1-\lambda)\mathbf{c}_b' R \mathbf{c}_b + \left[(1-\lambda)^2 - (1-\lambda)\right]\mathbf{c}_b' R \mathbf{c}_b - 2\lambda \mathbf{c}_a'\mathbf{x} -$$

$$2(1-\lambda)\mathbf{c}_b'\mathbf{x} + 1 \overset{?}{\leq} \lambda \left[\mathbf{c}_a' R \mathbf{c}_a - 2\mathbf{c}_a'\mathbf{x} + 1\right] + (1-\lambda)\left[\mathbf{c}_b' R \mathbf{c}_b - 2\mathbf{c}_b'\mathbf{x} + 1\right].$$

Canceling terms and dividing by $\lambda(1-\lambda)$ produces the following relation

$$-\mathbf{c}_a' R \mathbf{c}_a + \mathbf{c}_b' R \mathbf{c}_a + \mathbf{c}_a' R \mathbf{c}_b - \mathbf{c}_b' R \mathbf{c}_b \leq 0$$

or,

$$(\mathbf{c}_a - \mathbf{c}_b)' R (\mathbf{c}_a - \mathbf{c}_b) \geq 0. \tag{8.A.2}$$

But from (8.15d) we know that R is positive semidefinite; that is, for any vector $\mathbf{z} = \mathbf{c}_a - \mathbf{c}_b$

$$\mathbf{z}' R \mathbf{z} \geq 0\ , \tag{8.A.3}$$

Thus (8.A.2) is satisfied, and we have demonstrated that the MSE is a convex function of the tap weights.

APPENDIX 8B ASYMPTOTIC EIGENVALUE DISTRIBUTION FOR THE CORRELATION MATRIX OF SYNCHRONOUS AND FRACTIONALLY-SPACED EQUALIZERS

In this appendix, we describe the eigenvalues of the correlation matrix for infinitely long synchronous and fractionally-spaced equalizers.

8B.1 Synchronous Equalizer

From the definition $R = \mathbf{r}_n \mathbf{r}_n'$, we observe that R is a Toeplitz matrix whose eigenvalue equation is given by

$$\sum_{l=-N}^{N} R_{k-l} P_l = \lambda P_k, \quad -N \le k \le N, \tag{8.B.1}$$

where R_{k-l} is the kth element of R, λ is an eigenvalue, and $\mathbf{p}' = (p_{-N}, \ldots, p_0, \ldots, p_N)$ is the associated eigenvector. As $N \to \infty$, taking the Fourier transform of both sides of (8.B.1) yields

$$R(f)P(f) = \lambda P(f), \quad |f| \le \frac{1}{2T}, \tag{8.B.2}$$

where

$$R(f) = \sum_k |X(f+k/T)|^2 + \sigma^2 = |X_{eq}(f)|^2 + \sigma^2, \quad |f| \le \frac{1}{2T}. \tag{8.B.3}^*$$

For very large N, the discrete asymptotic approximation to (8.B.2) is

$$R(f_i)P(f_i) = \lambda P(f_i), \quad f_i = \frac{i}{(2N+1)T}, \quad -N \le i \le N. \tag{8.B.4}$$

Unless $R(f_i)$ has the same value for two or more values of the index i, the only way (8.B.4) can be satisfied is for $P(f)$ to be concentrated at a single frequency, i.e., be a sinusoid. Repeated values of $R(f_i)$ correspond to repeated eigenvalues and an eigenvector subspace which can be spanned either by distinct sinusoids or by combinations of sinusoids. Then the solution to (8.B.2) is

$$\left.\begin{aligned} \lambda_i &= R(f_i) \\ P_i(f) &= \delta(f-f_i) \pm \delta(f+f_i) \end{aligned}\right\} \quad -N \le i \le N. \tag{8.B.5}$$

---

* The Nyquist-equivalent spectrum, $X_{eq}(f)$, is defined as $X_{eq}(f) = \sum_k X[f+k/T]$, where the receiver sampling phase is incorporated into $X(f)$.

Thus for a synchronous equalizer, the asymptotic ($N \to \infty$) eigenvalues uniformly sample the folded-channel-plus-noise spectrum, and the eigenvectors are the corresponding sinusoids.

8B.2 Fractionally-Spaced Equalizers

Here the channel-correlation matrix while symmetric is not Toeplitz; thus Fourier transform techniques do not yield the eigenvalues and eigenvectors in the above short order. For convenience, we consider the noiseless situation, where the eigenvalue equation is

$$\sum_{l=-N}^{N} R(kT', lT')p(lT') = \lambda p(kT') \quad -N \le k \le N, \tag{8.B.6}$$

where the kth element of R is given by

$$R(kT', lT') = \sum_{m} x(mT - kT')x(mT - lT'). \tag{8.B.7}$$

With $T' = T/2$, we write (8.B.6) for even and odd values of k.

$$\sum_{l\ even} R\left(k\frac{T}{2}, l\frac{T}{2}\right)p\left(l\frac{T}{2}\right) + \sum_{l\ odd} R\left(k\frac{T}{2}, l\frac{T}{2}\right)p\left(l\frac{T}{2}\right) = \lambda p\left(k\frac{T}{2}\right), \quad k \text{ even} \tag{8.B.8}$$

$$\sum_{l\ even} R\left(k\frac{T}{2}, l\frac{T}{2}\right)p\left(l\frac{T}{2}\right) + \sum_{l\ odd} R\left(k\frac{T}{2}, l\frac{T}{2}\right)p\left(l\frac{T}{2}\right) = \lambda p\left(k\frac{T}{2}\right), \quad k \text{ odd}. \tag{8.B.9}$$

Now (8.B.8) and (8.B.9) can be written as

$$\sum_{l=-N/2}^{N/2} R(kT, lT)p(lT) + \sum_{l=-N/2}^{N/2} R\left(kT, lT + \frac{T}{2}\right) p\left(lT + \frac{T}{2}\right) = \lambda p(kT), \quad -N/2 < k < N/2 \tag{8.B.10}$$

$$\sum_{\ell=-N/2}^{N/2} R\left(kT+\frac{T}{2},\ \ell T\right)p(\ell T) + \sum_{l=-N/2}^{N/2} R\left(kT+\frac{T}{2},\ lT+\frac{T}{2}\right)p\left(lT+\frac{T}{2}\right)$$

$$= \lambda p\left(kT+\frac{T}{2}\right),\quad -N/2 < k < N/2\ , \qquad (8.B.11)$$

respectively, where both equations hold for N integer values of k, and more importantly the various component matrices are now all Toeplitz.* If we consider the situation where $X(f)$ has less than 100 percent excess bandwidth, then it is useful to introduce the four spectra

$$X_{eq}(f) \triangleq X(f) + X\left(f-\frac{1}{T}\right) + X\left(f+\frac{1}{T}\right)$$

$$\tilde{X}_{eq}(f) \triangleq X(f) - X\left(f-\frac{1}{T}\right) - X\left(f+\frac{1}{T}\right),\quad |f| \le \frac{1}{2T} \qquad (8.B.12)$$

$$P_{eq}(f) \triangleq P(f) + P\left(f-\frac{1}{T}\right) + P\left(f+\frac{1}{T}\right)$$

$$\tilde{P}_{eq}(f) \triangleq P(f) - P\left(f-\frac{1}{T}\right) - P\left(f+\frac{1}{T}\right),\quad |f| \le \frac{1}{2T}\ , \qquad (8.B.13)$$

where the discrete Fourier transform of the eigenvector $p(l(T/2))$, $P(f)$, is given by

$$P(f) \triangleq \sum_{l=-N}^{N} P(l\frac{T}{2})e^{-jf_j l(T/2)} = P_{eq}(f_j) + e^{-j2\pi f(T/2)}\tilde{P}_{eq}(f_j),\ f_j$$
$$= (\frac{j}{N})\cdot(\frac{1}{2\pi}),\ -N \le j \le N\ . \qquad (8.B.14)$$

Taking the *synchronous* Fourier transform (i.e., with respect to the T second sampling interval) of (8.B.10) and (8.B.11) gives as an approximation (which become exact as $N \to \infty$)

$$|X_{eq}(f_j)|^2 P_{eq}(f_j) + X_{eq}(f_j)\tilde{X}^*_{eq}(f_j)\tilde{P}_{eq}(f_j) = \lambda P_{eq}(f_j)$$

$$\tilde{X}_{eq}(f_j)X^*_{eq}(f_j)P_{eq}(f_j) + |\tilde{X}_{eq}(f_j)|^2\tilde{P}_{eq}(f_j) = \lambda\tilde{P}_{eq}(f_j). \qquad (8.B.15)$$

Note that $p(kT+(T/2))$ has the discrete Fourier transform exp $(-j2\pi f(T/2)\ \tilde{P}_{eq}(f_j))$, where the synchronous transform of $p(kT)$ is $P_{eq}(f_j)$.

---

* For example, $R(kT, lT+(T/2)) = \sum_m x(mT-kT)x(mT-lT+(T/2)) = \sum_n x(nT)x(nT+(k-l)T+(T/2))$.

Arguing as we did for the synchronous equalizer, we see that the ith eigenvectors $P_i(f_j)$ and $\tilde{P}_i(f)$ must again be delta functions at $f_i = i/NT$. Consequently, the eigenvalues, λ_i, must satisfy the quadratic equation

$$\lambda_i^2 - \lambda_i[|X_{eq}(f_i)|^2 + |\tilde{X}_{eq}(f_i)|^2] = 0, \tag{8.B.16}$$

and the eigenvalues are

$$\lambda_i^{(1)} = 0$$

$$\lambda_i^{(2)} = |X_{eq}(f_i)|^2 + |\tilde{X}_{eq}(f_i)|^2 = \sum_k X\left[f_i + \frac{k}{T}\right]^2, \quad -\frac{N}{2} \le i \le \frac{N}{2}. \tag{8.B.17}$$

In contrast to (8.B.3), which applies to the synchronous equalizer, half the eigenvalues are exactly zero, while the other half are samples of the aliased magnitude-squared channel transfer function. Not surprisingly, the eigenvalues are independent of both the channel phase characteristics and the receiver sampling phase, and if $|X(f)|^2$ is Nyquist, then all the nonzero eigenvalues are unity. Since the eigenvalues have been determined, we can solve for the eigenvectors. Since $p(lT)$ has the transform $P_{eq}(f)$ and $p(lT+(T/2))$ has the transform $\tilde{P}_{eq}(f)e^{j\pi f}$, the eigenvectors associated with the zero eigenvalue are constructed as

$$p_i(\frac{nT}{2}) = \begin{cases} \tilde{X}_{eq}(f_i)e^{j\pi f_i nT}, & n \text{ odd} \\ -X_{eq}(f_i)e^{j\pi f_i nT}, & n \text{ even} \end{cases} \tag{8.B.18}$$

while the eigenvector associated with the nonzero eigenvalue is

$$p_i(\frac{nT}{2}) = \begin{cases} X_{eq}(f_i)e^{j\pi f_i nT)}, & n \text{ even} \\ \tilde{X}_{eq}(f_i)e^{j\pi f_i nT)}, & n \text{ odd} \end{cases} \tag{8.B.19}$$

When f_i is not in the rolloff region, then $X_{eq}(f_j) = \tilde{X}_{eq}(f_i)$, and (8.B.19) describes a sinusoid of frequency f_i, since the even and odd portions of $p_i(n(T/2))$ mesh together in a continuous manner (i.e., $p_i(n(T/2)) = X_{eq}(f_i)e^{j\pi f_i nT)}$). However, (8.B.18) describes a function which changes sign and oscillates almost a full cycle in T seconds. Consequently, $p_i(n(T/2))$, as given by (8.B.18), will have most of its spectral energy concentrated near $1/T$ Hz. When f_i is in the rolloff region, the frequency content of (8.B.18) will differ somewhat from the above extreme cases, but the general results will still be as above. Numerical evaluations have confirmed the above.

APPENDIX 8C DERIVATION OF THE MATRIX INVERSION LEMMA

If the matrix B_n satisfies the recursion

$$B_{n+1} = B_n + \mathbf{r}_{n+1}\mathbf{r}'_{n+1} \quad , \tag{8.C.1}$$

and if B_n is positive definite for all n, then the inverse matrix, B_n^{-1}, satisfies the matrix inversion lemma

$$B_{n+1}^{-1} = B_n^{-1} - \frac{B_n^{-1}\mathbf{r}_{n+1}\mathbf{r}'_{n+1}B_n^{-1}}{1+\mathbf{r}'_{n+1}B^{-1}\mathbf{r}_{n+1}} \quad . \tag{8.C.2}$$

PROOF:

Since B_{n+1} is positive definite, it has a positive definite square root $B_{n+1}^{1/2}$ and we can write

$$B_n + \mathbf{r}_{n+1}\mathbf{r}'_{n+1} = B_n^{1/2}\left[I+B_n^{-1/2}\ \mathbf{r}_{n+1}\mathbf{r}'_{n+1}B_n^{-1/2}\right]B_n^{1/2} \quad . \tag{8.C.3}$$

Defining the vector

$$y_{n+1} = B_n^{-1/2}\mathbf{r}_{n+1}, \tag{8.C.4}$$

we have from (8.C.1)

$$B_{n+1}^{-1} = \left[B_n+\mathbf{r}_{n+1}\mathbf{r}'_{n+1}\right]^{-1} = \left[B_n^{1/2}(I+\mathbf{y}_{n+1}\mathbf{y}'_{n+1})B_n^{1/2}\right]^{-1}$$

$$= B_n^{-1/2}\left(I+\mathbf{y}_{n+1}\mathbf{y}'_{n+1}\right)^{-1}B_n^{-1/2} \quad . \tag{8.C.5}$$

We use the fact that

$$(I+\mathbf{y}_{n+1}\mathbf{y}'_{n+1})^{-1} = I - \frac{\mathbf{y}_{n+1}\ \mathbf{y}'_{n+1}}{1+|\mathbf{y}_{n+1}|^2}, \tag{8.C.6}$$

which is shown by multiplying both sides by $I + \mathbf{y}_{n+1}\mathbf{y}'^{n+1}$.

Finally, substituting (8.C.6) and (8.C.4) into (8.C.5) gives

$$B_{n+1}^{-1} = B_n^{-1/2}\left[I-\frac{\mathbf{y}_{n+1}\mathbf{y}'_{n+1}}{1+|\mathbf{y}_{n+1}|^2}\right]B_n^{-1/2}$$

$$= B_n^{-1} - \frac{B_n^{-1/2}B_n^{-1/2}\mathbf{r}_{n+1}\mathbf{r}'_{n+1}B_n^{-1/2}B_n^{-1/2}}{1-\mathbf{r}'_{n+1}B_n^{-1/2}B_n^{-1/2}\mathbf{r}_{n+1}}$$

$$= B_n^{-1} - \frac{B_n^{-1}\mathbf{r}_{n+1}\mathbf{r}_n^1B_n^{-1}}{1+\mathbf{r}_{n+1}^1B_n^{-1}\mathbf{r}_{n+1}} \quad , \tag{8.C.7}$$

which proves the matrix inversion lemma.

APPENDIX 8D TRACKING PROPERTIES OF THE LMS AND RLS ALGORITHMS

In Chapter 8 we have described the rapid convergence of the Kalman (or RLS) algorithm in a stationary environment. In this appendix we compare the behavior of the LMS and RLS algorithms in a nonstationary environment, where the tracking capabilities of the algorithms are the quantities of interest. There are numerous applications where a tracking capability is desired, most typically to follow the drift in channel characteristics. As we have seen in Chapter 6, in connection with our study of phase-locked loops, there is generally a conflict between a narrow bandwidth (small step size in the LMS algorithm and a forgetting factor close to unity in the RLS algorithm) for achieving a small steady-state error and a larger bandwidth (large step size in the LMS algorithm and a forgetting factor not close to unity in the RLS algorithm) to provide a tracking capability. In general, the mean-squared error can be decomposed into these two components: an estimation and tracking (or lag) error. In this Appendix we assume a Markovian channel model, i.e., the channel vector $\mathbf{x}$, varies with time according to the following first-order difference equation.

$$\mathbf{x}_{n+1} = \mathbf{x}_n + \omega_n , \tag{8.D.1}$$

where the vectors ω_n are independent of each other, and where the components of each vector are also independent, identically distributed, zero-mean random variables with variance σ_ω^2.

The estimation and tracking errors will be derived with the aid of the following representation for the tap error vector ε_n,

$$\varepsilon_n = \mathbf{c}_n - \mathbf{c}_{opt,n} = \underbrace{\mathbf{c}_n - E\mathbf{c}_n}_{estimation\ error} + \underbrace{E\mathbf{c}_n - \mathbf{c}_{opt,n}}_{lag\ error} \tag{8.D.2}$$

In (8.D.2) note that the optimum weight vector, $\mathbf{c}_{opt,n}$, is a function of time, and that $\mathbf{c}_n - E\mathbf{c}_n$ represents the (noise-averaging) tap-weight estimation error and that $E\mathbf{c}_n - \mathbf{c}_{opt,n}$ represents the tracking (or lag) tap-weight error. The fundamental relationship that we will use in this appendix is the familiar expression relating the excess MSE and the minimum MSE,

$$\varepsilon_n = \varepsilon_{opt} + q_n , \tag{8.D.3}$$

where q_n is the excess MSE.

LMS TRACKING

We first consider the tracking capability of the LMS algorithm. From (8.D.3) we recall that

$$\mathbf{q}_n = < \varepsilon_n' R \varepsilon_n > ,$$

and with the channel model of (8.D.1), by superposition the excess MSE, q_n, is the *sum* of the excess MSEs due to estimation noise and lag error. In [32], Widrow *et al.* show that the

$$q_{lag}(LMS) = \frac{N\,\sigma_\omega^2}{4\beta}, \tag{8.D.4}$$

where N is the number of weights and β is the step size in the LMS algorithm. Recall from Chapter 8 that the steady-state estimation error, when all eigenvalues are equal and the input power is unity, is given by

$$\underset{estimation}{q}(LMS) = \frac{\beta}{2(1-\beta N)}\,N\,\varepsilon_{opt}\,; \tag{8.D.5}$$

thus the excess MSE for the LMS algorithm is given by

$$q_{LMS} = \frac{\beta}{2(1-\beta N)}\,N\,\varepsilon_{opt} + \frac{N}{4\beta}\,\sigma_\omega^2\,. \tag{8.D.6}$$

RLS TRACKING

Eleftheriou and Falconer [21] have shown that the estimation error for the RLS algorithm is given by

$$\underset{estimation}{q}(RLS) = \left[\frac{1-\alpha}{1+\alpha}\right] N\,\varepsilon_{opt}\,, \tag{8.D.7}$$

where α is the forgetting factor in the RLS algorithm. Note that if $\alpha = 1$, the estimation error is zero, as expected; however, with $\alpha \neq 1$ (as will be required for tracking), the RLS exhibits an excess estimation error that grows with the number of taps. The same authors have also shown that

$$q_{lag}(RLS) = \frac{N}{2(1-\alpha)}\,\sigma_\omega^2\,, \tag{8.D.8}$$

which again is similar to the lag error in the LMS algorithm; but note that as $\alpha \to 1$, the lag error becomes arbitrarily large (since an equal weighting of the errors implies that changes in the channel will not be sensed after the system has converged to its initial setting). Combining (8.D.7) and (8.D.8), we have

$$q_{RLS} = \left[\frac{1-\alpha}{1+\alpha}\right] N\varepsilon_{opt} + \frac{N}{2(1-\alpha)}\,\sigma_\omega^2\,. \tag{8.D.9}$$

Comparing (8.D.6) and (8.D.9) we observe that as β or $(1-\alpha)$ decreases, the noise (estimation) component of the error decreases, while the lag component increases. Similarly, an increase in β or $(1-\alpha)$ reduces the lag component, but makes the noise component worse. Thus optimum values of β and α exist [33]. Each of the excess MSE expressions can be optimized by differentiation.

Minimizing q_{LMS} with respect to β we find

$$q_{LMS}^{*} = N\ \varepsilon_{opt}\ \sigma_{\omega} + \frac{N^2}{2}\ \sigma_{\omega}^2 \tag{8.D.10a}$$

$$= N\ h\ \varepsilon_{opt} \left[1 + \frac{Nh}{2} \right] , \tag{8.D.10b}$$

where the parameter h is defined by

$$h^2 \equiv \frac{\sigma_{\omega}^2}{\varepsilon_{opt}} , \tag{8.D.11}$$

and the optimum asterisk denotes that the optimum step size has been chosen. For the RLS algorithm we find

$$q_{RLS}^{*} = \begin{cases} Nh\varepsilon_{opt} \left[1 + \dfrac{h}{4} \right], \ h < 2.5 \\ \\ N\varepsilon_{opt}(.6 + .54h^2), \ h \geq 2.5 \ . \end{cases} \tag{8.D.12}$$

If the time variation of the channel is small relative to the steady-state MSE, ε_{opt}, i.e., h is small, then the LMS and RLS algorithms have similar tracking

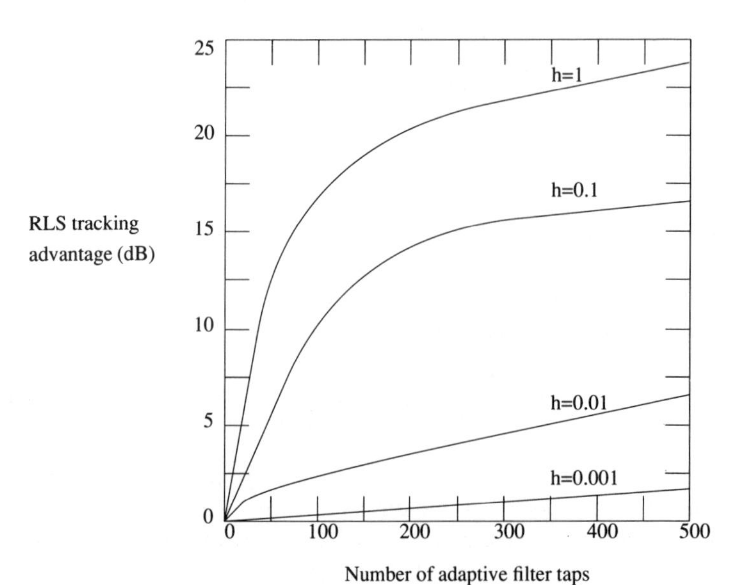

Fig. 8D.1 RLS/LMS tracking comparison.

errors given by

$$q_{LMS}^* = q_{RLS}^* = Nh\varepsilon_{opt} . \tag{8.D.13}$$

(Note that the optimized errors still grow with N, yet another reason for keeping N to a minimum.) On the other hand, when h is not small, the RLS algorithm will have a tracking advantage that grows with the number of taps (e.g., if $h = 1$ and $N = 10$. The advantage of the RLS algorithm is given in Figure 8.D.1. Note that for the *most common practical cases,* say $h = 0.001$, the *RLS algorithm does not* have a tracking advantage over the LMS algorithm. However, there have been applications reported [34] in fading dispersive channels where h is large, and the RLS algorithm does indeed have an advantage over the LMS algorithm.

REFERENCES

[1] R. W. Lucky, "Automatic equalization for digital communications," *Bell System Tech. J.*, Vol. 44, pp. 547–588, April 1965.

[2] B. Widrow and M. E. J. Hoff, "Adaptive Switching Circuits," *IRE 1960 Wescon Conv. Record*, pp. 563–587.

[3] R. W. Lucky, J. Salz, and E. J. Weldon, Jr., *Principles of Data Communication*, McGraw-Hill, 1968.

[4] R. D. Gitlin and S. B. Weinstein, "Fractionally-spaced equalization: An improved digital transversal equalizer," *Bell System Tech. J.*, Vol. 55, March 1976.

[5] J. Proakis, *Digital Communications,* McGraw-Hill, 1983.

[6] J. Salz, "Optimum Mean-Square Decision Feedback Equalization," *Bell System Tech. J.*, Vol. 52, pp. 1341–1373, October 1973.

[7] A. Gersho and T. L. Lim, "Adaptive Cancellation of Interference for Data Transmission," *Bell System Tech. J.*, Vol. 60, pp. 1997–2021, November 1981.

[8] R. D. Gitlin and T. L. Lim, "Equalization of Modulated Data Signals Utilizing Tentative and Final Decisions," U.S. Patent No. 4,615,038, September 30, 1988.

[9] D. Godard, "Channel equalization using a Kalman filter for fast data transmission," *IBM J. Research and Development*, pp. 267–273, May 1974.

[10] R. D. Gitlin and F. R. Magee, Jr., "Self Orthogonalizing Adaptive Equalization Algorithms," *IEEE Trans. on Communications*, Vol. COM-25, No. 7, pp. 666–672, July 1977.

[11] D. Falconer and L. Ljung, "Application of fast Kalman estimation to adaptive equalization," *IEEE Trans. on Communications*, Vol. COM-26, pp. 1439–1446, October 1978.

[12] C. A. Siller, Jr., "Multipath Propagation," *IEEE Communications Magazine*, February 1984.

[13] F. P. Duffy and T. N. Thatcher, Jr., "Analog Transmission Performance on the Switched Telecommunications Network," *Bell System Tech. J.*, Vol. 50, No. 4, pp. 1311–1347, April 1971.

[14] R. E. Bellman, *Matrix Analysis,* 2nd edition, McGraw-Hill, 1968.

[15] J. E. Mazo, "On the Independence Theory of Equalizer Convergence," *Bell System Tech. J.*, Vol. 58, pp. 963–993, May-June 1978.

[16] R. E. Kalman, "A New Approach to Linear and Prediction Problems," *J. Basic Engineering,* Vol. 82, pp. 34–45, March 1960.

[17] S. T. Alexander, *Adaptive Signal Processing: Theory and Applications,* Springer-Verlag, 1986.

[18] J. M. Cioffi and T. Kailath, "Fast Recursive Least-Squares Transversal Filters for Adaptive Filtering," *IEEE Trans. on Acoustics, Speech and Signal Processing,* Vol. ASSP-32, pp. 304–338, April 1984.

[19] D. T. M. Slock and T. Kailath, "Numerically Stable Fast Transversal Filters for Recursive Least-Squares Adaptive Filtering," *Proc. ICASSP '88 Conf.*, pp. 1365–1368, April 1988.

[20] E. H. Satorius and J. D. Pack, "Application of Least-Squares Lattice Algorithms to Adaptive Equalization," *IEEE Trans. on Communications,* Vol. COM-29, pp. 136–142, February 1981.

[21] E. Eleftheriou and D. D. Falconer, "Tracking Properties and Steady-State Performance of RLS Adaptive Filter Algorithms," *IEEE Trans. on Acoustics, Speech and Signal Processing,* Vol. ASSP-34, No. 5, October 1986.

[22] S. U. H. Qureshi, "Adaptive Equalization," *Proceedings of the IEEE* Vol. 73, No. 9, September 1985.

[23] K. A. Mueller and D. Spaulding, "Cyclic equalization — a new rapidly converging equalization technique for synchronous data communication," *Bell System Tech. J.*, Vol. 54, No. 2, February 1975.

[24] D. D. Falconer, "Jointly Adaptive Equalization and Carrier Recovery in Two-Dimensional Digital Communication Systems," *Bell System Tech. J., 55,* No. 3 pp. 317–334, March 1976.

[25] R. D. Gitlin, E. Y. Ho, and J. E. Mazo, "Passband Equalization for Differentially Phase-Modulated Data Signals," *Bell System Tech. J.*, *52*, No. 2 pp. 219–238, February 1973.

[26] D. D. Falconer, "Analysis of a Gradient Algorithm for Simultaneous Passband Equalization and Carrier Phase Recovery," *Bell System Tech. J.*, *55*, No. 4 pp. 409–428, April 1976.

[27] E. Biglieri, A. Gersho, R. D. Gitlin, and T. L. Lim, "Adaptive Cancellation of Nonlinear Intersymbol Interference for Voiceband Data Transmission," *IEEE J. on Selected Areas in Communications*, Vol. SAC-2, pp. 765–777.

[28] J. E. Mazo, "Analysis of Decision Directed Equalizer Convergence," *Bell System Tech. J.*, Vol. 59, No. 10, pp. 1857–1876, December 1980.

[29] J. R. Treichler, C. R. Johnson, Jr., and M. G. Larimore, *Theory and Design of Adaptive Filters,* New York, NY, Wiley, 1987.

[30] D. N. Godard, "Self-recovering equalization and carrier tracking in two-dimensional data communication systems," *IEEE Trans. Communications*, Vol. COM-28, No. 11, pp. 1867–1875, November 1980.

[31] D. N. Godard and P. E. Thirion, "Method and device for training an adaptive equalizer by means of an unknown data signal in a QAM transmission system," U.S. Patent 4 227 152, October 7, 1980.

[32] B. Widrow *et al.*, "Stationary and Nonstationary Learning Characteristics of the LMS Adaptive Filter," *Proc. IEEE,* Vol. 64, No. 8, pp. 1151–1162, August 1976.

[33] J. M. Cioffi, "When Do I Use an RLS Adaptive Filter?," *Proc. Asilomar Conf. on Circuits and Systems*, 1985.

[34] F. Ling and J. Proakis, "Lattice Decision-Feedback Equalizers and their Application to Fading Dispersive Channels," *IEEE Trans. on Communications,* Vol. 33, No. 4, April 1985.

[35] S. Hariharan and A. P. Clark, "HF Channel Estimation Using a Fast Transversal Filter Algorithm," *IEEE Trans. on Acoustics, Speech, and Signal Processing,* Vol. 38, No. 5, pp 1355–1362, August 1990.

EXERCISES

8.1: Consider a channel with binary transmission (± 1) and a (discrete-time) impulse response of $h_0 = 0.97$ and $h_1 = 0.25$ with additive noise having standard deviation $\sigma = 0.25$. Let r_n denote the channel output

(a) What is the matched filter bound on the error rate?

(b) Calculate the bit error rate for the above channel.

(c) Show that the filter shown below inverts the channel.

(d) Calculate the error rate for a receiver that follows q_n by a limiter.

(e) Calculate the error rate for a decision feedback equalizer (assuming correct decisions).

(f) What is the error rate for a maximum likelihood detector?

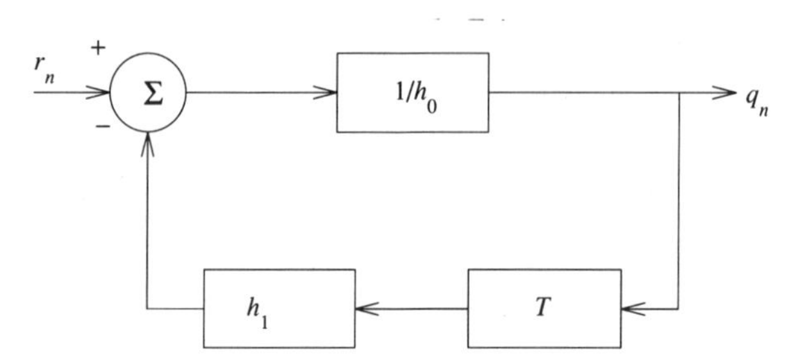

8.2: Show that the channel correlation matrix defined by (8.14) is Toeplitz for a synchronous equalizer. Is R Toeplitz for a fractionally spaced equalizer?

8.3: Show that the correlation matrix is positive definite if the channel does not have any spectral nulls.

8.4: Assume that the channel has a phase function $\theta(f) = b\sin 2\pi f T_0$. Show that the impulse response may be written as $\sum J_h(b)\, p(t - hT_0)$, where $p(t)$ is the transmitted pulse and $J_h(\cdot)$ is the Bessel function used in the expansion

$$e^{iz\sin\theta} = \sum_{h=-\infty}^{\infty} J_h(z) e^{ih\theta} .$$

8.5: (a) Derive equation (8.23).

(b) Derive a similar equation when the noise has power spectrum $N(f)$, and the input data has a correlation function f_n.

8.6: Derive equation (8.41).

8.7: Demonstrate that the MSE for a decision feedback equalizer is a convex function of the tap weights.

8.8: Show that the average tap error using the LMS algorithm is governed by equation (8.25b). What assumptions have you made?

8.9: Derive an LMS-type algorithm to minimize

(a) the absolute value of the output error.

(b) the fourth power of the output error.

(c) Give an intuitive explanation as to why these algorithms will converge under the same general conditions as the LMS algorithm.

8.10: Show that minimizing the mean-squared in-phase error (or the quadrature error) gives the same resulting complex mean-squared error (8.110).

8.11: How small do α and β of (8.126) have to be for the algorithm to converge?

8.12: Derive the equations corresponding to (8.127) and (8.128) for a passband decision feedback equalizer.

8.13: Prove that the filter W shown in Figure 8.33 is a matched filter.

8.14: Derive the LMS tap-update equation for the filters W and C in Figure 8.33.

8.15: Show that, for a fractionally spaced equalizer, if the channel does not have any nulls, then the correlation matrix R will be positive definite.

8.16: Derive the Kalman tap-update equations for a decision-feedback equalizer.

8.17: Derive the tap-update equation for the constant modulus blind equalization algorithm.

9

Echo Cancellation

9.0 INTRODUCTION

The simplest way to realize *full-duplex* data communication is by using completely separate transmission media for the two directions of transmission. But often, only a single, bilateral (simultaneous two-way) transmission medium is available rather than two one-way channels. For the ordinary telephone subscriber, served by a twisted-pair copper line from a central office, a dialed telephone circuit is a bilateral channel, usable for voice or data (at speeds up to about 14.4 kbps). It should be recognized that this circuit normally traverses interoffice facilities, in addition to the subscriber lines at each end, and that these facilities have a large influence on the characteristics of the end-to-end circuit.

There is a second type of bilateral channel becoming available to the telephone subscriber, and that is the channel offered by the twisted-pair copper line by itself, between the subscriber and the central office. For many subscriber locations, this subscriber line will, in the future, be dedicated to a full-duplex digital access channel to ISDN and other digital networks. The characteristics of the subscriber line alone are, of course, quite different from that of the dialed telephone line because of the absence of interoffice facilities and their restrictions, such as bandlimiting filters. Full-duplex signaling is carried out at 160 kbps (144 kbps information payload) for ISDN Basic access and faster for the high-speed digital subscriber line, as explained in Chapter 4.

There are several ways to use the available bandwidth to accomplish full-duplex data transmission over bilateral channels such as these, which can be summarized as *bandsplitting, time sharing, spectrum overlap,* and *same spectrum in both directions* (Figure 9.1). For bandsplitting, completely separate high-frequency and low-frequency subbands are used in each direction, with bandsplitting filters to avoid self-interference at each end of the circuit (Figure 9.2). The local transmitter can easily be 40 dB stronger in power than the

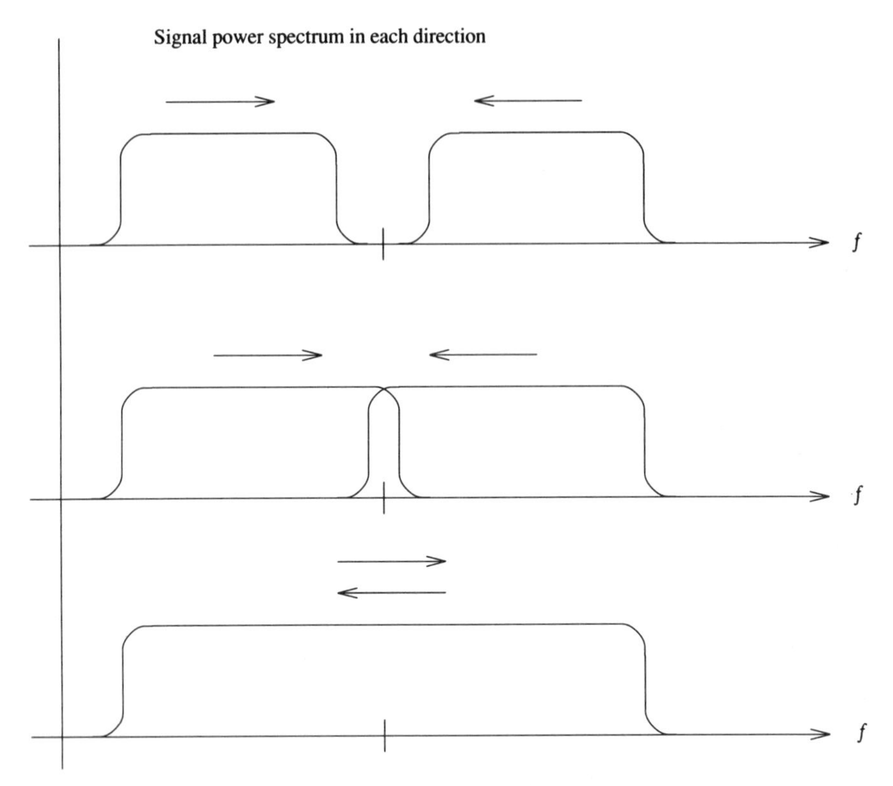

Fig. 9.1 Spectrum management for full-duplex communications: Bandsplitting, overlap, and same spectrum in both directions (used either simultaneously or alternately in time sharing).

desired signal arriving from the distant transmitter. A significant cost may be paid in waste of what is usually the least distorted part of the channel, at its center, and in degradations from band-edge distortions. Given a fixed total bandwidth and equal data rates in each direction as in the dialed telephone channel, the useful transmission rate in each direction is less than half of what the full

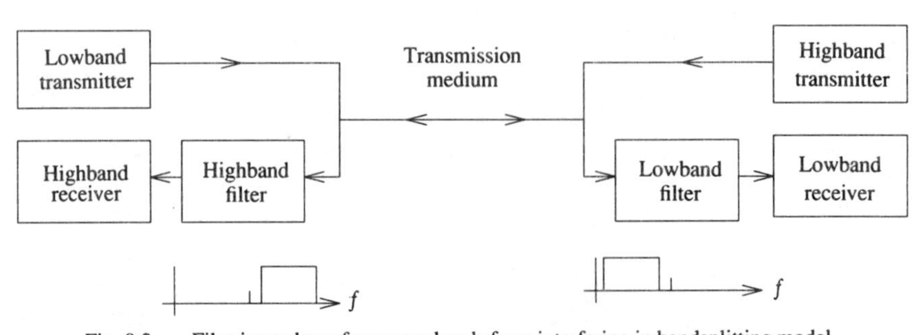

Fig. 9.2 Filtering to keep frequency bands from interfering in bandsplitting model.

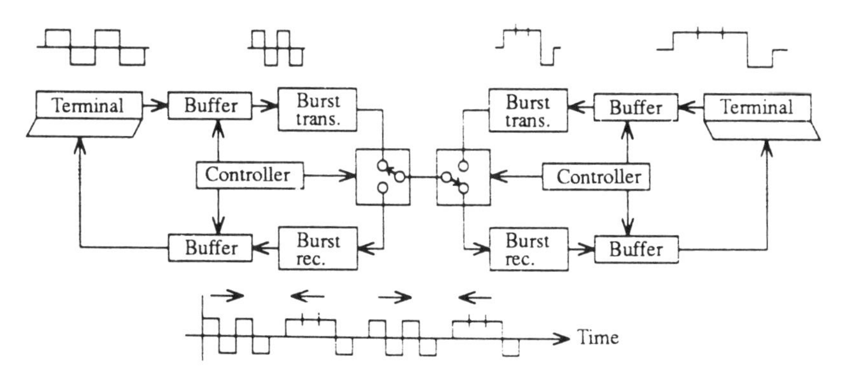

Fig. 9.3 Time-compression multiplexing (TCM), in which transmission is one way at a time in high-rate bursts.

channel would offer if it were used for one direction of transmission only. Bandsplitting is used in duplex voiceband modems up to 2.4 kbps and in two-way CATV systems (where the data rates in each direction are not equal).

Time compression multiplexing (TCM, Figure 9.3) is a time-domain sharing technique which is more or less equivalent to bandsplitting in its total bandwidth requirement, although each direction of transmission uses the same large spectrum. This technique alternates fast transmission bursts in each direction, saving up data (submitted to each transmitter at a lower rate) in buffers so that each of the end terminals has the illusion of a continuously available channel. If there were no propagation delay and hence no interval between bursts, the burst transmission rate required would be exactly twice the average data rate of each terminal and performance would be independent of buffer size (burst length). Realistically, there is propagation delay and the penalty must be paid in either buffer length (delay) or increased burst rate (bringing increased band-edge distortion), or in some reasonable combination (Exercise 9.1). TCM is practical for digital network access in most subscriber lines at rates up to about 80 kbps [1], but probably not for the 160 kbps required by ISDN.

For the spectrum overlap technique, filters suppress most of the self-interfering energy, and the remaining interfering energy, in the overlap region of the spectrum, is largely eliminated by the hybrid coupler/echo cancellation techniques explained below. Having partial rather than full overlap reduces the burden on the echo canceler, when bandwidth is available for this alternative.

When one is faced with the need to severely constrain bandwidth and achieve a high data rate, all with acceptable error rate, the best approach at the present time is the combination of hybrid coupler and echo canceler (Figure 9.4), which is the main subject of this chapter. The hybrid coupler, or "hybrid" for short, is a directional coupler realized as a passive 4-port [2]. Referring to Figure 9.4, if the impedances offered to the L and B ports are the same, and the impedances offered to the T and R ports are the same, then energy input to the T

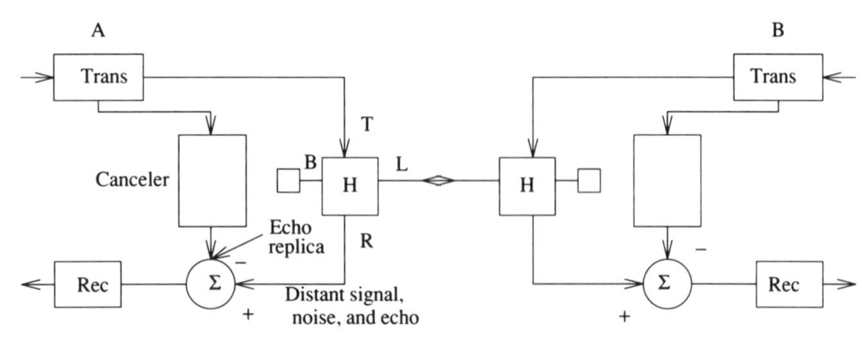

Fig. 9.4 Hybrid coupler and echo canceler for use of same spectrum in both directions.

port emerges from the L port (with 3 dB attenuation), and, ideally, none at all emerges from the R port. Similarly, energy arriving to the L port is totally directed (with 3 dB attenuation) to the R port. Thus the local transmitter does not interfere with the local receiver, and an ideal two-wire/four-wire interface is realized.

In practice, a compromise balancing impedance matches the typical two-wire line impedance. Also, the transmission circuit is not an ideal transmission line. Thus the reality is that considerable signal energy leaks from the T to the R port, and damaging "echoes" return from impedance mismatches at various points in the communication circuit. The hybrid coupler helps, but does not provide adequate separation of the transmitted and received signals.

Leakage through the hybrid coupler can be greatly reduced by isolating the line and balancing impedances (Figure 9.5) — a 50 dB echo suppression is possible [3] — but this presumes an ideal transmission line model for the two-wire line, with no returning echoes. There are also possibilities for adaptive hybrids that automatically generate a balancing impedance closely matched to the transmission line (to the extent that that line is an ideal transmission line).

In practice, it has been found more promising to use a compromise hybrid and depend on an echo canceler to attenuate local-transmitter energy that has

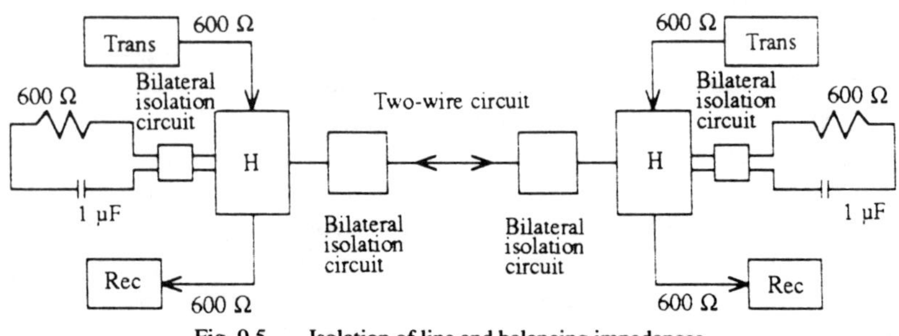

Fig. 9.5 Isolation of line and balancing impedances.

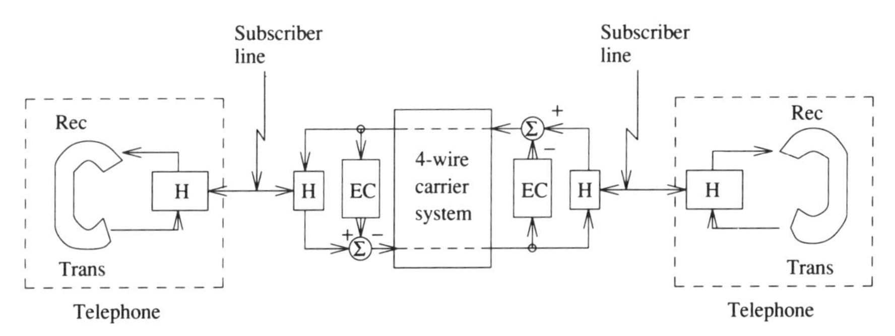

Fig. 9.6 Voice echo cancellation in the telephone network. Cancelers are provided only for "distant" echoes; nearer echoes are not disturbing to voice conversation.

leaked through the hybrid, to a level well below that of the desired distant signal. There is also a distant echo to be cancelled, as we shall see. Echo cancelers are also widely used within the telephone network (Figure 9.6) to attenuate distant echoes on voice connections with long delays, such as those including satellite links [4]. The original echo cancellation work was done in this voice echo cancellation context [5–7], and powerful integrated circuits were designed for deployment in the "tails" of long-distance telephone circuits [8]. But the differences between voice and data cancellation requirements, and in particular the fact that data may be transmitted in both directions simultaneously for long periods of time, unlike typical voice conversations, are such that voice echo cancelers in a long-distance telephone circuit (see Figure 9.6) are intentionally disabled for data communication sessions. A good overview is available in a paper by Messerschmitt [9].

A presumption is generally made here that the echo channel is linear and that the echo is free of timing, phase jitter, or frequency offset. In practice these impairments do occur, and later sections of this chapter suggest how these perturbations can be handled.

The bulk of this chapter describes structures and adaptation techniques for echo cancelers, seeking the best compromise among the generally conflicting goals of rapid adaptation, low residual cancellation error, and moderate implementation complexity. Some special considerations for operation of *digital* echo cancelers are described in Chapter 10. The next two sections of this chapter introduce system models for the two main applications of data echo cancellation in the telephone network, for full-duplex communication over dialed telephone circuits and via digital subscriber lines.

9.1 THE DIALED TELEPHONE CIRCUIT ECHO CANCELLATION MODEL

The application we are discussing is end-to-end full-duplex data communication over a dialed telephone circuit. Such connections are widely used for ter-

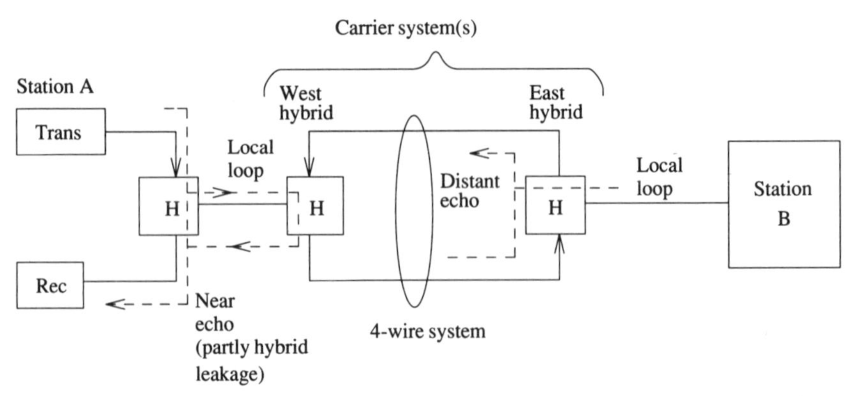

Fig. 9.7 Typical long-distance telephone circuit, showing the sources of most echoes returning to station A.

minal to host connections where digital data networks are unavailable, and as backups to four-wire private lines. They are also used for access to digital data networks when such access is not available at the nearest telephone office. We presume the standard 300–3300 Hz telephone channels described in Chapter 1.

As we shall see, the dialed telephone circuit typically consists of a concatenation of two-wire subscriber lines and four-wire circuits in carrier systems. In order to multiplex a number of voiceband circuits on a carrier system, the individual voiceband channels are, as described in Chapter 1, bandlimited to roughly 300–3300 Hz and either frequency-translated (FDM) or converted to digital format, usually with a nonlinear quantization characteristic, and time-division multiplexed (TDM). These operations introduce many of the limitations and distortions perturbing data communications in general and echo cancellation in particular. The model we develop here is applicable, with some minor changes, to other bandlimited channels, such as satellite circuits.

Figure 9.7 illustrates the origins of most echoes on dialed telephone channels. In general, a long-distance two-wire telephone connection, with a station at each end, will begin and end with two-wire facilities (i.e., the subscriber line) but will consist largely of four-wire carrier facilities. The two-wire–four-wire interfaces, in telephone offices, are provided by hybrid couplers with compromise balancing impedances. Depending on the characteristics of the two-wire transmission segment that the telephone switch has connected to the hybrid coupler, the hybrid may or may not be well balanced. Note that we are talking about a hybrid coupler inside the telephone network, using a compromise balancing impedance for the different two-wire subscriber lines it will be connected to. There may also be hybrid couplers within terminating station sets, as illustrated in Figure 9.7.

As a result of the mismatch, an echo returns through the carrier system, as illustrated by the "distant echo" dashed line in Figure 9.7. It returns with

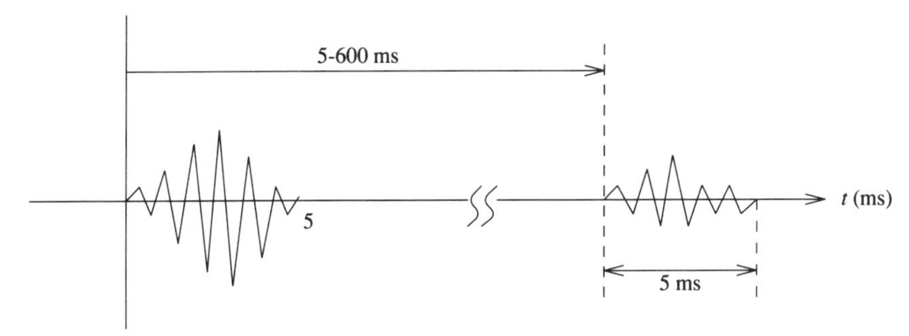

Fig. 9.8 Typical impulse response pattern of the baseband equivalent to a telephone echo channel. The echoes cluster in near and distant groups, each with a typical duration of perhaps 5 ms.

(mostly) linear distortion and at a delay ranging from 5 to 600 ms, depending on the electrical length of the circuit. The terrestrial fiber network has increased the maximum propagation delay in recent years. This round-trip signal is the "talker" distant echo. The distant echo returning to station A in Figure 9.7 results not only from leakage through the east hybrid coupler, but also from impedance mismatches all along the two-wire circuit between the east hybrid coupler and station B. It therefore exhibits dispersion, typically of the order of 5 ms (Figure 9.8).* It is possible for several distant echoes to return from several two-wire segments separated by four-wire transmission segments, but this is not usually the case for the public switched telephone network.

The distant echo may also return with phase perturbations contributed by the oscillators of FDM carrier systems. These are described, along with tracking methods for partially compensating them, in Section 9.5. Nonlinearities in the distant echo, contributed by companding devices (described in Chapter 1), cannot be canceled by the linear cancellation devices described in Section 9.3, but can be handled by the nonlinear "memory compensation" canceler defined in Section 9.4.

A near echo, composed of hybrid leakage at station A, reflections from impedance mismatches along the two-wire transmission path between station A and the two-wire–four-wire carrier system interface, and leakage through the west hybrid coupler, is also observed at the input to receiver A. This echo is ordinarily much larger, typically by 15 dB, than the distant echo, as suggested by Figure 9.8, although it, too, is dispersed over a relatively short interval. The separation between the near and distant echoes may, as already noted, be as large as 600 ms when there is a satellite in the four-wire system.

---

* It is important to note that the dispersion contributed by the transmitter filter, which constitutes part of the echo channel in the analysis of data driven echo cancelers, may contribute several additional milliseconds.

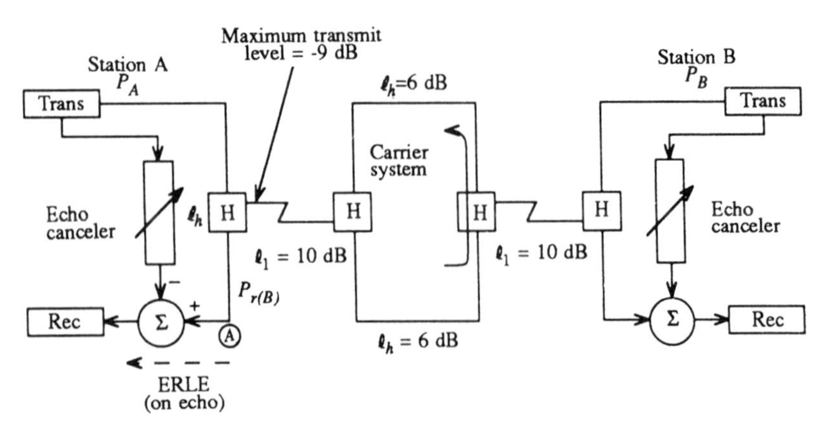

Definitions of terms:

ERLE	Echo return loss enhancement: the attenuation, in dB, of the echo, contributed by the echo canceler.
ℓ_h	Attenuation, in dB, of the hybrid leakage path from "transmit" to "receive" port. (Typical range equals 12 dB ± 4 dB.)
ℓ_1	Standard loss, in dB, of subscriber line. (10 dB would be close to the worst case.)
ℓ_2	Via net loss of carrier portion of transmission path.
P_A, P_B	Transmitter power, in dB.
P_{ea}	Power of echo, in dB. (-43 dBm is lowest level generally usable. Although unlikely to be that low on the public telephone network, it can happen on some private networks.)
$P_{r(B)}$	Power receiver from distant transmitter, in dB.
P_{near}	Power of near echo.
P_{dist}	Power of distant echo.

Fig. 9.9 Signal levels relevant to echo cancellation on a dialed telephone circuit. Noise is neglected. The pre-cancellation signal-to-echo ratio $SNR_{pre} = P_{r(b)} - P_{ea}$ is increased to $SNR_{post} = P_{r(B)} - (P_{ea} - ERLE)$ by the echo canceler.

The relative levels of received signal and total echo can vary widely. Nominally, the circuit losses can be roughly modeled as shown in Figure 9.9. The subscriber loops are often "padded out" to a standard loss ℓ_1 (dB) of around 10 dB, and the four-wire transmission plant has an intentional "via net loss" ℓ_2 of 4 dB on the average. The hybrid couplers both within the network and at station locations are presumed to have a leakage attenuation ℓ_h, between the incoming and outgoing lines of the four-wire circuit, in the range of 12 dB ± 4 dB. The attenuation is actually 3 dB greater, but in signal-to-echo calculations the 3 dB is canceled by the 3 dB loss in the normal signal path through the hybrid coupler, which is also neglected here.

With a local transmitted level P_A (in dB with respect to 1 mW into 600 ohms) and a distant-signal transmitted level of P_B, the received level of the desired signal at point A just before the canceler is (in dB)

$$P_{r(B)} = P_B - 2\ell_1 - 2\ell_2 . \tag{9.1}$$

At the same point, the received level of the near echo signal is

$$P_{near} = P_A - \ell_h , \tag{9.2}$$

while the level of the distant echo signal is

$$P_{dist} = P_A - 2\ell_1 - 2\ell_2 - \ell_h . \tag{9.3}$$

From these formulas, the distant echo is seen to be roughly $2(\ell_1 + \ell_2)$ dB below the near echo, a number with a nominal range of 10 to 22 dB. The signal-to-near-echo and signal-to-distant-echo ratios are, respectively,

$$\begin{aligned} P_{r(B)} - P_{near} &= P_B - 2\ell_1 - \ell_2 - P_A + \ell_h \\ &= (P_B - P_A) - 2\ell_1 - \ell_2 + \ell_h \end{aligned} \tag{9.4a}$$

and

$$\begin{aligned} P_{r(B)} - P_{dist} &= P_B - 2\ell_1 - \ell_2 - P_A + 2\ell_1 + \ell_2 + \ell_h \\ &= (P_B - P_A) + \ell_2 + \ell_h . \end{aligned} \tag{9.4b}$$

The ranges of these ratios are, for $P_A = P_B = -4$ dBm and $\ell_1 = 10$ dB,

$$-18 < P_{r(B)} - P_{near} < -6, \qquad 10 < P_{r(B)} - P_{dist} < 22.$$

It is fortunate that it is the near echo rather than the distant echo which requires a great deal of attenuation, since the near echo is more stable and generally less dispersed than the distant echo.

The improvement which the canceler must contribute is called the echo return loss enhancement (ERLE). For a 9600-bps QAM modem requiring a 25 dB SNR for an error rate below 10^{-7}, the required ERLE for the near echo ranges from 31 to 41 dB, and for the distant echo from 7 to 15 dB. On some dialed lines the maximum required ERLEs are actually more than these numbers suggest — perhaps more than 50 dB on the near echo — because of noise and excessive transmission loss. As a severe example, if $P_A = 0$ dBm, $\ell_h = 6$ dB, and $P_{r(B)} = 43$ dBm, then a total ERLE of 62 dB is required to achieve a 25 dB SNR.

In order for the echo canceler to work most effectively, the transmitted signals from the two ends of the circuit should be uncorrelated. This will become clear from later analysis, where the distant data signal, if present, will be regarded by the local adapting echo canceler as a form of interfering noise. To make the signals uncorrelated, data scramblers, with different scrambling patterns, must be provided between terminal and modem at each end.

Furthermore, to use data echo cancelers at station locations, echo suppressors and echo cancelers in the network have to be disabled, as has already been suggested. The echo suppressor, a network element in use since the late 1920s now largely replaced by the echo canceler, allows only one direction of transmission at a time and is obviously incompatible with full-duplex transmission. Network echo cancelers are incompatible to a lesser degree, going out of adjustment with certain spectrally degenerate data signals and with some signal power level scenarios in full-duplex transmission. These elements can be (temporarily) disabled with appropriate in-band signals.

9.2 THE ECHO CANCELLATION MODEL FOR DIGITAL SUBSCRIBER LINES

The copper twisted-pair lines used for subscriber access to telephone service at a telephone central office are modified for access to a digital network (Figure 9.10) by installing echo-canceling transceivers at the two ends, just as the two-way transmission on a telephone line. The major differences between this circuit and the dialed telephone line are (as described in Chapter 1):

(1) The digital subscriber line is of limited distance, typically under 6000 meters.

(2) It is baseband (it passes low-frequency components).

(3) Transmission is at rates of 160 kbps full duplex and above, revealing entirely different channel characteristics and introducing new problems such as crosstalk.

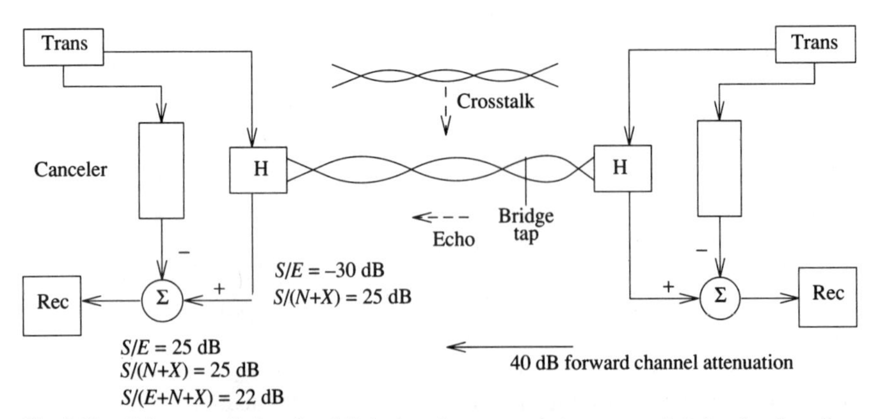

Fig. 9.10 Echo cancellation for full-duplex data transmission on a digital subscriber line. Crosstalk interference comes from other twisted pairs also carrying wideband data signals. Nominal ratios of signal to echo, signal to noise+crosstalk, and signal to echo+noise+crosstalk are shown for a channel requiring an echo cancellation (ERLE) of about 55 dB to achieve a final signal to "noise" ratio of about 22 dB.

(4) Transmission impairments characteristic of passband circuits in carrier systems (voice-band filtering, phase jitter, frequency offset) are absent.

The four impairments that must be traded off against one another are impulse noise, channel distortion, echo, and (mostly near-end) crosstalk (thermal noise is not an issue, usually). Echo comes from leakage of local transmitted signal through the hybrid to the local receiver, and from bridge taps and other impedance discontinuities in the subscriber line. Because of the limited distance, there is no distant echo; the echo is constrained to a total dispersion of perhaps a couple of milliseconds. The ratio of desired received signal to echo may be as low as –30 dB for a subscriber loop channel with 40 dB attenuation.

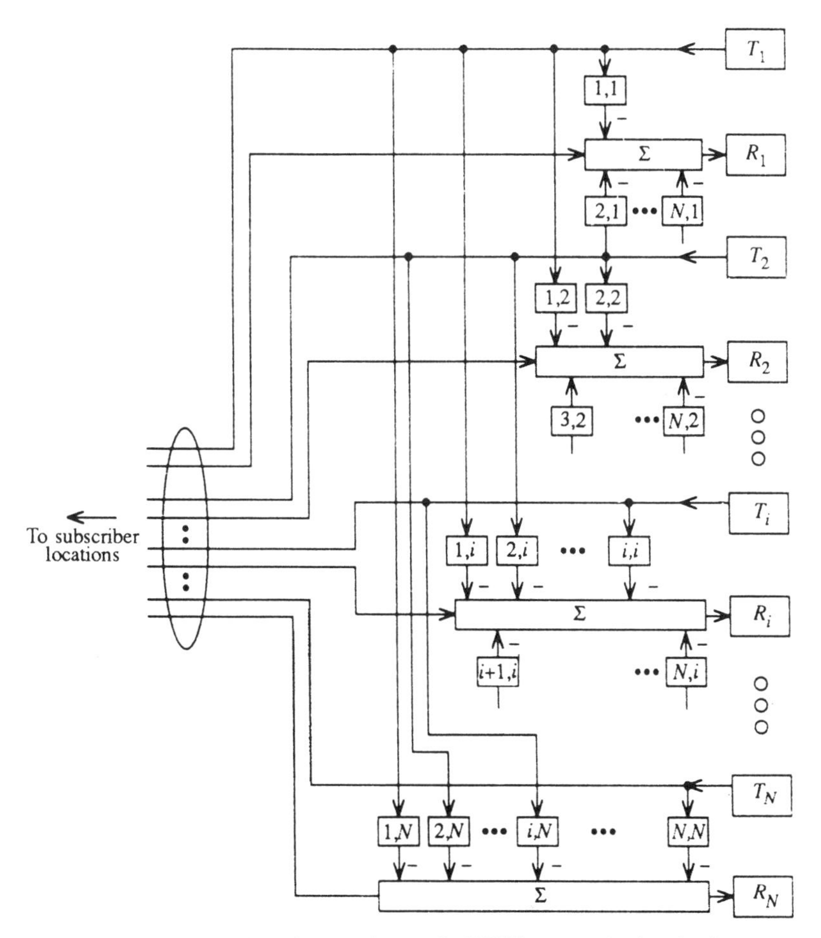

Fig. 9.11 Adaptive cancellation of near-end crosstalk (NEXT) at a serving location (e.g., a central office) for digital subscriber lines. The notation (i,j) within a box designates an adaptive filter replicating the crosstalk from transmitter i to receiver j.

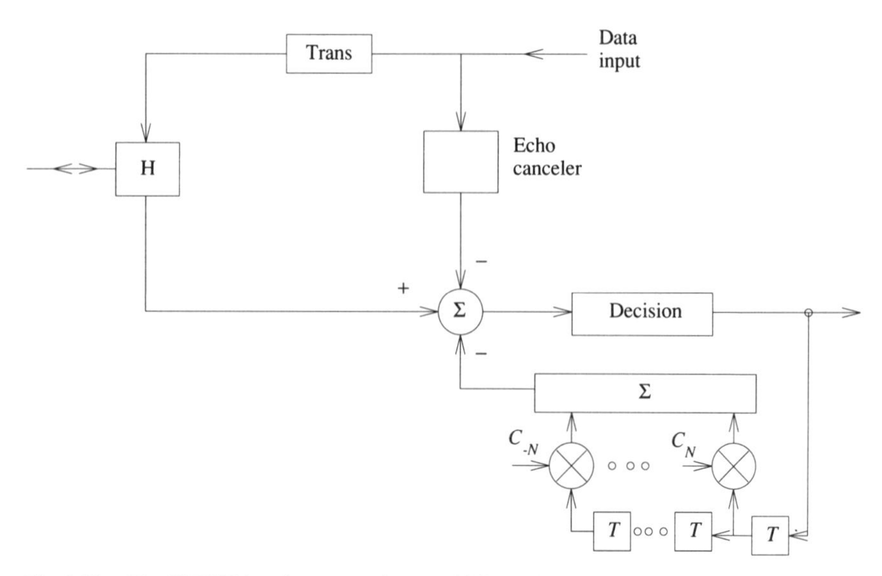

Fig. 9.12 The ISDN-U interface transceiver combining echo cancellation with decision-feedback equalization.

In order to realize a 25 dB signal-to-echo ratio, allowing a few dB margin for Gaussian noise and crosstalk, an ERLE of 55 dB is required [12].

Adaptive cancellation structures can, to a limited extent, be used to attenuate near-end crosstalk (NEXT) interference [11], an impairment characterized in Chapters 1 and 4. Consider the model of Figure 9.11 for the central office from which a bundle of digital subscriber lines (twisted pairs) radiate out to subscribers. The near-end crosstalk observed by one of the digital receivers comes from the other locally transmitted signals, which presumably are accessible and can be used to drive adaptive cancelers. Figure 9.11 shows a set of cancelers, one for each local transmitter contributing to the crosstalk interference, helping to clean up the signal supplied to the particular receiver. For N active line interfaces, N^2 adaptive cancelers would be required. Adaptation of this set of N^2 cancelers can be described as one large system adaptation [12]. There is also the possibility of NEXT cancellation at those subscriber locations, such as business offices, where subscriber terminations are clustered. In practice, it has proven difficult to access and utilize the locally transmitted signals to drive the N^2 cancelers.

A combination of techniques is required in the digital transceiver to bring all the distortions down to tolerable levels. For the ISDN Basic rate interface operating at 160 kbps, the preferred technique is to combine echo cancellation with decision-feedback equalization, as shown in Figure 9.12. As described in Chapter 7, the decision-feedback equalizer (DFE) can eliminate postcursor intersymbol interference but not precursor intersymbol interference. In the digital subscriber loop, the intersymbol interference tends to be mostly postcursor due to the ringing of bridge taps and the loss of energy near DC in the hybrids [11]. A DFE of reasonable length, such as 40 symbol intervals, is ordinarily adequate.*

9.3 FIR (TAPPED DELAY LINE) CANCELER STRUCTURES

Just as for the channel equalizers described in the last two chapters, the tapped delay line offers an easily described and analyzed structure for realizing an echo canceler. For echo cancellation, the tapped delay line realizes a replica, within the constraint of its finite length, of the echo, which is subtracted from the arriving echo. But other structures, with better performance under certain conditions, are also possible, including lattice filters and lookup tables (memory compensation). These are introduced in Section 9.4. We will also show, in the present section, that fractionally spaced tapped delay line echo cancelers offer the same advantages as fractionally spaced channel equalizers, in particular an output that can be interpolated to any sampling epoch, freeing the system from complex synchronization requirements. One important difference between equalization (which is basically channel inversion) and echo cancellation (which is system identification) is that for the latter system the number of tap weights required can be determined by inspection of the echo channel. As described in Chapter 8, it is not a straightforward operation to determine the required number of equalizer taps.

Assume, for the moment, a real baseband model, shown in Figure 9.13, for the two-way data communication system. We will later generalize to a passband model, which helps the discussion of echo cancellation for the dialed telephone channel, but the important principles are easiest to show in the real baseband mode. We assume, for the present discussion, that the echo channel is a linear filter $h_e(t)$; that is, the received signal at the input to the canceler of Figure 9.13 is

$$r_A(t) = s_B(t) + s_A(t)*h_e(t) + n(t), \tag{9.5}$$

where $s_A(t)$ is the local transmitted signal, $s_B(t)$ is the desired distant transmitted signal, and $n(t)$ is a combination of noise and crosstalk interference. The echo is the convolution of $s_A(t)$ with the echo channel. The canceler output is an echo replica, $q(t)$, which is subtracted from $r_A(t)$ to produce the receiver input.

9.3.1 VOICE-TYPE ECHO CANCELER

Figure 9.13 also illustrates the tapped delay line echo canceler, consisting of a transversal (FIR) filter with $2N+1$ taps, spaced at T' sec intervals, where

---

* For the 2B1Q (4-level PAM) line signaling described in Chapter 4, the baud for ISDN Basic access is 80,000 symbols/sec. Forty symbol-interval taps corresponds to a maximum dispersion of 0.5 ms. This may not correspond to all of the postcursor dispersion, but does include the great bulk of postcursor energy.

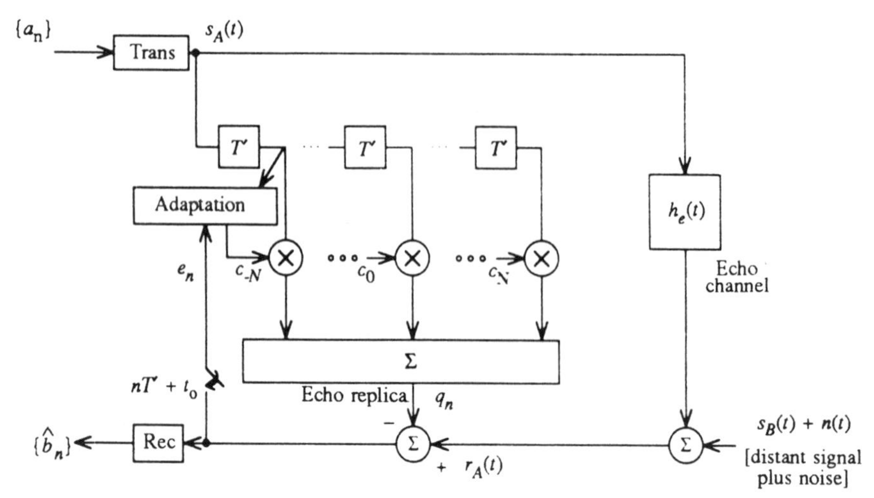

Fig. 9.13 Model for a "voice signal type" echo cancellation system. Adaptation is shown for one tap only.

$1/2T'$ is greater than the highest significant frequency in the received signal $r_A(t)$. The reader will recognize the maximum allowed value of T' as the Nyquist interval. The length $2NT'$ of the tapped delay line spans the maximum effective dispersion of the echo channel. The canceler as shown here is driven by the local transmitted line signal $s_A(t)$, and its tap weights, $c_{-N},...,c_0,...,c_N$ are adapted by a correlation between the tap voltages (at time nT') and the cancellation error e_n, as will be described.

The canceler of Figure 9.13 is a voice-type structure, identical to that widely used in the voice echo canceler of Figure 9.6. Although an echo canceler for data communication can be built this way [13], it has limitations which can be overcome by slightly different structures tailored to the data application. The major limitation is implementation complexity, requiring $2(2N+1)/T'$ high-resolution multiplications per second for filtering and tap updating in a digital implementation. Other problems concern convergence rate, stability, and residual error. *Data-driven* structures mitigating these problems are described in subsequent sections.

If the transversal filter's total length of $2NT'$ sec is greater than the dispersion of the echo channel, then any echo can theoretically be exactly replicated, and canceled. The fact that any echo can be replicated is seen in the transfer function

$$C_\infty(f) = \sum_{-\infty}^{\infty} c_m \, e^{-j2\pi m f T'} \tag{9.6}$$

of an infinite-tap transversal filter, which is a Fourier series which can generate an exact copy of any echo channel transfer function bandlimited to $1/2T'$ Hz.

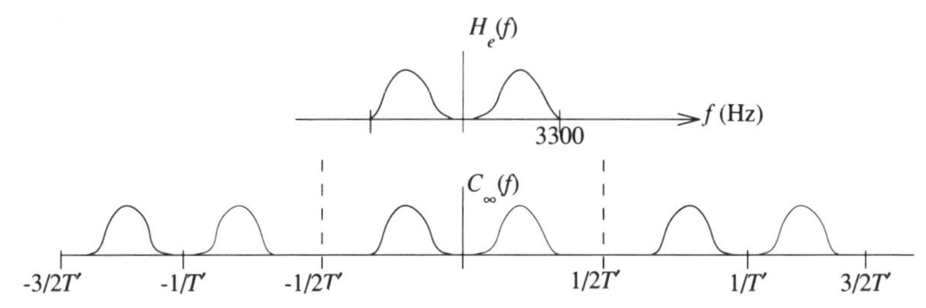

Fig. 9.14 Echo channel transfer function $H_e(f)$, and periodic transfer function generated by an ideal infinite-length transversal filter $C_\infty(f)$ with tap spacing T'.

As Figure 9.14 shows, the transfer function $C_\infty(f)$ is periodic with period T'. No aliasing distortion will result if there is no signal, echo, or noise energy at frequencies above $1/2T'$ Hz. The transversal filter limited to $2N+1$ taps, with transfer function

$$C_N(f) = \sum_{m=-N}^{N} c_m \, e^{-j2\pi mfT'} , \tag{9.7}$$

is all that is needed to generate any echo channel effectively bandlimited to $1/2T'$ Hz and with a dispersion effectively less than $2NT'$ sec. [Strictly speaking, $h_e(t)$ cannot be simultaneously bandlimited and time limited.] As mentioned above, the dispersion of an echo channel can be easily measured, allowing one to specify an appropriate number of taps. This is very different from the problem faced by a designer of a channel equalizer, where the delay spread determining the length of the transversal filter is not obvious from the transmission channel dispersion; it depends on the inverse of the channel characteristic in a complicated way, as derived in Chapter 8.

The active length of the echo canceler need not exceed the actual dispersion introduced by the echo channel. For the dialed telephone circuit, the distant echo may be delayed by as much as 600 ms, although the dispersion of the distant echo may be only a few milliseconds. Rather than make the tapped delay line 600 ms long, which could imply thousands of taps, the active taps can be provided in two shorter sections separated by a bulk delay, as shown in Figure 9.15. This matches the echo response suggested in Figure 9.8. The bulk delay must be determined by some separate means, such as an initial channel sounding (described below).

The voice-type echo canceler generates outputs not only at intervals T', but at any time instant, using the interpolation specified by the sampling theorem (Chapter 4). One could, in fact, construct a digital version of the canceler clocked to the sampling epoch τ of the local receiver A, so that clean symbol-interval samples at the canceler output would be available at the optimum times for sampling the desired distant signal $s_B(t)$. A slight mismatch in symbol rates between the transmitters at the two ends would be compensated by a continual

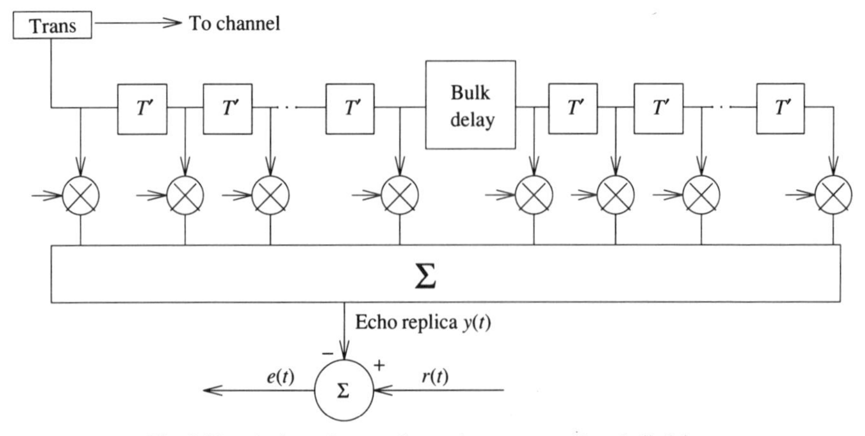

Fig. 9.15 Active echo canceler sections separated by a bulk delay.

tracking adaptation in the echo canceler. This synchronization mismatch is more difficult to handle in the data-driven cancelers described in subsequent sections.

By simple Fourier analysis (Exercise 9.2), the tap weights of the ideal echo canceler are the samples, at intervals T', of the impulse response $h_e(t)$ of the echo channel. One can easily show (Exercise 9.7) that this choice of tap weights minimizes the mean-square error at the canceler output provided the local and distant data trains are uncorrelated.

We have already noted that the distant signal $s_B(t)$ adds to the interfering noise at end A. This is equally true for the data-driven structures described in the rest of this chapter. For initial start-up, when the tap weights of the echo canceler may be far from their optimum settings, this "noise" can make convergence extremely slow. The common solution, which is specified, for example, in the V.32 standard for full-duplex modems on dialed telephone lines, is to initially adapt one end at a time. That is, transmitter A sends a training sequence and transmitter B is silent so that echo canceler A can adapt, and then transmitter B sends a training sequence and transmitter A is silent so that echo canceler B can adapt. The situation when both transmitters are simultaneously active (as in normal full-duplex data communication) is known as *double-talking*. Tracking of changes in the echo channels (one for each direction) is very slow during double talking, as implied above, but this is not a serious problem if the echo channels change very slowly and the cancelers are already adapted.

An alternative technique for startup is to measure the echo channel impulse response directly and "jam-set" the canceler tap weights as samples, at intervals T', of the measured impulse response. The impulse response of a fixed linear filter (which the echo channel is to a first approximation) is the expected value of the cross-correlation of a white-noise input with the filter output. Mathematically, if $n(t)$ is the white noise input with two-sided spectral power density equal

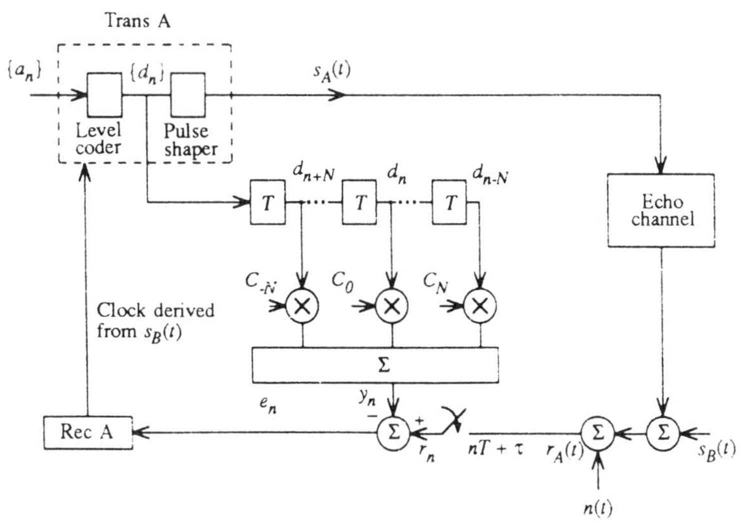

Fig. 9.16 Symbol-interval data-driven echo canceler. Adaptation, driven by top voltages and e_n, is not shown.

to 1 watt/Hz, $h_e(t)$ is the unknown echo channel impulse response, and $y(t)$ is the echo channel output, then the expected value of the cross-correlation of $n(t)$ with $y(t)$ is

$$E[n(t)\ y(t + \tau)] = \int_{-\infty}^{t+\tau} n(t)n(\beta)h_e(t + \tau - \beta)d\beta = h_e(\tau). \tag{9.8}$$

The expectation can be estimated by averaging a series of cross-correlations, and the initial tap values are simply these estimates at $\tau = mT'$, $-N \le m \le N$. An equivalent procedure is possible in a data-driven echo canceler [17].

In practice, a channel is sometimes used intermittently. In these cases, the tap weights from a previous period of activity can be saved and used as the initial settings for a new period of activity.

9.3.2 SYMBOL-INTERVAL DATA-DRIVEN ECHO CANCELER

We now begin consideration of *data-driven* echo cancelers [14], where the driving signal is the data stream at the input to the local transmitter rather than the line signal at its output. Our first model is the *symbol-interval data-driven canceler* of Figure 9.16. Here the transversal filter develops echo channel estimates at symbol intervals T, rather than the finer Nyquist intervals T', implying considerably fewer taps to span a given dispersion. For example, if the echo dispersion is 5 ms, a symbol-interval canceler with tap spacing $T = 1/2400$ sec requires 12 taps, while a canceler with Nyquist-interval taps of $T' = 125\ \mu s$

requires 40 taps. This substantially reduces the number of multiplications for tap updating. Also, each multiplication is much simpler, since the data value in the tapped delay line only takes on a few discrete values, e.g., four for four-level PAM.

Except for these important factors, the model remains the same as that of Figure 9.13. A linear echo channel $h_e(t)$ is estimated by the FIR filter, and the received signal $r_A(t)$ consists of the sum of echo, noise, and desired distant data signal.

It seems reasonable to build a canceler to cancel echoes at symbol intervals, since decisions are made by the local receiver only at symbol intervals, but this approach brings with it the difficult synchronization problems alluded to in the previous section. The times at which corrected samples must be generated for receiver A are the optimal sampling times for the data train modulating the received distant signal $s_B(t)$. The echo canceler serving receiver A is, on the other hand, driven by the data train modulating the local transmitted signal $s_A(t)$, and its outputs are necessarily synchronized with that data train, unlike the voice-type canceler, which has more flexibility. When the transmitters at the two ends of a dialed connection run on separate, unsynchronized clocks, as is normally the case, it is impossible to synchronize a symbol-interval canceler clocked by transmitter A with the local receiver clock derived from $s_B(t)$.

Even when transmitting at nominally the same rate, there is a drift between the data trains that requires special buffering and slippage correction. In contrast, the transmitters at the two ends of a subscriber line used for digital access (Section 9.2) are synchronized by a master clock arrangement, allowing symbol-interval cancellation, although even here there may be implementation conveniences from having canceler outputs at Nyquist intervals. Transmitter B provides the master clock, so that timing of transmitter A is controlled by the timing signal that receiver A derives from the data stream arriving from transmitter B.

Synchronization, when possible, offers a potentially large advantage in adaptive reference echo cancellation [21], described in Section 9.3.5. In this scheme, subtraction of an estimate of the distant data signal largely eliminates that signal as a component of the "noise" interfering with adaptation of the local echo canceler, leading to faster convergence and/or smaller residual cancellation errors.

9.3.3 GRADIENT ADAPTATION ALGORITHM

The "known gradient" (or steepest descent) algorithm, introduced in Chapter 8, gives insight into the behavior of a tapped delay line echo canceler. The stochastic least mean squares (LMS) adaptation algorithm is also applicable to echo cancelers. In this section, the known gradient algorithm for adaptation of the canceler tap weights is derived and analyzed. The algorithm could also be described for the voice-type canceler of Figure 9.13.

Referring to Figure 9.16, at time nT, define the tap-weight set as the column vector

$$\mathbf{c}_n = \left[c_{-N}(n), \ldots, c_N(n)\right]' , \tag{9.9}$$

and the tap voltage vector as

$$\mathbf{d}_n = \left[d_{n+N}, \ldots, d_{n-N}\right]' . \tag{9.10}$$

The constant and (for convenience) noncausal echo channel impulse response is

$$\mathbf{h}_e = \left[h_e(-NT + \tau), \ldots, h_e(NT + \tau)\right]' , \tag{9.11}$$

where τ is whatever sampling epoch is chosen by the receiver. Note that for different sampling epochs, there is effectively a different echo channel as far as the echo canceler is concerned. Note also that the transmitter pulse shaping is now included in $h_e(t)$. The sample at time nT of the received signal is

$$r_n = \mathbf{d}'_n \mathbf{h}_e + \nu_n , \tag{9.12}$$

where ν_n is the desired distant signal plus Gaussian noise at time nT (and is noise in its entirety from the point of view of canceling the echo signal). In (9.12) and throughout this book, the prime denotes transpose, so that $\mathbf{d}'_n$ is a row vector.

Whatever technique is used to specify the initial tap weight vector $\mathbf{c}_0$, we will try to adapt the tap weights to arrive at an eventual value $\mathbf{c}_{opt}$ which minimizes the mean-square error

$$MSE = E\, e_n^2 , \tag{9.13}$$

where the cancellation error at time nT is (from Figure 9.16)

$$e_n = \mathbf{r}_n - \mathbf{c}'_n\, \mathbf{d}_n = (\mathbf{h}_e - \mathbf{c}_n)'\, \mathbf{d}_n + \nu_n . \tag{9.14}$$

From (9.14) we have

$$\begin{aligned} MSE &= E\left\{(\mathbf{c}_n - \mathbf{h}_e)'\, \mathbf{d}_n \mathbf{d}'\, (\mathbf{c}_n - \mathbf{h}_e) - 2\nu_n\, (\mathbf{c}_n - \mathbf{h}_e)'\, \mathbf{d}_n + \nu_n^2\right\} \\ &= (\mathbf{c}_n - \mathbf{h}_e)'\, R(\mathbf{c}_n - \mathbf{h}_e) - 2(\mathbf{c}_n - \mathbf{h}_e)'\, \mathbf{p} + \sigma_n^2 , \end{aligned} \tag{9.15}$$

where

$$R = E\, [\mathbf{d}_n \mathbf{d}'_n] \tag{9.16}$$

is the reference signal covariance matrix. Note (Exercise 9.5) that if the data input levels to the tapped delay line are zero mean and uncorrelated, R is, except for a scale factor, the identity matrix. We shall, however, allow the possibility of a correlated sequence in the analysis below. That makes this analysis also applicable to the voice-type echo canceler, driven by line signal samples.

We further define the noise-data correlation vector,

$$\mathbf{p} = E\nu_n\, \mathbf{d}_n \,, \tag{9.17}$$

and the noise variance $\sigma^2 = E\, \nu_n^2$. From Chapter 8, recall that the quadratic form (9.15) is minimized by selecting tap weights

$$\mathbf{c}_{opt} = \mathbf{h}_e + R^{-1}\mathbf{p} \,. \tag{9.18}$$

If the "noise" ν_n (including the distant desired signal) is uncorrelated with the signal giving rise to the echo, i.e., $\mathbf{p} = 0$, the optimum tap weight setting is the echo channel impulse response, as has already been suggested.

As with the adaptive equalization discussion in Chapter 8, the known gradient algorithm adjusts the tap weight vector in the direction of the negative gradient of the MSE. We begin with

$$MSE = E\, e_n^2 = E\left[r_n - \sum_{m=-N}^{N} c_m\, d_{n-m} \right]^2 , \tag{9.19}$$

where $c_m = c_m(n)$ and d_{n-m} are understood as elements of $\mathbf{c}_n$ and $\mathbf{d}_n$, and r_n is the sample at time nT of the received signal (including echo) at the subtractor input. The MSE can be shown to be a convex function of the tap weights just as was shown in Chapter 8 for channel equalization.

The partial derivative of MSE with respect to c_m is $-2\, E(e_n d_{n-m})$, so the gradient with respect to $\mathbf{c}_n$ is $-2\, E(e_n \mathbf{d}_n)$ and the adaptation algorithm, moving in the negative gradient direction, is

$$\mathbf{c}_{n+1} = \mathbf{c}_n + \beta\, E(e_n \mathbf{d}_n) \,, \tag{9.20}$$

where β is the step size, which is arbitrary at this point.

To examine convergence, define the tap-weight error vector as

$$\xi(n+1) = \mathbf{c}_{n+1} - \mathbf{c}_{opt} \,.$$

Then the expected tap-weight error vector is

$$\mathbf{E}_{n+1} = E(\xi_{n+1}) = E\{\mathbf{c}_n + \beta\, e_n \mathbf{d}_n - \mathbf{c}_{opt}\} = [I - \beta R]\, \mathbf{E}_n \,, \tag{9.21}$$

where we have assumed that the tap voltage vector is uncorrelated with the tap weight vector. Thus

$$\mathbf{E}_n = [I - \beta R]^n\, \mathbf{E}_0 \,, \tag{9.22}$$

and diagonalizing R,

$$R = \mathbf{P}\, \Lambda\, \mathbf{P}' \,, \tag{9.23}$$

where $\mathbf{P}$ is the (orthonormal) matrix of (column) eigenvectors of R and Λ is the (diagonal) eigenvalue matrix, we have (Exercise 9.6)

$$(I - \beta R)^n = \mathbf{P}(I - \beta\Lambda)^n \mathbf{P}' \,. \tag{9.24}$$

Because R is symmetric and positive definite, it is invertible and has positive real eigenvalues. Its inverse R^{-1} is also symmetric. Placing (9.24) into (9.22) gives an iterative equation for the tap error:

$$\mathbf{E}_n = \mathbf{P}(1 - \beta\Lambda)^n \mathbf{P}' \mathbf{E} . \tag{9.25}$$

For $\mathbf{E}_n$ to converge to zero, each element must converge to zero. The fastest converging (and least stable) element corresponds to $\lambda_{\max}$; all elements will converge if this one does, i.e., if

$$\left| 1 - \beta\lambda_{\max} \right| < 1 ,$$

or

$$0 < \beta < 2/\lambda_{\max} . \tag{9.26}$$

For $\beta = 1/\lambda_{\max}$, an appropriate value for rapid convergence, the slowest-converging element of $\mathbf{E}_n$ is the factor

$$\left[1 - \beta\lambda_{\min}\right]^n = \left[1 - \frac{\lambda_{\min}}{\lambda_{\max}}\right]^n . \tag{9.27}$$

The results (9.26) and (9.27) put a cap on the step size and show how the rate of convergence is decreased with increasing eigenvalue spread.

EXAMPLE 1 Convergence of ideal gradient algorithm for uncorrelated and correlated driving signals

1. Uncorrelated driving signal samples $\{d_m\}$, and a 2-tap canceler. Assume that the $\{d_m\}$ are uncorrelated random variables. Then if each element has variance A,

$$R = E\mathbf{d}_n\mathbf{d}_n' = \begin{bmatrix} A & 0 \\ 0 & A \end{bmatrix}, \quad \Lambda = \begin{bmatrix} A & 0 \\ 0 & A \end{bmatrix}.$$

   Let $\beta = 1/\lambda_{\max} = 1/A$. Then $I - \beta\Lambda = 0$, implying $\mathbf{E}_1 = 0$, independent of the starting value $\mathbf{c}_0$ of the tap-weight vector. Convergence, in this degenerate example, occurs in one iteration! This will not be the case for the stochastic adaptation algorithm, even when, as is frequently the case, the elements of $\mathbf{d}_n$ are uncorrelated and R is a diagonal matrix, with unity eigenvalue ratio.

2. Correlated signal samples $\{d_n\}$. Assume $\mathbf{h}_e' = [1, -1]$. Then $\mathbf{c}_{opt}' = [1\ -1]$.

Assume further that the correlation of the signal samples $\{d_n\}$ is such that

$$R = E\mathbf{d}_n\mathbf{d}_n' = A\begin{bmatrix} 2 & -1 \\ -1 & 2 \end{bmatrix}.$$

The eigenvalues of R are found from the determinant equation

$$0 = |R - \lambda I| = \begin{vmatrix} 2-\lambda & -1 \\ -1 & 2-\lambda \end{vmatrix} = \lambda^2 - 4\lambda + 3\,.$$

Thus the eigenvalue and eigenvector matrices are

$$\Lambda = \begin{bmatrix} 3 & 0 \\ 0 & 1 \end{bmatrix}, \quad P = \begin{bmatrix} 1 & 1 \\ -1 & 1 \end{bmatrix}$$

Let $\beta = 1/\lambda_{\max} = 1/3$. Then

$$I - \beta\Lambda = \begin{bmatrix} 1 - \beta\lambda_1 & 0 \\ 0 & 1 - \beta\lambda_2 \end{bmatrix} = \begin{bmatrix} 0 & 0 \\ 0 & 1/3 \end{bmatrix}$$

and

$$P(1 - \beta\Lambda)^n P' = \begin{bmatrix} 1 & 1 \\ -1 & 1 \end{bmatrix} \begin{bmatrix} 0 & 0 \\ 0 & 1/3 \end{bmatrix} \begin{bmatrix} 1 & -1 \\ 1 & 1 \end{bmatrix} = (1/3)^n \begin{bmatrix} 1 & 1 \\ 1 & 1 \end{bmatrix}.$$

Thus equation (9.25) implies that both elements of the tap-weight errors vector decrease as $(1/3)^n$.

For the idealized model we have been considering, in which the gradient of the mean-square error is assumed known, the expected tap-weight error converges to zero, as (9.25) indicates. We will see in the next section that for stochastic adaptation, which is a practical approximation to the gradient algorithm, a residual error remains for any finite step size.

9.3.4 STOCHASTIC (LMS) ADAPTATION ALGORITHM

When ensemble expectations are unknown, time averages must suffice. That is the essential motivation for the stochastic LMS adaptation algorithm here and in Chapter 8. This algorithm adapts in the general direction of the negative gradient of the mean-square error through a long series of small random movements whose time-averaged value is in the correct direction.

Explicitly, the least mean-square (LMS) algorithm utilizes the gradient of the *instantaneous* squared error, rather than the mean-square error. This gradient is

$$\partial e_n^2 / \partial \mathbf{c}(n) = e_n [\partial e_n / \partial \mathbf{c}(n)] = -2e(n)\mathbf{d}(n) , \tag{9.28}$$

so that the stochastic adaptation algorithm is

$$\mathbf{c}(n+1) = \mathbf{c}(n) + \beta e(n)\, \mathbf{d}(n) . \tag{9.29}$$

The properties of (9.29) can be evaluated through examination of either the mean-square tap-weight error or the mean-square cancellation error at the *n*th iteration. We choose the latter, but the analysis begins with definition of the tap-weight error vector. This is, assuming the local data signal is uncorrelated with the distant data signal and the additive noise so that the ideal echo canceler tap weights are the echo channel samples,

$$\xi(n) = \mathbf{c}(n) - \mathbf{c}_{opt} = \mathbf{c}(n) - \mathbf{h}_e . \tag{9.30}$$

From (9.29) and (9.30)

$$\xi(n+1) = \xi(n) + \beta e(n)\mathbf{d}_n . \tag{9.31}$$

Since $\nu(n)$ denotes the combination of received distant signal and noise in Figure 9.16, the error at the subtractor output is

$$e(n) = [\mathbf{h}_e' - \mathbf{c}'(n)]\, \mathbf{d}(n) + \nu(n) = -\xi'(n)\, \mathbf{d}(n) + \nu(n) . \tag{9.32}$$

Then the MSE is

$$E\, e_n^2 = E\left\{(-\xi_n'\mathbf{d}_n + \nu(n))(-\mathbf{d}_n'\, \xi_n + \nu(n))\right\}$$

$$= E\left\{\xi_n'\mathbf{d}_n\mathbf{d}_n'\xi_n\right\} - 2\, E\left\{\nu(n)\xi_n'\mathbf{d}_n\right\} + E\nu^2(n) . \tag{9.33}$$

With the assumption, nonrigorous but in practice, that the tap weights are essentially uncorrelated with the current data vector, the first expectation is

$$E\{\xi_n'\mathbf{d}_n\mathbf{d}_n'\xi_n\} = E\{\xi_n' E\, R\xi_n\} ,$$

where the matrix $R = \mathbf{d}_n\mathbf{d}_n'$ (prior to taking the expected value) has elements

$$r_{jk} = d_{n-j}d_{n-j} . \tag{9.34}$$

The expectation of R is the diagonal matrix AI, where $A = E\{d^2(n+i)\}$ is the power in any data symbol, and recalling that the data symbols are mutually uncorrelated.

The second term on the right hand side of (9.33) is zero, since the noise (including desired distant signal) is uncorrelated with the echo or its replica.

Thus (9.33) becomes

$$E\, e_n^2 = A\, E\{\xi_n'\, \xi_n\} + \sigma^2 \,, \tag{9.35}$$

where $\sigma^2 = E\{\nu_n^2\}$.

At the $(n+1)$th iteration,

$$e_{n+1} = \xi_{n+1}' \mathbf{d}(n+1) + \nu_{n+1} \,, \tag{9.36}$$

so that (9.35) becomes

$$\begin{aligned} E\{e_{n+1}^2\} &= A\, E\{\xi_{n+1}' \xi_{n+1}\} + \sigma^2 \\ &= \left[1 - 2\beta A + \beta^2 (2N+1) A^2\right] E\{e_n^2\} + 2\beta A\, \sigma^2 \,. \end{aligned} \tag{9.37}$$

Following Mueller [15] and Werner [16], the solution for this recurrence relationship is

$$\begin{aligned} E\{e_n^2\} &= \left[1 - 2\beta A + \beta^2 (2N+1) A^2\right]^n E\{e_0^2\} \\ &+ \frac{1 - \left[1 - 2\beta A + \beta^2 (2N+1) A^2\right]^n}{1 - \left[1 - 2\beta A + \beta^2 (2N+1) A^2\right]} [2\beta A \sigma^2] \,. \end{aligned} \tag{9.38}$$

The expected square error decreases if

$$\left|1 - 2\beta A + \beta^2 (2N+1) A^2\right| < 1 \,,$$

i.e.,

$$0 < \beta < 2A/(2N+1)A^2 \,. \tag{9.39}$$

As with adaptive equalization, the step size is inversely proportional to the number of taps. Intuitively, the more taps contributing random fluctuations, the slower one must adapt the tap weights by the stochastic algorithm. It is this consideration, not the eigenvalue ratio (which was the most prominent factor in determining convergence rate for the ideal gradient algorithm), which limits the convergence rate for the stochastic algorithm.

The step size providing the fastest convergence rate may be derived by equating to zero the derivative of the bracketed term in (9.38), yielding

$$\beta = 1/\left[(2N+1)A\right] . \tag{9.40}$$

Substituting (9.40) into the first term of (9.38) shows that convergence proceeds as

$$E\{e_n^2\} = \left[1 - 1/(2N+1)\right]^n E\{e_0^2\} \,. \tag{9.41}$$

The residual error after an infinite number of iterations ($n \to \infty$) is found from

(9.38) to be

$$E\{e_\infty^2\} = \sigma^2 / \left[1 - (2N + 1)\beta A/2\right], \tag{9.42}$$

which is the unavoidable "noise" variance, σ^2, increased by a factor which is greater than one for any $\beta \neq 0$. Since the "noise" with variance σ^2 is largely the desired distant signal, a high received SNR (ratio of desired distant signal to echo plus other noise) requires that the excess in $E\{e_\infty^2\}$ above σ^2 be much smaller than σ^2.

It is important to note that convergence for the LMS algorithm *cannot* occur in one step, which *is* possible for the known-gradient algorithm. Even with a unity eigenvalue ratio, for independent data symbols, the structure of the LMS algorithm results in exponential convergence, as per (9.41).

EXAMPLE 2 Suppose that an echo canceler must provide a ratio of desired distant signal to echo + noise of 20 dB. There is a 33-tap FIR filter in the canceler, a symbol power $A = 1$, and additive Gaussian noise 26 dB below the desired distant signal. What is the maximum asymptotic step size β_∞?

ANSWER: $\sigma^2 = P_B + \sigma_n^2$, where P_B is the power in the desired distant signal and σ_n^2 is the power in the Gaussian noise. Since $10 \log_{10}(P_B/\sigma_n^2) = 26$, $P_B = 400\ \sigma_n^2$, and $\sigma^2 = 1.0025\ P_B$.

For a ratio of desired signal to echo + noise of 20 dB, $E\{e_\infty^2\} = 1.01\ P_B$, i.e., the excess MSE is 1/100 (20 dB below) P_B. Thus $1.01\ P_B = 1.0025\ P_B/[1 - 33\beta/2]$, or $\beta = 0.00045$.

Figure 9.17 illustrates some experimental results [16] for SNR vs. step size, with a 32-tap filter, an (uncancelled) echo-to-far-signal ratio of 20 dB, and with and without additive Gaussian noise 24 dB below the distant signal. Convergence characteristics are shown in later sections.

9.3.5 LEAST-SQUARES (KALMAN) ADAPTATION ALGORITHM

The rate of convergence of the stochastic adaptation (LMS) algorithm is such that even with optimum step sizes, the number of iterations required to converge can be an order of magnitude greater than the number of taps to be adapted. As described in Chapter 8, least squares algorithms, minimizing a weighted sum of a number of most recent errors, can, in contrast, be adapted in a number of iterations not much greater than the number of taps to be adapted. This advantage comes with a price, consisting partly of increased complexity and partly of instability under certain conditions. The complexity price can be considerably reduced by the (computationally) fast Kalman algorithm, which brings the complexity down to the order of the gradient algorithm.

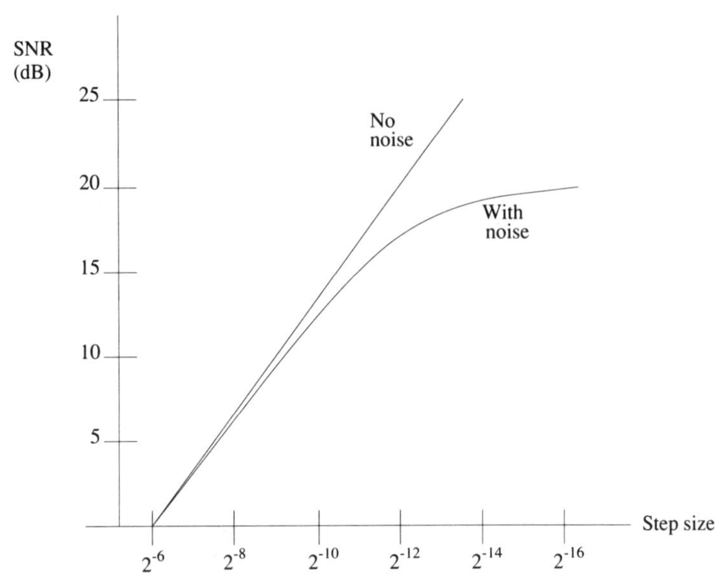

Fig. 9.17 SNR (desired distant signal to echo+noise) vs. asymptotic step size [16].

After $(2N+1)$ adaptation steps, the $(2N+1)$-tap transversal filter has had $(2N+1)$ successive tap voltage vectors $\mathbf{d}_n$ which, with high probability, are (algebraically) linearly independent. They span a $(2N+1)$-dimensional signal space and can be used as a basis to represent any vector in that space. The Kalman algorithm is an (implicit) Gram-Schmitt orthogonalization that (implicitly) builds a set of basis vectors that can represent a $\mathbf{c}_{opt}$ minimizing the accumulated squared error. If the $(2N+1)$ tap voltage vectors are not linearly independent, then "wait a little longer," and convergence is in mean-square for an arbitrary "random" input.

Following the development in Section 8.5 and referring to the notation of Figure 9.16, we assume a transversal (tapped delay line) filter with $(2N + 1)$ taps. At the nth iteration, the tap weight vector is $\mathbf{c}_n$, with components c_{nj}, and the output of the canceler filter is

$$y_n = \mathbf{c}_n' \mathbf{d}_n = \sum_{k=-N}^{N} c_{nk} d_{n-k} . \tag{9.43}$$

At this iteration, the least squares estimate of the echo channel $\mathbf{h}_e$ is that which selects the tap weight vector $\mathbf{c}_i$ such that

$$S_n = \sum_{j=0}^{n} w_j e_j^2 \tag{9.44}$$

is minimized, where the $\{ w_j \}$ are a set of weights, ordinarily heavier for recent than for old errors. A typical weighting is

$$w_j = \Delta^{n-j} , \tag{9.45}$$

where Δ is a positive number slightly less than one. The error e_j in (9.44) is given by

$$e_j = r_j - \mathbf{c}'_n \mathbf{d}_j , \tag{9.46}$$

where r_j is the received signal sample, including the echo component. Note that the tap weight vector used to define e_j is the latest one, not the tap weight vector that was available at the jth iteration. Thus e_j would have to be recomputed, for all j, $0 \le j \le n$, at each iteration, if it were used in the tap adjustment algorithm, which it is not.

The minimum of S is found from the $(2N + 1)$ equations resulting from equating the partial derivatives of S (with respect to the tap weights c_{nk}, $k = -N,..., N$) to zero:

$$0 = \partial S / \partial c_{nk} = 2 \sum_{j=0}^{n} w_j d_{j-k} \left[r_j - \sum_{\ell=-N}^{N} c_{n\ell} d_{j-\ell} \right] , \quad -N \le k \le N ,$$

or

$$\sum_{j=0}^{n} \Delta^{n-j} r_j d_{j-k} = \sum_{\ell=-N}^{N} c_{n\ell} \sum_{j=0}^{n} \Delta^{n-j} d_{j-\ell} d_{j-k} , \quad -N \le k \le N . \tag{9.47a}$$

In vector format, (9.47a) may be written as

$$\sum_{j=0}^{n} \Delta^{n-j} r_j \mathbf{d}_j = \left[\sum_{j=0}^{n} \Delta^{n-j} \mathbf{d}_j \mathbf{d}'_j \right] \mathbf{c}_n . \tag{9.47b}$$

This is a set of n linear equations for the $(2N+1)$ elements of $\mathbf{c}_n$. When n reaches $2N+1$, and if the linear independence property holds, it should be possible to unambiguously represent $\mathbf{c}_{opt}$. Solving for $\mathbf{c}_n$,

$$\mathbf{c}_n = R_n^{-1} \sum_{j=0}^{n} \Delta^{n-j} r_j \mathbf{d}_j , \tag{9.48}$$

where

$$R_n = \sum_{j=0}^{n} \Delta^{n-j} \mathbf{d}_j \mathbf{d}'_j . \tag{9.49}$$

As in Chapter 8, the object is to use a recursive algorithm for $\mathbf{c}_n$ in order not to have to calculate $\mathbf{c}_n$ from (9.48) each iteration. Noting that

$$R_n = \Delta R_{n-1} + \mathbf{d}_n \mathbf{d}'_n \tag{9.50}$$

and using the matrix inversion lemma, and Kalman gain vector given by equations (8.88) and (8.89) of Chapter 8 (with α replaced by Δ and $\mathbf{r}_n$ replaced by $\mathbf{d}_n$), we arrive at the recursion relationship

$$\mathbf{c}_n = \mathbf{c}_{n-1} - e_n \mathbf{K}_n \tag{9.51}$$

by the steps outlined in Chapter 8 to derive (8.95). Here the Kalman gain vector is

$$\mathbf{K}_n = \frac{R_{n-1}^{-1}\, \mathbf{d}_n}{\Delta + \mathbf{d}_n'\, R_{n-1}^{-1}\, \mathbf{d}_n}\,. \tag{9.52}$$

At each iteration, the Kalman gain is recomputed (using the matrix inversion lemma (8.88) to update R_{n-1}^{-1}) and used in (9.51) to generate the new $\mathbf{c}_n$.

Some results using the Kalman algorithm are given in Section 9.4.1.

9.3.6 DATA-DRIVEN ECHO CANCELLATION: THE FRACTIONALLY-SPACED CANCELER

We noted earlier that echo canceler output samples are needed at Nyquist intervals rather than at symbol intervals if the local receiver is unsynchronized with the local data transmitter. If the echo canceler, locked to the local transmitter's clock, can produce outputs at Nyquist intervals, an interpolator can derive samples at the sampling times, locked to the distant transmitter, demanded by the local receiver.

A line-signal-driven echo canceler satisfies this requirement, but at a formidable cost in complexity. In a digital implementation, if there are M taps, at Nyquist intervals T', to span the echo channel dispersion, the filtering operation $\mathbf{c}_n'\, \mathbf{s}_A(n)$ requires (for a real baseband canceler) M multiplications and additions each T' seconds. The filtering multiplications, if done digitally, will involve tap weights c_{nm} of about 12 bits precision and tap voltages of about 8 bits precision (see Chapter 10).

The updating operation in the stochastic adaptation algorithm,

$$\mathbf{c}_{n+1} = \mathbf{c}_n + \beta e_n \mathbf{s}_A(n), \tag{9.53}$$

similarly requires M multiplications and additions, so that there is a total of $2M$ multiplications and additions per Nyquist interval. If the variables used above represent an equivalent (complex) baseband model of a passband canceler, as would actually be the case in practice, the number of real multiplications is four times as large. Thus there could be $8M$ high-precision multiplications per T' interval. If T' were one-fourth the symbol interval T, as it may be in practice, there would be $32M$ multiplications per symbol interval T, or $32M/T$ multiplications per second. With $N = 128$ taps and $T = 1/2400$ sec, the requirement would be close to 10 million high-precision multiplications per second.

The severity of this requirement stimulated development of *data-driven* echo cancelers, as described in the previous section, in which outputs are computed only at symbol intervals and the multiplications are simpler because tap voltages are restricted to a very few values, e.g., two if the data are binary [18]. The echo canceler replicates the combination of transmitter filter and echo channel. But this solution does not provide samples at Nyquist intervals, and requires synchronization of the two transmitters. Fortunately techniques have

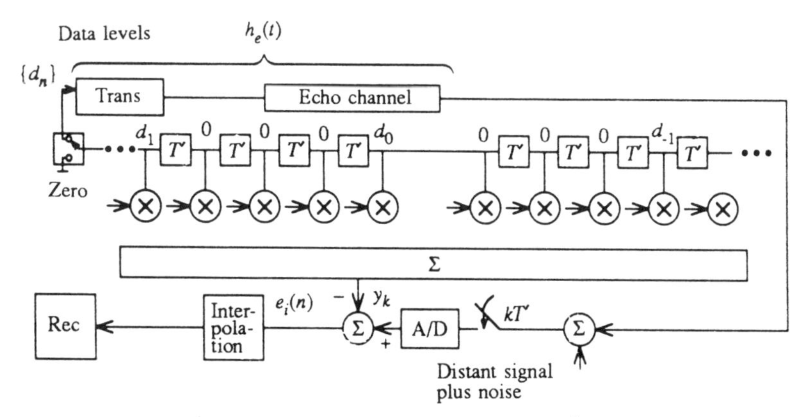

Fig. 9.18 Fractionally-spaced echo cancellation, illustrated for $T' = T/4$. Adaptation is not shown.

been found to simultaneously obtain most of the advantages of both data-driven and unsynchronized (Nyquist-interval) cancellation techniques [16,19]. The number of multiplications, as well as the complexity of each multiplication, is substantially reduced from the requirements of a straightforward line-signal-driven Nyquist-interval canceler.

As shown in Figure 9.18, the canceler is similar to that of Figure 9.13 (for the line-signal-driven model) in that it updates tap weights at Nyquist intervals T' that are a fraction of T, but the reference signal driving the canceler is the data train feeding the local transmitter rather than the line signal produced by the local transmitter. Since data occur only at symbol intervals T, the reference signal driving the canceler at intervals T' is nonzero only part of the time. If $T' = T/4$, then only one in four reference signal inputs to the canceler is nonzero. We will show that this structure produces exactly the same echo estimates as the line-signal-driven canceler which had nonzero inputs every T' seconds. The Nyquist-interval outputs, locked to the local data transmitter, are interpolated, as described above, to produce samples at the times desired by the local receiver.

This fractionally-spaced canceler retains the computational simplicity of the symbol-interval data-driven echo canceler, with arithmetic using a few data levels, and most reference inputs equal to zero. In fact, as we will show, it consists of ℓ interleaved symbol-interval cancelers, where $\ell = T/T'$. It allows complete independence of operation of the stations at the two ends of the line, just as with the line-signal-driven canceler.

To derive the structure of Figure 9.18, assume a received echo signal

$$r_e(t) = \sum_m d_m h_e(t - mT) , \tag{9.54}$$

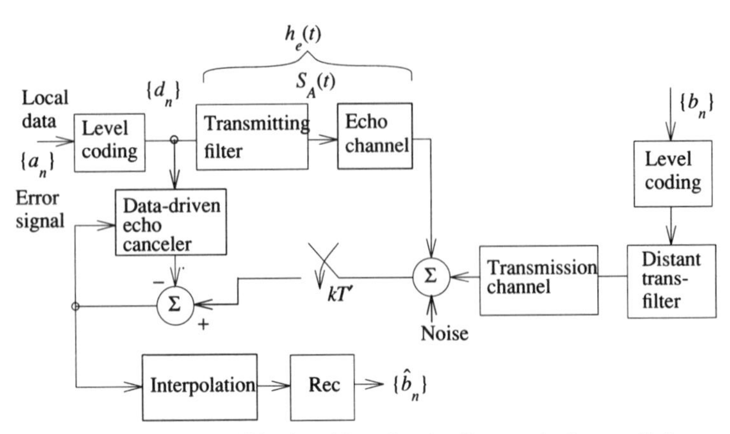

Fig. 9.19 System model for data-driven, fractionally-spaced echo cancellation.

where $\{ d_m \}$ is the data train (modulation levels) feeding the local transmitter's pulse generator, and $h_e(t)$ is the overall echo channel, including the local transmitter filter, as shown in Figure 9.19.

At intervals T', samples of the received echo signal are

$$r_e(kT') = \sum_m d_m h_e(kT' - mT) \ . \tag{9.55}$$

Assuming $T' < 1/2f_{\max}$, where $f_{\max}$ is the highest frequency in $r_e(t)$, the sample train is sufficient to determine $r_e(t)$ for *any* value of t.

Suppose that

$$T' = T/\ell \ , \tag{9.56}$$

where ℓ is an integer, e.g., $\ell = 4$. Then

$$r_e(kT') = \sum_m d_m h\left[(k - m\ell)\,T'\right] , \tag{9.57a}$$

and defining $h_k = h_e(kT')$,

$$r_e(kT') = \sum_m d_m h_{k-m\ell} \ . \tag{9.57b}$$

This echo-channel convolution can be duplicated in an infinite transversal filter with tap weights ($c_i = h_i$). Note in (9.57) that only every ℓth tap weight is used. Different sets of tap weights, spaced ℓ taps apart in each case, are used for $k = 0, k = 1, ..., k = \ell - 1$. For $k + \ell$, $r_e((k + \ell)T')$ requires the same set of tap weights as for computing $r_e(kT')$, but with the data shifted to the right. These operations can be realized by the "zero stuffer" canceler shown in Figure 9.18.

What we effectively have is ℓ interleaved *symbol-interval* transversal filters corresponding to ℓ timing instants within a symbol interval. The ℓ cancelers are

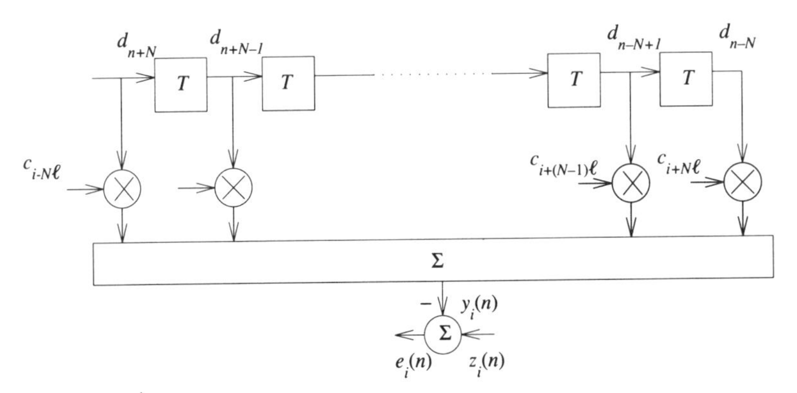

Fig. 9.20 Symbol-interval data-driven echo canceler, the ith subcanceler of the fractionally-spaced data-driven echo canceler.

operated and adapted independently. This becomes clearer with a further refinement of notation. Let

$$kT' = nT + iT' = (n\ell + i)T' , \tag{9.58}$$

where n is the largest multiple of T contained in kT', and $i = k \bmod(\ell)$. Let the echo replica be

$$y_i(n) = y(kT') = \sum_m d_m c_{k-m\ell}(n)$$

$$= \sum_{j=-N}^{N} d_{n-j} c_{i+j\ell}(n), \text{ all } n, i = 0,1,\ldots,\ell - 1 . \tag{9.59}$$

For each of the ℓ values of i, (9.59) describes a symbol-interval transversal filter (Figure 9.20) with $(2N + 1)$ tap weights.

Defining $\mathbf{c}_i(n)$ as the tap weight set that generates $y_i(n)$, with jth element $c_{i+j\ell}(n)$, and $\mathbf{d}(n)$ as the vector $(d_{n+N}, \ldots, d_{n-N})$ of data levels, (9.59) becomes

$$y_i(n) = \mathbf{c}_i'(n)\, \mathbf{d}(n) . \tag{9.60}$$

Each of the ℓ interleaved cancelers is updated once each symbol interval, according to the stochastic algorithm described in the previous section. Explicitly, the error in the output of the ith canceler at the nth symbol interval is

$$e_i(n) = \underbrace{\mathbf{h}_i'\mathbf{d}(n)}_{\substack{\textit{echo sample at}\\ t\,=\,(n\ell+i)T'}} + \underbrace{\mathbf{g}_i'\mathbf{b}(n)}_{\substack{\textit{received distant}\\ \textit{signal at } t\,=\,(n\ell+i)T'}} + \underbrace{\eta_{n\ell+i}}_{\textit{noise}} - \underbrace{\mathbf{c}_i'(n)\mathbf{d}(n)}_{\textit{echo replica}} , \tag{9.61}$$

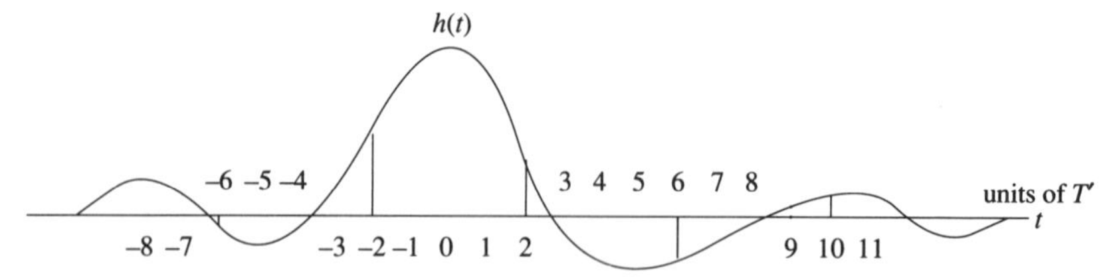

Fig. 9.21 Interleaved vectors of echo-channel samples for $\ell = 4$. h_i samples shown for $i = 2$.

where $\mathbf{g}_i$ represents the direct transmission channel from the distant transmitter to the local receiver, $\mathbf{b}(n)$ is the data vector at the distant transmitter, and

$$\mathbf{h}_i = h_e(iT' - NT), \ldots, h_e(iT' + NT), \quad i = 0,1,\ldots,(\ell - 1) \qquad (9.62)$$

is the ith of ℓ interleaved vectors of echo channel impulse response samples at symbol intervals (Figure 9.21).

The tap updating formula is, from (9.29),

$$\mathbf{c}_i(n+1) = \mathbf{c}_i(n) + \beta e_i(n)\mathbf{d}(n), \quad i = 0,\ldots,\ell-1.$$

Thus each symbol interval, each of the ℓ symbol-interval subcancelers, is updated according to the usual LMS algorithms. As with channel equalizers, it may be desirable to use a larger step size β at first for rapid convergence and a smaller one later for small residual error.

The convergence characteristics of the interleaved symbol-interval cancelers may vary to some extent, because the covariance matrices of the ℓ symbol-interval sampled echo channels may differ significantly. However, in practice, the convergence characteristics are similar. Figure 9.22 illustrates convergence from simulation experiments using a typical echo channel characteristic.

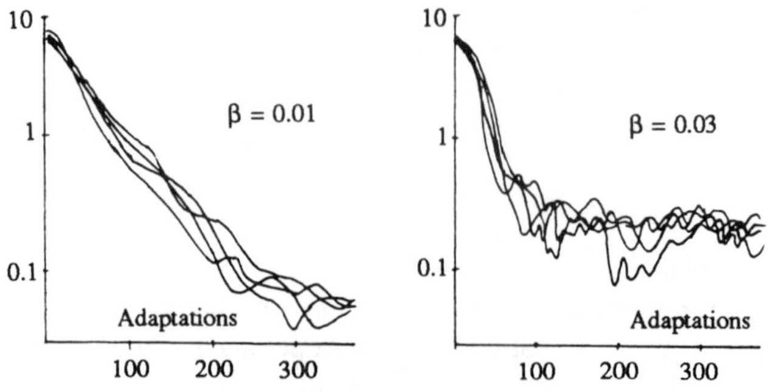

Fig. 9.22 Simulation runs, at two step sizes, for convergence of the four interleaved symbol-interval subcancelers of a fractionally-spaced echo canceler with two 10-tap "near" and "distant" echo sections, operating on the echo channel $h_e(t) = h_0(t-t_a) + h_0(t-t_b)$. Noise was present at a level 10 dB below the total echo power.

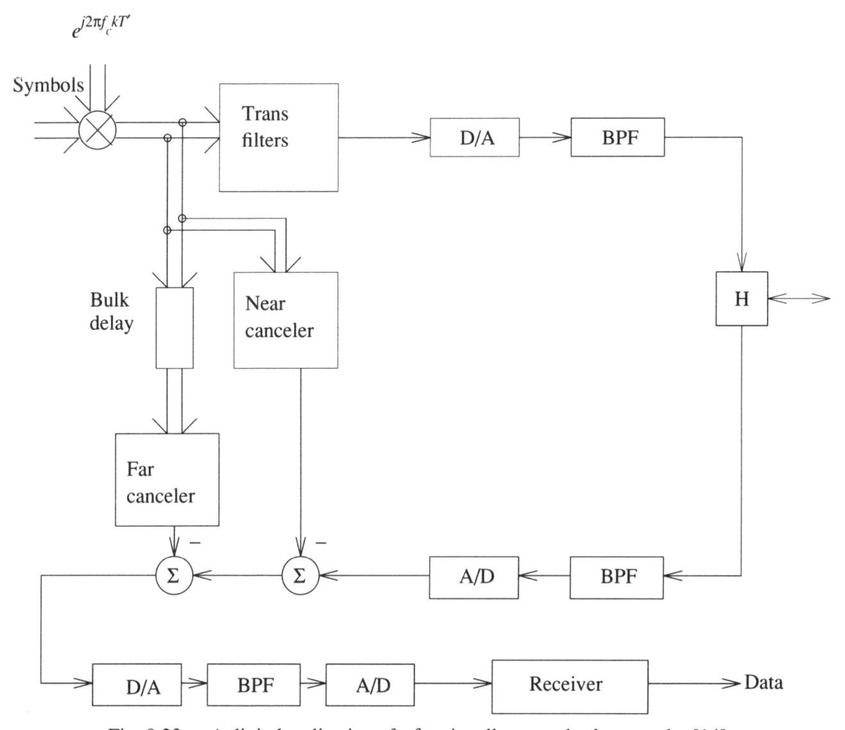

Fig. 9.23 A digital realization of a fractionally-spaced echo canceler [14].

For digital implementations, digital resolution and high accuracy in analog to digital converters are essential. Figure 9.23 shows an implementation [16] in which passband cancellation and receiver operations are done in the digital domain. However, the interpolation between the Nyquist-interval samples generated at the canceler output, clocked by the local transmitted data stream, and the symbol-interval samples required by the receiver, which is clocked by the distant transmitted data stream, is achieved by D/A conversion of the clean Nyquist-interval sample stream, conversion to a continuous analog waveform in a bandpass filter, and sampling by the receiver with the timing epoch recovered from the received data. It has been found [20] that considerably more digital resolution is required in the tap weights than in the tap voltages, in order to allow tap updating sufficient to reduce the mean-square cancellation error to an acceptable level. These and related considerations of digital resolution are described in Chapter 10.

9.3.7 ADAPTIVE REFERENCE ECHO CANCELLATION

We have noted that the desired distant signal is a component of the noise as far as adaptation of the local echo canceler is concerned. It is included in the

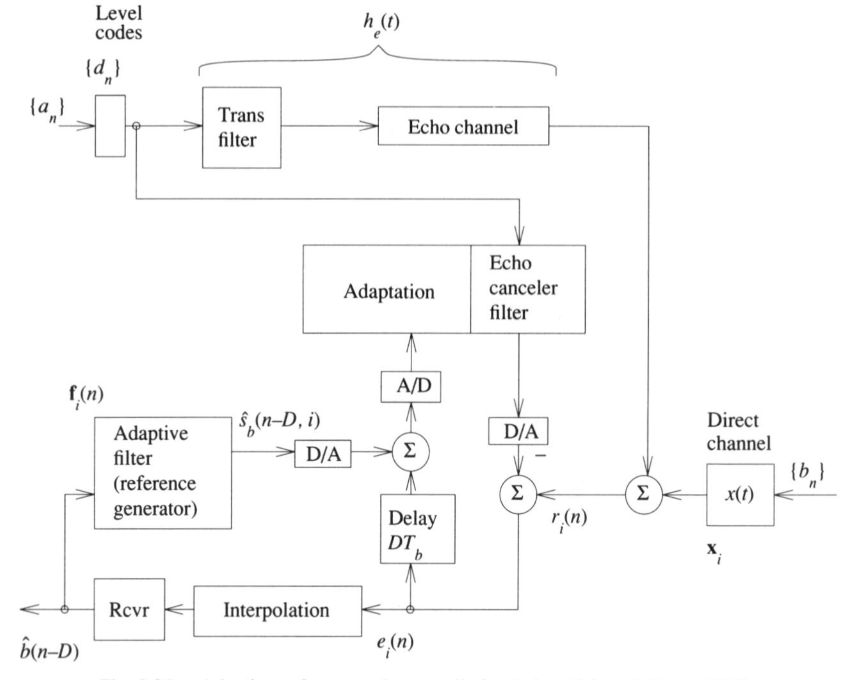

Fig. 9.24 Adaptive-reference echo cancellation (adapted from Falconer [21]).

"noise" variance of expression (9.42) for the residual mean-square cancellation error. This is a major performance disadvantage, but cancelers which treat the desired distant signal as part of the noise do have the advantage of avoiding all coordination between the near and distant transmitters (other than assuring that their signals are uncorrelated).

In principle, however, it should be possible to subtract out an estimate of the desired distant signal from the error signal $e(n)$ used to adapt the echo canceler, so that the echo canceler does not have to work against the large "noise" represented by the desired distant signal. In a symbol interval canceler, with adequate synchronization capabilities, this is possible with an adjunct decision feedback equalizer, generating a replica at symbol intervals of the precursor interference. If a delay in canceler adaptation is accepted, the "precursors" can include almost all significant pulse samples, allowing almost total elimination of the distant signal from the canceler error signal.

D. D. Falconer [21] extended these ideas to Nyquist-interval cancellation, yielding the advantages of adaptive-reference echo cancellation without requiring synchronization of the two transmitters.

To see how it works, consider the system diagram of Figure 9.24. Assume, as in the last section, an echo channel with $2NT$ dispersion represented by vectors

$$\mathbf{h}_i = h_e\left([i/\ell]T - NT\right), ..., h_e\left([i/\ell]T + NT\right), \quad i = 0, ..., (\ell - 1) , \tag{9.63}$$

where the index i corresponds to one of ℓ sampling phases in a symbol interval T. This echo channel is replicated in the echo cancellation filters $\mathbf{c}_i$ in Figure 9.24. Similarly, the direct transmission channel $x(t)$, represented by $(2N + 1)$-dimensional vectors

$$\mathbf{x}_i' = x\left([i/\ell]T - NT_b\right) , ..., x\left([i/\ell]T + NT_b\right) , \quad i = 0, ..., (\ell - 1) , \tag{9.64}$$

is replicated in the reference generator filters $\mathbf{f}_i$. Here T_b is used instead of T to make clear that the symbol rate T_b at the distant transmitter may differ slightly from the symbol interval T at the local transmitter.

The subtraction of the reference signal emerging from the $\mathbf{f}_i$ operations from the cancellation error signal is done in the analog domain with interpolated waveforms, thereby avoiding any concern over lack of synchronization (including a possible small difference of T_b from T) between the two transmitters. Analog subtraction also avoids high-resolution requirements for digital subtraction in cases where the distant signal has considerably lower energy than the echo signal.

The delay of D symbol intervals for the error signal $e_i(n)$, before the reference signal is subtracted from it and the result is used to adapt the echo canceler, corresponds to the processing delay in the local receiver. That is, the estimate of the distant data symbol b_n is not available until DT_b seconds after the pulse on which it is modulated appears at the receiver input. If the echo channel is constant or slowly varying, this delay in adaptation of the echo canceler is generally not a problem. (See Chapter 10 for a quantitative discussion of the effect of delay on the LMS algorithm.)

The reference signal output sample at time $t = (n + i/\ell)T_b$ is given by

$$\hat{s}_b(n - D,i) = \sum_{k=-N}^{N} b(n - D - k)\, f_i(k) . \tag{9.65}$$

The delayed error signal is

$$e_i(n) = \mathbf{b}'(n - D)\ \mathbf{V}_i(n) + \mathbf{d}'(n - D)\ \mathbf{U}_i(n - D) + \nu_i(n - D) , \tag{9.66}$$

where

$$\mathbf{v}_i(n) = \mathbf{x}_i - \mathbf{f}_i(n) \tag{9.67}$$

is the error in the filter replicating the direct channel, and

$$\mathbf{U}_i(n) = \mathbf{h}_i - \mathbf{c}_i(n) \tag{9.68}$$

is the error in the filter replicating the echo channel. The index n is actually different in (9.67) and (9.68), clocked to T_b for the former and T for the latter, but it is not a critical distinction in actual operation, implying only that updatings are made at slightly different times in $\mathbf{c}_i$ and $\mathbf{f}_i$.

A stochastic adaptation algorithm follows immediately from these considerations. Squaring $e_i(n)$ and taking the derivatives with respect to elements of the echo cancellation and reference generation filters, and using these as estimates of the gradient, the adaptation algorithm is

$$\begin{aligned} \mathbf{f}_i(n+1) &= \mathbf{f}_i(n) + \beta_1 \, e_i(n) \, \mathbf{b}(n-D) \\ \mathbf{c}_i(n+1) &= \mathbf{c}_i(n) + \beta_2 \, e_i(n) \, \mathbf{d}(n-D) \ , \end{aligned} \tag{9.69}$$

where β_1 and β_2 are step sizes. Algorithm (9.69) presumes, just as for data-driven channel equalization (Chapter 8), that the decisions $b(n)$ are correct, but occasional errors will not seriously degrade adaptation.

An examination of convergence of one of the symbol-interval filter pairs $(\mathbf{f}_i, \mathbf{c}_i)$ is sufficient to describe convergence of the entire fractionally-spaced structure. It is shown in [21] that the mean tap weight error $E\{\mathbf{U}_i(n)\}$ converges exponentially to zero with n for $0 < \beta_2 < 1$ and $D = 0$, and shows a sampled oscillatory convergence to zero for moderate values of β_2 and nonzero integer values of the delay D. A stable value for the mean tap-weight error vector is achieved for $0 < \beta_2 < 0.2$ and $0 \leq D \leq 8$. The mean-square error, $E\{|\mathbf{U}_i(n)|^2\}$, converges at a rate independent of the sampling phase i and of the channel and echo impulse responses, just as for the zero reference echo cancellation systems of previous sections. The rate of convergence is related in a complicated way to the step sizes, delay D, and dispersion lengths of the direct and echo channels. The asymptotic (steady-state) mean-square error is

$$E\left\{ |\mathbf{U}_i(n)|^2 \right\} = \frac{\beta_2 N_e \sigma_v^2}{2 - \left[\beta_1 N_h + \beta_2 N_e\right]} \tag{9.70}$$

where N_e is the length (symbol-interval samples) of the echo channel, and N_h is the length of the direct channel. Figure 9.25 compares the convergence and steady-state error performance of zero reference and adaptive reference cancellation algorithms respectively in a typical system, assuming $N_e = N_h$.

9.3.8 FAST STARTUP ECHO CANCELLATION

We have seen, in the algorithms described in earlier sections, that the desired distant signal is part of the noise perturbing echo canceler operation and significantly slows convergence. This is a particularly severe problem during startup, when the tap weight errors are initially far away from their optimum values and convergence time may not be acceptable even with a large initial step

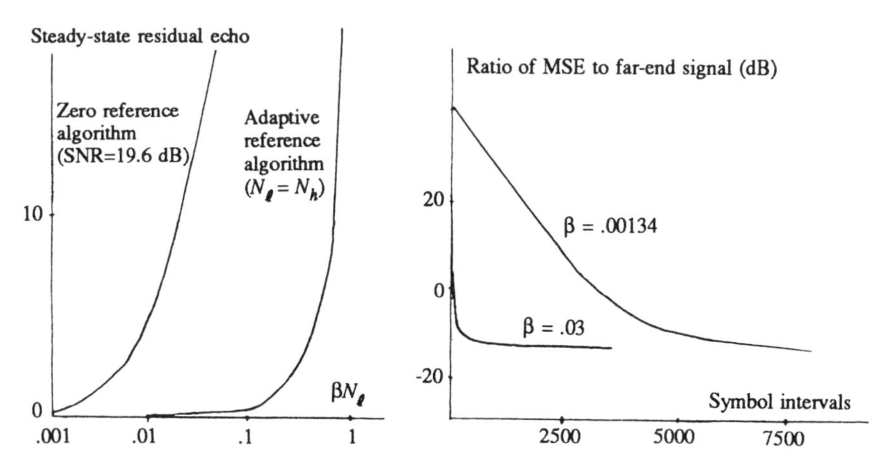

Fig. 9.25 (a) Relative steady-state residual echo vs. step size β for zero reference and adaptive reference cancelers respectively (SNR = 19.6 dB). (b) Convergence characteristics for zero reference and adaptive reference cancelers. (SNR = 15 dB, $N_e = 10$) (from Falconer [21]).

size. For this reason, a commonly used procedure for starting up an echo canceler is to do an initial adaptation in the absence of double-talking, i.e., with the distant transmitter quiet. Another practical procedure is to sound the echo channel and correlate the output with the input, as described in Section 9.3.1.

However, even with these procedures, startup is slower than desired in many applications. One approach to speeding it up is with the least squares adjustment algorithm of Section 9.3.5, which has the complexity costs described in that section. Alternatively, a modified stochastic adaptation algorithm discovered by Farrow and Salz exhibits performance comparable to the least squares algorithm, that is, adaptation in a number of iterations comparable to the length of the canceler filter, but with lower complexity comparable to the LMS algorithm.

Salz [31] motivates the procedure, with a pseudorandom (PN) sequence driving the transmitter filter and echo canceler of Figure 9.16. Recall (9.48), from Section 9.3.5, for the best estimator $\mathbf{c}_n$ of the tap weight set at the nth iteration of the least-squares adaptation algorithm. With the weighting $\Delta = 1$ in that equation, we have

$$\sum_{j=0}^{n} \mathbf{d}_j r_j = \left[\sum_{j=0}^{n} \mathbf{d}_j \mathbf{d}_j' \right] \mathbf{c}_n \ , \tag{9.71}$$

or

$$R_n \mathbf{c}_n = \mathbf{u}_n , \tag{9.72}$$

where

$$R_n = \sum_{j=0}^{n} \mathbf{d}_j \mathbf{d}_j' \text{ and } \mathbf{u}_n = \sum_{j=0}^{n} \mathbf{d}_j r_j . \tag{9.73}$$

As in earlier sections, all vectors are of length $2N+1$, the span of the echo canceler. If the $\{d_k\}$ are random ± 1 data, and if one makes an assumption of ergodicity, so that time averages approach ensemble averages, then as $n \to \infty$, and recalling (9.12) for r_n,

$$\mathbf{u}_n \to n\, E\{\mathbf{d}_j r_j\} = n\, E\{\mathbf{d}_j \mathbf{d}_j'\} \mathbf{h}_e = n \mathbf{h}_e \tag{9.74a}$$

$$R_n \to n\, E\{\mathbf{d}_j \mathbf{d}_j'\} = nI . \tag{9.74b}$$

Thus from (9.72),

$$\mathbf{c}_n = R_n^{-1} \mathbf{u}_n \underset{n \to \infty}{\to} \mathbf{h}_e , \tag{9.75}$$

the optimum tap weight vector yielding exact echo cancellation.

But for finite n, the statistical behavior of $\mathbf{c}_n$, and thus the progress of convergence, is hard to determine. The conditions on the random data sequence that would guarantee the existence of R_n^{-1} are not known. At a minimum, n has to be greater than $(2N + 1)$, for if it were not, the zero elements of $\mathbf{d}_n$ not yet occupied with data would result in all-zero rows and columns of R_n. The zero elements could be filled in with fake data, but there would be little point in adaptation with such a reference.

If an inverse of R_n exists, we can show that the tap weight estimator is unbiased. Specifically, from (9.14), (9.72), and (9.73), $r_j = \mathbf{d}_j' \mathbf{h}_e + \nu_j$,

$$\mathbf{c}_n = R_n^{-1} \mathbf{u}_n = R_n^{-1} \left[\sum_{j=0}^{n} \mathbf{d}_j \mathbf{d}_j' \right] \mathbf{h}_e + R_n^{-1} \sum_{j=0}^{n} \mathbf{d}_j \nu_j$$

$$= \mathbf{h}_e + R_n^{-1} \sum_{j=0}^{n} \mathbf{d}_j \nu_j , \tag{9.76}$$

so that, with the data and noise uncorrelated,

$$E\mathbf{c}_n = \mathbf{h}_e . \tag{9.77}$$

To guarantee the existence of the inverse of R_n, we seek appropriate *fixed* data sequences. Assuming that such a sequence $\{d_j\}$ has been found, we can define an *error matrix*

$$M = E\left\{ (\mathbf{c}_n - \mathbf{h}_e)\, (\mathbf{c}_n - \mathbf{h}_e)' \right\} . \tag{9.78}$$

Substituting (9.76) into (9.78), and recalling that the noise variance is $E\nu_j^2 = \sigma_\nu^2$,

$$M = E\left\{ R_n^{-1} \left[\sum_{j,k=0}^{n} \mathbf{d}_j \nu_j \nu_k \mathbf{d}_k' \right] R_n^{-1} \right\} = \sigma_\nu^2 R_n^{-1} . \tag{9.79}$$

The variance of the vector estimator $\mathbf{c}_n$ is the trace (sum of the diagonal terms) of the error matrix:

$$E||\mathbf{c}_n - \mathbf{h}_e||^2 = \text{trace } M = \sigma_\nu^2 \text{ trace} \left[\sum_{j=0}^{n} \mathbf{d}_j \mathbf{d}_j' \right]^{-1} . \tag{9.80}$$

As $n \to \infty$, the trace approaches zero. For any finite n, and any chosen data sequence, the variance $E||\mathbf{c}_n - \mathbf{h}_e||^2$ can be evaluated.

As candidates for a good fixed data sequence, consider two possible $(2N + 1)$-length $\pm$ data sequences, one orthogonal and the other pseudorandom. For the orthogonal sequence, which by definition has zero correlation with any time shift of itself, and $n = 2N + 1$, the ikth element of R_{2N+1} is

$$\left[\sum_{j=0}^{2N+1} \mathbf{d}_j \mathbf{d}_j' \right]_{ik} = \sum_{j=0}^{2N+1} d_{j-i} d_{j-k} = \begin{cases} 2N+1, & i = k \\ 0, & i \neq k \end{cases} \tag{9.81}$$

implying

$$E||\mathbf{c}_n - \mathbf{h}_e||^2 = \sigma_\nu^2 . \tag{9.82}$$

This is the best possible result, but it is not known how to construct an orthogonal sequence for any $(2N + 1)$. Pseudorandom (PN) sequences will be shown to do almost as well. A ± 1 PN sequence of length $(2N+1)$ P has the property

$$\left[R_{2N+1} \right]_{ik} = \sum_{j=0}^{2N} d_{j-i} d_{j-k} = \begin{cases} 2N+1, & i = k \\ -1, & i \neq k \end{cases} \tag{9.83a}$$

i.e., its correlation with itself is, for any cyclical time shift, equals to -1. The inverse matrix R_{2N+1}^{-1} has elements

$$\left[R_{2N+1}^{-1} \right]_{ik} = \begin{cases} \dfrac{2}{(2N + 1) + 1}, & i = j \\[2ex] \dfrac{1}{(2N + 1) + 1}, & i \neq j \end{cases} \tag{9.83b}$$

Then from (9.80)

$$E||\mathbf{c}_n - \mathbf{h}_e||^2 = \sigma_v^2 \; \frac{2[2N+1]}{[2N+1]+1} \; . \tag{9.84}$$

For large $(2N+1)$, this is a factor of two worse than (9.82) for the orthogonal sequence, but it can be reduced by using more than $(2N+1)$ samples.

To converge the echo canceler in exactly $(2N+1)$ iterations we use the modified LMS algorithm

$$\mathbf{c}_{n+1} = \mathbf{c}_n + \beta\,(\mathbf{d}_n + \mathbf{1})\; e_n, \tag{9.85}$$

where $\mathbf{1}$ is a column vector whose elements are all equal to one, and β is a step size equal to $1/2N+1$. We will use the property of a *PN* sequence that

$$\mathbf{d}_n' \; \mathbf{1} = 1, \tag{9.86}$$

i.e., the sum of the elements of a *PN* sequence (which has almost, but not quite, an equal number of +1 and – 1 elements) is one.

The proof that (9.85) converges in $(2N+1)$ steps is an inductive one suggested by Werner. First, recall (rewriting (9.32)) that the cancellation error is

$$e_n = d_n' \; (\mathbf{h}_e - \mathbf{c}_n) + \nu_n \; . \tag{9.87}$$

Assume an initial tap weight vector $\mathbf{c}_0 = \mathbf{0}$. Then from (9.85) and (9.87)

$$\mathbf{c}_1 = \beta(\mathbf{d}_0 + \mathbf{1})\;(\mathbf{d}_0'\mathbf{h}_e + \nu_0) \tag{9.88a}$$

$$\begin{aligned}
\mathbf{c}_2 &= \mathbf{c}_1 + \beta(\mathbf{d}_1 + \mathbf{1})\; e_1 \\
&= \beta(\mathbf{d}_0 + \mathbf{1})\;(\mathbf{d}_0'\mathbf{h}_e + \nu_0) + \beta(\mathbf{d}_1 + \mathbf{1}) \left[\mathbf{d}_1'(\mathbf{h}_e - \mathbf{c}_1) + \nu_1\right] \\
&= \beta\left\{(\mathbf{d}_0 + \mathbf{1})\;(\mathbf{d}_0'\mathbf{h}_e + \nu_0) + (\mathbf{d}_1 + \mathbf{1})\;(\mathbf{d}_1'\mathbf{h}_e + \nu_1)\right\} \\
&\quad - \beta^2 \left\{(\mathbf{d}_1 + \mathbf{1})\;\mathbf{d}_1'\;(\mathbf{d}_0 + \mathbf{1})\;[\mathbf{d}_0'\mathbf{h}_e + \nu_0]\right\} .
\end{aligned}$$

By property (9.83a), $\mathbf{d}_1\mathbf{d}_0' = -1$, since $\mathbf{d}_0$ is merely a cyclical shift, by one, of $\mathbf{d}_1$. By property (9.86), $\mathbf{d}_1'\mathbf{1} = 1$; thus

$$\mathbf{d}_1'\;(\mathbf{d}_0 + \mathbf{1}) = 0,$$

implying that the β^2 term of $\mathbf{c}_2$ is zero, so that

$$\mathbf{c}_2 = \beta\left[(\mathbf{d}_0 + \mathbf{1})(\mathbf{d}_0' h_e + \nu_0) + (\mathbf{d}_1 + \mathbf{1})(\mathbf{d}_1'\mathbf{h}_e + \nu_1)\right] . \tag{9.88b}$$

Continuing to iterate in this way, we finally arrive at

$$\mathbf{c}_{2N+1} = \alpha \left[\sum_{j=0}^{2N} (\mathbf{d}_j + \mathbf{1})(\mathbf{d}_j' \mathbf{h}_e + \nu_j) \right] . \qquad (9.88c)$$

But from (9.83a)

$$[R_{2N+1}]_{ik} = \left[\sum_{j=0}^{2N} \mathbf{d}_j \mathbf{d}_j' \right]_{ik} = \begin{cases} 2N+1, & i=k \\ -1, & i \neq k \end{cases} \qquad (9.89a)$$

and from (9.86)

$$\left[\sum_{j=0}^{2N} \mathbf{1} \cdot \mathbf{d}_j' \right]_{ik} = 1 \quad \text{(a matrix of all ones)} .$$

Thus

$$\left[\sum_{j=0}^{2N} (\mathbf{d}_j + 1)\, \mathbf{d}_j' \right]_{ik} = \begin{cases} 2N+2, & i=k \\ 0, & i \neq k \end{cases} \qquad (9.89b)$$

and with $\beta = 1/2N+1$, (9.88c) and (9.89b) yield

$$\mathbf{c}_{2N+1} = \mathbf{h}_e + \frac{1}{2N+1} \sum_{j=0}^{2N} (\mathbf{d}_j + \mathbf{1})\, \nu_0 .$$

Each element of $\mathbf{c}_{2N+1}$ has converged to the corresponding echo channel element plus a Gaussian random variable of mean zero and variance

$$E \left[\frac{1}{2N+1} \sum_{j=0}^{2N} (d_{jm} + 1) \nu_j \right]^2 = \frac{1}{(2N+1)^2} \sum_{j=0}^{2N} E(d_{jm} + 1)^2\, \sigma_\nu^2$$

$$= \frac{2\sigma_\nu^2}{2N+1}\, 1 ,$$

where d_{jm} is the ± 1 data random variable in the mth position of the jth data vector. The algorithm (9.85) has in fact converged in $(2N+1)$ steps, the size of the tap-weight vector, but with residual noise. Further adaptation, according to the stochastic adaptation algorithm (9.29) with a smaller step size β, can reduce the residual tap-weight fluctuation. Although fast startup by (9.85) has not yet been seen in practice because of other practical difficulties, such as restrictions on the length of the echo canceler, it is an excellent illustration of the opportunities for rapid convergence offered by special testing sequences.

It should be noted that there are other techniques for fast startup. Cioffi [27] describes a sounding technique using the discrete Fourier transform (DFT), and Ling and Long [17] propose white periodic complex training sequences with orthogonal real and imaginary parts, correlated with the (real) echo to generate an echo channel estimate.

9.4 OTHER CANCELER STRUCTURES

The tapped delay line is not the only possible adaptive filter that could be utilized in an echo canceler. Although the FIR filter and the LMS adaptation algorithm are extremely robust, they are unable to handle nonlinear echo channel characteristics, and convergence is slower than that possible in some alternative structures. In this section, we will describe the lattice filter and memory canceler structures that can offer improved performance in some circumstances.

9.4.1 LATTICE FILTER CANCELER

An adaptive lattice filter is used for mean-square estimation of values of a correlated time sequence [22]. Its salient virtue is that of orthogonalizing successive driving inputs, which increases the rate of convergence. It is an obvious candidate for the adaptive filter of an echo canceler, although the advantages it might have for a line-signal-driven canceler, where successive Nyquist-interval samples of the driving waveform are correlated, are not evident for the data-driven cancelers we have been considering, where the driving sequence consists of uncorrelated data values. One can, of course, conceive of data-driven applications, such as natural language and facsimile data streams, where the data are in fact correlated. We will describe how the lattice filter canceler works.

Assume, as shown in Figure 9.26, a sampled-data echo cancellation system. A reference sequence $\{,..., s_{i-2}, s_{i-1}, s_i, ...\}$ drives a "black box" adaptive filter with output q_i which is subtracted from the incoming echo plus noise sampler r_i. The objective, as described by Honig [23], is to configure the black box to minimize the mean-square error $E[e_i^2]$.

Consider first predicting a reference sequence sample s_i as a linear combination of n past sequence samples. The error in the prediction is the nth-order forward prediction residual

$$e_f(i|n) = s_i - \mathbf{f}'(i|n)\mathbf{s}(i-1|n) \tag{9.90}$$

where

$$\mathbf{s}'(i|n) = \left[s_{i-n+1}, ..., s_i\right] \quad (9.91$$

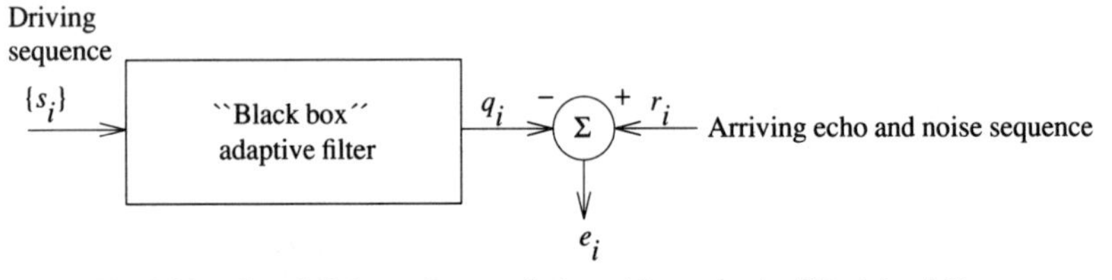

Fig. 9.26 Sampled-data echo cancellation with an adaptive "black box" filter.

is the vector of the most recent n samples of the reference sequence, and

$$\mathbf{f}'(i|n) = \left[f_{1|n}, \ldots, f_{n|n}\right] \tag{9.92}$$

is the vector of forward prediction coefficients. Similarly, one can predict a past sequence value s_{i-n} from n subsequent samples, yielding an nth-order backward prediction residual

$$e_b(i|n) = s_{i-n} - \mathbf{b}'(i|n)\mathbf{s}(i|n) , \tag{9.93}$$

where

$$\mathbf{b}'(i|n) = \left[b_{1|n}, \ldots, b_{n|n}\right] \tag{9.94}$$

is the vector of backward prediction coefficients. If the second-order statistics of the reference sequence are known, optimal values of the forward and backward prediction coefficients, minimizing $Ee_f^2(i|n)$ and $Ee_b^2(i|n)$ respectively, can be computed. It can be shown [22] that at any instant, the backward residuals are uncorrelated, i.e.,

$$E\left[e_b(i|m)e_b(i|n)\right] = \delta_{mn}\, E\left[e_b^2(i|n)\right] , \tag{9.95}$$

and that the optimum forward and backward prediction coefficients obey the recursive (order, not time) relationships

$$\begin{aligned} e_f(i|n) &= e_f(i|n-1) - K_n(i)e_b(i-1, n-1) \\ e_b(i|n) &= e_b(i-1|n-1) - K_n(i)e_f(i, n-1) , \end{aligned} \tag{9.96}$$

where $1 \leq n \leq N$ = the order of the filter, and the "parcor" (partial correlation) coefficients are defined as

$$K_n(i) \triangleq E\left[e_f(i|n-1)e_b(i-1|n-1)\right] \Big/ E\left[e_f^2(i|n-1)\right] . \tag{9.97}$$

The recursion relationships (9.96) are graphically represented by the lattice of Figure 9.27. The lattice begins (at the left) by estimating the residuals $e_f(i|1)$

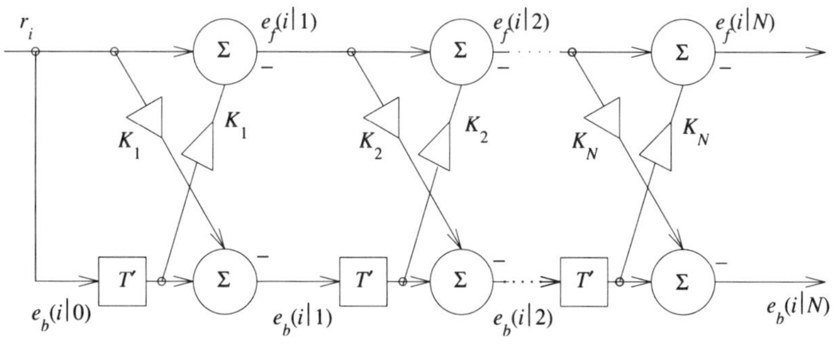

Fig. 9.27 Lattice structure for order-recursive generation of reference sequence prediction coefficients.

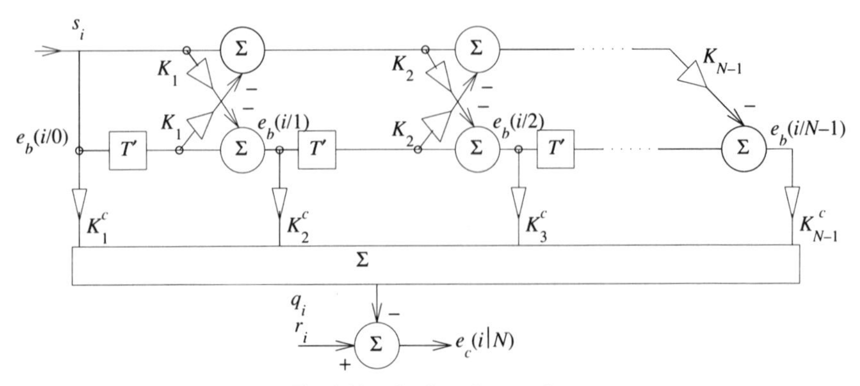

Fig. 9.28 Lattice echo canceler.

and $e_b(i|1)$ from the current (r_i) and most recent past (r_{i-1}) received samples.

The structure of Figure 9.27 yields prediction residuals for a driving signal consisting of one scalar value, s_i, at the ith iteration. For the echo cancellation application, however, the echo estimate q_i at the ith iteration, which in the tapped delay line structure is formed from the (possibly correlated) N-dimensional input sample vector,

$$\mathbf{s}'(i|N) = \left[s_{i-N+1}, \ldots, s_i\right], \tag{9.98}$$

is generated as a linear combination of the (uncorrelated) backward prediction residuals $[e_b(i|0), \ldots, e_b(i|N-1)]$. That is,

$$q_i = \sum_{n=0}^{N-1} K_{n+1}^c \, e_b(i|n) . \tag{9.99}$$

Thus, as shown in Figure 9.28, the post-cancellation error is

$$e_i = r_i - q_i = r_i - \sum_{n=0}^{N-1} K_{n+1}^c \, e_b(i|n) . \tag{9.100}$$

The coefficients $\{K_n^c\}$ which minimize $E\left[e_i^2\right]$ are readily derived (Exercise 9.5), yielding

$$\begin{aligned} K_n^c &= E\left[r_i e_b(i|n-1)\right] \Big/ E\left[e_b^2(i|n-1)\right] \\ &= E\left[e_c(i|n-1)e_b(i|n-1)\right] \Big/ E\left[e_b^2(i|n-1)\right] , \end{aligned} \tag{9.101}$$

where $e_c(i|n-1)$ has been defined as an $(n-1)$th-order filter residual of the cancellation error e_i, such that $e_c(i|N) = e_i$. Note that the coefficients are the normalized correlation of the (output) error and the voltage at the corresponding

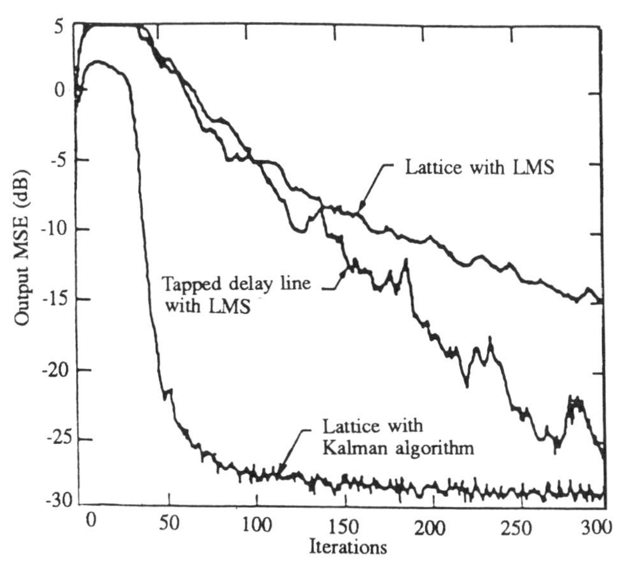

Fig. 9.29 Relative performance of lattice and tapped delay line cancelers with $N = 30$ taps, on an echo channel with exponential impulse response ($r_i = 0.96^i$) and SNR = 40 dB. The stochastic adaptation (LMS) algorithem was used in these simulations. Operation of the lattice with the Kalman adaptation algorithm is also shown. From Honig [23].

weight. The residual obeys, from (9.100), the recursion

$$e_c(i|n) = e_c(i|n-1) - K_n^c e_b(i|n-1) \, . \tag{9.102}$$

As we have noted, the performance of the lattice structure for an uncorrelated driving sequence is not superior to that of the tapped delay line, as shown in Figure 9.29. In fact it is worse, when adapted via the LMS algorithm, because of statistical fluctuations of the forward and backward prediction coefficients, which, in the case of an uncorrelated driving sequence, should be equal to zero. It has, however, been shown to be superior in cases where the driving sequence was correlated [23,24], as well as when "fast" adaptation algorithms, such as the Kalman algorithm described in Chapter 8 for channel equalization, are applied to echo cancellation (see Figure 9.29). Note that Figure 9.29 shows the ability of the Kalman algorithm to significantly improve convergence relative to the LMS algorithm, even when the inputs to the canceler are uncorrelated. As with equalizers, the Kalman and fast Kalman algorithms can be applied to both tapped delay line and lattice structures.

9.4.2 MEMORY COMPENSATION STRUCTURES

The transversal filter and lattice echo cancelers described so far, with multiple tap weights adapted to minimize mean-square error, have several disadvan-

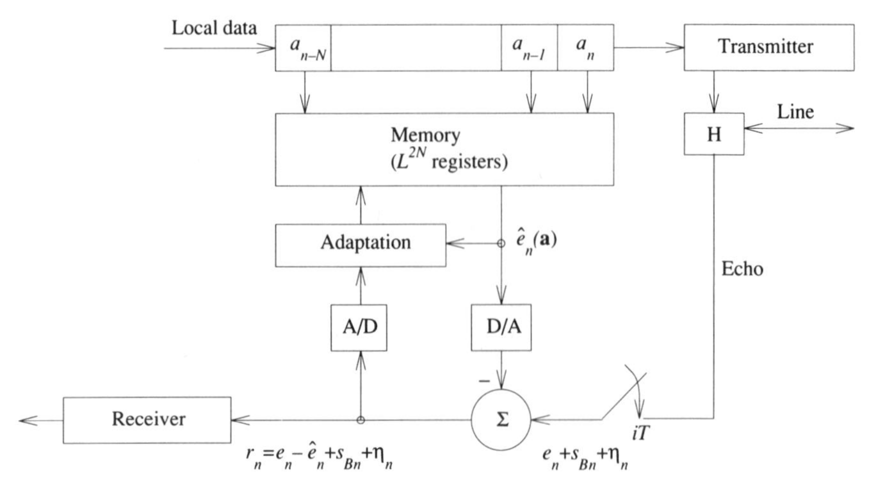

Fig. 9.30 Principle of memory echo canceler.

tages. Even when data-driven, the updating and filtering operations are computationally intensive, requiring high-speed logic. But most significantly, the echo canceler cannot replicate a nonlinear echo channel.

If the number L of pulse levels is small, and the echo channel memory, denoted as N symbol intervals, is not too long, there is a very simple and elegant way to realize a nonlinear canceler [25]. It is based on the observation that e_n (Figure 9.30), the echo sample at the receiver input at the nth symbol-interval sampling time nT, is completely determined by the N most recent local data values, denoted as the vector $\mathbf{a}(n)$, and the echo channel.

Let $\mathbf{a}_k$ denote the kth of the L^N values of this data vector. As Figure 9.30 illustrates, a memory simply stores all L^N possible values of $\hat{e}_n$. The contents of the register addressed by the current value of the data vector $\mathbf{a}(n)$ is simply read out as the echo estimate $\hat{e}_n$ and fed to the subtracter.

The memory echo canceler works only under special circumstances, fortunately the case for most digital subscriber lines (but not most dialed telephone circuits):

1. The echo channel changes only very slowly.

2. The clocks of the transmitters at the two ends of the subscriber line are locked together, so that the sampling times $nT + t_0$, with the sampling epoch t_0 chosen for best reception of the distant signal, are at a fixed delay t_0' from the transitions in the local data train $\{a_i\}$.

Condition (1) assures that the weak tracking capability of the memory canceler is not a problem, and condition (2) guarantees that there is no drift in the sampling time with respect to the local data train. A drift in the sampling time would require a change in the stored value $\hat{e}_n$ corresponding to a particular data vector **a**.

Given these conditions, a simple stochastic adaptation algorithm is one that minimizes the power

$$\left[r_n\right]^2 = \left[e_n - \hat{e}_n(\mathbf{a}) + s_{Bn} + \eta_n\right]^2 \tag{9.103}$$

of each symbol-interval sample of the post-cancellation error signal, where a particular N-element reference data sequence **a** happens to be in the shift register of Figure 9.30 at time nT. The effect, over many data symbols, is to minimize the average error-signal power; since the far-end signal and noise are independent of e_n and $\hat{e}_n$, this means that $\hat{e}_n \rightarrow e_n$.

The vector **a** addresses a particular location in the memory, where an echo estimate $\hat{e}_n(\mathbf{a})$ is stored at time nT, and it is this particular stored value, and it only, which is adapted when **a** is the reference input. No other stored data are adapted. Adaptation is according to a negative gradient algorithm:

$$\begin{aligned} \hat{e}_{n+1}(\mathbf{a}) &= \hat{e}_n(\mathbf{a}) - (\alpha/2)\partial r_n^2/\partial \hat{e}_n(\mathbf{a}) \\ &= \hat{e}_n(\mathbf{a}) + \alpha r_n \,. \end{aligned} \tag{9.104}$$

The value of α is, as usual, a compromise between fast convergence and small residual error after convergence. If $s_{Bn} + \nu_n$ in Figure 9.30 were zero, making $r_n = e_n - \hat{e}_n$, α could be made equal to one and the correction would make the echo estimate equal to the echo.

The complexity of the canceler of Figure 9.30 can be reduced by using a sign algorithm, allowing removal of the A/D converter and resulting in the adaptation algorithm

$$\hat{e}_{n+1}(\mathbf{a}) = \hat{e}_n(\mathbf{a}) + \Delta \operatorname{sgn}(r_n + \xi_n) \,, \tag{9.105}$$

where ξ_n is a uniformly distributed "dithering" noise, over a range comparable to the amplitude range of r_n, which reduces quantization error on the average, and Δ is the quantization step size of the memory registers, normalized by the RMS value of $r_n + \xi_n$. As a penalty for its computational simplicity, the sign algorithm tracks much more slowly than (9.104).

9.5 PASSBAND CONSIDERATIONS

We have so far presumed a baseband data communication system and examined echo cancelers for attenuation of baseband echos. But for many echo cancellation requirements, the echo is a passband modulated signal, as suggested in Figure 9.31. If one uses a data-driven echo canceler, some means must be found to synthesize a passband echo from a baseband data stream. There are several realization alternatives, both baseband and passband, described in the

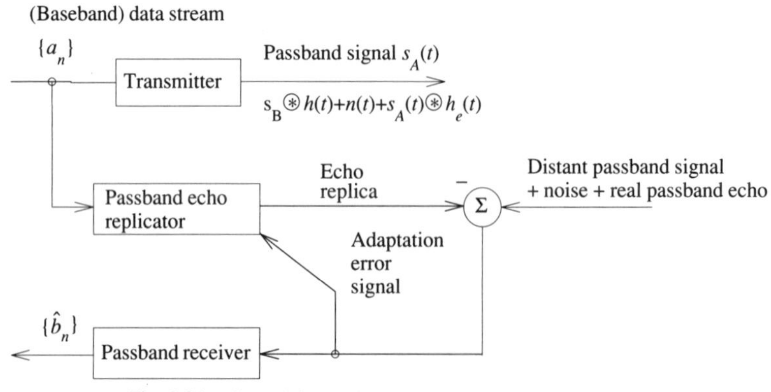

Fig. 9.31 Data-driven echo cancellation for passband echo.

following two subsections. There are also the transmission impairments of phase jitter and frequency offset on the distant echo, for which compensation can be arranged, as described in Section 9.5.3.

9.5.1 COMPLEX NOTATION FORMULATION

Referring to Figure 9.32, the complex transmitted signal is

$$\tilde{s}_A(t) = \sum_n \tilde{d}_n\, p(t-nT)\, e^{j2\pi f_c t}, \tag{9.106}$$

where the $\{\tilde{d}_n\}$ are, as in Section 5.2.1 of Chapter 5, the complex data levels corresponding to points in a two-dimensional signal constellation. The echo

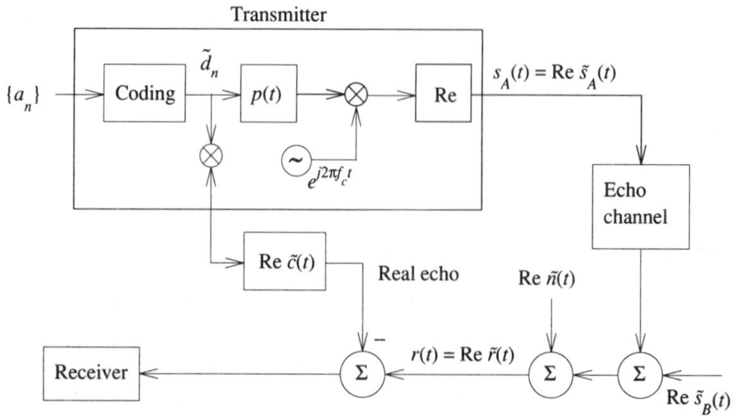

Fig. 9.32 Complex terminology. The echo characteristic $h_e(t) = \text{Re}\,\tilde{h}_e(t)$ is composed of $p(t)$ and the echo channel.

channel satisfies the relationships

$$h_e(t) = R_e\, \tilde{h}_e(t)$$

and (9.107)

$$\tilde{h}_e(t) = \tilde{h}_{eBB}(t)\, e^{j2\pi f_c t},$$

where $\tilde{h}_e(t)$ is the complex analytic echo channel and $\tilde{h}_{eBB}$ is the equivalent complex baseband echo channel. We assume that $h_e(t)$ includes the transmitter pulse filter $p(t)$. Thus the received signal just before the canceler of Figure 6.32 is

$$r(t) = \text{Re}\, \tilde{r}(t) = \text{Re}\left[\sum_n \tilde{d}_n\, \tilde{h}_e[t-nT]\, e^{j2\pi f_c t} + \tilde{n}(t) + \tilde{s}_B(t) * \tilde{h}(t)\right], \tag{9.108}$$

where $\tilde{n}(t)$ is the complex analytic noise, $\tilde{s}_B(t)$ is the complex analytic distant signal, and $\tilde{h}(t)$ is the complex analytic transmission channel from the distant transmitter.

The echo canceler and receiver will ordinarily be implemented as digital signal processors operating at a sampling rate at least as large as the Nyquist rate. Thus the canceler will deal with received signal samples

$$r(nT') = \text{Re}\left[\sum_n \tilde{d}_n \tilde{h}_e(nT'-nT) e^{j2\pi f_c nT'} + \tilde{n}(nT') + \tilde{s}_B(t) * \tilde{h}(t)\,\Big|_{nT'}\right]. \tag{9.109}$$

The canceler will generate echo replica samples at these same time instants nT', as described in Section 9.3.6, but not all samples may have to be generated.

The signal processing in the echo canceler can also be represented in complex notation. The canceler is attempting to create a replica of $\tilde{h}_e(t)$ or its real part. We will let $\tilde{\mathbf{c}}_i(n)$ denote the (complex) tap weight vector, at time $nT+iT'$, of the ith symbol-interval subcancelers. Similarly, $\tilde{e}_i(n)$ denotes the complex error after subtraction of the echo replica. If $A = E\, a_n^2 = E b_n^2$, then

$$E|\tilde{d}_n|^2 = 2A, \tag{9.110}$$

and formula (9.38) for the mean-square error becomes [16]

$$E|\tilde{e}_i(n)|^2 = [1-2\beta A+2\beta^2(2N+1)A^2]\; E|\tilde{e}_i(0)|^2 + \frac{1-(1-2\beta A+2\beta^2(2N+1)A^2)^2}{1-(1+2\alpha A+2\alpha^2(2N+1)A^2)} \cdot 2\beta A\sigma^2 . \tag{9.111}$$

9.5.2 COMPLEX CANCELER ALTERNATIVES

As we shall see, the signal processing in the echo canceler can also be presented in complex analytic notation. The canceler is attempting to create a

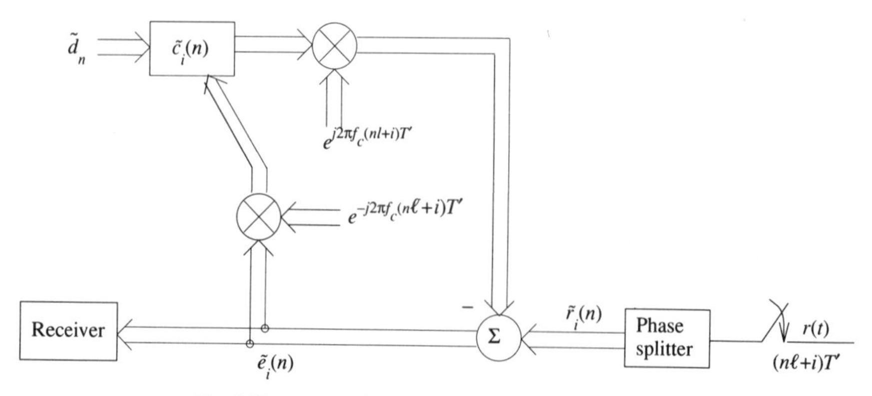

Fig. 9.33 A complex canceler with baseband adaptation.

replica of $\tilde{h}_e(t)$, and it can do this in separate steps of adaptive filtering and translation to passband, as described in this section, or, under certain conditions, by direct inband signal generation.

The various configurations differ in whether adaptive filtering is done in baseband or passband, in the complexity of implementation, and in the speed of convergence. A comparison of some of the alternatives is made in the paper by Im et al. [26].

Figure 9.33 pictures one of the alternatives, a complex adaptive baseband filter, followed by a multiplication by $\exp[j2\pi f_c nT']$ that does the translation to passband, followed by a complex subtracter [19] The algorithm is described below, but the convergence analyses, similar to the real baseband version described in Section 9.3.6, are not reproduced here.

The subtraction of the echo replica from the complex analytic received signal requires generation of the complex analytic received signal in a phase splitter that includes a Hilbert filter (phase splitter), as illustrated in Figure 5.13 of Chapter 5. However, no one has been able to build a Hilbert filter, of acceptable delay and complexity, that is accurate enough for 60 dB ERLE in the complex subtracter. For this reason, and complexity considerations, designers have looked for modification of this basic fractionally-spaced design. The complexity has already been characterized in Section 9.3.6 as $8(2N+1)/T'$ real multiplications per second. All of these multiplications are, however, relatively simple, using as factors values of $\{d_n\}$, which take on only a few distinct values.

With complex processing (9.60) and (9.63) for filtering and updating respectively take on the forms

$$\left.\begin{aligned} \tilde{y}_{n\ell+i} &= \left[\tilde{\mathbf{c}}_i'(n)\tilde{\mathbf{d}}(n)\right] e^{j2\pi f_c(n\ell+i)T'} \\ \tilde{\mathbf{c}}_i(n+1) &= \tilde{\mathbf{c}}_i(n) + \beta\left[\tilde{e}_i(n)e^{-j\pi f_c(n\ell+i)T'}\right]\tilde{\mathbf{d}}(n) \end{aligned}\right\} i = 0,\ldots,\ell-1 \,. \quad (9.112)$$

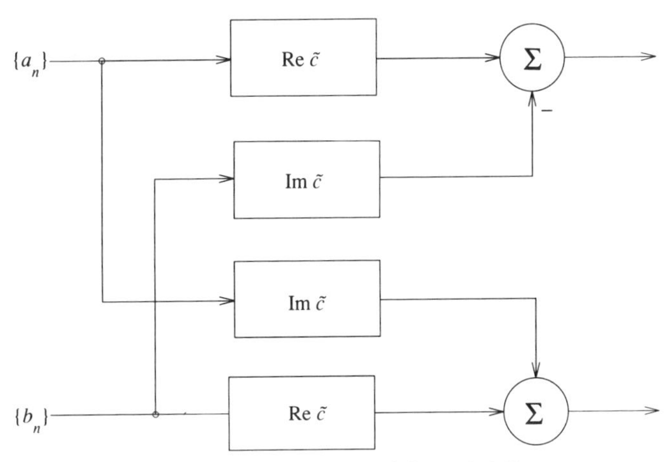

Fig. 9.34 The cross-coupled (full complex) filter.

Where, as in (9.56), $\ell = T/T'$, n is the time index in units of the symbol interval T, and i is the index of a particular symbol-interval subcancelers. The complex passband error sample $\tilde{e}_i(n)$, which is the error sample at time $t = (n\ell + i)T'$, is translated to baseband by the exponential with negative exponent.

To get around the phase splitter problem, the algorithm can be modified, at some cost in rate of convergence, by doing the subtraction in real variables and using the real error to drive the adaptation algorithm

$$\tilde{\mathbf{c}}_i(n+1) = \tilde{\mathbf{c}}_i(n) + \beta \text{ Re}\left[\tilde{e}_i(n)\right] e^{-j2\pi f_c (n\ell+i)T'} \tilde{\mathbf{d}}(n) . \quad (9.113)$$

It should be noted that (9.111) describes a *cross-coupled* structure. The product $\tilde{\mathbf{c}}_i'(n)\tilde{\mathbf{d}}(n)$ is, dropping the indices for convenience and recalling that a data symbol has real and imaginary components a and b respectively,

$$\begin{aligned} \tilde{\mathbf{c}}'\tilde{\mathbf{d}} &= \left[\text{Re}(\tilde{\mathbf{c}}) + i \text{ Im}(\tilde{\mathbf{c}})\right]' \left[\mathbf{a} + j\mathbf{b}\right] \\ &= \text{Re}(\tilde{\mathbf{c}})' \mathbf{a} - \text{Im}(\tilde{\mathbf{c}})'\mathbf{b} + j\left[\text{Re}(\tilde{\mathbf{c}})'\mathbf{b} + \text{Im}(\tilde{\mathbf{c}})'\mathbf{a}\right] . \end{aligned} \quad (9.114)$$

These four inner products are indicated in Figure 9.34.

Rather than generate a filter characteristic first and modulate to passband second, an echo canceler can modulate the data to passband first and adapt a *passband* filter second, as shown in Figure 9.35. The algorithm is expressed as

$$\begin{aligned} \tilde{y}_{n\ell+i} &= \tilde{\omega}_i'(n) \left[\tilde{\mathbf{d}}(n)\ e^{j2\pi f_c (n\ell+i)T'}\right] \\ \tilde{\omega}_i(n+1) &= \tilde{\omega}_i(n) + \beta\tilde{e}_i(n) \left[\tilde{\mathbf{d}}_{(n)}\ e^{j2\pi f_c (n\ell+i)T'}\right], \end{aligned} \quad (9.115)$$

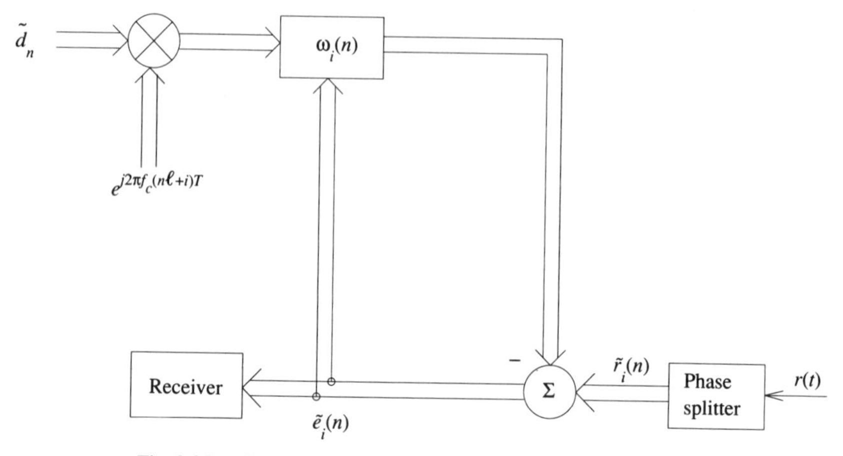

Fig. 9.35 Complex echo canceler with passband adaptive filtering.

where $\tilde{\omega}$ represents the passband filter tap weights. If $f_c T'$ equals a rational number, there are only a finite number of different phase rotations of $\tilde{\mathbf{d}}(n)$, analogous to what was described earlier as an option for the canceler with a baseband filter. Here, however, the cross-coupled complex filter can be reduced to a pair of real filters, since that is all that is needed to produce the real echo replica. This is shown in Figure 9.36. Cancellation with real signals avoids the phase-splitter problem alluded to earlier. The cost, of course, is in rate of convergence.

This non-cross-coupled structure is sometimes used for near-echo cancellation, with the cross-coupled structure reserved for the more difficult distant echo, which may suffer phase perturbations. The arrangement is suggested in Figure 9.37. With much of the near echo canceled, the ERLE expected of the far-echo canceler is modest and a phase splitter becomes practical.

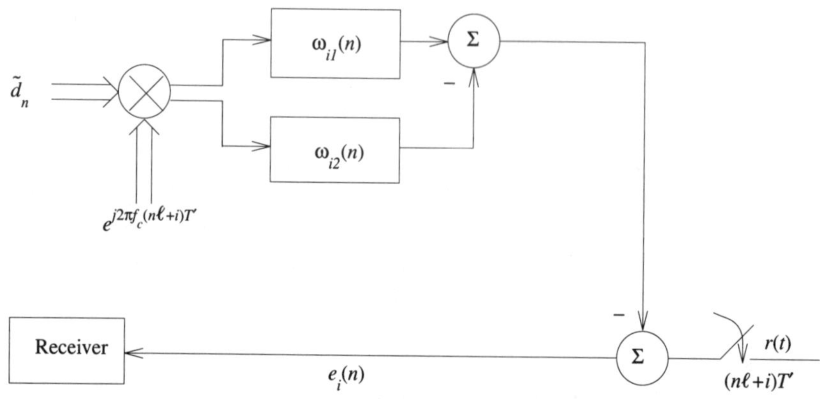

Fig. 9.36 Complex echo canceler with simplified passband adaptive filtering for real output only.

9.5.3 PHASE JITTER/FREQUENCY OFFSET COMPENSATION

The distant echo, having in some cases transversed FDM carrier systems, may arrive back at the source station with phase perturbation introduced in carrier systems, as explained in Chapter 1. A transversal filter operating an echo replica can, theoretically, generate the phase notations caused by phase jitter frequency offset, but in practice it cannot adapt fast enough to track most phase perturbations.

The situation is very similar to phase tracking on the forward channel, described in detail in Chapter 6. If (using complex notation) a phase perturbation represented by $e^{j\theta(t)}$ is present in the echo channel following the linear distortion, as illustrated in Figure 9.38, it can be partly canceled by a phase factor $e^{-j\hat{\theta}(t)}$. To achieve this, a first-order phase tracking loop can be derived as follows. The complex error following echo cancellation is

$$\tilde{e}_i(n) = \tilde{r}_i(n) - e^{\hat{\theta}_i(n)} \, \tilde{\mathbf{c}}_i(n)' \tilde{\mathbf{d}}(n) \, . \tag{9.116}$$

The gradient adaptation algorithm is

$$\hat{\theta}_i(n) = \hat{\theta}_i(n-1) - \frac{\lambda}{2} \frac{\partial |\tilde{e}_i(n)|^2}{\partial \theta} \, , \tag{9.117}$$

and since

$$\frac{\partial |\tilde{e}_i(n)|^2}{\partial \theta} = 2 \, \mathrm{Im} \left[\tilde{e}_i(n) e^{-j\hat{\theta}_i(n)} \, \tilde{\mathbf{c}}_i^*(n)' \tilde{\mathbf{d}}^*(n) \right] , \tag{9.118}$$

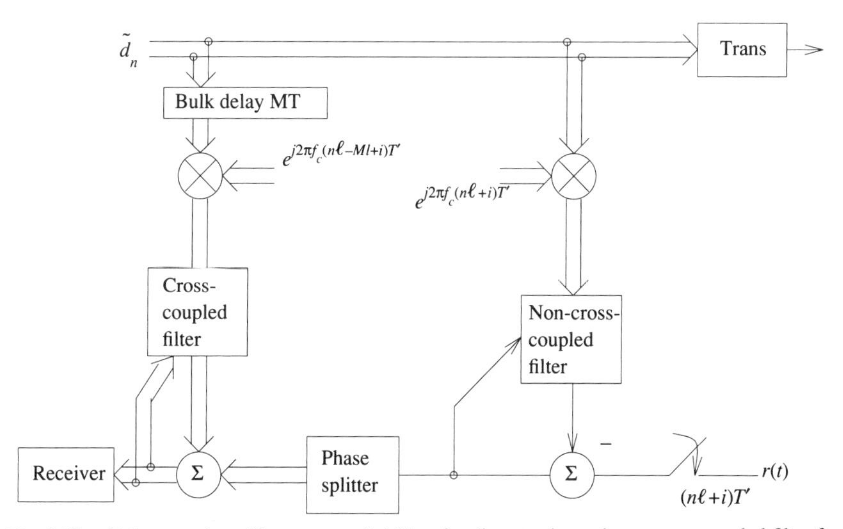

Fig. 9.37 Echo canceler with cross-coupled filter for distant echo and non-cross-coupled filter for near echoes.

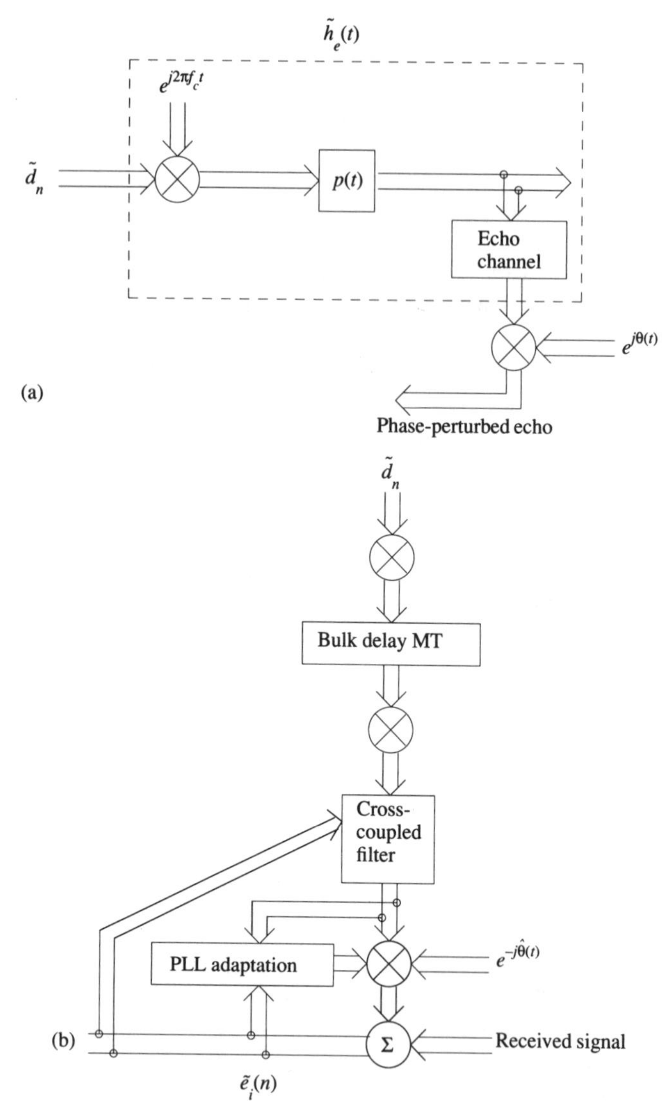

Fig. 9.38 Phase tracking for distant echo. (a) Introduction of phase perturbation $\theta(t)$ in echo channel. (b) Phase-locked tracking loop.

we have

$$\hat{\theta}_i(n) = \hat{\theta}_i(n-1) + \lambda \operatorname{Im} \left[\tilde{e}_i(n) e^{-j\hat{\theta}_i(n)} \tilde{\mathbf{c}}_i^*(n)' \tilde{\mathbf{d}}^*(n) \right]. \tag{9.119}$$

Further analysis similar to that in Chapter 6 shows that, in the absence of linear distortion and noise,

$$E\left[\tilde{e}_i(n)\, e^{-j\hat{\theta}_i(n)}\ \tilde{\mathbf{c}}_i^*(n)'\tilde{\mathbf{d}}^*(n)\right] \sim \sin\left[\theta_i(n)-\hat{\theta}_i(n)\right], \qquad (9.120)$$

so that (9.119) does indeed approximate a first-order phase-locked loop.

The phase perturbation is often largely a linearly increasing phase

$$\theta(t) = 2\pi f_{off} t, \qquad (9.121)$$

where f_{off} is the *frequency offset*. The frequency offset is rarely greater than 1 Hz. A second-order phase-locked loop might be considered, requiring minimum tracking bandwidth, not really true because of the need to track under double-talking conditions so that the step size γ in (9.119) has to be made small. In practice, even for small frequency offset, it is found that a second-order loop is required (see Harman, Wang, and Werner, [28]) but a first-order loop is adequate, since the frequency offset is small enough and may be combined with phase jitter.

9.5.4 PERFORMANCE WITHOUT AND WITH PHASE TRACKING

Wang and Werner [28,30] derive expressions for the echo return loss enhancement in non-cross-coupled and cross-coupled symbol interval echo cancelers perturbed by frequency offset, without any phase tracking. In the first case, the ERLE is

$$ERLE = 10\log\left[\frac{1-\beta A}{1-\beta(2N+1)A}\cdot\frac{\Delta^2}{\beta^2 A+(1-\beta A)\Delta^2}\right], \qquad (9.122a)$$

where $\Delta = 2\pi f_{OFF} T$. In the second case,

$$ERLE = -10\log\left[\frac{1-\beta A}{1-\beta(2N+1)A}\cdot\frac{\Delta^2}{4\beta^2 A^2+(1-2\beta A)\Delta^2}\right]. \qquad (9.122b)$$

For practical values of the parameters, the cross-coupled structure performs about 6 dB better.

REFERENCES

[1] A. Brosio, U. DeJulio, V. Lazzari, R. Ravaglia, and A. Topanelli, "A comparison of digital subscriber line transmission systems employing different line codes," *IEEE Trans. on Communications*, Vol. COM-29, No. 11, pp. 1581–1588, November 1981.

[2] J. Bingham, *Theory and Practice of Modem Design*, Wiley, 1988.

[3] S. Yoneda and Y. Fukui, "Suppression of impedance fluctuations in hybrid couplers," *IEEE Trans. on Communications*, Vol. COM-27, No. 1, pp. 102–105, January 1984.

[4] S. B. Weinstein, "Echo cancellation in the telephone network," *IEEE Communications Magazine*, Vol. 15, No. 1, pp. 9–15, January 1987.

[5] F. K. Becker, and H. R. Rudin, "Application of automatic transversal filters to the problem of echo suppression," *Bell System Tech. J.*, Vol. 45, No. 12, pp. 1947–1850, December 1966.

[6] M. M. Sondhi, and A. J. Presti, "A self-adaptive echo canceler," *Bell System Tech. J.*, Vol. 45, No. 10, pp. 1851–1854, December 1966.

[7] M. M. Sondhi, "An adaptive echo canceler," *Bell System Tech. J.*, Vol. 46, No. 3, pp. 497–511, March 1967.

[8] D. L. Duttweiler, "A twelve-channel digital echo canceler," *IEEE Trans. on Communications*, Vol. COM-26, No. 5, pp. 647–653, May 1978.

[9] D. G. Messerschmitt, "Echo cancellation in speech and data transmission," *IEEE Journal on Selected Areas in Communications*, SAC-2, No. 2, March 1984.

[10] N. S. Lin and C. P. J. Tzeng, "Full-duplex data transmission over local loops," *IEEE Communications Magazine*, Vol. 26, No. 2, pp. 31–47, February 1988.

[11] D. G. Messerschmitt, "Design issues in the ISDN U-interface transceiver," *IEEE Journal on Selected Areas in Communications*, Vol. SAC-4, pp. 1281–1293, November 1986.

[12] M. Honig, K. Steiglitz, and B. Goppuath, "Multichannel signal processing for data communications in the presence of crosstalk," IEEE Trans. on Communications, Vol. 38, No. 4, pp. 551–558, April 1990.

[13] V. G. Koll and S. B. Weinstein, "Simultaneous two-way data transmission over a two-wire circuit," *IEEE Trans. on Communications*, Vol. COM-21, No. 2, pp. 143–147, February 1973.

[14] D. D. Falconer and K. H. Mueller, "Adaptive echo cancellation/AGC structures for two-wire full-duplex transmission on two-wire circuits," *Bell System Tech. J.*, Vol. 58, pp. 1593–1616, September 1979.

[15] K. H. Mueller, "A new digital echo canceler for two-wire full-duplex data transmission," *IEEE Trans. on Communications*, Vol. COM-24, No. 9, pp. 956–967, September 1976.

[16] J. J. Werner, "An echo-cancellation-based 4800 bit/s full-duplex DDD

modem," *IEEE Journal on Selected Areas in Communications*, Vol. SAC-2, No. 5, pp. 722–730, September 1984.

[17] F. Ling and G. Long, "Correlation-based fast training of data-driven Nyquist in-band echo cancelers," *Proceedings ICC '90*, Atlanta, pp. 1280–1284, April 1990.

[18] N. A. M. Verhoeckx, H. C. Van den Elzen, F. A. M. Snijders, and P. J. Van Gerwen, "Digital echo cancellation for baseband data transmission," *IEEE Trans. on Acoustics, Speech, and Signal Processing*, Vol. ASSP-27, No. 6, December 1979.

[19] S. B. Weinstein, "A passband data-driven echo canceler for full-duplex transmission on two-wire circuits," *IEEE Trans. on Communications*, Vol. COM-26, No. 7, pp. 654–666, July 1977.

[20] R. D. Gitlin and S. B. Weinstein, "The effects of large interference on the tracking capability of digitally echo cancelers," *IEEE Trans. on Communications*, pp. 833–839, June 1978.

[21] D. D. Falconer, "Adaptive reference echo cancellation," *IEEE Trans. on Communications*, Vol. COM-30, No. 9, pp. 2083–2094, September 1982.

[22] B. Friedlander, "Lattice filters for adaptive processing," *Proc. IEEE*, Vol. 70, pp. 829–867, August 1982.

[23] M. Honig, "Echo cancellation of voiceband data signals using recursive least squares and stochastic gradient algorithms," *IEEE Trans. on Communications*, Vol. COM-33, No. 1, pp. 65–73, January 1985.

[24] E. H. Satarius and J. D. Pack, "Application of least squares lattice algorithms to adaptive equalization," *IEEE Trans. on Communications*, Vol. COM-29, pp. 136–142, February 1981.

[25] N. Holte and S. Stueflotten, "A new digital echo canceler for two-wire subscriber lines," *IEEE Trans. on Communications*, Vol. COM-29, No. 11, pp. 1573–1581, November 1986.

[26] G. H. Im, L. K. Un, and J. C. Lee, "Performance of a class of adaptive data-driven echo cancelers," *IEEE Trans. on Communications*, Vol. COM-37, No. 12, pp. 1254–1263, December 1989.

[27] J. Cioffi, "A fast echo canceler initialization method for the CCITT V.32 modem," *Proc. Globecom '87*, Tokyo, pp. 1950–1954, November 1987.

[28] D. Harman, J. D. Wang, and J. J. Werner, "Frequence Offset Compensation Techniques for Echo-Cancellation Based Modems," *Conference Record Globecom '87*, Tokyo, Japan.

[29] J. J. Werner, "Effects of channel impairments on the performance of an in-band data-driven echo canceler," *AT&T Tech. J.*, Vol. 64, No. 31, pp. 91–113, January 1985.

[30] J. D. Wang, and J. J. Werner, "Performance analysis of an echo-cancellation arrangement that compensates for frequency offset in the far echo," *IEEE Trans. on Communications*, Vol. 36, No. 3, pp. 364–372, March 1988.

[31] J. Salz, "On the start-up problem in digital echo canceler," *Bell System Tech. J.*, Vol. 60, No. 10, pp. 2345–2358, July-August 1983.

EXERCISES

9.1: A time-compression multiplexing system is to provide continuous full-duplex data communication at 100 kbps in each direction. Assume that the one-way, end-to-end propagation time is 10 ms, and a station does not start a transmission burst until it has observed the end of the transmission burst from the other direction

(a) What value of the buffer capacity yields the minimum transmission burst rate? What is that minimum burst rate? What burst rate yields the minimum buffer size? What is that size? [Hint: These are extreme cases.]

(b) For a realistic buffer capacity of 10,000 bits, what is the burst transmission rate? The end-to-end transmission delay (including buffer delay)?

(c) For this burst transmission rate, and if four-level baseband PAM with 50% rolloff is used for the line signal in each direction, what is the required bandwidth?

9.2: Prove that for a linear echo channel $h_e(t)$, the echo canceler incorporating an infinite-length transversal filter with taps spaced at $T' < 1/2f_{\max}$, where $f_{\max}$ is the highest frequency component of the bandlimited echo signal, cancels perfectly, and has tap weights equal to samples of the echo channel.

9.3: Starting from the relationship

$$E\, e_n^2 = A\, E\{\xi_n'\, \xi_n\} + \sigma^2 \,, \tag{9.35}$$

$$e_{n+1} = \xi_{n+1}' \mathbf{d}(n+1) + \nu_{n+1} \,, \tag{9.36}$$

where $\underline{\xi}_n$ is the tap-weight error vector, ν_n is noise uncorrelated with the data, and $\sigma^2 = E\, \nu_n^2$, and $(2N+1)$ is the number of taps, show that

$$\left[1 - 2\beta A + \beta^2 (2N+1) A^2\right] E\{e_n^2\} + 2\beta A\, \sigma^2 \,. \tag{9.37}$$

9.4: For 4-level PAM where the transmitted pulse level dz_n at time nT takes on the values [–3, –1, 1, 3] with equal probability, and with the $\{d_n\}$ mutually uncorrelated, derive the reference signal covariance matrix

$$R = E\{\mathbf{d}_n \mathbf{d}'_n\} .$$

9.5: Determine the values of the coefficients K_n^c, $n = 0, \ldots, N-1$, which minimize $E(e_i^2)$ where e_i is the echo estimation error, given by (9.100), for the lattice filter.

9.6: Let P be the matrix of column eigenvectors of the covariance matrix R, i.e., $Rp_j = \lambda_j p_j$, p_j the jth column of P.

(a) Show that $PP' = I$, i.e., P is orthonormal.

(b) Show that $[I - \beta R]^n = P[1 - \beta]^n P'$.

9.7: Show that an echo canceler tap-weight set equal to samples, at Nyquist interval T', of the echo channel $h_e(t)$ minimizes the mean-squared cancellation error, provided the local and distant data trains are uncorrelated.

10

Topics in Digital Communications

10.0 INTRODUCTION

In this chapter we discuss several advanced topics in digital communications. These concepts are advanced from two perspectives: (1) they represent a synthesis and/or an extension of material that has been discussed earlier in this text; (2) the work is very timely in the technical literature and for product applications, and has not yet appeared in texts. With regard to the first item, the foundation technologies that have been described in earlier chapters are the basic building blocks used for the design and analysis of communications systems. Although several of the subjects we consider could have been treated in earlier chapters, we have chosen to collect them here so that we can draw on the foundation that has been developed.

We begin this chapter with a discussion of the effects of digital implementation on adaptive systems — with emphasis on equalization. As we demonstrate, adaptation with limited precision is not as straightforward as designing a fixed (nonadaptive) digital filter. We also describe the affects of biases and latency on adaptive systems. The second topic is the adaptive phase-locked loop, which is capable of tracking an arbitrary phase jitter spectrum. This significantly extends the capabilities of the data-directed PLL of Chapter 6 by using the adaptivity concepts described in Chapter 8. Finally, we describe the potential applicability of (coded) modulation and equalization to emerging fiber-optic-based communication systems. As fiber-optic systems increase in rate, it is expected that designers will be faced with the same sorts of issues that have faced those designing systems for telephone, radio, twisted pair, and other media.

10.1 EFFECT OF DIGITAL IMPLEMENTATION ON THE PERFORMANCE OF ADAPTIVE EQUALIZERS

As described in Chapter 8, state-of-the-art adaptive equalizers are generally digitally implemented and strive to minimize the equalized mean-square error. An important consideration in assessing the complexity of such an adaptive digital equalizer is the number of bits required to represent the stored signal samples and the equalizer tap weights. In this section we will show that the precision required for successful adaptive operation, via the LMS algorithm, can be significantly greater than that required for static or fixed equalization. We will also determine the precision required in the tap-updating circuitry so that the equalizer mean-square error can be reduced to an acceptable level. The effects of biases in adaptive digital systems, particularly fractionally spaced equalizers, will also be discussed.

10.1.1 THE LMS ALGORITHM WITH LIMITED PRECISION

In the LMS (or estimated-gradient) tap adjustment algorithm,

$$\mathbf{c}_{n+1} = \mathbf{c}_n - \Delta e_n \mathbf{r}_n , \tag{10.1}$$

the tap weights, $\mathbf{c}_n$, are incremented by a term proportional to the product of the instantaneous output error, e_n, and the voltage stored in the corresponding delay element, r_n. Generally, it is assumed that the algorithm will settle and continue to fluctuate about the optimum tap weights. However, when the *correction term is less than half a tap-weight quantization interval, the algorithm ceases to make any further substantive adjustment.* [1] To determine the minimum number of bits needed to achieve an acceptable performance level (mean-square error), an appropriate proportionality constant, or step size, must be determined for use in the algorithm. From pure analog, or infinite precision considerations, a relatively large step size is desirable to accelerate initial convergence, while a small step size is needed to reduce the residual mean-square error (that part of the error in excess of the minimum attainable mean-square error). If the channel is stationary, then in the converged mode the analog algorithm should use a vanishingly small step size to provide almost no fluctuation about the minimum obtainable mean-square error. In a digitally implemented algorithm, *a decrease in the step size may actually degrade performance,* unless there is a compensating increase in the precision of the tap weights. This occurs when the error is so small that an increased number of bits are needed in order that the proportionately smaller corrections be "seen" by the equalizer.

A useful compromise will be to choose a step size that provides a slight increase in the steady-state mean-square error predicted by infinite precision considerations to a level which can be attained by a digital equalizer of reasonable precision. This precludes the choice of an unrealistically small step size — with its concomitant requirement of excessive precision — and provides a

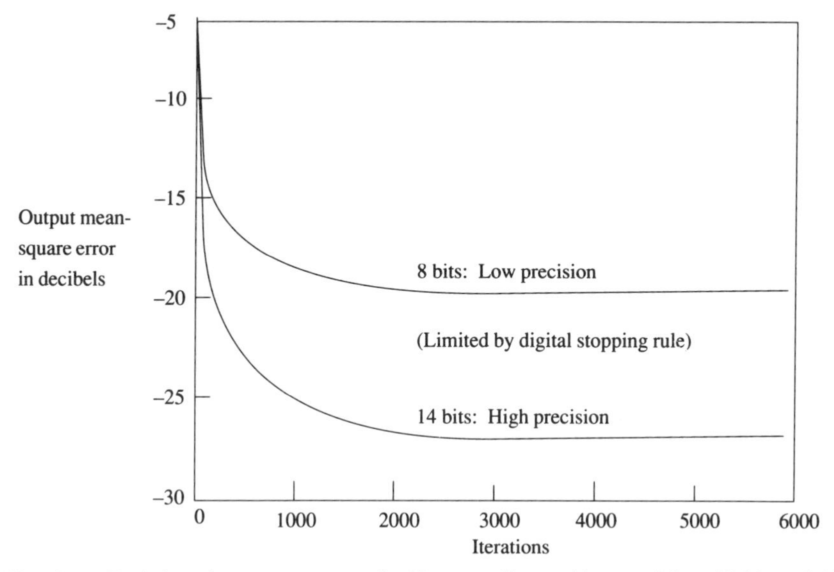

Fig. 10.1 Evolution of mean-square error for 32-tap equalizers with tap weights of 8 bits and 14 bits resolution. Steady state step size = 0.00021.

mechanism for the determination of the necessary level of precision. We first describe the effects of digital implementation on the LMS estimated-gradient tap-adjustment algorithm, and we then combine our previous analog results with these digital results to compute the minimum precision necessary to achieve acceptable performance.

In Figure 10.1 we sketch the evolution of the mean-square error in high- and low-precision equalizers. Note the substantially increased steady-state MSE with limited precision. In the low-precision equalizer, the steady-state, mean-square error is constrainted by the impossibility of changing a tap weight when the correction term in the LMS algorithm, given by (8.48) decreases below half a quantization interval. Some corrections will be possible until the peaks of the correction terms fall below the critical level, i.e., until for some tap

$$|\Delta e_n r_n| \leq \tfrac{1}{2}\,\delta = 2^{-B}\alpha \tag{10.1b}$$

where e_n is the equalizer output error, r_n is the signal sample in the delay line, δ is the interval between quantization levels, B is the number of bits (including sign) used to represent the equalizer tap weights, and $(-\alpha,\alpha)$ is the range covered by the (uniform) quantizer. Suppose (10.1) is satisfied for the ith tap, and as the particular input sample propagates down the equalizer, the error and step size will further decrease in magnitude, thus ensuring that this sample will turn off all the taps "down the line." Because of this observation, it is reasonable to assume that all the taps stop adapting at the same time. To a first approximation, it is also reasonable to replace $|r_n|$ by its rms value, R_{rms}, giving the following relation for the rms error when adaptation stops:

$$|e_n| \le \frac{\alpha 2^{-B}}{\Delta \cdot R_{rms}} \equiv e_d(\Delta). \tag{10.2}$$

We call $e_d(\Delta)$ the rms digital residual error (DRE). It is apparent from (10.2) that this error is *inversely* proportional to the step size. This is completely counter to the analog (or infinite precision) analysis of the LMS algorithm, where decreasing the step size always decreases the steady state error; as (10.2) shows, *when the LMS algorithm is dominated by digital considerations, decreasing the step size may increase the steady-state error.*

The above stopping condition can be approximated by replacing the magnitude in (10.2) by its peak value, which is assumed to be $\sqrt{2}$ times its rms value, i.e., adaptation continues if

$$\sqrt{2}\ \Delta \cdot \sqrt{\langle e_n^2 \rangle} \cdot \sqrt{\langle r_n^2 \rangle} \ge 2^{-B}\alpha\ , \tag{10.3}$$

where the MSE which satisfies (10.3), with equality, is the DRE. In a passband equalizer for which (10.2) applies separately to the in-phase and quadrature parts of the tap increment, the condition that is equivalent to (10.3) is $\Delta\sqrt{\langle |e_n|^2 \rangle\ \langle |r_n|^2 \rangle /2} \ge 2^{-B}\alpha$.

To illustrate the practical significance of (10.3), consider a 17-tap equalizer with tap weights quantized to 12 bits, an input data stream having an rms value of unity, and with the step size fixed at 0.07. For this example, the DRE will be about 0.35×10^{-2}. Compare this with the rms error that would be expected if the only source of error were the quantization of the desired tap weights to 12 bits; that is, we assume that the ideal error-free equalizer has 17 taps, each of infinite precision. In Appendix 10A we show that the quantization error (QE) for the equalizer, when the taps are set (perhaps by a genie) to the best 12-bit values, can be approximated by

$$\varepsilon_Q \approx (2N+1)^{1/2} \cdot LSD \cdot X_{rms}\ . \tag{10.4}$$

For the numerical example described above, the QE is about 10^{-3}. Thus the DRE, due to the failure of the algorithm to find the best coefficient, is roughly 3.5 times worse than the rms error would be if the best 12-bit coefficients had been found. As explained in Appendix 10A, the ratio of the DRE to the QE is proportional to $N^{1/2}$. Thus the residual error phenomenon will be even more pronounced in longer digital equalizers.

The ratio of the DRE to the QE demonstrates one of the manifestations of digital implementation. The tap weights are, to a first approximation, trying to approach the quantized versions of the optimum settings; however, when the tap weights get close to the optimum setting, the mean-square error and step size have decreased appreciably, and by the nature of the algorithm the taps try to approach the optimum setting by using very small correction terms. Once the

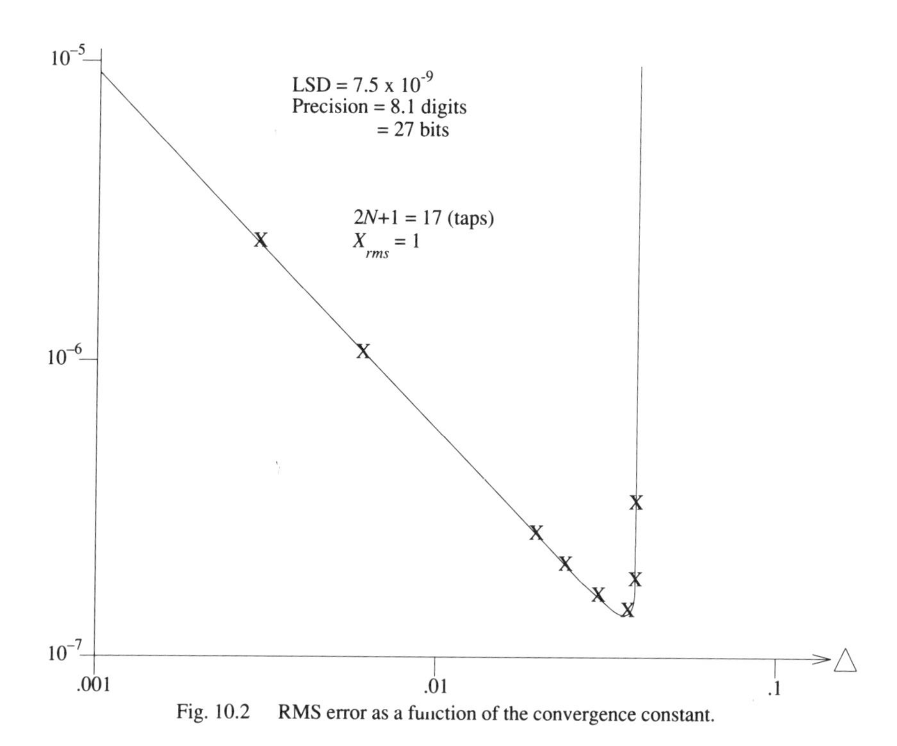

Fig. 10.2 RMS error as a function of the convergence constant.

correction term becomes smaller than the least significant digit (LSD), adaptation stops, and, as shown in Appendix 10A, the algorithm terminates while the taps are appreciably further away from the optimum setting than one LSD. The net effect is that the quantization is enhanced and produces a relatively large DRE.

To see how the above observations are manifested in practice we show, in Figure 10.2, the results of a computer experiment on a 17-tap digital adaptive equalizer. The experiment consisted of sending the same input stream into both the adaptive equalizer and the desired equalizer, a 17-tap equalizer with fixed optimum coefficients. The adaptive equalizer was adjusted by minimizing the mean-square difference between their outputs. The final rms error is plotted for various values of constant step size Δ. This error is normalized in the sense that the input data stream has an rms value of unity and the largest tap weight is one. As Δ increases, the stability limit is approached; as Δ is *decreased* we see the profound effects of the *digital stopping rule* as the RMS error *increases*.

There are several important consequences of the above discussion:

1) If the DRE exceeds the steady-state MSE attainable with infinite precision, any attempt to make Δ arbitrarily small, will ultimately *increase* the steady-state MSE so that (10.3) is satisfied.

2) The ratio of the mean-square error of an *adaptive* digital equalizer to the MSE due to quantizing the tap weights $\{c_n\}$ to within a LSB of their optimum values grows linearly with the number of taps and the "effective" eigenvalue ratio. (See Appendix 10A.) In other words, considerably more precision is required for adaptation of the tap weights than for filtering the received signal (i.e., performing the equalizer convolution).

3) The number of adaptive parameters should be kept to a minimum consistent with achieving the desired steady-state MSE, since any increase in the number of taps calls for a decreased Δ, which in turn increases the precision required.

4) The excess MSE (associated with a finite step size) evaluated in Chapter 8 (see (8.74)) can be traded against the required precision.

5) Highly dispersive channels have a larger eigenvalue spread than do moderately dispersive channels and thus require more precision to achieve the same MSE. However, the increased precision will be shown typically to be only 1 bit for channels of moderate distortion.

6) We will show that the digital word size will be restricted to a reasonable value if the digital equalizer is designed to allow an appropriate excess MSE on the order of the minimum steady-state MSE.

10.1.2 REQUIRED PRECISION

With the above observations, we let the steady-state step size Δ, determined from (8.70), be used in a *digitally implemented* equalizer. The analog parameters are assumed such that the steady-state MSE is equal to the digitally limited MSE. Thus performance will be determined by the available precision. The digital word length, B bits, needed in the tap weights to achieve E_∞ can be determined by substituting (8.70) in the digital stopping condition (10.3):

$$2\frac{\bar{\lambda}}{\lambda_M}\,\frac{\gamma}{1+\gamma}\,\frac{1}{(2N+1)}\cdot\frac{1}{\langle r_n^2\rangle}\cdot\sqrt{\langle e_n^2\rangle\langle r_n^2\rangle}\geq 2^{-B}\alpha, \tag{10.5}$$

which reduces to

$$2^{-B}\alpha\leq 2\frac{\bar{\lambda}}{\lambda_M}\,\frac{\gamma}{1+\gamma}\,\frac{1}{(2N+1)}\cdot\frac{1}{\sqrt{SNR\cdot\rho}}\,, \tag{10.6}$$

where

$$SNR=\frac{S_{out}}{\langle e_n^2\rangle} \tag{10.7}$$

is the (equalized) output signal-to-noise ratio, and where

$$\rho=\frac{\langle r_n^2\rangle}{S_{out}} \tag{10.8}$$

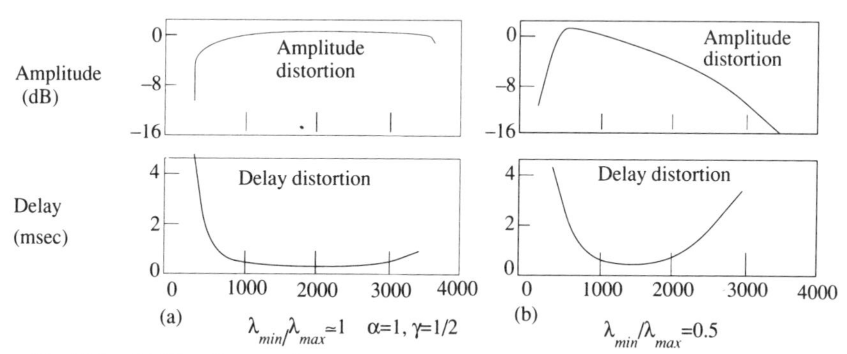

Fig. 10.3 (a) "Good" and (b) "distorted" channels used in analytical and simulation examples.

is the ratio of input signal power to output (baseband) signal power.* Using (8.74), the equalized mean-square error appearing in (10.7) can be written as $\langle e_n^2 \rangle = \xi_{opt} + q_\infty = (1 + \gamma)\, \xi_{opt}$. Thus for a given (or known) channel, all the terms in (10.7) can be readily computed. We may use the above equations to estimate the required digital word length under typical channel conditions. [2]†

i) *Operation on "Good" Channel* (Figure 10.3a). A good channel is one for which $\bar{\lambda}/\lambda_M \cong 1$, and the number of equalizer taps is quite small. In practice, however, a synchronously spaced equalizer will have a fixed number of taps, typically 32. With a passband equalizer having 32 complex tap pairs, the effective eigenvalue ratio is 0.93 for the "good" channel, using a near-optimum sampling epoch. We assume that the maximum quantization level is $\alpha = 1$, that $\gamma = ½$ (corresponding to a 1.1 dB degradation in output SNR ratio), and an output SNR ratio of 25.7 dB is observed (down 1.1 dB from that observed with effectively infinite resolution and a vanishing small step size). An AGC setting is assumed such that $\rho = 2$. We find from (10.7) that $B \approx 11$ bits.

ii) *Operation on a Severely Distorted Channel* (Figure 10.3b). This severely distorted channel has an effective eigenvalue ratio $\bar{\lambda}/\lambda_M$ (for the best sampling phase) of 0.5. Using this latter value and with the parameters of the previous sample, except for an output SNR ratio of 23.4 dB associated with the distorted channel, we find that $B \approx 11.5$ bits.

---

* For the complex passband equalizer, the factor on the right-hand size of (10.7) is $\sqrt{2}$ instead of 2, and $\langle |r_n|^2 \rangle$ replaces $\langle r_n^2 \rangle$ in (10.8).

† It should be noted that the word lengths derived in the following examples are longer than the precision indicated from using $\sigma^2 = \delta^2/12$ for the quantization error in each tap weight.

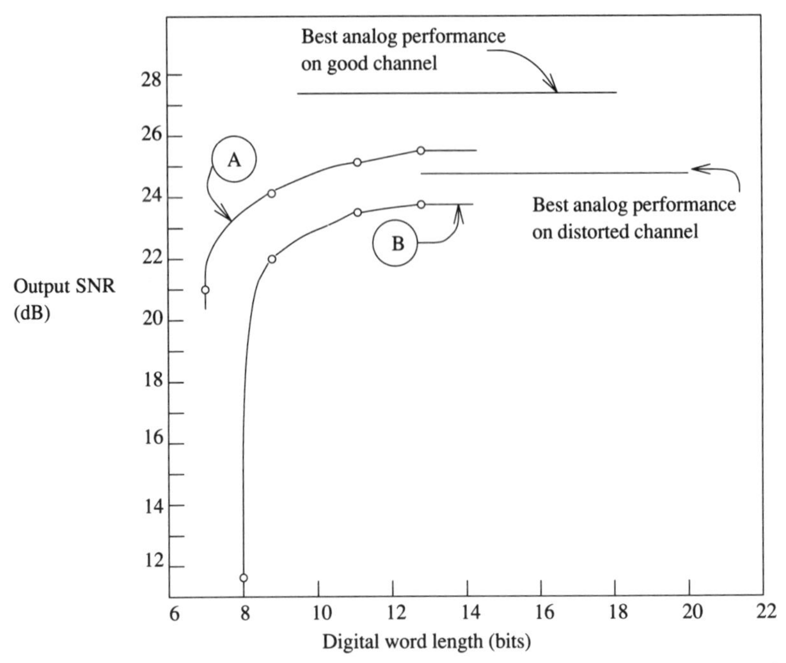

Fig. 10.4 Measured output signal-to-noise ratio vs. digital word length (of equalizer tap weights) for a "good" channel (Figure 10.3.A) (eigenvalue ratio = 0.93), $\Delta = 0.001$; and a distorted channel (Figure 10.3.B) (eigenvalue ratio = 0.5), $\Delta = 0.0005$. A 32-tap synchronous equalizer was used in both cases. Simulation results are indicated by the dotted points.

iii) *Operation with a Fractionally-Spaced Equalizer on the Distorted Channel.* With channel samples taken at $T/2$ intervals, where T is a symbol interval, a 64-tap equalizer is appropriate. The effective eigenvalue ratio* is 0.58, and find that $B \approx 12$ bits for the same 1.1 dB degradation used in the above examples.

The precision predicted by the above formulas was compared with results obtained using a simulation program for a QAM data communication system operating at 9600 bps with a baud of 2400 per second. Only the tap weights were quantized (rounded to the nearest quantization level); all other variables had the IBM 370 single-precision resolution of roughly 24 bits. The magnitude of the largest real tap weight was about 0.5 in a full quantization range of $(-1, 1)$. Timing and phase references were ideal and not subject to statistical fluctuation. The equalizer was either a 32 (complex) tap synchronous (tap spac-

* The "average" tap weight was the average over the 26 tap weights which collectively contained 95 percent of the tap-weight mass.

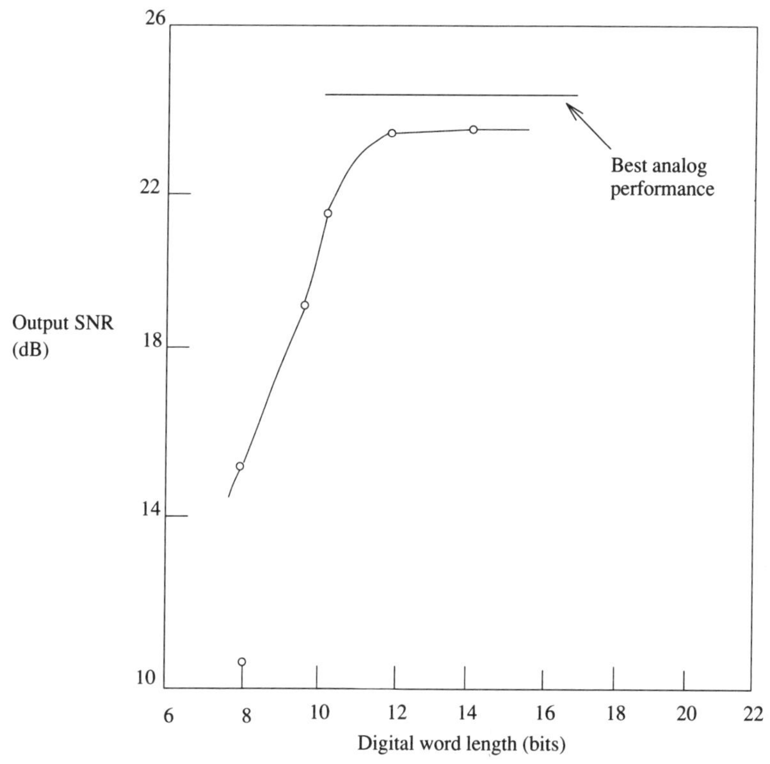

Fig. 10.5 Output signal-to-noise ratio vs. digital word length (of equalizer tap weights) for a 64 $T/2$ tap equalizer operating on the severely distorted channel (with step size = 0.003). Simulation results are indicated by the dotted points.

ing = T = symbol interval) structure, or a 64 (complex) tap $T/2$ structure. Some simulation runs were made with a gear-shifting sequence of adaptation step sizes to reach convergence within a reasonable number of iterations. However, great care was taken to reach the smallest (steady-state) step size well before complete convergence. This is because, with a larger step size, digital equalizer performance can possibly be better for a transient period than that corresponding to the chosen steady-state value. Deterioration of this "good" performance, once it is achieved, depends upon large signal and/or noise values, and may not be observed over the short duration of a simulation run.

Curve A in Figure 10.4, for operation with the synchronous equalizer on the "good" channel, indicates a degradation of about 1.5 dB for the step size of 0.001 calculated from (8.70). This is not far from the 1.1 dB (γ = 0.5) used in that formula; however, the signal-to-noise ratio degrades another 0.6 dB when the predicted digital word size of 11 bits is used instead of infinite resolution. This additional degradation is probably due to quantization noise and the slowed

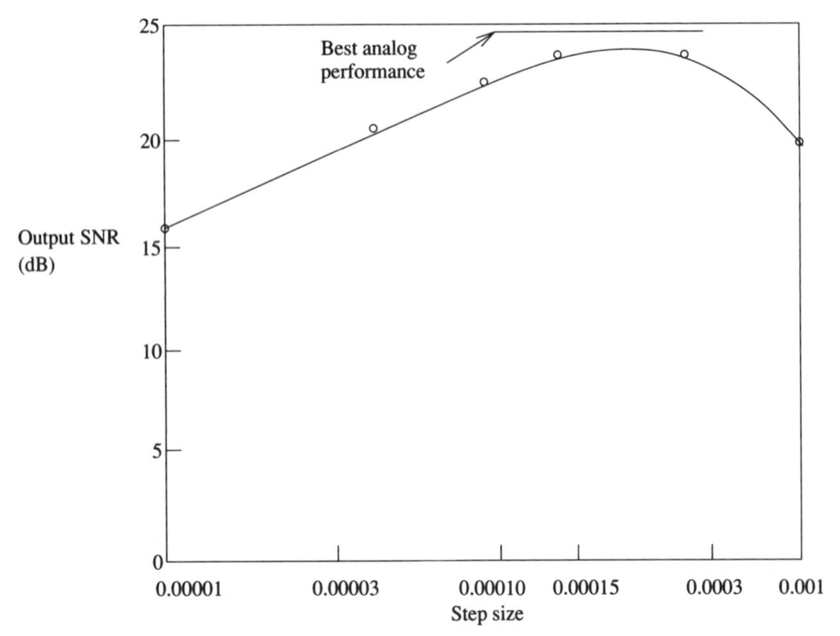

Fig. 10.6 Output s/n ratio vs. adaptation step size for 64 $T/2$ tap equalizer, tap weight, quantized to 12 bits, and severely distorted basic-conditioned channel.

rate of adaptation, just before the digital stopping condition prevails. Curve B, for operation on the distorted channel, is down 1.1 dB for a digital word length very close to the predicted 11.5 bits.

Figure 10.5 illustrates a similar curve for operation on the distorted channel with the 64-tap $T/2$ equalizer. The digital resolution of about 12 bits predicted for 1.1 dB degradation is consistent with the simulation results. Figure 10.6 presents another view of the performance of the $T/2$ equalizer on the distorted channel. This is a curve of output SNR ratio vs. adaptation step size for 12-bit resolution. The step size of 0.0003 used in the experiments, and represented as points on the curves of Figure 10.4, corresponds to a near-peak value of output signal-to-noise ratio. This supports the prior analysis as providing a useful formula for deciding on the steady-state step size to be used in a digitally implemented equalizer.

10.1.3 MORE ON FRACTIONALLY-SPACED EQUALIZERS: STABLE OPERATION OF A SYSTEM WITH TOO MANY DEGREES OF FREEDOM

In the previous section we discussed the affect of limited precision, or quantization, on synchronous and fractionally-spaced equalizers. In Chapter 8 we described how, in the frequency domain, the FSE spans a bandwidth greater than that of the input signal. Thus there is a region of the spectrum where there is no input power, but where the FSE can synthesize a response. Since each frequency band of $1/NT$ Hz corresponds to a degree of freedom (or dimension), the FSE has more degrees of freedom than it needs to equalize the channel. In this section we describe the operation of the FSE in light of the above observation, and examine what changes have to be made to the algorithm and structures described above to insure successful operation of FSEs.

In laboratory experiments [3] with a digitally implemented FSE it was noticed that after an extended period of operation some of the equalizer tap weights would invariably become large, while the mean-square error (MSE) remained at a satisfactory level. The taps generally would become so large that one or more registers, which computes partial sums of the equalizer output, would overflow, and the modem performance was then substantially degraded. This phenomenon is a consequence of the fact that an FSE, in contrast to a conventional synchronous equalizer, generally has many sets of tap values that correspond to roughly the same MSE. Included in the set of tap values that correspond to the minimum MSE are some tap coefficients of relatively large magnitude. These large tap values can be attained because of the cumulative effect of noise or any bias in the digital circuitry that performs the equalizer updating. Even though the value of the MSE is satisfactory, some of these tap values will be large enough to cause occasional overflow of the partial sums computed to form the equalizer output. The primary purpose of this section is to explain why a fractionally spaced equalizer has so many apparently "good" sets of tap values, and to indicate how equalizer operation can be stabilized by modifying the conventional estimated-gradient tap-adjustment algorithm. A natural question to ask is: does a fractionally spaced equalizer have a unique optimum setting?

To answer the above question, we refer to the simplified baseband data transmission system shown in Figure 10.7.

For the purposes of exploring the phenomenon of large-tap buildup, this baseband model will suffice. Referring to the figure: $\{a_n\}$ are the discrete-valued multilevel data symbols, $1/T$ is the symbol rate, $p(t)$ is the band-limited transmitter pulse (whose spectrum is shown in Figure 10.7b), $f(t)$ is the channel impulse response, and $v(t)$ is the additive background noise. Note that the receiving filter output, $r(t)$, is sampled at the rate $1/T'$, and the samples are then passed through the tapped delay-line equalizer (shown in Figure 10.7c) having $2N + 1$ delay elements spaced $T'(<T)$ seconds apart and tap weights $\{c_n\}$. The FSE output

$$q(nT) = \sum_{m=-N}^{N} c_m r(nT - mT'), \quad n = 1,2,... \tag{10.9}$$

is computed at the symbol rate and quantized (sliced) to provide the data decision, $\hat{a}_n$. The transmitted pulse spectrum will generally be band-limited to

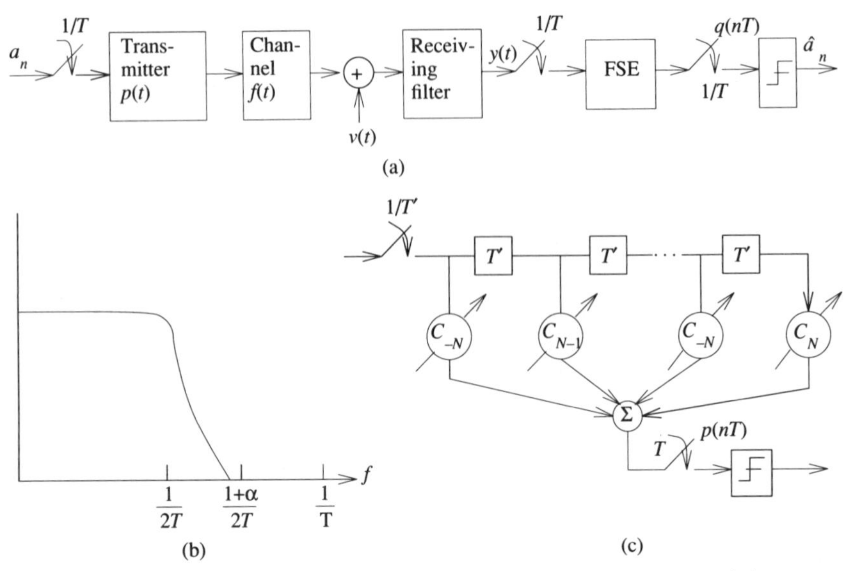

Fig. 10.7 Simplified (baseband) pulse amplitude modulation (PAM) data transmission system incorporating an FSE. (a) Transmission system. (b) Spectrum of the transmitted pulse. (c) A fractionally-spaced equalizer.

$(1 + \alpha)/2T$ Hz/second, where the rolloff factor, α, varies between 0 and 1. An FSE with tap spacing T' seconds will have a transfer function, $C_{T'}(f)$, with period $1/T'$, and, as shown in Figure 10.8 if $T' < T/(1 + \alpha)$, the transfer function of the FSE will span the entire spectral range of the transmitted signal. As described in Chapter 8, this enables the equalizer to exert complete control over the amplitude and delay distortion present in the region $0 < |f| \leq (1+\alpha)/2T$, and,

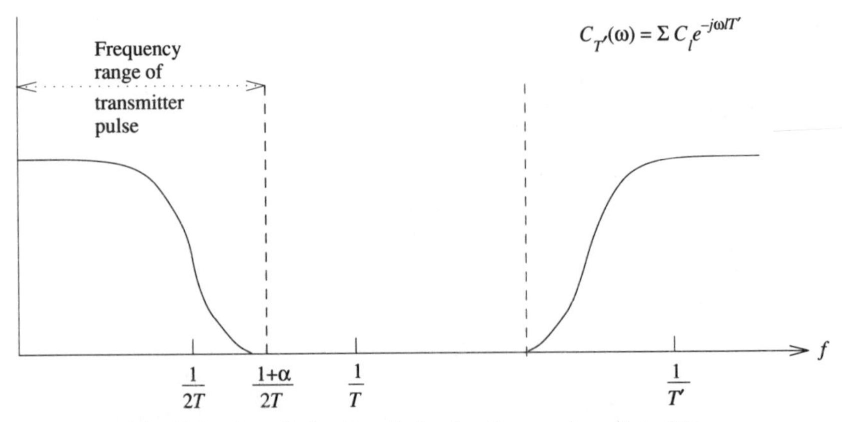

Fig. 10.8 Transfer function of a fractionally-spaced equalizer (FSE).

as we have said before, the FSE can compensate for these distortions directly, rather than filtering the aliased (folded) spectrum, as is done by the conventional equalizer.

From Figure 10.8 it should also be evident that when the noise becomes vanishingly small, there is concern as to what function(s) the equalizer will synthesize in the region, $(1+\alpha)/2T < |f| < 1/T$, where there is no signal energy. As we have shown previously, for an infinitely long equalizer, the composite transfer characteristic that minimizes the MSE is

$$C'_T(f) = \frac{X^*(f)}{\sum_{\ell} \left| X(f + \ell/T) \right|^2 + N_0} , \quad |f| \leq \frac{1}{T}, \tag{7.81}$$

where the asterisk denotes the complex conjugate, $X(f)$ is the transfer function of the received pulse, $x(t)$, presented to the equalizer input, and N_0 is the noise spectral density. As long as $N_0 \neq 0$, the equalizer function is zero whenever the received signal has no power; however, as $N_0 \to 0$ (7.81) approaches 0/0 in the region $\frac{(1+\alpha)}{2/T} < |f| < 1/2T$. As the noise vanishes, an infinitely long FSE can synthesize the required channel characteristic in the region $0 < |f| < \frac{(1+\alpha)}{2/T}$, and an arbitrary — and nonunique — characteristic in the remaining frequency band. An interesting question, then, is what happens to the optimum tap setting for the finite length FSE as the noise becomes vanishingly small.

10.1.4 UNIQUENESS OF SOLUTION FOR FINITE LENGTH FSE AS THE NOISE VANISHES

In Chapter 8, the MSE is readily evaluated as the quadratic form

$$\xi = \mathbf{c}' A \mathbf{c} - 2\mathbf{c}'\mathbf{x} + \langle a_n^2 \rangle , \tag{10.10}$$

where the prime denotes the transposed vector, $\mathbf{c}'$ is the tap vector $(c_{-N}, \ldots, c_0, \ldots, c_N)$, x' is the truncated impulse-response vector $|x(NT'), \ldots, x(-NT')|$, and A is the channel-correlation matrix whose *kl*th element is given by

$$A_{kl} = \sum_{m=-\infty}^{\infty} x(mT - kT')x(mT - lT') + \sigma^2 \delta_{k-l} . \tag{10.11}$$

Note that A is not a Toeplitz matrix, as it would be for a synchronous equalizer $(T' = T)$.

The matrix A is the sum of two matrices. The channel-dependent component of A is always positive semidefinite. Since the other component of the channel-correlation matrix, $\sigma^2 I$, is positive definite, then A will also be positive definite, and we can conclude that when there is noise present, the optimum tap setting is unique.

We now consider the situation as the noise becomes vanishingly small; the optimum tap setting will be unique if, and only if, A is nonsingular. A sufficient condition for A to be nonsingular is the ñonvanishing of the quadratic form $\mathbf{u}'A\mathbf{u}$, for any *nonzero* test vector u with components $[u_l]$. Using (10.11),

$$\mathbf{u}'A\mathbf{u} = \sum_{m,n=-N}^{N} u_m A_{mn} u_n = \sum_{l=-\infty}^{m} \left[\sum_{m=-N}^{N} u_m x(lT - mT') \right]^2 \geq 0 . \qquad (10.12)$$

The above inequality establishes the positive semidefinite nature of the matrix A and in [3] it is shown that *for a finite-length FSE* with an excess bandwidth of less than 100 percent, *even as the noise becomes vanishingly small, the A matrix is nonsingular and there is a unique optimum tap setting.*

10.1.5 THE TAP-WANDERING PHENOMENON

We have shown that for a finite-length fractionally-spaced equalizer in the practical range of interest—where the excess bandwidth is on the order of 10 to 50 percent—even with vanishingly small noise there will always be a unique best tap setting. One issue of interest is the "closeness" (or ill-conditioning) of the A matrix to a singular matrix. This is important for two reasons. First, the distribution of the eigenvalues of A, which is a measure of the ill-conditioning of the matrix, influences the rate of convergence of the equalizer taps to their optimum setting. Second, more importantly, from Chapter 8 we know that the contours of equal MSE are elliptical, and the eccentricity of these contours is directly related to the eigenvalue distribution. In Appendix 8B it was shown that for an infinitely long equalizer with $T' = T/2$, half the eigenvalues are zero; in Figure 10.9 we illustrate some constant MSE contours for a finite-length $T/2$ equalizer, whose optimum tap setting is denoted by $\mathbf{c}_{opt}$. Recall that even with an infinite-precision analog implementation, the use of a finite step size in the conventional estimated-gradient tap-adjustment algorithm results in a steady-state MSE that exceeds E_{opt}. This is depicted in Figure 10.9, where the inner contour is the MSE that can be attained with the chosen step size. Owing to the random component in the algorithm's correction term, the taps will wander along the constant MSE contour, and there will be a certain probability that the taps will become so large that one or more registers will saturate.* Thus, even in an analog implementation, random tap wandering can, in principle, lead to degraded performance.

It has been observed in laboratory experiments with a digitally implemented FSE, that under control of the conventional LMS tap adjustment algorithm,

$$\mathbf{c}_{n+1} = \mathbf{c}_n - \Delta[e_n \mathbf{r}_n], \quad n = 1, 2, 3, \ldots \qquad (10.12)$$

* For a synchronous equalizer the ellipses will not be very eccentric, and the tap wandering will not do any damage.

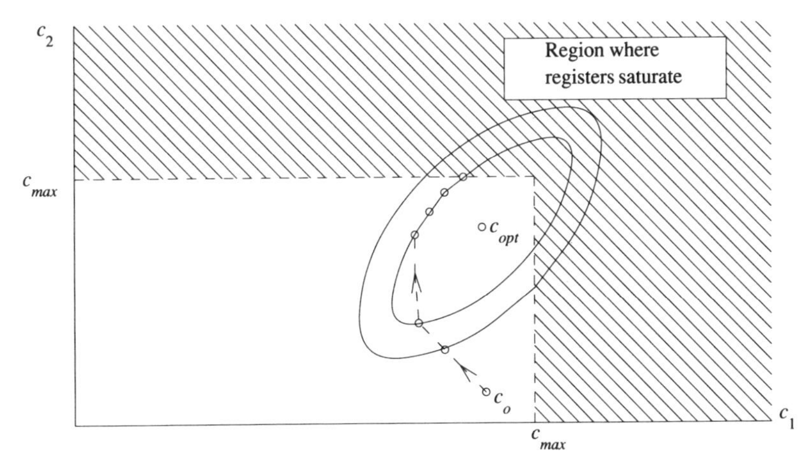

Fig. 10.9 Contours of equal MSE showing tap convergence.

the equalizer taps inevitably drift close to the shaded (large tap) region of Figure 10.9. In (10.12), c_n is the tap vector at $t = nT$, Δ is the step-size, which as we know from Chapter 8, influences both the convergence rate of the equalizer and the steady-state MSE; the brackets around $e_n r_n$ indicate that this increment is quantized to a specified number of bits. The term $[e_n r_n]$ will have a deterministic component proportional to the desired correction term, and a random component owing to both the manner in which the digital quantization is performed and the influence of the noise and data-dependent terms. Generally, $[e_n r_n]$ will also possess a deterministic component since a bias inevitably present in a digital implementation. A typical mechanism for such a bias is the two's complement type of quantizing characteristic shown in Figure 10.10. To quantify our discussion, we denote the bias by a time-invariant vector, **b**, and model the adjustment algorithm as

$$\mathbf{c}_{n+1} = \mathbf{c}_n - \Delta(e_n \mathbf{r}_n + \mathbf{b}), \tag{10.13}$$

where the bias vector has equal components, and b has a magnitude of less than half a quantization. Equation (10.13) ignores, except for the bias, the other effects of limited precision on the algorithm (including those discussed earlier in this chapter). Since it has been observed in the laboratory that tap wandering results in a systematic buildup of some tap values, the above model should be useful in relating the magnitude of the bias to the other system parameters. It is the bias component that can drive, in a *deterministic* manner,* the tap vector toward the tap region corresponding to large tap values. Since the equalizer output is generally formed as a series of partial sums, of the form $\sum_m c_m r_{n-m}$, it

* As opposed to the random wandering associated with the self-noise of the algorithm.

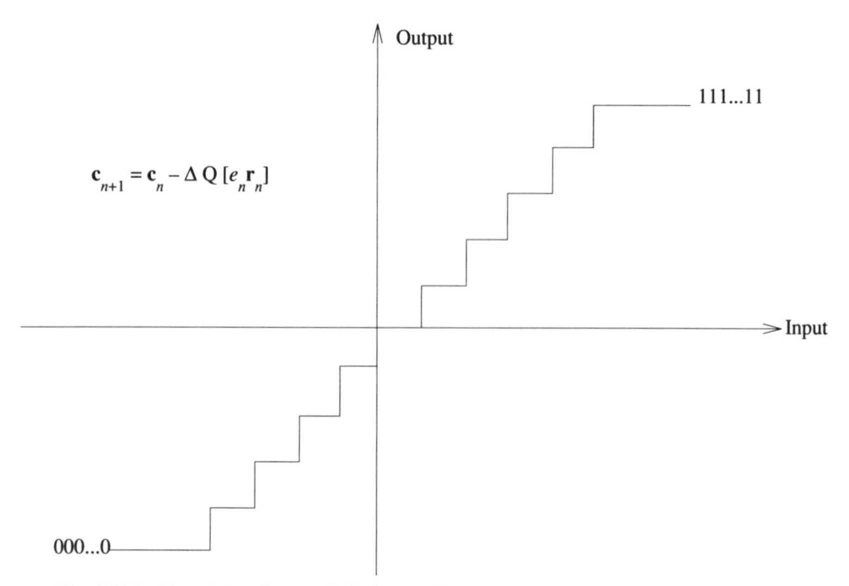

Fig. 10.10 Quantizing characteristic that produces a bias in the tap-adjustment algorithm.

is clear that large-tap values can lead either to an overflow of a partial sum or saturation of a tap. As the taps grow, occasional register overflows begin to occur, resulting in noise-like "hits" on the equalizer output. The occurrence of such a "hit" is a function of the specific pattern of data samples contained in the equalizer. Continued growth of the taps increases the frequency of the "hits", as more data patterns can produce these events. The error rate becomes very high, relative to what constitutes acceptable performance, but the frequency of occurrence of "hits" is still low enough that the MSE is almost unaffected. Growth continues until the tap coefficients themselves begin to saturate. At this point severe degradation occurs as the degrees of freedom of the system are reduced. In fact, with an experimental digitally implemented FSE, with a symbol rate of 2400 symbols/second, overflows typically begin to occur within several minutes of operation [3]. The overflows produce noise-like hits and a degraded equalizer output.

10.1.6 THE MEAN TAP ERROR AND THE MEAN-SQUARED ERROR

To assess the effects of the bias on the tap-error vector,

$$\boldsymbol{\varepsilon}_n \equiv \mathbf{c}_n - \mathbf{c}_{opt}, \tag{10.14}$$

we use the above model to write the tap-error evolution as

$$\boldsymbol{\varepsilon}_{n+1} = \boldsymbol{\varepsilon}_n - \Delta(e_n \mathbf{r}_n + \mathbf{b}) . \tag{10.15}$$

From (10.15) we have

$$\langle \boldsymbol{\varepsilon}_{n+1} \rangle = \langle \boldsymbol{\varepsilon}_n \rangle - \Delta \langle e_n \mathbf{r}_n \rangle - \Delta \mathbf{b} = (I - \Delta A) \langle \boldsymbol{\varepsilon}_n \rangle - \Delta \mathbf{b} , \quad (10.16)$$

and the steady-state mean tap error satisfies

$$\langle \boldsymbol{\varepsilon} \rangle = A^{-1} \mathbf{b}, \quad (10.17)$$

where λ_i and p_i, respectively, denote the ith eigenvalue and eigenvector of A.

$$\langle \boldsymbol{\varepsilon} \rangle = \sum_{-N}^{N} \frac{\mathbf{p}_i' \mathbf{b}}{\lambda_i} \mathbf{p}_i , \quad (10.18a)$$

If there is a small eigenvalue whose eigenvector is *not* orthogonal to b, then the steady-state tap error can be quite large. It is interesting to note that for a small number of taps, one would expect that the eigenvalues not be small, i.e., the equalizer would not possess enough degress of freedom to realize a somewhat arbitrary transfer function beyond the rolloff region. For a one-tap equalizer (or automatic gain control) where $A = \langle r^2(nT) \rangle$, and consequently

$$\langle \boldsymbol{\varepsilon} \rangle = \frac{\mathbf{b}}{\langle r^2(nT) \rangle} . \quad (10.18b)$$

Thus, for a one-tap equalizer, the tap error is directly proportional to the magnitude of the bias, and the buildup of a large tap value is prohibited. However, as the number of taps becomes infinite, it has been shown in Appendix 8B that with $T' = T/2$, that half the eigenvalues are zero, while the other half tend to uniformly sample the aliased (with respect to the symbol rate) squared magnitude of the channel transfer function. Moreover, the eigenvectors corresponding to the zero eigenvalues have most of their energy concentrated near $1/T$ Hz (and are thus close to being orthogonal to b), while the ith eigenvector corresponding to the nonzero eigenvalues approaches a sinusoid of radian frequency $\omega_i = i/N \cdot \pi/T$. For practical, finite-length equalizers, these limiting conditions will only be approximated, and there will be small eigenvalues whose corresponding eigenvector is not orthogonal to b. Consequently, $<\boldsymbol{\varepsilon}>$ can become as large as the largest ratio $[\mathbf{p}_i' \mathbf{B}]/\lambda_i$, and the steady-state tap error would then be biased away from the optimum value.

The effect of the bias term on the equalized MSE is determined by the size of the residual MSE. An exact analysis (or tight bounds) of the behavior of q_n is an extremely difficult problem; however, by assuming that the sequence $\{\mathbf{r}_n\}$ is independent, and that when the taps are at or near their optimum settings, the squared output error is relatively insensitive to the transmitted data pattern, it is possible to establish simple, but useful, relationships between the relevant system parameters. The residual MSE satisfies the recursion

$$q_{n+1} = \langle \boldsymbol{\varepsilon}_{n+1}' A \boldsymbol{\varepsilon}_{n+1} \rangle + \langle [\boldsymbol{\varepsilon}_n' - \Delta(e_n \mathbf{r}_n' + \mathbf{b}')] A [\boldsymbol{\varepsilon}_n - \Delta(e_n \mathbf{r}_n + \mathbf{b})] \rangle, \quad (10.19)$$

and by using the above assumptions we have

$$q_{N+1} \cong [1-2\Delta\bar{\lambda}+\Delta^2\lambda_M(2N+1)\langle r_n^2\rangle]q_n+\Delta^2\lambda_M(2N+1)(\langle r_n^2\rangle E_{opt}+b^2), \tag{10.20}$$

where λ_M is the maximum eigenvalue of A, and where

$$\bar{\lambda} = \frac{1}{2N+1}\sum_{-N}^{N}\lambda_i \tag{10.21}$$

is the average eigenvalue. Thus, the steady-state fluctuation about the minimum MSE is

$$q_{in} \cong \frac{\Delta\lambda_M(2N+1)[\langle r^2(nT)\rangle E_{opt} + b^2]}{2\bar{\lambda} - \Delta\lambda_M(2N+1)\langle r^2(nT)\rangle}. \tag{10.22}$$

To assess the effect of the bias on q_∞, we note that the bias "seen" by a tap component, will be approximately $2^{-B}c_{\max}$, where B is the number of bits used to represent the tap weights and $c_{\max}$ is the maximum tap value. If the equalizer signal is assumed to have unity power, then $c_{\max} \approx 1/[(2N+1)\langle r_n^2\rangle]^{1/2}$ and α is typically on the order of $1/|(2N+1)\cdot\langle r_n^2\rangle|$. Thus, $b^2/|r^2(nT)|$ will be on the order of $2^{-2B}(2N+1)$, and when the equalized output signal power is unity, the minimum MSE is roughly the inverse of the output signal-to-noise ratio. With typically parameters like $(2N+1) = 60$, $B = 12$, and $E_{opt} = 0.001$, it is clear that the effect of the bias on q_∞ is negligible. Thus, owing to the quantizing bias, there can be, on the average, a buildup of one or more large-tap weights, while the MSE is relatively unaffected. In other words, under the influence of a bias the taps would still remain on the boldface MSE contour of Figure 10.8, but would spend most of the time near the shaded region, and the system would be subject to random overflows, or hits. This phenomenon has been repeatedly observed experimentally, and in the next subsection we will describe a simple means of controlling the tap wandering.

10.1.7 THE TAP-LEAKAGE EQUALIZER ADJUSTMENT ALGORITHM

As we have discussed in the previous section, some or all of the tap weights in an FSE can reach unacceptably large values when the conventional tap adjustment algorithm, is used. A simple means of controlling large-tap buildup is by minimizing either of the augmented cost functions

$$J_1 = E + \mu\sum_{i=-N}^{N} c_i^2, \tag{10.23}$$

$$J_2 = E + \mu\sum_{i=-N}^{N} |c_i|, \tag{10.24}$$

where E is the MSE and μ is a suitably chosen (small) constant. The cost function J_1 ascribes a quadratic penalty to the magnitude of the tap vector, while J_2 provides a magnitude penalty, i.e., the cost function is penalized whenever the tap vectors builds up excessively. Since the taps are to be adjusted adaptively, we cannot interpret μ as a Lagrange multiplier. The use of a Lagrange multiplier would be appropriate if we were actually able to minimize J in a deterministic manner by using the true gradient. However, since the gradient of E with respect to c, $A\mathbf{c} - \mathbf{x}$, is not available, we must implement a stochastic algorithm analogous to the LMS algorithm. Thus, μ must be chosen beforehand by using some prior knowledge of the system parameters.

Let us first consider the degradation in the minimum attainable steady-state MSE caused by choosing $\mathbf{c}$ to minimize J_1 instead of E. Note that, for the moment, we are neglecting the bias and only assessing the increased MSE caused by minimizing the augmented cost function, J_1, via the true gradient algorithm. For binary transmission, that the taps will attempt to minimize the modified criterion:

$$J_1 = \mathbf{c}'(A + \mu_I)\mathbf{c} - 2\mathbf{c}'\mathbf{x} + 1 \tag{10.25}$$

$$= \mathbf{c}'B\mathbf{c} - 2\mathbf{c}'\mathbf{x} + 1, \tag{10.26}$$

where $B = A + \mu I$. The matrix B has the same eigenvalues as A, while the eigenvalues of B are $\lambda + \mu$. Note that the contours of equal values of J_1 are still ellipses but the maximum-to-minimum eigenvalue ratio governing the tap wandering is now $(\lambda_{\max} + \mu)/(\lambda_{\min} + \mu)$, where $\lambda_{\max}$ and $\lambda_{\min}$ are the maximum and minimum eigenvalues of A, respectively. Thus, by choosing μ properly the eccentricity can be controlled, and the equalizer tap vector is now determined as if the noise power were increased from σ^2 to $\sigma^2 + \mu$. Of course, the use of a tap vector selected on the basis of the pseudo-noise power, $\sigma^2 + \mu$, will increase the steady-state MSE. The steady-state tap vector will satisfy

$$\mathbf{c}(\mu) = B^{-1}\mathbf{x} = (A + \mu I)^{-1}\mathbf{x}, \tag{10.27}$$

where $\mathbf{c}(\mu)$ denotes the steady-state tap vector corresponding to the chosen value of μ. The excess MSE, $E(\mu) - E_{opt}$, is

$$E(\mu) - E_{opt} = [\mathbf{c}(\mu) - \mathbf{c}_{opt}]'A[\mathbf{c}(\mu) - \mathbf{c}_{opt}]. \tag{10.28}$$

To make a more detailed evaluation of the increase in MSE, we let

$$\boldsymbol{\varepsilon}(\mu) = \mathbf{c}(\mu) - \mathbf{c}_{opt} \tag{10.29}$$

denote the tap-error vector, and by diagonalizing A we have

$$\varepsilon(\mu) = [(A + \mu I)^{-1} - A^{-1}]\mathbf{x} = -\sum_i \frac{\mu}{\lambda_i(\lambda_i + \mu)} \mathbf{p}_i'\mathbf{x} \cdot \mathbf{p}_i \tag{10.30}$$

and

$$E(\mu) - E_{opt} = \mu^2 \sum_{i=-N}^{N} (\mathbf{p}_i'\mathbf{x})^2 \frac{1}{\lambda_i(\lambda_i + \mu)^2}. \tag{10.31}$$

Thus, to a first approximation, the increased MSE grows only as the square of the leakage parameter, μ, while the eigenvalue distribution—and the range in which the taps can wander—can be favorably altered, in a significant manner, by using even a very small value of μ.

10.1.8 THE TAP-LEAKAGE ALGORITHM

In a manner analogous to the LMS algorithm, the adaptive tap-leakage algorithms are constructed by minimizing the augmented instantaneous squared error, $e_n^2 + \mu \mathbf{c}_n' \mathbf{c}_n$, or, $e_n^2 + \mu \sum_{i=-N}^{N} |c_n^{(i)}|$. The first algorithm modifies the correction term in the LMS algorithm by a term proportional to the tap vector itself, giving the algorithm

$$\mathbf{c}_{n+1} = \mathbf{c}_n - \Delta[e_n \mathbf{r}_n + \mu \mathbf{c}_n] = (1 - \Delta\mu)\mathbf{c}_n - \Delta e_n \mathbf{r}_n , \qquad (10.32)$$

while the second algorithm is of the form

$$\mathbf{c}_{n+1} = \mathbf{c}_n - \Delta(e_n \mathbf{r}_n + \mu \text{ sgn } \mathbf{c}_n) = \mathbf{c}_n - \Delta\mu \text{ sgn } \mathbf{c}_n - \Delta e_n \mathbf{r}_n , \qquad (10.33)$$

where the sgn operation is applied individually to each component of the tap vector. Note that from an implementation point of view, the algorithm can be modified with almost no hardware change other than applying a systematic decrement to the magnitude of each tap. The latter algorithm has the practical advantage that adjustments will continue to be made no matter how small any tap weight becomes, while the first algorithm has the "advantage" of analytical tractability. When the former leakage algorithm is used, the algorithm evolves according to

$$\mathbf{c}_{n+1} = \mathbf{c}_n - \Delta[e_n \mathbf{r}_n + b + \mu \mathbf{c}_n] . \qquad (10.34)$$

Subtracting c_{opt} from both sides of (10.34), and solving for the steady-state average tap error we find

$$\langle \boldsymbol{\varepsilon} \rangle = \sum_{i=-N}^{N} \frac{\mathbf{p}_i' \mathbf{b}}{\lambda_i + \mu} \mathbf{p}_i + \mu \sum_{i=-N}^{N} \frac{\mathbf{p}_i' \mathbf{x}}{\lambda_i(\lambda_i + \mu)} \mathbf{p}_i . \qquad (10.35)$$

The first term, on the right-hand side is similar to (10.18), but note that the eigenvalues have been modified to eliminate the castrophic effects that can accompany vanishingly small eigenvalues. The second term, which is proportional to the leakage parameter, is similar to (10.31), and represents the increased tap error caused by the minimization of J_1 and not E. The leakage parameter, μ, must be chosen sufficiently large so that the first term is properly controlled in magnitude but not so large that the magnitude of the second term becomes appreciable. In general, the choice of μ is best done empirically.

To assess the effect of the tap-leakage algorithm on the steady-state MSE

$$E_n = E_{opt} + q_n ,$$

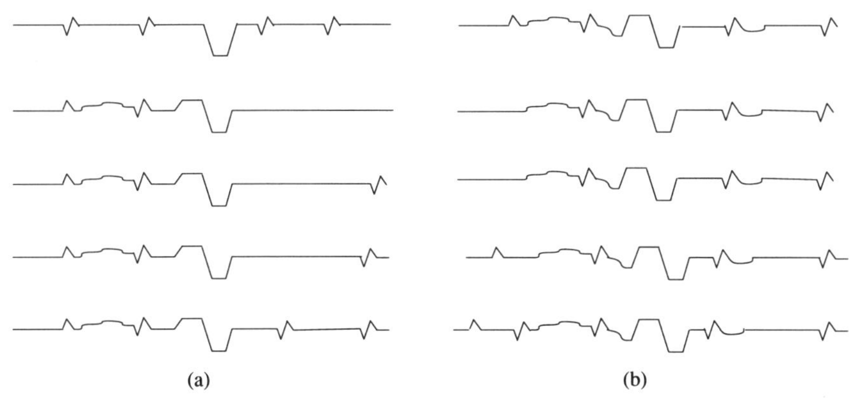

(a) (b)

Fig. 10.11 Tap buildup as a function of time. (a) Three-minute intervals. (b) Five-minute intervals. (9600 bps system, 16-QAM, T=1/2400 sec, tap quantization: 12 bits for update, 24 bits for coefficient.)

(where E_{opt} is the minimum MSE when both b and μ are zero) we find that the fluctuation about the minimum MSE is

$$q_{\infty} = \frac{\Delta\lambda M[(2N+1)\cdot\langle r^2(nT)\rangle\, e_{opt} + \mathbf{b}'\mathbf{b}] + 2\Delta\mu\mathbf{b}'\mathbf{x} + \mu^2\mathbf{x}'A^{-1}\mathbf{x}}{2\bar{\lambda} + \mu(1-\Delta\bar{\lambda}) - \Delta[\lambda_m(2N+1)\langle r^2(nT)\rangle + \mu^2]} \tag{10.36}$$

This expression is channel dependent, but when μ is chosen on the order of a small eigenvalue, then the fluctuation about the minimum MSE is a rather insensitive function of the leakage parameter, while the magnitude of the mean tap error, can be effectively controlled by the proper choice of μ.

For an experimental 9.6 kbps data transmission system using 16-point quadrature amplitude modulation, the tendency of the tap coefficients of a digitally implemented equalizer with T/2 sample spacing to drift is demonstrated by the waveforms of Figures 10.11a and 10.11b. These waveforms are analog representations of the values of one component of a set of complex tap coefficients associated with data samples taken T $(=1/2400)$ second apart. The equalizer is physically constructed by using two conventional T-spaced interleaved structures, requiring four tap coefficient distributions to totally describe the state of the equalizer.

The equalizer goes through a conventional start-up procedure, except that timing recovery and carrier phase adjustments are suppressed so that they will not change through interaction with either of these operations. The transmitted signal is fed back to the receiver, after passing through appropriate attenuators, thus avoiding any time varying channel characteristics, and assuring a low noise environment.

The first trace of Figure 10.11a illustrates the distribution of components among the particular collection of coefficients immediately after start-up. A single large negative component is noted, with all other components relatively insignificant. The coefficients are updated in the usual fashion, via the LMS algorithm, without the addition of a tap-leakage adjustment. Subsequent traces in Figure 10.11a are taken at 3-minute intervals. The traces of Figure 10.11b are a continuation of Figure 10.11a with separation in time extended to 5 minutes. A clear pattern of buildup in the amplitude of the taps, particularly those immediately preceding the original dominant tap, is demonstrated. A similar compensatory buildup occurs among those tap coefficient components not displayed, such that the MSE is essentially unchanged over the duration of the test. This deterministic growth of tap amplitudes will eventually lead to saturation of shift register accumulators used in forming the various components of the equalizer passband outputs. The observed output signal constellation will display frequent apparent noise-like hits of large amplitude and an unacceptable output error rate results.

The tap-coefficient components in the laboratory configuration are stored in 24-bit shift registers. The 12 most significant bits are used in the multiplication to form tap-product outputs. The remaining bits were to average out the effects of tap updating. The components of the updating quantity $-\Delta(e_n \mathbf{r}_n)$ — are stored in 12-bit words which are added to the 20 most significant bits of the coefficients during the normal steady-state mode of operation. To counteract bias in the arithmetic, the updating is changed to the tap-leakage algorithm

$$\mathbf{c}_{n+1} = \mathbf{c}_n - \beta \operatorname{sgn} \mathbf{c}_n - \Delta(e_n \mathbf{r}_n), \tag{10.37}$$

where $\beta = \Delta\mu$ of (10.33). In the experimental setup, a count of 1 is added or subtracted to the 23rd most significant bit of each component of each tap coefficient once each symbol interval. In steady-state operation, the 12-bit updating signal will typically show activity in a minimum of the five least significant bits. In general, therefore, the leakage term is quite small compared with the conventional updating term.

The effect of introducing this leakage is shown in the waveforms of Figure 10.12. The top trace shows the coefficient distribution 40 minutes after initial start-up, at the time the leakage is enabled. Subsequent traces were taken at 30-second intervals. Within two minutes the coefficient components had been virtually restored to the state that existed immediately after start-up.

The experimental arrangement allows the leakage to be scaled over a wide range. Viewing the 24-bit coefficient as an integer, the leakage increment can be made 2^r, $r = 0, 1, \ldots, 7$ ($r = 1$ is the case displayed). Since the coefficients are chosen to span an analog range of ± 4, this corresponds to β as defined in (10.48) ranging from 2^{-21} to 2^{-14}, with $\beta = 2^{-20}$ displayed. The value of α used in the experiment is $\alpha = 2^{-11}$. It is observed that $\beta = 2^{-21}$, the lowest possible level of continuous leakage, is adequate to suppress tap drift, indicating

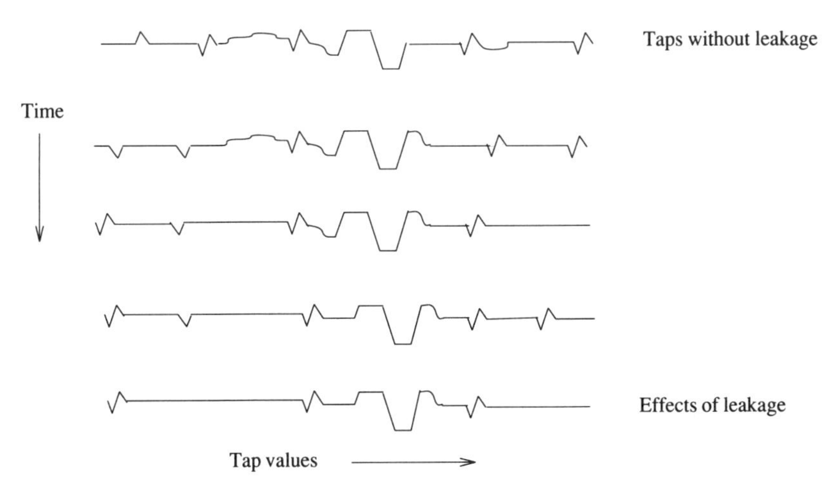

Fig. 10.12 The effect of the tap-leakage algorithm (30-second intervals between traces).

the extreme low level of the system bias to which the FSE updating algorithm appears susceptible.

The larger the value of β that is chosen, the more the leakage will degrade the equalizer performance, although the degradation is negligible for β less than 2^{-17}. In normal data-set operation, it has been observed that a substantially rapid shift in the sampling epoch, which may occur during the timing recovery operation, will greatly accelerate the buildup of tap coefficients. The choice of $\beta = 2^{-18}$ will allow rapid response to this situation without perceivable performance degradation.

Effective control of tap drifting for a fractionally spaced equalizer, has been demonstrated by employing the easily implemented tap-leakage algorithm. The tap-leakage algorithm, or some variation of it (such as injecting noise in the out-of-band region), might be appropriate for any digitally implemented adaptive system that has too many degrees of freedom and that exhibits coefficient wandering.

10.1.9 ADAPTIVE TRANSVERSAL FILTERS WITH DELAYED ADAPTATION

In practical applications of LMS adaptive filters, delay in updating the weighting coefficients is sometimes unavoidable due to the architecture of the signal processor which realizes the adaptive system. In this section we determine the affects of this latency on the choice of step size, convergence speed, and steady-state error, and precision. A typical application of these results would be to the design of adaptive systems using systolic arrays [7], where the desired symbol or the signal sample is not available for several symbol periods. Under these circumstances the LMS adaptation is governed by

$$\mathbf{c}_{n+1} = \mathbf{c}_n - \Delta e_{n-d} \mathbf{r}_{n-d} , \tag{10.38}$$

where the correction term is delayed by d symbol periods. The most significant impact of the delayed correction term is that the algorithm is now a $(d+1)$th-order system, as opposed to a first-order loop. Since adaptation stops when

$$\overline{e_{n-d} \mathbf{r}_{n-d}} = \overline{e_n \mathbf{r}_n} = (A\mathbf{c} - \mathbf{x}), \tag{10.39}$$

the delay in adaptation does not affect the optimum tap weights, but the delay in the algorithm will affect the fluctuation about this value and hence the steady-state MSE will be affected.

Following a development similar to that leading to (8.67) for the evolution of the excess MSE, it can be shown that q_n satisfies a $d+1$th-order recursion which is stable if [6,7]

$$0 < \Delta \leq \frac{2}{\left| \frac{\lambda_{\max}}{\bar{\lambda}} (2N+1) + 2d - 2 \right| \langle r^2 \rangle} . \tag{10.40}$$

Note that if $d = 1$ (as in the conventional LMS algorithm), that (10.40) is identical to (8.69). It can also be shown that choosing Δ equal to half the stability limit will result in the fastest possible convergence, as it would with the LMS algorithm. Unlike the LMS algorithm, the characteristic roots of the recursion are not necessarily real, and oscillatory behavior is possible. Thus the MSE does not always decay monotonically. Note that the affect of increasing d is to *decrease* the maximum step size (as well as the dominant root [6]) — this implies a slower speed of convergence, poorer tracking, and increased precision.

10.2 ADAPTIVE CARRIER RECOVERY SYSTEMS

In high-performance Quadrature Amplitude Modulated (QAM) digital data communication systems, errors associated with extreme amounts of phase jitter are an impediment to achieving the desired system performance. This section focuses on novel adaptive carrier recovery systems [8,9] which are more powerful than the nonadaptive, data-directed techniques (described in Chapter 6) which are a compromise between wide-band systems for jitter tracking and narrow-band systems for noise rejection. Incorporation of these novel adaptive carrier recovery systems may be necessary to achieve reliable operation of high-performance (~6–8 bps/Hz) data communications systems. These techniques may also be used in systems employing trellis-coded modulation. Furthermore, these adaptive structures can track jitter appearing at more than one frequency, and in the absence of jitter, the adaptive loop will strive to minimize the noise appearing at the loop output.

10.2.1 DECISION-DIRECTED PHASE-LOCKED LOOP (PLL)

We begin with a quick review of data-directed PLL for QAM systems. For ideal transmission, except for the phase errors associated with the carrier at the output of the adaptive equalizer, the QAM data communications receiver is shown in Figure 10.13. Referring to Figure 10.13,

$$x(n) = a(n)e^{j(2\pi f_c nT + \theta(n))} + v(n), \tag{10.41}$$

where $x(n)$ is the output of a passband equalizer, $a(n) = a_r(n) + ja_i(n)$ is the nth transmitted symbol (a discrete-valued complex number) representing the in-phase, $a_r(n)$, and quadrature, $a_i(n)$, data, f_c is the known transmitted carrier frequency, T is the symbol period, $\theta(n)$ is the uncompensated carrier phase at the receiver input ($\hat{\theta}(n)$ will be used to denote an estimate of $\theta(n)$), and $v(n)$ is additive white Gaussian noise.

The carrier error, $\theta(n)$, will generally have three components: phase jitter, frequency offset, and phase offset leading to the following mathematical model,

$$\theta(n) = 2\pi f_c nT + \sum_{j=0}^{J} A_j \sin 2\pi f_j T + \theta_0 . \tag{10.42}$$

The nonadaptive data-directed PLL (see Chapter 6) and Figure 10.14 commonly used in QAM receivers [4] uses the loop error signal

$$\text{Im}\left[\frac{x(n)a^*(n)}{|\,a(n)\,|^2}\; e^{-j(2\pi f_c nT + \hat{\theta}(n))}\right] = E(n), \tag{10.43}$$

where Im denotes the imaginary part of its argument. The error signal, $E(n)$, has a deterministic portion, sin $(\theta(n)-\hat{\theta}(n))$, which for small errors is proportional to the phase error, $\theta(n)-\hat{\theta}(n)$, and a random portion, $\mu(n)$, which is

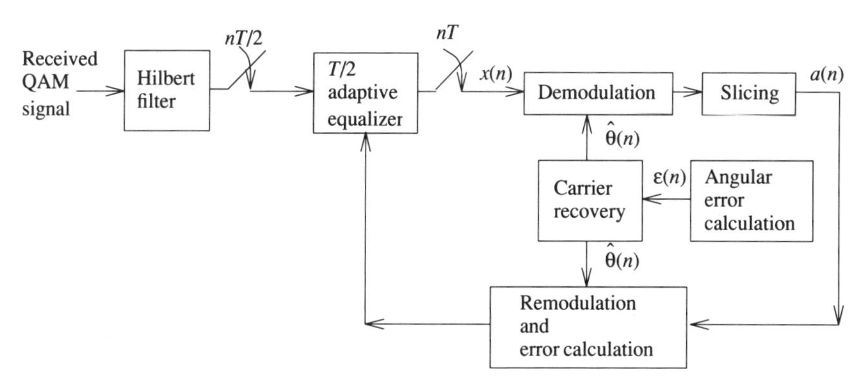

Fig. 10.13 QAM receiver block diagram.

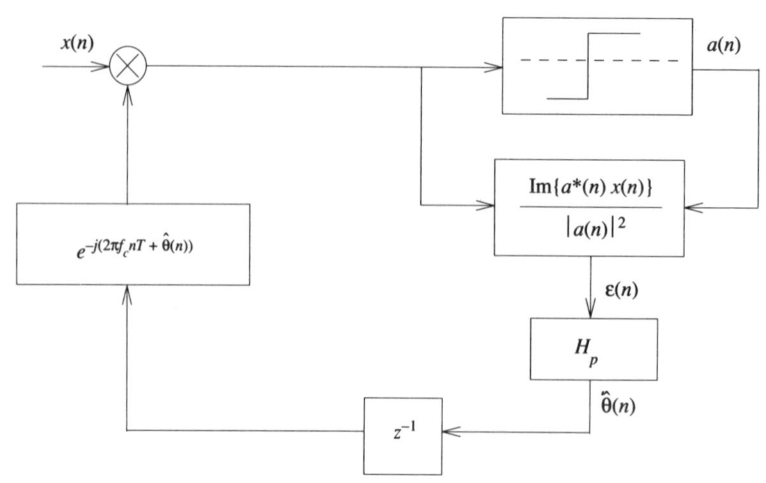

Fig. 10.14 Carrier recovery system using a phase locked loop (PLL) with a filter H_p.

given by $\mathrm{Im}\left[\dfrac{x(n)a^*(n)}{|a(n)|^2}\; e^{-j(2\pi f_c nT + \hat{\theta}(n))}\right]$. As shown in Figure 10.14, the error can be processed by the filter, H_p, to improve performance.

10.2.2 AN ADAPTIVE FIR PHASE PREDICTOR

As discussed in Chapter 6, the PLL parameters are chosen and fixed to achieve a compromise between a wide-band system for the purpose of tracking the phase jitter and a narrow-band system to minimize the noise enhancement. For a high-performance carrier recovery system, a structure is needed that will *adaptively* synthesize the jitter spectrum to reduce the effect of noise enhancement and which will have the capability to track jitter with multiple frequencies.

The essential idea behind this technique is to predict the next phase estimate by filtering several past phase estimates using a *finite impulse response adaptive line enhancer* (FIR-ALE) to realize H_p of Figure 10.14. In Figure 10.15, the structure used in the data-directed PLL of Figure 10.14, to derive the phase error of the pre-processor which provides the noisy phase error, which we now denote by $\psi(n)$ to avoid confusion with $E(n)$. Thus, $\psi(n)$ is the phase error of the predictive loop, while $E(n)$ is the phase error of the PLL. Ideally, a predictor would use prior carrier phase values to produce an estimate of the next phase angle. So far, we only have access to the phase error, $\psi(n) = \theta(n) - \hat{\theta}(n) + \mu(n)$. However, by adding $\hat{\theta}(n)$ *to* $\psi(n)$, we obtain $\theta(n) + \mu(n)$, a noisy estimate of the phase. This quantity will serve as the input to the predictor. With this approach, the predictor generates the phase estimate, $\hat{\theta}(n)$, as a *linear* combination of previous phase angles.

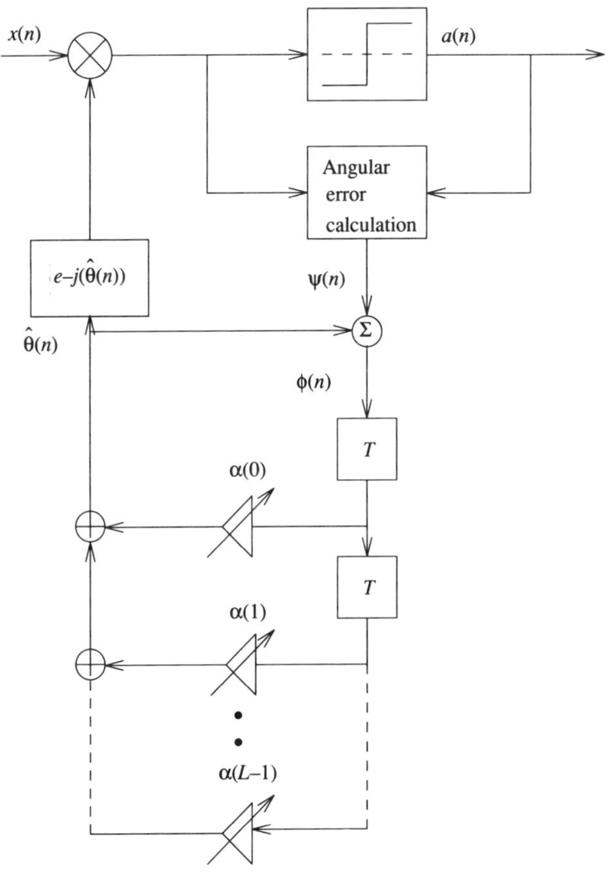

Fig. 10.15 FIR-ALE structure.

$$\hat{\theta}(n) = \sum_{k=0}^{L-1} \alpha(k)\phi(n-k-1) = \boldsymbol{\alpha}^T \boldsymbol{\Phi} , \tag{10.44}$$

where the superscript T denotes the transposed vector. The predictor has L weights $\boldsymbol{\alpha}^T = [\alpha_0, \alpha_1, ..., \alpha_{L-1}]$ and the inputs $\boldsymbol{\Phi}^T = [\phi(n-1), \phi(n-2), ..., \phi(n-L)]$, and where we have the defining relationship

$$\phi(n) = \hat{\theta}(n) + \psi(n) = \theta(n) + \mu(n) , \tag{10.45}$$

for the noisy estimate of $\theta(n)$. Since the phase estimate is a linear combination of prior phases, we avoid the problems associated with adapting IIR filters and a unimodal (i.e., a unique minimum performance surface may be formed so the optimum set of tap weights can be determined adaptively via the LMS algorithm). The performance measure we use is the observable mean-square phase error,

$$J = <\psi(n)^2> \quad . \tag{10.46}$$

Expanding J, using (10.44) and taking expectations gives the quadratic expression

$$J = \sum_{k=0}^{L-1} \sum_{k'=0}^{L-1} \alpha(k)\alpha(k') \left[r_{\theta\theta}(k-k') + \delta(k-k')\sigma_\mu^2 \right]$$

$$- 2 \sum_{k=0}^{L-1} \alpha(k) \left[r_{\theta\theta}(k-1) + \delta(k-1)\sigma_\mu^2 \right] + \sigma_\mu^2 + r_{\theta\theta}(0) \quad , \tag{10.47}$$

where $\delta(m)$ is the Kronecker delta function, σ_μ^2 is the variance of the noise component $\mu(n)$, and $r_{\theta\theta}(k)$ is the autocorrelation function of $\theta(n)$. Differentiating with respect to $\boldsymbol{\alpha}$, and setting the result equal to zero gives the Wiener solution for the tap weights:

$$\boldsymbol{\alpha}_{opt} = R^{-1}\mathbf{p} \ , \tag{10.48}$$

where

$$R = E[\boldsymbol{\Phi}(n)\boldsymbol{\Phi}(n)^T]$$

$$\mathbf{p} = E[\phi(n)\boldsymbol{\Phi}(n)] \quad . \tag{10.49}$$

Thus, there exists a *unique* optimum set of parameters, $\boldsymbol{\alpha}_{opt}$, and the LMS algorithm may be used to adapt the alpha parameters via

$$\boldsymbol{\alpha}(n+1) = \boldsymbol{\alpha}(n) + \Delta\psi(n)\boldsymbol{\Phi}(n) \quad , \tag{10.50}$$

where Δ is the step size of the adaptive algorithm. To determine Δ_{opt}, we follow the analysis of Chapter 8 for the evaluation of the mean-square phase error. For the case of sinusoidal jitter with a uniformly distributed random phase ρ_m, it can be shown that

$$0 < \Delta_{opt} < \frac{2}{L[\sum_{m=0}^{J} \frac{A_m^2}{2} + \sigma_\mu^2]} \quad . \tag{10.51}$$

We now calculate the optimal solution $\boldsymbol{\alpha}_{opt}$ due to the superposition of the phase jitter and noise. Consider the elements of the autocorrelation matrix of a single sinusoid in noise,

$$R = \frac{A_0^2}{2} \left[\mathbf{g}\ \mathbf{g}^T + \mathbf{h}\ \mathbf{h}^T \right] + \sigma_\mu^2 I \quad , \tag{10.52}$$

where $\mathbf{g}$ and $\mathbf{h}$ have elements:

$$g_m = \cos 2\pi f_0 mT \qquad\qquad m = 0,\ldots,L-1 \ .$$
$$h_m = \sin 2\pi f_0 mT$$

Using this notation (10.48) becomes:

$$\left[\frac{A_0^2}{2}\left[\mathbf{g}\mathbf{g}^T + \mathbf{h}\mathbf{h}^T\right] + \sigma_\mu^2 I\right]\boldsymbol{\alpha} = \frac{A_0^2}{2}\,\mathbf{g}\ . \tag{10.53}$$

If the filter length, L, is chosen such that exactly one-half the period of the unknown jitter frequency is spanned, i.e.,

$$L = \frac{k}{2f_0T}, \quad k = 1,2,\ldots, \tag{10.54}$$

some interesting results and insights can be obtained. The following orthogonality conditions will exist

$$\mathbf{g}^T\mathbf{h} = \sum_{j=0}^{L-1} \cos 2\pi f_0 jT \ \sin 2\pi f_0 jT = \frac{1}{2}\sum_{j=0}^{L-1} \sin\frac{2jk\pi}{L} = 0 \tag{10.55}$$

and

$$\mathbf{g}^T\mathbf{g} = \sum_{m=0}^{L-1} \cos^2 2\pi f_0 mT = \frac{L}{2} + \sum_{m=0}^{L-1} \cos\frac{2\pi km}{L} = \frac{L}{2} = \mathbf{h}^T\mathbf{h}\ . \tag{10.56}$$

To determine the optimal vector, $\boldsymbol{\alpha}_{opt}$, we make use of the identity

$$\left[I + \lambda(\mathbf{g}\,\mathbf{g}^T + \mathbf{h}\,\mathbf{h}^T)\right]^{-1} = I - \frac{\lambda(\mathbf{g}\,\mathbf{g}^T)}{1 + \lambda\mathbf{g}^T\mathbf{g}} - \frac{\lambda\mathbf{h}\,\mathbf{h}^T}{1 + \lambda\mathbf{h}^T\mathbf{h}}, \tag{10.57}$$

and after some manipulation, the optimum weight vector can be shown to be

$$\boldsymbol{\alpha}_{opt} = \left[\frac{1}{\dfrac{2\sigma_u^2}{A_0^2} + \dfrac{L}{2}}\right]\mathbf{g}\ . \tag{10.58}$$

Equation (10.58) shows the extent to which the level of the noise power will affect the optimal solution, and also that for sinusoidal phase jitter, the optimal coefficients are themselves samples of a sinusoid of frequency ω_0, the jitter frequency. Note from (10.58) that if the jitter vanishes, $\boldsymbol{\alpha}_{opt}$ will become zero and noise enhancement will *not* occur. This is in sharp contrast to the conventional data-directed PLL.

Applying the final value theorem to the observable angular error, for the case of noiseless sinusoidal phase jitter, for the structure of Figure 10.15, results in

$$\lim_{n\to\infty}\psi(n) = \lim_{z^{-1}\to 1}(1-z^{-1})\,\frac{(1 - H_F(z)z^{-1})A_0 \sin 2\pi f_0 T\, z^{-1}}{1-2\cos 2\pi f_0 Tz^{-1} + z^{-2}}, \tag{10.59}$$

where $H_F(z)$ is the transfer function of the FIR-ALE, i.e., $H_F(z) = \alpha_0$

$+ \alpha_1 z^{-1} + \alpha_2 z^{-2} + \cdots + \alpha_{L-1} z^{-(L-1)}$. Equation (10.59) shows that $\psi(\infty)$ is indeed equal to zero.

10.2.3 PERFORMANCE OF THE FIR PREDICTIVE PLL

The bias inherent in most digital implementations causes the effect of coefficient drifting, as discussed earlier in this chapter, for fractionally spaced equalizers. This has been noticed experimentally for the FIR-ALE. A small leakage [6] equal to 1/2 the LSB of the tap coefficients was used and the system was stabilized.

To select the number of tap weights, a sufficient condition for detection is to span 1/2 the period of the incoming sinusoid. However, (10.54) provides an upper bound on the requirements of L. Shorter lengths may suffice at a possible cost of less attenuation of the incoming jitter. Theoretically speaking, as L increases toward infinity without any background noise, the error approaches

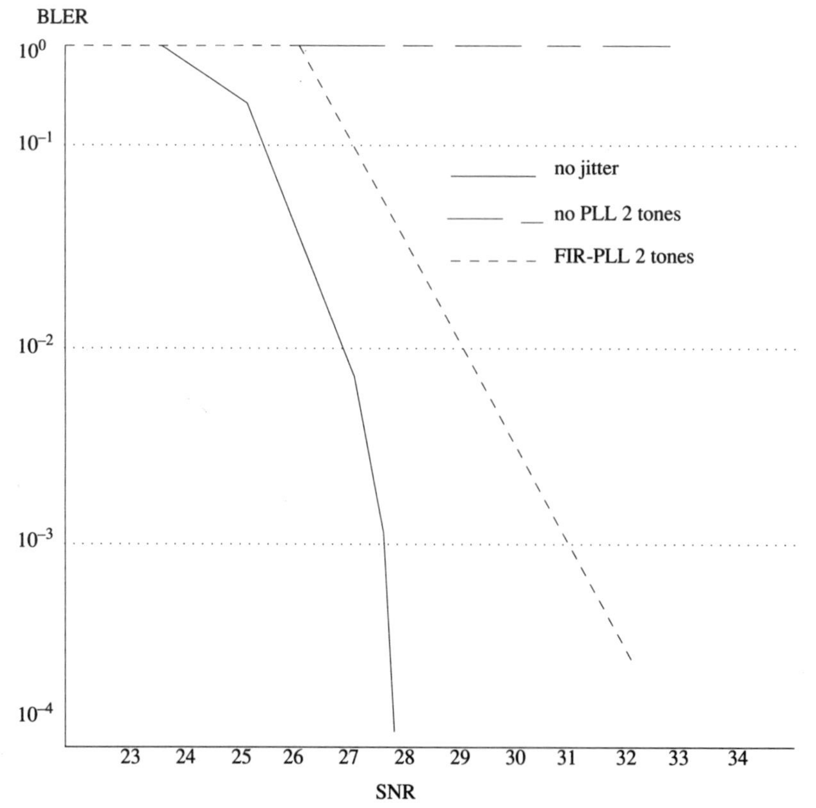

Fig. 10.16 19.2 kbps BLER of FIR-ALE and with 5 degrees of jitter at 60 Hz and 5 degrees of jitter at 120 Hz.

zero. However, practically speaking, an increase in L increases the misadjustment of the process. Also, there is a point of diminishing returns in performance. This point is seen somewhere between 20 to 30 taps for 60-Hz jitter. At a baud of 2400, 20 taps will span one-half of the period of a 60-Hz sinusoid. To compensate for lower frequencies, say 20 Hz, this requirement becomes 60 taps. The use of sparse taps, i.e., spacing greater than T, could then be applied. That is, use 30 taps and $2T$ spacing, to compensate for 20-Hz jitter. The sparse taps simply change the effective baud.

For experimental real-time results, a 19.2 kbps system using 7 bits per symbol and 2743 symbols/sec was used. In Figure 10.16 we show the block error rate (BLER) versus SNR for 5 degrees of jitter at both 60 and 120 Hz, before and after the predictive filter was introduced. Without the PLL, performance is unsatisfactory, but with the adaptive loop, the input jitter was reduced to 2.2 degrees.

10.3 SIGNAL PROCESSING FOR FIBER-OPTIC SYSTEMS

Presently almost all of the long-haul transmission facilities are fiber-optic based, and fiber optics will soon begin to penetrate the local loop. Apart from these replacement and/or (bit-rate) upgrade applications, several emerging local-area and metropolitan area network (LAN/MAN) standards are based on fiber-optic transmission. In this section we present the basic transmission principles for the operation of lightwave systems, and then we examine the prospects for the application of the digital communications technologies described in earlier chapters to such systems — from the perspective of assessing both the need for such techniques and the practicality of implementation. A summary of the transmission characteristics of optical fibers is presented in Section 1.6.4.

Most current long-haul digital fiber optic systems, as shown in Figure 10.17, use simple modulation and detection techniques such as on-off keying with matched-filter receiver techniques, for digital communications over single-mode fibers at ~1 Gbps rates between repeaters spaced 40 km apart. However,

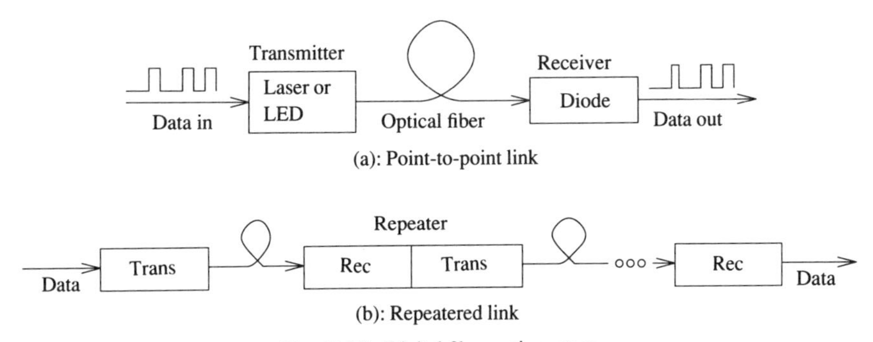

Fig. 10.17 Digital fiber-optic system.

more complex techniques such as multilevel signaling, equalization, and/or coding could conceivably be used in lightwave systems to significantly increase the data rate and/or reduce the effect of transmission impairments and improve performance. This section provides an overview of these techniques, which will probably become increasingly important as device technology matures, in which case substantial increases in performance (especially for installed systems) may only be achieved through these techniques. Furthermore, the introduction of optical amplifiers (as a cost-effective replacement for repeaters) will lengthen the distance between repeaters, and will result in systems where the accumulated impairments will have to be compensated for (in order to achieve a satisfactory error rate).

10.3.1 FUNDAMENTAL LIMITS ON LIGHTWAVE SYSTEMS: DIRECT AND COHERENT DETECTION*

In this section we describe some of the *fundamental* limits on the performance of idealized lightwave systems. The systems are idealized with regard to impairments in the sources, fiber, and detector. We consider two types of detection: direct and coherent. Much of this material is based on the survey article by Salz [11].

Direct Detection

Most commercial systems use direct detection, as shown in Figure 10.18a. In a direct detection system the input current modulates a laser or a light-emitting diode (LED) and a photodetector (either a PIN diode or avalanche detector) at the receiver converts light energy into an electrical signal — where the (average) current in the electrical signal is proportional to the power of the incident light wave. The received electrical current, $r(t)$, is actually formed by a photon-counting process which is subject to statistical fluctuations. The photon-counting process is a time-varying Poisson process whose intensity function $\lambda(t)$ is the average rate of photons arriving at the photodetector input. To obtain the fundamental limits on the performance of the direct detector, we will assume that only one bit is being transmitted† by on-off keying (OOK) so that

$$s(t) = \begin{cases} 0 & \text{"0" sent} \\ 1 & \text{"1" sent} \end{cases} \quad 0<t<T \ , \tag{10.60}$$

With a nondistorting fiber the intensity function will equal the transmitted electrical signal, i.e.,

$$\lambda(t) = s(t) \ . \tag{10.61}$$

---

* A summary of the transmission characteristics of optical fiber is contained in Section 1.6.4.

† Later in this chapter we will discuss the reception of a pulse train and indicate the conditions for superposition to hold.

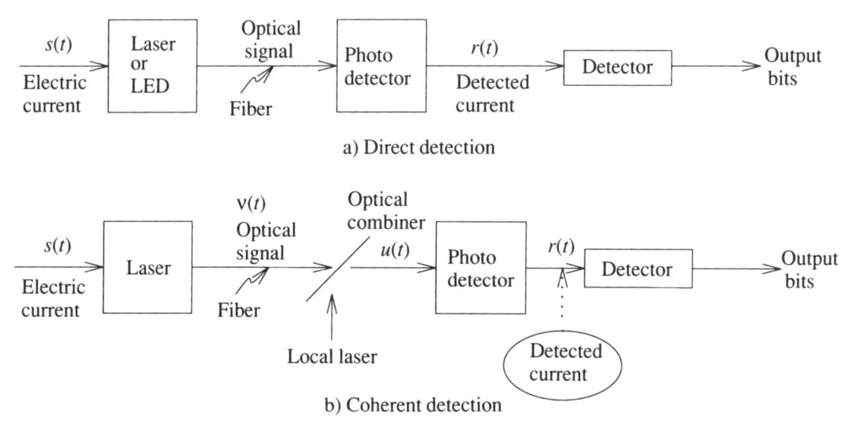

Fig. 10.18 Direct and coherent detection.

The detected current, $r(t)$, is given by

$$r(t) = \sum_n g_n \, w(t-t_n) \quad , \tag{10.62}$$

where g_n is the gain of an avalanche detector ($g_n = 1$ for a PIN diode), $w(t-t_n)$ is the pulse response of the photodetector to a photon arriving at $t=t_n$, and the photon arrival times are a Poisson process* with intensity $\lambda(t)$. For this simple model, the bit may be detected by integrating $r(t)$ over a bit period (T). This procedure counts the number of photons, n, in a bit period, and compares the output with a threshold. If the threshold is exceeded a "one" is declared, otherwise a "zero" is declared. For an ideal photodetector, the average output is given by $\int_0^T \lambda(t)\, dt = \int_0^T s(t)\, dt = S$ when a "one" is sent and zero when a "zero" is sent. Thus in the absence of any other impairment, such as additive noise, it makes sense to set the threshold to zero (see Example 2 in Section 2.2.1), and an error will only be made when a "one" is sent (with probability 1/2) and no photons are detected. When a "one" is transmitted, the number of photons counted by the integrator is a Poisson process with distribution

$$p(n) = \frac{S^n \, e^{-S}}{n!} \quad , \tag{10.63}$$

so that the probability of error is $p(n=0)/2$ or

$$P_e = \frac{1}{2} \, e^{-S} \quad . \tag{10.64}$$

---

* The probability of k pulses arriving in an interval Δ is given by $\lambda^k(t)\Delta e^{-\lambda(t)\Delta}/k!$ This assumes that Δ is small enough for $\lambda(t)$ to be essentially constant.

Since the average transmitted optical energy is

$$P = \frac{1}{2} S , \tag{10.65}$$

the error rate is given by the *quantum limit*

$$P_e = \frac{1}{2} e^{-2P} . \tag{10.66}$$

The *quantum limit* states that to achieve an error rate of 10^{-9} with on-off keying, about 10 photons/bit are required. In practical systems, the ideal performance of the quantum limit cannot be approached, because the thermal noise introduced in the preamplifier masks the extremely weak signal produced by 10 photons. Practical direct-detection systems require a received optical power of 10 to 20 dB higher (or 200 to 2000 photons/bit) [12]. As we will see in the next section, coherent detection offers the promise of a practical system with performance close to the quantum limit (as do direct detection systems with optical amplifiers before the detector).

Coherent Detection

Coherent detection for lightwave systems has a broader meaning than we have used for other media. Some forms of *coherent* detection actually use *incoherent* receivers. We begin with homodyne (true coherent) detection, where the electromagnetic wave, $v(t)$, at the laser output of Figure 10.18b is represented by

$$v(t) = A \cos 2\pi f_0 t , \tag{10.67}$$

where $2\pi f_0$ is the optical radian frequency and the optical power is proportional to A^2. Consider phase modulation of this wave where

$$v(t) = \begin{cases} -A \cos 2\pi f_0 t, \text{ "0" sent} \\ A \cos 2\pi f_0 t, \text{ "1" sent} \end{cases} \quad 0 \le t \le T . \tag{10.68}$$

Suppose the local laser in Figure 10.18b adds (via an optical combiner) a coherent local carrier of the same amplitude, frequency, and phase to the received signal, giving as the input to the photodetector

$$u(t) = (A \pm A) \cos 2\pi f_0 t . \tag{10.69}$$

After detection by the photodetector and integration for T seconds, the average number of counts is $4A^2T$ (for a "one") or zero (for a "zero"). The transmitted optical power is $P = A^2T$, so the probability of a bit error is

$$P_e = \frac{1}{2} e^{-4P} . \tag{10.70}$$

Note that this is a 3 dB improvement over the quantum limit, and is often referred to as the *super quantum limit*. This improvement is due to the fact that binary phase modulation (when coherently detected) is antipodal signaling, which, as we have seen in Section 2.2.5, has a 3 dB advantage over on-off keying (which is orthogonal signaling). As a practical matter, phase coherence is very difficult to achieve, but even if phase coherence could be achieved, the weak detected signal might very well be masked by the receiver noise.

If we consider local lasers with a very large amplitude B (but still assuming frequency and phase coherence), then

$$u(t) = (B \pm A) \cos 2\pi f_0 t \ , \tag{10.71}$$

and the average number of counts after integration is $(B \pm A)^2 T$. To overcome the limitation of receiver noise, we make $B >> A$, so that the optical power incident on the photodetector is much larger than the thermal noise. To compute the bit error rate we use the fact that a shot noise process with a large arrival rate may be approximated by a "white Gaussian" noise process. Since the local laser amplitude, B, is under the receiver's control, the average number of photons $(B^2 + A^2 \pm 2AB)T$ can be made arbitrarily larger and the approximation is valid.

If the data-signal independent (and presumed known) bias term is subtracted from the count, then we have the antipodal signal $\pm 2ABT$. Thus the received signal can be modeled as

$$r(t) = \pm 2AB + n(t) \ , \ 0 \le t \le T \ , \tag{10.72}$$

where the noise is Gaussian with (double-sided) spectral density B^2, and the bit error rate is

$$P_e = \sim e^{-2A^2 T} = e^{-2P} \ , \tag{10.73}$$

which is identical (to within a multiplicative factor of two) to the quantum limit. Thus a phase-coherent homodyne detector (using antipodal signaling) achieves the same performance as the ideal direct detector (for ON-OFF orthogonal) signaling.

With the relatively immature technology of laser sources, phase coherence is generally not attainable in practical systems, and researchers have proposed *heterodyne* detection, where phase coherence is *not* assumed and detection is achieved by translating the incident optical signal to an intermediate frequency (IF), rather than to baseband. With heterodyne detection, if the frequency of the local laser is denoted by ω_l and the incoming optical frequency by ω_0, with the IF frequency $f_i \triangleq f_0 - f_\ell$, then the incident optical signal is given by

$$u(t) = \pm A \cos 2\pi f_0 t + B \cos 2\pi f_\ell t \ , \ 0 \le t \le T \ . \tag{10.74}$$

If we express $u(t)$ in terms of the envelope and phase about ω_l we have

$$u(t) = E(t) \cos(2\pi f_\ell t + \beta(t)) \ , \tag{10.75}$$

where

$$E(t) = \sqrt{(B \pm A\cos 2\pi f_i t)^2 + A^2 \sin^2 2\pi f_i t} \tag{10.76}$$

and

$$\beta(t) = \tan^{-1} \frac{\pm A \sin 2\pi f_i t}{B \pm A\cos 2\pi f_i t} . \tag{10.77}$$

As before, the photodetector responds to the intensity (or squared envelope) of $u(t)$, producing a shot-noise process, $r(t)$, with intensity equal to the squared envelope*

$$\lambda_0(t) = B^2 + A^2 \pm 2AB\cos 2\pi f_i t . \tag{10.78}$$

Subtracting the bias term, $B^2 + A^2$, and using the same reasoning (i.e., shot noise approaching Gaussian noise), we have the detection problem

$$r(t) = \pm 2AB\cos 2\pi f_i t + n(t), \ 0 \le t \le T , \tag{10.79}$$

where the noise is white Gaussian with two-sided spectral density $\cong B^2$. Equation (10.79) describes an elementary detection problem, and since operation is at the IF band, the phase of the $\cos\omega_i t$ can be tracked so that the optimum detector can be used. This detector multiplies $r(t)$ of (10.79) by the coherent IF reference $\cos\omega_i t$ and integrates for T seconds. Neglecting double-frequency terms, the decision statistic

$$\pm ABT + \int_0^T n(t)\cos 2\pi f_i t \, dt \tag{10.80}$$

is compared to zero. The variance of the second term is $B^2 T/2$, so the error rate is given by

$$P_e = Q(\sqrt{2A^2 T}) . \tag{10.81a}$$

For high SNR we use the bound of (2.34b) to obtain

$$P_e \sim e^{-\frac{A^2B^2T^2}{B^2T}} = e^{-A^2T} = e^{-P} . \tag{10.81b}$$

Observe that the exponent is a factor of 2 smaller than for homodyne detection, so heterodyne detection is inferior, by 3 dB, to the ideal quantum-limited detector. The performance of various receivers is summarized in Table 10.1.

If these detectors are compared on a *peak power* basis (lasers are peak power limited), the quantum limit becomes e^{-P}, which is the performance level that a heterodyne detector can achieve — an important observation.

To summarize, coherent detection:

(1) Offers the possibility that a practical receiver can approach the quantum limit (direct detectors lose 10~20 dB to overcome receiver thermal noise).

---

* Thus heterodyne (coherent) detection is really incoherent envelope detection (in conventional digital communications terminology).

Table 10.1 Ideal Performance (Bit Error Rate)

1.	Direct Detection (on–off keyed)	$P_e \sim .5e^{-2P}$
2.	Superhomodyne detection (PSK)	$P_e \sim e^{-4P}$
3.	Homodyne detection (PSK)	$P_e \sim e^{-2P}$
4.	Heterodyne (PSK)	$P_e \sim e^{-P}$
	$P = A^2T$ = Energy/bit	

(2) Lends itself to high-capacity frequency division multiplexed system, with channel spacings on the order of ~0.1 GHz (as opposed to the 100 GHz required for wavelength division multiplexing with noncoherent systems).

(3) Allows the use of (linear) electronic equalization to compensate for linear dispersion in the optical fiber (as discussed later in this section) because of the linearity of the detected signal.

10.3.2 OVERVIEW OF TRANSMISSION IMPAIRMENTS IN SINGLE-MODE FIBER LIGHTWAVE SYSTEMS [26]

Having described the performance of (for the most part) idealized lightwave systems, we now describe the impairments generally encountered in single-mode fiber lightwave systems. These impairments may degrade performance well below the levels described in the last section. The transmission impairments in lightwave systems may be segmented into four categories: 1) signal dispersion, 2) thermal and shot noise, 3) laser phase noise, and 4) intermodulation or crosstalk (including echo). The two detection techniques we consider are direct and coherent detection.

Signal dispersion refers to those impairments that broaden the width of pulses, resulting in intersymbol interference that limits the maximum bit rate [13]. The dominant impairment is chromatic dispersion in the fiber which causes a delay in the received signal that varies with frequency. The loss and delay characteristics are shown in Figure 10.19 for signal-mode fibers [14]. The delay distortion is linear with about –17 psec/km/nm at 1.55 μm (the dispersion minimum is at 1.3 μm), and quadratic with about 0.1 psec/km/nm^2 at the dispersion minimum. Thus, the delay variation is linear with distance, but is fixed for a given length of fiber, i.e., it does not vary significantly with time. A second source of dispersion is polarization dispersion [15,16]. Polarization dispersion is generated by signal delays that are both polarization and frequency dependent. These delays increase with distance and also vary slowly with time. At a given frequency, the different polarizations in the fiber have two different delays. Thus, if the delay is comparable to, or larger than, a bit period, a pulse with a

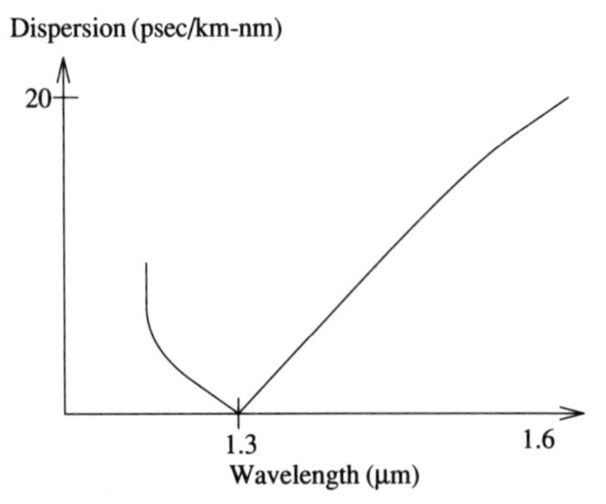

Fig. 10.19a Typical chromatic dispersion in silica fiber. Shown is the magnitude of the dispersion; the direction of the dispersion actually reverses at the zero-crossing.

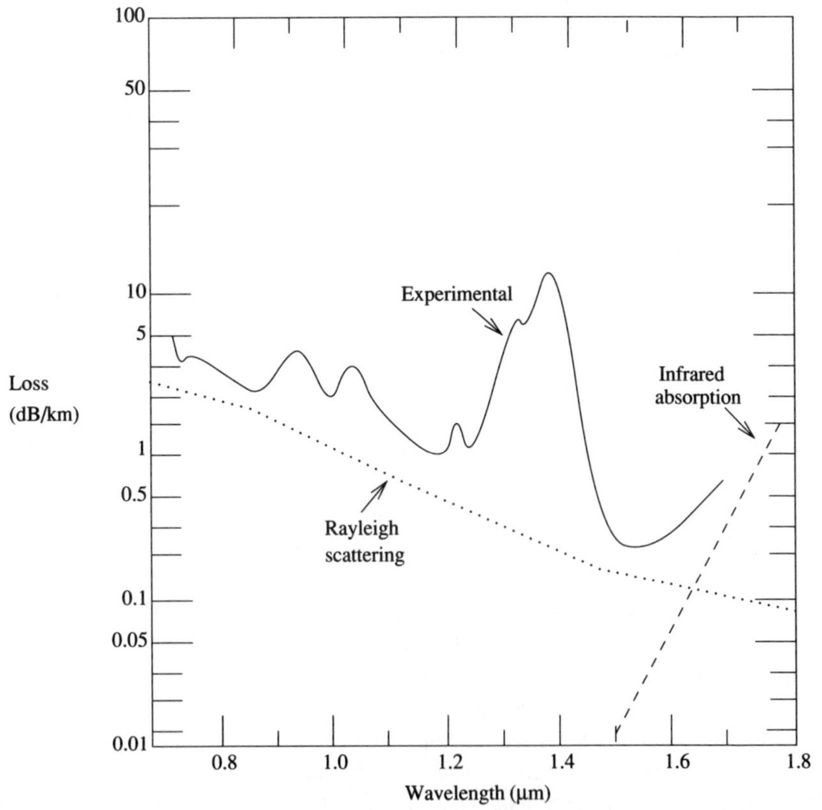

Fig. 10.19b Observed loss spectrum of an ultra-low-loss germanosilicate single mode fiber together with the loss due to intrinsic material effects.

sufficiently narrow frequency spectrum is received as two pulses (i.e., intersymbol interference is introduced) with a time delay between them. A pulse with a sufficiently wide frequency spectrum can be received as a broadened pulse. In general, both these effects will occur. More distortion arises from limited laser bandwidth and nonlinearities, as well as receiver bandwidth limitations. Another possible source of dispersion is a semiconductor optical amplifier [17]. These amplifiers can have phase and amplitude nonlinearities which distort the transmitted signal.

The above impairments all increase in severity with the signal bandwidth. This bandwidth is lower bounded by the data rate, but may be much larger than this because of other factors. For a single-mode (or frequency) laser, phase noise and other impairments, such as chirping or relaxation oscillation with direct modulation of the laser, increase the bandwidth of the signal. Chirp is the variation in carrier frequency of the laser with changes in drive signal amplitude. Chirp can be eliminated by the use of an external modulator.

The second class of transmission impairments to be considered is noise at the receiver. This noise can be shot noise or thermal noise and, with optical amplifiers, amplified spontaneous emission (ASE) noise. As discussed above, shot noise is the quantum noise due to the fact that the received signal is actually a series of photons. The number of photons received during each symbol interval has a Poisson distribution, and therefore the received signal level varies randomly from symbol to symbol. Thermal noise is introduced by the receiver preamplifier, and is usually assumed to be additive, white Gaussian noise. ASE is additive Gaussian noise in the optical signal that increases with the gain of the amplifiers. Without optical amplifiers, it can be shown [18] that in practical receivers, thermal noise is the major limitation with direct detection, while shot noise is the major limitation with coherent detection if the local oscillator power is large enough. With large local oscillator power, however, the high-intensity shot noise can also be modeled as additive, white Gaussian noise. Thus from a digital communications viewpoint, the noise sources in both direct and coherent detectors are well modeled by Gaussian noise.

A third transmission impairment is phase noise, which is the random variation in phase of the transmitting laser. The main parameter of interest with phase noise is the width of the phase noise spectrum relative to the data rate. Wider spectra (or linewidths) result in more signal dispersion as discussed above. Also, wider linewidths require wider receive filters (if all the signal energy in the received signal is to be detected), which results in higher thermal noise. Consider the effect of phase noise on the direct, detection and coherent-detection receivers shown in Figure 10.18. With direct-detection, the electrical signal current is proportional to the intensity of the received optical signal. If the linewidth is much greater than the data rate, then the received optical signals added *incoherently* [19], and the detected current is proportional to the sum of the intensities of the individual received signals (i.e., power superposition of the signals). If the linewidth is much less than the data rate, then the received optical signals add *coherently*, and the output current is proportional to the squared coherent sum of the received signals. For the cases between these two extremes, the current is proportional to neither the sum of intensities nor the coherent sum of the fields; the received signal varies between these extremes with a random component that varies from symbol-to-symbol (see Section 10.3.3 for a more detailed discussion of linearity). With coherent detection, the electrical signal current is always proportional to the received optical signal voltage. If the linewidth is much greater than the data rate, coherent detection is usually not used since the system has no advantages over direct detection. If the linewidth is comparable or somewhat less than the data rate, modest amounts of phase noise can seriously degrade performance [20], and it is important to reduce the laser's linewidth or to choose modulation formats that are relatively insensitive to phase noise [21]. If the linewidth is comparable to the data rate, then coherent detection has better performance than direct detection only if the electrical signal is filtered and sampled at bandwidths (and rates) greater than the data bandwidth (rate) [21].

The final class of impairments is intermodulation or crosstalk. This occurs with multiple signals and can be due to nonlinearities in the fiber [22] or in optical amplifiers. In addition, in full-duplex systems, echo also can degrade performance. There are, of course, other impairments in fiber-optic systems, but we have just surveyed the major ones in order to assess the prospects for various signal-processing techniques in a wide range of cases.

10.3.3 MODELING OF FIBER-OPTIC COMMUNICATIONS SYSTEMS

Here, we provide a quantitative description of some of the impairments described above, with emphasis on direct-detection, long-haul, fiber-optic systems, such as that shown in Figure 10.20. Our discussion builds upon the models used earlier in this section.

The nonreturn-to-zero (NRZ) input data stream, $x_1(t)$, is filtered [by the transmit filter with frequency response $H_T(f)$] and the filtered signal, $x_2(t)$, modulates a signal-frequençy laser. Alternatively, the data stream can be used, as in the dashed path, to externally modulate the optical power (to avoid laser nonlinearity), and the transmitted electric field (optical signal) is given by

$$E = \left[\mathbf{a}_x E_x + \mathbf{a}_y E_y e^{j\theta}\right] e^{-j\beta z} x(t) \quad , \tag{10.82}$$

where $\mathbf{a}_x$ and $\mathbf{a}_y$ are unit vectors in the x and y directions respectively, θ is the phase angle that determines the signal polarization, β is the propagation constant, and z is the direction of propagation. For direct modulation,

$$x(t) = \sqrt{P(t)}\, e^{j(\phi_c(t) \angle x_2(t))}\, e^{j2\pi f_c t} \quad , \tag{10.83}$$

where ω_c is the radian lightwave frequency, $P(t)$ is the optical power, and

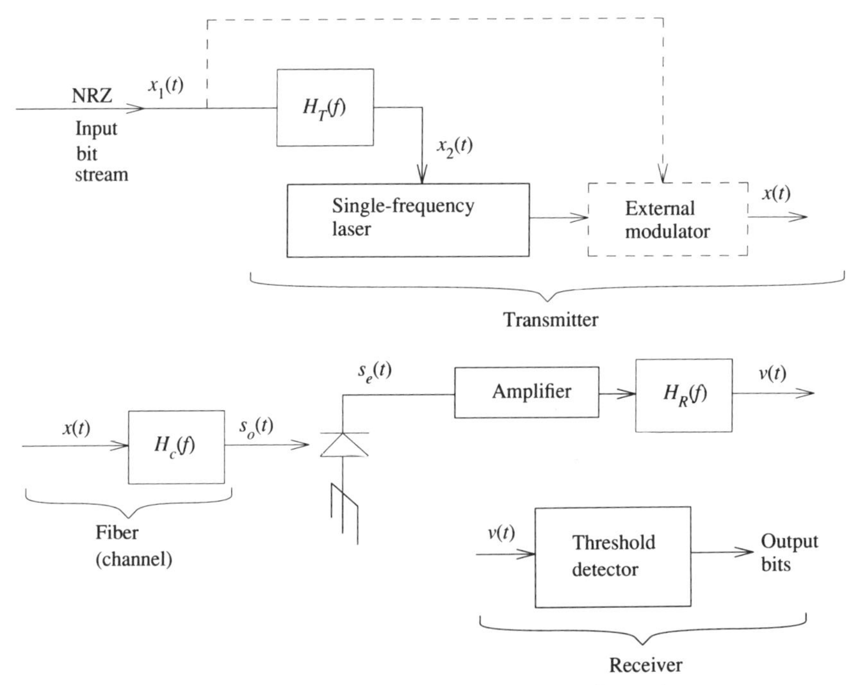

Fig. 10.20 Block diagram of a long-haul, direct-detection fiber-optic system.

$\angle x_2(t)$ is the phase angle of $x_2(t)$. The phase variation $d\phi_c(t)/dt$ is referred to as chirp, while the amplitude variation in $P(t)$ versus $\sqrt{|x_2(t)|}$ (since the laser output power is proportional to the input current) is referred to as relaxation oscillation. We refer to these two effects as laser nonlinearities. With external modulation,

$$x(t) = e^{j(2\pi f_c t + \phi(t))} \sqrt{|x_1(t)|}\ e^{j\angle x_1(t)} \quad , \tag{10.84}$$

where $\phi(t)$ is the phase noise of the laser. Note that the laser power, $|x(t)|^2$, is proportional to the electrical signal $x_1(t)$. The transmitted optical signal passes through a fiber with frequency response $H_C(f)$, which may have chromatic and polarization dispersion. Chromatic dispersion will have a significant impact on system performance if the laser frequency is different from the dispersion minimum of the fiber (which is 1.3 μm in a standard fiber). In this case, the main portion of the chromatic dispersion is linear delay distortion, and the frequency response of the fiber is given by

$$H_C(f) = e^{j\alpha f^2} \quad , \quad \alpha = \pi D \frac{\lambda^2}{c} L \quad , \tag{10.85}$$

where L is the fiber length and D is the linear delay coefficient, and λ is the wavelength ($= c/f$). For example, for a 1.55 μm signal in a standard fiber, $D = 17$ psec/km/nm. At the receiver the optical signal, $s_0(t) = h_C(t) * x(t)$ [where $h_C(t)$ is the Fourier transform of $H_C(f)$] is converted to an electrical signal by a photodiode detector (usually an APD).

With direct detection, as shown in Figure 10.20, the electrical signal $s_e(t)$ is proportional to $|s_0(t)|^2$. Alternatively, recall that with coherent detection the received optical signal is mixed with a local oscillator optical signal (at approximately the same frequency as the received signal) to generate an IF electrical signal whose *envelope* is proportional to $s_0(t)$.

Polarization dispersion can be characterized in terms of first and higher order (in frequency) effects. The first-order effect is a delay in the signal in one polarization relative to the delay in the signal in the other polarization. Thus, with first-order polarization dispersion and direct detection,

$$s_e(t) = \alpha_1 \left[|s_0(t)|^2 + \alpha |s_0(t+\tau)|^2 \right] , \tag{10.86}$$

where α_1 is the conversion constant between the optical and electrical signals, α is the ratio of the signal strengths in the two polarizations, and τ is the time delay between propagation in the two polarizations. With coherent detection, the receiver is polarization sensitive, and the effect of polarization dispersion depends on the receiver technique used. Because of the variety of coherent system receiver techniques that work around the polarization problem, e.g. [23], we note only that the signals in the two different polarizations add *powerwise* and we will not further discuss polarization dispersion in coherent systems.

Next, the electrical signal is amplified and filtered to increase the signal-to-noise ratio (where the receiver frequency response $H_R(f)$ is due to both the frequency characteristics of the receiver filter and the amplifier), producing an output signal $v(t) = h_R(t) * s_e(t)$. This signal is detected by comparing the signal level, during a short period of time at the peak opening of the eye of the signal (e.g., 60 psec for an 8-Gbps data rate), to a decision threshold.

Intersymbol interference in the detected signal can be caused by laser nonlinearity, chromatic dispersion, polarization dispersion, and nonideal receiver frequency response. The laser nonlinearity and receiver frequency response will vary among devices and polarization dispersion will vary slowly over time (e.g., on the order of hours). Chromatic dispersion is reasonably fixed for a given length of fiber, but its effect on system performance depends on laser nonlinearity and receiver frequency response, which can vary.

A key issue in the effectiveness of intersymbol interference compensation techniques is the linearity of the intersymbol interference. Linear distortions include the receiver frequency response and polarization dispersion. Chromatic dispersion is a linear distortion in the optical fiber; however, whether this distortion is linear at the detector (i.e., $v(t)$), depends on the transmitter and receiver techniques used. Specifically, chromatic dispersion appears as linear distortion in a coherent receiver, but as long as the laser linewidth is smaller than the data rate, this form of dispersion appears as *nonlinear* distortion in a direct detection system. Below, we discuss this critical point for external and direct modulation systems.

With external modulation in a direct detection system, the transmitted optical signal is given by

$$x(t) = \sqrt{|x_1(t)|\ e^{j\angle x_1(t)}}\ e^{j(\phi(t)+2\pi f t)} \quad ,$$

where $\phi(t)$ is the phase variation in laser signal, which, for the case of external modulation, is the phase noise of the laser. With chromatic dispersion, the received signal is given by

$$s_0(t) = x(t) * h_C(t) \tag{10.87}$$

and, after direct detection, the electrical signal is given by

$$s_e(t) = |x(t) * h_C(t)|^2 \ . \tag{10.88}$$

After the receiver filter, the signal is given by

$$v(t) = |x(t) * h_C(t)|^2 * h_R(t) \quad , \tag{10.89}$$

or, in baseband notation,

$$v(t) = |\sqrt{|x_1(t)|}\ e^{j\angle x_1(t)}\ e^{j\phi(t)} * h_C(t)|^2 * h_R(t) \ . \tag{10.90}$$

Since in representative long-haul systems the data rate is on the order of several Gbps while the phase noise is below 50 MHz, $e^{j\phi(t)}$ is approximately constant over the memory of the channel pulse response, and

$$v(t) \doteq |\sqrt{|x_1(t)|}\ e^{j\angle x_1(t)} * h_C(t)|^2 * h_R(t) \ . \tag{10.91}$$

Under these circumstances, the intersymbol interference in the electrical signal at the receiver is the *square* of the sum of the intersymbol interference in the optical signal, and the distortion is *nonlinear*. However, even with this nonlinearity, linear equalization can be partially effective in reducing intersymbol interference (see Appendix 10B for a detailed explanation).

With direct modulation of the laser, ignoring nonlinear amplitude variations in the laser output signal, the transmitted signal is

$$x(t) = \sqrt{|x_2(t)|}\ e^{j\angle x_2(t)}\ e^{j(\phi(t)+\omega t)} \quad , \tag{10.92}$$

where $e^j\phi(t)$ is mainly due to chirp, with the chirp bandwidth of current lasers roughly five times that of the data signal. From (10.89) and (10.92),

$$v(t) = |\sqrt{|x_2(t)|}\ e^{j\angle x_2(t)}\ e^{j\phi(t)} * h_C(t)|^2 * h_R(t) \ . \tag{10.93}$$

Now, for *random* phase variations with a bandwidth of five times the data rate it can be shown that the output is, to a good approximation,†

$$v(t) = |x_2(t)| * |h_C(t)|^2 * h_R(t) \quad . \tag{10.94}$$

Thus, the intersymbol interference in the electrical signal at the receiver is the sum of the squares of the intersymbol interference in the optical signal. The distortion described by (10.94) may be viewed as arising from a *linear* system with input $|x_2(t)|$ and pulse response $|h_C(t)|^2$. Note that $v(t)$ has twice the bandwidth of the original distortion. However, the chirp is caused by signal level changes and, therefore, does not have high bandwidth over the entire symbol duration. Furthermore, symbols surrounded by identical symbols may experience no chirp. Thus, in general, with today's lasers using direct modulation, over a single bit period the intersymbol interference is a combination of linear (10.94) and nonlinear (10.91) distortion. Therefore, in direct-detection systems with direct modulation, linear equalization can only be partially effective in reducing intersymbol interference.

10.3.4 COMPENSATION TECHNIQUES

A. Linear Equalization

To compensate for linear distortion, a linear equalizer (transversal filter) as described in Chapter 8, can be used between the receiver filter and the detector [27-28]. If we consider the familiar tapped delay line, the equalizer with N taps produces the equalizer output signal, $y(t)$, given by

$$y(t) = \sum_{j=1}^{N} c_j \, v\Big[t-(j-1)T\Big] \quad . \tag{10.95}$$

Since the receiver filter bandwidth is usually much less than the signal bandwidth (to reduce receiver noise), the tap spacing (T) need only be the symbol period, i.e., the equalizer is a synchronous linear equalizer (a fractionally spaced equalizer is not needed in this case to reduce ISI, although such an equalizer can reduce the equalizer noise enhancement [27]). At high data rates, symbol delays can be implemented by a short transmission line (e.g., less than 4 cm at 8 Gbps), and the weights can be implemented by a variable-gain amplifier. Since in many cases most of the intersymbol interference is due the symbols proceeding the detected symbol, there may be more taps for the precursor symbols than for postcursor symbols.

The two celebrated adaptive algorithms described in Chapter 8, the least-mean-square (LMS) algorithm of Widrow and the zero-forcing algorithm of

† That is, since $\phi(t)$ is rapidly varying, the expression $v(t) = \int\int \sqrt{|x_2(\tau_1)|}\sqrt{|x_2^*(\tau_2)|}\, e^{j\angle x_2(\tau_1)}\, e^{j\angle x_2(\tau_2)}\, e^{j\phi(\tau_1)}\, e^{j\phi(\tau_2)}\, h_C(t-\tau_1)\, h_C^*(t-\tau_2)\, d\tau_1\, d\tau_2$ is approximately zero unless $\tau_1 = \tau_2$, or $v(t) = \int |x_2(\tau_1)|\, |h_C(t-\tau_1)|^2 d\tau_1 * h_R(t)$, since $x_2(t)$ is confined to a symbol interval.

Lucky, may be used to adjust the equalizer tap weights. Recall that the mean-square-error algorithm minimizes the variance of the error signal, while the zero-forcing algorithm minimizes the peak distortion. The zero-forcing algorithm is only effective at minimizing peak distortion when the unequalized eye is open. With modest amounts of linear ISI and thermal noise, both equalizers produce distortion-free outputs and provide performance close to the optimum.

At the high data rates of long-haul systems, generating analog samples of signals is very costly and hence may not be practical. Thus, single-bit accuracy samples (± 1) must be used where ever possible. Two algorithms that provide single-bit accuracy are the sgn-sgn mean-square-error algorithm given by

$$\mathbf{c}_{k+1} = \mathbf{c}_k + \Delta \text{sgn } \varepsilon_k \text{ sgn } \mathbf{v}_k \ , \tag{10.96}$$

and the modified zero-forcing algorithm

$$\mathbf{c}_{k+1} = \mathbf{c}_k + \Delta \text{sgn } \varepsilon_k \ \mathbf{a}_k \tag{10.97}$$

where ε_k is the error in the equalizer output signal. Quantizing the signal samples reduces the rate of convergence of the algorithms (which is not a major concern, since channel impairments should change very slowly with time — on the order of hours or longer). Of more concern, especially for the sgn-sgn LMS algorithm, is the steady-state weights. Since the direction of the correction term ($\text{sgn } \mathbf{v}_k$) is different than for the LMS algorithm ($\mathbf{v}_k$), the weights may settle at a different value. However, for the quantized version of the zero-forcing algorithm, the direction of the correction term ($\mathbf{a}_k$) is the same as for the unquantized algorithm, and the algorithm converges to the same weights as the continuous version when the eye of the received signal is open (neither algorithm works when the eye is closed).

Recall that the (unquantized) mean-square-error algorithm gives better performance (i.e., lower bit error rate or higher system margin) than the zero-forcing algorithm because it minimizes both the intersymbol interference and thermal noise, while the zero-forcing algorithm only minimizes the intersymbol interference. Furthermore, unlike the zero-forcing algorithm, when supplied with a training sequence the mean-square-error algorithm also works when the eye of the received signal is closed. However, the implementation practicalities of Gbps operation preclude analog sampling of the signal, and, as mentioned above, the quantized version of the mean-square-error algorithm is not guaranteed to converge to the same weights as the continuous version. In fact, with the eye open, the quantized version of the mean-square-error algorithm is identical to the quantized version of the zero-forcing algorithm, since $a_k = \text{sgn}(v_k)$. However, the quantized mean-square-error algorithm requires samples of the received signal (before equalization) in addition to the samples of the error signal required by the zero-forcing algorithm, when the eye is open. Thus, the quantized version of the mean-square error algorithm requires twice as many detectors, yet gives the same performance as the quantized zero-forcing algorithm when the eye is open.

The performance criterion we use for our numerical results is the optical signal power penalty due to intersymbol interference (the penalty is defined as the increase in received optical signal power required to maintain the same eye opening with intersymbol interference, i.e., the same bit error rate with impairments in the presence of receiver noise), which can be derived from the minimum eye opening over all input bit sequences. The minimum eye opening is the minimum sampled signal value for a "1" minus the maximum sampled signal value for a "0" with no noise at the receiver. Thus, if the difference between the signal levels for a "1" and a "0" without ISI is Y, the minimum eye opening (in percent) is given by

$$\textit{eye opening} = \frac{\underset{k}{\text{Min}}\ (y_k | a_k = 1) - \underset{k}{\text{Max}}\ (y_k | a_k = 0)}{Y} \cdot 100 \ . \tag{10.98}$$

The optical power penalty is given by

$$\textit{penalty} = \begin{cases} 10 \log_{10} (\textit{eye opening}/100)\,\text{dB} & \text{for direct detection} \\ 20 \log_{10} (\textit{eye opening}/100)\,\text{dB} & \text{for coherent detection} \end{cases} \tag{10.99}$$

since the received current is proportional to the optical power with direct detection, and the received current is proportional to the optical voltage (i.e., the magnitude of the optical field) with coherent detection.

B. Nonlinear Cancellation

As shown above, there are cases when the distortion is nonlinear, and therefore nonlinear techniques must be used to significantly reduce the output distortion. Using knowledge of previously detected bits and perhaps estimates of bits to be detected, the decision threshold in the detector may be adjusted up or down to be halfway between the expected signal levels for each bit to be detected. If the adjustment is a linear sum of the previous and estimated bits, then the technique is linear cancellation (or decision-feedback equalization if only previously detected bits are used), as described in Chapter 8. Otherwise, a lookup table such as used in Chapter 9 for a RAM-based echo canceler may be used with 2^{N-1} entries for $N-1$ previously detected and estimated bits (a total of N bits are used to determine the data bit), to provide nonlinear cancellation (see also [29]).

Since estimating the bits to be detected requires an additional detector (or even additional interference reduction techniques if the eye is closed), the most practical technique is to adjust the decision threshold based on previously detected bits only, and use a tapped delay line to reduce distortion caused by postcursor bits. Such a nonlinear canceler is shown in Figure 10.21 where N_1 previous (precursor) bits are used for nonlinear cancellation and a tapped delay

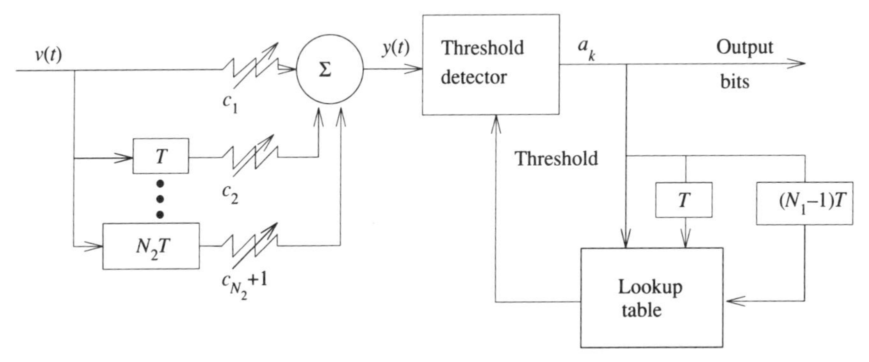

Fig. 10.21 Nonlinear canceler using N_1 previous bits and a tapped-delay line for reduction of intersymbol interference from N_2 postcursor bits.

line with N_2+1 taps is used for the N_2 postcursor bits plus the data bit to be detected. Thus, at the detector,

$$y(t) = \sum_{i=1}^{N_2+1} c_i v(t-(i-1)T) \ , \tag{10.100}$$

and the decision rule is

$$a_k = \begin{cases} 1, & \text{if } y(t_0) > y_0 + f(a_{k-1}, \ldots, a_{k-N_1}) \\ 0, & \text{otherwise}, \end{cases} \tag{10.101}$$

where y_0 is the unadjusted decision threshold, t_0 is the sampling time, and $f(\cdot)$ is the threshold appearing at output of the lookup table and provides an estimate of the (nonlinear) intersymbol interference. Such a situation is shown in Figure 10.22, where the curve labeled "10" is the voltage when the previous bit was "1" and the current bit is "0". The level F is the threshold halfway between the voltages corresponding to a current bit of "0" and a "1". Observe the proximity of F to the "01" curve; this indicates a rather small margin to noise. In the nonlinear canceler, the threshold f_i is used if the prior bit was detected as i (i = 0 or 1). Note the increased distances from the signal traces to the thresholds.

The performance of nonlinear cancellation is given by the minimum distance between sampled signal values for a "1" and a "0", given the same N_1 precursor and N_2 postcursor bits. Nonlinear cancellation will fail (i.e., the error rate will be nonzero even without noise at the receiver) if the levels for "1"'s and "0"'s overlap for some set of N_1 precursor and N_2 postcursor bits. To see where this could occur, consider the simple example of direct detection where the sample values of the optical signal with intersymbol interference are given by (using baseband notation)

$$s_0(kT) = a_k - 0.5\, a_{k-1} \ . \tag{10.102}$$

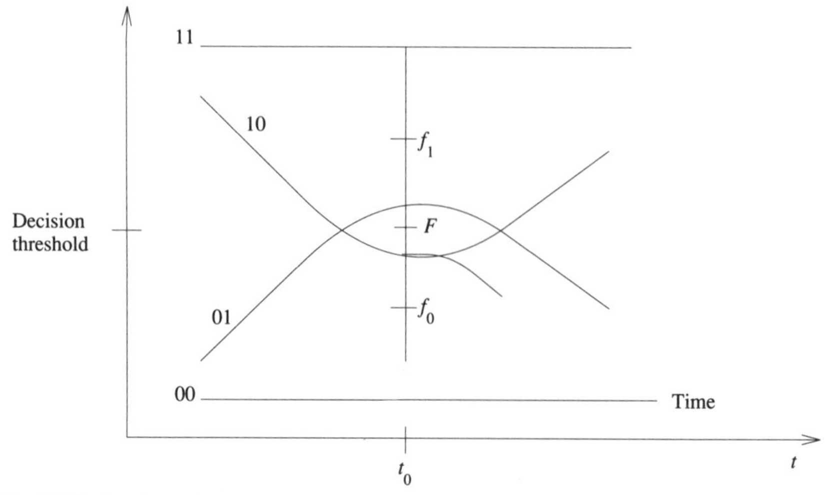

Fig. 10.22 Received signal eye with ISI: F is the threshold used by an equalizer and f_i are the thresholds used by the nonlinear canceler when the previous bit is i ($i = 0$ or 1).

Thus, neglecting the receiver filter and any noise,

$$y_k = s_e(kT) = |s_0(kT)|^2 = a_k^2 - a_k\, a_{k-1} + 0.25\, a_{k-1}^2 \ , \tag{10.103}$$

or

$$y_k = \begin{cases} 0\,, & a_{k-1} = 0, \quad a_k = 0, \\ 1\,, & a_{k-1} = 0, \quad a_k = 1, \\ 0.25, & a_{k-1} = 1, \quad a_k = 0,1, \end{cases} \tag{10.104}$$

Note that when $a_{k-1} = 1$, the signal level y_k is independent of a_k, and therefore the output levels cannot be separated and nonlinear cancellation will not work. However, we can see that a_k can be determined from y_{k+1} (with a three-level detector) which leads us to maximum likelihood detection in the next section.

C. Maximum Likelihood Detection

As we know from Chapter 7, maximum likelihood sequence detection (MLSD) is the optimum detection technique in that it minimizes the error probability for determining a bit (or bit sequence), given *N* received signal samples. It can be complex to implement for large *N*, but is nevertheless useful as a bound on performance. There are techniques to implement simplified, approximate versions of maximum likelihood detection, however, that may be practical at high data rates, if *N* is small [30]. As before, we assume that the received signal levels and intersymbol interference are deterministic, and that the only source of noise is additive Gaussian noise (thermal noise with direct detection and high-intensity shot noise with coherent detection).

In the simplified block-oriented maximum likelihood detector, we determine the bits in a block of length N, based on N consecutive signal samples, where the signal is corrupted by additive Gaussian noise. Specifically, we compare blocks of N consecutive received signal samples to each of 2^N possible (stored) signal sample vectors (corresponding to the 2^N possible bit sequences). With the additive Gaussian noise model, the detected bits correspond to the vector that has the closest Euclidean distance to the received vector. That is, we find the closest signal vector that maximizes the correlation between the received vector and the allowable (transmitted) signal vectors,

$$\underset{l}{\text{Max}} \left\{ \sum_{j=1}^{N} \left(2d_{lj}\, y_{k-j} - d_{lj}^2\right) \right\}, \tag{10.105}$$

where l is the set of the 2^N stored signal vectors and d_{lj} is the jth value of the lth signal vector. (Note that, for $N=1$, the technique is just bit-by-bit detection.) The stored signal vectors (the d_{lj}'s) are chosen to minimize the error rate, a task which can become very complicated with large N and severe intersymbol interference.

The performance of this detector can be determined from the minimum Euclidean distance between received signal vectors with different transmitted bits. The performance of this detector is degraded by the edge effects of ISI. That is, since the MLD makes decisions on block of N bits from N signal samples of these bits, bits at the edge of the block are more likely to be detected in error because of ISI from bits outside the block.

D. Coding

As we saw in Chapter 3, coding can be used to increase system margin by allowing higher raw (before decoding) bit error rates while maintaining the same output bit error rate. The performance improvement with coding (i.e., increase in margin) depends on the decrease in system margin versus the increase in data rate due to coding. Specifically, the increase in data rate (for the binary systems we are discussing) due to coding must result in a smaller penalty than the coding gain in order for coding to be useful. Thus, as a first step in studying the effect of coding, we must determine the increase in penalty with data rate. For example, in [24] it was shown that coding, which increased the data rate by 4%, reduced the error rate to $311P_e^2$, or from 2×10^{-6} to 10^{-9}, corresponding to an increase in system margin of 1.1 dB at a 10^{-9} bit error rate. Therefore, in this example, the decrease in system margin (increase in optical power penalty) with a 4% increase in data rate must be less than 1.1 dB for coding to be useful. It should be noted that the coding gain applies to the required electrical signal-to-noise ratio. The optical power penalty decrease is only half the electrical signal-to-noise ratio improvement (in dB) for direct detection systems, but equal to the electrical signal-to-noise ratio improvement for coherent detection systems. Thus, we would expect that coding is more likely to be useful in coherent systems.

E. Multilevel Signaling

Multilevel signaling can also be used to increase system margin when intersymbol interference is present. As compared to on-off keying (with two-levels), multilevel signaling with M levels decreases the symbol rate by a factor of $\log_2 M$, which reduces the amount of dispersion and decreases the noise power in the detector by $\log_2 M$. However, multilevel signaling, with the same average power, also decreases the eye opening by a factor of at least $M-1$, even without intersymbol interference. Thus, without intersymbol interference, for a given symbol error rate, multilevel signaling introduces an optical power penalty of $10\log_{10}\ (M-1)/\sqrt{\log_2 M}$ dB in direct detection systems and $20\log_{10}\ (M-1)/\sqrt{\log_2 M}$ dB in coherent systems. Thus, multilevel signals ($M>2$) can only be useful in reducing the optical power penalty due to intersymbol interference when the optical power penalty with 2-level signaling exceeds the above values. In these cases, we need to compare the optical power penalty with 2-level signaling to that with M-level signaling at $1/\log_2 M$ times the symbol rate. For example, 4-level signaling could only reduce the optical power penalty of a 2-level signaling system if the penalty exceeded 3.3 and 6.5 dB in direct detection and coherent systems, respectively. Thus, we would expect that multilevel signaling is more likely to be useful in direct detection systems. Note that this is opposite to the conclusion with coding.

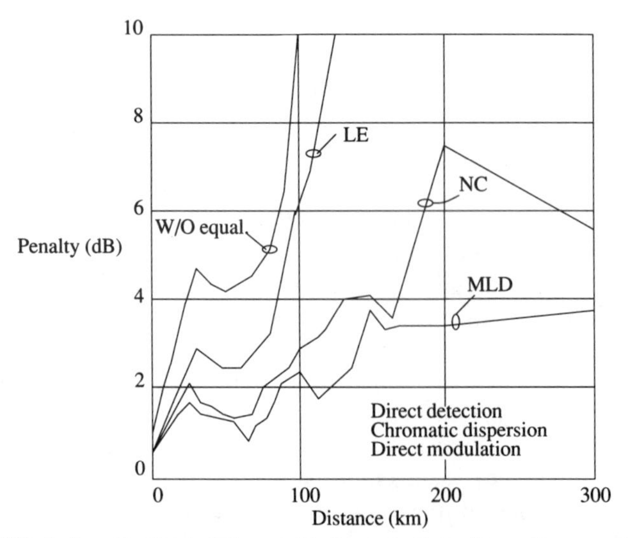

Fig. 10.23 Effect of equalization techniques with chromatic dispersion and laser nonlinearity: optical power penalty versus distance with LE, NLC, and MLD for $N = 6$.

10.3.5 NUMERICAL RESULTS

We conclude with some numerical results that were generated, via computer simulation, as follows. The data rate was 8 Gbps, and the transmit filter was a low-pass RC filter with a 3 dB bandwidth of 4 GHz for the cases with chromatic dispersion, and 32 GHz (i.e., the transmit filter can be neglected) in the other cases. The programs use a repetitive pseudorandom data stream of length 64, which contains all bit sequences of length 6. Thus, the results should be accurate as long as the intersymbol interference extends over only a few bit periods.

Figure 10.23 shows the effect of equalization techniques with chromatic dispersion and laser nonlinearity. Results are shown for the linear equalizer (LE), nonlinear canceler (NLC), and maximum likelihood detector (MLD), $N = 6$, optimum sampling time, and a Butterworth receiver filter. Note that the combined effect of chromatic dispersion and laser nonlinearity produces a dip in penalty above 40 km. Although chromatic dispersion produces nonlinear intersymbol interference at the receiver, the LE still reduces the power penalty somewhat — by at least 1.5 dB for distances above 40 km (or equivalently increases the maximum distance for a given penalty by about 20%). NLC and MLD, however, decrease the penalty by more than 3 dB for distances above 40 km and substantially increase the dispersion-limited transmission distance (to beyond 300 km). Note that these curves are jagged because the results were generated at discrete distances (at least 10 km between points), and because the penalty

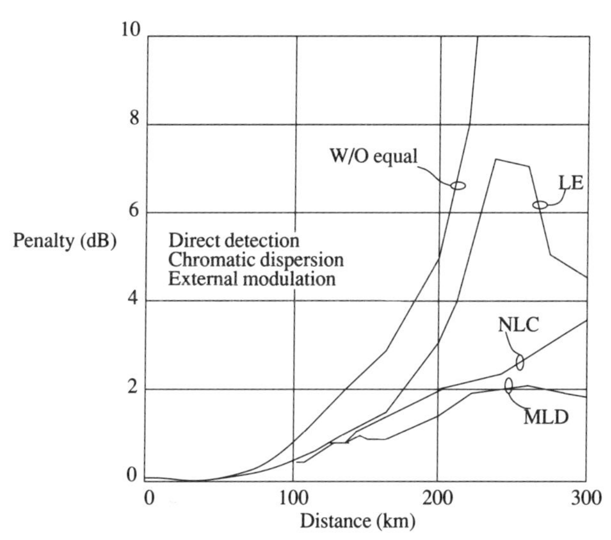

Fig. 10.24 Effect of equalization techniques with chromatic dispersion without laser nonlinearity in direct detection systems.

fluctuates with the level of intersymbol interference in the signal samples. Here, as with nonideal receiver filtering, the best N consecutive samples included at most one postcursor bit sample, which simplified implementation of NLC.

Figure 10.24 shows the effect of equalization techniques with chromatic dispersion when laser nonlinearity is not present (i.e., with external modulation). As before, LE decreases the penalty (by more than 1.5 dB above 160 km) or, alternatively, increases the dispersion-limited distance (by 25% for a 3 dB penalty). Although with LE the penalty dips above 240 km, this effect is highly dependent on the transmitter pulse shape and receiver frequency characteristics and therefore may not necessarily be present in systems with slightly different characteristics. On the other hand, NLC and MLD do not have such variations. These techniques greatly reduce the penalty above 200 km, increasing the dispersion-limited distance for a 3 dB penalty to 270 and 400 km with NLC and MLD, respectively.

APPENDIX 10A A COMPARISON OF THE QUANTIZATION ERROR (QE) OF A FIXED EQUALIZER WITH THE ACHIEVABLE DIGITAL RESIDUAL ERROR (DRE) OF AN ADAPTIVE EQUALIZER

In this appendix we first derive an expression for the rms quantization error (QE) of a *fixed* (nonadaptive) equalizer. We then compare the digital residual error (DRE) of the *fixed* equalizer to the digital residual error (DRE) that can be achieved for an adaptive equalizer implemented with limited precision. Let us start by considering the following two equalizers shown in Figure 10.A.1. The first has tap weights $\{c_j\}$ that are assumed to be of infinite precision. The output of this equalizer is given by

$$y(i) = \sum_{j=-N}^{N} c_j x(i-j) . \tag{10.A.1}$$

The tap weights in the second equalizer $\{\hat{c}_j\}$ are found by truncating the corresponding tap weights in the first filter. The output of the second equalizer equals

$$\hat{y}(i) = \sum_{j=-N}^{N} \hat{c}_j x(i-j) . \tag{10.A.2}$$

The QE is the mean-square difference between the two outputs or

$$\xi(i) = y(i) - \hat{y}(i) = \sum_{j=-N}^{N} (c_j - \hat{c}_j)\cdot x(i-j) \tag{10.A.3}$$

and the mean-squared QE equals

$$\varepsilon_Q^2 = E[\xi_i^2(i)] = E\left[\sum_{j=-N}^{N} \sum_{l=-N}^{N} (c_j - \hat{c}_j)\cdot(c_l - \hat{c}_l)\cdot x(i-j)\cdot x(i-l)\right]$$

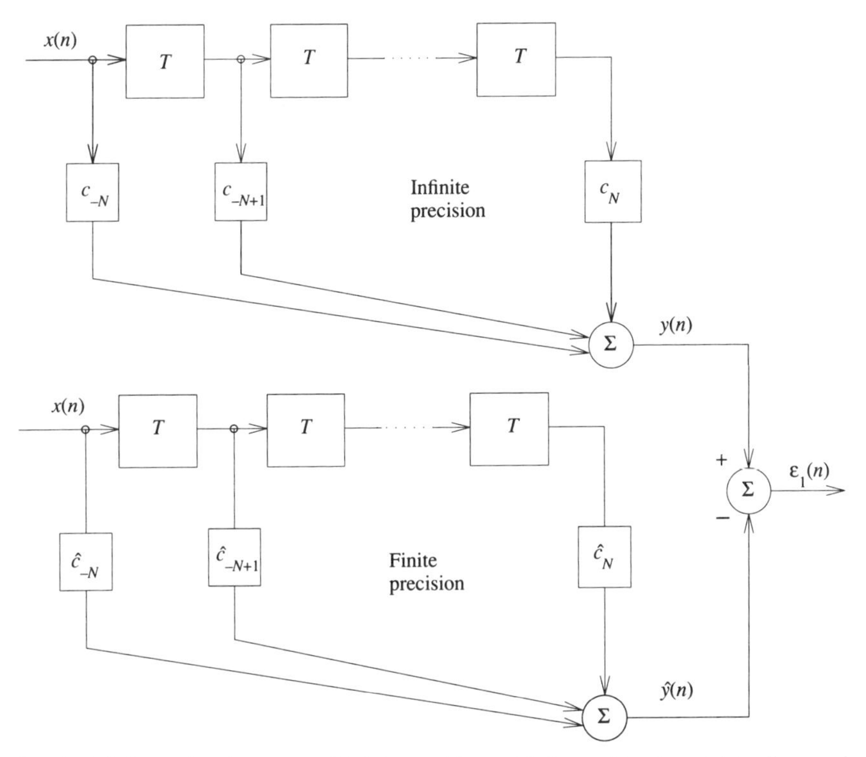

Fig. 10.A.1 Comparing the outputs of infinite precision and finite precision tapped delay line equalizer.

Let us assume that the inputs $\{x_i\}$ are independent random variables with rms value equal to X_{rms}. Then

$$\varepsilon_Q^2 = \sum_{j=-N}^{N} (c_j - \hat{c}_j)^2 X_{rms}^2 \le (2N+1)\cdot LSD^2 \cdot X_{rms}^2 \ , \qquad (10.A.4)$$

where LSD is the value of the least significant digit in the truncated tap weights. The above expressions are exact for the case of statistically independent inputs and provide useful approximate results when the inputs are dependent. The term $(c_j - \hat{c}_j)^2$ could be treated as a random variable, rather than bounded. In this case, if we assume that the tap error is a uniform random variable over the quantization region $(-\Delta/2,\ \Delta/2)$, we would arrive at the well-known $\Delta^2/12$ term as the mean-squared quantization error, rather than the worst case bound appearing above.

As explained in Section 10.1, the rms digital error (DRE) using the LMS algorithm can be approximated by

$$e_d \approx \frac{LSD}{\Delta\cdot X_{rms}} \ . \qquad (10.A.5)$$

In this appendix, the rms QE was found to be upper bounded by

$$\varepsilon_Q \leq (2N+1)^{1/2} \cdot LSD \cdot x_{rms} . \tag{10.A.6}$$

To relate these two errors we must express Δ as a function of N. This can be accomplished by means of the sufficient condition (8.70) for the convergence of the LMS algorithm, namely,

$$0 \leq \Delta \leq \frac{\lambda_{\min}}{\lambda_{\max}} \frac{1}{(2N+1)X_{rms}^2} , \tag{10.A.7}$$

which is a lower bound on the maximum step size, where X_{rms} is the RMS value of the signal at the input to the equalizer in terms of the maximum and minimum eigenvalues of the channel correlation matrix ($\lambda_{\max}$ and $\lambda_{\min}$ respectively).

Combining the above, we can lower bound the ratio of DRE to QE as follows:

$$\frac{e_d}{\varepsilon_Q} \geq \frac{LSD \cdot (2N+1)X_{rms}^2}{X_{rms} \frac{\lambda_{\min}}{\lambda_{\max}} (2N+1)^{1/2} \cdot LSD \cdot X_{rms}} = \frac{\lambda_{\min}}{\lambda_{\max}} (2N+1)^{1/2} , \tag{10.A.8}$$

a number that grows with the length of the equalizer and the eccentricity of the input signal correlation. This is yet another reason for minimizing the number of tap weights.

APPENDIX 10B THE EFFECT OF LINEAR EQUALIZATION ON QUADRATIC DISTORTION FOR LIGHTWAVE SYSTEMS

Here we consider the effect of linear equalization on quadratic (nonlinear) distortion. In general, linear equalization will not improve performance with quadratic distortion. However, in lightwave systems, if the signaling is binary, on–off keying and the major portion of the intersymbol interference (ISI) is from a single adjacent symbol, then linear equalization can be effective in reducing the ISI. Let us consider an 8-Gbps systems using a DFB laser [25] where the extinction ratio of the laser modulation is 5, i.e., the level for a "1" is approximately 5 times that for a "0". Thus, with ISI from one adjacent bit, the received electrical signal at time kT is given by

$$\begin{aligned} s_e(kT) &= |s_0(kT)|^2 \approx |0.45+0.55a_k+p(T)a_{k+1}|^2 \\ &\approx 0.2+0.3a_k^2+0.5a_k+p^2(T)a_{k-1}^2+(0.9+1.1a_k)a_{k-1}p(T) , \end{aligned} \tag{10.B.1}$$

where $p(t)$ is the received pulse response of the system (assumed to be maximum at $t=0$). Note that if $p(T)=0$, $s_e(kT)$ is 0.2 for $a_k=0$ and 1 for $a_k=1$

(for an extinction ratio of 5). The linear component* of the ISI (in terms of squared data symbols) is $p^2(T)a_{k-1}^2$, and if the ISI is reasonably small, then this component will be much less than the nonlinear component $(0.9+1.1I_k)a_{k-1}p(T)$. In this case, cancellation of the linear component (or partial cancellation through linear equalization) will have a negligible impact.

However, for binary, on–off keying, the effect of the nonlinear component *can* be reduced by linear equalization. First, since $a_k^2 = a_k$, (10.B.1) can be expressed as

$$s_e(kT) = 0.2+0.8I_k+(p^2(T)+0.9)a_{k-1}+1.1p(T)a_k a_{k-1} \quad (10.B.2)$$

or

$$s_e(kT) = \begin{cases} 0.2 & \text{if } a_{k-1}=0 \text{ and } a_k=0 \\ 1 & \text{if } a_{k-1}=0 \text{ and } a_k=1 \\ 0.2+(p^2(T)+0.9 & \text{if } a_{k-1}=1 \text{ and } a_k=0 \\ 1+(p^2(T)+0.9)+1.1p(T) & \text{if } a_{k-1}=1 \text{ and } a_k=1\,. \end{cases} \quad (10.B.3)$$

The linear component of the ISI is now $p^2(T)+0.9$ and the nonlinear component is $1.1p(T)a_k a_{k-1}$. Thus, in contrast to (B.10.1), the linear component is always larger (and is much larger with small ISI) than the nonlinear component, and linear equalization can be effective. Furthermore, the effect of the smaller nonlinear term can also be somewhat reduced by linear equalization. When $p(T)<0$, the nonlinear term $1.1p(T)a_k a_{k-1}$ reduces the level for a "1" when $a_{k-1}=1$. (The nonlinear term increases the eye opening for $p(T)>0$.) In this case, when the ISI is small, we can increase the minimum distance to the threshold (set at 0.6) for $p(T)<0$, by using a linear equalizer to add the term $0.55p(T)s_e((k-1)T)$ (which is approximately $0.55p(T)a_{k-1}$ with small ISI) to $s_e(kT)$.

REFERENCES

[1] R. D. Gitlin, J. E. Mazo, and M. G. Taylor, "On the Design of Gradient Algorithms for Digitally Implemented Adaptive Filters," *IEEE Trans. on Circuit Theory,* Vol. CT-20, No. 2, pp. 125–136, March 1973.

[2] R. D. Gitlin and S. B. Weinstein, "On the Required Tap-Weight Precision for Digitally Implemented, Adaptive, Mean-Squared Equalizers," *Bell System Tech. J.*, Vol. 58, No. 2, pp. 301–321, February 1979.

---

* In terms of the above notation, the system would be linear, in terms of the squared data symbols, if $s_e(kT) = \sum_j a_{k-j}^2 p^2(jT)$.

[3] R. D. Gitlin, H. C. Meadors, and S. B. Weinstein, "The Tap Leakage Algorithm: An Algorithm for the Stable Operation of a Digitally Implemented Fractionally Spaced Adaptive Equalizer," *Bell System Tech. J.*, Vol. 61, No. 8, pp. 1817–1839, October 1982.

[4] J. M. Cioffi, "Limited Precision Effects in Adaptive Filtering," *IEEE Trans. on Circuits and Systems,* Vol. CAS-34, No. 7, July 1987.

[5] S. Haykin, *Adaptive Filter Theory*, Prentice-Hall, 1986.

[6] G. Long, F. Ling, and J. G. Proakis, "Adaptive Transversal Filters with Delayed Coefficient Adaptation," *Acoustics, Speech and Signal Processing Symposium Conference Record,* April 1987.

[7] P. Kabal, "The Stability of Adaptive Minimum Mean Square Error Equalizers Using Delayed Adjustment," *IEEE Trans. on Communications,* Vol. COM-31, No. 3, pp. 430–432, March 1983.

[8] R. D. Gitlin and R. L. Cupo, "Adaptive Carrier Recovery Systems for Digital Data Communications Receivers," *IEEE Journal on Selected Areas in Communication,* December 1989.

[9] R. P. Gooch and M. J. Reddy, "An Adaptive Phase Lock Loop for Phase Jitter Tracking," *Conference Record* 21st Asilomar Conference on Signals, Systems and Computers, November 2–4, 1987.

[10] D. D. Falconer, "Jointly Adaptive Equalization and Carrier Recovery in Two-Dimensional Digital Communication Systems," *Bell System Tech. J.,* Vol. 55, No. 3, pp. 317–334, March 1976.

[11] J. Salz, "Modulation and Detection for Coherent Lightwave Communications," *IEEE Communications Magazine,* 24(6), June 1986.

[12] J. C. Campbell, A. G. Dentai, W. S. Holden, and B. L. Kasper, "High Performance Avalanche Photodiode with Separate Absorption, Grading, and Multiplication Regions," *Elec. Lett. 19,* pp. 818–819, September 29, 1983.

[13] P. S. Henry, "Introduction to Lightwave Transmission," *IEEE Communications Magazine,* pp. 12–16, May 1985.

[14] A. F. Elrefaie, R. E. Wagner, D. A. Atlas, and D. G. Daut, "Chromatic Dispersion Limitations in Coherent Lightwave Transmission Systems," *Journal of Lightwave Technology,* Vol. 6, pp. 704–709, May 1988.

[15] R. E. Wagner and A. F. Elrefaie, "Polarization-Dispersion Limitations in Lightwave Systems," in *Technical Digest, Optical Fiber Communications Conference,* New Orleans, LA, p. 37, January 25–28, 1988.

[16] C. D. Poole and C. R. Giles, "Polarization-Dependent Pulse Compression and Broadening Due to Polarization Dispersion in Dispersion-Shifted Fiber," *Optics Letters,* Vol. 13, pp. 155–157, February 1988.

[17] A. A. M. Saleh, "Nonlinear Models of Traveling-Wave Optical Amplifiers," *Electronics Letters,* Vol. 24, No. 14, July 7, 1988.

[18] U. Timor and R. A. Linke, "A Comparison of Sensitivity Degradations for Optical Homodyne vs. Direct Detection of On-Off Keyed Signals," *Journal of Lightwave Technology,* Vol. 6, pp. 1782–1788, November 1988.

[19] D. G. Messerschmitt, "Minimum MSE Equalization of Digital Fiber Optic Systems," *IEEE Trans. on Communications,* Vol. COM-16, pp. 1110–1118, July 1978.

[20] J. Salz, "Coherent Lightwave Communications," *AT&T Tech. J.,* Vol. 64, No. 10, December 1985.

[21] G. J. Foschini, L. J. Greenstein, and G. Vannucci, "Noncoherent Detection of Coherent Lightwave Signals Corrupted by Phase Noise," *IEEE Trans. Commun.,* Vol. COM-36, pp. 306–314, March 1988.

[22] W. J. Tomlinson and R. H. Stolen, "Nonlinear Phenomena in Optical Fibers," *IEEE Communications Magazine,* Vol. 26, pp. 36–44, April 1988.

[23] I. M. I. Habbab and L. J. Cimini, Jr., "Polarization-Switching Techniques for Coherent Optical Communications," *Journal of Lightwave Technology,* Vol. 6, No. 10, October 1988.

[24] W. D. Grover, "Forward Error Correction in Dispersion-Limited Lightwave Systems," *Journal of Lightwave Technology,* Vol. 6, pp. 643–654, May 1988.

[25] T. L. Koch and P. J. Corvini, "Semiconductor Laser Chirping-Induced Dispersion Distortion in High-Bit-Rate Optical Fiber Communications Systems," *Proc of the IEEE International Conference on Communications '88,* pp. 19.4.1–4, June 12–15, 1988.

[26] J. H. Winters and R. D. Gitlin, "Electrical Signal Processing Techniques for Fiber Optic Communications Systems," *IEEE Trans. on Communications,* September 1990.

[27] J. H. Winters, "Equalization in Coherent Lightwave Systems Using a Fractionally-Spaced Equalizer," *Journal of Lightwave Technology,* October 1990.

[28] J. H. Winters and M. Santoro, "Experimental Equalization of Polarization Dispersion," *Photonics Technology Letters,* Vol. 2, pp. 591–593, August 1990.

[29] S. Kasturia and J. H. Winters, "Techniques for High-Speed Implementation of Nonlinear Cancellation," *IEEE Journal on Selected Areas in Communications,* Vol. 9, pp. 711–717, June 1991.

[30] J. H. Winters and S.. Kasturia, "Constrained Maximum-Likelihood Detection for High-Speed Fiber-Optic Systems," *Proc. of GLOBECOM '91,* Paper 44.1, December 2–6, 1991.

Index